To the Student

As authors our highest goal has been to write a solid mathematics textbook that will truly help you succeed in your mathematics course. After years of refining our textbooks, we have now developed a feature that we think will help ensure your success in mathematics. It is called the *How Am I Doing? Guide to Math Success.*

The *How Am I Doing? Guide to Math Success* shows how you can effectively use this textbook to succeed in your mathematics course. This clear path for you to follow is based upon how our successful students have utilized the textbook in the past. Here is how it works:

EXAMPLES and PRACTICE PROBLEMS: When you study an Example, you should immediately do the Practice Problem that follows to make sure you understand each step in solving a particular problem. The worked-out solution to every Practice Problem can be found in the back of the text starting at page SP-1 so you can check your work and receive immediate guidance in case you need to review.

EXERCISE SETS—Practice, Practice, Practice: You learn math by *doing* math. The best way to learn math is to *practice, practice, practice.* The exercise sets provide the opportunity for this practice. Be sure that you complete every exercise your instructor assigns as homework. In addition, check your answers to the odd-numbered exercises in the back of the text to see whether you have correctly solved each problem.

HOW AM I DOING? MID-CHAPTER REVIEW: This feature allows you to check if you understand the important concepts covered to that point in a particular chapter. Many students find that halfway through a chapter is the point of greatest need because so many different types of problems have been covered. This review covers each of the types of problems from the first half of the chapter. Do these problems and check your answers at the back of the text. If you need to review any of these problems, simply refer back to the section and objective indicated next to the answer. Before you go further into the chapter, it is important to understand what has been learned so far.

HOW AM I DOING? CHAPTER TEST: This test (found at the end of every chapter) provides you with an excellent opportunity to both practice and review for any test you will take in class. Take this test to see how much of the chapter you have mastered. By checking your answers, you can once again refer back to the section and the objective of any exercise you want to review further. This allows you to see at once what has been learned and what still needs more study as you prepare for your test or exam.

HOW AM I DOING? CHAPTER TEST PREP VIDEO CD: If you need to review any of the exercises from the *How Am I Doing? Chapter Test,* this video CD found at the back of the text provides worked-out solutions to each of the exercises. Simply insert the CD into a computer and watch a math instructor solve each of the Chapter Test exercises in detail. By reviewing these problems, you can study any points of difficulty and better prepare yourself for your upcoming test or exam.

These steps provide a clear path you can follow in order to successfully complete your math course. More importantly, the ***How Am I Doing? Guide to Math Success*** is a tool to help you achieve an understanding of mathematics. We encourage you to take advantage of this new feature.

John Tobey
Jeffrey Slater
North Shore Community College

MORE TOOLS FOR SUCCESS

In addition to the ***How Am I Doing? Guide to Math Success,*** your Tobey/Slater textbook is filled with other tools and features to help you succeed in your mathematics course. These include:

Blueprint for Problem Solving

The Mathematics Blueprint for Problem Solving provides you with a consistent outline to organize your approach to problem solving. You will not need to use the blueprint to solve every problem, but it is available when you are faced with a problem with which you are not familiar, or when you are trying to figure out where to begin solving a problem.

Chapter Organizers

The key concepts and mathematical procedures covered in each chapter are reviewed at the end of the chapter in a unique Chapter Organizer. This device not only lists the key concepts and methods, but provides a completely worked-out example for each type of problem. The Chapter Organizer should be used in conjunction with the ***How Am I Doing? Chapter Test*** and the ***How Am I Doing? Chapter Test Prep Video CD*** as a study aid to help you prepare for tests.

Developing Your Study Skills

These notes appear throughout the text to provide you with suggestions and techniques for improving your study skills and succeeding in your math course.

RESOURCES FOR SUCCESS

In addition to the textbook, Prentice Hall offers a wide range of materials to help you succeed in your mathematics course. These include:

Student Study Pack

Includes the *Student Solutions Manual* (fully worked-out solutions to odd-numbered exercises), access to the *Prentice Hall Tutor Center,* and the CD *Lecture Series Videos* that accompany the text.

MyMathLab

MyMathLab offers the entire textbook online with links to video clips and practice exercises in addition to tutorial exercises, homework, and tests. *MyMathLab* also offers a personalized Study Plan for each student based on student test results. The Study Plan links directly to unlimited tutorial exercises for the areas you need to study and re-test so you can practice until you have mastered the skills and concepts.

Beginning Algebra

SIXTH EDITION

John Tobey

North Shore Community College
Danvers, Massachusetts

Jeffrey Slater

North Shore Community College
Danvers, Massachusetts

PEARSON
Prentice Hall

Upper Saddle River, NJ 07458

Library of Congress Cataloging-in-Publication Data

Tobey, John
Beginning algebra.—6th ed. / John Tobey, Jeffrey Slater.
p. cm.
Includes index.
ISBN 0-13-148287-4
    1.   Algebra. I. Slater, Jeffery, II. Title
QA152.3.T63 2006
512—dc22   2004044785

Senior Acquisitions Editor: *Paul Murphy*
Project Managers: *Elaine Page and Dawn Nuttall*
Editor in Chief: *Christine Hoag*
Production Editor: *Lynn Savino Wendel*
Vice President/Director of Production and Manufacturing: *David W. Riccardi*
Senior Managing Editor: *Linda Mihatov Behrens*
Executive Managing Editor: *Kathleen Schiaparelli*
Assistant Manufacturing Manager/Buyer: *Michael Bell*
Executive Marketing Manager: *Eilish Collins Main*
Development Editor: *Tony Palermino*
Marketing Assistant: *Rebecca Alimena*
Editor in Chief, Development: *Carol Trueheart*
Media Project Manager, Developmental Math: *Audra J. Walsh*
Media Production Editor: *Zachary Hubert*
Print Supplements Editor: *Christina Simoneau*
Art Director: *Jonathan Boylan*
Interior and Cover Designer: *Susan Anderson-Smith*
Editorial Assistant: *Mary Burket*
Art Editor: *Thomas Benfatti*
Director of Creative Services: *Paul Belfanti*
Director, Image Resource Center: *Melinda Reo*
Manager, Rights and Permissions: *Zina Arabia*
Manager, Visual Research: *Beth Brenzel*
Manager, Cover Visual Research & Permissions: *Karen Sanatar*
Image Permission Coordinator: *Cynthia Vincenti*
Photo Researcher: *Kathy Ringrose*
Art Studio: *Scientific Illustrators* and *Laserwords*
Compositor: *Interactive Composition Corporation*
Cover Photo Credits: *Ron Chapple/Thinkstock/Alamy*

Photo credits appear on page P-1, which constitutes a continuation of the copyright page.

Pearson Education Ltd., London
Pearson Education Australia Pty. Limited, Sydney
Pearson Education Singapore Pte. Ltd.
Pearson Education North Asia, Ltd, Hong Kong
Pearson Education Canada, Ltd., Toronto
Pearson Educación de Mexico, S.A., de C.V.
Pearson Education, Japan, Tokyo
Pearson Education Malaysia, Pte. Ltd.

This book is dedicated to the memory of Lexie Tobey and John Tobey, Sr.
They have left a legacy of love, a memory of four decades of faithful teaching,
and a sense of helping others that will influence generations to come.
For their grandchildren they have left an inspiring model of a loving family,
true character, and service to God and community.

Contents

CHAPTER 0

A Brief Review of Arithmetic Skills   1

CHAPTER 1

Real Numbers and Variables   67

CHAPTER 2

Equations and Inequalities   133

CHAPTER 3

Solving Applied Problems   191

CHAPTER 4

Exponents and Polynomials   251

CHAPTER 5

Factoring   303

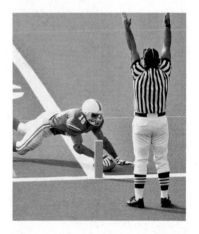

CHAPTER 6

Rational Expressions and Equations   355

CHAPTER 10

Quadratic Equations   569

Preface

TO THE INSTRUCTOR

We share a partnership with you. For over thirty-three years we have taught mathematics courses at North Shore Community College. Each semester we join you in the daily task of sharing the knowledge of mathematics with students who often struggle with this subject. We enjoy teaching and helping students—and we are confident that you share these joys with us.

Mathematics instructors and students face many challenges today. *Beginning Algebra* was written with these needs in mind. This textbook explains mathematics slowly, clearly, and in a way that is relevant to everyday life for the college student. As with previous editions, special attention has been given to problem solving in the sixth edition. This text is written to help students organize the information in any problem-solving situation, to reduce anxiety, and to provide a guide that enables students to become confident problem solvers.

One of the hallmark characteristics of *Beginning Algebra* that makes the text easy to learn and teach from is the building-block organization. Each section is written to stand on its own, and each homework set is completely self-testing. Exercises are paired and graded and are of varying levels and types to ensure that all skills and concepts are covered. As a result, the text offers students an effective and proven learning program suitable for a variety of course formats—including lecture-based classes; discussion-oriented classes; distance learning centers; modular, self-paced courses; mathematics laboratories; and computer-supported centers. The book has been written to be especially helpful in online classes. The authors teach at least one course online each semester.

Beginning Algebra is part of a series that includes the following:

Tobey/Slater, *Basic College Mathematics,* Fifth Edition

Tobey/Slater, *Essentials of Basic College Mathematics*

Blair/Tobey/Slater, *Prealgebra,* Third Edition

Tobey/Slater, *Beginning Algebra,* Sixth Edition

Tobey/Slater/Blair, *Beginning Algebra: Early Graphing*

Tobey/Slater, *Intermediate Algebra,* Fifth Edition

Tobey/Slater/Blair, *Beginning and Intermediate Algebra,* Second Edition

We have visited and listened to teachers across the country and have incorporated a number of suggestions into this edition to help you with the particular learning delivery system at your school. The following pages describe the key continuing features and changes in the sixth edition.

NEW! AND IMPROVED FEATURES IN THE SIXTH EDITION

How Am I Doing? Guide to Math Success

The ***How Am I Doing? Guide to Math Success*** shows how students can effectively use this textbook to succeed in their mathematics course. This clear path for them to follow is based upon how successful students have utilized this textbook in the past.

EXAMPLES and PRACTICE PROBLEMS The design of the text has been enhanced so the Examples and Practice Problems are clearly connected in a cohesive unit. This has been done to encourage students to immediately do the Practice Problem associated with each Example to make sure they understand each step in solving a particular problem. The worked-out solution to every Practice Problem can be found in the back of the text starting at page SP-1 so students can check their work and receive immediate guidance in case they need to review.

Enhanced Exercise Sets

➤ First, *each* exercise set has been:

- enhanced to have a **better progression from easy to medium to challenging** problems, with appropriate quantities of each.
- restructured to have **better matching of even and odd** problems.
- rewritten to ensure that **all concepts are fully represented** with every example from the section covered by a group of exercises.

➤ In addition, the **number and frequency of *Mixed Practice* problems have been increased,** when appropriate, throughout the exercise sets. These *Mixed Practice* problems require the students to identify the type of problem and the best method they should use to solve it. **More *Verbal and Writing Skills* exercises have also been added,** where appropriate, to allow students more time to interact with new concepts and to explain them fully in their own words.

➤ Lastly, throughout the text the **application exercises have been updated and labeled** to better indicate the scope and relevance of each real-world and real-data problem at-a-glance. These applications relate to everyday life, global issues beyond the borders of the United States, and other academic disciplines. Many include source citations. Roughly 30 percent of the applications have been contributed by actual students based on scenarios they have encountered in their home or work lives.

How Am I Doing? Mid-Chapter Review

This feature allows you to check if your students understand the important concepts covered to that point in a particular chapter. Many students find that halfway through a chapter is the point of greatest need because so many different types of problems have been covered. This review covers each of the types of problems from the first half of the chapter. Specific section and objective references are provided with each answer to indicate where a student should look for further review.

How Am I Doing? Chapter Test

This test (found at the end of every chapter) provides your students with an excellent opportunity to both practice and review for any test they will take in class. Encourage your students to take this test to see how much of the material they have mastered and then check their answers in the back of the text. Specific section and objective references are provided with each answer to indicate where a student should look for further review.

How Am I Doing? Chapter Test Prep Video CD

If students need to review any of the exercises from the ***How Am I Doing? Chapter Test,*** this video CD found at the back of the text provides worked-out solutions to each of the exercises. Students can simply insert the CD into a computer and watch a math instructor solve each of the Chapter Test exercises in detail. By reviewing these problems, students can study through any points of difficulty and better prepare for an upcoming test or exam.

How Am I Doing? Chapter Test & TestGen

For this new edition of the text, we have provided a new file in each chapter of the *TestGen* program specific to each Chapter Test in the text. The *TestGen* Chapter Test file provides algorithms specific to the problems in the corresponding textbook Chapter Test. This provides a convenient way to create similar tests for practice or assessment purposes.

KEY FEATURES IN THE SIXTH EDITION

Developing Problem-Solving Abilities

We are committed as authors to producing a textbook that emphasizes mathematical reasoning and problem-solving techniques as recommended by AMATYC, NCTM, AMS, NADE, MAA, and other bodies. To this end, the problem sets are built on a wealth of real-life and real-data applications. Unique problems have been developed and incorporated into the exercise sets that help train students in data interpretation, mental mathematics, estimation, geometry and graphing, number sense, critical thinking, and decision making.

Mathematics Blueprint for Problem Solving

The successful Mathematics Blueprint for Problem Solving strengthens problem-solving skills by providing a consistent and interactive outline to help students organize their approach to problem solving. Once students fill in the blueprint, they can refer back to their plan as they do what is needed to solve the problem. Because of its flexibility, this feature can be used with single-step problems, multi-step problems, applications, and nonroutine problems that require problem solving strategies. Students will not need to use the blueprint to solve every problem. It is available for those faced with a problem with which they are not familiar, to alleviate anxiety, to show them where to begin, and to assist them in the steps of reasoning.

Developing Your Study Skills

This highly successful feature has been retained in the new edition. The boxed notes are integrated throughout the text to provide students with techniques for improving their study skills and succeeding in math courses.

Putting Your Skills to Work

This highly successful feature has been revised and expanded in the sixth edition. There are eleven new Putting Your Skills to Work applications in the new edition. These nonroutine application problems challenge students to synthesize the knowledge they have gained and apply it to a totally new area. Each problem is specifically arranged for independent and cooperative learning or group investigation of mathematical problems that pique student interest. Students are given the opportunity to help one another discover mathematical solutions to extended problems. The investigations feature open-ended questions and extrapolation of data to areas beyond what is normally covered in such a course.

Increased Integration and Emphasis on Geometry

Due to the emphasis on geometry on many statewide exams, geometry problems are integrated throughout the text. The new edition contains an increased number of geometry problems. Additionally, examples and exercises that incorporate a principle of geometry are marked with a triangle icon for easy identification.

Graphs, Charts, and Tables

When students encounter mathematics in real-world publications, they often encounter data represented in a graph, chart, or table and are asked to make a reasonable conclusion based on the data presented. This emphasis on graphical interpretation is a continuing trend with today's expanding technology. The number of mathematical problems based on charts, graphs, and tables has been significantly increased in this edition. Students are asked to make simple interpretations, to solve medium-level problems, and to investigate challenging applied problems based on the data shown in a chart, graph, or table.

MASTERING MATHEMATICAL CONCEPTS

Text features that develop the mastery of concepts include the following:

Learning Objectives

Concise learning objectives listed at the beginning of each section allow students to preview the goals of that section.

To Think About

These critical thinking questions may follow examples in the text and appear in the exercise sets. They extend the concept being taught, providing the opportunity for all students to stretch their minds, to look for patterns, and to make conclusions based on their previous experience. The number of these exercises has significantly increased in this edition.

Cumulative Review

Almost every exercise set concludes with a section of cumulative review problems. These problems review topics previously covered, and are designed to assist students in retaining the material. Many additional applied problems have been added to the cumulative review sections.

Calculator Problems

Calculator boxes are placed in the margin of the text to alert students to a scientific or graphing calculator application. In the exercise section, icons are used to indicate problems that are designed for solving with a graphing or scientific calculator.

REVIEWING MATHEMATICAL CONCEPTS

At the end of each chapter we have included problems and tests to provide your students with several different formats to help them review and reinforce the ideas that they have learned. This assists them not only with that specific chapter, but reviews previously covered topics as well.

Chapter Organizers

The concepts and mathematical procedures covered are reviewed at the end of each chapter in a unique chapter organizer. This device has been extremely popular with faculty and students alike. It not only lists concepts and methods, but provides a completely worked-out example for each type of problem. Students find that preparing a similar chapter organizer on their own in higher-level math courses becomes an invaluable way to master the content of a chapter of material.

Chapter Review Problems

These problems are grouped by section as a quick refresher at the end of the chapter. They can also be used by the student as a quiz of the chapter material.

How Am I Doing? Chapter Tests

Found at the end of the chapter, the chapter test is a representative review of the material from that particular chapter that simulates an actual testing format. This provides the students with a gauge to their preparedness for the actual examination.

Cumulative Tests

At the end of each chapter is a cumulative test. One-half of the content of each cumulative test is based on the math skills learned in previous chapters. By completing these tests for each chapter, the students build confidence that they have mastered not only the contents of the chapter but those of previous chapters as well.

RESOURCES FOR THE INSTRUCTOR
Printed Resources
Annotated Instructor's Edition (ISBN: 0-13-148288-2)

- Complete student text
- Answers appear in place on the same text page as exercises
- **Teaching Tips** placed in the margin at key points where students historically need extra help
- Answers to all exercises in the section sets, mid-chapter reviews, chapter reviews, chapter tests, cumulative tests, and practice final

Instructor's Solutions Manual (ISBN: 0-13-148291-2)

- Detailed step-by-step solutions to the even-numbered section exercises
- Solutions to every exercise (odd and even) in the mid-chapter reviews, chapter reviews, chapter tests, cumulative tests, and practice final
- Solution methods reflect those emphasized in the text

Instructor's Resource Manual with Tests (ISBN: 0-13-148290-4)

- New! One *Mini-Lecture* is provided per section. These include key learning objectives, classroom examples, and teaching notes.
- New! One *Skill Builder* is provided per section. These include concept rules, explained examples, and extra problems for students. All answers included.
- New! Two *Activities* per chapter provide short group activities in a convenient ready-to-use handout format. All answers included.
- Twenty *Additional Exercises* are provided per section for added test exercises or worksheets. All answers included.
- *Tests* provide additional suggested testing materials:
  - two *Chapter Pretests* per chapter (1 free response, 1 multiple choice)
  - six *Chapter Tests* per chapter (3 free response, 3 multiple choice)
  - two *Cumulative Tests* per even-numbered chapter (1 free response, 1 multiple choice)
  - two *Final Exams* (1 free response, 1 multiple choice)
  - answers to all items

Media Resources
TestGen with QuizMaster (Windows/Macintosh) (ISBN: 0-13-148294-7)

- Algorithmically driven, text-specific testing program—covers all objectives of the text
- New! Chapter Test file for each chapter provides algorithms specific to exercises in each *How Am I Doing? Chapter Test* from the text
- Edit and add own questions with the built-in question editor, which also allows you to create graphs, import graphics, and insert math notation
- Create a nearly unlimited number of tests and worksheets as well as assorted reports and summaries
- Networkable for administering tests and capturing grades online or on a local area network

CD Lecture Series Videos—Lab Pack (ISBN: 0-13-148293-9)

- Lecture-based instruction for Key Concepts, Examples, and Practice Problems from each text section

- Excellent support for students who want to review classroom lectures when doing homework or studying for a test
- Convenient anytime access to video tutorial support when provided for students as part of the Student Study Pack

New! MyMathLab (instructor)

MyMathLab is an all-in-one, online tutorial, homework, assessment, course management tool with the following features.

- **Powered by CourseCompass**™—Pearson Education's online teaching and learning environment
- **Tutorial powered by *MathXL*®**—our online homework, tutorial, and assessment system
- Rich and flexible set of course materials, **featuring free-response exercises** algorithmically generated for unlimited practice and mastery.
- **The entire textbook online with links to multimedia resources**—video clips, practice exercises, and animations—that are correlated to the textbook examples and exercises.
- **Homework and test managers** to select and assign online exercises correlated directly to the text
- **A personalized Study Plan** generated based on student test results. The Study Plan links directly to unlimited tutorial exercises for the areas students need to study and re-test, so they can practice until they have mastered the skills and concepts.
- **Easy to use tracking** in the *MyMathLab* gradebook of all of the online homework, tests, and tutorial work.
- **Import *TestGen* tests**
- **Online gradebook**—designed specifically for mathematics—automatically tracks students' homework and test results and provides grade control.
- http://www.mymathlab.com

New! MathXL®

MathXL® is a powerful online homework, tutorial, and assessment system. With *MathXL* instructors can:

- Create, edit, and assign online homework and tests using algorithmically generated exercises correlated at the objective level to your textbook.
- Track student work in *MathXL*'s online gradebook.

RESOURCES FOR THE STUDENT

New! Student Study Pack (ISBN: 0-13-171162-8)

Includes the *Student Solutions Manual,* access to the *Prentice Hall Tutor Center,* and the *CD Lecture Series Videos* that accompany the text.

Printed Resources

Student Solutions Manual (ISBN: 0-13-153060-7)

- Solutions to all odd-numbered section exercises
- Solutions to every (even and odd) exercises found in the mid-chapter reviews, chapter reviews, chapter tests, and cumulative reviews
- Solution methods reflect those emphasized in the text
- Ask your bookstore about ordering

MEDIA RESOURCES

New! Chapter Test Prep Video CD (ISBN: 0-13-153053-4)

Provides step-by-step video solutions to each problem in each *How Am I Doing? Chapter Test* in the textbook.

New! MyMathLab (student)

MyMathLab is an all-in-one, online tutorial, homework, assessment, course management tool with the following student features.

- **The entire textbook online with links to multimedia resources**—video clips, practice exercises, and animations—that are correlated to the textbook examples and exercises.
- **Online tutorial, homework, and tests**
- **A personalized Study Plan** generated based on student test results. The Study Plan links directly to unlimited tutorial exercises for the areas students need to study and re-test, so they can practice until they have mastered the skills and concepts.
- http://www.mymathlab.com

New! MathXL®

MathXL® is a powerful online homework, tutorial, and assessment system. With *MathXL* students can:

- Take chapter tests and receive a personalized study plan based on their test results.
- See diagnosed weaknesses and link directly to tutorial exercises for the objectives they need to study and retest.
- Access supplemental animations and video clips directly from selected exercises.

New! MathXL® Tutorials on CD (ISBN: 0-13-153063-1)

This interactive tutorial CD-ROM provides:

- Algorithmically-generated practice exercises correlated at the objective level.
- Practice exercises accompanied by an example and a guided solution.
- Tutorial video clips within the exercise to help students visualize concepts.
- Easy-to-use tracking of student activity and scores and printed summaries of students' progress.

New! Interact Math www.interactmath.com

The power of the *MathXL* text-specific tutorial exercises available for unlimited practice online, without an access code, and without tracking capabilities.

Prentice Hall Tutor Center www.prenhall.com/tutorcenter (ISBN: 0-13-064604-0)

- Tutorial support via phone, fax or email staffed by developmental math faculty.
- Available Sunday—Thursday 5pm EST to midnight—5 days a week, 7 hours a day
- Accessed through a registration number that may be bundled with a new text as part of the Student Study Pack or purchased separately with a used book. Comes automatically within *MyMathLab*.

ACKNOWLEDGMENTS

This book is the product of many years of work and many contributions from faculty and students across the country. We would like to thank the many reviewers and participants in focus groups and special meetings with the authors in preparation of previous editions.

Our deep appreciation to each of the following:

George J. Apostolopoulos, *DeVry Institute of Technology*

Katherine Barringer, *Central Virginia Community College*

Mark Billiris, *St. Petersburg Junior College*

Jamie Blair, *Orange Coast College*

Larry Blevins, *Tyler Junior College*

Connie Buller, *Metropolitan Community College*

Robert Christie, *Miami-Dade Community College*

Nelson Collins, *Joliet Junior College*

Mike Contino, *California State University at Heyward*

Judy Dechene, *Fitchburg State University*

Floyd L. Downs, *Arizona State University*

Robert Dubuc, Jr., *New England Institute of Technology*

Barbara Edwards, *Portland State University*

Janice F. Gahan-Rech, *University of Nebraska at Omaha*

Colin Godfrey, *University of Massachusetts, Boston*

Mary Beth Headlee, *Manatee Community College*

Doug Mace, *Baker College*

Carl Mancuso, *William Paterson College*

James Matovina, *Community College of Southern Nevada*

Janet McLaughlin, *Montclair State College*

Beverly Meyers, *Jefferson College*

Nancy Meyers, *University of Southern Indiana*

Gloria Mills, *Tarrant County Junior College*

Norman Mittman, *Northeastern Illinois University*

Wayne L. Miller, *Lee College*

Sharon L. Morrison, *St. Petersburg Junior College*

Jim Osborn, *Baker College*

Linda Padilla, *Joliet Junior College*

Cathy Panik, *Manatee Community College*

Elizabeth A. Polen, *County College of Morris*

Joel Rappaport, *Miami-Dade Community College*

Jose Rico, *Laredo Community College*

Ronald Ruemmler, *Middlesex County College*

Dennis Runde, *Manatee Community College*

Sally Search, *Tallahassee Community College*

Carolyn Gigi Smith, *Armstrong Atlantic State University*

Lee Ann Spahr, *Durham Technical Community College*

Richard Sturgeon, *University of Southern Maine*

Ara B. Sullenberger, *Tarrant County Community College*

Margie Thrall, *Manatee Community College*

Michael Trappuzanno, *Arizona State University*

Jerry Wisnieski, *Des Moines Community College*

In addition, we want to thank the following individuals for providing splendid insight and suggestions for this new edition.

Carla Ainsworth, *Salt Lake Community College*

Mary Lou Baker, *Columbia State Community College*

Jana Barnard, *Angelo State University*

Linda Beattie, *Western New Mexico University*

Lydia Botsford, *Heald Business College,* Milpitas

Bonnie Brooks, *Sussex County Community College*

David Brown, *Heald Business College,* Vacaville

John Buckley, *Delaware Technical and Community College*

John Close, *Salt Lake Community College*

Gene Coco, *Heald Business College,* Roseville

Kevin Cooper, *National American University*

Gregory Daubenmire, *Las Positas College*

Robert Dubuc, *New England Institute of Technology*

Mahmoud El-Hashash, *Bridgewater State College*

Naomi Gibbs, *Pitt Community College*

Mark Glines, *Salt Lake Community College*

Edna Greenwood, *Tarrant County College, Northwest Campus*

Cynthia Gubitose, *Southern Connecticut State University*

Renu Gupta, *Louisiana State University,* Alexandria

Philip Hammersley, *Ivy Tech State College*

Autumn Hoover, *Angelo State University*

Laura Kaufmann, *Orange Coast College*

Carolyn Krause, *Delaware Technical and Community College*

Euniz Lochte, *Florida Metropolitan University,* Orlando North

Elizabeth Lucas, *North Shore Community College*

Jim Lynn, *Florida Metropolitan University,* Orlando North

Valerie Maley, *Cape Fear Community College*

Mary Marlin, *West Virginia Northern Community College*

Todd Mattson, *DeVry University, DuPage Campus*

Jim McKee, *Florida Metropolitan University,* Tampa

Maria Mendez, *Laredo Community College*

Kathy Morgan, *Coastal Carolina Community College*

Christina Morian, *Lincoln University*

Joe Nagengast, *Florida Career College*

Charlcie Neal, *Clayton College and State University*

Armando Perez, *Laredo Community College*

Larry Pontaski, *Pueblo Community College*

Juan Rivera, *Florida Metropolitan University,* Orlando South

Graciela Rodriguez, *Laredo Community College*

Jeff Simmons, *Ivy Tech State College*

Jed Soifer, *Atlantic Cape Community College*

Gwen Terwilliger, *University of Toledo*

Sharon Testone, *Onondaga Community College*

Georgina Vastola, *Middlesex Community College*

Jonathan Weissman, *Essex County College*

Alice Williamson, *Sussex County Community College*

Mike Wilson, *Florida Metropolitan University,* Tampa

Mary Jane Wolfe, *University of Rio Grande*

Michelle Younker, *Terra Community College*

Peter Zimmer, *West Chester University*

We have been greatly helped by a supportive group of colleagues who not only teach at North Shore Community College but have also provided a number of ideas as well as extensive help on all of our mathematics books. Also, a special word of thanks to Hank Harmeling, Tom Rourke, Wally Hersey, Bob McDonald, Judy Carter, Bob Campbell, Rick Ponticelli, Russ Sullivan, Kathy LeBlanc, Lora Connelly, Sharyn Sharaf, Donna Stefano, and Nancy Tufo. Joan Peabody has done an excellent job of typing various materials for the manuscript and her help is gratefully acknowledged. Sue Ellen Robinson provided new problems, new ideas, and great mental energy. She greatly assisted us with error checking. Her excellent help was much appreciated. A special word of thanks to Richard Semmler, Jenny Crawford, and Ron Salzman for their excellent work in accuracy reviewing page proofs.

Each textbook is a combination of ideas, writing, and revisions from the authors and wise editorial direction and assistance from the editors. We want to thank our Prentice Hall editor, Paul Murphy, for his helpful insight and perspective on each phase of the revision of the textbook. Paul is a man of ideas, wisdom, and energy. He has that rare ability to look at the big picture and consider all the possibilities and then make a wise plan to accomplish the goal in the best possible way. He has found the secret of how to listen to a multitude of ideas and distill them down to a few excellent concepts upon which to improve a textbook. It has been a joy to work with him in the planning, writing, and revising of the new sixth edition of *Beginning Algebra*. Elaine Page, our project manager, provided daily support and encouragement as the book progressed. Elaine is a woman of insight, organization, and helpful suggestions. She has an amazing ability to take hundreds of pages of ideas and condense them into a manageable collection of concrete suggestions. From time to time, as we rolled up our sleeves and went to work on the actual revision and preparation of the manuscript for production, we found that we had much to do and we asked Elaine for help. In every case, she came to our assistance and we accomplished the task together. Tony Palermino, our developmental editor, sifted through mountains of material and offered excellent suggestions for improvement and change.

For the final phases of production of this book, Dawn Nuttall has been our project manager. Dawn is a woman with remarkable perspective and thoughtful analysis. She possesses a wonderful ability to see through problems and find solutions that not only solve the difficulty at hand, but prevent further difficulties from occurring at some later point. When we encountered a wall blocking the way, Dawn always found a clear path to take us around the obstacle. It has been a delight to work with Dawn.

Our mathematics production editor, Lynn Savino Wendel, has been a wonderful help as we finalized the manuscript and checked over the page proofs. Lynn kept things moving on schedule and adjusted tasks in order to make production more efficient. Lynn knows textbook production like the back of her hand and her knowledge proved most helpful throughout the entire process. Special thanks to our art director, Jonathan Boylan, and designer, Susan Anderson-Smith, who coordinated the attractive design and layout of the text and cover for all books in this series.

Nancy Tobey retired from teaching and joined the team as our administrative assistant. Mailing, editing, photocopying, collating, and taping were cheerfully done each day. A special thanks goes to Nancy. We could not have finished the book without you.

Book writing is impossible for us without the loyal support of our families. Our deepest thanks and love to Nancy, Johnny, Melissa, Marcia, Shelley, Rusty, and Abby. Your understanding, your love and help, and your patience have been a source of great encouragement. Finally, we thank God for the strength and energy to write and the opportunity to help others through this textbook.

We have spent more than 33 years teaching mathematics. Each teaching day, we find that our greatest joy is helping students learn. We take a personal interest in ensuring that each student has a good learning experience in taking this course. If you have some personal comments, suggestions, or ideas for future editions of this textbook, please write to us at:

Prof. John Tobey and Prof. Jeffrey Slater
Prentice Hall Publishing
Office of the College Mathematics Editor
Room 300
75 Arlington Street
Boston, MA 02116

or e-mail us at

jtobey@northshore.edu

We wish you success in this course and in your future life!

George

John Tobey
Jeffrey Slater

1. _____

2. _____

3. _____

4. _____

5. _____

6. _____

7. _____

8. _____

9. _____

10. _____

11. _____

12. _____

13. _____

14. _____

15. _____

16. _____

17. _____

18. _____

19. _____

20. _____

21. _____

22. _____

23. _____

Follow the directions for each problem. Simplify each answer.

Chapter 0

1. Add. $3\frac{1}{4} + 2\frac{3}{5}$

2. Multiply. $\left(1\frac{1}{6}\right)\left(2\frac{2}{3}\right)$

3. Divide. $\frac{15}{4} \div \frac{3}{8}$

4. Multiply. $(1.63)(3.05)$

5. Divide. $120 \div 0.0006$

6. Find 7% of 64,000.

Chapter 1

7. Add. $-3 + (-4) + (+12)$

8. Subtract. $-20 - (-23)$

9. Combine. $5x - 6xy - 12x - 8xy$

10. Evaluate. $2x^2 - 3x - 4$ when $x = -3$.

11. Remove the grouping symbols. $2 - 3\{5 + 2[x - 4(3 - x)]\}$

12. Evaluate. $-3(2 - 6)^2 + (-12) \div (-4)$

Chapter 2

In questions 13–16, solve each equation for x.

13. $40 + 2x = 60 - 3x$

14. $7(3x - 1) = 5 + 4(x - 3)$

15. $\frac{2}{3}x - \frac{3}{4} = \frac{1}{6}x + \frac{21}{4}$

16. $\frac{4}{5}(3x + 4) = 20$

17. Solve for p. $A = \frac{1}{2}(3p - 4f)$

18. Solve for x and graph the result. $42 - 18x < 48x - 24$

Chapter 3

19. The length of a rectangle is 7 meters longer than twice the width. The perimeter is 46 meters. Find the dimensions.

20. One side of a triangle is triple the second side. The third side is 3 meters longer than double the second side. Find each side of the triangle if the perimeter of the triangle is 63 meters.

21. Hector has four test scores of 80, 90, 83, and 92. What does he need to score on the fifth test to have an average of 86 on the five tests?

22. Marcia invested $6000 in two accounts. One earned 5% interest, while the other earned 7% interest. After one year she earned $394 in interest. How much did she invest in each account?

23. Melissa has three more dimes than nickels. She has twice as many quarters as nickels. The value of the coins is $4.20. How many of each coin does she have?

24. The drama club put on a play for Thursday, Friday, and Saturday nights. The total attendance for the three nights was 6210. Thursday night had 300 fewer people than Friday night. Saturday night had 510 more people than Friday night. How many people came each night?

Chapter 4

25. Multiply. $(-2xy^2)(-4x^3y^4)$ **26.** Divide. $\dfrac{36x^5y^6}{-18x^3y^{10}}$

27. Raise to the indicated power. $(-2x^3y^4)^5$

28. Evaluate. $(-3)^{-4}$

29. Multiply. $(3x^2 + 2x - 5)(4x - 1)$

30. Divide. $(x^3 + 6x^2 - x - 30) \div (x - 2)$

Chapter 5

Factor completely.

31. $5x^2 - 5$ **32.** $x^2 - 12x + 32$

33. $8x^2 - 2x - 3$ **34.** $3ax - 8b - 6a + 4bx$

Solve for x.

35. $x^3 + 7x^2 + 12x = 0$ **36.** $16x^2 - 24x + 9 = 0$

Chapter 6

37. Simplify. $\dfrac{x^2 + 3x - 18}{2x - 6}$

38. Multiply. $\dfrac{6x^2 - 14x - 12}{6x + 4} \cdot \dfrac{x + 3}{2x^2 - 2x - 12}$

39. Divide and simplify.

$$\dfrac{x^2}{x^2 - 4} \div \dfrac{x^2 - 3x}{x^2 - 5x + 6}$$

40. Add.

$$\dfrac{3}{x^2 - 7x + 12} + \dfrac{4}{x^2 - 9x + 20}$$

41. Solve for x.

$$2 - \dfrac{5}{2x} = \dfrac{2x}{x + 1}$$

42. Simplify.

$$\dfrac{3 + \dfrac{1}{x}}{\dfrac{9}{x} + \dfrac{3}{x^2}}$$

Chapter 7

43. Graph $y = 2x - 4$.

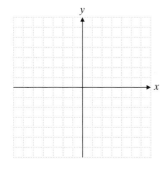

44. Graph $3x + 4y = -12$.

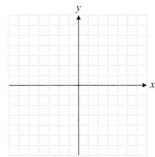

45. What is the slope of a line passing through $(6, -2)$ and $(-3, 4)$?

46. If $f(x) = 2x^2 - 3x + 1$, find $f(3)$.

24. _____

25. _____

26. _____

27. _____

28. _____

29. _____

30. _____

31. _____

32. _____

33. _____

34. _____

35. _____

36. _____

37. _____

38. _____

39. _____

40. _____

41. _____

42. _____

43. _____

44. _____

45. _____

46. _____

47. _____

48. _____

49. _____

50. _____

51. _____

52. _____

53. _____

54. _____

55. _____

56. _____

57. _____

58. _____

59. _____

60. _____

61. _____

62. _____

63. _____

64. _____

65. _____

66. _____

47. Graph the region. $y \geq -\dfrac{1}{3}x + 2$

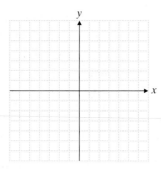

48. Find the equation of a line with a slope of $\dfrac{3}{5}$ that passes through the point $(-1, 3)$.

Chapter 8 _Solve each system by the appropriate method._

49. Substitution method
$$x + y = 17$$
$$2x - y = -5$$

50. Addition method
$$-5x + 4y = 8$$
$$2x + 3y = 6$$

51. Any method
$$2(x - 2) = 3y$$
$$6x = -3(4 + y)$$

52. Any method
$$x + \dfrac{1}{3}y = \dfrac{10}{3}$$
$$\dfrac{3}{2}x + y = 8$$

53. Is $(2, -3)$ a solution for the system
$$3x + 5y = -9$$
$$2x - 3y = 13?$$

54. A man bought three pairs of gloves and four scarves for \$53. A woman bought two pairs of the same-priced gloves and three scarves of the same-priced scarves for \$38. How much did each item cost?

Chapter 9

55. Evaluate. $\sqrt{121}$

56. Simplify. $\sqrt{125x^3y^5}$

57. Multiply and simplify.
$$\left(\sqrt{2} + \sqrt{6}\right)\left(2\sqrt{2} - 3\sqrt{6}\right)$$

58. Rationalize the denominator.
$$\dfrac{\sqrt{5} - \sqrt{3}}{\sqrt{6}}$$

59. In the right triangle with sides a, b, and c, find side c if side $a = 4$ and side $b = 6$.

60. y varies directly with x. When $y = 56$, then $x = 8$. Find y when $x = 11$.

Chapter 10 _In questions 61–64, solve for x._

61. $14x^2 + 21x = 0$

62. $2x^2 + 1 = 19$

63. $2x^2 - 4x - 5 = 0$

64. $x^2 - x + 8 = 5 + 6x$

65. Graph the equation.
$$y = x^2 + 8x + 15$$

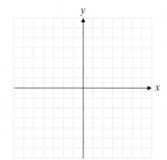

66. A rectangular box is 4 inches longer in length than in width. The area of the box is 96 square inches. Find the length and the width of the box.

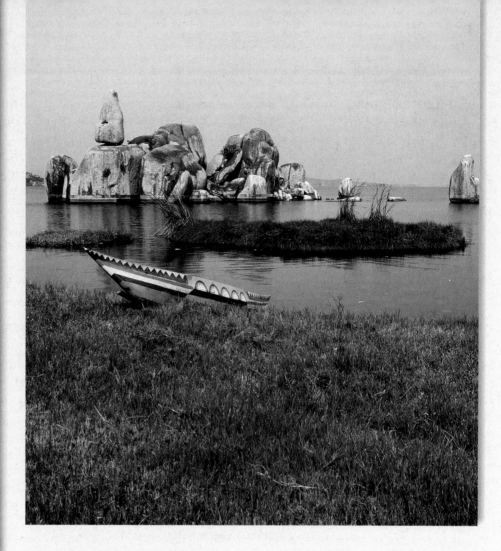

CHAPTER

0

Have you ever thought about the large lakes that cover the earth? What is the largest lake that you have visited? Some of them are more than 30,000 square miles in area. Turn to page 59 and see if you can solve some mathematical problems involving some of the larger lakes.

A Brief Review of Arithmetic Skills

0.1 SIMPLIFYING FRACTIONS

Chapter 0 is designed to give you a mental "warm-up." In this chapter you'll be able to step back a bit and tone up your math skills. This brief review of arithmetic will increase your math flexibility and give you a good running start into algebra.

1 Understanding Basic Mathematical Definitions

Whole numbers are the set of numbers 0, 1, 2, 3, 4, 5, 6, 7, ... They are used to describe whole objects, or entire quantities.

Fractions are a set of numbers that are used to describe parts of whole quantities. In the object shown in the figure there are four equal parts. The *three* of the *four* parts that are shaded are represented by the fraction $\frac{3}{4}$. In the fraction $\frac{3}{4}$ the number 3 is called the **numerator** and the number 4, the **denominator.**

$$\frac{3}{4}$$

$3 \leftarrow$ *Numerator* is on the top
$4 \leftarrow$ *Denominator* is on the bottom

The *denominator* of a fraction shows the number of equal parts in the whole and the *numerator* shows the number of these parts being talked about or being used.

Numerals are symbols we use to name numbers. There are many different numerals that can be used to describe the same number. We know that $\frac{1}{2} = \frac{2}{4}$. The fractions $\frac{1}{2}$ and $\frac{2}{4}$ both describe the same number.

Usually, we find it more useful to use fractions that are simplified. A fraction is considered to be in **simplest form** or **reduced form** when the numerator (top) and the denominator (bottom) can both be divided exactly by no number other than 1, and the denominator is greater than 1.

$$\frac{1}{2} \text{ is in simplest form.}$$

$$\frac{2}{4} \text{ is } not \text{ in simplest form, since the numerator and the denominator can both be divided by 2.}$$

If you get the answer $\frac{2}{4}$ to a problem, you should state it in simplest form, $\frac{1}{2}$. The process of changing $\frac{2}{4}$ to $\frac{1}{2}$ is called **simplifying** or **reducing** the fraction.

2 Simplifying Fractions to Lowest Terms Using Prime Numbers

Natural numbers or **counting numbers** are the set of whole numbers excluding 0. Thus the natural numbers are the numbers 1, 2, 3, 4, 5, 6, ...

When two or more numbers are multiplied, each number that is multiplied is called a **factor.** For example, when we write $3 \times 7 \times 5$, each of the numbers 3, 7, and 5 is called a factor.

Prime numbers are all natural numbers greater than 1 whose only natural number factors are 1 and itself. The number 5 is prime. The only natural number factors of 5 are 5 and 1.

$$5 = 5 \times 1$$

The number 6 is not prime. The natural number factors of 6 are 3 and 2 or 6 and 1.

$$6 = 3 \times 2 \qquad 6 = 6 \times 1$$

The first 15 prime numbers are

$$2, 3, 5, 7, 11, 13, 17, 19, 23, 29, 31, 37, 41, 43, 47.$$

Any natural number greater than 1 either is prime or can be written as the product of prime numbers. For example, we can take each of the numbers 12, 30, 14, 19, and 29 and either indicate that they are prime or, if they are not prime, write them as the product of prime numbers. We write as follows:

$$12 = 2 \times 2 \times 3 \qquad 30 = 2 \times 3 \times 5 \qquad 14 = 2 \times 7$$

19 is a prime number.   29 is a prime number.

To reduce a fraction, we use prime numbers to factor the numerator and the denominator. Write each part of the fraction (numerator and denominator) as a product of prime numbers. Note any *factors* that appear in both the *numerator* (top) and *denominator* (bottom) of the fraction. If we divide numerator and denominator by these values we will obtain an equivalent fraction in *simplest form*. When the new fraction is simplified, it is said to be in **lowest terms.** Throughout this text, to *simplify* a fraction will always mean to simplify the fraction to lowest terms.

EXAMPLE 1   Simplify each fraction.   **(a)** $\dfrac{14}{21}$ **(b)** $\dfrac{15}{35}$ **(c)** $\dfrac{20}{70}$

Solution

(a) $\dfrac{14}{21} = \dfrac{\cancel{7} \times 2}{\cancel{7} \times 3} = \dfrac{2}{3}$ We factor 14 and factor 21. Then we divide numerator and denominator by 7.

(b) $\dfrac{15}{35} = \dfrac{\cancel{5} \times 3}{\cancel{5} \times 7} = \dfrac{3}{7}$ We factor 15 and factor 35. Then we divide numerator and denominator by 5.

(c) $\dfrac{20}{70} = \dfrac{2 \times \cancel{2} \times \cancel{5}}{7 \times \cancel{2} \times \cancel{5}} = \dfrac{2}{7}$ We factor 20 and factor 70. Then we divide numerator and denominator by both 2 and 5.

Practice Problem 1   Simplify.   **(a)** $\dfrac{10}{16}$ **(b)** $\dfrac{24}{36}$ **(c)** $\dfrac{36}{42}$

Sometimes when we simplify a fraction, all the prime factors in the top (numerator) are divided out. When this happens, we must remember that a 1 is left in the numerator.

EXAMPLE 2   Simplify each fraction.   **(a)** $\dfrac{7}{21}$ **(b)** $\dfrac{15}{105}$

Solution   **(a)** $\dfrac{7}{21} = \dfrac{\cancel{7} \times 1}{\cancel{7} \times 3} = \dfrac{1}{3}$ **(b)** $\dfrac{15}{105} = \dfrac{\cancel{5} \times \cancel{3} \times 1}{7 \times \cancel{5} \times \cancel{3}} = \dfrac{1}{7}$

Practice Problem 2   Simplify.   **(a)** $\dfrac{4}{12}$ **(b)** $\dfrac{25}{125}$ **(c)** $\dfrac{73}{146}$

NOTE TO STUDENT: Fully worked-out solutions to all of the Practice Problems can be found at the back of the text starting at page SP-1

If all the prime numbers in the bottom (denominator) are divided out, we do not need to leave a 1 in the denominator, since we do not need to express the answer as a fraction. The answer is then a whole number and is not usually expressed as a fraction.

EXAMPLE 3   Simplify each fraction.   **(a)** $\dfrac{35}{7}$ **(b)** $\dfrac{70}{10}$

Solution   **(a)** $\dfrac{35}{7} = \dfrac{5 \times \cancel{7}}{\cancel{7} \times 1} = 5$ **(b)** $\dfrac{70}{10} = \dfrac{7 \times \cancel{5} \times \cancel{2}}{\cancel{5} \times \cancel{2} \times 1} = 7$

Practice Problem 3   Simplify.   **(a)** $\dfrac{18}{6}$ **(b)** $\dfrac{146}{73}$ **(c)** $\dfrac{28}{7}$

Sometimes the fraction we use represents how many of a certain thing are successful. For example, if a baseball player was at bat 30 times and achieved 12 hits, we could say that he had a hit $\frac{12}{30}$ of the time. If we reduce the fraction, we could say he had a hit $\frac{2}{5}$ of the time.

EXAMPLE 4   Cindy got 48 out of 56 questions correct on a test. Write this as a fraction.

Solution   Express as a fraction in simplest form the number of correct responses out of the total number of questions on the test.

$$48 \text{ out of } 56 \;\rightarrow\; \frac{48}{56} = \frac{6 \times \cancel{8}}{7 \times \cancel{8}} = \frac{6}{7}$$

Cindy answered the questions correctly $\frac{6}{7}$ of the time.

Practice Problem 4   The major league pennant winner in 1917 won 56 games out of 154 games played. Express as a fraction in simplest form the number of games won in relation to the number of games played.

NOTE TO STUDENT: Fully worked-out solutions to all of the Practice Problems can be found at the back of the text starting at page SP-1

The number *one* can be expressed as $1, \frac{1}{1}, \frac{2}{2}, \frac{6}{6}, \frac{8}{8}$, and so on, since

$$1 = \frac{1}{1} = \frac{2}{2} = \frac{6}{6} = \frac{8}{8}.$$

We say that these numerals are *equivalent ways* of writing the number *one* because they all express the same quantity even though they appear to be different.

SIDELIGHT: The Multiplicative Identity
When we simplify fractions, we are actually using the fact that we can multiply any number by 1 without changing the value of that number. (Mathematicians call the number 1 the **multiplicative identity** because it leaves any number it multiplies with the same identical value as before.)
    Let's look again at one of the previous examples.

$$\frac{14}{21} = \frac{7 \times 2}{7 \times 3} = \frac{7}{7} \times \frac{2}{3} = 1 \times \frac{2}{3} = \frac{2}{3}$$

So we see that

$$\frac{14}{21} = \frac{2}{3}$$

When we simplify fractions, we are using this property of multiplying by 1.

3 **Converting Between Improper Fractions and Mixed Numbers**

If the numerator is less than the denominator, the fraction is a **proper fraction**. A proper fraction is used to describe a quantity smaller than a whole.
    Fractions can also be used to describe quantities larger than a whole. The following figure shows two bars that are equal in size. Each bar is divided into 5 equal pieces. The first bar is shaded in completely. The second bar has 2 of the 5 pieces shaded in.
    The shaded-in region can be represented by $\frac{7}{5}$ since 7 of the pieces (each of which is $\frac{1}{5}$ of a whole box) are shaded. The fraction $\frac{7}{5}$ is called an improper fraction. An **improper fraction** is one in which the numerator is larger than or equal to the denominator.

The shaded-in region can also be represented by 1 whole added to $\frac{2}{5}$ of a whole, or $1 + \frac{2}{5}$. This is written as $1\frac{2}{5}$. The fraction $1\frac{2}{5}$ is called a mixed number. A **mixed number** consists of a whole number added to a proper fraction (the numerator is smaller than the denominator). The addition is understood but not written. When we write $1\frac{2}{5}$, it represents $1 + \frac{2}{5}$. The numbers $1\frac{7}{8}$, $2\frac{3}{4}$, $8\frac{1}{3}$, and $126\frac{1}{10}$ are all mixed numbers. From the preceding figure it seems clear that $\frac{7}{5} = 1\frac{2}{5}$. This suggests that we can change from one form to the other without changing the value of the fraction.

From a picture it is easy to see how to *change improper fractions to mixed numbers*. For example, suppose we start with the fraction $\frac{11}{3}$ and represent it by the following figure (where 11 of the pieces, each of which is $\frac{1}{3}$ of a box, are shaded). We see that $\frac{11}{3} = 3\frac{2}{3}$, since 3 whole boxes and $\frac{2}{3}$ of a box are shaded.

Changing Improper Fractions to Mixed Numbers You can follow the same procedure without a picture. For example, to change $\frac{11}{3}$ to a mixed number, we can do the following:

$$\frac{11}{3} = \frac{3}{3} + \frac{3}{3} + \frac{3}{3} + \frac{2}{3} \qquad \text{By the rule for adding fractions (which is discussed in detail in Section 0.2)}$$

$$= 1 + 1 + 1 + \frac{2}{3} \qquad \text{Write 1 in place of } \frac{3}{3}, \text{ since } \frac{3}{3} = 1.$$

$$= 3 + \frac{2}{3} \qquad \text{Write 3 in place of } 1 + 1 + 1.$$

$$= 3\frac{2}{3} \qquad \text{Use the notation for mixed numbers.}$$

Now that you know how to change improper fractions to mixed numbers and why the procedure works, here is a shorter method.

> ## TO CHANGE AN IMPROPER FRACTION TO A MIXED NUMBER
>
> 1. Divide the denominator into the numerator.
> 2. The result is the whole-number part of the mixed number.
> 3. The remainder from the division will be the numerator of the fraction. The denominator of the fraction remains unchanged.

We can write the fraction as a division statement and divide. The arrows show how to write the mixed number.

$$\frac{7}{5} \qquad \begin{array}{r} 1 \\ 5\overline{)7} \\ \underline{5} \\ 2 \end{array} \text{ Remainder}$$

Whole-number part $\longrightarrow 1\frac{2}{5} \longleftarrow$ Numerator of fraction

Thus, $\frac{7}{5} = 1\frac{2}{5}$.

$$\frac{11}{3} \qquad \begin{array}{r} 3 \\ 3\overline{)11} \\ \underline{9} \\ 2 \end{array} \text{ Remainder}$$

Whole-number part $\longrightarrow 3\frac{2}{3} \longleftarrow$ Numerator of fraction

Thus, $\frac{11}{3} = 3\frac{2}{3}$.

Sometimes the remainder is 0. In this case, the improper fraction changes to a whole number.

EXAMPLE 5 Change to a mixed number or to a whole number.

(a) $\dfrac{7}{4}$ **(b)** $\dfrac{15}{3}$

Solution

(a) $\dfrac{7}{4} = 7 \div 4$

$$\begin{array}{r} 1 \\ 4{\overline{\smash{)}7}} \\ \underline{4} \\ 3 \end{array}$$ Remainder

Thus $\dfrac{7}{4} = 1\dfrac{3}{4}$

(b) $\dfrac{15}{3} = 15 \div 3$

$$\begin{array}{r} 5 \\ 3{\overline{\smash{)}15}} \\ \underline{15} \\ 0 \end{array}$$ Remainder

Thus $\dfrac{15}{3} = 5$

NOTE TO STUDENT: *Fully worked-out solutions to all of the Practice Problems can be found at the back of the text starting at page SP-1*

Practice Problem 5 Change to a mixed number or to a whole number.

(a) $\dfrac{12}{7}$ **(b)** $\dfrac{20}{5}$

Changing Mixed Numbers to Improper Fractions It is not difficult to see how to change mixed numbers to improper fractions. Suppose that you wanted to write $2\frac{2}{3}$ as an improper fraction.

$$2\dfrac{2}{3} = 2 + \dfrac{2}{3} \qquad \text{The meaning of mixed number notation}$$

$$= 1 + 1 + \dfrac{2}{3} \qquad \text{Since } 1 + 1 = 2$$

$$= \dfrac{3}{3} + \dfrac{3}{3} + \dfrac{2}{3} \qquad \text{Since } 1 = \frac{3}{3}$$

When we draw a picture of $\frac{3}{3} + \frac{3}{3} + \frac{2}{3}$, we have this figure:

If we count the shaded parts, we see that

$$\dfrac{3}{3} + \dfrac{3}{3} + \dfrac{2}{3} = \dfrac{8}{3}. \qquad \text{Thus} \quad 2\dfrac{2}{3} = \dfrac{8}{3}.$$

Now that you have seen how this change can be done, here is a shorter method.

TO CHANGE A MIXED NUMBER TO AN IMPROPER FRACTION

1. Multiply the whole number by the denominator.
2. Add this to the numerator. The result is the new numerator. The denominator does not change.

EXAMPLE 6   Change to an improper fraction.

(a) $3\dfrac{1}{7}$

(b) $5\dfrac{4}{5}$

Solution

(a) $3\dfrac{1}{7} = \dfrac{(3 \times 7) + 1}{7} = \dfrac{21 + 1}{7} = \dfrac{22}{7}$

(b) $5\dfrac{4}{5} = \dfrac{(5 \times 5) + 4}{5} = \dfrac{25 + 4}{5} = \dfrac{29}{5}$

Practice Problem 6   Change to an improper fraction.

(a) $3\dfrac{2}{5}$ **(b)** $1\dfrac{3}{7}$ **(c)** $2\dfrac{6}{11}$ **(d)** $4\dfrac{2}{3}$

④ Changing a Fraction to an Equivalent Fraction with a Given Denominator

Fractions can be changed to an equivalent fraction with a different denominator by multiplying both numerator and denominator by the same number.

$$\frac{5}{6} = \frac{5}{6} \times 1 = \frac{5}{6} \times \frac{2}{2} = \frac{5 \times 2}{6 \times 2} = \frac{10}{12} \qquad \frac{3}{7} = \frac{3}{7} \times 1 = \frac{3}{7} \times \frac{3}{3} = \frac{3 \times 3}{7 \times 3} = \frac{9}{21}$$

So $\dfrac{5}{6}$ is equivalent to $\dfrac{10}{12}$. $\dfrac{3}{7}$ is equivalent to $\dfrac{9}{21}$.

We often multiply in this way to obtain an equivalent fraction with a *particular denominator*.

EXAMPLE 7   Find the missing number.

(a) $\dfrac{3}{5} = \dfrac{?}{25}$

(b) $\dfrac{3}{7} = \dfrac{?}{21}$

(c) $\dfrac{2}{9} = \dfrac{?}{36}$

Solution

(a) $\dfrac{3}{5} = \dfrac{?}{25}$ Observe that we need to multiply the denominator by 5 to obtain 25. So we multiply the numerator 3 by 5 also.

$\dfrac{3 \times 5}{5 \times 5} = \dfrac{15}{25}$ The desired numerator is 15.

(b) $\dfrac{3}{7} = \dfrac{?}{21}$ Observe that $7 \times 3 = 21$. We need to multiply the numerator by 3 to get the new numerator.

$\dfrac{3 \times 3}{7 \times 3} = \dfrac{9}{21}$ The desired numerator is 9.

(c) $\dfrac{2}{9} = \dfrac{?}{36}$ Observe that $9 \times 4 = 36$. We need to multiply the numerator by 4 to get the new numerator.

$\dfrac{2 \times 4}{9 \times 4} = \dfrac{8}{36}$ The desired numerator is 8.

Practice Problem 7   Find the missing number.

(a) $\dfrac{3}{8} = \dfrac{?}{24}$ **(b)** $\dfrac{5}{6} = \dfrac{?}{30}$ **(c)** $\dfrac{2}{7} = \dfrac{?}{56}$

0.1 EXERCISES

Student Solutions Manual | CD/Video | PH Math Tutor Center | MathXL®Tutorials on CD | MathXL® | MyMathLab® | Interactmath.com

Verbal and Writing Skills

1. In the fraction $\frac{12}{13}$, what number is the numerator?

2. In the fraction $\frac{13}{17}$, what is the denominator?

3. What is a factor? Give an example.

4. Give some examples of the number 1 written as a fraction.

5. Draw a diagram to illustrate $2\frac{2}{3}$.

6. Draw a diagram to illustrate $3\frac{3}{4}$.

Simplify each fraction.

7. $\frac{18}{24}$ **8.** $\frac{20}{35}$ **9.** $\frac{12}{36}$ **10.** $\frac{12}{48}$ **11.** $\frac{60}{12}$ **12.** $\frac{45}{15}$

Change to a mixed number.

13. $\frac{17}{6}$ **14.** $\frac{19}{5}$ **15.** $\frac{111}{9}$ **16.** $\frac{124}{8}$ **17.** $\frac{38}{7}$ **18.** $\frac{41}{6}$

Change to an improper fraction.

19. $3\frac{1}{5}$ **20.** $2\frac{6}{7}$ **21.** $6\frac{3}{5}$ **22.** $5\frac{3}{8}$ **23.** $\frac{72}{9}$ **24.** $\frac{78}{6}$

Find the missing numerator.

25. $\frac{3}{11} = \frac{?}{44}$ **26.** $\frac{5}{7} = \frac{?}{28}$ **27.** $\frac{3}{5} = \frac{?}{35}$ **28.** $\frac{5}{9} = \frac{?}{45}$ **29.** $\frac{4}{13} = \frac{?}{39}$ **30.** $\frac{13}{17} = \frac{?}{51}$

Applications

Solve.

31. *Basketball* Charles Barkley of the Phoenix Suns once scored 1560 points during 68 games played during the season. Express as a mixed number in simplified form how many points he averaged per game.

32. *Kentucky Derby* In 2002, 417 horses were nominated to compete in the Kentucky Derby. Only 18 horses were actually chosen to compete in the Derby. What simplified fraction shows what portion of nominated horses actually competed in the Derby?

33. *Income Tax* Last year, my parents had a combined income of $64,000. They paid $13,200 in federal income taxes. What simplified fraction shows how much my parents spent on their federal taxes?

34. *Transfer Students* The University of California system accepted 14,664 out of 18,330 California Community College students who applied as transfers in the fall of 2003. What simplified fraction shows what portion of applications were accepted?

Trail Mix   *The following chart gives the recipe for a trail mix.*

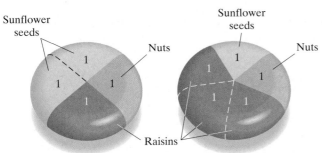

35. What part of the premium blend is nuts?

36. What part of the high-energy blend is raisins?

37. What part of the premium blend is not sunflower seeds?

38. What part of the high-energy blend does not contain nuts?

Baseball   *The following chart provides some statistics for the North Andover Knights.*

39. Determine how often each player hit the ball based on the number of times at bat. Write each answer as a reduced fraction.

40. Which player was the best hitter?

Fish Catch   *The following chart provides some statistics about the fish catch in the United States.*

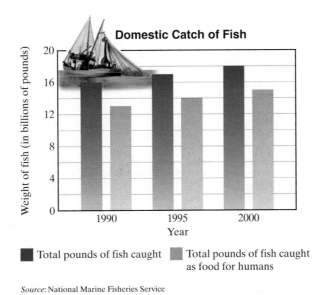

Source: National Marine Fisheries Service

41. What fractional part of the fish catch in 1990 was used as food for humans?

42. What fractional part of the fish catch in 2000 was used as food for humans?

0.2 ADDING AND SUBTRACTING FRACTIONS

Student Learning
Objectives

After studying this section, you
will be able to:

1. Add or subtract fractions
   with a common
   denominator.

2. Use prime factors to find
   the least common
   denominator of two or
   more fractions.

3. Add or subtract fractions
   with different
   denominators.

4. Add or subtract mixed
   numbers.

1 Adding or Subtracting Fractions with a Common Denominator

If fractions have the same denominator, the numerators may be added or subtracted. The denominator remains the same.

TO ADD OR SUBTRACT TWO FRACTIONS WITH A COMMON DENOMINATOR

1. Add or subtract the numerators.
2. Keep the same (common) denominator.
3. Simplify the answer whenever possible.

EXAMPLE 1

Add the fractions. Simplify your answer whenever possible.

(a) $\dfrac{5}{7} + \dfrac{1}{7}$ (b) $\dfrac{2}{3} + \dfrac{1}{3}$ (c) $\dfrac{1}{8} + \dfrac{3}{8} + \dfrac{2}{8}$ (d) $\dfrac{3}{5} + \dfrac{4}{5}$

Solution

(a) $\dfrac{5}{7} + \dfrac{1}{7} = \dfrac{5+1}{7} = \dfrac{6}{7}$

(b) $\dfrac{2}{3} + \dfrac{1}{3} = \dfrac{2+1}{3} = \dfrac{3}{3} = 1$

(c) $\dfrac{1}{8} + \dfrac{3}{8} + \dfrac{2}{8} = \dfrac{1+3+2}{8} = \dfrac{6}{8} = \dfrac{3}{4}$

(d) $\dfrac{3}{5} + \dfrac{4}{5} = \dfrac{3+4}{5} = \dfrac{7}{5} = 1\dfrac{2}{5}$

Practice Problem 1 Add.

(a) $\dfrac{3}{6} + \dfrac{2}{6}$ (b) $\dfrac{3}{11} + \dfrac{8}{11}$ (c) $\dfrac{1}{8} + \dfrac{2}{8} + \dfrac{1}{8}$ (d) $\dfrac{5}{9} + \dfrac{8}{9}$

EXAMPLE 2

Subtract the fractions. Simplify your answer whenever possible.

(a) $\dfrac{9}{11} - \dfrac{2}{11}$ (b) $\dfrac{5}{6} - \dfrac{1}{6}$

Solution

(a) $\dfrac{9}{11} - \dfrac{2}{11} = \dfrac{9-2}{11} = \dfrac{7}{11}$

(b) $\dfrac{5}{6} - \dfrac{1}{6} = \dfrac{5-1}{6} = \dfrac{4}{6} = \dfrac{2}{3}$

Practice Problem 2 Subtract.

(a) $\dfrac{11}{13} - \dfrac{6}{13}$ (b) $\dfrac{8}{9} - \dfrac{2}{9}$

NOTE TO STUDENT: Fully worked-out solutions to all of the Practice Problems can be found at the back of the text starting at page SP-1

Although adding and subtracting fractions with the same denominator is fairly simple, most problems involve fractions that do not have a common denominator. Fractions and mixed numbers such as halves, fourths, and eighths are often used. To add or subtract such fractions, we begin by finding a common denominator.

Using Prime Factors to Find the Least Common Denominator of Two or More Fractions

Before you can add or subtract fractions, they must have the same denominator. To save work, we select the smallest possible common denominator. This is called the **least common denominator** or LCD (also known as the *lowest common denominator*).

The LCD of two or more fractions is the smallest whole number that is exactly divisible by each denominator of the fractions.

EXAMPLE 3   Find the LCD. $\frac{2}{3}$ and $\frac{1}{4}$

Solution   The numbers are small enough to find the LCD by inspection. The LCD is 12, since 12 is exactly divisible by 4 and by 3. There is no smaller number that is exactly divisible by 4 and 3.

Practice Problem 3   Find the LCD. $\frac{1}{8}$ and $\frac{5}{12}$

In some cases, the LCD cannot easily be determined by inspection. If we write each denominator as the product of prime factors, we will be able to find the LCD. We will use $(\cdot)$ to indicate multiplication. For example, $30 = 2 \cdot 3 \cdot 5$. This means $30 = 2 \times 3 \times 5$.

PROCEDURE TO FIND THE LCD USING PRIME FACTORS

1. Write each denominator as the product of prime factors.

2. The LCD is a product containing each different factor.

3. If a factor occurs more than once in any one denominator, the LCD will contain that factor repeated the greatest number of times that it occurs in any one denominator.

EXAMPLE 4   Find the LCD of $\frac{5}{6}$ and $\frac{1}{15}$ by this new procedure.

Solution

$$6 = 2 \cdot 3$$
$$15 = \quad 3 \cdot 5$$

$$LCD = 2 \cdot 3 \cdot 5$$
$$LCD = 2 \cdot 3 \cdot 5 = 30$$

Write each denominator as the product of prime factors.

The LCD is a product containing each different prime factor. The different factors are 2, 3, and 5, and each factor appears at most once in any one denominator.

Practice Problem 4   Use prime factors to find the LCD of $\frac{8}{35}$ and $\frac{6}{15}$.

Great care should be used to determine the LCD in the case of repeated factors.

EXAMPLE 5 Find the LCD of $\frac{4}{27}$ and $\frac{5}{18}$.

Solution

$$27 = 3 \cdot 3 \cdot 3$$

Write each denominator as the product of prime factors. We observe that the factor 3 occurs three times in the factorization of 27.

$$18 = \quad 3 \cdot 3 \cdot 2$$
$$LCD = 3 \cdot 3 \cdot 3 \cdot 2$$
$$LCD = 3 \cdot 3 \cdot 3 \cdot 2 = 54$$

The LCD is a product containing each different factor. The factor 3 *occurred most* in the factorization of 27, where it occurred *three* times. Thus the LCD will be the product of *three* 3's and *one* 2.

NOTE TO STUDENT: *Fully worked-out solutions to all of the Practice Problems can be found at the back of the text starting at page SP-1*

Practice Problem 5 Find the LCD of $\frac{5}{12}$ and $\frac{7}{30}$.

EXAMPLE 6 Find the LCD of $\frac{5}{12}$, $\frac{1}{15}$, and $\frac{7}{30}$.

Solution

$$12 = 2 \cdot 2 \cdot 3$$

Write each denominator as the product of prime factors. Notice that the only repeated factor is 2, which occurs twice in the factorization of 12.

$$15 = \quad 3 \cdot 5$$
$$30 = \quad 2 \cdot 3 \cdot 5$$
$$LCD = 2 \cdot 2 \cdot 3 \cdot 5$$
$$LCD = 2 \cdot 2 \cdot 3 \cdot 5 = 60$$

The LCD is the product of each different factor with the factor 2 appearing twice since it occurred twice in one denominator.

Practice Problem 6 Find the LCD of $\frac{2}{27}$, $\frac{1}{18}$, and $\frac{5}{12}$.

③ Adding or Subtracting Fractions with Different Denominators

Before you can add or subtract them, fractions must have the same denominator. Using the LCD will make your work easier. First you must find the LCD. Then change each fraction to a fraction that has the LCD as the denominator. Sometimes one of the fractions will already have the LCD as the denominator. Once all the fractions have the same denominator, you can add or subtract. Be sure to simplify the fraction in your answer if this is possible.

TO ADD OR SUBTRACT FRACTIONS THAT DO NOT HAVE A COMMON DENOMINATOR

1. Find the LCD of the fractions.
2. Change each fraction to an equivalent fraction with the LCD for a denominator.
3. Add or subtract the fractions.
4. Simplify the answer whenever possible.

Let us return to the two fractions of Example 3. We have previously found that the LCD is 12.

EXAMPLE 7   Bob picked $\frac{2}{3}$ of a bushel of apples on Monday and $\frac{1}{4}$ of a bushel of apples on Tuesday. How much did he have in total?

Solution   To solve this problem we need to add $\frac{2}{3}$ and $\frac{1}{4}$, but before we can do so, we must change $\frac{2}{3}$ and $\frac{1}{4}$ to fractions with the same denominator. We change each fraction to an equivalent fraction with a common denominator of 12, the LCD.

$$\frac{2}{3} = \frac{?}{12} \qquad \frac{2}{3} \times \frac{4}{4} = \frac{8}{12} \quad \text{so} \quad \frac{2}{3} = \frac{8}{12}$$

$$\frac{1}{4} = \frac{?}{12} \qquad \frac{1}{4} \times \frac{3}{3} = \frac{3}{12} \quad \text{so} \quad \frac{1}{4} = \frac{3}{12}$$

Then we rewrite the problem with common denominators and add.

$$\frac{2}{3} + \frac{1}{4} = \frac{8}{12} + \frac{3}{12} = \frac{8+3}{12} = \frac{11}{12}$$

In total Bob picked $\frac{11}{12}$ of a bushel of apples.

Practice Problem 7   Carol planted corn in $\frac{5}{12}$ of the farm fields at the Old Robinson Farm. Connie planted soybeans in $\frac{1}{8}$ of the farm fields. What fractional part of the farm fields of the Old Robinson Farm was planted in corn or soybeans?

Sometimes one of the denominators is the LCD. In such cases the fraction that has the LCD for the denominator will not need to be changed. If every other denominator divides into the largest denominator, the largest denominator is the LCD.

EXAMPLE 8   Find the LCD and then add. $\frac{3}{5} + \frac{7}{20} + \frac{1}{2}$

Solution   We can see by inspection that both 5 and 2 divide exactly into 20. Thus 20 is the LCD. Now add.

$$\frac{3}{5} + \frac{7}{20} + \frac{1}{2}$$

We change $\frac{3}{5}$ and $\frac{1}{2}$ to equivalent fractions with a common denominator of 20, the LCD.

$$\frac{3}{5} = \frac{?}{20} \qquad \frac{3}{5} \times \frac{4}{4} = \frac{12}{20} \quad \text{so} \quad \frac{3}{5} = \frac{12}{20}$$

$$\frac{1}{2} = \frac{?}{20} \qquad \frac{1}{2} \times \frac{10}{10} = \frac{10}{20} \quad \text{so} \quad \frac{1}{2} = \frac{10}{20}$$

Then we rewrite the problem with common denominators and add.

$$\frac{3}{5} + \frac{7}{20} + \frac{1}{2} = \frac{12}{20} + \frac{7}{20} + \frac{10}{20} = \frac{12+7+10}{20} = \frac{29}{20} \quad \text{or} \quad 1\frac{9}{20}$$

Practice Problem 8   Find the LCD and add. $\frac{3}{5} + \frac{4}{25} + \frac{1}{10}$

Now we turn to examples where the selection of the LCD is not so obvious. In Examples 9 through 11 we will use the prime factorization method to find the LCD.

EXAMPLE 9 Add. $\dfrac{7}{18} + \dfrac{5}{12}$

Solution First we find the LCD.

$$18 = 3 \cdot 3 \cdot 2$$
$$12 = \ \ 3 \cdot 2 \cdot 2$$
$$\text{LCD} = 3 \cdot 3 \cdot 2 \cdot 2 = 9 \cdot 4 = 36$$

Now we change $\dfrac{7}{18}$ and $\dfrac{5}{12}$ to equivalent fractions that have the LCD.

$$\frac{7}{18} = \frac{?}{36} \qquad \frac{7}{18} \times \frac{2}{2} = \frac{14}{36}$$

$$\frac{5}{12} = \frac{?}{36} \qquad \frac{5}{12} \times \frac{3}{3} = \frac{15}{36}$$

Now we add the fractions.

$$\frac{7}{18} + \frac{5}{12} = \frac{14}{36} + \frac{15}{36} = \frac{29}{36} \qquad \text{This fraction cannot be simplified.}$$

NOTE TO STUDENT: Fully worked-out solutions to all of the Practice Problems can be found at the back of the text starting at page SP-1

Practice Problem 9 Add. $\dfrac{1}{49} + \dfrac{3}{14}$

EXAMPLE 10 Subtract. $\dfrac{25}{48} - \dfrac{5}{36}$

Solution First we find the LCD.

$$48 = 2 \cdot 2 \cdot 2 \cdot 2 \cdot 3$$
$$36 = \ \ \ \ 2 \cdot 2 \cdot 3 \cdot 3$$
$$\text{LCD} = 2 \cdot 2 \cdot 2 \cdot 2 \cdot 3 \cdot 3 = 16 \cdot 9 = 144$$

Now we change $\dfrac{25}{48}$ and $\dfrac{5}{36}$ to equivalent fractions that have the LCD.

$$\frac{25}{48} = \frac{?}{144} \qquad \frac{25}{48} \times \frac{3}{3} = \frac{75}{144}$$

$$\frac{5}{36} = \frac{?}{144} \qquad \frac{5}{36} \times \frac{4}{4} = \frac{20}{144}$$

Now we subtract the fractions.

$$\frac{25}{48} - \frac{5}{36} = \frac{75}{144} - \frac{20}{144} = \frac{55}{144} \qquad \text{This fraction cannot be simplified.}$$

Practice Problem 10 Subtract. $\dfrac{1}{12} - \dfrac{1}{30}$

EXAMPLE 11   Combine. $\dfrac{1}{5} + \dfrac{1}{6} - \dfrac{3}{10}$

Solution   First we find the LCD.

$$
\begin{aligned}
5 &= 5 \\
6 &= \quad 2 \cdot 3 \\
10 &= 5 \cdot 2 \\
\end{aligned}
$$
$$\downarrow \ \downarrow \ \downarrow$$
$$\text{LCD} = 5 \cdot 2 \cdot 3 = 10 \cdot 3 = 30$$

Now we change $\frac{1}{5}, \frac{1}{6}$, and $\frac{3}{10}$ to equivalent fractions that have the LCD for a denominator.

$$\frac{1}{5} = \frac{?}{30} \qquad \frac{1}{5} \times \frac{6}{6} = \frac{6}{30}$$

$$\frac{1}{6} = \frac{?}{30} \qquad \frac{1}{6} \times \frac{5}{5} = \frac{5}{30}$$

$$\frac{3}{10} = \frac{?}{30} \qquad \frac{3}{10} \times \frac{3}{3} = \frac{9}{30}$$

Now we combine the three fractions.

$$\frac{1}{5} + \frac{1}{6} - \frac{3}{10} = \frac{6}{30} + \frac{5}{30} - \frac{9}{30} = \frac{2}{30} = \frac{1}{15}$$

Note the important step of simplifying the fraction to obtain the final answer.

Practice Problem 11   Combine. $\dfrac{2}{3} + \dfrac{3}{4} - \dfrac{3}{8}$

4 Adding or Subtracting Mixed Numbers

If the problem you are adding or subtracting has mixed numbers, change them to improper fractions first and then combine (add or subtract). As a convention in this book, if the original problem contains mixed numbers, express the result as a mixed number rather than as an improper fraction.

EXAMPLE 12   Combine. Simplify your answer whenever possible.

(a) $5\dfrac{1}{2} + 2\dfrac{1}{3}$ **(b)** $2\dfrac{1}{5} - 1\dfrac{3}{4}$ **(c)** $1\dfrac{5}{12} + \dfrac{7}{30}$

Solution
(a) First we change the mixed numbers to improper fractions.

$$5\frac{1}{2} = \frac{5 \times 2 + 1}{2} = \frac{11}{2} \qquad 2\frac{1}{3} = \frac{2 \times 3 + 1}{3} = \frac{7}{3}$$

Next we change each fraction to an equivalent form with a common denominator of 6.

$$\frac{11}{2} = \frac{?}{6} \qquad \frac{11}{2} \times \frac{3}{3} = \frac{33}{6}$$

$$\frac{7}{3} = \frac{?}{6} \qquad \frac{7}{3} \times \frac{2}{2} = \frac{14}{6}$$

Finally, we add the two fractions and change our answer to a mixed number.

$$\frac{33}{6} + \frac{14}{6} = \frac{47}{6} = 7\frac{5}{6}$$

Thus $5\frac{1}{2} + 2\frac{1}{3} = 7\frac{5}{6}$.

(b) First we change the mixed numbers to improper fractions.

$$2\frac{1}{5} = \frac{2 \times 5 + 1}{5} = \frac{11}{5} \qquad 1\frac{3}{4} = \frac{1 \times 4 + 3}{4} = \frac{7}{4}$$

Next we change each fraction to an equivalent form with a common denominator of 20.

$$\frac{11}{5} = \frac{?}{20} \qquad \frac{11}{5} \times \frac{4}{4} = \frac{44}{20}$$

$$\frac{7}{4} = \frac{?}{20} \qquad \frac{7}{4} \times \frac{5}{5} = \frac{35}{20}$$

Now we subtract the two fractions.

$$\frac{44}{20} - \frac{35}{20} = \frac{9}{20}$$

Thus $2\frac{1}{5} - 1\frac{3}{4} = \frac{9}{20}$.

Note: It is not necessary to use these exact steps to add and subtract mixed numbers. If you know another method and can use it to obtain the correct answers, it is all right to continue to use that method throughout this chapter.

(c) Now we add $1\frac{5}{12} + \frac{7}{30}$.

The LCD of 12 and 30 is 60. Why? Change the mixed number to an improper fraction. Then change each fraction to an equivalent form with a common denominator.

$$1\frac{5}{12} = \frac{17}{12} \times \frac{5}{5} = \frac{85}{60} \qquad \frac{7}{30} \times \frac{2}{2} = \frac{14}{60}$$

Then add the fractions, simplify, and write the answer as a mixed number.

$$\frac{85}{60} + \frac{14}{60} = \frac{99}{60} = \frac{33}{20} = 1\frac{13}{20}$$

Thus $1\frac{5}{12} + \frac{7}{30} = 1\frac{13}{20}$.

NOTE TO STUDENT: Fully worked-out solutions to all of the Practice Problems can be found at the back of the text starting at page SP-1

Practice Problem 12   Combine.

(a) $1\frac{2}{3} + 2\frac{4}{5}$

(b) $5\frac{1}{4} - 2\frac{2}{3}$

▲ **EXAMPLE 13**   Manuel is enclosing a triangular-shaped exercise yard for his new dog. He wants to determine how many feet of fencing he will need. The sides of the yard measure $20\frac{3}{4}$ feet, $15\frac{1}{2}$ feet, and $18\frac{1}{8}$ feet. What is the perimeter of (total distance around) the triangle?

Solution

Understand the problem. Begin by drawing a picture.

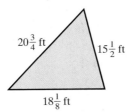

We want to add up the lengths of all three sides of the triangle. This distance around the triangle is called the **perimeter.**

$$20\frac{3}{4} + 15\frac{1}{2} + 18\frac{1}{8} = \frac{83}{4} + \frac{31}{2} + \frac{145}{8}$$
$$= \frac{166}{8} + \frac{124}{8} + \frac{145}{8} = \frac{435}{8} = 54\frac{3}{8} \text{ feet}$$

He will need $54\frac{3}{8}$ feet of fencing.

▲ **Practice Problem 13**   Find the perimeter of a rectangle with sides of $4\frac{1}{5}$ cm and $6\frac{1}{2}$ cm. Begin by drawing a picture. Label the picture by including the measure of *each* side.

Developing Your Study Skills

Class Attendance

A student of mathematics needs to get started in the right direction by choosing to attend class every day, beginning with the first day of class. Statistics show that class attendance and good grades go together. Classroom activities are designed to enhance learning and you must be in class to benefit from them. Vital information and explanations that can help you in understanding concepts are given each day. Do not be deceived into thinking that you can just find out from a friend what went on in class. There is no good substitute for firsthand experience. Give yourself a push in the right direction by developing the habit of going to class every day.

Class Participation

People learn mathematics through active participation, not through observation from the sidelines. If you want to do well in this course, be involved in classroom activities. Sit near the front where you can see and hear well and where your focus is on the instruction process and not on the students around you. Ask questions, be ready to contribute toward solutions, and take part in all classroom activities. Your contributions are valuable to the class and to yourself. Class participation requires an investment of yourself in the learning process, which you will find pays huge dividends.

Verbal and Writing Skills

1. Explain why the denominator 8 is the least common denominator of $\frac{3}{4}$ and $\frac{5}{8}$.

2. What must you do before you add or subtract fractions that do not have a common denominator?

Find the LCD (least common denominator) of each pair of fractions. Do not combine the fractions; only find the LCD.

3. $\frac{7}{15}$ and $\frac{11}{21}$

4. $\frac{13}{25}$ and $\frac{29}{40}$

5. $\frac{7}{10}$ and $\frac{1}{4}$

6. $\frac{3}{16}$ and $\frac{1}{24}$

7. $\frac{5}{18}$ and $\frac{7}{54}$

8. $\frac{7}{24}$ and $\frac{5}{48}$

9. $\frac{1}{2}, \frac{1}{18},$ and $\frac{13}{30}$

10. $\frac{5}{8}, \frac{3}{14},$ and $\frac{11}{16}$

Combine. Be sure to simplify your answer whenever possible.

11. $\frac{3}{8} + \frac{2}{8}$

12. $\frac{2}{9} + \frac{5}{9}$

13. $\frac{5}{14} - \frac{1}{14}$

14. $\frac{9}{20} - \frac{3}{20}$

15. $\frac{3}{8} + \frac{5}{6}$

16. $\frac{7}{10} + \frac{8}{15}$

17. $\frac{5}{7} - \frac{2}{9}$

18. $\frac{7}{8} - \frac{2}{3}$

19. $\frac{1}{3} + \frac{2}{5}$

20. $\frac{2}{7} + \frac{1}{3}$

21. $\frac{7}{18} + \frac{1}{12}$

22. $\frac{3}{10} + \frac{3}{25}$

23. $\frac{2}{3} + \frac{7}{12} + \frac{1}{4}$

24. $\frac{4}{7} + \frac{7}{9} + \frac{1}{3}$

25. $\frac{5}{36} + \frac{7}{9} - \frac{5}{12}$

26. $\frac{5}{24} + \frac{3}{8} - \frac{1}{3}$

27. $4\frac{1}{3} + 3\frac{2}{5}$

28. $3\frac{1}{8} + 2\frac{1}{6}$

29. $1\frac{5}{24} + \frac{5}{18}$

30. $6\frac{2}{3} + \frac{3}{4}$

31. $7\frac{1}{6} - 2\frac{1}{4}$

32. $7\frac{2}{5} - 3\frac{3}{4}$

33. $8\frac{5}{7} - 2\frac{1}{4}$

34. $7\frac{8}{15} - 2\frac{3}{5}$

35. $2\frac{1}{8} + 3\frac{2}{3}$

36. $3\frac{1}{7} + 4\frac{1}{3}$

37. $11\frac{1}{7} - 6\frac{5}{7}$

38. $17\frac{1}{5} - 10\frac{3}{5}$

39. $2\frac{1}{8} + 6\frac{3}{4}$

40. $3\frac{1}{4} + 4\frac{5}{8}$

Mixed Practice

41. $\frac{7}{9} + \frac{5}{6}$

42. $\frac{7}{15} + \frac{3}{20}$

43. $2\frac{1}{7} + 3\frac{11}{14}$

44. $4\frac{1}{6} + 5\frac{11}{12}$

45. $\frac{16}{21} - \frac{2}{7}$

46. $\frac{15}{24} - \frac{3}{8}$

47. $5\frac{1}{5} - 2\frac{1}{2}$

48. $6\frac{1}{3} - 4\frac{1}{4}$

Applications

49. ***Inline Skating*** Jenny and Laura went inline skating. They skated $3\frac{1}{8}$ miles on Monday, $2\frac{2}{3}$ miles on Tuesday, and $4\frac{1}{2}$ miles on Wednesday. What was their total distance for those three days?

50. ***Marathon Training*** Jomo and Eskinder were training for the New York Marathon. Their coach scheduled them to run $11\frac{1}{2}$ miles Monday, $13\frac{2}{3}$ miles Wednesday, and $21\frac{1}{8}$ miles on Friday, with Tuesday and Thursday reserved for weight training. How many miles did they run this week?

51. ***MTV Video*** Sheryl has $8\frac{1}{2}$ hours this weekend to work on her new video. She estimates that it will take $2\frac{2}{3}$ hours to lip sync the new song and $1\frac{3}{4}$ hours to learn the new dance steps. How much time will she have left over for her MTV interview?

52. ***Aquariums*** Carl bought a 20 gallon aquarium. He put $17\frac{3}{4}$ gallons of water into the aquarium, but it looked too low, so he added $1\frac{1}{4}$ more gallons of water. He then put in the artificial plants and the gravel but now the water was too high, so he siphoned off $2\frac{2}{3}$ gallons of water. How many gallons of water are now in the aquarium?

To Think About

Carpentry *Carpenters use fractions in their work. The picture below is a diagram of a spice cabinet. The symbol " means inches. Use the picture to answer exercises 53 and 54.*

53. Before you can determine where the cabinet will fit, you need to calculate the height, *A,* and the width, *B.* Don't forget to include the $\frac{1}{2}$-inch thickness of the wood where needed.

$\frac{1}{2}''$ hardwood for all framing and drawer fronts

$\frac{1}{4}''$ plywood for all drawer sides, back and bottom

54. Look at the close-up of the drawer. The width is $4\frac{9}{16}''$. In the diagram, the width of the opening for the drawer is $4\frac{5}{8}''$. What is the difference?
Why do you think the drawer is smaller than the opening?

55. ***Putting Green Care*** The Falmouth Country Club maintains the putting greens with a grass height of $\frac{7}{8}$ inch. The grass on the fairways is maintained at $2\frac{1}{2}$ inches. How much must the mower blade be lowered by a person mowing the fairways if that person will be using the same mowing machine on the putting greens?

56. ***Fairway Care*** The director of facilities maintenance at the club discovered that due to slippage in the adjustment lever that the lawn mower actually cuts the grass $\frac{1}{16}$ of an inch too long or too short on some days. What is the maximum height that the fairway grass could be after being mowed with this machine? What is the minimum height that the putting greens could be after being mowed with this machine?

Cumulative Review

57. Simplify. $\dfrac{36}{44}$

58. Change to an improper fraction. $26\dfrac{3}{5}$

1 Multiplying Fractions, Whole Numbers, and Mixed Numbers

Multiplying Fractions During a recent snowstorm, the runway at Beverly Airport was plowed. However, the plow cleared only $\frac{3}{5}$ of the width and $\frac{2}{7}$ of the length. What fraction of the total runway area was cleared? To answer this question, we need to multiply $\frac{3}{5} \times \frac{2}{7}$.

The answer is that $\frac{6}{35}$ of the total runway area was cleared.

The multiplication rule for fractions states that to multiply two fractions, we multiply the two numerators and multiply the two denominators.

> **TO MULTIPLY ANY TWO FRACTIONS**
>
> 1. Multiply the numerators.
> 2. Multiply the denominators.

EXAMPLE 1 Multiply.

(a) $\frac{3}{5} \times \frac{2}{7}$ **(b)** $\frac{1}{3} \times \frac{5}{4}$ **(c)** $\frac{7}{3} \times \frac{1}{5}$ **(d)** $\frac{6}{5} \times \frac{2}{3}$

Solution

(a) $\frac{3}{5} \times \frac{2}{7} = \frac{3 \cdot 2}{5 \cdot 7} = \frac{6}{35}$ **(b)** $\frac{1}{3} \times \frac{5}{4} = \frac{1 \cdot 5}{3 \cdot 4} = \frac{5}{12}$

(c) $\frac{7}{3} \times \frac{1}{5} = \frac{7 \cdot 1}{3 \cdot 5} = \frac{7}{15}$ **(d)** $\frac{6}{5} \times \frac{2}{3} = \frac{6 \cdot 2}{5 \cdot 3} = \frac{12}{15} = \frac{4}{5}$

Note that we must simplify this fraction.

Practice Problem 1 Multiply.

(a) $\frac{2}{7} \times \frac{5}{11}$ **(b)** $\frac{1}{5} \times \frac{7}{10}$ **(c)** $\frac{9}{5} \times \frac{1}{4}$ **(d)** $\frac{8}{9} \times \frac{3}{10}$

It is possible to avoid having to simplify a fraction at the last step. In many cases we can divide by a value that appears as a factor in both a numerator and a denominator. Often it is helpful to write a number as a product of prime factors in order to do this.

EXAMPLE 2 Multiply. **(a)** $\frac{3}{5} \times \frac{5}{7}$ **(b)** $\frac{4}{11} \times \frac{5}{2}$ **(c)** $\frac{15}{8} \times \frac{10}{27}$

Solution **(a)** $\frac{3}{5} \times \frac{5}{7} = \frac{3 \cdot 5}{5 \cdot 7} = \frac{3 \cdot \overset{1}{\cancel{5}}}{7 \cdot \cancel{5}} = \frac{3}{7}$ Note that here we divided numerator and denominator by 5.

If we factor each number, we can see the common factors.

(b) $\frac{4}{11} \times \frac{5}{2} = \frac{2 \cdot \overset{1}{\cancel{2}}}{11} \times \frac{5}{\underset{1}{\cancel{2}}} = \frac{10}{11}$ **(c)** $\frac{15}{8} \times \frac{10}{27} = \frac{\overset{1}{\cancel{3}} \cdot 5}{2 \cdot 2 \cdot \underset{1}{\cancel{2}}} \times \frac{5 \cdot \overset{1}{\cancel{2}}}{\underset{1}{\cancel{3}} \cdot 3 \cdot 3} = \frac{25}{36}$

After dividing out common factors, the resulting multiplication problem involves smaller numbers and the answers are in simplified form.

Practice Problem 2 Multiply. **(a)** $\frac{3}{5} \times \frac{4}{3}$ **(b)** $\frac{9}{10} \times \frac{5}{12}$

SIDELIGHT: Dividing Out Common Factors
Why does this method of dividing out a value that appears as a factor in both numerator and denominator work? Let's reexamine one of the examples we have solved previously.

$$\frac{3}{5} \times \frac{5}{7} = \frac{3}{\cancel{5}_{1}} \times \frac{\cancel{5}^{1}}{7} = \frac{3}{7}$$

Consider the following steps and reasons.

$$\frac{3}{5} \times \frac{5}{7} = \frac{3 \cdot 5}{5 \cdot 7} \qquad \text{Definition of multiplication of fractions.}$$

$$= \frac{5 \cdot 3}{5 \cdot 7} \qquad \begin{array}{l}\text{Change the order of the factors in the numerator, since}\\ 3 \cdot 5 = 5 \cdot 3. \text{ This is called the commutative property of}\\ \text{multiplication.}\end{array}$$

$$= \frac{5}{5} \cdot \frac{3}{7} \qquad \text{Definition of multiplication of fractions.}$$

$$= 1 \cdot \frac{3}{7} \qquad \text{Write 1 in place of } \frac{5}{5}, \text{ since 1 is another name for } \frac{5}{5}.$$

$$= \frac{3}{7} \qquad \begin{array}{l}1 \cdot \frac{3}{7} = \frac{3}{7}, \text{ since any number can be multiplied by 1 without}\\ \text{changing the value of the number.}\end{array}$$

Think about this concept. It is an important one that we will use again when we discuss rational expressions.

Multiplying a Fraction by a Whole Number   Whole numbers can be named using fractional notation. $3, \frac{9}{3}, \frac{6}{2},$ and $\frac{3}{1}$ are ways of expressing the number *three*. Therefore,

$$3 = \frac{9}{3} = \frac{6}{2} = \frac{3}{1}.$$

When we multiply a fraction by a whole number, we merely express the whole number as a fraction whose denominator is 1 and follow the multiplication rule for fractions.

EXAMPLE 3   Multiply.

(a) $7 \times \frac{3}{5}$

(b) $\frac{3}{16} \times 4$

Solution

(a) $7 \times \frac{3}{5} = \frac{7}{1} \times \frac{3}{5} = \frac{21}{5} = 4\frac{1}{5}$

(b) $\frac{3}{16} \times 4 = \frac{3}{16} \times \frac{4}{1} = \frac{3}{4 \cdot \cancel{4}} \times \frac{\cancel{4}}{1} = \frac{3}{4}$

Notice that in (b) we did not use *prime* factors to factor 16. We recognized that $16 = 4 \cdot 4$. This is a more convenient factorization of 16 for this problem. Choose the factorization that works best for each problem. If you cannot decide what is best, factor into primes.

Practice Problem 3   Multiply.

(a) $4 \times \frac{2}{7}$

(b) $12 \times \frac{3}{4}$

Multiplying Mixed Numbers   When multiplying mixed numbers, we first change them to improper fractions and then follow the multiplication rule for fractions.

EXAMPLE 4 How do we find the area of a rectangular field $3\frac{1}{3}$ miles long by $2\frac{1}{2}$ miles wide?

Solution To find the area, we multiply length times width.

$$3\frac{1}{3} \times 2\frac{1}{2} = \frac{10}{3} \times \frac{5}{2} = \frac{\cancel{2} \cdot 5}{3} \times \frac{5}{\cancel{2}} = \frac{25}{3} = 8\frac{1}{3}$$

The area is $8\frac{1}{3}$ square miles.

NOTE TO STUDENT: *Fully worked-out solutions to all of the Practice Problems can be found at the back of the text starting at page SP-1*

▲ **Practice Problem 4** Delbert Robinson had a farm that had a rectangular field that measured $5\frac{3}{5}$ miles long and $3\frac{3}{4}$ miles wide. What was the area of that field?

EXAMPLE 5 Multiply. $2\frac{2}{3} \times \frac{1}{4} \times 6$

Solution
$$2\frac{2}{3} \times \frac{1}{4} \times 6 = \frac{8}{3} \times \frac{1}{4} \times \frac{6}{1} = \frac{\cancel{4} \cdot 2}{\cancel{3}} \times \frac{1}{\cancel{4}} \times \frac{2 \cdot \cancel{3}}{1} = \frac{4}{1} = 4$$

Practice Problem 5 Multiply. $3\frac{1}{2} \times \frac{1}{14} \times 4$

② Dividing Fractions, Whole Numbers, and Mixed Numbers

Dividing Fractions To divide two fractions, we invert the second fraction (that is, the divisor) and then multiply the two fractions.

> **TO DIVIDE TWO FRACTIONS**
> 1. Invert the second fraction (that is, the divisor).
> 2. Now multiply the two fractions.

EXAMPLE 6 Divide.

(a) $\frac{1}{3} \div \frac{1}{2}$ (b) $\frac{2}{5} \div \frac{3}{10}$ (c) $\frac{2}{3} \div \frac{7}{5}$

Solution

(a) $\frac{1}{3} \div \frac{1}{2} = \frac{1}{3} \times \frac{2}{1} = \frac{2}{3}$ Note that we always invert the *second* fraction.

(b) $\frac{2}{5} \div \frac{3}{10} = \frac{2}{5} \times \frac{10}{3} = \frac{2}{\cancel{5}} \times \frac{\cancel{5} \cdot 2}{3} = \frac{4}{3} = 1\frac{1}{3}$ (c) $\frac{2}{3} \div \frac{7}{5} = \frac{2}{3} \times \frac{5}{7} = \frac{10}{21}$

Practice Problem 6 Divide.   (a) $\frac{2}{5} \div \frac{1}{3}$ (b) $\frac{12}{13} \div \frac{4}{3}$

Dividing a Fraction and a Whole Number The process of inverting the second fraction and then multiplying the two fractions should be done very carefully when one of the original values is a whole number. Remember, a whole number such as 2 is equivalent to $\frac{2}{1}$.

EXAMPLE 7   Divide.

(a) $\frac{1}{3} \div 2$

(b) $5 \div \frac{1}{3}$

Solution

(a) $\frac{1}{3} \div 2 = \frac{1}{3} \div \frac{2}{1} = \frac{1}{3} \times \frac{1}{2} = \frac{1}{6}$

(b) $5 \div \frac{1}{3} = \frac{5}{1} \div \frac{1}{3} = \frac{5}{1} \times \frac{3}{1} = \frac{15}{1} = 15$

Practice Problem 7   Divide.

(a) $\frac{3}{7} \div 6$

(b) $8 \div \frac{2}{3}$

SIDELIGHT: Number Sense

Look at the answers to the problems in Example 7. In part (a), you will notice that $\frac{1}{6}$ is less than the original number $\frac{1}{3}$. Does this seem reasonable? Let's see. If $\frac{1}{3}$ is divided by 2, it means that $\frac{1}{3}$ will be divided into two equal parts. We would expect that each part would be less than $\frac{1}{3}$. $\frac{1}{6}$ is a reasonable answer to this division problem.

In part (b), 15 is greater than the original number 5. Does this seem reasonable? Think of what $5 \div \frac{1}{3}$ means. It means that 5 will be divided into thirds. Let's think of an easier problem. What happens when we divide 1 into thirds? We get *three* thirds. We would expect, therefore, that when we divide 5 into thirds, we would get 5×3 or 15 thirds. 15 is a reasonable answer to this division problem.

Complex Fractions

Sometimes division is written in the form of a **complex fraction** with one fraction in the numerator and one fraction in the denominator. It is best to write this in standard division notation first; then complete the problem using the rule for division.

EXAMPLE 8   Divide.

(a) $\dfrac{\frac{3}{7}}{\frac{3}{5}}$

(b) $\dfrac{\frac{2}{9}}{\frac{5}{7}}$

Solution

(a) $\dfrac{\frac{3}{7}}{\frac{3}{5}} = \frac{3}{7} \div \frac{3}{5} = \frac{\cancel{3}}{7} \times \frac{5}{\cancel{3}} = \frac{5}{7}$

(b) $\dfrac{\frac{2}{9}}{\frac{5}{7}} = \frac{2}{9} \div \frac{5}{7} = \frac{2}{9} \times \frac{7}{5} = \frac{14}{45}$

Practice Problem 8   Divide.

(a) $\dfrac{\frac{3}{11}}{\frac{5}{7}}$

(b) $\dfrac{\frac{12}{5}}{\frac{8}{15}}$

SIDELIGHT: Invert and Multiply

Why does the method of "invert and multiply" work? The division rule really depends on the property that any number can be multiplied by 1 without changing the value of the number. Let's look carefully at an example of division of fractions:

$$\frac{2}{5} \div \frac{3}{7} = \frac{\frac{2}{5}}{\frac{3}{7}}$$ We can write the problem using a complex fraction.

$$= \frac{\frac{2}{5}}{\frac{3}{7}} \times 1$$ We can multiply by 1, since any number can be multiplied by 1 without changing the value of the number.

$$= \frac{\frac{2}{5}}{\frac{3}{7}} \times \frac{\frac{7}{3}}{\frac{7}{3}}$$ We write 1 in the form $\frac{\frac{7}{3}}{\frac{7}{3}}$, since any nonzero number divided by itself equals 1. We choose this value as a multiplier because it will help simplify the denominator.

$$= \frac{\frac{2}{5} \times \frac{7}{3}}{\frac{3}{7} \times \frac{7}{3}}$$ Definition of multiplication of fractions.

$$= \frac{\frac{2}{5} \times \frac{7}{3}}{1}$$ The product in the denominator equals 1.

$$= \frac{2}{5} \times \frac{7}{3}.$$

Thus we have shown that $\frac{2}{5} \div \frac{3}{7}$ is equivalent to $\frac{2}{5} \times \frac{7}{3}$ and have shown some justification for the "invert and multiply rule."

Dividing Mixed Numbers This method for division of fractions can be used with mixed numbers. However, we first must change the mixed numbers to improper fractions and then use the rule for dividing fractions.

EXAMPLE 9 Divide.   **(a)** $2\frac{1}{3} \div 3\frac{2}{3}$ **(b)** $\dfrac{2}{3\frac{1}{2}}$

Solution

(a) $2\frac{1}{3} \div 3\frac{2}{3} = \frac{7}{3} \div \frac{11}{3} = \frac{7}{\cancel{3}} \times \frac{\cancel{3}}{11} = \frac{7}{11}$

(b) $\dfrac{2}{3\frac{1}{2}} = 2 \div 3\frac{1}{2} = \frac{2}{1} \div \frac{7}{2} = \frac{2}{1} \times \frac{2}{7} = \frac{4}{7}$

NOTE TO STUDENT: *Fully worked-out solutions to all of the Practice Problems can be found at the back of the text starting at page SP-1*

Practice Problem 9 Divide.

(a) $1\frac{2}{5} \div 2\frac{1}{3}$ **(b)** $4\frac{2}{3} \div 7$ **(c)** $\dfrac{1\frac{1}{5}}{1\frac{2}{7}}$

EXAMPLE 10   A chemist has 96 fluid ounces of a solution. She pours the solution into test tubes. Each test tube holds $\frac{3}{4}$ fluid ounce. How many test tubes can she fill?

Solution   We need to divide the total number of ounces, 96, by the number of ounces in each test tube, $\frac{3}{4}$.

$$96 \div \frac{3}{4} = \frac{96}{1} \div \frac{3}{4} = \frac{96}{1} \times \frac{4}{3} = \frac{\cancel{3} \cdot 32}{1} \times \frac{4}{\cancel{3}} = \frac{128}{1} = 128$$

She will be able to fill 128 test tubes.

Pause for a moment to think about the answer. Does 128 test tubes filled with solution seem like a reasonable answer? Did you perform the correct operation?

Practice Problem 10   A chemist has 64 fluid ounces of a solution. He wishes to fill several jars, each holding $5\frac{1}{3}$ fluid ounces. How many jars can he fill?

Sometimes when solving word problems involving fractions or mixed numbers, it is helpful to solve the problem using simpler numbers first. Once you understand what operation is involved, you can go back and solve using the original numbers in the word problem.

EXAMPLE 11   A car traveled 301 miles on $10\frac{3}{4}$ gallons of gas. How many miles per gallon did it get?

Solution   Use simpler numbers: 300 miles on 10 gallons of gas. We want to find out how many miles the car traveled on 1 gallon of gas. You may want to draw a picture.

10 gallons

300 miles

Divide. $300 \div 10 = 30.$

Now use the original numbers given in the problem.

$$301 \div 10\frac{3}{4} = \frac{301}{1} \div \frac{43}{4} = \frac{301}{1} \times \frac{4}{43} = \frac{1204}{43} = 28$$

This is 28 miles to the gallon.

Practice Problem 11   A car can travel $25\frac{1}{2}$ miles on 1 gallon of gas. How many miles can a car travel on $5\frac{1}{4}$ gallons of gas? Check your answer to see if it is reasonable.

Developing Your Study Skills

Why Is Homework Necessary?

Mathematics is a set of skills that you learn by doing, not by watching. Your instructor may make solving a mathematics problem look very easy, but for you to learn the necessary skills, you must practice them over and over again, just as your instructor once had to do. There is no other way. Learning mathematics is like learning to play a musical instrument, to type, or to play a sport. No matter how much you watch someone else do it, how many books you may read on "how to" do it, or how easy it may seem to be, the key to success is to practice on a regular basis. Homework provides this practice.

0.3 EXERCISES

Student Solutions Manual · CD/Video · PH Math Tutor Center · MathXL®Tutorials on CD · MathXL® · MyMathLab® · Interactmath.com

Verbal and Writing Skills

1. Explain in your own words how to multiply two mixed numbers.

2. Explain in your own words how to divide two proper fractions.

Multiply. Simplify your answer whenever possible.

3. $\dfrac{36}{7} \times \dfrac{5}{9}$

4. $\dfrac{3}{8} \times \dfrac{24}{5}$

5. $\dfrac{17}{18} \times \dfrac{3}{5}$

6. $\dfrac{21}{22} \times \dfrac{11}{6}$

7. $\dfrac{4}{5} \times \dfrac{3}{10}$

8. $\dfrac{5}{6} \times \dfrac{7}{11}$

9. $\dfrac{24}{25} \times \dfrac{5}{2}$

10. $\dfrac{15}{24} \times \dfrac{8}{9}$

11. $9 \times \dfrac{2}{5}$

12. $\dfrac{8}{11} \times 3$

Divide. Simplify your answer whenever possible.

13. $\dfrac{8}{5} \div \dfrac{8}{3}$

14. $\dfrac{13}{9} \div \dfrac{13}{7}$

15. $\dfrac{3}{7} \div 3$

16. $\dfrac{7}{8} \div 4$

17. $10 \div \dfrac{5}{7}$

18. $18 \div \dfrac{2}{9}$

19. $\dfrac{\frac{7}{8}}{\frac{3}{4}}$

20. $\dfrac{\frac{5}{6}}{\frac{10}{13}}$

21. $\dfrac{\frac{5}{6}}{\frac{7}{9}}$

22. $\dfrac{\frac{3}{4}}{\frac{11}{12}}$

23. $1\dfrac{3}{7} \div 6\dfrac{1}{4}$

24. $4\dfrac{1}{2} \div 3\dfrac{3}{8}$

25. $3\dfrac{1}{3} \div 2\dfrac{1}{2}$

26. $5\dfrac{1}{2} \div 3\dfrac{3}{4}$

27. $\dfrac{\frac{7}{9}}{1\frac{1}{3}}$

28. $\dfrac{\frac{5}{8}}{1\frac{3}{4}}$

29. $\dfrac{\frac{2}{3}}{1\frac{1}{4}}$

30. $\dfrac{\frac{5}{6}}{2\frac{1}{2}}$

Mixed Practice

Perform the proper calculations. Reduce your answer whenever possible.

31. $\dfrac{6}{5} \times \dfrac{10}{12}$

32. $\dfrac{5}{24} \times \dfrac{18}{15}$

33. $\dfrac{5}{14} \div \dfrac{2}{7}$

34. $\dfrac{5}{6} \div \dfrac{11}{18}$

35. $10\dfrac{3}{7} \times 5\dfrac{1}{4}$

36. $10\dfrac{2}{9} \div 2\dfrac{1}{3}$

37. $2\dfrac{1}{8} \div \dfrac{1}{4}$

38. $4 \div 1\dfrac{7}{9}$

39. $6 \times 4\dfrac{2}{3}$

40. $5\dfrac{1}{2} \times 10$

41. $2\dfrac{1}{2} \times \dfrac{1}{10} \times \dfrac{3}{4}$

42. $3\dfrac{1}{3} \times \dfrac{1}{5} \times \dfrac{2}{3}$

Applications

43. *Shirt Manufacturing* A denim shirt at the Gap requires $2\dfrac{3}{4}$ yards of material. How many yards would be needed to make 26 shirts?

44. *Pullover Manufacturing* A fleece pullover requires $1\dfrac{5}{8}$ yds of material. How many yards will it take to make 18 pullovers?

45. *Mountain Biking* Jennifer rode her mountain bike for $4\dfrac{1}{5}$ miles after work. Two-thirds of the distance was over a mountain bike trail. How long is the mountain bike trail?

46. *Cross-Country Racing* Phil ran a cross-country race that was $3\dfrac{3}{8}$ miles long. One-third of that race was on a hilly nature preserve trail. How long is the nature preserve trail?

Cumulative Review

In exercises 47 and 48, simplify each fraction.

47. $\dfrac{116}{124}$

48. $\dfrac{33}{77}$

0.4 USING DECIMALS

1 Understanding the Meaning of Decimals

We can express a part of a whole as a fraction or as a decimal. A **decimal** is another way of writing a fraction whose denominator is 10, 100, 1000, and so on.

$$\frac{3}{10} = 0.3 \qquad \frac{5}{100} = 0.05 \qquad \frac{172}{1000} = 0.172 \qquad \frac{58}{10,000} = 0.0058$$

The period in decimal notation is known as the **decimal point.** The number of digits in a number to the right of the decimal point is known as the number of **decimal places** of the number. The place value of decimals is shown in the following chart.

Hundred-thousands	Ten-thousands	Thousands	Hundreds	Tens	ones	← Decimal point	Tenths	Hundredths	Thousandths	Ten-thousandths	Hundred-thousandths
100,000	10,000	1000	100	10	1	.	$\frac{1}{10}$	$\frac{1}{100}$	$\frac{1}{1000}$	$\frac{1}{10,000}$	$\frac{1}{100,000}$

EXAMPLE 1 Write each of the following decimals as a fraction. State the number of decimal places. Write out in words the way the number would be spoken.

(a) 0.6   **(b)** 0.29   **(c)** 0.527   **(d)** 1.38   **(e)** 0.00007

Solution

Decimal Form	Fraction Form	Number of Decimal Places	The Words Used to Describe the Number
(a) 0.6	$\frac{6}{10}$	one	six-tenths
(b) 0.29	$\frac{29}{100}$	two	twenty-nine hundredths
(c) 0.527	$\frac{527}{1000}$	three	five hundred twenty-seven thousandths
(d) 1.38	$1\frac{38}{100}$	two	one and thirty-eight hundredths
(e) 0.00007	$\frac{7}{100,000}$	five	seven hundred-thousandths

Practice Problem 1 Write each decimal as a fraction and in words.

(a) 0.9   **(b)** 0.09   **(c)** 0.731   **(d)** 1.371   **(e)** 0.0005

NOTE TO STUDENT: Fully worked-out solutions to all of the Practice Problems can be found at the back of the text starting at page SP-1

You have seen that a given fraction can be written in several different but equivalent ways. There are also several different equivalent ways of writing the decimal form of fractions. The decimal 0.18 can be written in the following equivalent ways:

Fractional form: $\frac{18}{100} = \frac{180}{1000} = \frac{1800}{10,000} = \frac{18,000}{100,000}$

Decimal form: $0.18 = 0.180 = 0.1800 = 0.18000.$

Thus we see that *any number of terminal zeros may be added to the right-hand side of a decimal* without changing its value.

$$0.13 = 0.1300 \qquad 0.162 = 0.162000$$

Similarly, *any number of terminal zeros may be removed from the right-hand side of a decimal* without changing its value.

2 Changing a Fraction to a Decimal

A fraction can be changed to a decimal by dividing the denominator into the numerator.

EXAMPLE 2   Write each of the following fractions as a decimal.

(a) $\dfrac{3}{4}$ **(b)** $\dfrac{21}{20}$ **(c)** $\dfrac{1}{8}$ **(d)** $\dfrac{3}{200}$

Solution

(a) $\dfrac{3}{4} = 0.75$ since
$$
\begin{array}{r}
0.75 \\
4\overline{)3.00} \\
\underline{28} \\
20 \\
\underline{20} \\
0
\end{array}
$$

(b) $\dfrac{21}{20} = 1.05$ since
$$
\begin{array}{r}
1.05 \\
20\overline{)21.00} \\
\underline{20} \\
100 \\
\underline{100} \\
0
\end{array}
$$

(c) $\dfrac{1}{8} = 0.125$ since
$$
\begin{array}{r}
0.125 \\
8\overline{)1.000} \\
\underline{8} \\
20 \\
\underline{16} \\
40 \\
\underline{40} \\
0
\end{array}
$$

(d) $\dfrac{3}{200} = 0.015$ since
$$
\begin{array}{r}
0.015 \\
200\overline{)3.000} \\
\underline{200} \\
1000 \\
\underline{1000} \\
0
\end{array}
$$

Practice Problem 2   Write as decimals.

(a) $\dfrac{3}{8}$ **(b)** $\dfrac{7}{200}$ **(c)** $\dfrac{33}{20}$

Calculator

 Fraction to Decimal

You can use a calculator to change $\frac{3}{5}$ to a decimal. Enter:

$$3 \;\boxed{\div}\; 5 \;\boxed{=}$$

The display should read

$$\boxed{0.6}$$

Try the following.

(a) $\dfrac{17}{25}$ **(b)** $\dfrac{2}{9}$

(c) $\dfrac{13}{10}$ **(d)** $\dfrac{15}{19}$

Answers:

(a) 0.68   **(b)** $0.\overline{2}$

(c) 1.3

(d) 0.7894737 (Rounded to seven decimal places.)

Sometimes division yields an infinite repeating decimal. We use three dots to indicate that the pattern continues forever. For example,

$$\frac{1}{3} = 0.3333\ldots$$
$$
\begin{array}{r}
0.333 \\
3\overline{)1.000} \\
\underline{9} \\
10 \\
\underline{9} \\
10 \\
\underline{9} \\
1
\end{array}
$$

An alternative notation is to place a bar over the repeating digit(s):

$$0.3333\ldots = 0.\overline{3} \qquad 0.575757\ldots = 0.\overline{57}$$

EXAMPLE 3 Write each fraction as a decimal.

(a) $\dfrac{2}{11}$

(b) $\dfrac{5}{6}$

Solution

(a) $\dfrac{2}{11} = 0.181818\ldots$ or $0.\overline{18}$

$$
\begin{array}{r}
0.1818 \\
11\overline{)2.0000} \\
\underline{11} \\
90 \\
\underline{88} \\
20 \\
\underline{11} \\
90 \\
\underline{88} \\
2
\end{array}
$$

(b) $\dfrac{5}{6} = 0.8333\ldots$ or $0.8\overline{3}$

$$
\begin{array}{r}
0.8333 \\
6\overline{)5.0000} \\
\underline{48} \\
20 \\
\underline{18} \\
20 \\
\underline{18} \\
20 \\
\underline{18} \\
2
\end{array}
$$

Note that the 8 does not repeat. Only the digit 3 is repeating.

Practice Problem 3 Write each fraction as a decimal.

(a) $\dfrac{1}{6}$

(b) $\dfrac{5}{11}$

NOTE TO STUDENT: Fully worked-out solutions to all of the Practice Problems can be found at the back of the text starting at page SP-1

Sometimes division must be carried out to many places in order to observe the repeating pattern. This is true in the following example:

$$\frac{2}{7} = 0.285714285714285714\ldots \qquad \text{This can also be written as } \frac{2}{7} = 0.\overline{285714}.$$

It can be shown that the denominator determines the maximum number of decimal places that might repeat. So $\frac{2}{7}$ must repeat in the seventh decimal place or sooner.

3 Changing a Decimal to a Fraction

To convert from a decimal to a fraction, merely write the decimal as a fraction with a denominator of 10, 100, 1000, 10,000, and so on, and simplify the result when possible.

EXAMPLE 4 Write each decimal as a fraction.

(a) 0.2 **(b)** 0.35 **(c)** 0.516 **(d)** 0.74 **(e)** 0.138 **(f)** 0.008

Solution

(a) $0.2 = \dfrac{2}{10} = \dfrac{1}{5}$

(b) $0.35 = \dfrac{35}{100} = \dfrac{7}{20}$

(c) $0.516 = \dfrac{516}{1000} = \dfrac{129}{250}$

(d) $0.74 = \dfrac{74}{100} = \dfrac{37}{50}$

(e) $0.138 = \dfrac{138}{1000} = \dfrac{69}{500}$

(f) $0.008 = \dfrac{8}{1000} = \dfrac{1}{125}$

Practice Problem 4 Write each decimal as a fraction and simplify whenever possible.

(a) 0.8 **(b)** 0.88 **(c)** 0.45 **(d)** 0.148 **(e)** 0.612 **(f)** 0.016

All repeating decimals can also be converted to fractional form. In practice, however, repeating decimals are usually rounded to a few places. It will not be necessary, therefore, to learn how to convert $0.\overline{033}$ to $\frac{11}{333}$ for this course.

 ## Adding and Subtracting Decimals

Last week Bob spent $19.83 on lunches purchased at the cafeteria at work. During this same period, Sally spent $24.76 on lunches. How much did the two of them spend on lunches last week?

Adding or subtracting decimals is similar to adding and subtracting whole numbers, except that it is necessary to line up decimal points. To perform the operation $19.83 + 24.76$, we line up the numbers in column form and add the digits:

$$\begin{array}{r} 19.83 \\ + 24.76 \\ \hline 44.59 \end{array}$$

Thus Bob and Sally spent $44.59 on lunches last week.

ADDITION AND SUBTRACTION OF DECIMALS

1. Write in column form and line up decimal points.
2. Add or subtract the digits.

EXAMPLE 5   Perform the following operations.

(a) $3.6 + 2.3$

(b) $127.32 - 38.48$

(c) $3.1 + 42.36 + 9.034$

(d) $5.0006 - 3.1248$

Solution

(a)
$$\begin{array}{r} 3.6 \\ + 2.3 \\ \hline 5.9 \end{array}$$

(b)
$$\begin{array}{r} 127.32 \\ - 38.48 \\ \hline 88.84 \end{array}$$

(c)
$$\begin{array}{r} 3.1 \\ 42.36 \\ + 9.034 \\ \hline 54.494 \end{array}$$

(d)
$$\begin{array}{r} 5.0006 \\ - 3.1248 \\ \hline 1.8758 \end{array}$$

Practice Problem 5   Add or subtract.

(a) $3.12 + 5.08 + 1.42$

(b) $152.003 - 136.118$

(c) $1.1 + 3.16 + 5.123$

(d) $1.0052 - 0.1234$

SIDELIGHT: Adding Zeros to the Right-Hand Side of the Decimal

When we added fractions, we had to have common denominators. Since decimals are really fractions, why can we add them without having common denominators? Actually, we have to have common denominators to add any fractions, whether they are in decimal form or fraction form. However, sometimes the notation does not show this. Let's examine Example 5(c).

Original Problem   We are adding the three numbers:

$$\begin{array}{r} 3.1 \\ 42.36 \\ + 9.034 \\ \hline 54.494 \end{array}$$

$3\frac{1}{10} + 42\frac{36}{100} + 9\frac{34}{1000}$

$3\frac{100}{1000} + 42\frac{360}{1000} + 9\frac{34}{1000}$

$3.100 + 42.360 + 9.034$ This is the new problem.

Original Problem   *New Problem*

3.1	3.100
42.36	42.360
+ 9.034	+ 9.034
54.494	54.494

We notice that the results are the same. The only difference is the notation. We are using the property that any number of zeros may be added to the right-hand side of a decimal without changing its value.

This shows the convenience of adding and subtracting fractions in decimal form. Little work is needed to change the decimals so that they have a common denominator. All that is required is to add zeros to the right-hand side of the decimal (and we usually do not even write out that step except when subtracting).

As long as we line up the decimal points, we can add or subtract any decimal fractions.

In the following example we will find it useful to add zeros to the right-hand side of the decimal.

EXAMPLE 6   Perform the following operations.

(a) $1.0003 + 0.02 + 3.4$ **(b)** $12 - 0.057$

Solution   We will add zeros so that each number shows the same number of decimal places.

(a)

1.0003
0.0200
+ 3.4000
4.4203

(b)

12.000
− 0.057
11.943

Practice Problem 6   Perform the following operations.

(a) $0.061 + 5.0008 + 1.3$ **(b)** $18 - 0.126$

NOTE TO STUDENT: *Fully worked-out solutions to all of the Practice Problems can be found at the back of the text starting at page SP-1*

5 Multiplying Decimals

MULTIPLICATION OF DECIMALS

To multiply decimals, you first multiply as with whole numbers. To determine the position of the decimal point, you count the total number of decimal places in the two numbers being multiplied. This will determine the number of decimal places that should appear in the answer.

EXAMPLE 7   Multiply. 0.8×0.4

Solution

0.8	(one decimal place)
×0.4	(one decimal place)
0.32	(two decimal places)

Practice Problem 7   Multiply. 0.5×0.3

Note that you will often have to add zeros to the left of the digits obtained in the product so that you obtain the necessary number of decimal places.

EXAMPLE 8 Multiply. 0.123×0.5

Solution

$$
\begin{array}{r}
0.123 \quad (\text{three decimal places}) \\
\times \quad 0.5 \quad (\text{one decimal place}) \\
\hline
0.0615 \quad (\text{four decimal places})
\end{array}
$$

Practice Problem 8 Multiply. 0.12×0.4

Here are some examples that involve more decimal places.

EXAMPLE 9 Multiply.

(a) 2.56×0.003 **(b)** 0.0036×0.008

Solution

(a)
$$
\begin{array}{r}
2.56 \quad (\text{two decimal places}) \\
\times \quad 0.003 \quad (\text{three decimal places}) \\
\hline
0.00768 \quad (\text{five decimal places})
\end{array}
$$

(b)
$$
\begin{array}{r}
0.0036 \quad (\text{four decimal places}) \\
\times \quad 0.008 \quad (\text{three decimal places}) \\
\hline
0.0000288 \quad (\text{seven decimal places})
\end{array}
$$

Practice Problem 9 Multiply.

(a) 1.23×0.005 **(b)** 0.003×0.00002

SIDELIGHT: Counting the Number of Decimal Places
Why do we count the number of decimal places? The rule really comes from the properties of fractions. If we write the problem in Example 8 in fraction form, we have

$$
0.123 \times 0.5 = \frac{123}{1000} \times \frac{5}{10} = \frac{615}{10,000} = 0.0615.
$$

6 Dividing Decimals

When discussing division of decimals, we frequently refer to the three primary parts of a division problem. Be sure you know the meaning of each term.

The **divisor** is the number you divide into another.
The **dividend** is the number to be divided.
The **quotient** is the result of dividing one number by another.

In the problem $6 \div 2 = 3$ we represent each of these terms as follows:

When dividing two decimals, count *the number of decimal places* in the divisor. Then *move the decimal point to the right* that *same number of places* in both *the divisor* and *the dividend*. Mark that position with a caret ($_\wedge$). Finally, perform the division. Be sure to line up the decimal point in the quotient with the position indicated by the caret in the dividend.

EXAMPLE 10 Four friends went out for lunch. The total bill, including tax, was $32.68. How much did each person pay if they shared the cost equally?

Solution To answer this question, we must calculate $32.68 \div 4$.

$$
\begin{array}{r}
8.17 \\
4\overline{)32.68} \\
\underline{32} \\
6 \\
\underline{4} \\
28 \\
\underline{28} \\
0
\end{array}
$$

Since there are no decimal places in the divisor, we do not need to move the decimal point. We must be careful, however, to place the decimal point in the quotient directly above the decimal point in the dividend.

Thus $32.68 \div 4 = 8.17$, and each friend paid $8.17.

NOTE TO STUDENT: Fully worked-out solutions to all of the Practice Problems can be found at the back of the text starting at page SP-1

Practice Problem 10 Sally Keyser purchased 6 boxes of paper for the inkjet printer. The cost was $31.56. There was no tax since she purchased the paper for a charitable organization. How much did she pay for each box of paper?

Note that sometimes we will need to place extra zeros in the dividend in order to move the decimal point the required number of places.

EXAMPLE 11 Divide. $16.2 \div 0.027$

Solution

$$0.027\overline{)16.200}$$

There are **three** decimal places in the divisor, so we move the decimal point **three places** to **the right** in the **divisor** and **dividend** and mark the new position by a caret. Note that we must add two zeros to 16.2 in order to do this.

three decimal places

$$
\begin{array}{r}
600. \\
0.027\overline{)16.200} \\
\underline{16\,2} \\
000
\end{array}
$$

Now perform the division as with whole numbers. The decimal point in the answer is directly above the caret in the dividend.

Thus $16.2 \div 0.027 = 600$.

Practice Problem 11 Divide. $1800 \div 0.06$

Special care must be taken to line up the digits in the quotient. Note that sometimes we will need to place zeros in the quotient after the decimal point.

EXAMPLE 12 Divide. 0.04288 ÷ 3.2

Solution

$$3.2\overline{)0.0\overset{\wedge}{4}288}$$

There is **one** decimal place in the divisor, so we move the decimal point **one place** to **the right** in the **divisor** and **dividend** and mark the new position by a caret.

one decimal place

$$\begin{array}{r} 0.0134 \\ 3.2\overline{)0.04288} \\ \underline{32} \\ 108 \\ \underline{96} \\ 128 \\ \underline{128} \\ 0 \end{array}$$

Now perform the division as for whole numbers. The decimal point in the answer is directly above the caret in the dividend. Note the need for the initial zero after the decimal point in the answer.

Thus 0.04288 ÷ 3.2 = 0.0134.

Practice Problem 12 Divide. 0.01764 ÷ 4.9

SIDELIGHT: Dividing Decimals by Another Method

Why does this method of dividing decimals work? Essentially, we are using the steps we used in Section 0.1 to change a fraction to an equivalent fraction by multiplying both the numerator and denominator by the same number. Let's reexamine Example 12.

$$0.04288 \div 3.2 = \frac{0.04288}{3.2}$$

Write the original problem using fraction notation.

$$= \frac{0.04288 \times 10}{3.2 \times 10}$$

Multiply the numerator and denominator by 10. Since this is the same as multiplying by 1, we are not changing the fraction.

$$= \frac{0.4288}{32}$$

Write the result of multiplication by 10.

$$= 0.4288 \div 32$$

Rewrite the fraction as an equivalent problem with division notation.

Notice that we have obtained a new problem that is the same as the problem in Example 12 when we moved the decimal one place to the right in the divisor and dividend. We see that the reason we can move the decimal point as many places as necessary to the right in divisor and dividend is that this is the same as multiplying the numerator and denominator of a fraction by a power of 10 to obtain an equivalent fraction.

7 Multiplying and Dividing a Decimal by a Multiple of 10

When multiplying by 10, 100, 1000, and so on, a simple rule may be used to obtain the answer. For every zero in the multiplier, move the decimal point one place to the right.

EXAMPLE 13 Multiply.

(a) 3.24×10 **(b)** 15.6×100 **(c)** 0.0026×1000

Solution

(a) $3.24 \times 10 = 32.4$ One zero—move decimal point one place to the right.

(b) $15.6 \times 100 = 1560$ Two zeros—move decimal point two places to the right.

(c) $0.0026 \times 1000 = 2.6$ Three zeros—move decimal point three places to the right.

Practice Problem 13 Multiply.

(a) 0.0016×100 **(b)** 2.34×1000 **(c)** $56.75 \times 10,000$

NOTE TO STUDENT: Fully worked-out solutions to all of the Practice Problems can be found at the back of the text starting at page SP-1

The reverse rule is true for division. When dividing by 10, 100, 1000, 10,000, and so on, move the decimal point one place to the left for every zero in the divisor.

EXAMPLE 14 Divide.

(a) $52.6 \div 10$ **(b)** $0.0038 \div 100$ **(c)** $5936.2 \div 1000$

Solution

(a) $\dfrac{52.6}{10} = 5.26$ Move one place to the left.

(b) $\dfrac{0.0038}{100} = 0.000038$ Move two places to the left.

(c) $\dfrac{5936.2}{1000} = 5.9362$ Move three places to the left.

Practice Problem 14 Divide.

(a) $\dfrac{5.82}{10}$ **(b)** $123.4 \div 1000$ **(c)** $\dfrac{0.00614}{10,000}$

Developing Your Study Skills

Making a Friend in the Class

Attempt to make a friend in your class. You may find that you enjoy sitting together and drawing support and encouragement from one another. Exchange phone numbers so you can call each other whenever you get stuck in your study. Set up convenient times to study together on a regular basis, to do homework, and to review for exams.

You must not depend on a friend or fellow student to tutor you, do your work for you, or in any way be responsible for your learning. However, you will learn from one another as you seek to master the course. Studying with a friend and comparing notes, methods, and solutions can be very helpful. And it can make learning mathematics a lot more fun!

0.4 EXERCISES

Student Solutions Manual | CD/Video | PH Math Tutor Center | MathXL®Tutorials on CD | MathXL® | MyMathLab® | Interactmath.com

Verbal and Writing Skills

1. A decimal is another way of writing a fraction whose denominator is _____.

2. We write 0.42 in words as _____.

3. When dividing 7432.9 by 1000 we move the decimal point _____ places to the _____.

4. When dividing 96.3 by 10,000 we move the decimal point _____ places to the _____.

Write each fraction as a decimal.

5. $\dfrac{5}{8}$　　　**6.** $\dfrac{6}{25}$　　　**7.** $\dfrac{3}{15}$　　　**8.** $\dfrac{12}{15}$　　　**9.** $\dfrac{7}{11}$　　　**10.** $\dfrac{2}{3}$

Write each decimal as a fraction in simplified form.

11. 0.8　　　**12.** 0.15　　　**13.** 0.625　　　**14.** 0.08　　　**15.** 2.6　　　**16.** 1.8

Add or subtract.

17. $1.71 + 0.38$　　　**18.** $3.42 + 0.38$　　　**19.** $2.5 + 3.42 + 4.9$　　　**20.** $6.31 + 4.2 + 8.5$

21. $46.03 + 215.1 + 0.078$　　**22.** $33.01 + 0.38 + 175.401$　**23.** $147.18 - 15.39$　　**24.** $131.43 - 86.95$

Multiply or divide.

25. 7.21×4.2　　　**26.** 7.12×2.6　　　**27.** 4.23×0.025　　　**28.** 3.84×0.0017

29. $169{,}000 \times 0.0013$　　**30.** $368{,}000 \times 0.00021$　　**31.** $7.9728 \div 3.02$　　　**32.** $6.519 \div 2.05$

33. $0.5230 \div 0.002$　　**34.** $0.031 \div 0.005$　　**35.** $0.02056 \div 0.08$　　**36.** $0.03222 \div 0.09$

Multiply or divide by moving the decimal point.

37. 3.45×1000　　　**38.** 1.36×1000　　　**39.** $0.76 \div 100$　　　**40.** $175{,}318 \div 1000$

41. $7.36 \times 10{,}000$　　**42.** $0.00243 \times 100{,}000$　　**43.** $73{,}892 \div 100{,}000$　　**44.** $3.52 \div 1000$

Mixed Practice

Perform the indicated calculations.

45. 23.75×0.06

46. 1.824×0.004

47. $1.62 + 2.005 + 8.1007$

48. $1.5 + 3.06 + 4.209$

49. $0.05724 \div 0.027$

50. $77.136 \div 0.003$

51. 0.7683×1000

52. $34.72 \times 10,000$

Applications

53. *Measurement* In a recent chemistry lab, Jerome needed to change the measured data from inches to centimeters. If there are 2.54 cm in an inch, and the original measurement was 9.5 inches, what is the measured data in cms?

54. *Hybrid Cars* Hai Hung bought a hybrid electric-gas–powered car because of its great miles per gallon in the city. The car averages 46.2 miles per gallon in the city, and the tank holds 9.7 gallons of gas. How many miles can Hai travel in the city on a full tank of gas?

55. *Curtain Material* Lexi decided that she would need 18.5 yards of material to make curtains. If the material she chose cost $11.50 per yard, what did the material for her curtains cost?

56. *Turkey Weight* The Parkins family is going to order a turkey for Thanksgiving dinner from Ray's Turkey Farm. Mrs. Parkins estimates that she will need 12 pounds of meat to serve all her guests. The turkey farm owner claims that each turkey yields about 0.78 of its weight in meat. If Mrs. Parkins buys a 17.3-pound turkey, will there be enough meat for everyone?

57. *Drinking Water* The EPA standard for safe drinking water is a maximum of 1.3 milligrams of copper per liter of water. A water testing firm found 6.8 milligrams of copper in a 5-liter sample drawn from Jim and Sharon LeBlanc's house. Is the water safe or not? By how much does the amount of copper exceed or fall short of the maximum allowed?

58. *Wages* Harry has a part-time job at Stop and Shop. He earns $8.50 an hour. Last week he worked 19 hours. He had hoped to earn at least $150. Did he reach his goal or not? By how much did he exceed or fall short of his goal?

Cumulative Review

Perform each operation. Simplify all answers.

59. $3\frac{1}{2} \div 5\frac{1}{4}$

60. $\frac{3}{8} \cdot \frac{12}{27}$

61. $\frac{12}{25} + \frac{9}{20}$

62. $1\frac{3}{5} - \frac{1}{2}$

How are you doing with your homework assignments in Sections 0.1 to 0.4? Do you feel you have mastered the material so far? Do you understand the concepts you have covered? Before you go further in the textbook, take some time to do each of the following problems.

0.1

In exercises 1 and 2, simplify each fraction.

1. $\dfrac{15}{55}$

2. $\dfrac{46}{115}$

3. Write $\dfrac{15}{4}$ as a mixed number.

4. Change $4\dfrac{5}{7}$ to an improper fraction.

Find the missing number.

5. $\dfrac{3}{7} = \dfrac{?}{14}$

6. $\dfrac{7}{4} = \dfrac{?}{20}$

0.2

7. Find the LCD, but do not add. $\dfrac{3}{8}, \dfrac{5}{6},$ and $\dfrac{7}{15}$

Perform the calculation indicated. Write the answer in simplest form.

8. $\dfrac{3}{7} + \dfrac{2}{7}$

9. $\dfrac{5}{14} + \dfrac{2}{21}$

10. $2\dfrac{3}{4} + 5\dfrac{2}{3}$

11. $\dfrac{17}{18} - \dfrac{5}{9}$

12. $\dfrac{6}{7} - \dfrac{2}{3}$

13. $3\dfrac{1}{5} - 1\dfrac{3}{8}$

0.3

Perform the calculations indicated. Write the answer in simplest form.

14. $\dfrac{25}{7} \times \dfrac{14}{45}$

15. $2\dfrac{4}{5} \times 3\dfrac{3}{4}$

16. $4 \div \dfrac{8}{7}$

17. $2\dfrac{1}{3} \div 3\dfrac{1}{4}$

0.4

Change to a decimal. Use bar notation for repeating decimals.

18. $\dfrac{7}{8}$

19. $\dfrac{5}{9}$

20. $\dfrac{3}{200}$

Perform the calculations indicated.

21. $15.23 + 3.6 + 0.821$

22. 3.28×0.63

23. $3.015 \div 6.7$

24. $12.13 - 9.884$

Now turn to page SA-1 for the answer to each of these problems. Each answer also includes a reference to the objective in which the problem is first taught. If you missed any of these problems, you should stop and review the Examples and Practice Problems in the referenced objective. A little review now will help you master the material in the upcoming sections of the text.

1. _____

2. _____

3. _____

4. _____

5. _____

6. _____

7. _____

8. _____

9. _____

10. _____

11. _____

12. _____

13. _____

14. _____

15. _____

16. _____

17. _____

18. _____

19. _____

20. _____

21. _____

22. _____

23. _____

24. _____

0.5 USING PERCENT

Student Learning Objectives

After studying this section, you will be able to:

1 Understand the meaning of percent.

2 Change a decimal to a percent.

3 Change a percent to a decimal.

4 Find the percent of a given number.

5 Find the missing percent when given two numbers.

NOTE TO STUDENT: *Fully worked-out solutions to all of the Practice Problems can be found at the back of the text starting at page SP-1*

1 Understanding the Meaning of Percent

A **percent** is a fraction that has a denominator of 100. When you say "sixty-seven percent" or write 67%, you are just expressing the fraction $\frac{67}{100}$ in another way. The word *percent* is a shortened form of the Latin words *per centum*, which means "by the hundred." In everyday use, percent means per one hundred.

Russell Camp owns 100 acres of land in Montana. 49 of the acres are covered with trees. The rest of the land is open fields. We say that 49% of his land is covered with trees.

It is important to see that 49% means 49 parts out of 100 parts. It can also be written as a fraction, $\frac{49}{100}$, or as a decimal 0.49. Understanding the meaning of the notation allows you to change from one form to another. For example,

$$49\% = 49 \text{ out of } 100 \text{ parts} = \frac{49}{100} = 0.49.$$

Similarly, you can express a fraction with denominator 100 as a percent or a decimal.

$$\frac{11}{100} \text{ means 11 parts out of 100 or } 11\%. \qquad \text{So } \frac{11}{100} = 11\% = 0.11.$$

2 Changing a Decimal to a Percent

Now that we understand the concept, we can use some quick procedures to change from decimals to percent, and vice versa.

> **CHANGING A DECIMAL TO PERCENT**
>
> 1. Move the decimal point two places to the right.
> 2. Add the % symbol.

EXAMPLE 1 Change to a percent.

(a) 0.23   **(b)** 0.461   **(c)** 0.4

Solution We move the decimal point two places to the right and add the % symbol.

(a) 0.23 = 23%   **(b)** 0.461 = 46.1%   **(c)** 0.4 = 0.40 = 40%

Practice Problem 1 Change to a percent.

(a) 0.92   **(b)** 0.418   **(c)** 0.7

Be sure to follow the same procedure for percents that are less than 1%. Remember, 0.01 is 1%. Thus we would expect 0.001 to be less than 1%. 0.001 = 0.1% or one-tenth (0.1) of a percent.

EXAMPLE 2   Change to a percent.

(a) 0.0364   **(b)** 0.0026   **(c)** 0.0008

Solution   We move the decimal point two places to the right and add the % symbol.

(a) 0.0364 = 3.64%   **(b)** 0.0026 = 0.26%   **(c)** 0.0008 = 0.08%

Practice Problem 2   Change to a percent.

(a) 0.0019   **(b)** 0.0736   **(c)** 0.0003

Be sure to follow the same procedure for percents that are greater than 100%. Remember that 1 is 100%. Thus we would expect 1.5 to be greater than 100%. In fact, 1.5 = 150%.

EXAMPLE 3   Change to a percent.

(a) 1.48   **(b)** 2.938   **(c)** 4.5

Solution   We move the decimal point two places to the right and add the percent symbol.

(a) 1.48 = 148%   **(b)** 2.938 = 293.8%   **(c)** 4.5 = 4.50 = 450%

Practice Problem 3   Change to a percent.

(a) 3.04   **(b)** 5.186   **(c)** 2.1

  3 **Changing a Percent to a Decimal**

In this procedure we move the decimal point to the left and remove the % symbol.

> **CHANGING A PERCENT TO A DECIMAL**
>
> **1.** Move the decimal point two places to the left.
> **2.** Remove the % symbol.

EXAMPLE 4   Change to a decimal.

(a) 4%   **(b)** 3.2%   **(c)** 0.6%

Solution   First we move the decimal point two places to the left. Then we remove the % symbol.

(a) 4% = 4.% = 0.04

      ↑

    The unwritten decimal point is understood to be here.

(b) 3.2% = 0.032   **(c)** 0.6% = 0.006

Practice Problem 4   Change to a decimal.

(a) 7%   **(b)** 9.3%   **(c)** 0.2%

Calculator

Percent to Decimal

You can use a calculator to change 52% to a decimal. If your calculator has a $\boxed{\%}$ key, do the following:

Enter: 52 $\boxed{\%}$
The display should read:

$$\boxed{0.52}$$

If your calculator does not have a $\boxed{\%}$ key, divide the number by 100.
Enter:

$$52 \; \boxed{\div} \; 100 \; \boxed{=}$$

Try the following:

(a) 46%   **(b)** 137%
(c) 9.3%   **(d)** 6%

Note: The calculator divides by 100 when the percent key is pressed. If you do not have a $\boxed{\%}$ key, then you can divide by 100 instead.

EXAMPLE 5   Change to a decimal.

(a) 192%   **(b)** 254.8%   **(c)** 0.027%

Solution   First we move the decimal point two places to the left. Then we re-move the % symbol.

(a) 192% = 192.% = 1.92

↑

The unwritten decimal point is understood to be here.

(b) 254.8% = 2.548   **(c)** 0.027% = 0.00027

Practice Problem 5   Change to a decimal.

(a) 131%   **(b)** 301.6%   **(c)** 0.04%

NOTE TO STUDENT: Fully worked-out solutions to all of the Practice Problems can be found at the back of the text starting at page SP-1

4 Finding the Percent of a Given Number

How do we find 60% of 20? Let us relate it to a problem we did in Section 0.2. Consider the following problem.

What is $\dfrac{3}{5}$ of 20?

↓ ↓ ↓ ↓ ↓

$\boxed{?}$ = $\dfrac{3}{5}$ × 20

$\boxed{?}$ = $\dfrac{3}{\cancel{5}}$ × $\cancel{20}^{4}$ = 12   The answer is 12.

Since a percent is really a fraction, a percent problem is solved similarly to the way a fraction problem is solved. Since $\frac{3}{5} = \frac{6}{10} = 60\%$, we could write the problem as

What is 60% of 20?

↓ ↓ ↓ ↓ ↓

$\boxed{?}$ = 60% × 20

$\boxed{?}$ = 0.60 × 20

$\boxed{?}$ = 12.0   The answer is 12.

Thus we have developed the following rule.

FINDING THE PERCENT OF A NUMBER

To find the percent of a number, change the percent to a decimal and multiply the number by the decimal.

EXAMPLE 6   Find.

(a) 10% of 36   **(b)** 2% of 350   **(c)** 182% of 12   **(d)** 0.3% of 42

Solution

(a) 10% of 36 = 0.10 × 36 = 3.6   **(b)** 2% of 350 = 0.02 × 350 = 7

(c) 182% of 12 = 1.82 × 12 = 21.84   **(d)** 0.3% of 42 = 0.003 × 42 = 0.126

Practice Problem 6   Find.   **(a)** 18% of 50   **(b)** 4% of 64   **(c)** 156% of 35

Calculator

 Finding the Percent of a Number

You can use a calculator to find 12% of 48.
Enter:

12 $\boxed{\%}$ $\boxed{\times}$ 48 $\boxed{=}$

The display should read

$\boxed{5.76}$

If your calculator does not have a $\boxed{\%}$ key, do the following:

Enter: 0.12 $\boxed{\times}$ 48 $\boxed{=}$

What is 54% of 450?

There are many real-life applications for finding the percent of a number. When you go shopping in a store, you may find sale merchandise marked 35% off. This means that the sale price is 35% off the regular price. That is, 35% of the regular price is subtracted from the regular price to get the sale price.

EXAMPLE 7   A store is having a sale of 35% off the retail price of all sofas. Melissa wants to buy a particular sofa that normally sells for $595.

(a) How much will Melissa save if she buys the sofa on sale?

(b) What will the purchase price be if Melissa buys the sofa on sale?

Solution

(a) To find 35% of $595 we will need to multiply 0.35×595.

$$\begin{array}{r} 595 \\ \times \quad 0.35 \\ \hline 2975 \\ 1785 \\ \hline 208.25 \end{array}$$

Thus Melissa will save $208.25 if she buys the sofa on sale.

(b) The purchase price is the difference between the original price and the amount saved.

$$\begin{array}{r} 595.00 \\ - \ 208.25 \\ \hline 386.75 \end{array}$$

If Melissa buys the sofa on sale, she will pay $386.75.

Practice Problem 7   John received a 4.2% pay raise at work this year. He had previously earned $38,000 per year.

(a) What was the amount of his pay raise in dollars?

(b) What is his new salary?

5 Finding the Missing Percent When Given Two Numbers

Recall that we can write $\frac{3}{4}$ as $\frac{75}{100}$ or 75%. If we were asked the question, "What percent is 3 of 4?" we would say 75%. This gives us a procedure for finding what percent one number is of a second number.

> **FINDING THE MISSING PERCENT**
>
> 1. Write a fraction with the two numbers. The number *after* the word "of" is always the denominator, and the other number is the numerator.
> 2. Simplify the fraction (if possible).
> 3. Change the fraction to a decimal.
> 4. Express the decimal as a percent.

EXAMPLE 8   What percent of 24 is 15?

Solution   This can be solved quickly as follows.

Step 1 $\dfrac{15}{24}$ Write the relationship as a fraction. The number after "of" is 24, so 24 is the denominator.

Step 2 $\dfrac{15}{24} = \dfrac{5}{8}$ Simplify the fraction (when possible).

Step 3 $= 0.625$ Change the fraction to a decimal.

Step 4 $= 62.5\%$ Change the decimal to percent.

Practice Problem 8   What percent of 148 is 37?

The question in Example 8 can also be written as "15 is what percent of 24?" To answer the question, we begin by writing the relationship as $\frac{15}{24}$. Remember that "of 24" means 24 will be the denominator.

EXAMPLE 9

(a) 82 is what percent of 200?

(b) What percent of 16 is 3.8?   (c) $150 is what percent of $120?

Solution

(a) 82 is what percent of 200?

$$\frac{82}{200}$$ Write the relationship as a fraction with 200 in the denominator.

$$\frac{82}{200} = \frac{41}{100} = 0.41 = 41\%$$

(b) What percent of 16 is 3.8?

$$\frac{3.8}{16}$$ Write the relationship as a fraction.

You can divide to change the fraction to a decimal and then change the decimal to a percent.

$$16\overline{)3.8000} \quad 0.2375 \rightarrow 23.75\%$$

(c) $150 is what percent of $120?

$$\frac{150}{120} = \frac{5}{4}$$ Reduce the fraction whenever possible to make the division easier.

$$4\overline{)5.00} \quad 1.25 \rightarrow 125\%$$

Practice Problem 9

(a) What percent of 48 is 24?   (b) 4 is what percent of 25?

NOTE TO STUDENT: *Fully worked-out solutions to all of the Practice Problems can be found at the back of the text starting at page SP-1*

Calculator

Finding the Percent Given Two Numbers

You can use a calculator to find a missing percent. What percent of 95 is 19?

1. Enter as a fraction.

$$19 \;\boxed{\div}\; 95$$

2. Change to a percent.

$$19 \;\boxed{\div}\; 95 \;\boxed{\times}\; 100 \;\boxed{=}$$

The display should read

$$\boxed{20}$$

This means 20%.

What percent of 625 is 250?

EXAMPLE 10

Marcia made 29 shots on goal during the last high school field hockey season. She actually scored a goal eight times. What percent of her total shots were goals? Round your answer to the nearest whole percent.

Solution   Marcia scored a goal eight times out of 29 tries. We want to know what percent of 29 is 8.

Step 1 $\quad\dfrac{8}{29}$ Express the relationship as a fraction. The number after the word *of* is 29, so 29 appears in the denominator.

Step 2 $\quad\dfrac{8}{29} = \dfrac{8}{29}$ Note that this fraction cannot be reduced.

Step 3 $\quad = 0.2758\ldots$ The decimal equivalent of the fraction has many digits.

Step 4 $\quad = 27.58\ldots\%$ We change the decimal to a percent, which we round to the nearest whole percent.

$\qquad\qquad \approx 28\%$ (The $\approx$ symbol means approximately equal to.)

Therefore, Marcia scored a goal approximately 28% of the time she made a shot on goal.

Practice Problem 10   Roberto scored a basket 430 times out of 1256 attempts during his high school basketball career. What percent of the time did he score a basket? Round your answer to the nearest whole percent.

Verbal and Writing Skills

1. When you write 19%, what do you really mean? Describe the meaning in your own words.

2. When you try to solve a problem like "What percent of 80 is 30?" how do you know if you should write the fraction as $\frac{80}{30}$ or as $\frac{30}{80}$?

Change to a percent.

3. 0.28

4. 0.61

5. 0.568

6. 0.089

7. 0.076

8. 0.325

9. 2.39

10. 7.4

Change to a decimal.

11. 3%

12. 66%

13. 0.4%

14. 0.62%

15. 250%

16. 175%

17. 7.4%

18. 87.59%

Find the following.

19. What is 8% of 65?

20. What is 7% of 69?   4.83

21. What is 10% of 130?

22. What is 25% of 600?

23. What is 112% of 65?

24. What is 154% of 270?   415.8

25. 36 is what percent of 24?

26. 49 is what percent of 28?

27. What percent of 340 is 17?

28. 48 is what percent of 600?

29. 75 is what percent of 30?

30. What percent of 35 is 28?

Applications

31. *Exam Grades* Dave took an exam with 80 questions. He had 68 answered correctly. What was his grade for the exam? Write the grade as a percent.

32. *Soccer* Ronaldo, the great Brazilian soccer star, made 8 goals out of 21 attempts on goal in the 2002 World Cup Series. What percent of the times that he shoots does Ronaldo make a goal? (Round to the nearest whole percent.)

33. *Tipping* Diana and Russ ate a meal costing $32.80 when they went out to dinner. If they want to leave the standard 15% tip for their server, how much will they tip and what will their total bill be?

34. *CD Failure* Music CDs have a failure rate of 1.8%. (They skip or get stuck on a song.) If 36,000 CDs were manufactured last week, how many of them were defective?

35. *Food Budget* The Gonzalez family has a combined monthly income of $1850. Their food budget is $380/month. What percentage of their monthly income is budgeted for food? (Round your answer to the nearest whole percent.)

36. *Survey* In a local college survey, it was discovered that 137 out of 180 students had a grandparent not born in the United States. What percent of the students had a grandparent not born in the United States? (Round your answer to the nearest hundredth of a percent.)

37. *Gift Returns* Last Christmas season, the Jones Mill Outlet Store chain calculated that they sold 36,000 gift items. If they assume that there will be a 1.5% return rate after the holidays, how many gifts can they expect to be exchanged?

38. *Rent* The total cost of a downtown Boston apartment, including rent, heat, electricity, and water is $3690 per month. Sarah is renting this apartment with four friends. Based on a complicated formula, which takes into account the size of her bedroom and the furniture she brought to the apartment, she must pay 17% of the apartment cost each month. What is her monthly cost?

39. *Wages* Jim is earning $12.50 per hour as a cook. He has earned an 8% pay raise this year. What will his raise be, and what will his new hourly rate be?

40. *On-Time Arrivals* Eastern Shores charter airline had 2390 flights last year. Of these flights, 560 did not land on time. What percent of the flights landed on time? Round your answer to the nearest whole percent.

41. *Salary* Abdul sells computers for a local computer outlet. He gets paid $450 per month plus a commission of 3.8% on all the computer hardware he sells. Last year he sold $780,000 worth of computer hardware. (a) What was his sales commission on the $780,000 worth of hardware he sold? (b) What was his annual salary (that is, his monthly pay and commission combined)?

42. *Business Travel* Bruce sells medical supplies on the road. He logged 18,600 miles in his car last year. He declared 65% of his mileage as business travel. (a) How many miles did he travel on business last year? (b) If his company reimburses him 31 cents for each mile traveled, how much should he be paid for travel expenses?

To Think About

43. *Tipping* Fred and Nancy Adams go out once a week to a restaurant where their daughter is the chef. They are given a 33% discount on the cost of the meal. They always leave a tip of 20%. Fred thinks the tip should be based on the cost of the meal before the discount. Nancy thinks the tip should be based on the cost of the meal after the discount. If they normally purchase a meal each week that costs $45 for the two of them before the discount and use Nancy's method of calculating the tip, by how much will the tip increase if they switch to Fred's method?

44. *Tax Deduction* Dave Bagley traveled 24,500 miles last year. He is a salesperson and 74% of his mileage is for business purposes. He was planning to deduct 31 cents per business mile on his income tax return. However, his accountant told him he can deduct 35 cents per mile. By how much will his deduction increase if he uses the new larger amount?

Cumulative Review

45. *Checking Account Balance* Janeen had an opening balance of $45.50 in her checking account this month. She deposited her paycheck of $1189 and a refund check of $33.90. She was charged by the bank $1.50 for ATM fees. Then she made out checks for $98.00, $128.00, $56.89, and $445.88. What will be her final balance at the end of the month?

46. *Car Loan* Tally is buying a car and has a bank loan. He has to make 33 more monthly payments of $188.50. How much money will he pay to the bank over the next 33 months?

47. *Gas Mileage* Dan took a trip in his Ford Taurus. At the start of the trip his car odometer read 68,459.5 miles. At the end of the trip his car odometer read 69,229.5 miles. He used 35 gallons of gas on the trip. How many miles per gallon did his car achieve?

48. *Rainfall* In Hilo, Hawaii, Brad spent three months working in a restaurant for a summer job. In June he observed that 4.6 inches of rain fell. In July the rainfall was 4.5 inches and in August it was 2.9 inches. What was the average monthly rainfall that summer for those three months in Hilo?

0.6 ROUNDING AND ESTIMATING

1 Using Rounding to Estimate

As we begin this section, we will take some time to be sure you understand the idea of rounding a number. You will probably recall the following simple rule from your previous mathematics courses.

Student Learning Objective

After studying this section, you will be able to:

 Use rounding to estimate.

ROUNDING A NUMBER

If the first digit to the right of the round-off place is

1. less than 5, we make no change to the digit in the round-off place.
2. 5 or more, we increase the digit in the round-off place by 1.

To illustrate, 4689 rounded to the nearest hundred is 4700. Rounding 233,987 to the nearest ten thousand, we obtain 230,000. We will now use our experience in rounding as we discuss the general area of estimation.

Estimation is the process of finding an approximate answer. It is not designed to provide an exact answer. Estimation will give you a rough idea of what the answer might be. For any given problem, you may choose to estimate in many different ways.

ESTIMATION BY ROUNDING

1. Round each number so that there is one nonzero digit.
2. Perform the calculation with the rounded numbers.

EXAMPLE 1 Find an estimate of the product 5368×2864.

Solution

Step 1 Round 5368 to 5000.
Round 2864 to 3000.

Step 2 Multiply.
$$5000 \times 3000 = 15,000,000$$

An estimate of the product is 15,000,000.

Practice Problem 1 Find an estimate of the product $128,621 \times 378$.

NOTE TO STUDENT: Fully worked-out solutions to all of the Practice Problems can be found at the back of the text starting at page SP-1

EXAMPLE 2 The four walls of a college classroom are $22\frac{1}{4}$ feet long and $8\frac{3}{4}$ feet high. A painter needs to know the area of these four walls in square feet. Since paints is sold in gallons, an estimate will do. Estimate the area of the four walls.

Solution

Step 1 Round $22\frac{1}{4}$ feet to 20 feet.

Round $8\frac{3}{4}$ feet to 9 feet.

Step 2 Multiply 20×9 to obtain an estimate of the area of one wall.
Multiply $20 \times 9 \times 4$ to obtain an estimate of the area of all four walls.
$$20 \times 9 \times 4 = 720 \text{ square feet}$$

Our estimate for the painter is 720 square feet of wall space.

NOTE TO STUDENT: Fully worked-out solutions to all of the Practice Problems can be found at the back of the text starting at page SP-1

▲ **Practice Problem 2** Mr. and Mrs. Ramirez need to carpet two rooms of their house. One room measures $12\frac{1}{2}$ feet by $9\frac{3}{4}$ feet. The other room measures $14\frac{1}{4}$ feet by $18\frac{1}{2}$ feet. Estimate the number of square feet (square footage) in these two rooms.

EXAMPLE 3 Won Lin has a small compact car. He drove 396.8 miles in his car and used 8.4 gallons of gas.

(a) Estimate the number of miles he gets per gallon.

(b) Estimate how much it will cost him for fuel to drive on a cross-country trip of 2764 miles if gasoline usually costs $\$1.59\frac{9}{10}$ per gallon.

Solution

(a) Round 396.8 miles to 400 miles. Round 8.4 gallons to 8 gallons. Now divide.

$$8\overline{)400} \quad \frac{50}{}$$

Won Lin's car gets about 50 miles per gallon.

(b) We will need to use the information we found in part (a) to determine how many gallons of gasoline Won Lin will use on his trip. Round 2764 miles to 3000 miles and divide 3000 miles by 50 gallons.

$$50\overline{)3000} \quad \frac{60}{}$$

Won Lin will use about 60 gallons of gas for his cross-country trip.

To estimate the cost, we need to ask ourselves, "What kind of an estimate are we looking for?" It may be sufficient to round $\$1.59\frac{9}{10}$ to $\$2.00$ and multiply.

$$60 \times \$2.00 = \$120.00$$

Keep in mind that this is a broad estimate. You may want an estimate that will be closer to the exact answer. In that case round $\$1.59\frac{9}{10}$ to $\$1.60$ and multiply.

$$60 \times \$1.60 = \$96.00$$

CAUTION An estimate is only a rough guess. If we are estimating the cost of something, we may want to round the unit cost to a value that is closer to the actual amount so that our estimate is more accurate. When we estimate costs, it is a good idea to round to a value above the actual unit cost to make sure that we will have enough money for the expenditure. The actual cost to the nearest penny of the cross-country trip in Example 3 is $93.56.

Practice Problem 3 Roberta drove 422.8 miles in her truck and used 19.3 gallons of gas. Assume that gasoline costs $\$1.69\frac{9}{10}$ per gallon.

(a) Estimate the number of miles she gets per gallon.

(b) Estimate how much it will cost her to drive to Chicago and back, a distance of 3862 miles.

Other words that indicate estimation in word problems are *about* and *approximate*.

EXAMPLE 4   A local manufacturing company releases 1684 pounds of sulfur dioxide per day from its plant. The company operates 294 days per year and has been operating at this level for 32 years. Approximately how many pounds of sulfur dioxide have been released into the air by this company over the last 32 years?

Solution   The word *approximately* means we are looking for an estimate. We begin by rounding the given data so that there is only one nonzero digit in each number.

Round 1684 tons to 2000 tons.
Round 294 days to 300 days.
Round 32 years to 30 years.

Now determine the calculation you would perform to find the answer and calculate using the rounded numbers. We multiply the amount released in one day × the number of days in an operating year × years.

$$2000 \times 300 \times 30 = 18{,}000{,}000$$

Thus 18,000,000 tons is an approximation of the sulfur dioxide released by the company during its 32 years of operation.

Practice Problem 4   A space probe is sent from Earth at 43,300 miles per hour toward the planet Pluto, which has an average distance from Earth of 3,580,000,000 miles.

(a) About how many hours will it take for the space probe to travel from Earth to Pluto?

(b) About how many days will it take for the space probe to travel from Earth to Pluto?

EXAMPLE 5   Find an estimate for 2.68% of $54,361.92.

Solution   Begin by writing 2.68% as a decimal.

$$2.68\% = 0.0268$$

Next round 0.0268 to the nearest hundredth.

0.0268 is 0.03 rounded to the nearest hundredth.

Finally, round $54,361.92 to $50,000.
    Now multiply.

$0.03 \times 50{,}000 = 1500$ Remember that to find a percent of a number you multiply.

Thus we estimate the answer to be $1500.
    To get a better estimate of 2.68% of $54,361.92, you may wish to round $54,361.92 to the nearest thousand. Try it and compare this estimate to the previous one. Remember, an estimate is only an approximation. How close you should try to get to the exact answer depends on what you need the estimate for.

Practice Problem 5   Find an estimate for 56.93% of $293,567.12.

In exercises 1–10, follow the principles of estimation to find an approximate value. Round each number so that there is one nonzero digit. Do not find the exact value.

1. 693×307

2. 437×892

3. 2862×5986

4. 4893×6174

5. $14 + 73 + 80 + 21 + 56$

6. $318 + 494 + 613 + 243$

7. $41 \overline{)829{,}346}$

8. $16 \overline{)5{,}846{,}213}$

9. $\dfrac{2714}{31500}$

10. $\dfrac{53610}{786}$

11. Find 17% of $21,365.85.

12. Find 4.9% of $9321.88.

Applications

In exercises 13–22, determine an estimate of the exact answer. Use estimation by rounding. Do not find the exact value.

▲ **13.** *Carpeting* Estimate the cost of new carpeting that costs $22.50 per square yard, if you want to carpet a room that is 18.5 square yards in size.

▲ **14.** *Garden Design* Estimate the size of Mr. Brown's garden if it measures $15\frac{2}{3}$ feet by $11\frac{2}{3}$ feet.

15. *Checkout Sales* A typical customer at the local Piggly Wiggly supermarket spends approximately $82 at the checkout register. The store keeps four registers open, and each handles 22 customers per hour. Estimate the amount of money the store receives in one hour.

16. *Weekend Spending Money* The Westerly Community Credit Union in Rhode Island has found that on Fridays, the average customer withdraws $85 from the ATM for weekend spending money. Each ATM averages 19 customers per hour. There are five machines. Estimate the amount of money withdrawn by customers in one hour.

17. *Gas Mileage* Rod's trip from Salt Lake City to his cabin in the mountains was 117.7 miles. If his car used 3.8 gallons of gas for the trip, estimate the number of miles his car gets to the gallon.

18. *Book Transport* The Paltrows are transferring their stock of books from one store to another. To save money, they are moving the books themselves. If their car can transport 430 books per trip, and they have 11,900 books to move, estimate the number of trips it will take.

19. *Revenue* For fiscal year 2002–2003, Krispy Kreme Donuts had a total revenue of $491,549.00. If there were 279 stores systemwide, estimate the average revenue per store.

20. *Happy Meals Sales* The local McDonald's registered $437,690 in sales last month. They know from experience that 18% of their sales are Happy Meals. Estimate the amount of money spent on Happy Meals last month.

21. *Education Budget* In the year 2000, it was estimated that 26% of the California state budget of $217,970,000,000 was spent for education. Estimate the amount of money spent on education in California in 2000.

(*Source:* U.S. Census Bureau)

22. *Operations* A recent survey of local hospitals by a state medical board estimates that 6.23% of all operations performed in hospitals are unnecessary. If 28,364,122 operations were performed last year, estimate the number of unnecessary operations.

Income and Expenditures   *The data for exercises 23–26 was obtained by the U.S. Bureau of Labor Statistics.*

23. In 2002 it was found that the average consumer aged 25 to 34 years had an annual income of $38,945.89 and spent 14% of his/her income on food. Estimate the annual expenditure for food for a consumer in this age group.

24. In 2002 it was found that the average consumer aged 45 to 54 years had an annual income of $46,160.56 and spent 31% of his/her income on housing. Estimate the annual expenditure for housing for a consumer in this age group.

25. In 2002 it was found that the average consumer 65 years of age or older had an annual income of $26,533.67. It was found that this average consumer spent $3247.54 per year for health care. Estimate the percent of income spent for healthcare for consumers in this group.

26. In 2002 it was found that the average consumer aged 35 to 44 years of age had an annual income of $45,149.82. It was found that this average consumer spent $8702.47 per year for transportation. Estimate the percent of income spend for transportation for consumers in this group.

To Think About

27. ***Long Distance Phone Bills*** Laura's boyfriend, James, was transferred to San Diego. He averages 43 long-distance calls to her per month. The average cost of his long-distance calls is $3.24 per call. A new long-distance telephone company has promised James a 23% reduction in the cost of his phone bills. If that is true, estimate how much he will save in one year.

28. ***Commuting Costs*** Eddie usually drives his car to his job in Denver. His car costs 8 cents per mile to drive. He commutes 22 miles round-trip every day, and drives to work 280 days per year. He can buy an unlimited-use bus pass for one year for $300. Estimate how much he would save by taking the bus to work.

Cumulative Review

29. What is 0.6% of 350?

30. ***Basketball*** If Samuel plays basketball and scores 9 baskets out of 16 throws, what percent of the time does he score a basket?

31. ***Tipping*** Michael had breakfast at the Agawam Diner. The cost of his breakfast listed on the menu was $5.85. If he leaves a 15% tip, what will be the total amount he paid for his meal?

32. ***Tire Inspection*** In a recent tire inspection for Ford Explorers, local dealers found that 12% of the treads were defective. For each Explorer that was defective, the dealer offered immediate replacement of all five tires (including the spare). If the local dealers inspected the tires on 450 Explorers, how many new tires did they need?

1 Using the Mathematics Blueprint to Solve Real-life Problems

When a builder constructs a new home or office building, he or she often has a blueprint. This accurate drawing shows the basic form of the building. It also shows the dimensions of the structure to be built. This blueprint serves as a useful reference throughout the construction process.

Similarly, when solving real-life problems, it is helpful to have a "mathematics blueprint." This is a simple way to organize the information provided in the word problem, in a chart, or in a graph. You can record the facts you need to use. You can determine what it is you are trying to find and how you can go about actually finding it. You can record other information that you think will be helpful as you work through the problem.

As we solve real-life problems, we will use three steps.

Step 1 *Understand the problem.* Here we will read through the problem. Draw a picture if it will help, and use the Mathematics Blueprint as a guide to assist us in thinking through the steps needed to solve the problem.

Step 2 *Solve and state the answer.* We will use arithmetic or algebraic procedures along with problem-solving strategies to find a solution.

Step 3 *Check.* We will use a variety of techniques to see if the answer in step 2 is the solution to the word problem. This will include estimating to see if the answer is reasonable, repeating our calculation, and working backward from the answer to see if we arrive at the original conditions of the problem.

▲ EXAMPLE 1 Nancy and John want to install wall-to-wall carpeting in their living room. The floor of the rectangular living room is $11\frac{2}{3}$ feet wide and $19\frac{1}{2}$ feet long. How much will it cost if the carpet is $18.00 per square yard?

Solution

1. *Understand the problem.* First, read the problem carefully. Drawing a sketch of the living room may help you see what is required. The carpet will cover the floor of the living room, so we need to find the area. Now we fill in the Mathematics Blueprint.

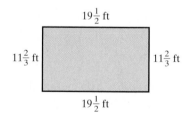

Mathematics Blueprint for Problem Solving

Gather the Facts	What Am I Solving For?	What Must I Calculate?	Key Points to Remember
The living room measures $11\frac{2}{3}$ ft by $19\frac{1}{2}$ ft. The carpet costs $18.00 per square yard.	(a) the area of the room in square feet (b) the area of the room in square yards (c) the cost of the carpet	(a) Multiply $11\frac{2}{3}$ ft by $19\frac{1}{2}$ ft to get area in square feet. (b) Divide the number of square feet by 9 to get the number of square yards. (c) Multiply the number of square yards by $18.00.	There are 9 square feet, 3 feet × 3 feet, in 1 square yard; therefore, we must divide the number of square feet by 9 to obtain square yards.

2. *Solve and state the answer.*

(a) To find the area of a rectangle, we multiply the length times the width.

$$11\frac{2}{3} \times 19\frac{1}{2} = \frac{35}{3} \times \frac{39}{2}$$

$$= \frac{455}{2} = 227\frac{1}{2} \text{ sq ft}$$

A minimum of $227\frac{1}{2}$ square feet of carpet will be needed. We say a minimum because some carpet may be wasted in cutting. Carpet is sold by the square yard. We will want to know the amount of carpet needed in square yards.

(b) To determine the area in square yards, we divide $227\frac{1}{2}$ by 9. (9 sq ft = 1 sq yd.)

$$227\frac{1}{2} \div 9 = \frac{455}{2} \div \frac{9}{1}$$

$$= \frac{455}{2} \times \frac{1}{9} = \frac{455}{18} = 25\frac{5}{18} \text{ sq yd}$$

A minimum of $25\frac{5}{18}$ square yards of carpet will be needed.

(c) Since the carpet costs $18.00 per square yard, we will multiply the number of square yards needed by $18.00.

$$25\frac{5}{18} \times 18 = \frac{455}{18} \times \frac{18}{1} = \$455$$

The carpet will cost a minimum of $455.00 for this room.

3. *Check.* We will estimate to see if our answers are reasonable.

(a) We will estimate by rounding each number to the nearest 10.

$$11\frac{2}{3} \times 19\frac{1}{2} \longrightarrow 10 \times 20 = 200 \text{ sq ft}$$

This is close to our answer of $227\frac{1}{2}$ sq ft. Our answer is reasonable. ✓

(b) We will estimate by rounding to one significant digit.

$$227\frac{1}{2} \div 9 \longrightarrow 200 \div 10 = 20 \text{ sq yd}$$

This is close to our answer of $25\frac{5}{18}$ sq yd. Our answer is reasonable. ✓

(c) We will estimate by rounding each number to the nearest 10.

$$25\frac{5}{18} \times 18 \longrightarrow 30 \times 20 = \$600$$

This is close to our answer of $455. Our answer seems reasonable. ✓

"Remember to estimate. It will save you time and money!"

▲ **Practice Problem 1**   Jeff went to help Abby pick out wall-to-wall carpet for her new house. Her rectangular living room measures $16\frac{1}{2}$ feet by $10\frac{1}{2}$ feet. How much will it cost to carpet the room if the carpet costs $20 per square yard?

NOTE TO STUDENT: Fully worked-out solutions to all of the Practice Problems can be found at the back of the text starting at page SP-1

Mathematics Blueprint for Problem Solving

Gather the Facts	What Am I Solving For?	What Must I Calculate?	Key Points to Remember

TO THINK ABOUT: Example 1 Follow-Up   Assume that the carpet in Example 1 comes in a standard width of 12 feet. How much carpet will be wasted if it is laid out on the living room floor in one strip that is $19\frac{1}{2}$ feet long? How much carpet will be wasted if it is laid in two sections side by side that are each $11\frac{2}{3}$ feet long? Assuming you have to pay for wasted carpet, what is the minimum cost to carpet the room?

EXAMPLE 2   The following chart shows the 2003 sales of Micropower Computer Software for each of the four regions of the United States. Use the chart to answer the following questions (round all answers to the nearest whole percent):

(a) What percent of the sales personnel are assigned to the Northeast?

(b) What percent of the volume of sales is attributed to the Northeast?

(c) What percent of the sales personnel are assigned to the Southeast?

(d) What percent of the volume of sales is attributed to the Southeast?

(e) Which of these two regions of the country has sales personnel that appear to be more effective in terms of the volume of sales?

Region of the U.S.	Number of Sales Personnel	Dollar Volume of Sales
Northeast	12	1,560,000
Southeast	18	4,300,000
Northwest	10	3,660,000
Southwest	15	3,720,000
Total	55	13,240,000

Solution

1. *Understand the problem.* We will only need to deal with figures from the Northeast region and the Southeast region.

Mathematics Blueprint for Problem Solving

Gather the Facts	What Am I Solving For?	What Must I Calculate?	Key Points to Remember

Personnel:
  12 Northeast
  18 Southeast
  55 total
Sales Volume:
  $1,560,000 NE
  $4,300,000 SE
  $13,240,000 Total
	(a) the percent of the total personnel in the Northeast
	(b) the percent of the total sales in the Northeast
	(c) the percent of the total personnel in the Southeast
	(d) the percent of the total sales in the Southeast
	(e) compare the percentages from the two regions
		(a) 12 of 55 is what percent? Divide. $12 \div 55$
		(b) 1,560,000 of 13,240,000 is what percent?
		  $1,560,000 \div 13,240,000$
		(c) $18 \div 55$
		(d) $4,300,000 \div 13,240,000$
			We do not need to use the numbers for the Northwest or the Southwest.

2. *Solve and state the answer.*

(a) $\dfrac{12}{55} = 0.21818\ldots$

   $\approx 22\%$

(b) $\dfrac{1{,}560{,}000}{13{,}240{,}000} = \dfrac{156}{1324} \approx 0.1178$

   $\approx 12\%$

(c) $\dfrac{18}{55} = 0.32727\ldots$

   $\approx 33\%$

(d) $\dfrac{4{,}300{,}000}{13{,}240{,}000} = \dfrac{430}{1324} \approx 0.3248$

   $\approx 32\%$

(e) We notice that 22% of the sales force in the Northeast made 12% of the sales. The percent of the sales compared to the percent of the sales force is about half (12% of 24% would be half) or 50%. 33% of the sales force in the Southeast made 32% of the sales. The percent of sales compared to the percent of the sales force is close to 100%. We must be cautious here. *If there are no other significant factors,* it would appear that the Southeast sales force is more effective. (There may be other significant factors affecting sales, such as a recession in the Northeast, new and inexperienced sales personnel, or fewer competing companies in the Southeast.)

3. *Check.* You may want to use a calculator to check the division in step 2, or you may use estimation.

(a) $\dfrac{12}{55} \rightarrow \dfrac{10}{60} \approx 0.17$

   $= 17\%$ ✓

(b) $\dfrac{1{,}560{,}000}{13{,}240{,}000} \rightarrow \dfrac{1{,}600{,}000}{13{,}000{,}000} \approx 0.12$

   $= 12\%$ ✓

(c) $\dfrac{18}{55} \rightarrow \dfrac{20}{60} \approx 0.33$

   $= 33\%$ ✓

(d) $\dfrac{4{,}300{,}000}{13{,}240{,}000} \rightarrow \dfrac{4{,}300{,}000}{13{,}000{,}000} \approx 0.33$

   $= 33\%$ ✓

Practice Problem 2   Using the chart for Example 2, answer the following questions. (Round all answers to the nearest whole percent.)

(a) What percent of the sales personnel are assigned to the Northwest?

(b) What percent of the sales volume is attributed to the Northwest?

(c) What percent of the sales personnel are assigned to the Southwest?

(d) What percent of the sales volume is attributed to the Southwest?

(e) Which of these two regions of the country has sales personnel that appear to be more effective in terms of volume of sales?

NOTE TO STUDENT: Fully worked-out solutions to all of the Practice Problems can be found at the back of the text starting at page SP-1

Mathematics Blueprint for Problem Solving

Gather the Facts	What Am I Solving For?	What Must I Calculate?	Key Points to Remember

TO THINK ABOUT: Example 2 Follow-Up   Suppose in 2006 the number of sales personnel (55) increases by 60%. What would the new number of sales personnel be? Suppose in 2006 that the number of sales personnel decreases by 60% from the number of sales personnel in 2003. What would the new number be? Why is this number not 55, since we have increased the number by 60% and then decreased the result by 60%? Explain.

Applications

Use the Mathematics Blueprint for Problem Solving to help you solve each of the following exercises.

▲ **1.** *Vinyl Flooring* Jocelyn wants to put new vinyl flooring in her kitchen. The kitchen measures $12\frac{3}{4}$ feet long by $9\frac{1}{2}$ feet wide. If the vinyl flooring she chose costs $20.00 per square yard, how much will the new flooring cost her? (Round your answer to the nearest cent.)

▲ **2.** *Decking Costs* The Carters need to replace the deck floor off their kitchen door. The deck is $11\frac{1}{2}$ feet by $20\frac{1}{2}$ feet. If the new decking costs $4.50 per square foot, how much will it cost them to replace the deck? (Round your answer to the nearest cent.)

▲ **3.** *Garden Design* Jenna White has room for a rectangular garden that can measure as much as $15\frac{1}{2}$ feet wide and $25\frac{2}{3}$ feet long.

   (a) She decides the garden should be as large as possible, but wants to fence it off to protect the vegetables from the neighborhood raccoons. How much fencing does she need?

   (b) She finds that she can buy a prepackaged 90 feet of fencing for $155.00 or she can get a cut-to-order length for $2.10 per foot. Which should she buy? How much money does she save?

▲ **4.** *Pond Design* The Lee family is installing a backyard pond for their prize koi. (A koi is a kind of fish bred in Japan. They are noted for their large size and variety of color.)

   (a) If the pond measures $11\frac{1}{2}$ feet by 7 feet, and is $3\frac{1}{2}$ feet deep, how much water will it take to fill the pond?

   (b) If each cubic foot is 7.5 gallons, how many gallons of water will the pond hold?

   (c) If at least 200 gallons of water is recommended for each koi, how many koi can the Lee family place in their pond?

Exercise Training   *The following directions are posted on the wall at the gym.*

Beginning exercise training schedule
On day 1, each athlete will begin the morning as follows:
Jog.................. $1\frac{1}{2}$ miles
Walk............. $1\frac{3}{4}$ miles
Rest............... $2\frac{1}{2}$ minutes
Walk............. 1 mile

5. Betty's athletic trainer told her to follow the beginning exercise training schedule on day 1. On day 2, she is to increase all distances and times by $\frac{1}{3}$ that of day 1. On day 3, she is to increase all distances and times by $\frac{1}{3}$ that of day 2. What will be her training schedule on day 3?

6. Melinda's athletic trainer told her to follow the beginning exercise training schedule on day 1. On day 2, she is to increase all distances and times by $\frac{1}{3}$ that of day 1. On day 3, she is to once again increase all distances and times by $\frac{1}{3}$ that of day 1. What will be her training schedule on day 3?

To Think About

Refer to exercises 5 and 6 in working exercises 7–10.

7. Who will have a more demanding schedule on day 3, Betty or Melinda? Why?

8. If Betty kept up the same type of increase day after day, how many miles would she be jogging on day 5?

9. If Melinda kept up the same type of increase day after day, how many miles would she be jogging on day 7?

10. Which athletic trainer would appear to have the best plan for training athletes if they used this plan for 14 days? Why?

11. *House Prices* In 1985, the average selling price of an existing single-family home in Atlanta, Georgia, was $66,200. Between 1985 and 1990, the average price increased by 30%. Between 1990 and 2005 the average price increased again, this time by 15%. What was the median house price in Atlanta in 2005?

12. *Egg Weight* Chicken eggs are classified by weight per dozen eggs. Large eggs weigh 24 ounces per dozen and medium eggs weigh 21 ounces per dozen.
  (a) If you do not include the shell, which is 12% of the total weight of an egg, how many ounces of eggs do you get from a dozen large eggs? From a dozen medium eggs?

  (b) At a local market, large eggs sell for $1.79 a dozen, and medium eggs for $1.39 a dozen. If you do not include the shell, which is a better buy, large or medium eggs?

Family Budgets   *For the following problems, use the chart below.*

The Johnson family has created the following budget:

Rent	20%	Clothing	10%
Food	31%	Medical	12%
Utilities	5%	Savings	10%
Entertainment	6%	Miscellaneous	6%

13. The income for this family comes from Mrs. Johnson's annual salary of $50,000.
  (a) If 28% of her salary is withheld for various taxes, how much money does the family have available for their budget?

  (b) How much of Mrs. Johnson's take home pay is budgeted for food?

14. With one child ready to head off to college, Mrs. Johnson is looking for ways to save money.
  (a) She figures out that if they plant a garden this year, the family can save 12% of their food costs. How much will she save in food costs? (This does not account for the cost of planting a garden.) What is the new food budget amount?

  (b) What percentage of Mrs. Johnson's budget is the new food amount?

Paycheck Stub   *Use the following information from a paycheck stub to solve exercises 15–18.*

TOBEY & SLATER INC.					Check Number	Payroll Period		Pay Date

| TOBEY & SLATER INC. 5000 Stillwell Avenue Queens, NY 10001 | | | | | Check Number 495885 | From Date 10-30-04 | To Date 11-30-04 | Pay Date 12-01-04 |
| Name Fred J. Gilliani | Social Security No. 012-34-5678 | I.D. Number 01 | File Number 1379 | Rate/Salary 1150.00 | Department 0100 | MS M | DEP 5 | Res NY |

	Current	Year to Date			Current	Year to Date
GROSS	1,150.00	6,670.00	STATE		67.76	388.45
FEDERAL	138.97	781.07	LOCAL		5.18	30.04
FICA	87.98	510.28	DIS-SUI		.00	.00
W-2 GROSS		6,670.00	NET		790.47	4,960.16

Earnings						Special Deductions		
No.	Type	Hours	Rate	Amount	Dept/Job No.	No.	Description	Amount
96	REGULAR			1,150.00	0100	82	Retirement	12.56
						75	Medical	36.28
						56	Union Dues	10.80

Gross pay is the pay an employee receives for his or her services before deductions. Net pay is the pay the employee actually gets to take home. You may round each amount to the nearest whole percent for exercises 15–18.

15. What percent of Fred's gross pay is deducted for federal, state, and local taxes?

16. What percent of Fred's gross pay is deducted for retirement and medical?

17. What percent of Fred's gross pay does he actually get to take home?

18. What percent of Fred's deductions are special deductions?

Developing Your Study Skills

Reading the Textbook

Your homework time each day should begin with the careful reading of the section(s) assigned in your textbook. Usually, much time and effort have gone into the selection of a particular text, and your instructor has decided that this is the book that will help you to become successful in this mathematics class. Textbooks are expensive, but they can be a wise investment if you take advantage of them by reading them.

Reading a mathematics textbook is unlike reading many other types of books that you may find in your literature, history, psychology, or sociology courses. Mathematics texts are technical books that provide you with exercises for practice. Reading a mathematics text requires slow and careful reading of each word, which takes time and effort.

Begin reading your textbook with a paper and pencil in hand. As you come across a new definition or concept, underline it in the text and/or write it down in your notebook. Whenever you encounter an unfamiliar term, look it up and make a note of it. When you come to an example, work through it step-by-step. Be sure to read each word and to follow directions carefully.

Notice the helpful hints that the author provides. They guide you to correct solutions and prevent you from making errors. Take advantage of these pieces of expert advice.

Be sure that you understand what you are reading. Make a note of any of those things that you do not understand and ask your instructor about them. Do not hurry through the material. Learning mathematics takes time.

Putting Your Skills to Work

Bodies of Water: The World's Largest Lakes

Did you know that any body of water surrounded by land is called a lake, even salty bodies of water? Thus, the Caspian Sea in Asia, the Aral Sea in Asia, the Great Salt Lake in the United States, and the Eyre in Australia are all considered lakes. Look at the following information on the largest *freshwater* lakes in the world.

Lake	Location	Area (in 1000 square miles)
Superior	US	31.8
Victoria	Africa	26.6
Huron	United States, Canada	23
Michigan	United States	22.4
Tanganyika	Africa	13.9
Great Bear	Canada	12.3
Lake Baikul	Russia	11.8

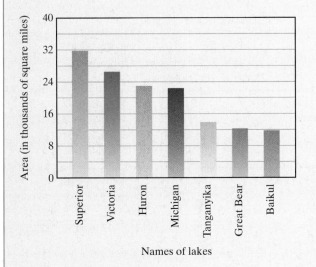

Problems for Individual Investigation and Analysis:

1. How much larger in area is Lake Victoria than Lake Tanganyika?

2. How much larger in area is Lake Superior than Lake Michigan?

Problems for Group Investigation and Cooperative Learning:

3. The smallest Great Lake is Lake Ontario with an area of 7540 sq. mi. About how many times would Lake Ontario fit into Lake Superior?

4. The largest lake in New Hampshire is Lake Winnipesaukee with an area of 72 square miles. About how many times would Lake Winnipesaukee fit into Lake Huron?

Topic	Procedure	Examples
Simplifying fractions, p. 3.	**1.** Write the **numerator** and **denominator** as a product of prime factors. **2.** Use the basic rule of fractions that $$\frac{a \times c}{b \times c} = \frac{a}{b}$$ for any factor that appears in both the numerator and the denominator. **3.** Multiply the remaining factors for the numerator and separately for the denominator.	$$\frac{15}{25} = \frac{\cancel{5} \cdot 3}{\cancel{5} \cdot 5} = \frac{3}{5}$$ $$\frac{36}{48} = \frac{\cancel{2} \cdot \cancel{2} \cdot 3 \cdot \cancel{3}}{\cancel{2} \cdot \cancel{2} \cdot 2 \cdot 2 \cdot \cancel{3}} = \frac{3}{4}$$ $$\frac{26}{39} = \frac{2 \cdot \cancel{13}}{3 \cdot \cancel{13}} = \frac{2}{3}$$
Changing improper fractions to mixed numbers, p. 5.	**1.** Divide the denominator into the numerator to obtain the whole-number part of the mixed fraction. **2.** The remainder from the division will be the numerator of the fraction. **3.** The denominator remains unchanged.	$\frac{14}{3} = 4\frac{2}{3}$ $\quad$ $\frac{19}{8} = 2\frac{3}{8}$ since $3\overline{)14}$ $\quad$ since $8\overline{)19}$ $ \frac{12}{2}$ $\qquad$ $ \frac{16}{3}$
Changing mixed numbers to improper fractions, p. 6.	**1.** Multiply the whole number by the denominator and add the result to the numerator. This will yield the new numerator. **2.** The denominator does not change.	$$4\frac{5}{6} = \frac{(4 \times 6) + 5}{6} = \frac{24 + 5}{6} = \frac{29}{6}$$ $$3\frac{1}{7} = \frac{(3 \times 7) + 1}{7} = \frac{21 + 1}{7} = \frac{22}{7}$$
Changing fractions to equivalent fractions with a given denominator, p. 7.	**1.** Divide the original denominator into the new denominator. This result is the value that we use for multiplication. **2.** Multiply the numerator and the denominator of the original fraction by that value.	$$\frac{4}{7} = \frac{?}{21}$$ $3 \leftarrow$ Use this to multiply $\frac{4 \times 3}{7 \times 3} = \frac{12}{21}$ $7\overline{)21}$
Finding the LCD (least common denominator) of two or more fractions, p. 11.	**1.** Write each denominator as the product of prime factors. **2.** The LCD is a product containing each different factor. **3.** If a factor occurs more than once in any one denominator, the LCD will contain that factor repeated the greatest number of times that it occurs in any one denominator.	Find the LCD of $\frac{4}{15}$ and $\frac{3}{35}$. $15 = 5 \cdot 3$ $35 = 5 \cdot 7$ LCD $= 3 \cdot 5 \cdot 7 = 105$ Find the LCD of $\frac{11}{18}$ and $\frac{7}{45}$. $18 = 3 \cdot 3 \cdot 2$ (factor 3 appears twice) $45 = 3 \cdot 3 \cdot 5$ (factor 3 appears twice) LCD $= 2 \cdot 3 \cdot 3 \cdot 5 = 90$
Adding and subtracting fractions that do not have a common denominator, p. 12.	**1.** Find the LCD. **2.** Change each fraction to an equivalent fraction with the LCD for a denominator. **3.** Add or subtract the fractions and simplify the answer if possible.	$$\frac{3}{8} + \frac{1}{3} = \frac{3 \cdot 3}{8 \cdot 3} + \frac{1 \cdot 8}{3 \cdot 8} = \frac{9}{24} + \frac{8}{24} = \frac{17}{24}$$ $$\frac{11}{12} - \frac{1}{4} = \frac{11}{12} - \frac{1 \cdot 3}{4 \cdot 3} = \frac{11}{12} - \frac{3}{12} = \frac{8}{12} = \frac{2}{3}$$
Adding and subtracting mixed numbers, p. 15.	**1.** Change the mixed numbers to improper fractions. **2.** Follow the rules for adding and subtracting fractions. **3.** If necessary, change your answer to a mixed number.	$$1\frac{2}{3} + 1\frac{3}{4} = \frac{5}{3} + \frac{7}{4} = \frac{5 \cdot 4}{3 \cdot 4} + \frac{7 \cdot 3}{4 \cdot 3}$$ $$= \frac{20}{12} + \frac{21}{12} = \frac{41}{12} = 3\frac{5}{12}$$ $$2\frac{1}{4} - 1\frac{3}{4} = \frac{9}{4} - \frac{7}{4} = \frac{2}{4} = \frac{1}{2}$$
Multiplying fractions, p. 20.	**1.** If there are no common factors, multiply the numerators. Then multiply the denominators. **2.** If possible, write the numerators and denominators as the product of prime factors. Use the basic rule of fractions to divide out any value that appears in both a numerator and a denominator. Multiply the remaining factors in the numerator. Multiply the remaining factors in the denominator.	$$\frac{3}{7} \times \frac{2}{13} = \frac{6}{91}$$ $$\frac{6}{15} \times \frac{35}{91} = \frac{2 \cdot \cancel{3}}{\cancel{3} \cdot \cancel{5}} \times \frac{\cancel{5} \cdot \cancel{7}}{\cancel{7} \cdot 13} = \frac{2}{13}$$ $$3 \times \frac{5}{8} = \frac{3}{1} \times \frac{5}{8} = \frac{15}{8} \text{ or } 1\frac{7}{8}$$

Topic	Procedure	Examples
Dividing fractions, p. 22.	1. Change the division sign to multiplication. 2. Invert the second fraction. 3. Multiply the fractions.	$\dfrac{4}{7} \div \dfrac{11}{3} = \dfrac{4}{7} \times \dfrac{3}{11} = \dfrac{12}{77}$ $\dfrac{5}{9} \div \dfrac{5}{7} = \dfrac{\cancel{5}}{9} \times \dfrac{7}{\cancel{5}} = \dfrac{7}{9}$
Multiplying and dividing mixed numbers, pp. 21 and 24.	1. Change each mixed number to an improper fraction. 2. Use the rules for multiplying or dividing fractions. 3. Change your answer to a mixed number.	$2\dfrac{1}{4} \times 3\dfrac{3}{5} = \dfrac{9}{4} \times \dfrac{18}{5}$ $= \dfrac{3 \cdot 3}{2 \cdot \cancel{2}} \times \dfrac{\cancel{2} \cdot 3 \cdot 3}{5} = \dfrac{81}{10} = 8\dfrac{1}{10}$ $1\dfrac{1}{4} \div 1\dfrac{1}{2} = \dfrac{5}{4} \div \dfrac{3}{2} = \dfrac{5}{2 \cdot \cancel{2}} \times \dfrac{\cancel{2}}{3} = \dfrac{5}{6}$
Changing fractional form to decimal form, p. 29.	Divide the denominator into the numerator.	$\dfrac{5}{8} = 0.625 \ \text{ since } \ 8\overline{)5.000}^{\,0.625}$
Changing decimal form to fractional form, p. 30.	1. Write the decimal as a fraction with a denominator of 10, 100, 1000, and so on. 2. Simplify the fraction, if possible.	$0.37 = \dfrac{37}{100} \qquad 0.375 = \dfrac{375}{1000} = \dfrac{3}{8}$
Adding and subtracting decimals, p. 31.	1. Carefully line up the decimal points as indicated for addition and subtraction. (Extra zeros may be added to the right-hand side of the decimals if desired.) 2. Add or subtract the appropriate digits.	Add. $\qquad$ Subtract. $1.236 + 7.825 \qquad 2 - 1.32$ $\begin{array}{r} 1.236 \\ + 7.825 \\ \hline 9.061 \end{array} \qquad \begin{array}{r} 2.00 \\ - 1.32 \\ \hline 0.68 \end{array}$
Multiplying decimals, p. 32.	1. First multiply the digits. 2. Count the total number of decimal places in the numbers being multiplied. 3. This number determines the number of decimal places in the answer.	$\begin{array}{r} 0.9 \ \text{(one place)} \\ \times 0.7 \ \text{(one place)} \\ \hline 0.63 \ \text{(two places)} \end{array} \quad \begin{array}{r} 0.009 \ \text{(three places)} \\ \times \ 0.07 \ \text{(two places)} \\ \hline 0.00063 \ \text{(five places)} \end{array}$
Dividing decimals, p. 33.	1. Count the number of decimal places in the divisor. 2. Move the decimal point to the right the same number of places in both the divisor and dividend. 3. Mark that position with a caret $(_\wedge)$. 4. Perform the division. Line up the decimal point in the quotient with the position indicated by the caret in the dividend.	Divide. $7.5 \div 0.6$. Move decimal point one place to the right. $\begin{array}{r} 12.5 \\ 0.6_\wedge \overline{)7.5_\wedge 0} \\ \underline{6} \\ 15 \\ \underline{12} \\ 30 \\ \underline{30} \\ 0 \end{array}$ $\quad$ Therefore, $7.5 \div 0.6 = 12.5$
Changing a decimal to a percent, p. 40.	1. Move the decimal point two places to the right. 2. Add the % symbol.	$0.46 = 46\%$ $0.002 = 0.2\% \qquad 1.59 = 159\%$ $0.013 = 1.3\% \qquad 0.0007 = 0.07\%$
Changing a percent to a decimal, p. 41.	1. Remove the % symbol. 2. Move the decimal point two places to the left.	$49\% = 0.49 \qquad 180\% = 1.8$ $59.8\% = 0.598 \qquad 0.13\% = 0.0013$
Finding a percent of a number, p. 42.	1. Convert the percent to a decimal. 2. Multiply the decimal by the number.	Find 12% of 86. $12\% = 0.12 \qquad 86 \times 0.12 = 10.32$ Therefore, 12% of $86 = 10.32$.
Finding what percent one number is of another number, p. 43.	1. Place the number after the word *of* in the denominator. 2. Place the other number in the numerator. 3. If possible, simplify the fraction. 4. Change the fraction to a decimal. 5. Express the decimal as a percent.	What percent of 8 is 7? $\dfrac{7}{8} = 0.875 = 87.5\%$ 42 is what percent of 12? $\dfrac{42}{12} = \dfrac{7}{2} = 3.5 = 350\%$

Topic	Procedure	Examples
Estimation, p. 47.	1. Round each number so that there is one nonzero digit. 2. Perform the calculation with the rounded numbers.	Estimate the number of square feet in a room that is 22 feet long and 13 feet wide. Assume that the room is rectangular. 1. We round 22 to 20. We round 13 to 10. 2. To find the area of a rectangle, we multiply length times width. $$20 \times 10 = 200.$$ We estimate that there are 200 square feet in the room.

Problem solving, p. 52.	In solving a real-life problem, you may find it helpful to complete the following steps. You will not use all of the steps all of the time. Choose the steps that best fit the conditions of the problem.
	1. Understand the problem.
	  (a) Read the problem carefully.
	  (b) Draw a picture if this helps you.
	  (c) Use the Mathematics Blueprint for Problem Solving.
	2. Solve and state the answer.
	3. Check.
	  (a) Estimate to see if your answer is reasonable.
	  (b) Repeat your calculation.
	  (c) Work backward from your answer. Do you arrive at the original conditions of the problem?
		Susan is installing wall-to-wall carpeting in her $10\frac{1}{2}$-ft-by-12-ft bedroom. How much will it cost at \$20 a square yard?
		1. Understand the problem.
		We need to find the area of the room in square yards. Then we can find the cost.
		2. Solve and state the answer.
		Area: $10\frac{1}{2} \times 12 = \frac{21}{2} \times \frac{12}{1} = 126$ sq ft
		$126 \div 9 = 14$ sq yd
		Cost: $14 \times 20 = \$280$
		The carpeting will cost \$280.
		3. Check.
		Estimate: $10 \times 12 = 120$ sq ft
		$120 \div 9 = 13\frac{1}{3}$ sq yd
		$13 \times 20 = \$260$
		Our answer is reasonable. ✓

Chapter 0 Review Problems

Section 0.1

In exercises 1–4, simplify.

1. $\dfrac{36}{48}$

2. $\dfrac{15}{50}$

3. $\dfrac{36}{82}$

4. $\dfrac{18}{30}$

5. Write $4\dfrac{3}{5}$ as an improper fraction.

6. Write $\dfrac{34}{5}$ as a mixed number.

7. Write $\dfrac{39}{6}$ as a mixed number.

Change each fraction to an equivalent fraction with the specified denominator.

8. $\dfrac{5}{8} = \dfrac{?}{24}$

9. $\dfrac{1}{7} = \dfrac{?}{35}$

10. $\dfrac{5}{9} = \dfrac{?}{72}$

11. $\dfrac{2}{5} = \dfrac{?}{55}$

Section 0.2

Combine.

12. $\dfrac{3}{5} + \dfrac{1}{4}$

13. $\dfrac{7}{12} + \dfrac{5}{8}$

14. $\dfrac{7}{20} - \dfrac{1}{12}$

15. $\dfrac{7}{10} - \dfrac{4}{15}$

16. $3\dfrac{1}{6} + 2\dfrac{3}{5}$

17. $1\dfrac{1}{4} + 2\dfrac{7}{10}$

18. $6\dfrac{2}{9} - 3\dfrac{5}{12}$

19. $3\dfrac{1}{15} - 1\dfrac{3}{20}$

Section 0.3

Multiply.

20. $6 \times \dfrac{5}{11}$ **21.** $2\dfrac{1}{3} \times 4\dfrac{1}{2}$ **22.** $1\dfrac{1}{8} \times 2\dfrac{1}{9}$ **23.** $\dfrac{4}{7} \times 5$

Divide.

24. $\dfrac{3}{8} \div 6$ **25.** $\dfrac{\dfrac{8}{3}}{\dfrac{5}{9}}$ **26.** $\dfrac{15}{16} \div 6\dfrac{1}{4}$ **27.** $2\dfrac{6}{7} \div \dfrac{10}{21}$

Section 0.4

Combine.

28. $1.634 + 3.007 + 2.560$ **29.** $24.831 - 17.094$ **30.** $47.251 - 17.69$ **31.** $1.9 + 2.53 + 0.006$

Multiply.

32. 0.007×5.35 **33.** 362.341×1000 **34.** $2.6 \times 0.03 \times 1.02$ **35.** $1.08 \times 0.06 \times 160$

Divide.

36. $0.186 \div 100$ **37.** $71.32 \div 1000$ **38.** $0.523 \div 0.4$ **39.** $1.35 \div 0.015$

40. $4.186 \div 2.3$ **41.** $0.19 \div 0.38$ **42.** Write as a decimal: $\dfrac{3}{8}$.

43. Write as a fraction in simplified form: 0.36.

Section 0.5

In exercises 44–47, write each percentage in decimal form.

44. 1.4%   **45.** 36.1%   **46.** 0.02%

47. 125.3%   **48.** What is 85% of 600?   **49.** Find 7.2% of 55.

50. 48 is what percent of 75?   **51.** What percent of 120 is 15?   **52.** What percent of 1250 is 750?

53. *Education Level* In 1997, 80.7% of California residents over the age of 25 reported that they had at least a 4-year high school education. If the population of California over the age of 25 was 20,229,000, how many people reported at least a high school education?   (*Source*: U.S. Census Bureau)

54. *Math Deficiency* In a given university, 720 of the 960 freshmen had a math deficiency. What percentage of the class had a math deficiency?

Section 0.6

In exercises 55–60, estimate. Do not find an exact value.

55. $234{,}897 \times 1{,}936{,}112$

56. $357 + 923 + 768 + 417$

57. $634{,}318 - 284{,}000$

58. $7\frac{1}{3} + 3\frac{5}{6} + 8\frac{3}{7}$

59. Find 18% of $56,297.

60. $12{,}482 \div 389$

61. *Salary* Estimate Juan's salary for the week if he makes $7.85 per hour, and worked 32.5 hours.

62. *Apartment Sharing* Estimate the monthly cost for each of three roommates who want to share an apartment that costs $923.50 per month.

Section 0.7

Solve. You may use the Mathematics Blueprint for Problem Solving.

▲ **63.** *Carpeting* Mr. and Mrs. Carr are installing wall-to-wall carpeting in a room that measures $12\frac{1}{2}$ ft by $9\frac{2}{3}$ ft. How much will it cost if the carpet is $26.00 per square yard?

64. *Population* The population of Falmouth was 34,000 in 1980 and 36,720 in 2000. What was the percent of increase in population?

Gas Mileage A six-passenger Piper Cub airplane has a gas tank that holds 240 gallons. Use this information to answer exercises 65 and 66.

65. When flying at cruising speed, the plane averages $7\frac{2}{3}$ miles per gallon. How far can the plane fly at cruising speed? If the pilot never plans to fly more than 80% of his maximum cruising distance, what is the longest trip he would plan to fly?

66. When flying at maximum speed, the plane averages $6\frac{1}{4}$ miles per gallon. How far can the plane fly at maximum speed? If the pilot never plans to fly more than 70% of his maximum flying distance when flying at full speed, what is the longest trip he would plan to fly at full speed?

▲ **67.** *Geometry* What is the volume of a matchbox toy box if it measures: $4''$ long $\times$ $2\frac{1}{2}''$ long $\times$ $\frac{3}{4}''$ high?

▲ **68.** *Geometry* Using your results from Exercise 67, if a packing crate measures $16''$ long $\times$ $12\frac{1}{2}''$ wide $\times$ $7\frac{1}{2}''$ high, how many matchbox toy boxes can be packed inside the crate?

69. *Sales Commission* Mike sells sporting goods on an 8% commission. During the first week in July, he sold goods worth $5785. What was his commission for the week?

70. *Package Weight* Sam took to the post office several packages that weighed 0.75 pounds each. The total weight of all the packages was 34.5 pounds. How many packages did he bring to the post office?

71. *Car Loan* Dick and Ann Wright purchased a new car. They took out a loan of $9214.50 to help pay for the car and paid the rest in cash. They paid off the loan with payments of $225 per month for four years. How much more did they pay back than the amount of the car loan? (This is the amount of interest they were charged for the car loan.)

72. *Wages* Frank works as a clerk at Wal-Mart. He is paid $6.40 an hour for a 40-hour week. For any additional time he gets paid 1.5 times the normal rate. Last week he worked 52 hours. How much did he get paid last week?

Remember to use your Chapter Test Prep Video CD to see the worked-out solutions to the test problems you want to review.

In exercises 1 and 2, simplify.

1. $\dfrac{16}{18}$

2. $\dfrac{48}{36}$

3. Write as an improper fraction. $6\dfrac{3}{7}$

4. Write as a mixed number: $\dfrac{105}{9}$.

In exercises 5–12, perform the operations indicated. Simplify answers whenever possible.

5. $\dfrac{2}{3} + \dfrac{5}{6} + \dfrac{3}{8}$

6. $1\dfrac{1}{8} + 3\dfrac{3}{4}$

7. $3\dfrac{2}{3} - 2\dfrac{5}{6}$

8. $\dfrac{5}{7} \times \dfrac{28}{15}$

9. $\dfrac{5}{18} \times \dfrac{3}{4}$

10. $\dfrac{7}{4} \div \dfrac{1}{2}$

11. $2\dfrac{1}{2} \times 3\dfrac{1}{4}$

12. $5\dfrac{3}{8} \div 2\dfrac{3}{4}$

In exercises 13–18, perform the calculations indicated.

13. $1.6 + 3.24 + 9.8$

14. $7.0046 - 3.0149$

15. 32.8×0.04

16. 0.07385×1000

17. $12.88 \div 0.056$

18. $26{,}325.9 \div 100$

1. _____

2. _____

3. _____

4. _____

5. _____

6. _____

7. _____

8. _____

9. _____

10. _____

11. _____

12. _____

13. _____

14. _____

15. _____

16. _____

17. _____

18. _____

19. _____

19. Write as a percent. 0.073   **20.** Write as a decimal. 196.5%

20. _____

21. What is 3.5% of 180?   **22.** What is 2% of 16.8?

21. _____

23. 39 is what percent of 650?   **24.** What percent of 460 is 138?

22. _____

25. A 4-inch stack of computer chips is on the table. Each computer chip is $\frac{2}{9}$ of an inch thick. How many computer chips are in the stack?

23. _____

In exercises 26–27, estimate. Round each number to one nonzero digit. Then calculate.

26. $52{,}344\overline{)4{,}678{,}987}$

24. _____

25. _____

27. $285.36 + 311.85 + 113.6$

26. _____

Solve. You may use the Mathematics Blueprint for Problem Solving.

27. _____

28. Allison is paid \$14,000 per year plus a sales commission of 3% of the value of her sales. Last year she sold \$870,000 worth of products. What percent of her total income was her commission?

28. _____

▲ **29.** Fred and Melinda are laying wall tile in the kitchen. Each tile covers $3\frac{1}{2}$ square inches of space. They plan to cover 210 square inches of wall space. How many tiles will they need?

29. _____

Do you have any idea how much the use of the Internet has grown in recent years? Do you think it has spread to farmers and fishermen? Do you have any idea how many people use a computer both at home and at work? Turn to the Putting Your Skills to Work on page 125 and see how good your estimates were.

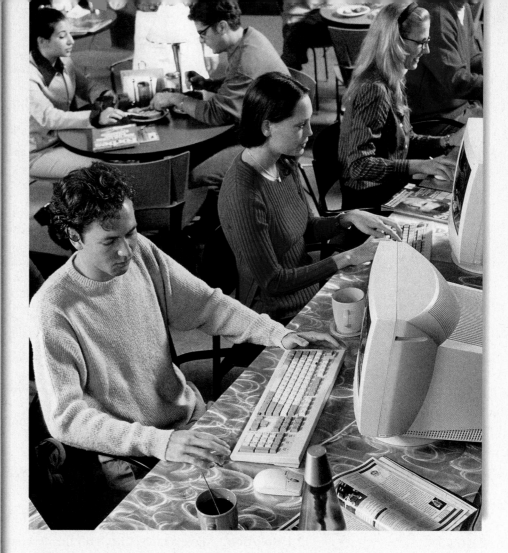

Real Numbers and Variables

1.2   SUBTRACTING REAL NUMBERS   79

1.3   MULTIPLYING AND DIVIDING REAL NUMBERS   84

1.4   EXPONENTS   94

1.5   THE ORDER OF OPERATIONS   98

       HOW AM I DOING? SECTIONS 1.1–1.5   102

1.6   USING THE DISTRIBUTIVE PROPERTY TO SIMPLIFY ALGEBRAIC EXPRESSIONS   103

1.7   COMBINING LIKE TERMS   108

1.8   USING SUBSTITUTION TO EVALUATE ALGEBRAIC EXPRESSIONS AND FORMULAS   113

1.9   GROUPING SYMBOLS   121

       CHAPTER 1 ORGANIZER   126

       CHAPTER 1 REVIEW PROBLEMS   127

       HOW AM I DOING? CHAPTER 1 TEST   131

Student Learning Objectives

After studying this section, you will be able to:

1. Identify different types of numbers.

2. Use real numbers in real-life situations.

3. Add real numbers with the same sign.

4. Add real numbers with opposite signs.

5. Use the addition properties for real numbers.

① Identifying Different Types of Numbers

Let's review some of the basic terms we use to talk about numbers.

Whole numbers are numbers such as $0, 1, 2, 3, 4, \ldots$

Integers are numbers such as $\ldots, -3, -2, -1, 0, 1, 2, 3, \ldots$.

Rational numbers are numbers such as $\frac{3}{2}, \frac{5}{7}, -\frac{3}{8}, -\frac{4}{13}, \frac{6}{1}$, and $-\frac{8}{2}$.

Rational numbers can be written as one integer divided by another integer (as long as the denominator is not zero!). Integers can be written as fractions ($3 = \frac{3}{1}$, for example), so we can see that all integers are rational numbers. Rational numbers can be expressed in decimal form. For example, $\frac{3}{2} = 1.5$, $-\frac{3}{8} = -0.375$, and $\frac{1}{3} = 0.333\ldots$ or $0.\overline{3}$. It is important to note that rational numbers in decimal form are either terminating decimals or repeating decimals.

Irrational numbers are numbers that cannot be expressed as one integer divided by another integer. The numbers π, $\sqrt{2}$, and $\sqrt[3]{7}$ are irrational numbers.

Irrational numbers can be expressed in decimal form. The decimal form of an irrational number is a nonterminating, nonrepeating decimal. For example, $\sqrt{2} = 1.414213\ldots$ can be carried out to an infinite number of decimal places with no repeating pattern of digits

Finally, **real numbers** are all the rational numbers and all the irrational numbers.

EXAMPLE 1 Classify as an integer, a rational number, an irrational number, and/or a real number.

(a) 5 **(b)** $-\dfrac{1}{3}$ **(c)** 2.85 **(d)** $\sqrt{2}$ **(e)** $0.777\ldots$

Solution Make a table. Check off the description of the number that applies.

	Number	Integer	Rational Number	Irrational Number	Real Number
(a)	5	✓	✓		✓
(b)	$-\frac{1}{3}$		✓		✓
(c)	2.85		✓		✓
(d)	$\sqrt{2}$			✓	✓
(e)	$0.777\ldots$		✓		✓

Practice Problem 1 Classify.

(a) $-\dfrac{2}{5}$ **(b)** $1.515151\ldots$ **(c)** -8 **(d)** π

NOTE TO STUDENT: Fully worked-out solutions to all of the Practice Problems can be found at the back of the text starting at page SP-1

Any real number can be pictured on a **number line.**

Positive numbers are to the right of 0 on the number line.

Negative numbers are to the left of 0 on the number line.

The **real numbers** include the positive numbers, the negative numbers, and zero.

② Using Real Numbers in Real-Life Situations

We often encounter practical examples of number lines that include positive and negative rational numbers. For example, we can tell by reading the accompanying thermometer that the temperature is 20° below 0. From the stock market report, we see that the stock opened at 36 and closed at 34.5, and the net change for the day was −1.5.

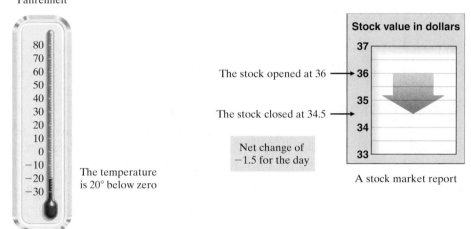

Temperature in degrees Fahrenheit

The temperature is 20° below zero

Stock value in dollars

The stock opened at 36

The stock closed at 34.5

Net change of −1.5 for the day

A stock market report

In the following example we use real numbers to represent real-life situations.

EXAMPLE 2   Use a real number to represent each situation.

(a) A temperature of 128.6°F below zero is recorded at Vostok, Antarctica.

(b) The Himalayan peak K2 rises 29,064 feet above sea level.

(c) The Dow gains 10.24 points.

(d) An oil drilling platform extends 328 feet below sea level.

Solution   A key word can help you to decide whether a number is positive or negative.

(a) 128.6°F *below* zero is −128.6.

(b) 29,064 feet *above* sea level is +29,064.

(c) A *gain* of 10.24 points is +10.24.

(d) 328 feet *below* sea level is −328.

Practice Problem 2   Use a real number to represent each situation.

(a) A population growth of 1259   **(b)** A depreciation of $763

(c) A wind-chill factor of minus 10

In everyday life we consider positive numbers the opposite of negative numbers. For example, a gain of 3 yards in a football game is the opposite of a loss of 3 yards; a check written for $2.16 on a checking account is the opposite of a deposit of $2.16.

Each positive number has an opposite negative number. Similarly, each negative number has an opposite positive number. **Opposite numbers,** also called **additive inverses,** have the same magnitude but different signs and can be represented on the number line.

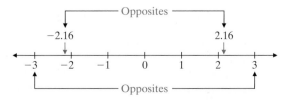

Opposites

−2.16   2.16

Opposites

EXAMPLE 3   Find the additive inverse (that is, the opposite).

(a) -7 **(b)** $\dfrac{1}{4}$

Solution

(a) The opposite of -7 is $+7$.   **(b)** The opposite of $\dfrac{1}{4}$ is $-\dfrac{1}{4}$.

Practice Problem 3   Find the additive inverse (the opposite).

(a) $+\dfrac{2}{5}$ **(b)** -1.92 **(c)** a loss of 12 yards on a football play

NOTE TO STUDENT: Fully worked-out solutions to all of the Practice Problems can be found at the back of the text starting at page SP-1

③ Adding Real Numbers with the Same Sign

To use a real number, we need to be clear about its sign. When we write the number three as $+3$, the sign indicates that it is a positive number. The positive sign can be omitted. If someone writes three (3), it is understood that it is a positive three ($+3$). To write a negative number such as negative three (-3), we must include the sign.

A concept that will help us add and subtract real numbers is the idea of absolute value. The **absolute value** of a number is the distance between that number and zero on the number line. The absolute value of 3 is written $|3|$.

Distance is always a positive number regardless of the direction we travel. This means that the absolute value of any number will be a positive value or zero. We place the symbols | and | around a number to mean the absolute value of the number.

The distance from 0 to 3 is 3, so $|3| = 3$. This is read "the absolute value of 3 is 3."

The distance from 0 to -3 is 3, so $|-3| = 3$. This is read "the absolute value of -3 is 3."

Some other examples are

$$|-22| = 22, \qquad |5.6| = 5.6, \qquad \text{and} \qquad |0| = 0.$$

Thus, the absolute value of a number can be thought of as the magnitude of the number, without regard to its sign.

EXAMPLE 4   Find the absolute value.

(a) $|-4.62|$ **(b)** $\left|\dfrac{3}{7}\right|$ **(c)** $|0|$

Solution

(a) $|-4.62| = 4.62$ **(b)** $\left|\dfrac{3}{7}\right| = \dfrac{3}{7}$ **(c)** $|0| = 0$

Practice Problem 4   Find the absolute value.

(a) $|-7.34|$ **(b)** $\left|\dfrac{5}{8}\right|$ **(c)** $\left|\dfrac{0}{2}\right|$

Now let's look at addition of real numbers when the two numbers have the same sign. Suppose that you are keeping track of your checking account at a local bank. When you make a deposit of 5 dollars, you record it as +5. When you write a check for 4 dollars, you record it as −4, as a debit. Consider two situations.

SITUATION 1: Total Deposit You made a deposit of 20 dollars on one day and a deposit of 15 dollars the next day. You want to know the total value of your deposits.
Your record for situation 1.

$$20 + 15 = \boxed{35}$$

The amount of the deposit on the first day added to the amount of the deposit on the second day is the total of the deposits made over the two days.

SITUATION 2: Total Debit You write a check for 25 dollars to pay one bill and two days later write a check for 5 dollars. You want to know the total value of debits to your account for the two checks.
Your record for situation 2.

$$-25 + (-5) = \boxed{-30}$$

The value of the first check added to the value of the second check is the total debit to your account.

In each situation we found that we added the absolute value of each number. (That is, we added the numbers without regarding their sign.) The answer always contained the sign that was common to both numbers.

We will now state these results as a formal rule.

ADDITION RULE FOR TWO NUMBERS WITH THE SAME SIGN

To add two numbers with the same sign, add the absolute values of the numbers and use the common sign in the answer.

EXAMPLE 5 Add.

(a) $14 + 16$ **(b)** $-8 + (-7)$

Solution

(a) $14 + \quad 16$ Add the absolute values of the numbers.
$\quad 14 + \quad 16 = 30$ Use the common sign in the answer. Here the common
$\quad 14 + \quad 16 = +30$ sign is the + sign.

(b) $-8 + \quad (-7)$ Add the absolute values of the numbers.
$\quad 8 + \quad\quad 7 = 15$ Use the common sign in the answer. Here the common
$\quad -8 + \quad (-7) = -15$ sign is the − sign.

Practice Problem 5 Add.

(a) $37 + 19$ **(b)** $-23 + (-35)$

NOTE TO STUDENT: *Fully worked-out solutions to all of the Practice Problems can be found at the back of the text starting at page SP-1*

EXAMPLE 6 Add. $\dfrac{2}{3} + \dfrac{1}{7}$

Solution

$$\dfrac{2}{3} + \dfrac{1}{7}$$

$$\dfrac{14}{21} + \dfrac{3}{21}$$ Change each fraction to an equivalent fraction with a common denominator of 21.

$$\dfrac{14}{21} + \dfrac{3}{21} = +\dfrac{17}{21} \text{ or } \dfrac{17}{21}$$ Add the absolute values of the numbers. Use the common sign in the answer. Note that if no sign is written, the number is understood to be positive.

Practice Problem 6 Add. $-\dfrac{3}{5} + \left(-\dfrac{4}{7}\right)$

EXAMPLE 7 Add. $-4.2 + (-3.94)$

Solution

$$-4.2 + (-3.94)$$
$$4.20 + 3.94 = 8.14$$ Add the absolute values of the numbers.
$$-4.20 + (-3.94) = -8.14$$ Use the common sign in the answer.

Practice Problem 7 Add. $-12.7 + (-9.38)$

The rule for adding two numbers with the same signs can be extended to more than two numbers. If we add more than two numbers with the same sign, the answer will have the sign common to all.

EXAMPLE 8 Add. $-7 + (-2) + (-5)$

Solution

$$-7 + (-2) + (-5)$$ We are adding three real numbers all with the same sign. We begin by adding the first two numbers.

$$= -9 + (-5)$$ Add $-7 + (-2) = -9$.
$$= -14$$ Add $-9 + (-5) = -14$.

Of course, this can be shortened by adding the three numbers without regard to sign and then using the common sign for the answer.

Practice Problem 8 Add. $-7 + (-11) + (-33)$

---

4 Adding Real Numbers with Opposite Signs

What if the signs of the numbers you are adding are different? Let's consider our checking account again to see how such a situation might occur.

SITUATION 3: Net Increase You made a deposit of 30 dollars on one day. On the next day you write a check for 25 dollars. You want to know the result of your two transactions.

Your record for situation 3.

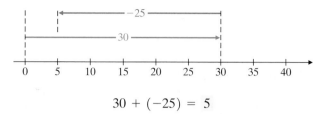

$$30 + (-25) = 5$$

A positive 30 for the deposit added to a negative 25 for the check, which is a debit, gives a net increase of 5 dollars in the account.

SITUATION 4: Net Decrease   You made a deposit of 10 dollars on one day. The next day you write a check for 40 dollars. You want to know the result of your two transactions.

Your record for situation 4.

$$10 + (-40) = -30$$

A positive 10 for the deposit added to a negative 40 for the check, which is a debit, gives a net decrease of 30 dollars in the account.

The result is a negative thirty (-30), because the check was larger than the deposit. If you do not have at least 30 dollars in your account at the start of Situation 4, you have overdrawn your account.

What do we observe from situations 3 and 4? In each case, first we found the difference of the absolute values of the two numbers. Then the sign of the result was always the sign of the number with the greater absolute value. Thus, in situation 3, 30 is larger than 25. The sign of 30 is positive. The sign of the answer (5) is positive. In situation 4, 40 is larger than 10. The sign of 40 is negative. The sign of the answer (-30) is negative.

We will now state these results as a formal rule.

ADDITION RULE FOR TWO NUMBERS WITH DIFFERENT SIGNS

1. Find the difference between the larger absolute value and the smaller one.

2. Give the answer the sign of the number having the larger absolute value.

EXAMPLE 9   Add. $8 + (-7)$

Solution

$8 + (-7)$	We are to add two numbers with opposite signs.
$8 - 7 = 1$	Find the difference between the two absolute values, which is 1.
$+8 + (-7) = +1$ or 1	The answer will have the sign of the number with the larger absolute value. That number is $+8$. Its sign is **positive**, so the answer will be $+1$.

Practice Problem 9   Add. $-9 + 15$

---

⑤ Using the Addition Properties for Real Numbers

It is useful to know the following three properties of real numbers.

> **1.** *Addition is commutative.*
> This property states that if two numbers are added, the result is the same no matter which number is written first. The order of the numbers does not affect the result.
>
> $$3 + 6 = 6 + 3 = 9$$
> $$-7 + (-8) = (-8) + (-7) = -15$$
> $$-15 + 3 = 3 + (-15) = -12$$
>
> **2.** *Addition of zero to any given number will result in that given number again.*
>
> $$0 + 5 = 5$$
> $$-8 + 0 = -8$$
>
> **3.** *Addition is associative.*
> This property states that if three numbers are added, it does not matter which two numbers are grouped by parentheses and added first.
>
> | | |
> |---|---|
> | $3 + (5 + 7) = (3 + 5) + 7$ | First combine numbers inside parentheses; then combine the remaining numbers. The results are the same no matter which numbers are grouped first. |
> | $3 + (12) = (8) + 7$ | |
> | $15 = 15$ | |

We can use these properties along with the rules we have for adding real numbers to add three or more numbers. We go from left to right, adding two numbers at a time.

EXAMPLE 10 Add. $\dfrac{3}{17} + \left(-\dfrac{8}{17}\right) + \dfrac{4}{17}$

Solution

$-\dfrac{5}{17} + \dfrac{4}{17}$ Add $\dfrac{3}{17} + \left(-\dfrac{8}{17}\right) = -\dfrac{5}{17}$.
The answer is negative since the larger of the two absolute values is negative.

$= -\dfrac{1}{17}$ Add $-\dfrac{5}{17} + \dfrac{4}{17} = -\dfrac{1}{17}$.
The answer is negative since the larger of the two absolute values is negative.

NOTE TO STUDENT: Fully worked-out solutions to all of the Practice Problems can be found at the back of the text starting at page SP-1

Practice Problem 10 Add. $-\dfrac{5}{12} + \dfrac{7}{12} + \left(-\dfrac{11}{12}\right)$

Sometimes the numbers being added have the same signs; sometimes the signs are different. When adding three or more numbers, you may encounter both situations.

EXAMPLE 11 Add. $-1.8 + 1.4 + (-2.6)$

Solution

$-0.4 + (-2.6)$ We take the difference of 1.8 and 1.4 and use the sign of the number with the larger absolute value.

$= -3.0$ Add $-0.4 + (-2.6) = -3.0$. The signs are the same; we add the absolute values of the numbers and use the common sign.

Practice Problem 11 Add. $-6.3 + (-8.0) + 3.5$

If many real numbers are added, it is often easier to add numbers with like signs in a column format. Remember that addition is commutative; therefore, real numbers can be added *in any order*. You do *not* need to combine the first two numbers as your first step.

EXAMPLE 12 Add. $-8 + 3 + (-5) + (-2) + 6 + 5$

Solution

$$
\begin{array}{r}
-8 \\
-5 \\
\underline{-2} \\
-15
\end{array}
$$
All the signs are the same. Add the three negative numbers to obtain -15.

$$
\begin{array}{r}
+3 \\
+6 \\
\underline{+5} \\
+14
\end{array}
$$
All the signs are the same Add the three positive numbers to obtain $+14$.

Add the two results.

$$-15 + 14 = -1$$

The answer is negative because the number with the larger absolute value is negative.

Practice Problem 12 Add. $-6 + 5 + (-7) + (-2) + 5 + 3$

A word about notation: The only time we really need to show the sign of a number is when the number is negative—for example, -3. The only time we need to show parentheses when we add real numbers is when we have two different signs preceding a number. For example, $-5 + (-6)$.

EXAMPLE 13 Add.

(a) $2.8 + (-1.3)$ **(b)** $-\dfrac{2}{5} + \left(-\dfrac{3}{4}\right)$

Solution

(a) $2.8 + (-1.3) = 1.5$

(b) $-\dfrac{2}{5} + \left(-\dfrac{3}{4}\right) = -\dfrac{8}{20} + \left(-\dfrac{15}{20}\right) = -\dfrac{23}{20}$ or $-1\dfrac{3}{20}$

Practice Problem 13 Add.

(a) $-2.9 + (-5.7)$ **(b)** $\dfrac{2}{3} + \left(-\dfrac{1}{4}\right)$

Student Solutions Manual   CD/Video   PH Math Tutor Center   MathXL®Tutorials on CD   MathXL®   MyMathLab®   Interactmath.cor

Verbal and Writing Skills

Check off any description of the number that applies.

	Number	Whole Number	Rational Number	Irrational Number	Real Number
1.	23				
2.	$-\frac{4}{5}$				
3.	π				
4.	2.34				
5.	$-6.666\ldots$				

	Number	Whole Number	Rational Number	Irrational Number	Real Number
6.	$-\frac{7}{9}$				
7.	$-2.3434\ldots$				
8.	14				
9.	$\sqrt{2}$				
10.	$3.232232223\ldots$				

Use a real number to represent each situation.

11. Jules Verne wrote a book with the title *20,000 Leagues under the Sea.*

12. The value of the dollar is up $0.07 with respect to the yen.

13. Ramona lost $37\frac{1}{2}$ pounds on Weight Watchers.

14. The scouts hiked from sea level to the top of a 3642-foot-high mountain.

15. The temperature rises 7°F.

16. Maya lost the game by 12 points.

Find the additive inverse (opposite).

17. 8

18. $-\frac{4}{5}$

19. -2.73

20. 85.4

Find the absolute value.

21. $|-1.3|$

22. $|-5.9|$

23. $\left|\frac{5}{6}\right|$

24. $\left|\frac{7}{12}\right|$

Add.

25. $-6 + (-5)$

26. $-13 + (-3)$

27. $-17 + (-14)$

28. $-12 + (-19)$

29. $-\frac{5}{16} + \frac{9}{16}$

30. $-\frac{2}{9} + \left(\frac{-4}{9}\right)$

31. $-\frac{2}{13} + \left(-\frac{5}{13}\right)$

32. $-\frac{5}{14} + \frac{2}{14}$

33. $-\frac{2}{5} + \frac{3}{7}$

34. $-\frac{2}{7} + \frac{3}{14}$

35. $-1.5 + (-2.3)$

36. $-1.8 + (-1.4)$

37. $0.6 + (-0.2)$

38. $-0.8 + 0.5$

39. $-5.26 + (-8.9)$

40. $-6.48 + (-3.7)$

41. $-8 + 5 + (-3)$ **42.** $7 + (-8) + (-4)$ **43.** $-3 + 5 + (-7)$ **44.** $-9 + 6 + (-12)$

45. $-\dfrac{4}{5} + \dfrac{8}{15}$ **46.** $-\dfrac{5}{6} + \dfrac{7}{18}$ **47.** $-7 + (-9) + 8$ **48.** $-4 + (-13) + 7$

Mixed Practice

Add.

49. $8 + (-11)$ **50.** $16 + (-24)$ **51.** $-83 + 142$ **52.** $-114 + 186$

53. $-\dfrac{4}{9} + \dfrac{5}{6}$ **54.** $-\dfrac{3}{5} + \dfrac{2}{3}$ **55.** $-\dfrac{1}{10} + \dfrac{1}{2}$ **56.** $-\dfrac{2}{3} + \left(-\dfrac{1}{4}\right)$

57. $4.36 + (-3.6)$ **58.** $4.79 + (-9.1)$ **59.** $4 + (-8) + 16$ **60.** $27 + (-11) + (-4)$

61. $34 + (-18) + 11 + (-27)$ **62.** $-23 + 4 + (-11) + 17$

63. $17.85 + (-2.06) + 0.15$ **64.** $23.17 + 5.03 + (-11.81)$

Applications

65. *Profit/Loss* Holly paid $47 for a vase at an estate auction. She resold it to an antiques dealer for $214. What was her profit or loss?

66. *Temperature* When we skied at Jackson Hole, Wyoming, yesterday, the temperature at the summit was $-12°$F. Today when we called the ski report, the temperature had risen $7°$F. What is the temperature at the summit today?

67. *Home Equity Line of Credit* Ramon borrowed $2300 from his home equity line of credit to pay off his car loan. He then borrowed another $1500 to pay to have his kitchen repainted. Represent how much Ramon owed on his home equity line of credit as a real number.

68. *Time Change* During the winter, New York City is on Eastern Standard Time (EST). Melbourne, Australia is 15 hours ahead of New York. If it is 11 P.M. in Melbourne, what time is it in New York?

69. *Football* On three successive football running plays, Jon gained 9 yards, lost 11 yards, and gained 5 yards. What was his total gain or loss?

70. *School Fees* Wanda's financial aid account at school held $643.85. She withdrew $185.50 to buy books for the semester. Does she have enough left in her account to pay the $475.00 registration fee for the next semester? If so, how much extra money does she have? If not, how much is she short?

71. *Butterfly Population* The population of a particular butterfly species was 8000. Twenty years later there were 3000 fewer. Today, there are 1500 fewer. Study the graph to the right. What is the new population?

72. *Credit Card Balance* Aaron owes $258 to a credit card company. He makes a purchase of $32 with the card and then makes a payment of $150 on the account. How much does he still owe?

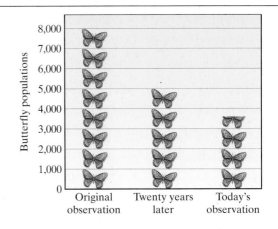

Profit/Loss   *During the first five months of 2005 a regional Midwest airline posted profit and loss figures for each month of operation, as shown in the accompanying bar graph.*

73. For the first three months of 2005, what were the total earnings of the airline?

74. For the first five months of 2005, what were the total earnings for the airline?

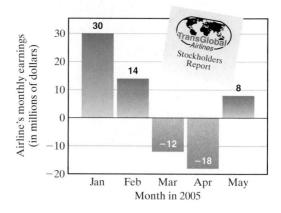

To Think About

75. What number must be added to -13 to get 5?

76. What number must be added to -18 to get 10?

77. ***Football*** Vern rushed for 8 yards to his team's 17-yard line in the last quarter of the home football game. On the next play, the quarterback was sacked and lost 4 yards. On the third down, the pass was incomplete. How many yards must the team make on the next play to make first down?

78. ***Discount Stores*** Filene's Basement Store in Boston is famous for its Automatic Markdown Policy. Each item is dated when it enters the store. If it does not sell after 14 days, the price is reduced by 25%. If it is still there after 21 days, the discount is now 50%. After 28 days, the discount becomes a whopping 75%. If the item has still not sold after 35 days, it is taken from the store and donated to charity. While shopping in Filene's Basement, Heather Bean found two blouses that she liked a lot. The red blouse was originally $87.99 and had been in the store for 22 days. The blue blouse was originally $57.99 but had only been in the store for 19 days. Which one was cheaper? Which one was a better value?

Cumulative Review

Perform the indicated calculations.

79. $\dfrac{3}{7} + \dfrac{5}{21}$

80. $\left(\dfrac{2}{5}\right)\left(\dfrac{20}{27}\right)$

81. $\dfrac{2}{15} - \dfrac{1}{20}$

82. $2\dfrac{1}{2} \div 3\dfrac{2}{5}$

83. $0.72 + 0.8$

84. $1.63 - 0.98$

85. $(1.63)(0.7)$

86. $0.208 \div 0.8$

① Subtracting Real Numbers with Like or Unlike Signs

So far we have developed the rules for adding real numbers. We can use these rules to subtract real numbers. Let's look at a checkbook situation to see how.

SITUATION 5: Subtract a Deposit and Add a Debit You have a balance of 20 dollars in your checking account. The bank calls you and says that a deposit of 5 dollars that belongs to another account was erroneously added to your account. They say they will correct the account balance to 15 dollars. The bank tells you that since they cannot take away the erroneous credit, they will add a debit to your account. You want to keep track of what's happening to your account.

Your record for situation 5.

$$20 - (+5) = 15$$

From your present balance subtract the deposit to give the new balance. This equation shows what needs to be done to your account. The bank tells you that because the error happened in the past they cannot "take it away." However, they can add to your account a debit of 5 dollars. Here is the equivalent addition.

$$20 + (-5) = 15$$

To your present balance add a debit to give the new balance. Subtracting a positive 5 has the same effect as adding a negative 5.

SUBTRACTION OF REAL NUMBERS

To subtract real numbers, add the opposite of the second number (that is, the number you are subtracting) to the first.

The rule tells us to do three things when we subtract real numbers. First, change subtraction to addition. Second, replace the second number by its opposite. Third, add the two numbers using the rules for addition of real numbers.

EXAMPLE 1 Subtract. $6 - (-2)$

Solution

$6 \qquad - \qquad (-2)$

| Change subtraction to addition. | Write the opposite of the second number. |

$= \quad 6 \quad + \quad (+2)$

| Add the two real numbers with the same sign. |

$= \qquad\qquad 8$

Practice Problem 1 Subtract. $9 - (-3)$

NOTE TO STUDENT: Fully worked-out solutions to all of the Practice Problems can be found at the back of the text starting at page SP-1

NOTE TO STUDENT: Fully worked-out solutions to all of the Practice Problems can be found at the back of the text starting at page SP-1

EXAMPLE 2   Subtract. $-8 - (-6)$

Solution

$$= \quad -8 \quad + \quad (+6)$$

Add the two real numbers with opposite signs.

$$= \quad -2$$

Practice Problem 2   Subtract. $-12 - (-5)$

EXAMPLE 3   Subtract.   **(a)** $\dfrac{3}{7} - \dfrac{6}{7}$ **(b)** $-\dfrac{7}{18} - \left(-\dfrac{1}{9}\right)$

Solution

(a) $\dfrac{3}{7} - \dfrac{6}{7} = \dfrac{3}{7} + \left(-\dfrac{6}{7}\right)$ Change the subtraction problem to one of adding the opposite of the second number. We note that the problem has two fractions with the same denominator.

$\qquad = -\dfrac{3}{7}$ Add two numbers with different signs.

(b) $-\dfrac{7}{18} - \left(-\dfrac{1}{9}\right) = -\dfrac{7}{18} + \dfrac{1}{9}$ Change subtracting to adding the opposite.

$\qquad = -\dfrac{7}{18} + \dfrac{2}{18}$ Change $\dfrac{1}{9}$ to $\dfrac{2}{18}$ since LCD $= 18$.

$\qquad = -\dfrac{5}{18}$ Add two numbers with different signs.

Practice Problem 3   Subtract.   **(a)** $\dfrac{5}{9} - \dfrac{7}{9}$ **(b)** $-\dfrac{5}{21} - \left(-\dfrac{3}{7}\right)$

EXAMPLE 4   Subtract. $-5.2 - (-5.2)$

Solution

$-5.2 - (-5.2) = -5.2 + 5.2$ Change the subtraction problem to one of adding the opposite of the second number.

$\qquad = 0$ Add two numbers with different signs.

Example 4 illustrates what is sometimes called the **additive inverse property.** When you add two real numbers that are opposites of each other, you will obtain zero. Examples of this are the following:

$$5 + (-5) = 0 \qquad -186 + 186 = 0 \qquad -\dfrac{1}{8} + \dfrac{1}{8} = 0.$$

Practice Problem 4   Subtract. $-17.3 - (-17.3)$

EXAMPLE 5 Calculate.

(a) $-8 - 2$ **(b)** $23 - 28$ **(c)** $5 - (-3)$ **(d)** $\dfrac{1}{4} - 8$

Solution

(a) $-8 - 2 = -8 + (-2)$ Notice that we are subtracting a positive 2. Change to addition.

$ = -10$ Add.

In a similar fashion we have

(b) $23 - 28 = 23 + (-28) = -5$

(c) $5 - (-3) = 5 + 3 = 8$

(d) $\dfrac{1}{4} - 8 = \dfrac{1}{4} + (-8) = \dfrac{1}{4} + \left(-\dfrac{32}{4}\right) = -\dfrac{31}{4}$

Practice Problem 5 Calculate.

(a) $-21 - 9$ **(b)** $17 - 36$ **(c)** $12 - (-15)$ **(d)** $\dfrac{3}{5} - 2$

EXAMPLE 6 A satellite is recording radioactive emissions from nuclear waste buried 3 miles below sea level. The satellite orbits the Earth at 98 miles above sea level. How far is the satellite from the nuclear waste?

Solution We want to find the difference between $+98$ miles and -3 miles. This means we must subtract -3 from 98.

$$98 - (-3) = 98 + 3$$
$$= 101$$

The satellite is 101 miles from the nuclear waste.

Practice Problem 6 A helicopter is directly over a sunken vessel. The helicopter is 350 feet above sea level. The vessel lies 186 feet below sea level. How far is the helicopter from the sunken vessel?

Developing Your Study Skills

Reading the Textbook

Begin reading your textbook with a paper and pencil in hand. As you come across a new definition or concept, underline it in the text and/or write it down in your notebook. Whenever you encounter an unfamiliar term, look it up and make a note of it. When you come to an example, work through it step-by-step. Be sure to read each word and follow directions carefully.

Notice the helpful hints the author provides. They guide you to correct solutions and prevent you from making errors. Take advantage of these pieces of expert advice.

Be sure that you understand what you are reading. Make a note of any of those things that you do not understand and ask your instructor about them. Do not hurry through the material. Learning mathematics takes time.

Student Solutions Manual | CD/Video | PH Math Tutor Center | MathXL®Tutorials on CD | MathXL® | MyMathLab® | Interactmath.cor

Verbal and Writing Skills

1. Explain in your own words how you would perform the necessary steps to find $-8 - (-3)$.

2. Explain in your own words how you would perform the necessary steps to find $-10 - (-15)$.

Subtract by adding the opposite.

3. $18 - 35$

4. $16 - 48$

5. $15 - 20$

6. $18 - 24$

7. $-14 - (-3)$

8. $-24 - (-7)$

9. $-52 - (-60)$

10. $-48 - (-80)$

11. $0 - (-5)$

12. $0 - (-7)$

13. $-18 - (-18)$

14. $-24 - (-24)$

15. $-11 - (-8)$

16. $-35 - (-10)$

17. $\dfrac{2}{5} - \dfrac{4}{5}$

18. $\dfrac{2}{9} - \dfrac{7}{9}$

19. $\dfrac{3}{4} - \left(-\dfrac{3}{5}\right)$

20. $-\dfrac{2}{3} - \dfrac{1}{4}$

21. $-\dfrac{3}{4} - \dfrac{5}{6}$

22. $-\dfrac{7}{10} - \dfrac{10}{15}$

23. $-0.6 - 0.3$

24. $-0.9 - 0.5$

25. $2.64 - (-1.83)$

26. $-0.03 - 0.06$

Mixed Practice

Calculate.

27. $\dfrac{3}{5} - 4$

28. $\dfrac{5}{6} - 3$

29. $-\dfrac{2}{7} + 6$

30. $-\dfrac{3}{8} + 5$

31. $34 - 87$

32. $19 - 76$

33. $-25 - 48$

34. $-74 - 11$

35. $2.3 - (-4.8)$

36. $8.4 - (-2.7)$

37. $8 - \left(-\dfrac{3}{4}\right)$

38. $\dfrac{2}{3} - (-6)$

39. $\dfrac{5}{6} - 7$

40. $9 - \dfrac{2}{3}$

41. $-\dfrac{3}{10} - \dfrac{3}{4}$

42. $-\dfrac{11}{12} - \dfrac{5}{18}$

43. $-135 - (-126.5)$

44. $-97.6 - (-146)$

45. $\dfrac{1}{5} - 6$

46. $\dfrac{2}{7} - (-3)$

47. $5 - (-3.162)$ **48.** $7 - (-6.183)$ **49.** $-3 - 2.047$ **50.** $-1.043 - 4$

51. Subtract -9 from -2.   **52.** Subtract -12 from 20.   **53.** Subtract 13 from -35.

One Step Further

Change each subtraction operation to "adding the opposite." Then combine the numbers.

54. $9 + 6 - (-5)$ **55.** $7 + (-6) - 3$ **56.** $8 + (-4) - 10$ **57.** $-10 + 6 - (-15)$

58. $18 - (-15) - 3$ **59.** $7 + (-42) - 27$ **60.** $-4.2 - (-3.8) + 1.5$ **61.** $-6.4 - (-2.7) + 5.3$

62. $-3 - (-12) + 18 + 15 - (-6)$ **63.** $42 - (-30) - 65 - (-11) + 20$

Applications

64. *Sea Rescue* A rescue helicopter is 300 feet above sea level. The captain has located an ailing submarine directly below it that is 126 feet below sea level. How far is the helicopter from the submarine?

65. *Checking Account Balance* Yesterday Jackie had $112 in her checking account. Today her account reads "balance $-\$37$." Find the difference in these two amounts.

66. *Temperature Change* On June 22, 1943, in Spearfish, South Dakota, the temperature was $-4°F$ at 7:30 A.M. At 7:32 A.M., the temperature was $+44.6°F$. Find the difference in temperature over those two minutes.

+300 feet

Sea level

−126 feet

67. *Elevation Difference* The highest point in Africa is Mt. Kilimanjaro in Tanzania, which is 5895 meters above sea level. The lowest point in Africa is Lake Assal in Djibouti, which is 156 meters below sea level. What is the difference in elevation between Mt. Kilimanjaro and Lake Assal?

68. *Federal Taxes* In 2004, Rachel had $1815 withheld from her paycheck in federal taxes. When she filed her tax return, she received a $265 refund. How much did she actually pay in taxes?

Cumulative Review

In exercises 69–73, perform the indicated operations.

69. $-37 + 16$ **70.** $-37 + (-14)$ **71.** $-3 + (-6) + (-10)$

72. *Temperature* What is the temperature after a rise of $13°C$ from a start of $-21°C$?

73. *Hiking* Sean and Khalid went hiking in the Blue Ridge Mountains. During their $8\frac{1}{3}$ mile hike, $\frac{4}{5}$ of the distance was covered with snow. How many miles were snow covered?

Student Learning Objectives

After studying this section, you will be able to:

1 Multiply real numbers.

2 Use the multiplication properties for real numbers.

3 Divide real numbers.

1 **Multiplying Real Numbers**

We are familiar with the meaning of multiplication for positive numbers. For example, $5 \times 90 = 450$ might mean that you receive five weekly checks of 90 dollars each and you gain \$450. Let's look at a situation that corresponds to $5 \times (-90)$. What might that mean?

SITUATION 6: Checking an Account Balance   You write a check for five weeks in a row to pay your weekly room rent of 90 dollars. You want to know the total impact on your checking account balance.

Your record for situation 6.

$(+5)$	$\times$	(-90)	$=$	-450
The number of checks you have written	times	negative 90 the value of each check that was a debit to your account,	gives	negative 450 dollars, a net debit to your account.

Note that a multiplication symbol is not needed between the $(+5)$ and the (-90) because the two sets of parentheses indicate multiplication. The multiplication $(5)(-90)$ is the same as repeated addition of five (-90)'s. Note that 5 multiplied by -90 can be written as $5(-90)$ or $(5)(-90)$.

$$(-90) + (-90) + (-90) + (-90) + (-90) = -450$$

This example seems to show that a positive number multiplied by a negative number is negative.

What if the negative number is the one that is written first? If $(5)(-90) = -450$, then $(-90)(5) = -450$ by the commutative property of multiplication. This is an example showing that *when two numbers with opposite signs* (one positive, one negative) *are multiplied, the result is negative.*

But what if both numbers are negative? Consider the following situation.

SITUATION 7: Renting a Room   Last year at college you rented a room at 90 dollars per week for 36 weeks, which included two semesters and summer school. This year you will not attend the summer session, so you will be renting the room for only 30 weeks. Thus the number of weekly rental checks will be six less than last year. You are making out your budget for this year. You want to know the financial impact of renting the room for six fewer weeks.

Your record for situation 7.

(-6)	$\times$	(-90)	$=$	540
The difference in the number of checks this year compared to last is -6, which is negative to show a decrease,	times	-90, the value of each check paid out,	gives	$+540$ dollars. The product is positive, because your financial situation will be 540 dollars better this year.

You could check that the answer is positive by calculating the total rental expenses.

	Dollars in rent last year	$(36)(90) =$	3240
(subtract)	Dollars in rent this year	$-(30)(90) =$	-2700
	Extra dollars available this year	$=$	$+540$

This agrees with our previous answer: $(-6)(-90) = +540$.

In this situation it seems reasonable that a negative number times a negative number yields a positive answer. We already know from arithmetic that a positive number times a positive number yields a positive answer. Thus we might see the general rule that *when two numbers with the same sign* (both positive or both negative) *are multiplied, the result is positive.*

We will now state our rule.

MULTIPLICATION OF REAL NUMBERS

To multiply two real numbers with **the same sign,** multiply the absolute values. The sign of the result is **positive.**

To multiply two real numbers with **opposite signs,** multiply the absolute values. The sign of the result is **negative.**

Note that negative 6 times -90 can be written as $-6(-90)$ or $(-6)(-90)$.

EXAMPLE 1   Multiply.

(a) $(3)(6)$ **(b)** $\left(-\dfrac{5}{7}\right)\left(-\dfrac{2}{9}\right)$ **(c)** $-4(8)$ **(d)** $\left(\dfrac{2}{7}\right)(-3)$

Solution

(a) $(3)(6) = 18$ When multiplying two numbers with the same sign, the result is a positive number.

(b) $\left(-\dfrac{5}{7}\right)\left(-\dfrac{2}{9}\right) = \dfrac{10}{63}$

(c) $-4(8) = -32$ When multiplying two numbers with opposite signs, the result is a negative number.

(d) $\left(\dfrac{2}{7}\right)(-3) = \left(\dfrac{2}{7}\right)\left(-\dfrac{3}{1}\right) = -\dfrac{6}{7}$

Practice Problem 1   Multiply.

(a) $(-6)(-2)$ **(b)** $(7)(9)$ **(c)** $\left(-\dfrac{3}{5}\right)\left(\dfrac{2}{7}\right)$ **(d)** $\left(\dfrac{5}{6}\right)(-7)$

NOTE TO STUDENT: Fully worked-out solutions to all of the Practice Problems can be found at the back of the text starting at page SP-1

To multiply more than two numbers, multiply two numbers at a time.

EXAMPLE 2   Multiply. $(-4)(-3)(-2)$

Solution

$(-4)(-3)(-2) = (+12)(-2)$ We begin by multiplying the first two numbers, (-4) and (-3). The signs are the same. The answer is positive 12.

$\qquad\qquad\qquad = -24$ Now we multiply $(+12)$ and (-2). The signs are different. The answer is negative 24.

Practice Problem 2   Multiply. $(-5)(-2)(-6)$

EXAMPLE 3   Multiply.

(a) $-3(-1.5)$ **(b)** $\left(-\dfrac{1}{2}\right)(-1)(-4)$ **(c)** $-2(-2)(-2)(-2)$

Solution   Multiply two numbers at a time. See if you find a pattern.

(a) $-3(-1.5) = 4.5$ Be sure to place the decimal point in your answer.

(b) $\left(-\dfrac{1}{2}\right)(-1)(-4) = +\dfrac{1}{2}(-4) = -2$

(c) $-2(-2)(-2)(-2) = +4(-2)(-2) = -8(-2) = +16$ or 16

What kind of answer would we obtain if we multiplied five negative numbers? If you guessed "negative," you probably see the pattern.

Practice Problem 3   Determine the sign of the product. Then multiply to check.

(a) $-2(-3)$ **(b)** $(-1)(-3)(-2)$ **(c)** $-4\left(-\dfrac{1}{4}\right)(-2)(-6)$

NOTE TO STUDENT: Fully worked-out solutions to all of the Practice Problems can be found at the back of the text starting at page SP-1

When you multiply two or more real numbers:

1. The result is always **positive** if there are an **even** number of negative signs.
2. The result is always **negative** if there are an **odd** number of negative signs.

2 Using the Multiplication Properties for Real Numbers

For convenience, we will list the properties of multiplication.

1. *Multiplication is commutative.*

   This property states that if two real numbers are multiplied, the order of the numbers does not affect the result. The result is the same no matter which number is written first.

   $$(5)(7) = (7)(5) = 35, \qquad \left(\dfrac{1}{3}\right)\left(\dfrac{2}{7}\right) = \left(\dfrac{2}{7}\right)\left(\dfrac{1}{3}\right) = \dfrac{2}{21}$$

2. *Multiplication of any real number by zero will result in zero.*

   $$(5)(0) = 0, \qquad (-5)(0) = 0, \qquad (0)\left(\dfrac{3}{8}\right) = 0, \qquad (0)(0) = 0$$

3. *Multiplication of any real number by 1 will result in that same number.*

   $$(5)(1) = 5, \qquad (1)(-7) = -7, \qquad (1)\left(-\dfrac{5}{3}\right) = -\dfrac{5}{3}$$

4. *Multiplication is associative.*

   This property states that if three real numbers are multiplied, it does not matter which two numbers are grouped by parentheses and multiplied first.

   $2 \times (3 \times 4) = (2 \times 3) \times 4$ First multiply the numbers in parentheses. Then multiply the remaining numbers.

   $2 \times (12) = (6) \times 4$ The results are the same no matter which numbers are grouped and multiplied first.

   $24 = 24$

3 **Dividing Real Numbers**

What about division? Any division problem can be rewritten as a multiplication problem.

We know that $20 \div 4 = 5$ because $4(5) = 20$.

Similarly, $-20 \div (-4) = 5$ because $-4(5) = -20$.

In both division problems the answer is positive 5. Thus we see that *when you divide two numbers with the same sign* (both positive or both negative), *the answer is positive.* What if the signs are different?

We know that $-20 \div 4 = -5$ because $4(-5) = -20$.

Similarly, $20 \div (-4) = -5$ because $-4(-5) = 20$.

In these two problems the answer is negative 5. So we have reasonable evidence to see that *when you divide two numbers with different signs* (one positive and one negative), *the answer is negative.*

We will now state our rule for division.

DIVISION OF REAL NUMBERS

To divide two real numbers with **the same sign,** divide the absolute values. The sign of the result is **positive.**

To divide two real numbers with **different signs,** divide the absolute values. The sign of the result is **negative.**

EXAMPLE 4   Divide.

(a) $12 \div 4$ **(b)** $(-25) \div (-5)$ **(c)** $\dfrac{-36}{18}$ **(d)** $\dfrac{42}{-7}$

Solution

(a) $12 \div 4 = 3$

(b) $(-25) \div (-5) = 5$

> When dividing two numbers with the same sign, the result is a positive number.

(c) $\dfrac{-36}{18} = -2$

(d) $\dfrac{42}{-7} = -6$

> When dividing two numbers with different signs, the result is a negative number.

Practice Problem 4   Divide.

(a) $-36 \div (-2)$ **(b)** $-49 \div 7$ **(c)** $\dfrac{50}{-10}$ **(d)** $\dfrac{-39}{13}$

EXAMPLE 5   Divide.   **(a)** $-36 \div 0.12$ **(b)** $-2.4 \div (-0.6)$

Solution   **(a)** $-36 \div 0.12$ Look at the problem to determine the sign. When dividing two numbers with different signs, the result will be a negative number.

We then divide the absolute values.

$$
\begin{array}{r}
3\ 00. \\
0.12_{\wedge}\overline{)36.00_{\wedge}} \\
\underline{36} \\
00
\end{array}
$$

Thus $-36 \div 0.12 = -300$. The answer is a negative number.

(b) $-2.4 \div (-0.6)$ Look at the problem to determine the sign. When dividing two numbers with the same sign, the result will be positive.

We then divide the absolute values.

$$
0.6_\wedge \overline{)2.4_\wedge} \\
\underline{2\ 4} \\
4.
$$

Thus $-2.4 \div (-0.6) = 4$. The answer is a positive number.

NOTE TO STUDENT: *Fully worked-out solutions to all of the Practice Problems can be found at the back of the text starting at page SP-1*

Practice Problem 5 Divide. **(a)** $-12.6 \div (-1.8)$ **(b)** $0.45 \div (-0.9)$

Note that the rules for multiplication and division are the same. When you **multiply** or **divide** two numbers with the **same** sign, you obtain **a positive** number. When you **multiply** or **divide** two numbers with **different** signs, you obtain a **negative** number.

EXAMPLE 6 Divide. $-\dfrac{12}{5} \div \dfrac{2}{3}$

Solution

$$= \left(-\frac{12}{5}\right)\left(\frac{3}{2}\right)$$ Divide two fractions. We invert the second fraction and multiply by the first fraction.

$$= \left(-\frac{\overset{6}{\cancel{12}}}{5}\right)\left(\frac{3}{\underset{1}{\cancel{2}}}\right)$$

$$= -\frac{18}{5} \quad \text{or} \quad -3\frac{3}{5}$$ The answer is negative since the two numbers divided have different signs.

Practice Problem 6 Divide. $-\dfrac{5}{16} \div \left(-\dfrac{10}{13}\right)$

Note that division can be indicated by the symbol $\div$ or by the fraction bar $-$. $\frac{2}{3}$ means $2 \div 3$.

EXAMPLE 7 Divide. **(a)** $\dfrac{\frac{7}{8}}{-21}$ **(b)** $\dfrac{-\frac{2}{3}}{-\frac{7}{13}}$

Solution

(a) $\dfrac{\frac{7}{8}}{-21}$

$$= \frac{7}{8} \div \left(-\frac{21}{1}\right)$$ Change -21 to a fraction. $-21 = -\frac{21}{1}$

$$= \frac{\overset{1}{\cancel{7}}}{8}\left(-\frac{1}{\underset{3}{\cancel{21}}}\right)$$ Change the division to multiplication. Cancel where possible.

$$= -\frac{1}{24}$$ Simplify.

(b) $\dfrac{-\dfrac{2}{3}}{-\dfrac{7}{13}} = -\dfrac{2}{3} \div \left(-\dfrac{7}{13}\right) = -\dfrac{2}{3}\left(-\dfrac{13}{7}\right) = \dfrac{26}{21}$ or $1\dfrac{5}{21}$

Practice Problem 7 Divide.   **(a)** $\dfrac{-12}{-\dfrac{4}{5}}$ **(b)** $\dfrac{-\dfrac{2}{9}}{\dfrac{8}{13}}$

---

1. *Division of 0 by any nonzero real number gives 0 as a result.*

$$0 \div 5 = 0, \qquad 0 \div \dfrac{2}{3} = 0, \qquad \dfrac{0}{5.6} = 0, \qquad \dfrac{0}{1000} = 0$$

You can divide zero by $5, \frac{2}{3}, 5.6, 1000$, or any number (except 0).

2. *Division of any real number by 0 is* **undefined.**

$$7 \div 0 \qquad\qquad \dfrac{64}{0}$$
$$\uparrow \qquad\qquad\qquad \uparrow$$

Neither of these operations is possible. **Division by zero is undefined.**

You may be wondering why division by zero is undefined. Let us think about it for a minute. We said that $7 \div 0$ is undefined. Suppose there were an answer. Let us call the answer a. So we assume for a minute that $7 \div 0 = a$. Then it would have to follow that $7 = 0(a)$. But this is impossible. Zero times any number is zero. So we see that if there were such a number, it would contradict known mathematical facts. Therefore there is no number a such that $7 \div 0 = a$. Thus we conclude that division by zero is undefined.

When combining two numbers, it is important to be sure you know which rule applies. Think about the concepts in the following chart. See if you agree with each example.

Operation	Two Real Numbers with the Same Sign	Two Real Numbers with Different Signs
Addition	Result may be positive or negative. $9 + 2 = 11$ $-5 + (-6) = -11$	Result may be positive or negative. $-3 + 7 = 4$ $4 + (-12) = -8$
Subtraction	Result may be positive or negative. $15 - 6 = 15 + (-6) = 9$ $-12 - (-3) = -12 + 3 = -9$	Result may be positive or negative. $-12 - 3 = -12 + (-3) = -15$ $5 - (-6) = 5 + 6 = 11$
Multiplication	Result is always positive. $9(3) = 27$ $-8(-5) = 40$	Result is always negative. $-6(12) = -72$ $8(-3) = -24$
Division	Result is always positive. $150 \div 6 = 25$ $-72 \div (-2) = 36$	Result is always negative. $-60 \div 10 = -6$ $30 \div (-6) = -5$

EXAMPLE 8   The Hamilton-Wenham Generals recently analyzed the
48 plays their team made while in the possession of the football during their last
game. The following bar graph illustrates the number of plays made in each
category. The team statistician
prepared the following chart
indicating the average number
of yards gained or lost during
each type of play.

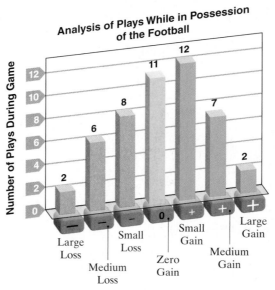

Analysis of Plays While in Possession of the Football

Type of Play	Average Yards Gained or Lost for Play
Large gain	+25
Medium gain	+15
Small gain	+5
Zero gain	0
Small loss	−5
Medium loss	−10
Large loss	−15

(a) How many yards were lost by the Generals in the plays that were considered
small losses?

(b) How many yards were gained by the Generals in the plays that were considered small gains?

(c) If the total yards gained in small gains were combined with the total yards
lost in small losses, what would be the result?

Solution

(a) We multiply the number of small losses by the average number of total yards
lost on each small loss:

$$8(-5) = -40.$$

The team lost approximately 40 yards with plays that were considered small
losses.

(b) We multiply the number of small gains by the average number of yards
gained on each small gain:

$$12(5) = 60.$$

The team gained approximately 60 yards with plays that were considered
small gains.

(c) We combine the results for (a) and (b):

$$-40 + 60 = 20.$$

A total of 20 yards was gained during the plays that were small losses and
small gains.

*NOTE TO STUDENT: Fully worked-out
solutions to all of the Practice Problems
can be found at the back of the text
starting at page SP-1*

Practice Problem 8   Using the information provided in Example 8, answer
the following:

(a) How many yards were lost by the Generals in the plays that were considered
medium losses?

(b) How many yards were gained by the Generals in the plays that were
considered medium gains?

(c) If the total yards gained in medium gains were combined with the total yards
lost in medium losses, what would be the result?

Verbal and Writing Skills

1. Explain in your own words the rule for determining the correct sign when multiplying two real numbers.

2. Explain in your own words the rule for determining the correct sign when multiplying three or more real numbers.

Multiply. Be sure to write your answer in the simplest form.

3. $5(-4)$

4. $8(-2)$

5. $0(-12)$

6. $0(-150)$

7. $16(1.5)$

8. $24(2.5)$

9. $(-1.32)(-0.2)$

10. $(-2.3)(-0.11)$

11. $0.7(-2.5)$

12. $0.6(-3.5)$

13. $\left(\dfrac{3}{8}\right)(-4)$

14. $(5)\left(-\dfrac{7}{10}\right)$

15. $\left(-\dfrac{3}{5}\right)\left(-\dfrac{15}{11}\right)$

16. $\left(-\dfrac{4}{9}\right)\left(-\dfrac{3}{5}\right)$

17. $\left(\dfrac{12}{13}\right)\left(\dfrac{-5}{24}\right)$

18. $\left(\dfrac{14}{17}\right)\left(-\dfrac{3}{28}\right)$

Divide.

19. $-36 \div (-9)$

20. $0 \div (-15)$

21. $-48 \div (-8)$

22. $-45 \div (9)$

23. $-220 \div (-11)$

24. $240 \div (-15)$

25. $156 \div (-13)$

26. $-0.6 \div 0.3$

27. $-9.1 \div (0.07)$

28. $8.1 \div (-0.03)$

29. $0.54 \div (-0.9)$

30. $-7.2 \div 8$

31. $-6.3 \div 7$

32. $\dfrac{2}{7} \div \left(-\dfrac{3}{5}\right)$

33. $\left(-\dfrac{1}{5}\right) \div \left(\dfrac{2}{3}\right)$

34. $\left(-\dfrac{5}{6}\right) \div \left(-\dfrac{7}{18}\right)$

35. $-\dfrac{5}{7} \div \left(-\dfrac{3}{28}\right)$

36. $\left(-\dfrac{4}{9}\right) \div \left(-\dfrac{8}{15}\right)$

37. $\left(-\dfrac{7}{12}\right) \div \left(-\dfrac{5}{6}\right)$

38. $\dfrac{12}{-\dfrac{2}{5}}$

39. $\dfrac{-6}{-\dfrac{3}{7}}$

40. $\dfrac{-\dfrac{3}{8}}{-\dfrac{2}{3}}$

41. $\dfrac{-\dfrac{2}{3}}{\dfrac{8}{15}}$

42. $\dfrac{\dfrac{5}{12}}{-\dfrac{7}{24}}$

43. $\dfrac{-\dfrac{7}{8}}{-\dfrac{14}{15}}$

Multiply. You may want to determine the sign of the product before you multiply.

44. $-6(2)(-3)(4)$

45. $-1(-2)(-3)(4)$

46. $-2(-1)(3)(-1)(-4)$

47. $-2(-2)(2)(-1)(-3)$

48. $-3(2)(-4)(0)(-2)$ **49.** $-3(-2)\left(\dfrac{1}{3}\right)(-4)(2)$ **50.** $60(-0.6)(-0.002)(0.5)$

51. $-3(-0.03)(0.001)(-2)$ **52.** $\left(\dfrac{3}{8}\right)\left(\dfrac{1}{2}\right)\left(-\dfrac{5}{6}\right)$ **53.** $\left(-\dfrac{4}{5}\right)\left(-\dfrac{6}{7}\right)\left(-\dfrac{1}{3}\right)$

54. $\left(-\dfrac{1}{2}\right)\left(\dfrac{4}{5}\right)\left(-\dfrac{7}{8}\right)\left(-\dfrac{2}{3}\right)$ **55.** $\left(-\dfrac{3}{4}\right)\left(-\dfrac{7}{15}\right)\left(-\dfrac{8}{21}\right)\left(-\dfrac{5}{9}\right)$

Mixed Practice

Take a minute to review the chart before Example 8. Be sure that you can remember the sign rules for each operation. Then do exercises 56–65. Perform the indicated calculations.

56. $-5 - (-2)$ **57.** $-36 \div (-4)$ **58.** $-3(-9)$ **59.** $5 + (-7)$

60. $18 \div (-6)$ **61.** $8 - (-9)$ **62.** $-6 + (-3)$ **63.** $6(-12)$

64. $18 \div (-18)$ **65.** $-37 \div 37$

Applications

66. ***Stock Trading*** Your favorite stock opened the day's trading at \$37.20 per share. When trading closed for the day, your stock was priced at \$27.30 per share. If you owned 75 shares, what was your profit or loss that day?

67. ***Equal Contributions*** Ed, Ned, Ted, and Fred went camping. They each contributed an equal share of money towards food. Fred did the shopping. When he returned from the store, he had \$17.60 left. How much money did Fred give back to each person?

68. ***Student Loans*** Ramon will pay \$6480 on his student loan over the next 3 years. If \$180 is automatically deducted from his bank account each month to pay the loan off, how much does he still owe after one year?

69. ***Car Payments*** Keith will pay the Ford dealer a total of \$15,768 to be paid off in 48 equal monthly installments. What is his monthly bill?

Football *The Beverly Panthers recently analyzed the 37 plays their team made while in the possession of the football during their last game. The team statistician prepared the following chart indicating the number of plays in each category and the average number of yards gained or lost during each type of play. Use this chart to answer exercises 70–77.*

Type of Play	Number of Plays	Average Yards Gained or Lost per Play
Large gain	1	+25
Medium gain	6	+15
Small gain	4	+5
Zero gain	5	0
Small loss	10	−5
Medium loss	7	−10
Large loss	4	−15

70. How many yards were lost by the Panthers in the plays that were considered small losses?

71. How many yards were gained by the Panthers in the plays that were considered small gains?

72. If the total yards gained in small gains were combined with the total yards lost in small losses, what would be the result?

73. How many yards were lost by the Panthers in the plays that were considered medium losses?

74. How many yards were gained by the Panthers in the plays that were considered medium gains?

75. If the total yards gained in medium gains were combined with the total yards lost in medium losses, what would be the result?

76. The game being studied was lost by the Panthers. The coach said that if two of the large-loss plays had been avoided and the number of small-gain plays had been doubled, they would have won the game. If those actions listed by the coach had happened, how many additional yards would the Panthers have gained? Assume each play would reflect the average yards gained or lost per play.

77. The game being studied was lost by the Panthers. The coach of the opposing team said that they could have scored two more touchdowns against the Panthers if they had caused three more large-loss plays and avoided four medium-gain plays. If those actions had happened, what effect would this have had on total yards gained by the Panthers? Assume each play would reflect the average yards gained or lost per play.

Cumulative Review

78. $-17.4 + 8.31 + 2.40$

79. $-\dfrac{3}{4} + \left(-\dfrac{2}{3}\right) + \left(-\dfrac{5}{12}\right)$

80. $-47 - (-32)$

81. $-37 - 51$

82. *Sneaker Production* A sneaker company needs $104\frac{1}{2}$ square yards of white leather, $88\frac{2}{3}$ square yards of neon-yellow leather, and $72\frac{5}{6}$ square yards of neon-red leather to make today's batch of designer sneakers. What is today's required total square yardage of leather?

Developing Your Study Skills

Steps Toward Success in Mathematics

Mathematics is a building process, mastered one step at a time. The foundation of this process consists of a few basic requirements. Those who are successful in mathematics realize the absolute necessity for building a study of mathematics on the firm foundation of these six minimum requirements.

1. Attend class every day.

2. Read the textbook.

3. Take notes in class.

4. Do assigned homework every day.

5. Get help immediately when needed.

6. Review regularly.

Student Learning Objectives

After studying this section, you will be able to:

1 Write numbers in exponent form.

2 Evaluate numerical expressions that contain exponents.

1 Writing Numbers in Exponent Form

In mathematics, we use exponents as a way to abbreviate repeated multiplication.

Long Notation		Exponent Form
$2 \cdot 2 \cdot 2 \cdot 2 \cdot 2 \cdot 2$	=	2^6

There are two parts to exponent notation: (1) the **base** and (2) the **exponent.** The **base** tells you what number is being multiplied and the **exponent** tells you how many times this number is used as a factor. (A *factor,* you recall, is a number being multiplied.)

$$2 \cdot 2 \cdot 2 \cdot 2 \cdot 2 \cdot 2 = 2^6$$

The *base* is 2 (the number being multiplied)

The *exponent* is 6 (the number of times 2 is used as a factor)

If the base is a *positive* real number, the exponent appears to the right and slightly above the level of the number as in, for example, 5^6 and 8^3. If the base is a *negative* real number, then parentheses are used around the number and the exponent appears outside the parentheses. For example, $(-2)(-2)(-2) = (-2)^3$.

In algebra, if we do not know the value of a number, we use a letter to represent the unknown number. We call the letter a **variable.** This is quite useful in the case of exponents. Suppose we do not know the value of a number, but we know the number is multiplied by itself several times. We can represent this with a variable base and a whole-number exponent. For example, when we have an unknown number, represented by the variable x, and this number occurs as a factor four times, we have

$$(x)(x)(x)(x) = x^4.$$

Likewise if an unknown number, represented by the variable w, occurs as a factor five times, we have

$$(w)(w)(w)(w)(w) = w^5.$$

EXAMPLE 1 Write in exponent form.

(a) $9(9)(9)$ **(b)** $13(13)(13)(13)$ **(c)** $-7(-7)(-7)(-7)(-7)$

(d) $-4(-4)(-4)(-4)(-4)(-4)$ **(e)** $(x)(x)$ **(f)** $(y)(y)(y)$

Solution

(a) $9(9)(9) = 9^3$ **(b)** $13(13)(13)(13) = 13^4$

(c) The -7 is used as a factor five times. The answer must contain parentheses. Thus $-7(-7)(-7)(-7)(-7) = (-7)^5$.

(d) $-4(-4)(-4)(-4)(-4)(-4) = (-4)^6$

(e) $(x)(x) = x^2$ **(f)** $(y)(y)(y) = y^3$

Practice Problem 1 Write in exponent form.

(a) $6(6)(6)(6)$ **(b)** $-2(-2)(-2)(-2)(-2)$ **(c)** $108(108)(108)$

(d) $-11(-11)(-11)(-11)(-11)(-11)$ **(e)** $(w)(w)(w)$ **(f)** $(z)(z)(z)(z)$

NOTE TO STUDENT: Fully worked-out solutions to all of the Practice Problems can be found at the back of the text starting at page SP-1

If the base has an exponent of 2, we say the base is **squared.**

If the base has an exponent of 3, we say the base is **cubed.**

If the base has an exponent greater than 3, we say the base is raised **to the (exponent)-th power.**

x^2 is read "x squared." y^3 is read "y cubed."

3^6 is read "three to the sixth power" or simply "three to the sixth."

2 Evaluating Numerical Expressions That Contain Exponents

EXAMPLE 2   Evaluate.   **(a)** 2^5 **(b)** $2^3 + 4^4$

Solution

(a) $2^5 = (2)(2)(2)(2)(2) = 32$

(b) First we evaluate each power.

$2^3 = 8$ $4^4 = 256$

Then we add. $8 + 256 = 264$

Practice Problem 2   Evaluate.   **(a)** 3^5 **(b)** $2^2 + 3^3$

If the base is negative, be especially careful in determining the sign. Notice the following:

$$(-3)^2 = (-3)(-3) = +9   (-3)^3 = (-3)(-3)(-3) = -27.$$

From Section 1.3 we know that when you multiply two or more real numbers, first you multiply their absolute values.

- The result is positive if there are an even number of negative signs.
- The result is negative if there are an odd number of negative signs.

SIGN RULE FOR EXPONENTS

Suppose that a number is written in exponent form and the base is negative. The result is **positive** if the exponent is **even.** The result is **negative** if the exponent is **odd.**

Be careful how you read expressions with exponents and negative signs.

$$(-3)^4 \text{ means } (-3)(-3)(-3)(-3) \text{ or } +81.$$
$$-3^4 \text{ means } -(3)(3)(3)(3) \text{ or } -81.$$

EXAMPLE 3   Evaluate.

(a) $(-2)^3$ **(b)** $(-4)^6$ **(c)** -3^6 **(d)** $-(5^4)$

Solution

(a) $(-2)^3 = -8$ The answer is negative since the base is negative and the exponent 3 is odd.

(b) $(-4)^6 = +4096$ The answer is positive since the exponent 6 is even.

(c) $-3^6 = -729$ The negative sign is not contained in parentheses. Thus we find 3 raised to the sixth power and then take the negative of that value.

(d) $-(5^4) = -625$ The negative sign is outside the parentheses.

Practice Problem 3   Evaluate.

(a) $(-3)^3$ **(b)** $(-2)^6$ **(c)** -2^4 **(d)** $-(3^6)$

Calculator

Exponents

You can use a calculator to evaluate 3^5. Press the following keys:

$$\boxed{3}\ \boxed{y^x}\ \boxed{5}\ \boxed{=}$$

The display should read

243

Try the following.

(a) 4^6 **(b)** $(0.2)^5$
(c) 18^6 **(d)** 3^{12}

Answers:

(a) 4096   **(b)** 0.00032
(c) 34,012,224   **(d)** 531,441

The steps needed to raise a number to a power are slightly different on some calculators.

EXAMPLE 4 Evaluate.

(a) $\left(\dfrac{1}{2}\right)^4$ **(b)** $(0.2)^4$ **(c)** $\left(\dfrac{2}{5}\right)^3$

(d) $(3)^3(2)^5$ **(e)** $2^3 - 3^4$

Solution

(a) $\left(\dfrac{1}{2}\right)^4 = \left(\dfrac{1}{2}\right)\left(\dfrac{1}{2}\right)\left(\dfrac{1}{2}\right)\left(\dfrac{1}{2}\right) = \dfrac{1}{16}$

(b) $(0.2)^4 = (0.2)(0.2)(0.2)(0.2) = 0.0016$

(c) $\left(\dfrac{2}{5}\right)^3 = \left(\dfrac{2}{5}\right)\left(\dfrac{2}{5}\right)\left(\dfrac{2}{5}\right) = \dfrac{8}{125}$

(d) First we evaluate each power.

$3^3 = 27 \qquad 2^5 = 32$

Then we multiply. $(27)(32) = 864$

(e) $2^3 - 3^4 = 8 - 81 = -73$

NOTE TO STUDENT: Fully worked-out solutions to all of the Practice Problems can be found at the back of the text starting at page SP-1

Practice Problem 4 Evaluate.

(a) $\left(\dfrac{1}{3}\right)^3$ **(b)** $(0.3)^4$ **(c)** $\left(\dfrac{3}{2}\right)^4$

(d) $(3)^4(4)^2$ **(e)** $4^2 - 2^4$

Verbal and Writing Skills

1. Explain in your own words how to evaluate 4^4.

2. Explain in your own words how to evaluate 9^2.

3. Explain how you would determine whether $(-5)^3$ is negative or positive.

4. Explain how you would determine whether $(-2)^5$ is negative or positive.

5. Explain the difference between $(-2)^4$ and -2^4. What answers do you obtain when you evaluate the expressions?

6. Explain the difference between $(-3)^4$ and -3^4. What answers do you obtain when you evaluate the expressions?

Write in exponent form.

7. $(5)(5)(5)(5)(5)(5)(5)$ **8.** $(7)(7)(7)(7)(7)$ **9.** $(w)(w)$ **10.** $(z)(z)(z)$

11. $(p)(p)(p)(p)$ **12.** $(x)(x)(x)(x)(x)$ **13.** $(3q)(3q)(3q)$ **14.** $(6x)(6x)(6x)(6x)$

Evaluate.

15. 3^3 **16.** 4^2 **17.** 3^4 **18.** 8^3 **19.** 6^3 **20.** 15^2

21. $(-3)^3$ **22.** $(-2)^3$ **23.** $(-4)^2$ **24.** $(-5)^4$ **25.** -5^2 **26.** -4^2

27. $\left(\dfrac{1}{4}\right)^2$ **28.** $\left(\dfrac{1}{2}\right)^3$ **29.** $\left(\dfrac{2}{5}\right)^3$ **30.** $\left(\dfrac{2}{3}\right)^4$ **31.** $(1.1)^2$ **32.** $(1.2)^2$

33. $(0.2)^4$ **34.** $(0.7)^3$ **35.** $(-16)^2$ **36.** $(-7)^4$ **37.** -16^2 **38.** -7^4

Evaluate.

39. $5^3 + 6^2$ **40.** $7^2 + 6^3$ **41.** $5^3 - 3^2$ **42.** $4^3 - 2^5$ **43.** $(-3)^3 - (8)^2$

44. $(-2)^3 - (-5)^4$ **45.** $2^5 - (-3)^2$ **46.** $8^2 - (-2)^3$ **47.** $(-4)^3(-3)^2$

48. $(-7)^3(-2)^4$ **49.** $8^2(-2)^3$ **50.** $9^2(-3)^3$ **51.** 4^{12} **52.** 6^{11}

To Think About

53. What number to the third power equals -343?

54. What number to the eighth power equals 256?

Cumulative Review

Evaluate.

55. $(-11) + (-13) + 6 + (-9) + 8$ **56.** $\dfrac{3}{4} \div \left(-\dfrac{9}{20}\right)$ **57.** $-17 - (-9)$ **58.** $(-2.1)(-1.2)$

59. Amanda decided to invest her summer job earnings of $1600 in a bank certificate of deposit, at an annual interest rate of 6% for nine months. How much money did Amanda have at the end of the nine months when her certificate of deposit matured?

Student Learning Objective

After studying this section, you will be able to:

 Use the order of operations to simplify numerical expressions.

① Using the Order of Operations to Simplify Numerical Expressions

It is important to know *when* to do certain operations as well as how to do them. For example, to simplify the expression $2 - 4 \cdot 3$, should we subtract first or multiply first?

Also remember that multiplication can be written several ways. Thus $4 \cdot 3$, 4×3, $4(3)$, and $(4)(3)$ all indicate that we are multiplying 4 times 3.

The following list will assist you. It tells which operations to do first: the correct **order of operations.** You might think of it as a *list of priorities*.

> **ORDER OF OPERATIONS FOR NUMBERS**
>
> Follow this order of operations:
>
> Do first
> 1. Do all operations inside parentheses.
> 2. Raise numbers to a power.
> 3. Multiply and divide numbers from left to right.
>
> Do last
> 4. Add and subtract numbers from left to right.

Let's return to the problem $2 - 4 \cdot 3$. There are no parentheses or numbers raised to a power, so multiplication comes next. We do that first. Then we subtract since this comes last on our list.

$$2 - 4 \cdot 3 = 2 - 12 \quad \text{Follow the order of operations by first multiplying } 4 \cdot 3 = 12.$$
$$= -10 \quad \text{Combine } 2 - 12 = -10.$$

EXAMPLE 1

Evaluate. $8 \div 2 \cdot 3 + 4^2$

Solution

$$8 \div 2 \cdot 3 + 4^2 = 8 \div 2 \cdot 3 + 16 \quad \text{Evaluate } 4^2 = 16 \text{ because the highest priority in this problem is raising to a power.}$$
$$= 4 \cdot 3 + 16 \quad \text{Next multiply and divide from left to right. So } 8 \div 2 = 4 \text{ and } 4 \cdot 3 = 12.$$
$$= 12 + 16$$
$$= 28 \quad \text{Finally, add.}$$

Practice Problem 1 Evaluate. $25 \div 5 \cdot 6 + 2^3$

Note: Multiplication and division have equal priority. We do not do multiplication first. Rather, we work from left to right, doing any multiplication or division that we encounter. Similarly, addition and subtraction have equal priority.

EXAMPLE 2

Evaluate. $(-3)^3 - 2^4$

Solution The highest priority is to raise the expressions to the appropriate powers.

$$(-3)^3 - 2^4 = -27 - 16 \quad \text{In } (-3)^3 \text{ we are cubing the number } -3 \text{ to obtain } -27.$$
$$\text{Be careful; } -2^4 \text{ is not } (-2)^4!$$
$$\text{Raise 2 to the fourth power and take the negative of the result.}$$
$$= -43 \quad \text{The last step is to add and subtract from left to right.}$$

Practice Problem 2 Evaluate. $(-4)^3 - 2^6$

NOTE TO STUDENT: Fully worked-out solutions to all of the Practice Problems can be found at the back of the text starting at page SP-1

EXAMPLE 3   Evaluate. $2 \cdot (2 - 3)^3 + 6 \div 3 + (8 - 5)^2$

Solution

$2 \cdot (2 - 3)^3 + 6 \div 3 + (8 - 5)^2$ Combine the numbers inside the parentheses.

$= 2 \cdot (-1)^3 + 6 \div 3 + 3^2$

$= 2 \cdot (-1) + 6 \div 3 + 9$ Next, raise to a power. Note that we need parentheses for -1 because of the negative sign, but they are not needed for 3.

$= -2 + 2 + 9$ Next, multiply and divide from left to right.

$= 9$ Finally, add and subtract from left to right.

Practice Problem 3   Evaluate. $6 - (8 - 12)^2 + 8 \div 2$

EXAMPLE 4   Evaluate. $\left(-\frac{1}{5}\right)\left(\frac{1}{2}\right) - \left(\frac{3}{2}\right)^2$

Solution   The highest priority is to raise $\frac{3}{2}$ to the second power.

$\left(\frac{3}{2}\right)^2 = \left(\frac{3}{2}\right)\left(\frac{3}{2}\right) = \frac{9}{4}$ Next we multiply.

$\left(-\frac{1}{5}\right)\left(\frac{1}{2}\right) - \left(\frac{3}{2}\right)^2 = \left(-\frac{1}{5}\right)\left(\frac{1}{2}\right) - \frac{9}{4}$

$= -\frac{1}{10} - \frac{9}{4}$

$= -\frac{1 \cdot 2}{10 \cdot 2} - \frac{9 \cdot 5}{4 \cdot 5}$ We need to write each fraction as an equivalent fraction with the LCD of 20.

$= -\frac{2}{20} - \frac{45}{20}$

$= -\frac{47}{20}$ Add.

Practice Problem 4   Evaluate. $\left(-\frac{1}{7}\right)\left(-\frac{14}{5}\right) + \left(-\frac{1}{2}\right) \div \left(\frac{3}{4}\right)$

Calculator

Order of Operations

Use your calculator to evaluate $3 + 4 \cdot 5$.
Enter

3 [+] 4 [×] 5 [=]

If the display is [23], the correct order of operations is built in. If the display is not 23, you will need to modify the way you enter the problem. You should use

4 [×] 5 [+] 3 [=]

Try $6 + 3 \cdot 4 - 8 \div 2$.

Answer:

14

Developing Your Study Skills

Previewing New Material

Part of your study time each day should consist of looking over the sections in your text that are to be covered the following day. You do not necessarily need to study and learn the material on your own, but a survey of the concepts, terminology, diagrams, and examples will help the new ideas seem more familiar as the instructor presents them. You can look for concepts that appear confusing or difficult and be ready to listen carefully for your instructor's explanations. You can be prepared to ask the questions that will increase your understanding. Previewing new material enables you to see what is coming and prepares you to be ready to absorb it.

Verbal and Writing Skills

Game Points *You have lost a game of UNO and are counting the points left in your hand. You announce that you have three fours and six fives.*

1. Write this as a number expression.

2. How many points have you in your hand?

3. What answer would you get for the number expression if you simplified it by
(a) performing the operations from left to right?

(b) following the order of operations?

4. Which procedure in exercise 3 gives the correct number of total points?

Evaluate.

5. $(2 - 5)^2 \div 3 \times 4$

6. $(3 - 7)^2 \div 2 \times 5$

7. $2(3 - 5 + 6) + 5$

8. $3(9 - 2 + 3) + 7$

9. $8 - 2^3 \cdot 5 + 3$

10. $6 - 3^2 \cdot 6 + 4$

11. $4 + 42 \div 3 \cdot 2 - 8$

12. $7 + 36 \div 12 \cdot 3 - 14$

13. $3 \cdot 5 + 7 \cdot 3 - 5 \cdot 3$

14. $2 \cdot 6 + 5 \cdot 3 - 7 \cdot 4$

15. $8 - 5(2)^3 \div (-8)$

16. $11 - 3(4)^2 \div (-6)$

17. $3(5 - 7)^2 - 6(3)$

18. $-2(3 - 6)^2 - (-2)$

19. $5 \cdot 6 - (3 - 5)^2 + 8 \cdot 2$

20. $(-3)^2 \cdot 6 \div 9 + 4 \cdot 2$

21. $\dfrac{1}{2} \div \dfrac{2}{3} + 6 \cdot \dfrac{1}{4}$

22. $\dfrac{5}{6} \div \dfrac{2}{3} - 6 \cdot \left(\dfrac{1}{2}\right)^2$

23. $0.8 + 0.3(0.6 - 0.2)^2$

24. $0.05 + 1.4 - (0.5 - 0.7)^3$

25. $\dfrac{3}{4}\left(-\dfrac{2}{5}\right) - \left(-\dfrac{3}{5}\right)$

26. $-\dfrac{2}{3}\left(\dfrac{3}{5}\right) + \dfrac{5}{7} \div \dfrac{5}{3}$

Mixed Practice

27. $(3 - 7)^2 \div 8 + 3$

28. $\left(\dfrac{3}{5}\right)\left(\dfrac{5}{6}\right) - \dfrac{3}{4} \div 6$

29. $\left(\dfrac{3}{4}\right)^2(-16) + \dfrac{4}{5} \div \dfrac{-8}{25}$

30. $\left(2\dfrac{4}{7}\right) \div \left(-1\dfrac{1}{5}\right)$

31. $-6.3 - (-2.7)(1.1) + (3.3)^2$

32. $4.35 + 8.06 \div (-2.6) - (2.1)^2$

33. $\left(\dfrac{1}{2}\right)^3 + \dfrac{1}{4} - \left(\dfrac{1}{6} - \dfrac{1}{12}\right) - \dfrac{2}{3} \cdot \left(\dfrac{1}{4}\right)^2$

34. $(2.4 \cdot 1.2)^2 - 1.6 \cdot 2.2 \div 4.0 - 3.6$

Applications

Tiger Woods just completed a round of golf, and created the following scoring table based on his score per hole, compared to par:

Score on a Hole	Number of Times the Score Occurred
Eagle (-2)	3
Birdie (-1)	9
Par (0)	5
Bogey ($+1$)	1

35. Write his score as the sum of eagles, birdies, pars, and bogeys.

36. What was his final score for the round, when compared to par?

37. What answer do you get if you do the arithmetic left to right?

38. Explain why the answers in #36 and #37 do not match.

Cumulative Review

Simplify.

39. $(0.5)^3$

40. $-\dfrac{3}{4} - \dfrac{5}{6}$

41. -1^{20}

42. $3\dfrac{3}{5} \div 6\dfrac{1}{4}$

43. *Health Drink* An Olympic weight lifter has been told by his trainer to increase his daily consumption of protein by 15 grams. The trainer suggested a health drink that provides 2 grams of protein for every 6 ounces consumed. How many ounces of this health drink would the weight lifter need to consume daily to reach his goal of 15 additional grams per day?

44. $(0.76)(0.9)$

45. $0.324 \div 0.6$

1. _____

2. _____

3. _____

4. _____

5. _____

6. _____

7. _____

8. _____

9. _____

10. _____

11. _____

12. _____

13. _____

14. _____

15. _____

16. _____

17. _____

18. _____

19. _____

20. _____

21. _____

22. _____

How are you doing with your homework assignments in Sections 1.1 to 1.5? Do you feel you have mastered the material so far? Do you understand the concepts you have covered? Before you go further in the textbook, take some time to do each of the following problems.

Simplify each of the following. If the answer is a fraction, be sure to leave it in reduced form.

1.1

1. $3 + (-12)$

2. $-\dfrac{5}{6} + \left(-\dfrac{7}{8}\right)$

3. $0.34 + 0.9$

4. $-14 + 3 + (-2.5) + 6.4$

1.2

5. $-23 - (-34)$

6. $-\dfrac{4}{5} - \dfrac{1}{3}$

7. $4.5 - (-7.8)$

8. $-4 - (-5) + 9$

1.3

9. $(-3)(-8)(2)(-2)$

10. $\left(-\dfrac{6}{11}\right)\left(-\dfrac{5}{3}\right)$

11. $-0.072 \div 0.08$

12. $\dfrac{5}{8} \div \left(-\dfrac{17}{16}\right)$

1.4

Evaluate.

13. $(0.7)^3$

14. $(-4)^4$

15. -2^8

16. $\left(\dfrac{2}{3}\right)^3$

17. $5^3 + (-2)^4$

1.5

18. $12 \div 6(2) + 3$

19. $15 + 3 - 2 + (-6)$

20. $(9 - 13)^2 + 15 \div (-3)$

21. $-0.12 \div 0.6 + (-3)(1.2) - (-0.5)$

22. $\left(\dfrac{3}{4}\right)\left(-\dfrac{2}{5}\right) + \left(-\dfrac{1}{2}\right)\left(\dfrac{4}{5}\right) + \left(\dfrac{1}{2}\right)^2$

Now turn to page SA-3 for the answer to each of these problems. Each answer also includes a reference to the objective in which the problem is first taught. If you missed any of these problems, you should stop and review the Examples and Practice Problems in the referenced objective. A little review now will help you master the material in the upcoming sections of the text.

1.6 USING THE DISTRIBUTIVE PROPERTY TO SIMPLIFY ALGEBRAIC EXPRESSIONS

Using the Distributive Property to Simplify Algebraic Expressions

As we learned previously, we use letters called *variables* to represent unknown numbers. If a number is multiplied by a variable we do not need any symbol between the number and variable. Thus, to indicate $(2)(x)$, we write $2x$. To indicate $3 \cdot y$, we write $3y$. If one variable is multiplied by another variable, we place the variables next to each other. Thus, $(a)(b)$ is written ab. We use exponent form if an unknown number (a variable) is used several times as a factor. Thus, $x \cdot x \cdot x = x^3$. Similarly, $(y)(y)(y)(y) = y^4$.

In algebra, we need to be familiar with several definitions. We will use them throughout the remainder of this book. Take some time to think through how each of these definitions is used.

An **algebraic expression** is a quantity that contains numbers and variables, such as $a + b, 2x - 3$, and $5ab^2$. In this chapter we will be learning rules about adding and multiplying algebraic expressions. A **term** is a number, a variable, or a product of numbers and variables. $17, x, 5xy$, and $22xy^3$ are all examples of terms. We will refer to terms when we discuss the distributive property.

An important property of algebra is the **distributive property.** We can state it in an equation as follows:

> ### DISTRIBUTIVE PROPERTY
>
> For all real numbers a, b, and c,
> $$a(b + c) = ab + ac.$$

A numerical example shows that it does seem reasonable.

$$5(3 + 6) = 5(3) + 5(6)$$
$$5(9) = 15 + 30$$
$$45 = 45$$

We can use the distributive property to multiply any term by the sum of two or more terms. In Section 1.4, we defined the word *factor*. Two or more algebraic expressions that are multiplied are called **factors.** Consider the following examples of multiplying algebraic expressions.

EXAMPLE 1 Multiply. **(a)** $5(a + b)$ **(b)** $-3(3x + 2y)$

Solution

(a) $5(a + b) = 5a + 5b$ Multiply the factor $(a + b)$ by the factor 5.

(b) $-3(3x + 2y) = -3(3x) + (-3)(2y)$ Multiply the factor $(3x + 2y)$ by the
$$= -9x - 6y$$ factor (-3).

Practice Problem 1 Multiply. **(a)** $3(x + 2y)$ **(b)** $-a(a - 3b)$

If the parentheses are preceded by a negative sign, we consider this to be the product of (-1) and the expression inside the parentheses.

NOTE TO STUDENT: Fully worked-out solutions to all of the Practice Problems can be found at the back of the text starting at page SP-1

EXAMPLE 2   Multiply. $-(a - 2b)$

Solution

$$-(a - 2b) = (-1)(a - 2b) = (-1)(a) + (-1)(-2b) = -a + 2b$$

NOTE TO STUDENT: Fully worked-out solutions to all of the Practice Problems can be found at the back of the text starting at page SP-1

Practice Problem 2   Multiply. $-(-3x + y)$

In general, we see that in all these examples we have multiplied each term of the expression in the parentheses by the expression in front of the parentheses.

EXAMPLE 3   Multiply.   **(a)** $\dfrac{2}{3}(x^2 - 6x + 8)$ **(b)** $1.4(a^2 + 2.5a + 1.8)$

Solution

(a) $\dfrac{2}{3}(x^2 - 6x + 8) = \left(\dfrac{2}{3}\right)(1x^2) + \left(\dfrac{2}{3}\right)(-6x) + \left(\dfrac{2}{3}\right)(8)$

$$= \dfrac{2}{3}x^2 + (-4x) + \dfrac{16}{3}$$

$$= \dfrac{2}{3}x^2 - 4x + \dfrac{16}{3}$$

(b) $1.4(a^2 + 2.5a + 1.8) = 1.4(1a^2) + (1.4)(2.5a) + (1.4)(1.8)$

$$= 1.4a^2 + 3.5a + 2.52$$

Practice Problem 3   Multiply.

(a) $\dfrac{3}{5}(a^2 - 5a + 25)$ **(b)** $2.5(x^2 - 3.5x + 1.2)$

There are times we multiply a variable by itself and use exponent notation. For example, $(x)(x) = x^2$ and $(x)(x)(x) = x^3$. In other cases there will be numbers and variables multiplied at the same time.

We will see problems like $(2x)(x) = (2)(x)(x) = 2x^2$. Some expressions will involve multiplication of more than one variable. We will see problems like $(3x)(xy) = (3)(x)(x)(y) = 3x^2y$. There will be times when we use the distributive property and all of these methods will be used. For example,

$$2x(x - 3y + 2) = 2x(x) + (2x)(-3y) + (2x)(2)$$
$$= 2x^2 + (-6)(xy) + 4(x)$$
$$= 2x^2 - 6xy + 4x.$$

We will discuss this type of multiplication of variables with exponents in more detail in Section 4.1. At that point we will expand these examples and other similar examples to develop the general rule for multiplication $(x^a)(x^b) = x^{a+b}$.

EXAMPLE 4   Multiply. $-2x(3x + y - 4)$

Solution

$$-2x(3x + y - 4) = -2(x)(3)(x) + (-2)(x)(y) + (-2)(x)(-4)$$
$$= -2(3)(x)(x) + (-2)(xy) + (-2)(-4)(x)$$
$$= -6x^2 - 2xy + 8x$$

Practice Problem 4   Multiply. $-4x(x - 2y + 3)$

The distributive property can also be presented with the a on the right.

$$(b + c)a = ba + ca$$

The a is "distributed" over the b and c inside the parentheses.

EXAMPLE 5   Multiply. $(2x^2 - x)(-3)$

Solution

$$(2x^2 - x)(-3) = 2x^2(-3) + (-x)(-3)$$
$$= -6x^2 + 3x$$

Practice Problem 5   Multiply.
$$(3x^2 - 2x)(-4)$$

EXAMPLE 6   A farmer has a rectangular field that is 300 feet wide. One portion of the field is $2x$ feet long. The other portion of the field is $3y$ feet long. Use the distributive property to find an expression for the area of this field.

Solution   First we draw a picture of a field that is 300 feet wide and $2x + 3y$ feet long.

To find the area of the field, we multiply the width times the length.

$$300(2x + 3y) = 300(2x) + 300(3y) = 600x + 900y$$

Thus the area of the field in square feet is $600x + 900y$.

Practice Problem 6   A farmer has a rectangular field that is 400 feet wide. One portion of the field is $6x$ feet long. The other portion of the field is $9y$ feet long. Use the distributive property to find an expression for the area of this field.

Developing Your Study Skills

How to Do Homework

As you begin your homework assignments, read the directions carefully. You need to understand what is being asked for. Concentrate on each exercise, taking time to solve it accurately. Rushing through your work usually causes errors. Check your answers with those given in the back of the textbook. If your answer is incorrect, check to see that you are doing the right exercise. Redo the exercise, watching for little errors. If it is still wrong, check with a friend. Perhaps the two of you can figure it out.

Verbal and Writing Skills

In exercises 1 and 2, complete each sentence by filling in the blank.

1. A _____ is a symbol used to represent an unknown number.

2. When we write an expression with numbers and variables such as $7x$, it indicates that we are _____ 7 by x.

3. Explain in your own words how we multiply a problem like $(4x)(x)$.

4. Explain why you think the property $a(b + c) = ab + ac$ is called the distributive property. What does distribute mean?

5. Does the following distributive property work? $a(b - c) = ab - ac$ Why or why not? Give an example.

6. Susan tried to use the distributive property and wrote

$$-5(x + 3y - 2) = -5x - 15y - 10.$$

What did she do wrong?

Multiply. Use the distributive property.

7. $3(x - 2y)$

8. $4(3x - y)$

9. $-2(4a - 3b)$

10. $-3(2a - 5b)$

11. $3(3x + y)$

12. $2(4x + y)$

13. $5(-2a - 3b)$

14. $6(-4a - 2b)$

15. $-(x - 3y)$

16. $-(-4x + y)$

17. $-9(9x - 5y + 8)$

18. $-5(3x + 9 - 7y)$

19. $2(-5x + y - 6)$

20. $3(2x - 6y - 5)$

21. $\frac{5}{6}(12x^2 - 24x + 18)$

22. $\frac{2}{3}(-27a^4 + 9a^2 - 21)$

23. $\frac{x}{5}(x + 10y - 4)$ $\left(\text{Hint: } \frac{x}{5} = \frac{1}{5}x\right)$

24. $\frac{y}{3}(3y - 4x - 6)$ $\left(\text{Hint: } \frac{y}{3} = \frac{1}{3}y\right)$

25. $5x(x + 2y + z)$

26. $3a(2a + b - c)$

27. $(2x - 3)(-2)$

28. $(5x + 1)(-4)$

29. $(6x + y - 1)(3x)$

30. $(3x - 3y + 4)(2x)$

Mixed Practice

Multiply. Use the distributive property.

31. $(3x + 2y - 1)(-xy)$

32. $(4a - 2b - 1)(-ab)$

33. $(2x + 3y - 2)3xy$

34. $(-2x + y - 3)4xy$

35. $2.3(4.1x^2 - 2.3x + 0.4)$

36. $3.1(2.5x^2 - 3.1x + 0.7)$

37. $-0.3x(-1.2x^2 - 0.3x + 0.5)$

38. $-0.9q(2.1q - 0.2r - 0.8s)$

39. $0.5x(0.6x + 0.8y - 5)$

Applications

▲ **40.** *Geometry* Blaine Johnson has a rectangular field that is 700 feet wide. One portion of the field is $12x$ feet long. The other portion of the field is $8y$ feet long. Use the distributive property to find an expression for the area of this field.

▲ **41.** *Geometry* Kathy DesMaris has a rectangular field that is 800 feet wide. One portion of the field is $5x$ feet long. The other portion of the field is $14y$ feet long. Use the distributive property to find an expression for the area of this field.

▲ **42.** *Mural Painting* Marc Adler was in charge of painting a mural on the side of the school gym. The gym was 85 feet long. If the distance from the floor to a banner rail is $3x$ feet and it is $2y$ feet from the banner rail to the ceiling, what was the area that the mural had to cover?

43. *Cell Phone Sales* The price of a cell phone was $2x$. A manager's special reduced the price by $5.00. If the store sold $4y$ cell phones, use the distributive property to find the value of the cell phones sold.

To Think About

▲ **44.** *Athletic Field* The athletic field at Baxford College is $3x$ feet wide. It used to be 1500 feet long. However, due to the construction of a new dormitory, the field length was decreased by $4y$ feet. Use the distributive property to find an expression for the area of the new field after construction of the dormitory.

▲ **45.** *Airport Runway* The Beverly Airport runway is $4x$ feet wide. The airport was supposed to have a 3000-foot-long runway. However, some of the land was wetland, so a runway could not be built on all of it. Therefore, the length of the runway was decreased by $2y$ feet. Use the distributive property to find an expression for the area of the final runway.

Cumulative Review

In exercises 46–48, evaluate.

46. $-18 + (-20) + 36 + (-14)$

47. $(-2)^6$

48. $-27 - (-41)$

49. $25 \div 5(2) + (-6)$

50. $(12 - 10)^2 + (-3)(-2)$

51. *Soccer Practice* The San Rafael Junior Soccer League would practice 365 days per year if it could. Last year because of the weather, the players could practice only 205 days. What percent of the days last year could players practice? (Round to the nearest whole percent.)

52. *Soccer Practice* The coach of the team in exercise 51 predicts that this year the team will be able to practice 60% of the year. How many more days of practice will be possible this year than were possible last year?

1.7 COMBINING LIKE TERMS

Student Learning Objectives

After studying this section, you will be able to:

1 Identify like terms.

2 Combine like terms.

1 Identifying Like Terms

We can add or subtract quantities that are *like quantities*. This is called **combining like quantities.**

$$5 \text{ inches} + 6 \text{ inches} = 11 \text{ inches}$$
$$20 \text{ square inches} - 16 \text{ square inches} = 4 \text{ square inches}$$

However, we cannot combine things that are not the same.

$$16 \text{ square inches} - 4 \text{ inches (Cannot be done!)}$$

Similarly, in algebra we can **combine like terms.** This means to add or subtract like terms. Remember, we cannot combine terms that are not the same. Recall that a *term* is a number, a variable, or a product of numbers and variables. **Like terms** are terms that have identical variables and exponents. In other words, like terms must have exactly the same letter parts.

> ### EXAMPLE 1
> List the like terms of each expression.
>
> **(a)** $5x - 2y + 6x$ **(b)** $2x^2 - 3x - 5x^2 - 8x$
>
> **Solution**
>
> **(a)** $5x$ and $6x$ are like terms. These are the only like terms.
>
> **(b)** $2x^2$ and $-5x^2$ are like terms.
>
> $-3x$ and $-8x$ are like terms.
>
> Note that x^2 and x are not like terms.

NOTE TO STUDENT: Fully worked-out solutions to all of the Practice Problems can be found at the back of the text starting at page SP-1

> **Practice Problem 1** List the like terms of each expression.
>
> **(a)** $5a + 2b + 8a - 4b$ **(b)** $x^2 + y^2 + 3x - 7y^2$

Do you really understand what a term is? A term is a number, a variable, or a product of numbers and variables. Terms are the parts of an algebraic expression separated by plus or minus signs. The sign in front of the term is considered part of the term.

2 Combining Like Terms

It is important to know how to combine like terms. Since

$$4 \text{ inches} + 5 \text{ inches} = 9 \text{ inches},$$

we would expect in algebra that $4x + 5x = 9x$.

Why is this true? Let's take a look at the distributive property.

> Like terms may be added or subtracted by using the distributive property:
>
> $$ab + ac = a(b + c) \quad \text{and} \quad ba + ca = (b + c)a.$$

For example,

$$-7x + 9x = (-7 + 9)x = 2x$$
$$5x^2 + 12x^2 = (5 + 12)x^2 = 17x^2.$$

EXAMPLE 2 Combine like terms.

(a) $-4x^2 + 8x^2$ **(b)** $5x + 3x + 2x$

Solution

(a) Notice that each term contains the factor x^2. Using the distributive property, we have

$$-4x^2 + 8x^2 = (-4 + 8)x^2 = 4x^2.$$

(b) Note that each term contains the factor x. Using the distributive property, we have

$$5x + 3x + 2x = (5 + 3 + 2)x = 10x.$$

Practice Problem 2 Combine like terms.

(a) $16y^3 + 9y^3$ **(b)** $5a + 7a + 4a$

In this section, the direction *simplify* means to remove parentheses and/or combine like terms.

EXAMPLE 3 Simplify. $5a^2 - 2a^2 + 6a^2$

Solution

$$5a^2 - 2a^2 + 6a^2 = (5 - 2 + 6)a^2 = 9a^2$$

Practice Problem 3 Simplify. $-8y^2 - 9y^2 + 4y^2$

After doing a few problems, you will find that it is not necessary to write out the step of using the distributive property. We will omit this step for the remaining examples in this section.

EXAMPLE 4 Simplify.

(a) $5.6a + 2b + 7.3a - 6b$ **(b)** $3x^2y - 2xy^2 + 6x^2y$

(c) $2a^2b + 3ab^2 - 6a^2b^2 - 8ab$

Solution

(a) $5.6a + 2b + 7.3a - 6b = 12.9a - 4b$ We combine the a terms and the b terms separately.

(b) $3x^2y - 2xy^2 + 6x^2y = 9x^2y - 2xy^2$ **Note:** x^2y and xy^2 are not like terms because of different powers.

(c) $2a^2b + 3ab^2 - 6a^2b^2 - 8ab$ These terms cannot be combined; there are no like terms in this expression.

Practice Problem 4 Simplify.

(a) $1.3x + 3a - 9.6x + 2a$ **(b)** $5ab - 2ab^2 - 3a^2b + 6ab$

(c) $7x^2y - 2xy^2 - 3x^2y - 4xy^2 + 5x^2y$

The two skills in this section that a student must practice are identifying like terms and correctly adding and subtracting like terms. If a problem involves many terms, you may find it helpful to rearrange the terms so that like terms are together.

EXAMPLE 5   Simplify. $3a - 2b + 5a^2 + 6a - 8b - 12a^2$

Solution   There are three pairs of like terms.

$$\underbrace{3a + 6a}_{a \text{ terms}} \underbrace{- 2b - 8b}_{b \text{ terms}} + \underbrace{5a^2 - 12a^2}_{a^2 \text{ terms}}$$ You can rearrange the terms so that like terms are together, making it easier to combine them.

$$= 9a - 10b - 7a^2$$ Combine like terms.

Because of the commutative property, the order of terms in an answer to this problem is not significant. These three terms can be rearranged in a different order. $-10b + 9a - 7a^2$ and $-7a^2 + 9a - 10b$ are also correct. However, we usually write polynomials in order of descending exponents. $-7a^2 + 9a - 10b$ would be the preferred way to write the answer.

NOTE TO STUDENT: *Fully worked-out solutions to all of the Practice Problems can be found at the back of the text starting at page SP-1*

Practice Problem 5   Simplify. $5xy - 2x^2y + 6xy^2 - xy - 3xy^2 - 7x^2y$

Use extra care with fractional values.

EXAMPLE 6   Simplify. $\dfrac{3}{4}x^2 - 5y - \dfrac{1}{8}x^2 + \dfrac{1}{3}y$

Solution   We need a least common denominator for the x^2 terms, which is 8. Change $\dfrac{3}{4}$ to eighths by multiplying the numerator and denominator by 2.

$$\frac{3}{4}x^2 - \frac{1}{8}x^2 = \frac{3 \cdot 2}{4 \cdot 2}x^2 - \frac{1}{8}x^2 = \frac{6}{8}x^2 - \frac{1}{8}x^2 = \frac{5}{8}x^2$$

The least common denominator for the y terms is 3. Change 5 to thirds.

$$-\frac{5}{1}y + \frac{1}{3}y = \frac{-5 \cdot 3}{1 \cdot 3}y + \frac{1}{3}y = \frac{-15}{3}y + \frac{1}{3}y = -\frac{14}{3}y$$

Thus, our solution is $\dfrac{5}{8}x^2 - \dfrac{14}{3}y$.

Practice Problem 6   Simplify. $\dfrac{1}{7}a^2 - \dfrac{5}{12}b + 2a^2 - \dfrac{1}{3}b$

EXAMPLE 7   Simplify. $6(2x + 3xy) - 8x(3 - 4y)$

Solution   First remove the parentheses; then combine like terms.

$$6(2x + 3xy) - 8x(3 - 4y) = 12x + 18xy - 24x + 32xy$$ Use the distributive property.

$$= -12x + 50xy$$ Combine like terms.

Practice Problem 7   Simplify. $5a(2 - 3b) - 4(6a + 2ab)$

Verbal and Writing Skills

1. Explain in your own words the mathematical meaning of the word *term*.

2. Explain in your own words the mathematical meaning of the phrase *like terms*.

3. Explain which terms are like terms in the expression $5x - 7y - 8x$.

4. Explain which terms are like terms in the expression $12a - 3b - 9a$.

5. Explain which terms are like terms in the expression $7xy - 9x^2y - 15xy^2 - 14xy$.

6. Explain which terms are like terms in the expression $-3a^2b - 12ab + 5ab^2 + 9ab$.

Combine like terms.

7. $-14b^2 - 11b^2$

8. $-17x^5 + 3x^5$

9. $10x^4 + 8x^4 + 7x^2$

10. $3a^3 - 6a^2 + 5a^3$

11. $3x + 2y - 8x - 7y$

12. $4a - 3b - 2a - 8b$

13. $1.3x - 2.6y + 5.8x - 0.9y$

14. $3.1a - 0.2b - 0.8a + 5.3b$

15. $1.6x - 2.8y - 3.6x - 5.9y$

16. $1.9x - 2.4b - 3.8x - 8.2b$

17. $3p - 4q + 2p + 3 + 5q - 21$

18. $6x - 5y - 3y + 7 - 11x - 5$

19. $2ab + 5bc - 6ac - 2ab$

20. $7ab - 3bc - 12ac + 8ab$

21. $2x^2 - 3x - 5 - 7x + 8 - x^2$

22. $5x + 7 - 6x^2 + 6 - 11x + 4x^2$

23. $2y^2 - 8y + 9 - 12y^2 - 8y + 3$

    $-10y^2 + 16y + 12$

24. $3y^2 + 9y - 12 - 4y^2 - 6y + 2$

25. $\frac{1}{3}x - \frac{2}{3}y - \frac{2}{5}x + \frac{4}{7}y$

    $\frac{-1}{15}x - \frac{2}{21}y$

26. $\frac{2}{5}s - \frac{3}{8}t - \frac{4}{15}s - \frac{5}{12}t$

27. $\frac{3}{4}a^2 - \frac{1}{3}b - \frac{1}{5}a^2 - \frac{1}{2}b$

28. $\frac{2}{5}y - \frac{3}{4}x^2 - \frac{1}{3}y + \frac{7}{8}x^2$

29. $ab - 7a - 9ab + 4a - 6b$

30. $ab + 3a - 4ab + 2a - 8b$

Simplify. Use the distributive property to remove parentheses; then combine like terms.

31. $5(2a - b) - 3(5b - 6a)$

32. $8(3x - 2y) + 4(3y - 5x)$

33. $-3b(5a - 3b) + 4(-3ab - 5b^2)$

34. $2x(x - 3y) - 4(-3x^2 - 2xy)$

35. $-3(x^2 + 3y) + 5(-6y - x^2)$

36. $-3(7xy - 11y^2) - 2y(-2x + 3y)$

37. $4(2 - x) - 3(-5 - 12x)$

38. $7(3 - x) - 6(8 - 13x)$

To Think About

▲ **39.** *Fencing in a Pool* Mr. Jimenez has a pool behind his house that needs to be fenced in. The backyard is an odd quadrilateral shape. The four sides are $3a$, $2b$, $4a$, and $7b$ in length. How much fencing (the length of the perimeter) would he need to enclose the pool?

▲ **40.** *Framing a Masterpiece* The new Degas masterpiece purchased by the Museum of Fine Arts in Boston needs to be reframed. If the rectangular picture measures $6x - 3$ wide by $8x - 7$ high, what is the perimeter of the painting?

▲ **41.** *Geometry* A triangle has sides of length $2a$ centimeters, $7b$ centimeters, and $5a + 3$ centimeters. What is the perimeter of the triangle?

▲ **42.** *Geometry* A rectangle has sides of length $7x - 2$ meters and $3x + 4$ meters. What is the perimeter of the rectangle?

▲ **43.** *Geometry* A square has a side of length $9x - 2$ inches. Each side is shortened by 3 inches. What is the perimeter of the new smaller square?

▲ **44.** *Geometry* A triangle has sides of length $4a - 5$ feet, $3a + 8$ feet, and $9a + 2$ feet. Each side is doubled in length. What is the perimeter of the new enlarged triangle?

Cumulative Review

Evaluate.

45. $-\dfrac{1}{3} - \left(-\dfrac{1}{5}\right)$

46. $\left(-\dfrac{5}{3}\right)\left(\dfrac{1}{2}\right)$

47. $\dfrac{4}{5} + \left(-\dfrac{1}{25}\right) + \left(-\dfrac{3}{10}\right)$

48. $\left(\dfrac{5}{7}\right) \div \left(-\dfrac{14}{3}\right)$

49. *Serving Size* A 2-quart container of orange juice produces 9.5 average servings. The container holds approximately 1.9 liters of orange juice. How many liters are in one serving?

1.8 USING SUBSTITUTION TO EVALUATE ALGEBRAIC EXPRESSIONS AND FORMULAS

1 Evaluating an Algebraic Expression for a Specified Value

You will use the order of operations to **evaluate** variable expressions. Suppose we are asked to evaluate

$$6 + 3x \text{ for } x = -4.$$

In general, x represents some unknown number. Here we are told x has the value -4. We can replace x with -4. Use parentheses around -4. Note that we always put replacement values in parentheses.

$$6 + 3(-4) = 6 + (-12) = -6$$

When we replace a variable by a particular value, we say we have **substituted** the value for the variable. We then evaluate the expression (that is, find a value for it).

Student Learning Objectives

After studying this section, you will be able to:

1 Evaluate an algebraic expression for a specified value.

2 Evaluate a formula by substituting values.

EXAMPLE 1 Evaluate $\dfrac{2}{3}x - 5$ for $x = -6$.

Solution

$$\frac{2}{3}x - 5 = \frac{2}{3}(-6) - 5 \quad \text{Substitute } x = -6. \text{ Be sure to enclose the } -6 \text{ in parentheses.}$$

$$= -4 - 5 \quad \text{Multiply } \left(\frac{2}{3}\right)\left(-\frac{6}{1}\right) = -4.$$

$$= -9 \quad \text{Combine.}$$

Practice Problem 1   Evaluate $4 - \dfrac{1}{2}x$ for $x = -8$.

NOTE TO STUDENT: Fully worked-out solutions to all of the Practice Problems can be found at the back of the text starting at page SP-1

Compare parts (a) and (b) in the next example. The two parts illustrate that you must be careful what value you raise to a power. *Note:* In part (b) we will need parentheses within parentheses. To avoid confusion, we use brackets [ ] to represent the outside parentheses.

EXAMPLE 2   **(a)** Evaluate $2x^2$ for $x = -3$.

(b) Evaluate $(2x)^2$ for $x = -3$.

Solution

(a) Here the value x is squared.

$$2x^2 = 2(-3)^2$$

$$= 2(9) \quad \text{First square } -3.$$

$$= 18 \quad \text{Then multiply.}$$

(b) Here the value $(2x)$ is squared.

$$(2x)^2 = [(2)(-3)]^2$$

$$= (-6)^2 \quad \text{First multiply the numbers inside the parentheses.}$$

$$= 36 \quad \text{Then square } -6.$$

Practice Problem 2   Evaluate for $x = -3$.   **(a)** $4x^2$ **(b)** $(4x)^2$

Carefully study the solutions to Example 2(a) and Example 2(b). You will find that taking the time to see *how* and *why* they are different is a good investment of study time.

EXAMPLE 3 Evaluate $x^2 + 3x$ for $x = -4$.

Solution

$$
\begin{aligned}
x^2 + 3x &= (-4)^2 + 3(-4) && \text{Replace } x \text{ by } -4 \text{ in the original expression.} \\
&= 16 + (3)(-4) && \text{Raise to a power.} \\
&= 16 - 12 && \text{Multiply.} \\
&= 4 && \text{Finally, add.}
\end{aligned}
$$

Practice Problem 3 Evaluate $2x^2 - 3x$ for $x = -2$.

---

② Evaluating a Formula by Substituting Values

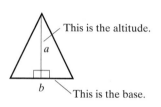

This is the altitude.

a

b

This is the base.

We can *evaluate a formula* by substituting values for the variables. For example, the area of a triangle can be found using the formula $A = \frac{1}{2}ab$, where b is the length of the base of the triangle and a is the altitude of the triangle (see figure). If we know values for a and b, we can substitute those values into the formula to find the area. The units for area are *square units*.

Because some of the examples and exercises in this section involve geometry, it may be helpful to review this topic.

However, before we proceed to do any examples, it may be helpful to review a little geometry. The following information is very important. If you have forgotten some of this material (or if you have never learned it), please take the time to learn it completely now. Throughout the entire book we will be using this information in solving applied problems.

Perimeter is the distance around a plane figure. Perimeter is measured in linear units (inches, feet, centimeters, miles). **Area** is a measure of the amount of surface in a region. Area is measured in square units (square inches, square feet, square centimeters).

In our sketches we will show angles of 90° by using a small square (⌐⌐). This indicates that the two lines are at right angles. All angles that measure 90° are called **right angles.** An **altitude** is perpendicular to the base of a figure. That is, the altitude forms right angles with the base. The small corner square in a sketch helps us identify the altitude of the figure.

The following box provides a handy guide to some facts and formulas you will need to know. Use it as a reference when solving word problems involving geometric figures.

GEOMETRIC FORMULAS: TWO-DIMENSIONAL FIGURES

A **parallelogram** is a four-sided figure with opposite sides parallel. In a parallelogram, opposite sides are equal and opposite angles are equal.

b

c a = altitude c

b

Perimeter = the sum of all four sides

Area = ab

A **rectangle** is a parallelogram with all interior angles measuring 90°.

l

w w = width

l = length

Perimeter = $2l + 2w$

Area = lw

A **square** is a rectangle with all four sides equal.

$s = $ side

Perimeter $= 4s$
Area $= s^2$

A **trapezoid** is a four-sided figure with two sides parallel. The parallel sides are called the *bases* of the trapezoid.

$b_1 = $ base one
d $a = $ altitude e
$b_2 = $ base two

Perimeter $=$ the sum of all four sides

Area $= \dfrac{1}{2}a(b_1 + b_2)$

A **triangle** is a closed plane figure with three sides.

$a = $ altitude
$b = $ base

Perimeter $=$ the sum of the three sides

Area $= \dfrac{1}{2}ab$

A **circle** is a plane curve consisting of all points at an equal distance from a given point called the center.
Circumference is the distance around a circle.

$r = $ radius

$d = $ diameter $= 2r$

Circumference $= 2\pi r$
Area $= \pi r^2$

π (the number *pi*) is a constant associated with circles. It is an irrational number that is approximately 3.141592654. We usually use 3.14 as a sufficiently accurate approximation. Thus we write $\pi \approx 3.14$ for most of our calculations involving π.

EXAMPLE 4   Find the area of a triangle with a base of 16 centimeters (cm) and a height of 12 centimeters (cm).

Solution   Use the formula $A = \dfrac{1}{2}ab$.

Substitute 12 centimeters for a and 16 centimeters for b.

$A = \dfrac{1}{2}(12 \text{ centimeters})(16 \text{ centimeters})$

$\quad = \dfrac{1}{2}(12)(16)(\text{cm})(\text{cm})$ If you take $\frac{1}{2}$ of 12 first, it will make your calculation easier.

$\quad = (6)(16)(\text{cm})^2 = 96$ square centimeters

The area of the triangle is 96 square centimeters.

▲ **Practice Problem 4**   Find the area of a triangle with an altitude of 3 meters and a base of 7 meters.

NOTE TO STUDENT: Fully worked-out solutions to all of the Practice Problems can be found at the back of the text starting at page SP-1

The area of a circle is given by

$$A = \pi r^2.$$

We will use 3.14 as an approximation for the *irrational number* π.

▲ **EXAMPLE 5**   Find the area of a circle if the radius is 2 inches.

Solution

$A = \pi r^2 \approx (3.14)(2 \text{ inches})^2$ Write the formula and substitute the given values for the letters.

$\quad = (3.14)(4)(\text{in.})^2$ Raise to a power. Then multiply.

$\quad = 12.56 \text{ square inches}$

▲ **Practice Problem 5**   Find the area of a circle if the radius is 3 meters.

The formula $C = \frac{5}{9}(F - 32)$ allows us to find the Celsius temperature if we know the Fahrenheit temperature. That is, we can substitute a value for F in degrees Fahrenheit into the formula to obtain a temperature C in degrees Celsius.

EXAMPLE 6   What is the Celsius temperature when the Fahrenheit temperature is $F = -22°$?

Solution   Use the formula.

$$C = \frac{5}{9}(F - 32)$$

$$= \frac{5}{9}((-22) - 32) \quad \text{Substitute } -22 \text{ for } F \text{ in the formula.}$$

$$= \frac{5}{9}(-54) \quad\quad\quad \text{Combine the numbers inside the parentheses.}$$

$$= (5)(-6) \quad\quad\quad\; \text{Simplify.}$$

$$= -30 \quad\quad\quad\quad\;\; \text{Multiply.}$$

The temperature is $-30°$ Celsius.

Practice Problem 6   What is the Celsius temperature when the Fahrenheit temperature is $F = 68°$? Use the formula $C = \frac{5}{9}(F - 32)$.

NOTE TO STUDENT: Fully worked-out solutions to all of the Practice Problems can be found at the back of the text starting at page SP-1

When driving in Canada or Mexico, we must observe speed limits posted in kilometers per hour. A formula that converts r (miles per hour) to k (kilometers per hour) is $k \approx 1.61r$. Note that this is an approximation.

EXAMPLE 7 You are driving on a highway in Mexico. It has a posted maximum speed of 100 kilometers per hour. You are driving at 61 miles per hour. Are you exceeding the speed limit?

Solution Use the formula.

$$k \approx 1.61r$$
$$= (1.61)(61) \quad \text{Replace } r \text{ by } 61.$$
$$= 98.21 \quad \text{Multiply the numbers.}$$

You are driving at approximately 98 kilometers per hour. You are not exceeding the speed limit.

Practice Problem 7 You are driving behind a heavily loaded truck on a Canadian highway. The highway has a posted minimum speed of 65 kilometers per hour. When you travel at exactly the same speed as the truck ahead of you, you observe that the speedometer reads 35 miles per hour. Assuming that your speedometer is accurate, determine whether the truck is violating the minimum speed law.

Developing Your Study Skills

Problems with Accuracy

Strive for accuracy. Mistakes are often made as a result of human error rather than from lack of understanding. Such mistakes are frustrating. A simple arithmetic or sign error can lead to an incorrect answer.

These five steps will help you cut down on errors.

1. Work carefully and take your time. Do not rush through a problem just to get it done.
2. Concentrate on one problem at a time. Sometimes problems become mechanical, and your mind begins to wander. You can become careless and make a mistake.
3. Check your problem. Be sure that you copied it correctly from the book.
4. Check your computations from step to step. Check the solution in the problem. Does it work? Does it make sense?
5. Keep practicing new skills. Remember the old saying, "Practice makes perfect." An increase in practice results in an increase in accuracy. Many errors are due simply to lack of practice.

There is no magic formula for eliminating all errors, but these five steps will be a tremendous help in reducing them.

Evaluate.

1. $-2x + 1$ for $x = 3$ **2.** $-4x - 2$ for $x = 5$ **3.** $\frac{2}{3}x - 5$ for $x = -9$ **4.** $\frac{3}{4}x + 8$ for $x = -8$

5. $5x + 10$ for $x = \frac{1}{2}$ **6.** $7x + 20$ for $x = -\frac{1}{2}$ **7.** $2 - 4x$ for $x = 7$ **8.** $3 - 5x$ for $x = 8$

9. $3.5 - 2x$ for $x = 2.4$ **10.** $6.3 - 3x$ for $x = 2.3$ **11.** $9x + 13$ for $x = -\frac{3}{4}$ **12.** $5x + 7$ for $x = -\frac{2}{3}$

13. $x^2 - 3x$ for $x = -2$ **14.** $x^2 + 3x$ for $x = 4$ **15.** $3x^2$ for $x = -1$ **16.** $4x^2$ for $x = -1$

17. $-3x^3$ for $x = 2$ **18.** $-7x^2$ for $x = 5$ **19.** $-5x^2$ for $x = -2$ **20.** $-2x^2$ for $x = -3$

21. $2x^2 + 3x$ for $x = -3$ **22.** $18 + 3x^2$ for $x = -3$ **23.** $(2x)^2 + x$ for $x = 3$ **24.** $2 - x^2$ for $x = -2$

25. $2 - (-x)^2$ for $x = -2$ **26.** $2x - 3x^2$ for $x = -4$

27. $7x + (2x)^2$ for $x = -3$ **28.** $5x + (3x)^2$ for $x = -2$

29. $3x^2 - 5x$ for $x = -3$ **30.** $7 - 2x^2$ for $x = -4$

31. $x^2 - 7x + 3$ for $x = 3$ **32.** $4x^2 - 3x + 9$ for $x = 2$

33. $\frac{1}{2}x^2 - 3x + 9$ for $x = -4$ **34.** $\frac{1}{3}x^2 + 2x - 5$ for $x = -3$

35. $x^2 - 2y + 3y^2$ for $x = -3$ and $y = 4$ **36.** $2x^2 - 3xy + 2y$ for $x = 4$ and $y = -1$

37. $a^3 + 2abc - 3c^2$ for $a = 5, b = 9,$ and $c = -1$ **38.** $a^2 - 2ab + 2c^2$ for $a = 3, b = 2,$ and $c = -4$

39. $\dfrac{a^2 + ab}{3b}$ for $a = -1$ and $b = -2$ **40.** $\dfrac{x^2 - 2xy}{2y}$ for $x = -2$ and $y = -3$

Applications

▲ **41.** *Geometry* A sign is made in the shape of a parallelogram. The base measures 22 feet. The altitude measures 16 feet. What is the area of the sign?

▲ **42.** *Geometry* A field is shaped like a parallelogram. The base measures 92 feet. The altitude measures 54 feet. What is the area of the field?

▲ **43.** *TV Parts* A square support unit in a television is made with a side measuring 3 centimeters. A new model being designed for next year will have a larger square with a side measuring 3.2 centimeters. By how much will the area of the square be increased?

▲ **44.** *Computer Chips* A square computer chip for last year's computer had a side measuring 23 millimeters. This year the computer chip has been reduced in size. The new square chip has a side of 20 millimeters. By how much has the area of the chip decreased?

▲ **45.** *Carpentry* A carpenter cut out a small trapezoid as a wooden support for the front step. It has an altitude of 4 inches. One base of the trapezoid measures 9 inches and the other base measures 7 inches. What is the area of this support?

▲ **46.** *Signal Tower* The Comcast signal tower has a small trapezoid frame on the top of the tower. The frame has an altitude of 9 inches. One base of the trapezoid is 20 inches and the other base measures 17 inches. What is the area of this small trapezoidal frame?

▲ **47.** *Geometry* Bradley Palmer State Park has a triangular piece of land on the border. The altitude of the triangle is 400 feet. The base of the triangle is 280 feet. What is the area of this piece of land?

▲ **48.** *Roofing* The ceiling in the Madisons' house has a leak. The roofer exposed a triangular region that needs to be sealed and then reroofed. The region has an altitude of 14 feet. The base of the region is 19 feet. What is the area of the region that needs to be reroofed?

▲ **49.** *Geometry* The radius of a circular opening of a chemistry flask is 4 cm. What is the area of the opening?

▲ **50.** *Geometry* An ancient outdoor sundial has a radius of 5 meters. What is its area?

Temperature   *For exercises 51 and 52, use the formula* $C = \dfrac{5}{9}(F - 32)$ *to find the Celsius temperature.*

51. Dry ice is solid carbon dioxide. Dry ice does not melt, it goes directly from the solid state to the gaseous state. Dry ice changes from a solid to a gas at $-109.3°F$. What is this temperature in C?

52. Ivana was an exchange student from Russia, attending school in Montgomery, Alabama. Her host family told her to take a jacket to school because the temperature was not supposed to rise above $50°$. Since Ivana was used to temperatures in Celsius, this made no sense at all, for $50°C$ is incredibly warm. The host family realized they needed to change the temperature into Celsius for Ivana. What was the temperature in Celsius?

Solve.

▲ **53.** *Sail Dimensions* Find the total cost of making a triangular sail that has a base dimension of 12 feet and a height of 20 feet if the price for making the sail is $19.50 per square foot.

▲ **54.** *Window Coating* A semicircular window of radius 15 inches is to be laminated with a sunblock coating that costs $0.85 per square inch to apply. What is the total cost of coating the window, to the nearest cent? (Use $\pi \approx 3.14$.)

55. *Temperature Tolerance* Some new computers can be exposed to extreme temperatures (as high as 60°C and as low as −50°C). What is the temperature range in Fahrenheit that these computers can be exposed to? (Use the formula $F = \frac{9}{5}C + 32$.)

56. *Temperature Tolerance* To deal with extreme temperatures while doing research at the South Pole, scientists have developed accommodations that can comfortably withstand an outside temperature of −60°C with no wind blowing, or −30°C with wind gusts of up to 50 miles per hour. What is the corresponding Fahrenheit temperature range? (Use the formula $F = \frac{9}{5}C + 32$.)

57. *Sea Level* Bruce becomes exhausted while on a bicycle trip in Canada. He reads on the map that his present elevation is 2.3 kilometers above sea level. How many miles to the nearest tenth above sea level is he? Why is he so tired? Use the formula $r = 0.62k$ where r is the number of miles and k is the number of kilometers.

58. *Bicycle Travel* While biking down the Pacific coast of Mexico, you see on the map that it is 20 kilometers to the nearest town. Approximately how many miles is it to the nearest town? Use the formula $r = 0.62k$ where r is the number of miles and k is the number of kilometers.

Cumulative Review

In exercises 59–60, simplify.

59. $(-2)^4 - 4 \div 2 - (-2)$

60. $3(x - 2y) - (x^2 - y) - (x - y)$

61. *CD Recording* A 93-minute-long recordable compact disc is used to record 15 songs. Express in decimal form the average number of minutes available per song.

62. *Homework Grading* The economics class lecture session had 214 students. If there were 4 teaching assistants to grade homework, approximately how many homework papers were given to each teaching assistant? (Give in decimal form; then round to the nearest whole number.)

1.9 GROUPING SYMBOLS

 Simplifying Algebraic Expressions by Removing Grouping Symbols

Student Learning Objective

After studying this section, you will be able to:

1 Simplify algebraic expressions by removing grouping symbols.

Many expressions in algebra use **grouping symbols** such as parentheses, brackets, and braces. Sometimes expressions are inside other expressions. Because it can be confusing to have more than one set of parentheses, brackets and braces are also used. How do we know what to do first when we see an expression like $2[5 - 4(a + b)]$?

To simplify the expression, we start with the innermost grouping symbols. Here is a set of parentheses. We first use the distributive law to multiply.

$$2[5 - 4(a + b)] = 2[5 - 4a - 4b]$$

We use the distributive law again.

$$= 10 - 8a - 8b$$

There are no like terms, so this is our final answer.

Notice that we started with two sets of grouping symbols, but our final answer has none. So we can say we *removed* the grouping symbols. Of course, we didn't just take them away; we used the distributive law and the rules for real numbers to simplify as much as possible. Although simplifying expressions like this involves many steps, we sometimes say "remove parentheses" as a shorthand direction. Sometimes we say "simplify."

Remember to remove the innermost parentheses first. Keep working from the inside out.

EXAMPLE 1 Simplify. $3[6 - 2(x + y)]$

Solution We want to remove the innermost parentheses first. Therefore, we first use the distributive property to simplify $-2(x + y)$.

$$3[6 - 2(x + y)] = 3[6 - 2x - 2y]$$ Use the distributive property.
$$= 18 - 6x - 6y$$ Use the distributive property again.

Practice Problem 1 Simplify. $5[4x - 3(y - 2)]$

You recall that a negative sign in front of parentheses is equivalent to having a coefficient of negative 1. You can write the -1 and then multiply by -1 using the distributive property.

$$-(x + 2y) = -1(x + 2y) = -x - 2y$$

Notice that this has the effect of removing the parentheses. Each term in the result now has its sign changed.

Similarly, a positive sign in front of parentheses can be viewed as multiplication by $+1$.

$$+(5x - 6y) = +1(5x - 6y) = 5x - 6y$$

If a grouping symbol has a positive or negative sign in front, we mentally multiply by $+1$ or -1, respectively.

Fraction bars are also considered grouping symbols. Later in this book we will encounter problems where our first step will be to simplify expressions above and below fraction bars. This type of operation will have some similarities to the operation of removing parentheses.

NOTE TO STUDENT: Fully worked-out solutions to all of the Practice Problems can be found at the back of the text starting at page SP-1

EXAMPLE 2 Simplify. $-2[3a - (b + 2c) + (d - 3e)]$

Solution

$= -2[3a - b - 2c + d - 3e]$ Remove the two innermost sets of parentheses. Since one is not inside the other, we remove both sets at once.

$= -6a + 2b + 4c - 2d + 6e$ Now we remove the brackets by multiplying each term by -2.

NOTE TO STUDENT: Fully worked-out solutions to all of the Practice Problems can be found at the back of the text starting at page SP-1

Practice Problem 2 Simplify. $-3[2a - (3b - c) + 4a]$

EXAMPLE 3 Simplify. $2[3x - (y + w)] - 3[2x + 2(3y - 2w)]$

Solution

$= 2[3x - y - w] - 3[2x + 6y - 4w]$ In each set of brackets, remove the inner parentheses.

$= 6x - 2y - 2w - 6x - 18y + 12w$ Remove each set of brackets by multiplying by the appropriate number.

$= -20y + 10w$ or $10w - 20y$ Combine like terms. (Note that $6x - 6x = 0x = 0$.)

Practice Problem 3 Simplify. $3[4x - 2(1 - x)] - [3x + (x - 2)]$

You can always simplify problems with many sets of grouping symbols by the method shown. Essentially, you just keep removing one level of grouping symbols at each step. Finally, at the end you add up the like terms if possible.

Sometimes it is possible to combine like terms at each step.

EXAMPLE 4 Simplify. $-3\{7x - 2[x - (2x - 1)]\}$

Solution

$= -3\{7x - 2[x - 2x + 1]\}$ Remove the inner parentheses by multiplying each term within the parentheses by -1.

$= -3\{7x - 2[-x + 1]\}$ Combine like terms by combining $+x - 2x$.

$= -3\{7x + 2x - 2\}$ Remove the brackets by multiplying each term within them by -2.

$= -3\{9x - 2\}$ Combine the x-terms.

$= -27x + 6$ Remove the braces by multiplying each term by -3.

Practice Problem 4 Simplify. $-2\{5x - 3x[2x - (x^2 - 4x)]\}$

Developing Your Study Skills

The Night Before Your Exam

With adequate preparation, you can spend the night before an exam pulling together the final details.

1. Look over each section to be covered in the exam. Review the steps needed to solve each type of problem.

2. Review your list of terms, rules, and formulas that you are expected to know for the exam.

3. Take the Practice Test at the end of the chapter just as though you were taking the actual exam. Do not look in your text or get help in any way. Time yourself so that you know how long it takes you to complete the test.

4. Check the Practice Test. Redo the problems you missed.

5. Be sure you have ready the necessary supplies for taking your exam.

1.9 EXERCISES

Student Solutions Manual · CD/Video · PH Math Tutor Center · MathXL®Tutorials on CD · MathXL® · MyMathLab® · Interactmath.com

Verbal and Writing Skills

1. Rewrite the expression $-3x - 2y$ using a negative sign and parentheses.

2. Rewrite the expression $-x + 5y$ using a negative sign and parentheses.

3. To simplify expressions with grouping symbols, we use the _____ property.

4. When an expression contains many grouping symbols, remove the _____ parentheses first.

Simplify. Remove grouping symbols and combine like terms.

5. $6x - 3(x - 2y)$

6. $-4x - 2(y - 3x)$

7. $2(a + 3b) - 3(b - a)$

8. $4(x - y) - 2(3x + y)$

9. $-3(x + 3y) + 2(2x + y)$

10. $-4(a + 2b) + 5(2b - a)$

11. $2x[4x^2 - 2(x - 3)]$

12. $4y[-3y^2 + 2(4 - y)]$

13. $2[5(x + y) - 2(3x - 4y)]$

14. $-3[2(3a + b) - 5(a - 2b)]$

15. $2(x - 2y) - [3 - 2(x - y)]$

16. $3(x + 2y) - [4 - 2(x + y)]$

17. $5[3a - 2a(3a + 6b) + 6a^2]$

18. $3[x - y(3x + y) + y^2]$

19. $6a(2a^2 - 3a - 4) - a(a - 2)$

20. $7b(3b^2 - 2b - 5) - 2b(4 - b)$

21. $3a^2 - 4[2b - 3b(b + 2)]$

22. $2b^2 - 3[5b + 2b(2 - b)]$

23. $6b - \{5a - 2[a + (b - 2a)]\}$

24. $2a - \{6b - 4[a - (b - 3a)]\}$

25. $3\{3b^2 + 2[5b - (2 - b)]\}$

26. $2\{3x^2 + 4[2x - (3 - x)]\}$

27. $-4\{3a^2 - 2[4a^2 - (b + a^2)]\}$

28. $-2\{x^2 - 3[x - (x - 2x^2)]\}$

To Think About

29. *Robot Success* The job of a four-wheeled robot called Nomad, built by the Robotics Institute at Carnegie Mellon University, is to find samples of meteorites in Antarctica. If Nomad is successful in finding meteorites 2.5% of the time, out of four tries per day for six years, estimate the number of successful and unsuccessful meteorite search attempts. Round to the nearest whole number.

30. *Tipping* Grandma and Grandpa Tobey had a tradition of eating out once a week. The average cost of the meal was $20. In Massachusetts there is a 5% state sales tax that is added to the cost of the meal. The Tobeys always left a 15% tip. They continued this pattern for ten years. Mrs. Tobey felt that they should calculate the tip on the total of the cost of the meal including the sales tax. Mr. Tobey felt they should calculate the tip on the cost of the meal alone. How much difference would this make over the ten-year period?

Cumulative Review

31. *Temperature* Use $F = 1.8C + 32$ to find the Fahrenheit temperature equivalent to 36.4° Celsius.

▲ **32.** *Geometry* Use 3.14 as an approximation for π to compute the area covered by a circular irrigation system with radial arm of length 380 feet. $A = \pi r^2$.

▲ **33.** *Marble Flooring* The base of an office building is in the shape of a trapezoid. The altitude of the trapezoid is 400 feet. The bases of the trapezoid are 700 feet and 800 feet, respectively. What is the area of the base of the office building? If the base has a marble floor that cost $55 per square foot, what was the cost of the marble floor?

▲ **34.** *Signal Paint* The Global Media Tower has a triangular signal tester at the top of the tower. The altitude of the triangle is 3.5 feet and the base is 6.5 feet. What is the area of the triangular signal tester? If the signal tester is coated with a special metallic surface paint that costs $122 per square foot, what was the cost of the amount of paint needed to coat one side of the triangle?

35. *Dog Weight* An average Great Dane weighs between 120 and 150 pounds. Express the range of weight for a Great Dane in kilograms. Use the formula $k = 2.205p$ (where k = kilograms, p = pounds).

36. *Dog Weight* An average Miniature Pinscher weighs between 9 and 14 pounds. Express the range of weight for a Miniature Pinscher in kilograms. Use the formula $k = 2.205p$ (where k = kilograms, p = pounds.)

Putting Your Skills to Work

Looking at Computer Usage

NTIA and the Economics and Statistics Administration have published *A Nation Online: How Americans Are Expanding Their Use of the Internet.* This report is based on the September 2001 U.S. Census Bureau's Current Population Survey—a survey of approximately 57,000 households and more than 137,000 individuals across the United States. As such, the data in this study are among the most broad-based and reliable data sets that have been gathered on Internet, broadband, and computer connectivity.

Computer use is expanding ever quicker both at work and at home. Look at the following information gathered for this government report.

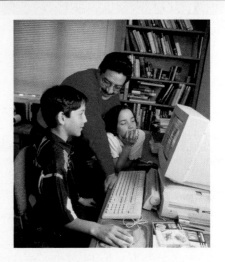

Computer Use at Work by Occupation (2001)

| Occupation | Employed | | Uses computer at main job | | Main computer uses (by percent) | | | |
	Total (in 1000s)	Percent Women	Total (in 1000s)	Percent of employed	E-mail	Word processing	Spreadsheet database	Calendar, schedule
TOTAL	**115,065**	**46.3**	**65,190**	**56.7**	**41.7**	**38.6**	**35.9**	**30.4**
Professional Managerial	39,412	50.2	31,723	80.5	66.8	63.2	56.5	48.8
Support Staff	31,482	62.9	22,205	70.5	49.2	45.5	43.1	34.7
Precision Production	13,083	8.4	4,152	31.7	19.0	14.8	16.6	14.6
Service	13,678	61.6	3,478	25.4	13.9	14.3	11.8	12.3
Laborers	14,504	24.3	3,006	20.7	9.2	7.6	8.7	7.0
Farming, forestry, fishing	2,905	20.3	625	21.5	14.6	13.0	13.2	9.2

Source: Statistical Abstract of the United States, 2002

Problems for Individual Investigation and Analysis

1. The data shows that 66.8% of computer using professionals use e-mail on a regular basis. How many professionals in the United States use e-mail at their main job?

2. How many more professionals than support staff use computers at their main job?

3. The NTIA estimates that 77% of Americans who use computers at home also use computers at work. How many American do you estimate used computers at home as of September 2001?

4. If computer use grows by 17.5% per year, how many people would you expect used computers at work in 2002, the year after this survey?

Problems for Group Investigation and Cooperative Learning

A number of business professionals feel that the number of people using computers at work in the future can be predicted in the near future by the formula

$$N = 65,190,000 + 4,500,000x,$$

where N is the number of people using computers at work and x is the number of years since 2001. For example, in the year 2003 the value of x would be 2. In the year 2004 the value of x would be 3.

5. Use this formula to estimate the number of people who will use computers at work in the year 2007.

6. Use this formula to estimate the number of people who will use computers at work in the year 2010.

Chapter 1 Organizer

Topic	Procedure	Examples
Absolute value, p. 70.	The absolute value of a number is the distance between that number and zero on the number line. The absolute value of any number will be positive or zero.	$\lvert 3 \rvert = 3$ $\lvert -2 \rvert = 2$ $\lvert 0 \rvert = 0$ $\left\lvert -\dfrac{5}{6} \right\rvert = \dfrac{5}{6}$ $\lvert -1.38 \rvert = 1.38$
Adding real numbers with the same sign, p. 71.	If the signs are the same, add the absolute values of the numbers. Use the common sign in the answer.	$-3 + (-7) = -10$
Adding real numbers with opposite signs, p. 73.	If the signs are different: **1.** Find the difference between the larger and the smaller absolute value. **2.** Give the answer the sign of the number having the larger absolute value.	$(-7) + 13 = 6$ $7 + (-13) = -6$
Adding several real numbers, p. 74.	When adding several real numbers, separate them into two groups by sign. Find the sum of all the positive numbers and the sum of all the negative numbers. Combine these two subtotals by the method described above.	$-7 + 6 + 8 + (-11) + (-13) + 22$ $\begin{array}{rr} -7 & +6 \\ -11 & +8 \\ \underline{-13} & \underline{+22} \\ -31 & +36 \end{array}$ $-31 + 36 = 5$ The answer is positive since 36 is positive.
Subtracting real numbers, p. 79.	Change the sign of the second number (the number you are subtracting) and then add.	$-3 - (-13) = -3 + (+13) = 10$
Multiplying and dividing real numbers, p. 85 and p. 87.	**1.** If the two numbers have the same sign, multiply (or divide) the absolute values. The result is positive. **2.** If the two numbers have different signs, multiply (or divide) the absolute values. The result is negative.	$-5(-3) = +15$ $-36 \div (-4) = +9$ $28 \div (-7) = -4$ $-6(3) = -18$
Exponent form, p. 94.	The base tells you what number is being multiplied. The exponent tells you how many times this number is used as a factor.	$2^5 = 2 \cdot 2 \cdot 2 \cdot 2 \cdot 2 = 32$ $4^3 = 4 \cdot 4 \cdot 4 = 64$ $(-3)^4 = (-3)(-3)(-3)(-3) = 81$
Raising a negative number to a power, p. 95.	When the base is negative, the result is positive for even exponents, and negative for odd exponents.	$(-3)^3 = -27$ but $(-2)^4 = 16$
Order of operations, p. 98.	Remember the proper order of operations: **1.** Perform operations inside parentheses. **2.** Raise to powers. **3.** Multiply and divide from left to right. **4.** Add and subtract from left to right.	$3(5 + 4)^2 - 2^2 \cdot 3 \div (9 - 2^3)$ $= 3 \cdot 9^2 - 4 \cdot 3 \div (9 - 8)$ $= 3 \cdot 81 - 12 \div 1$ $= 243 - 12 = 231$
Removing parentheses, p. 103.	Use the distributive property to remove parentheses. $a(b + c) = ab + ac$	$3(5x + 2) = 15x + 6$ $-4(x - 3y) = -4x + 12y$

Topic	Procedure	Examples
Combining like terms, **p. 108.**	Combine terms that have identical letters and exponents.	$7x^2 - 3x + 4y + 2x^2 - 8x - 9y = 9x^2 - 11x - 5y$
Substituting into variable expressions, p. 113.	**1.** Replace each letter by the numerical value given for it. **2.** Follow the order of operations in evaluating the expression.	Evaluate $2x^3 + 3xy + 4y^2$ for $x = -3$ and $y = 2$. $$2(-3)^3 + 3(-3)(2) + 4(2)^2$$ $$= 2(-27) + 3(-3)(2) + 4(4)$$ $$= -54 - 18 + 16$$ $$= -56$$
Using formulas, p. 115.	**1.** Replace the variables in the formula by the given values. **2.** Evaluate the expression. **3.** Label units carefully.	Find the area of a circle with radius 4 feet. Use $A = \pi r^2$, with π as approximately 3.14. $$A = (3.14)(4 \text{ feet})^2$$ $$= (3.14)(16 \text{ feet}^2)$$ $$= 50.24 \text{ feet}^2$$ The area of the circle is approximately 50.24 square feet.
Removing grouping symbols, p. 121.	**1.** Remove innermost grouping symbols first. **2.** Then remove remaining innermost grouping symbols. **3.** Continue until all grouping symbols are removed. **4.** Combine like terms.	$5\{3x - 2[4 + 3(x - 1)]\}$ $$= 5\{3x - 2[4 + 3x - 3]\}$$ $$= 5\{3x - 8 - 6x + 6\}$$ $$= 15x - 40 - 30x + 30$$ $$= -15x - 10$$

Chapter 1 Review Problems

Section 1.1

Add.

1. $-6 + (-2)$

2. $-12 + 7.8$

3. $5 + (-2) + (-12)$

4. $3.7 + (-1.8)$

5. $\dfrac{1}{2} + \left(-\dfrac{5}{6}\right)$

6. $-\dfrac{3}{11} + \left(-\dfrac{1}{22}\right)$

7. $\dfrac{3}{4} + \left(-\dfrac{1}{12}\right) + \left(-\dfrac{1}{2}\right)$

8. $-\dfrac{4}{15} + \dfrac{12}{5} + \left(-\dfrac{2}{3}\right)$

Section 1.2

Add or subtract.

9. $5 - (-3)$

10. $-2 - (-15)$

11. $-30 - (+3)$

12. $8 - (-1.2)$

13. $-\dfrac{7}{8} + \left(-\dfrac{3}{4}\right)$

14. $-\dfrac{3}{14} + \dfrac{5}{7}$

15. $-20.8 - 1.9$

16. $-151 - (-63)$

Section 1.3

17. $87 \div (-29)$

18. $-5(-6) + 4(-3)$

19. $\dfrac{-24}{-\dfrac{3}{4}}$

20. $-\dfrac{1}{2} \div \left(\dfrac{3}{4}\right)$

21. $\dfrac{5}{7} \div \left(-\dfrac{5}{25}\right)$

22. $-6(3)(4)$

23. $-1(-2)(-3)(-5)$

24. $(-5)\left(-\dfrac{1}{2}\right)(4)(-3)$

Mixed Practice

Sections 1.1–1.3

Perform the operations indicated. Simplify all answers.

25. $-5 + (-2) - (-3)$ **26.** $6 - (-4) + (-2) + 8$ **27.** $-16 + (-13)$ **28.** $-11 - (-12)$

29. $-\dfrac{4}{3} + \dfrac{2}{3} + \dfrac{1}{6}$ **30.** $-\dfrac{6}{7} + \dfrac{1}{2} + \left(-\dfrac{3}{14}\right)$ **31.** $-3(-2)(-5)$ **32.** $-6 + (-2) - (-3)$

33. $3.5(-2.6)$ **34.** $-5.4 \div (-6)$ **35.** $5 - (-3.5) + 1.6$ **36.** $-8 + 2 - (-4.8)$

37. $17 + 3.4 + (-16) + (-2.5)$ **38.** $37 + (-44) + 12.5 + (-6.8)$

Solve.

39. *Football* The Dallas Cowboys football team had three plays in which they lost 8 yards each time. What was the total yardage lost?

40. *Temperature Change* The low temperature in Anchorage, Alaska, last night was $-34°F$. During the day the temperature rose $12°F$. What was the temperature during the day?

41. *Elevation Levels* A mountain peak is 6895 feet above sea level. A location in Death Valley is 468 feet below sea level. What is the difference in height between these two locations?

42. *Stock Prices* During January 2000, IBM stock rose $1\frac{1}{2}$ points on Monday, dropped $3\frac{1}{4}$ points on Tuesday, rose 2 points on Wednesday, and dropped $2\frac{1}{2}$ points on Thursday. What was the total gain or loss on the value of the stock over this four-day period?

Section 1.4

Evaluate.

43. $(-3)^5$ **44.** $(-2)^7$ **45.** $(-5)^4$ **46.** $\left(\dfrac{2}{3}\right)^3$

47. -9^2 **48.** $(0.6)^2$ **49.** $\left(\dfrac{5}{6}\right)^2$ **50.** $\left(\dfrac{3}{4}\right)^3$

Section 1.5

Simplify using the order of operations.

51. $5(-4) + 3(-2)^3$ **52.** $20 - (-10) - (-6) + (-5) - 1$ **53.** $(7 - 9)^3 + -6(-2) + (-3)$

Section 1.6

Use the distributive property to multiply.

54. $5(3x - 7y)$ **55.** $2x(3x - 7y + 4)$ **56.** $-(7x^2 - 3x + 11)$ **57.** $(2xy + x - y)(-3y)$

Section 1.7

Combine like terms.

58. $3a^2b - 2bc + 6bc^2 - 8a^2b - 6bc^2 + 5bc$ **59.** $9x + 11y - 12x - 15y$

60. $4x^2 - 13x + 7 - 9x^2 - 22x - 16$

61. $-x + \dfrac{1}{2} + 14x^2 - 7x - 1 - 4x^2$

Section 1.8

Evaluate for the given value of the variable.

62. $7x - 6$ for $x = -7$

63. $7 - \dfrac{3}{4}x$ for $x = 8$

64. $x^2 + 3x - 4$ for $x = -3$

65. $-3x^2 - 4x + 5$ for $x = 2$

66. $-3x^3 - 4x^2 + 2x + 6$ for $x = -2$

67. $vt - \dfrac{1}{2}at^2$ for $v = 24$, $t = 2$, and $a = 32$

68. $\dfrac{nRT}{V}$ for $n = 16$, $R = -2$, $T = 4$, and $V = -20$

Solve.

69. *Simple Interest* Find the simple interest on a loan of $6000 at an annual interest rate of 18% per year for $\frac{3}{4}$ of a year. Use $I = prt$, where p = principal, r = rate per year, and t = time in years.

70. *Temperature* Find the Fahrenheit temperature if a radio announcer in Mexico City says that the high temperature today was 30°C. Use the formula
$$F = \frac{9C + 160}{5}.$$

▲ **71.** *Sign Painting* How much will it cost to paint a circular sign with a radius of 15 meters if the painter charges $3 per square meter? Use $A = \pi r^2$, where π is approximately 3.14.

72. *Profit* Find the daily profit P at a furniture factory if the initial cost of setting up the factory $C = \$1200$, rent $R = \$300$, and sales of furniture $S = \$56$. Use the profit formula $P = 180S - R - C$.

▲ **73.** *Parking Lot Sealer* A parking lot is in the shape of a trapezoid. The altitude of the trapezoid is 200 feet, and the bases of the trapezoid are 300 feet and 700 feet. What is the area of the parking lot? If the parking lot had a sealer applied that costs $2 per square foot, what was the cost of the amount of sealer needed for the entire parking lot?

▲ **74.** *Signal Paint* The Green Mountain Telephone Company has a triangular signal tester at the top of a communications tower. The altitude of the triangle is 3.8 feet and the base is 5.5 feet. What is the area of the triangular signal tester? If the signal tester is painted with a special metallic surface paint that costs $66 per square foot, what was the cost of the amount of paint needed to paint one side of the triangle?

Section 1.9

Simplify.

75. $5x - 7(x - 6)$

76. $3(x - 2) - 4(5x + 3)$

77. $2[3 - (4 - 5x)]$

78. $-3x[x + 3(x - 7)]$

79. $2xy^3 - 6x^3y - 4x^2y^2 + 3(xy^3 - 2x^2y - 3x^2y^2)$

80. $-5(x + 2y - 7) + 3x(2 - 5y)$

81. $2\{x - 3(y - 2) + 4[x - 2(y + 3)]\}$

82. $-5\{2a - b[5a - b(3 + 2a)]\}$

83. $-3\{2x - [x - 3y(x - 2y)]\}$

84. $2\{3x + 2[x + 2y(x - 4)]\}$

Mixed Practice

Simplify the following.

85. $-6.3 + 4$

86. $4 + (-8) + 12$

87. $-\dfrac{2}{3} - \dfrac{4}{5}$

88. $-\dfrac{7}{8} - \left(-\dfrac{3}{4}\right)$

89. $3 - (-4) + (-8)$

90. $-1.1 - (-0.2) + 0.4$

91. $\left(-\dfrac{3}{5}\right)\left(-2\dfrac{1}{2}\right)$

92. $(-4.2) \div (-0.7)$

93. $-14.4 \div (-0.06)$

94. $(-8.2)(3.1)$

95. *Jeopardy* A Jeopardy quiz show contestant began the second round (Double Jeopardy) with $400. She buzzed in on the first two questions, answering a $1000 question correctly, but then giving the incorrect answer to a $800 question. What was her score?

Simplify the following.

96. $(-0.3)^4$

97. -0.5^4

98. $9(5) - 5(2)^3 + 5$

99. $3.8x - 0.2y - 8.7x + 4.3y$

100. *Evaluate.* $\dfrac{2p + q}{3q}$ for $p = -2$ and $q = 3$

101. *Evaluate.* $\dfrac{4s - 7t}{s}$ for $s = -3$ and $t = -2$

102. *Dog Body Temperature* The normal body temperature of a dog is 38.6°C. Your dog has a temperature of 101.1°F. Does your dog have a fever? Use the formula $F = \dfrac{9}{5}C + 32$ to convert normal temperature in Fahrenheit.

103. $-7(x - 3y^2 + 4) + 3y(4 - 6y)$

104. $-2\{6x - 3[7y - 2y(3 - x)]\}$

Remember to use your Chapter Test Prep Video CD to see the worked-out solutions to the test problems you want to review

Simplify.

1. $-2.5 + 6.3 + (-4.1)$

2. $-5 - (-7)$

3. $\left(-\dfrac{2}{3}\right)(7)$

4. $-5(-2)(7)(-1)$

5. $-12 \div (-3)$

6. $-1.8 \div (0.6)$

7. $(-4)^3$

8. $(1.6)^2$

9. $\left(\dfrac{2}{3}\right)^4$

10. $(0.2)^2 - (2.1)(-3) + 0.46$

11. $3(4 - 6)^3 + 12 \div (-4) + 2$

12. $-5x(x + 2y - 7)$

13. $-2ab^2(-3a - 2b + 7ab)$

14. $6ab - \dfrac{1}{2}a^2b + \dfrac{3}{2}ab + \dfrac{5}{2}a^2b$

15. $2.3x^2y - 8.1xy^2 + 3.4xy^2 - 4.1x^2y$

16. $3(2 - a) - 4(-6 - 2a)$

17. $5(3x - 2y) - (x + 6y)$

1. _____

2. _____

3. _____

4. _____

5. _____

6. _____

7. _____

8. _____

9. _____

10. _____

11. _____

12. _____

13. _____

14. _____

15. _____

16. _____

17. _____

18. _____

In questions 18–20, evaluate for the value of the variable indicated.

18. $x^3 - 3x^2y + 2y - 5$ for $x = 3$ and $y = -4$

19. $3x^2 - 7x - 11$ for $x = -3$

19. _____

20. $2a - 3b$ for $a = \dfrac{1}{3}$ and $b = -\dfrac{1}{2}$

20. _____

21. If you are traveling 60 miles per hour on a highway in Canada, how fast are you traveling in kilometers per hour? (Use $k = 1.61r$, where $r =$ rate in miles per hour and $k =$ rate in kilometers per hour.)

21. _____

▲ **22.** A field is in the shape of a trapezoid. The altitude of the trapezoid is 120 feet and the bases of the trapezoid are 180 feet and 200 feet. What is the area of the field?

22. _____

▲ **23.** Jeff Slater's garage has a triangular roof support beam. The support beam is covered with a sheet of plywood. The altitude of the triangular region is 6.8 feet and the base is 8.5 feet. If the triangular piece of plywood was painted with paint that cost $0.80 per square foot, what was the cost of the amount of paint needed to coat one side of the triangle?

23. _____

▲ **24.** You wish to apply blacktop sealer to your driveway, but do not know how much to buy. If your rectangular driveway measures 60 feet long by 10 feet wide, and a can of blacktop sealer claims to cover 200 square feet, how many cans should you buy?

24. _____

25. _____

Simplify.

25. $3\left[x - 2y\left(x + 2y\right) - 3y^2\right]$

26. _____

26. $-3\left\{a + b\left[3a - b\left(1 - a\right)\right]\right\}$

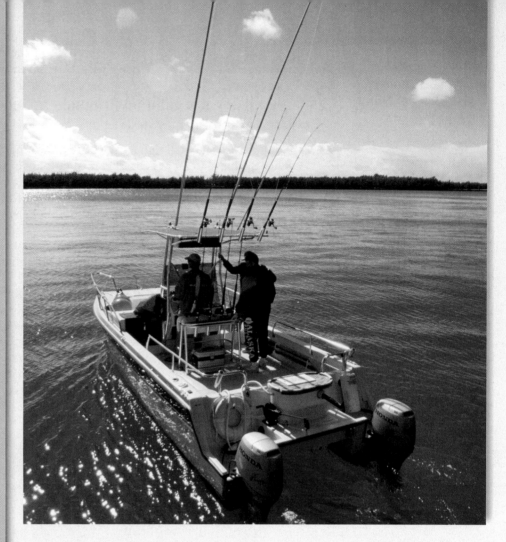

CHAPTER

2

People who own sport fishing boats and charter them out for fishing are often concerned with the cost of fuel. Many outboard boats are equipped with either one or two motors. Having two motors increases realibility and generates a higher speed. However, the added fuel cost may present a problem. Can you analyze exactly how much more fuel will be used if a boat is equipped with two outboard motors instead of one? Turn to the Putting Your Skills to Work on page 183 to find out.

Equations and Inequalities

Student Learning Objective

After studying this section, you will be able to:

 Use the addition principle to solve equations of the form $x + b = c$.

Using the Addition Principle to Solve Equations of the Form $x + b = c$

When we use an equal sign (=), we are indicating that two expressions are equal in value. Such a statement is called an **equation.** For example, $x + 5 = 23$ is an equation. A **solution** of an equation is a number that when substituted for the variable makes the equation true. Thus 18 is a solution of $x + 5 = 23$ because $18 + 5 = 23$. Equations that have exactly the same solutions are called **equivalent equations.** By following certain procedures, we can often transform an equation to a simpler equivalent one that has the form $x =$ some number. Then this number is a solution of the equation. The process of finding all solutions of an equation is called **solving the equation.**

One of the first procedures used in solving equations has an application in our everyday world. Suppose that we place a 10-kilogram box on one side of a seesaw and a 10-kilogram stone on the other side. If the center of the box is the same distance from the balance point as the center of the stone, we would expect the seesaw to balance. The box and the stone do not look the same, but their weights are equal. If we add a 2-kilogram lead weight to the center of weight of each object at the same time, the seesaw should still balance. The weights are still equal.

There is a similar principle in mathematics. We can state it in words as follows.

> **THE ADDITION PRINCIPLE**
>
> If the same number is added to both sides of an equation, the results on both sides are equal in value.

We can restate it in symbols this way.

> For real numbers a, b, and c,   if $a = b$,   then $a + c = b + c$.

Here is an example.   If $3 = \dfrac{6}{2}$,   then $3 + 5 = \dfrac{6}{2} + 5$.

Since we added the same amount, 5, to both sides, the sides remain equal to each other.

$$3 + 5 = \frac{6}{2} + 5$$

$$8 = \frac{6}{2} + \frac{10}{2}$$

$$8 = \frac{16}{2}$$

$$8 = 8$$

We can use the addition principle to solve certain equations.

EXAMPLE 1   Solve for x. $x + 16 = 20$

Solution $x + 16 + (-16) = 20 + (-16)$ Use the addition principle to add -16 to both sides.

$x + 0 = 4$ Simplify.

$x = 4$ The value of x is 4.

We have just found a solution of the equation. A **solution** is a value for the variable that makes the equation true. We then say that the value 4 in our example **satisfies** the equation. We can easily verify that 4 is a solution by substituting this value into the original equation. This step is called **checking** the solution.

Check. $x + 16 = 20$

$4 + 16 \overset{?}{=} 20$

$20 = 20$ ✓

When the same value appears on both sides of the equal sign, we call the equation an **identity.** Because the two sides of the equation in our check have the same value, we know that the original equation has been solved correctly. We have found a solution, and since no other number makes the equation true, it is the only solution.

Practice Problem 1   Solve for x and check your solution. $x + 14 = 23$

NOTE TO STUDENT: *Fully worked-out solutions to all of the Practice Problems can be found at the back of the text starting at page SP-1*

Notice that when you are trying to solve these types of equations, you must add a particular number to both sides of the equation. What is the number to choose? Look at the number that is on the same side of the equation with x, that is, the number added to x. Then think of the number that is **opposite in sign.** This is called the **additive inverse** of the number. The additive inverse of 16 is -16. The additive inverse of -3 is 3. The number to add to both sides of the equation is precisely this additive inverse.

It does not matter which side of the equation contains the variable. The x-term may be on the right or left. In the next example the x-term will be on the right.

EXAMPLE 2   Solve for x. $14 = x - 3$

Solution $14 + 3 = x - 3 + 3$ Notice that -3 is being added to x in the original equation. Add 3 to both sides, since 3 is the additive inverse of -3. This will eliminate the -3 on the right and isolate x.

$17 = x + 0$ Simplify.

$17 = x$ The value of x is 17.

Check. $14 = x - 3$

$14 \overset{?}{=} 17 - 3$ Replace x by 17.

$14 = 14$ ✓   Simplify. It checks. The solution is 17.

Practice Problem 2   Solve for x and check your solution. $17 = x - 5$

Before you add a number to both sides, you should always simplify the equation. The following example shows how combining numbers by addition—separately, on both sides of the equation—simplifies the equation.

EXAMPLE 3   Solve for x. $1.5 + 0.2 = 0.3 + x + 0.2$

Solution

$1.7 = x + 0.5$	Simplify by adding.
$1.7 + (-0.5) = x + 0.5 + (-0.5)$	Add the value -0.5 to both sides, since -0.5 is the additive inverse of 0.5.
$1.2 = x$	Simplify. The value of x is 1.2.

Check.

$1.5 + 0.2 = 0.3 + x + 0.2$	
$1.5 + 0.2 \overset{?}{=} 0.3 + 1.2 + 0.2$	Replace x by 1.2 in the original equation.
$1.7 = 1.7$ ✓	It checks.

Practice Problem 3   Solve for x and check your solution. $0.5 - 1.2 = x - 0.3$

In Example 3 we added -0.5 to each side. You could subtract 0.5 from each side and get the same result. In Chapter 1 we discussed how subtracting a 0.5 is the same as adding a negative 0.5. Do you see why?

Just as it is possible to add the same number to both sides of an equation, it is also possible to subtract the same number from both sides of an equation. This is so because any subtraction problem can be rewritten as an addition problem. For example, $17 - 5 = 17 + (-5)$. Thus the addition principle tells us that we can subtract the same number from both sides of the equation.

We can determine whether a value is the solution to an equation by following the same steps used to check an answer. Substitute the value to be tested for the variable in the original equation. We will obtain an identity if the value is the solution.

EXAMPLE 4   Is 10 the solution to the equation $-15 + 2 = x - 3$? If it is not, find the solution.

Solution   We substitute 10 for x in the equation and see if we obtain an identity.

$-15 + 2 = x - 3$	
$-15 + 2 \overset{?}{=} 10 - 3$	
$-13 \neq 7$	The values are not equal. The statement is not an identity.

Thus, 10 is not the solution. Now we take the original equation and solve to find the solution.

$-15 + 2 = x - 3$	
$-13 = x - 3$	Simplify by adding.
$-13 + 3 = x - 3 + 3$	Add 3 to both sides. 3 is the additive inverse of -3.
$-10 = x$	

Check to see if -10 is the solution. The value 10 was incorrect because of a sign error. We must be especially careful to write the correct sign for each number when solving equations.

NOTE TO STUDENT: Fully worked-out solutions to all of the Practice Problems can be found at the back of the text starting at page SP-1

Practice Problem 4   Is -2 the solution to the equation $x + 8 = -22 + 6$? If it is not, find the solution.

EXAMPLE 5   Find the value of x that satisfies the equation

$$\frac{1}{5} + x = -\frac{1}{10} + \frac{1}{2}.$$

Solution   To be combined, the fractions must have common denominators. The least common denominator (LCD) of the fractions is 10.

$$\frac{1 \cdot 2}{5 \cdot 2} + x = -\frac{1}{10} + \frac{1 \cdot 5}{2 \cdot 5} \qquad \text{Change each fraction to an equivalent fraction with a denominator of 10.}$$

$$\frac{2}{10} + x = -\frac{1}{10} + \frac{5}{10} \qquad \text{This is an equivalent equation.}$$

$$\frac{2}{10} + x = \frac{4}{10} \qquad \text{Simplify by adding.}$$

$$\frac{2}{10} + \left(-\frac{2}{10}\right) + x = \frac{4}{10} + \left(-\frac{2}{10}\right) \qquad \text{Add the additive inverse of } \frac{2}{10} \text{ to each side. You could also say that you are subtracting } \frac{2}{10} \text{ from each side.}$$

$$x = \frac{2}{10} \qquad \text{Add the fractions.}$$

$$x = \frac{1}{5} \qquad \text{Simplify the answer.}$$

Check.   We substitute $\frac{1}{5}$ for x in the original equation and see if we obtain an identity.

$$\frac{1}{5} + x = -\frac{1}{10} + \frac{1}{2}$$

$$\frac{1}{5} + \frac{1}{5} \overset{?}{=} -\frac{1}{10} + \frac{1}{2} \qquad \text{Substitute } \frac{1}{5} \text{ for } x.$$

$$\frac{2}{5} \overset{?}{=} -\frac{1}{10} + \frac{5}{10}$$

$$\frac{2}{5} \overset{?}{=} \frac{4}{10}$$

$$\frac{2}{5} = \frac{2}{5} \quad \checkmark \qquad \text{It checks.}$$

Practice Problem 5   Find the value of x that satisfies the equation

$$\frac{1}{20} - \frac{1}{2} = x + \frac{3}{5}.$$

Developing Your Study Skills

Why Study Mathematics?

In our present-day, technological world, it is easy to see mathematics at work. Many vocational and professional areas—such as the fields of business, statistics, economics, psychology, finance, computer science, chemistry, physics, engineering, electronics, nuclear energy, banking, quality control, and teaching—require a certain level of expertise in mathematics. Those who want to work in these fields must be able to function at a given mathematical level. Those who cannot will not make it. So if your field of study requires you to take higher-level mathematics courses, be sure to master the topics of this course. Then you will be ready for the next one.

Verbal and Writing Skills

In exercises 1–3, fill in the blank with the appropriate word.

1. When we use the _____ sign, we indicate two expressions are _____ in value.

2. If the _____ _____ is added to both sides of an equation, the results on each side are equal in value.

3. The _____ of an equation is a value of the variable that makes the equation true.

4. What is the additive inverse of -20?

5. Why do we add the additive inverse of a to each side of $x + a = b$ to solve for x?

6. What is the additive inverse of a?

Solve for x. Check your answers.

7. $x + 11 = 15$

8. $x + 12 = 18$

9. $17 = 5 + x$

10. $23 = x + 16$

11. $x - 3 = 14$

12. $x - 11 = 5$

13. $0 = x + 5$

14. $0 = x - 7$

15. $x - 6 = -19$

16. $x - 11 = -13$

17. $-12 + x = 50$

18. $-18 + x = 48$

19. $3 + 5 = x - 7$

20. $8 - 2 = x + 5$

21. $32 - 17 = x - 6$

22. $27 - 12 = x - 9$

23. $4 + 8 + x = 6 + 6$

24. $18 - 6 + x = 15 - 3$

25. $8 - 23 + 7 = 1 + x - 2$

26. $3 - 17 + 8 = 8 + x - 3$

27. $-12 + x - 3 = 15 - 18 + 9$

28. $-19 + x - 7 = 20 - 42 + 10$

In exercises 29–36, determine whether the given solution is correct. If it is not, find the solution.

29. Is $x = 5$ the solution to $-7 + x = 2$?

30. Is $x = 7$ the solution to $-13 + x = 4$?

31. Is -3 a solution to $-18 - 2 = x - 7$?

32. Is -5 a solution to $-16 + 5 = x - 6$?

33. Is -33 the solution to $x - 23 = -56$?

34. Is -8 the solution to $-39 = x - 47$?

35. Is 35 the solution to $15 - 3 + 20 = x - 3$?

36. Is -12 the solution to $x + 8 = 12 - 19 + 3$?

Find the value of x that satisfies each equation.

37. $2.5 + x = 0.7$

38. $4.2 + x = 1.3$

39. $2.7 + x - 1.4 = 3.8$

40. $4.3 + x - 2.6 = 3.4$

41. $x - \dfrac{1}{4} = \dfrac{3}{4}$

42. $x + \dfrac{1}{3} = \dfrac{2}{3}$

43. $\dfrac{2}{3} + x = \dfrac{1}{6} + \dfrac{1}{4}$

44. $\dfrac{2}{5} + x = \dfrac{1}{2} - \dfrac{3}{10}$

Mixed Practice

Solve for x.

45. $3 + x = -12 + 8$

46. $12 + x = -7 + 20$

47. $5\dfrac{1}{6} + x = 8$

48. $7\dfrac{1}{8} = -20 + x$

49. $\dfrac{5}{12} - \dfrac{5}{6} = x - \dfrac{3}{2}$

50. $\dfrac{4}{15} - \dfrac{3}{5} = x - \dfrac{4}{3}$

51. $1.6 + 4x - 3.2 = -2x + 5.6 + 5x$

52. $0.7 + 3x - 4.2 = 9x + 3.6 - 7x$

 53. $x + 0.7513 = 2.2419$

 54. $x - 0.2314 = -4.0144$

Cumulative Review

Simplify by adding like terms.

55. $x + 3y - 5x - 7y + 2x$

56. $y^2 + y - 12 - 3y^2 - 5y + 16$

57. *Quality Control* The Perception Toy Company wishes to cut costs and decides that improving quality control will help. It can do this by reducing the amount of rejected frisbees to 3%. In a recent week, a quality control inspector found that 8 out of 413 frisbees were rejected. Did they meet Perception's goal?

58. *Checking Account Balance* Trevor pays his monthly computer lease bill for $49.99, but forgets to look at his checking account balance before doing so. When he gets his checking account statement at the local ATM, his balance reads −$35.07. How much was in his account before he wrote the check?

59. *Radar Picture* A 90-meter-wide radar picture is taken of a swamp in northern Australia. The radar detects a rock outcrop that is 90 feet above sea level, and a vein of opal (a semiprecious stone) 27 feet below sea level. How far is the top of the rock from the location of the opal?

90 feet

−27 feet

1 Solving Equations of the Form $\frac{1}{a}x = b$

The addition principle allows us to add the same number to both sides of an equation. What would happen if we multiplied each side of an equation by the same number? For example, what would happen if we multiplied each side of an equation by 3?

To answer this question, let's return to our simple example of the box and the stone on a balanced seesaw. If we triple the weight on each side (that is, multiply the weight on each side by 3), the seesaw should still balance. The weight values of both sides remain equal.

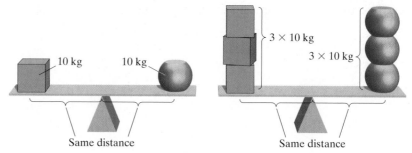

In words we can state this principle thus.

MULTIPLICATION PRINCIPLE

If both sides of an equation are multiplied by the same nonzero number, the results on both sides are equal in value.

In symbols we can restate the multiplication principle this way.

For real numbers a, b, and c with $c \neq 0$, if $a = b$, then $ca = cb$.

It is important that we say $c \neq 0$. We will explore this idea in the To Think About exercises.

Let us look at an equation where it would be helpful to multiply each side by 3.

EXAMPLE 1   Solve for x. $\frac{1}{3}x = -15$

Solution   We know that $(3)\left(\frac{1}{3}\right) = 1$. We will multiply each side of the equation by 3 because we want to isolate the variable x.

$$3\left(\frac{1}{3}x\right) = 3(-15) \qquad \text{Multiply each side of the equation by 3 since } (3)\left(\frac{1}{3}\right) = 1.$$

$$\left(\frac{3}{1}\right)\left(\frac{1}{3}\right)(x) = -45$$

$$1x = -45 \qquad \text{Simplify.}$$

$$x = -45 \qquad \text{The solution is } -45.$$

Check. $\frac{1}{3}(-45) \stackrel{?}{=} -15 \qquad \text{Substitute } -45 \text{ for } x \text{ in the original equation.}$

$-15 = -15 \checkmark \qquad \text{It checks.}$

Practice Problem 1   Solve for x. $\frac{1}{8}x = -2$

Note that $\frac{1}{5}x$ can be written as $\frac{x}{5}$. To solve the equation $\frac{x}{5} = 3$, we could multiply each side of the equation by 5. Try it. Then check your solution.

② Solving Equations of the Form $ax = b$

We can see that using the multiplication principle to multiply each side of an equation by $\frac{1}{2}$ is the same as dividing each side of the equation by 2. Thus, it would seem that the multiplication principle would allow us to divide each side of the equation by any nonzero real number. Is there a real-life example of this idea?

Let's return to our simple example of the box and the stone on a balanced seesaw. Suppose that we were to cut the two objects in half (so that the amount of weight of each was divided by 2). We then return the objects to the same places on the seesaw. The seesaw would still balance. The weight values of both sides remain equal.

In words we can state this principle thus.

DIVISION PRINCIPLE

If both sides of an equation are divided by the same nonzero number, the results on both sides are equal in value.

Note: We put a restriction on the number by which we are dividing. We cannot divide by zero. We say that expressions like $\frac{2}{0}$ are not defined. Thus we restrict our divisor to *nonzero* numbers. We can restate the division principle this way.

For real numbers a, b, and c where $c \neq 0$,   if $a = b$,   then $\dfrac{a}{c} = \dfrac{b}{c}$.

EXAMPLE 2   Solve for x. $5x = 125$

Solution
$$\frac{5x}{5} = \frac{125}{5} \qquad \text{Divide both sides by 5.}$$
$$x = 25 \qquad \text{Simplify. The solution is 25.}$$

Check.
$$5x = 125$$
$$5(25) \overset{?}{=} 125 \qquad \text{Replace } x \text{ by 25.}$$
$$125 = 125 \quad \checkmark \quad \text{It checks.}$$

Practice Problem 2   Solve for x. $9x = 72$

For equations of the form $ax = b$ (a number multiplied by x equals another number), we solve the equation by choosing to divide both sides by a particular number. What is the number to choose? We look at the side of the equation that contains x. We notice the number that is multiplied by x. We divide by that number.

The division principle tells us that we can still have a true equation provided that we divide by that number *on both sides* of the equation.

The solution to an equation may be a proper fraction or an improper fraction.

EXAMPLE 3   Solve for x. $4x = 38$

Solution

$$\frac{4x}{4} = \frac{38}{4} \quad \text{Divide both sides by 4.}$$

$$x = \frac{19}{2} \quad \text{Simplify. The solution is } \tfrac{19}{2}.$$

If you leave the solution as a fraction, it will be easier to check that solution in the original equation.

Check.

$$4x = 38$$

$$\overset{2}{\cancel{4}}\left(\frac{19}{\cancel{2}}\right) \overset{?}{=} 38 \qquad \text{Replace } x \text{ by } \tfrac{19}{2}.$$

$$38 = 38 \quad \checkmark \quad \text{It checks.}$$

Practice Problem 3   Solve for x. $6x = 50$

NOTE TO STUDENT: Fully worked-out solutions to all of the Practice Problems can be found at the back of the text starting at page SP-1

In Examples 2 and 3 we *divided by the number multiplied by x*. This procedure is followed regardless of whether the sign of that number is positive or negative. In equations of the form $ax = b$ the **coefficient** of x is a. A coefficient is a multiplier.

EXAMPLE 4   Solve for x. $-3x = 48$

Solution

$$\frac{-3x}{-3} = \frac{48}{-3} \quad \text{Divide both sides by } -3.$$

$$x = -16 \quad \text{The solution is } -16.$$

Check.   Can you check this solution?

Practice Problem 4   Solve for x. $-27x = 54$

The coefficient of x may be 1 or -1. You may have to rewrite the equation so that the coefficient of 1 or -1 is obvious. With practice you may be able to recognize the coefficient without actually rewriting the equation.

EXAMPLE 5   Solve for x. $-x = -24$.

Solution $-1x = -24$ Rewrite the equation. $-1x$ is the same as $-x$.
Now the coefficient of -1 is obvious.

$$\frac{-1x}{-1} = \frac{-24}{-1} \quad \text{Divide both sides by } -1.$$

$$x = 24 \quad \text{The solution is 24.}$$

Check.   Can you check this solution?

Practice Problem 5   Solve for x. $-x = 36$

The variable can be on either side of the equation. The equation $-78 = -3x$ can be solved in exactly the same way as $-3x = -78$.

EXAMPLE 6   Solve for x. $-78 = -3x$

Solution
$$\frac{-78}{-3} = \frac{-3x}{-3} \qquad \text{Divide both sides by } -3.$$
$$26 = x \qquad \text{The solution is 26.}$$

Check.
$$-78 = -3x$$
$$-78 \overset{?}{=} -3(26) \qquad \text{Replace } x \text{ by 26.}$$
$$-78 = -78 \quad \checkmark \quad \text{It checks.}$$

Practice Problem 6   Solve for x. $-51 = -6x$

There is a mathematical concept that unites what we have learned in this section. The concept uses the idea of a multiplicative inverse. For any nonzero number a, the **multiplicative inverse** of a is $\frac{1}{a}$. Likewise, for any nonzero number a, the multiplicative inverse of $\frac{1}{a}$ is a. So to solve an equation of the form $ax = b$, we say that we need to multiply each side by the multiplicative inverse of a. Thus to solve $5x = 45$, we would multiply each side of the equation by the multiplicative inverse of 5, which is $\frac{1}{5}$. In similar fashion, if we wanted to solve the equation $\left(\frac{1}{6}\right)x = 4$, we would multiply each side of the equation by the multiplicative inverse of $\frac{1}{6}$, which is 6. In general, all the problems we have covered so far in this section can be solved by multiplying both sides of the equation by the multiplicative inverse of the coefficient of x.

EXAMPLE 7   Solve for x. $31.2 = 6.0x - 0.8x$

Solution

$31.2 = 6.0x - 0.8x \qquad$ There are like terms on the right side.
$31.2 = 5.2x \qquad$ Collect like terms.
$\dfrac{31.2}{5.2} = \dfrac{5.2x}{5.2} \qquad$ Divide both sides by 5.2 (which is the same as multiplying both sides by the multiplicative inverse of 5.2).
$6 = x \qquad$ The solution is 6.

Note: Be sure to place the decimal point in the quotient directly above the caret ($_\wedge$) when performing the division.

$$\begin{array}{r} 6. \\ 5.2_\wedge \overline{)31.2_\wedge} \\ \underline{31.2} \\ 0 \end{array}$$

Check.   The check is up to you.

Practice Problem 7   Solve for x. $16.2 = 5.2x - 3.4x$

Developing Your Study Skills

Getting Help

Getting the right kind of help at the right time can be a key ingredient in being successful in mathematics. When you have gone to class on a regular basis, taken careful notes, methodically read your textbook, and diligently done your homework—all of which means making every effort possible to learn the mathematics—you may find that you are still having difficulty. If this is the case, then you need to seek help. Make an appointment with your instructor to find out what help is available to you. The instructor, tutoring services, a mathematics lab, videotapes, and computer software may be among the resources you can draw on.

Once you discover the resources available in your school, you need to take advantage of them. Do not put it off, or you will find yourself getting behind. You cannot afford that. When studying mathematics, you must keep up with your work.

Verbal and Writing Skills

1. To solve the equation $6x = -24$, divide each side of the equation by _____.

2. To solve the equation $-7x = 56$, divide each side of the equation by _____.

3. To solve the equation $\frac{1}{7}x = -2$, multiply each side of the equation by _____.

4. To solve the equation $\frac{1}{9}x = 5$, multiply each side of the equation by _____.

Solve for x. Be sure to reduce your answer. Check your solution.

5. $\frac{1}{9}x = 4$

6. $\frac{1}{7}x = 6$

7. $\frac{1}{3}x = -9$

8. $\frac{1}{4}x = -20$

9. $\frac{x}{5} = 16$

10. $\frac{x}{10} = 8$

11. $-3 = \frac{x}{5}$

12. $\frac{x}{3} = -12$

13. $13x = 52$

14. $15x = 60$

15. $56 = 7x$

16. $46 = 2x$

17. $-16 = 6x$

18. $-35 = 21x$

19. $1.5x = 75$

20. $2x = 0.36$

21. $-15 = -x$

22. $32 = -x$

23. $-112 = 16x$

24. $-108 = -18x$

25. $0.4x = 0.08$

26. $2.1x = 0.3$

27. $-3.9x = -15.6$

28. $-4.7x = -14.1$

Determine whether the given solution is correct. If it is not, find the correct solution.

29. Is 7 the solution for $-3x = 21$?

30. Is 8 the solution for $5x = -40$?

31. Is -6 the solution for $-11x = 66$?

32. Is -20 the solution for $-x = 20$?

Mixed Practice

Find the value of the variable that satisfies the equation.

33. $7y = -0.21$

34. $-3y = 0.42$

35. $-56 = -21t$

36. $34 = -51q$

37. $4.6y = -3.22$

38. $-2.8y = -3.08$

39. $4x + 3x = 21$

40. $5x + 4x = 36$

41. $2x - 7x = 20$

42. $3x - 9x = 18$

43. $-6x - 3x = -7$

44. $y - 11y = 7$

45. $12 - 19 = -7x$ **46.** $36 - 22 = -2x$ **47.** $6x = -18 + 36$ **48.** $11x = -20 + 42$

49. $\frac{2}{3}x = 18$ **50.** $\frac{3}{5}x = 39$ 📱 **51.** $3.6172x = -19.026472$ 📱 **52.** $-4.0518x = 14.505444$

To Think About

53. We have said that if $a = b$ and $c \neq 0$, then $ac = bc$. Why is it important that $c \neq 0$? What would happen if we tried to solve an equation by multiplying both sides by zero?

54. We have said that if $a = b$ and $c \neq 0$ then $\frac{a}{c} = \frac{b}{c}$. Why is it important that $c \neq 0$? What would happen if we tried to solve an equation by dividing both sides by zero?

Cumulative Review

Evaluate using the correct order of operations. (Be careful to avoid sign errors.)

55. $(-6)(-8) + (-3)(2)$ **56.** $(-3)^3 + (-20) \div 2$ **57.** $5 + (2 - 6)^2$

58. *Discount Merchandise* An off-price clothing store chain specializes in last year's merchandise. Their contact in Hong Kong has purchased 12,000 famous designer men's sport coats for the stores. When the shipment is unloaded, 800 sport coats have no left sleeve. What percent of the shipment is acceptable?

59. *Investments* In January, Keiko invested $600 in a certain stock. In February, the stock gained $82.00. In March, the stock lost $47.00. In April, the stock gained $103.00. In May, the stock lost $106.00. What was Keiko's stock holding worth after the May loss?

60. *Whale Calf Population* In 1995, the humpback whale calf population at Stellwagen Bank, near Gloucester, Massachusetts, was estimated at 12 calves. The population grew by 21 calves in 1996, 18 calves in 1997, and 51 calves in 1998. In 1999, the number of whale calves decreased by 4, and in 2000 it increased by 6. What was the whale calf population at the end of 2000?

61. *Earthquakes* In an average year, worldwide, there are 20 earthquakes of magnitude 7 on the Richter scale. If next year is predicted to be an exceptional year, and the number of earthquakes of magnitude 7 is expected to increase by 35%, about how many earthquakes of magnitude 7 can be expected? (Round off to the nearest whole number.)

2.3 USING THE ADDITION AND MULTIPLICATION PRINCIPLES TOGETHER

Student Learning Objectives

After studying this section, you will be able to:

1 Solve equations of the form $ax + b = c$.

2 Solve equations with the variable on both sides of the equation.

3 Solve equations with parentheses.

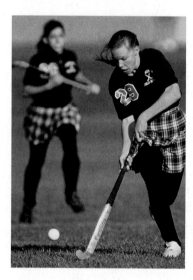

NOTE TO STUDENT: Fully worked-out solutions to all of the Practice Problems can be found at the back of the text starting at page SP-1

1 Solving Equations of the Form $ax + b = c$

Jenny Crawford scored several goals in field hockey during April. Her teammates scored three more than five times the number she scored. Together the team scored 18 goals in April. How many did Jenny score? To solve this problem, we need to solve the equation $5x + 3 = 18$.

To solve an equation of the form $ax + b = c$, we must use both the addition principle and the multiplication principle.

EXAMPLE 1 Solve for x to determine how many goals Jenny scored and check your solution.

$$5x + 3 = 18$$

Solution We first want to isolate the variable term.

$5x + 3 + (-3) = 18 + (-3)$	Use the addition principle to add -3 to both sides.
$5x = 15$	Simplify.
$\dfrac{5x}{5} = \dfrac{15}{5}$	Use the division principle to divide both sides by 5.
$x = 3$	The solution is 3. Thus Jenny scored 3 goals.

Check.
$$5(3) + 3 \overset{?}{=} 18$$
$$15 + 3 \overset{?}{=} 18$$
$$18 = 18 \checkmark \qquad \text{It checks.}$$

Practice Problem 1 Solve for x and check your solution. $9x + 2 = 38$

2 Solving Equations with the Variable on Both Sides of the Equation

In some cases the variable appears on both sides of the equation. We would like to rewrite the equation so that all the terms containing the variable appear on one side. To do this, we apply the addition principle to the variable term.

EXAMPLE 2 Solve for x. $9x = 6x + 15$

Solution

$9x + (-6x) = 6x + (-6x) + 15$	Add $-6x$ to both sides. Notice $6x + (-6x)$ eliminates the variable on the right side.
$3x = 15$	Combine like terms.
$\dfrac{3x}{3} = \dfrac{15}{3}$	Divide both sides by 3.
$x = 5$	The solution is 5.

Check. The check is left to the student.

Practice Problem 2 Solve for x. $13x = 2x - 66$

In many problems the variable terms and constant terms appear on both sides of the equations. You will want to get all the variable terms on one side and all the constant terms on the other side.

EXAMPLE 3   Solve for x and check your solution. $9x + 3 = 7x - 2$.

Solution   First we want to isolate the variable term.

$9x + (-7x) + 3 = 7x + (-7x) - 2$ Add $-7x$ to both sides of the equation.

$2x + 3 = -2$ Combine like terms.

$2x + 3 + (-3) = -2 + (-3)$ Add -3 to both sides.

$2x = -5$ Simplify.

$\dfrac{2x}{2} = \dfrac{-5}{2}$ Divide both sides by 2.

$x = -\dfrac{5}{2}$ The solution is $-\frac{5}{2}$.

Check. $9x + 3 = 7x - 2$

$9\left(-\dfrac{5}{2}\right) + 3 \overset{?}{=} 7\left(-\dfrac{5}{2}\right) - 2$ Replace x by $-\frac{5}{2}$.

$-\dfrac{45}{2} + 3 \overset{?}{=} -\dfrac{35}{2} - 2$ Simplify.

$-\dfrac{45}{2} + \dfrac{6}{2} \overset{?}{=} -\dfrac{35}{2} - \dfrac{4}{2}$ Change to equivalent fractions with a common denominator.

$-\dfrac{39}{2} = -\dfrac{39}{2}$ ✓   It checks. The solution is $-\frac{5}{2}$.

Practice Problem 3   Solve for x and check your solution. $3x + 2 = 5x + 2$

In our next example we will study equations that need simplifying before any other steps are taken. Where it is possible, you should first collect like terms on one or both sides of the equation. The variable terms can be collected on the right side or the left side. In this example we will collect all the x-terms on the right side.

EXAMPLE 4   Solve for x. $5x + 26 - 6 = 9x + 12x$

Solution

$5x + 20 = 21x$ Combine like terms.

$5x + (-5x) + 20 = 21x + (-5x)$ Add $-5x$ to both sides.

$20 = 16x$ Combine like terms.

$\dfrac{20}{16} = \dfrac{16x}{16}$ Divide both sides by 16.

$\dfrac{5}{4} = x$ Don't forget to reduce the resulting fraction.

Check.   The check is left to the student.

Practice Problem 4   Solve for z. $-z + 8 - z = 3z + 10 - 3$

Do you really need all these steps? No. As you become more proficient you will be able to combine or eliminate some of these steps. However, it is best to write each step in its entirety until you are consistently obtaining the correct solution. It is much better to show every step than to take a lot of shortcuts and possibly obtain a wrong answer. This is a section of the algebra course where working neatly and accurately will help you—both now and as you progress through the course.

③ Solving Equations with Parentheses

The equations that you just solved are simpler versions of equations that we will now discuss. These equations contain parentheses. If the parentheses are first removed, the problems then become just like those encountered previously. We use the distributive property to remove the parentheses.

EXAMPLE 5   Solve for x and check your solution.

$$4(x + 1) - 3(x - 3) = 25$$

Solution $4(x + 1) - 3(x - 3) = 25$

$4x + 4 - 3x + 9 = 25$	Multiply by 4 and -3 to remove parentheses. Be careful of the signs. Remember that $(-3)(-3) = 9$.

After removing the parentheses, it is important to collect like terms on each side of the equation. Do this before going on to isolate the variable.

$x + 13 = 25$	Collect like terms.
$x + 13 - 13 = 25 - 13$	Add -13 to both sides to isolate the variable.
$x = 12$	The solution is 12.

Check.

$4(12 + 1) - 3(12 - 3) \overset{?}{=} 25$	Replace x by 12.
$4(13) - 3(9) \overset{?}{=} 25$	Combine numbers inside parentheses.
$52 - 27 \overset{?}{=} 25$	Multiply.
$25 = 25$ ✓	Simplify. It checks.

Practice Problem 5   Solve for x and check your solution.

$$4x - (x + 3) = 12 - 3(x - 2)$$

NOTE TO STUDENT: Fully worked-out solutions to all of the Practice Problems can be found at the back of the text starting at page SP-1

EXAMPLE 6   Solve for x. $3(-x - 7) = -2(2x + 5)$

Solution

$-3x - 21 = -4x - 10$	Remove parentheses. Watch the signs carefully.
$-3x + 4x - 21 = -4x + 4x - 10$	Add $4x$ to both sides.
$x - 21 = -10$	Simplify.
$x - 21 + 21 = -10 + 21$	Add 21 to both sides.
$x = 11$	The solution is 11.

Check.   The check is left to the student.

Practice Problem 6   Solve for x. $4(-2x - 3) = -5(x - 2) + 2$

In problems that involve decimals, great care should be taken. In some steps you will be multiplying decimal quantities, and in other steps you will be adding them.

EXAMPLE 7 Solve for x. $0.3(1.2x - 3.6) = 4.2x - 16.44$

Solution

$0.36x - 1.08 = 4.2x - 16.44$	Remove parentheses.
$0.36x - 0.36x - 1.08 = 4.2x - 0.36x - 16.44$	Subtract $0.36x$ from both sides.
$-1.08 = 3.84x - 16.44$	Collect like terms.
$-1.08 + 16.44 = 3.84x - 16.44 + 16.44$	Add 16.44 to both sides.
$15.36 = 3.84x$	Simplify.
$\dfrac{15.36}{3.84} = \dfrac{3.84x}{3.84}$	Divide both sides by 3.84.
$4 = x$	The solution is 4.

Check. The check is left to the student.

Practice Problem 7 Solve for x. $0.3x - 2(x + 0.1) = 0.4(x - 3) - 1.1$

EXAMPLE 8 Solve for z and check. $2(3z - 5) + 2 = 4z - 3(2z + 8)$

Solution

$6z - 10 + 2 = 4z - 6z - 24$	Remove parentheses.
$6z - 8 = -2z - 24$	Collect like terms.
$6z - 8 + 2z = -2z + 2z - 24$	Add $2z$ to each side.
$8z - 8 = -24$	Simplify.
$8z - 8 + 8 = -24 + 8$	Add 8 to each side.
$8z = -16$	Simplify.
$\dfrac{8z}{8} = \dfrac{-16}{8}$	Divide each side by 8.
$z = -2$	Simplify. The solution is -2.

Check.

$2[3(-2) - 5] + 2 \overset{?}{=} 4(-2) - 3[2(-2) + 8]$	Replace z by -2.
$2[-6 - 5] + 2 \overset{?}{=} -8 - 3[-4 + 8]$	Multiply.
$2[-11] + 2 \overset{?}{=} -8 - 3[4]$	Simplify.
$-22 + 2 \overset{?}{=} -8 - 12$	
$-20 = -20$ ✓	It checks.

Practice Problem 8 Solve for z and check.

$$5(2z - 1) + 7 = 7z - 4(z + 3)$$

Student Solutions Manual   CD/Video   PH Math Tutor Center   MathXL®Tutorials on CD   MathXL®   MyMathLab®   Interactmath.com

Find the value of the variable that satisfies the equation in exercises 1–22. Check your solution. Answers that are not integers may be left in fractional form or decimal form.

1. $4x + 13 = 21$

2. $7x + 4 = 53$

3. $4x - 11 = 13$

4. $5x - 11 = 39$

5. $7x - 18 = -46$

6. $6x - 23 = -71$

7. $-4x + 17 = -35$

8. $-6x + 25 = -83$

9. $2x + 3.2 = 9.4$

10. $4x + 4.6 = 9.2$

11. $\frac{1}{5}x - 2 = 6$

12. $\frac{1}{3}x - 7 = 4$

13. $\frac{1}{3}x + 5 = -4$

14. $\frac{1}{8}x - 3 = -9$

15. $8x = 48 + 2x$

16. $5x = 22 + 3x$

17. $-6x = -27 + 3x$

18. $-7x = -26 + 6x$

19. $63 - x = 8x$

20. $56 - 3x = 5x$

21. $54 - 2x = -8x$

22. $72 - 4x = -12x$

23. Is 2 the solution for $2y + 3y = 12 - y$?

24. Is 4 the solution for $5y + 2 = 6y - 6 + y$?

25. Is 11 a solution for $7x + 6 - 3x = 2x - 5 + x$?

26. Is -12 a solution for $9x + 2 - 5x = -8 + 5x - 2$?

Solve for the variable. You may move the variable terms to the right or to the left.

27. $14 - 2x = -5x + 11$

28. $8 - 3x = 7x + 8$

29. $x - 6 = 8 - x$

30. $-x + 12 = -4 + x$

31. $0.8y - 0.4 = 0.9 - 0.5y$

32. $0.7y - 0.5 = 1.1 - 0.1y$

33. $5x - 9 = 3x + 23$

34. $9x - 5 = 7x + 43$

To Think About

First collect like terms on each side of the equation. Then solve for y by getting all the y-terms on the left. Then solve for y by getting all the y-terms on the right. Which approach is better?

35. $-3 + 10y + 6 = 15 + 12y - 18$

36. $7y + 21 - 5y = 5y - 7 + y$

Remove the parentheses and solve for the variable. Check your solution. Answers that are not integers may be left in fractional form or decimal form.

37. $5(x + 3) = 35$

38. $6(x + 2) = 42$

39. $6(3x + 2) - 8 = -2$

40. $4(2x + 1) - 7 = 6 - 5$

41. $7x - 3(5 - x) = 10$

42. $8x - 2(4 - x) = 14$

43. $0.5x - 0.3(2 - x) = 4.6$

44. $0.4x - 0.2(3 - x) = 1.8$

45. $5(x - 3) + 5 = 3(x + 2)$

46. $3(x - 2) + 2 = 2(x - 4)$

47. $-2(x + 3) + 4 = 3(x + 4) + 2$

48. $-3(x + 5) + 2 = 4(x + 6) - 9$

49. $-3(y - 3y) + 4 = -4(3y - y) + 6 + 13y$

50. $2(4x - x) + 6 = 2(2x + x) + 8 - x$

Mixed Practice

Solve for the variable.

51. $5.7x + 3 = 4.2x - 3$

52. $4x - 3.1 = 5.3 - 3x$

53. $5z + 7 - 2z = 32 - 2z$

54. $8 - 7z + 2z = 20 + 5z$

55. $-0.3a + 1.4 = -1.2 - 0.7a$

56. $-0.7b + 1.6 = -1.7 - 1.5b$

57. $6x + 8 - 3x = 11 - 12x - 13$ **58.** $4 - 7x - 13 = 8x - 3 - 5x$ **59.** $-3.5x + 1.3 = -2.7x + 1.5$

60. $2.8x - 0.9 = 5.2x - 3.3$ **61.** $5(4 + x) = 3(3x - 1) - 9$ **62.** $6(3x - 1) = 4(2x + 5) - 6$

63. $4x + 3.2 - 1.9x = 0.3x - 4.9$ **64.** $3x + 2 - 1.7x = 0.6x + 31.4$

65. $3(x + 4) - 5(3x - 2) = 8$ **66.** $3(2z - 4) - 4(z + 5) = 6$

Solve for x. Round your answer to the nearest hundredth.

 67. $1.63x - 9.23 = 5.71x + 8.04$ **68.** $-2.21x + 8.65 = 3.69x - 7.78$

Cumulative Review

Simplify.

69. $2x(3x - y) + 4(2x^2 - 3xy)$ **70.** $2\{x - 3[4 + 2(3 + x)]\}$

71. *Investments* On March 30, 2000, Nancy owned three different stocks: Coca Cola, General Motors and Alcoa. Her portfolio contained the following:

4.0 shares of Coca Cola stock valued at $\$52\frac{1}{8}$,
3.2 shares of General Motors stock valued at $\$81\frac{7}{8}$, and
5.2 shares of Alcoa stock valued at $\$71\frac{7}{8}$.

Find the market value of Nancy's stock holdings on March 30, 2000. (*Note:* In 2000 stock quotes were given in fractional form. Now decimals are used.)

72. *Employee Discount* Bea works for a large department store and obtains a 10% discount on anything she buys from the store. One item Bea wishes to purchase costs $140 and is on sale at a 25% discount.

(a) What is the price if Bea has a total discount of 35%? (Disregard sales tax.)

(b) What is the price if Bea has a 10% discount on the 25% discount price? (Disregard sales tax.)

2.4 SOLVING EQUATIONS WITH FRACTIONS

 Solving Equations with Fractions

Equations with fractions can be rather difficult to solve. This difficulty is simply due to the extra care we usually have to use when computing with fractions. The actual equation-solving procedures are the same, with fractions or without. To avoid unnecessary work, we transform the given equation with fractions to an equivalent equation that does not contain fractions. How do we do this? We multiply each side of the equation by the least common denominator of all the fractions contained in the equation. We then use the distributive property so that the LCD is multiplied by each term of the equation.

Student Learning Objective

After studying this section, you will be able to:

1 Solve equations with fractions.

EXAMPLE 1 Solve for x. $\dfrac{1}{4}x - \dfrac{2}{3} = \dfrac{5}{12}x$

Solution First we find that the LCD = 12.

$$12\left(\frac{1}{4}x - \frac{2}{3}\right) = 12\left(\frac{5}{12}x\right) \qquad \text{Multiply each side by 12.}$$

$$\left(\frac{12}{1}\right)\left(\frac{1}{4}\right)(x) - \left(\frac{12}{1}\right)\left(\frac{2}{3}\right) = \left(\frac{12}{1}\right)\left(\frac{5}{12}\right)(x) \qquad \text{Use the distributive property.}$$

$$3x - 8 = 5x \qquad \text{Simplify.}$$

$$3x + (-3x) - 8 = 5x + (-3x) \qquad \text{Add } -3x \text{ to each side.}$$

$$-8 = 2x \qquad \text{Simplify.}$$

$$\frac{-8}{2} = \frac{2x}{2} \qquad \text{Divide each side by 2.}$$

$$-4 = x \qquad \text{Simplify.}$$

Check.

$$\frac{1}{4}(-4) - \frac{2}{3} \overset{?}{=} \frac{5}{12}(-4)$$

$$-1 - \frac{2}{3} \overset{?}{=} -\frac{5}{3}$$

$$-\frac{3}{3} - \frac{2}{3} \overset{?}{=} -\frac{5}{3}$$

$$-\frac{5}{3} = -\frac{5}{3} \quad \checkmark \qquad \text{It checks.}$$

Practice Problem 1 Solve for x. $\dfrac{3}{8}x - \dfrac{3}{2} = \dfrac{1}{4}x$

NOTE TO STUDENT: Fully worked-out solutions to all of the Practice Problems can be found at the back of the text starting at page SP-1

In Example 1 we multiplied each side of the equation by the LCD. However, most students prefer to go immediately to the second step and multiply each term by the LCD. This avoids having to write out a separate step using the distributive property.

EXAMPLE 2   Solve for x and check your solution. $\dfrac{x}{3} + 3 = \dfrac{x}{5} - \dfrac{1}{3}$

Solution

$$15\left(\dfrac{x}{3}\right) + 15\,(3) = 15\left(\dfrac{x}{5}\right) - 15\left(\dfrac{1}{3}\right)$$

The LCD is 15. Use the multiplication principle to multiply each term by 15.

$$5x + 45 = 3x - 5$$ Simplify.

$$5x - 3x + 45 = 3x - 3x - 5$$ Add $-3x$ to both sides.

$$2x + 45 = -5$$ Combine like terms.

$$2x + 45 - 45 = -5 - 45$$ Add -45 to both sides.

$$2x = -50$$ Simplify.

$$\dfrac{2x}{2} = \dfrac{-50}{2}$$ Divide both sides by 2.

$$x = -25$$ The solution is -25.

Check.

$$\dfrac{-25}{3} + 3 \overset{?}{=} \dfrac{-25}{5} - \dfrac{1}{3}$$

$$-\dfrac{25}{3} + \dfrac{9}{3} \overset{?}{=} -\dfrac{5}{1} - \dfrac{1}{3}$$

$$-\dfrac{16}{3} \overset{?}{=} -\dfrac{15}{3} - \dfrac{1}{3}$$

$$-\dfrac{16}{3} = -\dfrac{16}{3} \;\checkmark$$

Practice Problem 2   Solve for x and check your solution.

$$\dfrac{5x}{4} - 1 = \dfrac{3x}{4} + \dfrac{1}{2}$$

NOTE TO STUDENT: Fully worked-out solutions to all of the Practice Problems can be found at the back of the text starting at page SP-1

EXAMPLE 3   Solve for x. $\dfrac{x + 5}{7} = \dfrac{x}{4} + \dfrac{1}{2}$

Solution

$$\dfrac{x}{7} + \dfrac{5}{7} = \dfrac{x}{4} + \dfrac{1}{2}$$

First we rewrite the left side as two fractions. This is actually multiplying $\frac{1}{7}(x + 5) = \frac{x}{7} + \frac{5}{7}$.

$$28\left(\dfrac{x}{7}\right) + 28\left(\dfrac{5}{7}\right) = 28\left(\dfrac{x}{4}\right) + 28\left(\dfrac{1}{2}\right)$$

We observe that the LCD is 28, so we multiply each term by 28.

$$4x + 20 = 7x + 14$$ Simplify.

$$4x - 4x + 20 = 7x - 4x + 14$$ Add $-4x$ to both sides.

$$20 = 3x + 14$$ Combine like terms.

$$20 - 14 = 3x + 14 - 14$$ Add -14 to both sides.

$$6 = 3x$$ Combine like terms.

$$\dfrac{6}{3} = \dfrac{3x}{3}$$ Divide both sides by 3.

$$2 = x$$ The solution is 2.

Check.   The check is left to the student.

Practice Problem 3   Solve for x $\dfrac{x + 6}{9} = \dfrac{x}{6} + \dfrac{1}{2}$

If a problem contains both parentheses and fractions, it is best to remove the parentheses first. Many students find it is helpful to have a written procedure to follow in solving these more involved equations.

PROCEDURE TO SOLVE EQUATIONS

1. Remove any parentheses.
2. If fractions exist, multiply all terms on both sides by the least common denominator of all the fractions.
3. Combine like terms if possible.
4. Add or subtract terms on both sides of the equation to get all terms with the variable on one side of the equation.
5. Add or subtract a constant value on both sides of the equation to get all terms not containing the variable on the other side of the equation.
6. Divide both sides of the equation by the coefficient of the variable.
7. Simplify the solution (if possible).
8. Check your solution.

Let's use each step in solving the next example.

EXAMPLE 4   Solve for x and check your solution.

$$\frac{1}{3}(x - 2) = \frac{1}{5}(x + 4) + 2$$

Solution

Step 1 $\dfrac{x}{3} - \dfrac{2}{3} = \dfrac{x}{5} + \dfrac{4}{5} + 2$ Remove parentheses.

Step 2 $15\left(\dfrac{x}{3}\right) - 15\left(\dfrac{2}{3}\right) = 15\left(\dfrac{x}{5}\right) + 15\left(\dfrac{4}{5}\right) + 15\,(2)$ Multiply by the LCD, 15.

$5x - 10 = 3x + 12 + 30$ Simplify.

Step 3 $5x - 10 = 3x + 42$ Combine like terms on each side.

Step 4 $5x - 3x - 10 = 3x - 3x + 42$ Add $-3x$ to both sides.

$2x - 10 = 42$ Simplify.

Step 5 $2x - 10 + 10 = 42 + 10$ Add 10 to both sides.

$2x = 52$ Simplify.

Step 6 $\dfrac{2x}{2} = \dfrac{52}{2}$ Divide both sides by 2.

Step 7 $x = 26$ Simplify the solution.

Step 8   *Check.* $\dfrac{1}{3}(26 - 2) \overset{?}{=} \dfrac{1}{5}(26 + 4) + 2$ Replace x by 26.

$\dfrac{1}{3}(24) \overset{?}{=} \dfrac{1}{5}(30) + 2$ Combine values within parentheses.

$8 \overset{?}{=} 6 + 2$ Simplify.

$8 = 8$ ✓   The solution is 26.

NOTE TO STUDENT: Fully worked-out solutions to all of the Practice Problems can be found at the back of the text starting at page SP-1

Practice Problem 4 Solve for x and check your solution.

$$\frac{1}{3}(x - 2) = \frac{1}{4}(x + 5) - \frac{5}{3}$$

Remember that not every step will be needed in each problem. You can combine some steps as well, *as long as you are consistently obtaining the correct solution.* However, you are encouraged to write out every step as a way of helping you to avoid careless errors.

It is important to remember that when we write decimals these numbers are really fractions written in a special way. Thus, $0.3 = \frac{3}{10}$ and $0.07 = \frac{7}{100}$. It is possible to take an equation containing decimals and to multiply each term by the appropriate value to obtain integer coefficients.

EXAMPLE 5 Solve for x. $0.2(1 - 8x) + 1.1 = -5(0.4x - 0.3)$

Solution

$0.2 - 1.6x + 1.1 = -2.0x + 1.5$	Remove parentheses.
$10(0.2) - 10(1.6x) + 10(1.1) = 10(-2.0x) + 10(1.5)$	Multiply each term by 10.
$2 - 16x + 11 = -20x + 15$	Multiplying by 10 moves the decimal point one place to the right.
$-16x + 13 = -20x + 15$	Simplify.
$-16x + 20x + 13 = -20x + 20x + 15$	Add $20x$ to each side.
$4x + 13 = 15$	Simplify.
$4x + 13 + (-13) = 15 + (-13)$	Add -13 to each side.
$4x = 2$	Simplify.
$\dfrac{4x}{4} = \dfrac{2}{4}$	Divide each side by 4.
$x = \dfrac{1}{2}$ or 0.5	Simplify.

Check.

$$0.2[1 - 8(0.5)] + 1.1 \overset{?}{=} -5[0.4(0.5) - 0.3]$$

$$0.2[1 - 4] + 1.1 \overset{?}{=} -5[0.2 - 0.3]$$

$$0.2[-3] + 1.1 \overset{?}{=} -5[-0.1]$$

$$-0.6 + 1.1 \overset{?}{=} 0.5$$

$$0.5 = 0.5 \checkmark$$

Practice Problem 5 Solve for x. $2.8 = 0.3(x - 2) + 2(0.1x - 0.3)$

TO THINK ABOUT: Does Every Equation Have One Solution?   Actually, no. There are some rare cases where an equation has no solution at all. Suppose we try to solve the equation

$$5(x + 3) = 2x - 8 + 3x.$$

If we remove the parentheses and collect like terms we have

$$5x + 15 = 5x - 8.$$

If we add $-5x$ to each side, we obtain

$$15 = -8.$$

Clearly this is impossible. There is no value of x for which these two numbers are equal. We would say this equation has **no solution.**

One additional surprise may happen. An equation may have an infinite number of solutions. Suppose we try to solve the equation

$$9x - 8x - 7 = 3 + x - 10.$$

If we combine like terms on each side, we have the equation

$$x - 7 = x - 7.$$

If we add $-x$ to each side, we obtain

$$-7 = -7.$$

Now this statement is always true, no matter what the value of x. We would say this equation has **an infinite number of solutions.**

In the To Think About exercises in this section, we will encounter some equations that have no solution or an infinite number of solutions.

Developing Your Study Skills

Taking Notes in Class

An important part of studying mathematics is taking notes. To take meaningful notes, you must be an active listener. Keep your mind on what the instructor is saying, and be ready with questions whenever you do not understand something.

If you have previewed the lesson material, you will be prepared to take good notes. The important concepts will seem somewhat familiar. If you frantically try to write all that the instructor says or copy all the examples done in class, you may find your notes nearly worthless when you are home alone. Write down *important* ideas and examples as the instructor lectures, making sure that you are listening and following the logic. Include any helpful hints or suggestions that your instructor gives you or refers to in your text.

In exercises 1–16, solve for the variable and check your answer. Noninteger answers may be left in fractional form or decimal form.

1. $\frac{1}{2}x + \frac{2}{3} = \frac{1}{6}$

2. $\frac{1}{3} + \frac{5}{12}x = \frac{3}{4}$

3. $\frac{2}{3}x = \frac{1}{15}x + \frac{3}{5}$

4. $\frac{5}{21}x = \frac{2}{3}x - \frac{1}{7}$

5. $\frac{x}{2} + \frac{x}{5} = \frac{7}{10}$

6. $\frac{x}{5} - \frac{x}{3} = \frac{8}{15}$

7. $20 - \frac{1}{3}x = \frac{1}{2}x$

8. $15 - \frac{1}{2}x = \frac{1}{4}x$

9. $2 + \frac{y}{2} = \frac{3y}{4} - 3$

10. $\frac{x}{3} - 1 = -\frac{1}{2} - x$

11. $\frac{x - 3}{5} = 1 - \frac{x}{3}$

12. $\frac{y - 5}{4} = 1 - \frac{y}{5}$

13. $\frac{x + 3}{4} = \frac{x}{2} + \frac{1}{6}$

14. $\frac{x + 5}{6} = \frac{x}{2} + \frac{3}{4}$

15. $0.6x + 5.9 = 3.8$

16. $1.2x - 2.2 = 5.6$

17. Is 4 a solution to $\frac{1}{2}(y - 2) + 2 = \frac{3}{8}(3y - 4)$?

18. Is 2 a solution to $\frac{1}{5}(y + 2) = \frac{1}{10}y + \frac{3}{5}$?

19. Is $\frac{5}{8}$ a solution to $\frac{1}{2}\left(y - \frac{1}{5}\right) = \frac{1}{5}(y + 2)$?

20. Is $\frac{13}{3}$ a solution to $\frac{y}{2} - \frac{7}{9} = \frac{y}{6} + \frac{2}{3}$?

Remove parentheses first. Then collect like terms. Solve for the variable. Noninteger answers may be left in fractional form or decimal form.

21. $\frac{3}{4}(3x + 1) = 2(3 - 2x) + 1$

22. $\frac{1}{4}(3x + 1) = 2(2x - 4) - 8$

23. $2(x - 2) = \frac{2}{5}(3x + 1) + 2$

24. $2(x - 4) = \frac{5}{6}(x + 6) - 6$

25. $0.3x - 0.2(3 - 5x) = -0.5(x - 6)$

26. $0.3(x - 2) + 0.4x = -0.2(x - 6)$

27. $-5(0.2x + 0.1) - 0.6 = 1.9$

28. $0.3x + 1.7 = 0.2x - 0.4(5x + 1)$

Mixed Practice

Solve. Noninteger answers may be left in fractional form or decimal form.

29. $\frac{1}{3}(y + 2) = 3y - 5(y - 2)$

30. $\frac{1}{4}(y + 6) = 2y - 3(y - 3)$

31. $\frac{1 + 2x}{5} + \frac{4 - x}{3} = \frac{1}{15}$

32. $\frac{1 + 3x}{2} + \frac{2 - x}{3} = \frac{5}{6}$

33. $\frac{1}{5}(x + 3) = 2x - 3(2 - x) - 3$

34. $\frac{2}{3}(x + 4) = 6 - \frac{1}{4}(3x - 2) - 1$

35. $\frac{1}{3}(x - 2) = 3x - 2(x - 1) + \frac{16}{3}$

36. $\frac{3}{4}(x - 2) + \frac{3}{5} = \frac{1}{5}(x + 1)$

37. $\frac{4}{5}x - \frac{2}{3} = \frac{3x + 1}{2}$

38. $\frac{4}{7}x + \frac{1}{3} = \frac{3x - 2}{14}$

39. $0.2(x + 3) = 4(0.5x - 0.03)$

40. $0.6(x + 0.1) = 2(0.4x - 0.2)$

To Think About

Solve. Be careful to examine your work to see if the equation may have no solution or an infinite number of solutions.

41. $-1 + 5(x - 2) = 12x + 3 - 7x$

42. $x + 3x - 2 + 3x = -11 + 7(x + 2)$

43. $9(x + 3) - 6 = 24 - 2x - 3 + 11x$

44. $7(x + 4) - 10 = 3x + 20 + 4x - 2$

45. $7x + 6 = 2(3x - 1) + 8$

46. $9x + 10 = 5(4x - 1) + 15$

47. $3(4x + 1) - 2x = 2(5x - 3)$

48. $5(-3 + 4x) = 4(2x + 4) + 12x$

Cumulative Review

49. Add. $\dfrac{3}{7} + 1\dfrac{5}{10}$

50. Subtract. $3\dfrac{1}{5} - 2\dfrac{1}{4}$

51. Multiply: $\left(-3\dfrac{1}{4}\right)\left(5\dfrac{1}{3}\right) =$

52. Divide. $5\dfrac{1}{2} \div 1\dfrac{1}{4}$

53. *Falcon Population* In 1975 there were 40 nesting pairs of the American peregrine falcon nesting in the United States and Canada. Between 1975 and 1985, the peregrine falcon population increased by 20 percent. Between 1985 and 2000, the population increased by 450 percent. How many pairs of peregrine falcons were thriving in 2000? (Source: U.S. Department of the Interior.)

54. *Auditorium Seating* The seating area of an auditorium is shaped like a trapezoid, with front and back sides parallel. The front of the auditorium measures 88 feet across, the back of the auditorium measures 150 feet across, and the auditorium is 200 feet from front to back. If each seat requires a space that is 2.5 feet wide by 3 feet deep, how many seats will the auditorium hold? (This will only be an approximation because of the angled side walls. Round off to the nearest whole number.)

88 feet

200 feet

150 feet

▲ **55.** *Vent Grill* The Newbury Elementary School needs a new air vent drilled into the wall of the maintenance room. The circular hole that is necessary will have a radius of 6 inches. The stainless steel grill that will cover the vent costs $2 per square inch. How much will the vent cost? Use $\pi \approx 3.14$.

▲ **56.** *Sail Material* Tom Rourke needs to replace the sail on his sailboat. It is in the shape of a triangle with an altitude of 9 feet and a base of 8 feet. The material to make the sail costs $3 per square foot. How much will the material cost to make a new sail for the boat?

How are you doing with your homework assignments in Sections 2.1 to 2.4? Do you feel you have mastered the material so far? Do you understand the concepts you have covered? Before you go further in the textbook, take some time to do each of the following problems.

Solve for x. If the solution is not an integer, you may express your answer as a fraction or as a decimal.

2.1

1. $5 - 8 + x = -12$

2. $3.6 + x = -7.3$

2.2

3. $-45 = -5x$

4. $12x - 6x = -48$

2.3

5. $-1.2x + 3.5 = 2.7$

6. $9x - 3 = -17x + 4$

7. $14x + 2(7 - 2x) = 20$

8. $0.5(1.2x - 3.4) = -1.4x + 5.8$

9. $3(x + 6) = -2(4x - 1) + x$

2.4

10. $\dfrac{x}{5} + \dfrac{x}{4} = \dfrac{2}{5}$

11. $\dfrac{1}{4}(x + 3) = 4x - 2(x - 3)$

12. $\dfrac{1}{2}(x - 1) + 2 = 3(2x - 1)$

13. $\dfrac{1}{7}(7x - 14) - 2 = \dfrac{1}{3}(x - 2)$

14. $0.2(x - 3) = 4(0.2x - 0.1)$

Now turn to page SA-6 for the answer to each of these problems. Each answer also includes a reference to the objective in which the problem is first taught. If you missed any of these problems, you should stop and review the Examples and Practice Problems in the referenced objective. A little review now will help you master the material in the upcoming sections of the text.

1. _____
2. _____
3. _____
4. _____
5. _____
6. _____
7. _____
8. _____
9. _____
10. _____
11. _____
12. _____
13. _____
14. _____

1 Solving a Formula for a Specified Variable

Formulas are equations with one or more variables that are used to describe real-life situations. The formula describes the relationship that exists among the variables. For example, in the formula $d = rt$, distance (d) is equal to the rate of speed (r) multiplied by the time (t). We can use this formula to find distance if we know the rate and time. Sometimes, however, we are given the distance and the rate, and we are asked to find the time.

EXAMPLE 1   American Airlines recently scheduled a non-stop flight in a new Boeing 777 jet from Chicago to London. The approximate air distance traveled on the flight was 3975 miles. The average speed of the aircraft on the trip was 530 miles per hour. How many hours did it take the Boeing 777 to fly this trip? (*Source:* www.aa.com)

Solution

$$d = rt \qquad \text{Use the distance formula.}$$

$$3975 = 530t \qquad \text{Substitute the known values for the variables.}$$

$$\frac{3975}{530} = \frac{530t}{530} \qquad \text{Divide both sides of the equation by 530 to solve for } t.$$

$$7.5 = t \qquad \text{Simplify.}$$

It took the Boeing 777 about 7.5 hours to fly this trip from Chicago to London.

Practice Problem 1   The airlines are planning a nonstop flight from Chicago to Prague. This distance is approximately 4565 miles and the time of the flight is 8.3 hours. Find the average rate of speed for the flight.

If we have many problems that ask us to find the time given the distance and rate, it may be worthwhile to rewrite the formula in terms of time.

EXAMPLE 2   Solve for t. $d = rt$

Solution

$$\frac{d}{r} = \frac{rt}{r} \qquad \text{We want to isolate } t. \text{ Therefore we divide both sides of the equation by the coefficient of } t, \text{ which is } r.$$

$$\frac{d}{r} = t \qquad \text{We have solved for the variable indicated.}$$

Practice Problem 2   Einstein's equation relating energy E to mass m and the speed of light c is $E = mc^2$. Solve it for m.

A straight line can be described by an equation of the form $Ax + By = C$ where A, B, and C are real numbers and A and B are not both zero. We will study this in later chapters. Often it is useful to solve such an equation for the variable y in order to make graphing the equation easier.

EXAMPLE 3 Solve for y. $3x - 2y = 6$

Solution

$$-2y = 6 - 3x$$ We want to isolate the term containing y, so we subtract $3x$ from both sides.

$$\frac{-2y}{-2} = \frac{6 - 3x}{-2}$$ Divide both sides by the coefficient of y.

$$y = \frac{6}{-2} + \frac{-3x}{-2}$$ Rewrite the fraction on the right side as two fractions.

$$y = \frac{3}{2}x - 3$$ Simplify and reorder the terms on the right.

This is known as the slope–intercept form of the equation of a line.

Practice Problem 3 Solve for y. $8 - 2y + 3x = 0$

Our procedure for solving an equation can be rewritten to give us a procedure for solving a formula for a specified variable.

PROCEDURE TO SOLVE A FORMULA FOR A SPECIFIED VARIABLE

1. Remove any parentheses.
2. If fractions exist, multiply all terms on both sides by the LCD of all the fractions.
3. Combine like terms on each side if possible.
4. Add or subtract terms on both sides of the equation to get all terms with the desired variable on one side of the equation.
5. Add or subtract the appropriate quantities to get all terms that do *not* have the desired variable on the other side of the equation.
6. Divide both sides of the equation by the coefficient of the desired variable.
7. Simplify if possible.

EXAMPLE 4 A trapezoid is a four-sided figure with two parallel sides. If the parallel sides are a and b and the altitude is h, the area is given by

$$A = \frac{h}{2}(a + b).$$

In some houses, two of the sides of the roof are in the shape of a trapezoid. Solve this equation for a.

Solution $$A = \frac{h}{2}(a + b)$$

$$A = \frac{ha}{2} + \frac{hb}{2}$$ Remove the parentheses.

$$2(A) = 2\left(\frac{ha}{2}\right) + 2\left(\frac{hb}{2}\right)$$ Multiply all terms by the LCD of 2.

$$2A = ha + hb$$ Simplify.

$$2A - hb = ha$$

We want to isolate the term containing *a*. Therefore, we subtract *hb* from both sides.

$$\frac{2A - hb}{h} = \frac{ha}{h}$$

Divide both sides by *h* (the coefficient of *a*).

$$\frac{2A - hb}{h} = a$$

The solution is obtained.

Note: Although the solution is in simple form, it could be written in an alternative way. Since

$$\frac{2A - hb}{h} = \frac{2A}{h} - \frac{hb}{h} = \frac{2A}{h} - b,$$

we could also have written $\dfrac{2A}{h} - b = a$.

NOTE TO STUDENT: Fully worked-out solutions to all of the Practice Problems can be found at the back of the text starting at page SP-1

▲ **Practice Problem 4** The relationship between the circumference *C* of a circle and the circle's diameter *d* is described by the equation $C = \pi d$. Solve it for *d*.

Developing Your Study Skills

Keep Trying

You may be one of those students who have had much difficulty with mathematics in the past and who are sure that you cannot do well in this course. Perhaps you are thinking, "I have never been any good at mathematics," or "I have always hated mathematics," or "Math always scares me," or "I have not had any math for so long that I have forgotten it all." You may have even picked up the label "math anxiety" and attached it to yourself. That is most unfortunate, and it is time for you to reprogram your thinking. Replace those negative thoughts with more positive ones. You need to say things like, "I will give this math class my best shot," or "I can learn mathematics if I work at it," or "I will try to do better than I have done in previous math classes." You will be pleasantly surprised at the difference a positive attitude makes!

We live in a highly technical world, and you cannot afford to give up on the study of mathematics. Dropping mathematics may prevent you from entering certain career fields that you may find interesting. You may not have to take math courses as high-level as calculus, but such courses as intermediate algebra, finite math, college algebra, and trigonometry may be necessary. Learning mathematics can open new doors for you.

Learning mathematics is a process that takes time and effort. You will find that regular study and daily practice are necessary to strengthen your skills and to help you to grow academically. This process will lead you toward success in mathematics. Then, as you become more successful, your confidence in your ability to do mathematics will grow.

2.5 EXERCISES

Student Solutions Manual | CD/ Video | PH Math Tutor Center | MathXL®Tutorials on CD | MathXL® | MyMathLab® | Interactmath.com

Verbal and Writing Skills

1. *Temperature* The formula for calculating the temperature in degrees Fahrenheit when you know the temperature in degrees Celsius is $F = \frac{9}{5}C + 32$. Explain in your own words how you would solve this equation for C.

▲ **2. *Geometry*** The formula for finding the area of a trapezoid with an altitude of 9 meters and bases of b meters and c meters is given by the equation $A = \frac{9}{2}(b + c)$. Explain in your own words how you would solve this equation for b.

Applications

▲ **3. *Geometry*** The formula for the area of a triangle is $A = \frac{1}{2}ab$, where b is the *base* of the triangle and a is the *altitude* of the triangle.

    (a) Use this formula to find the base of a triangle that has an area of 60 square meters and an altitude of 12 meters.

    (b) Use this formula to find the altitude of a triangle that has an area of 88 square meters and a base of 11 meters.

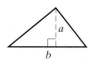

4. *Simple Interest* The formula for calculating simple interest is $I = Prt$, where P is the *principal* (amount of money invested), r is the *rate* at which the money is invested, and t is the *time*.

    (a) Use this formula to find how long it would take to earn \$720 in interest on an investment of \$3000 at the rate of 6%.

    (b) Use this formula to find the rate of interest if \$5000 earns \$400 interest in 2 years.

    (c) Use this formula to find the amount of money invested if the interest earned was \$120 and the rate of interest was 5% over 3 years.

5. The equation $4x + 3y = 18$ describes a line and is written in standard form.
    (a) Solve for the variable y.

    (b) Use this result to find y with $x = -3$.

6. The equation $5y - 3x = 15$ describes a line and is written in standard form.
    (a) Solve for the variable x.

    (b) Use this result to find x with $y = -6$.

In each formula or equation, solve for the variable indicated.

Area of a triangle

▲ **7.** $A = \frac{1}{2}bh$ Solve for b.

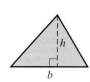

▲ **8.** $A = \frac{1}{2}bh$ Solve for h.

Simple interest formula

9. $I = Prt$ Solve for P.

10. $I = Prt$ Solve for r.

Slope–intercept form of a line

11. $y = mx + b$ Solve for m.

12. $y = mx + b$ Solve for b.

Standard form of a line

13. $8x - 12y = 24$.   Solve for y.

14. $3y - 5x = 9$.   Solve for y.

Slope–intercept form of a line

15. $y = -\dfrac{2}{3}x + 4$.   Solve for x.

16. $y = \dfrac{3}{8}x - 9$.   Solve for x.

Standard form of a line

17. $ax + by = c$ Solve for y.

18. $ax + by = c$ Solve for x.

Area of a circle

▲ **19.** $A = \pi r^2$ Solve for r^2.

Surface area of a sphere

▲ **20.** $s = 4\pi r^2$ Solve for r^2.

Distance of a falling object

21. $S = \dfrac{1}{2}gt^2$ Solve for g.

22. $S = \dfrac{1}{2}gt^2$ Solve for t^2.

Simple interest formula

23. $A = P(1 + rt)$ Solve for t.

Area of a trapezoid

▲ **24.** $A = \dfrac{1}{2}a(b_1 + b_2)$ Solve for b_1.

Surface area of a right circular cylinder

▲ **25.** $S = 2\pi rh + 2\pi r^2$ Solve for h.

26. $H = 5as + 10a^2$ Solve for s.

Volume of a right circular cylinder

▲ **27.** $V = \pi r^2 h$ Solve for h.

▲ **28.** $V = \pi r^2 h$ Solve for r^2.

Volume of a rectangular prism

▲ **29.** $V = LWH$ Solve for L.

▲ **30.** $V = LWH$ Solve for H.

Volume of a cone

▲ **31.** $V = \dfrac{1}{3}\pi r^2 h$ Solve for r^2.

▲ **32.** $V = \dfrac{1}{3}\pi r^2 h$ Solve for h.

Perimeter of a rectangle

▲ **33.** $P = 2L + 2W$ Solve for W.

Finding the nth term of an arithmetic sequence

34. $N = F + d(n - 1)$.   Solve for n.

Pythagorean theorem

▲ **35.** $c^2 = a^2 + b^2$ Solve for a^2.

▲ **36.** $c^2 = a^2 + b^2$ Solve for b^2.

Temperature conversion formulas

37. $F = \dfrac{9}{5}C + 32$ Solve for C.

38. $C = \dfrac{5}{9}(F - 32)$ Solve for F.

Ohm's law

Density formula (density, mass, and volume)

39. $P = \dfrac{E^2}{R}$. Solve for R.

40. $d = \dfrac{m}{v}$. Solve for v.

Area of a sector of a circle

▲ **41.** $A = \dfrac{\pi r^2 S}{360}$ Solve for S.

▲ **42.** $A = \dfrac{\pi r^2 S}{360}$ Solve for r^2.

Applications

▲ **43.** *Geometry* Use the result you obtained in exercise 33 to solve the following problem. A farmer has a rectangular field with a perimeter of 5.8 miles and a length of 2.1 miles. Find the width of the field.

44. Use the result you obtained in exercise 34 to solve the following problem. If you are given an arithmetic sequence where the first term (F) is 6, the difference (d) is 3, and nth term (N) is 24, what number term (n) is it?

▲ **45.** *Geometry* Use the result you obtained in exercise 29 to solve the following problem. The foundation of a house is in the shape of a rectangular solid. The volume held by the foundation is 5940 cubic feet. The height of the foundation is 9 feet and the width is 22 feet. What is the length of the foundation?

▲ **46.** *Geometry* Use the result you obtained in exercise 30 to solve the following problem. The fish tank at the Mandarin Danvers Restaurant is in the shape of a rectangular solid. The volume held by the tank is 3024 cubic inches. The length of the tank is 18 inches while the width of the tank is 14 inches. What is the height of the tank?

47. *Tourism* The number of foreign visitors measured in thousands (V) admitted to the United States for a pleasure trip for any given year can be predicted by the equation $V = 1100x + 7050$, where x is the number of years since 1985. For example, if $x = 20$ (this would be the year 2005), the predicted number of visitors in thousands would be $1100(20) + 7050 = 29{,}050$. Thus we would predict that in 2005, a total of 29,050,000 visitors came to the United States for a pleasure trip. (Source: U.S. Immigration and Naturalization Service.)

(a) Solve this equation for x.

(b) Use the result of your answer in (a) to find the year in which the number of visitors will be predicted to be 25,750,000. (*Hint:* Let $V = 25{,}750$ in your answer for (a).)

48. *Tourism* The number of foreign visitors from Europe measured in thousands (E) admitted to the United States from Europe for a pleasure trip for any given year can be predicted by the equation $E = 480x + 2400$, where x is the number of years since 1985. For example, if $x = 25$ (this would be the year 2010), the predicted number of visitors in thousands would be $480(25) + 2400 = 14{,}400$. Thus we would predict that in 2010, a total of 14,400,000 visitors from Europe came to the United States for a pleasure trip. (*Source:* U.S. Immigration and Naturalization Service.)

(a) Solve this equation for x.

(b) Use the result of your answer in (a) to find the year in which the number of visitors from Europe will be predicted to be 11,520,000. (*Hint:* Let $E = 11{,}520$ in your answer for (a).)

To Think About

▲ **49.** In the formula $A = \dfrac{1}{2}ab$, if b doubles, what is the effect on A?

▲ **50.** In the formula $A = \dfrac{1}{2}ab$, if both a and b double, what is the effect on A?

▲ **51.** In $A = \pi r^2$, if r doubles, what is the effect on A?

▲ **52.** In $A = \pi r^2$, if r is halved, what is the effect on A?

Cumulative Review

53. Find 12% of 260.

54. What is 0.2% of 48?

55. *Corral Fencing* Greg Boyd recently replaced 500 feet of corral fencing. He estimated that he replaced 40% of the total amount of fencing aroung the corral. How much fencing was around the corral?

56. *Stereo Defects* A major auto company received a shipment of car stereos. Four stereos out of 160 are defective. What percent of the shipment is defective?

57. *Electronic Games* A very popular handheld electronic game requires $3\frac{1}{4}$ square feet of a certain type of durable plastic in the manufacturing process. How many square feet of durable plastic does this company need to make 12,000 handheld games?

58. *Spotlight Rental* The Superstar Lighting Company rents out giant spotlights that shine up in the sky to mark the location of special events, such as the opening of a movie, a major sports play-off game, or a huge sales event at an auto dealership. A giant spotlight was used for $4\frac{1}{3}$ hours on Saturday, $2\frac{3}{4}$ hours on Tuesday, and $3\frac{1}{2}$ hours on Wednesday. What was the total number of hours that the spotlight was in use?

2.6 WRITING AND GRAPHING INEQUALITIES

1 Interpreting Inequality Statements

We frequently speak of one value being greater than or less than another value. We say that "5 is less than 7" or "9 is greater than 4." These relationships are called **inequalities.** We can write inequalities in mathematics by using symbols. We use the symbol < to represent the words "**is less than.**" We use the symbol > to represent the words "**is greater than.**"

Statement in Words	Statement in Algebra
5 is less than 7.	$5 < 7$
9 is greater than 4.	$9 > 4$

Note: "5 is less than 7" and "7 is greater than 5" have the same meaning. Similarly, $5 < 7$ and $7 > 5$ have the same meaning. They represent two equivalent ways of describing the same relationship between the two numbers 5 and 7.

We can better understand the concept of inequality if we examine a number line.

We say that one number is greater than another if it is to the right of the other on the number line. Thus $7 > 5$, since 7 is to the right of 5.

What about negative numbers? We can say "-1 is greater than -3" and write it in symbols as $-1 > -3$ because we know that -1 lies to the right of -3 on the number line.

EXAMPLE 1 In each statement, replace the question mark with the symbol < or >.

(a) $3 ? -1$ **(b)** $-2 ? 1$ **(c)** $-3 ? -4$ **(d)** $0 ? 3$ **(e)** $-3 ? 0$

Solution

(a) $3 > -1$ Use >, since 3 is to the right of -1 on the number line.

(b) $-2 < 1$ Use <, since -2 is to the left of 1. (Or equivalently, we could say that 1 is to the right of -2.)

(c) $-3 > -4$ Note that -3 is to the right of -4.

(d) $0 < 3$

(e) $-3 < 0$

Practice Problem 1 In each statement, replace the question mark with the symbol < or >.

(a) $7 ? 2$ **(b)** $-3 ? -4$ **(c)** $-1 ? 2$ **(d)** $-8 ? -5$ **(e)** $0 ? -2$ **(f)** $\dfrac{2}{5} ? \dfrac{3}{8}$

---

2 Graphing an Inequality on a Number Line

Sometimes we will use an inequality to express the relationship between a variable and a number. $x > 3$ means that x could have the value of *any number* greater than 3.

Any number that makes an inequality true is called a **solution** of the inequality. The set of all numbers that make the inequality true is called the **solution set.** A picture that represents all of the solutions of an inequality is called a **graph** of the inequality. The inequality $x > 3$ can be graphed on the number line as follows:

Case 1

Note that all of the points to the right of 3 are shaded. The open circle at 3 indicates that we do not include the point for the number 3.

Similarly, we can graph $x < -2$ as follows:

Case 2

Note that all of the points to the left of -2 are shaded.

Sometimes a variable will be either greater than or equal to a certain number. In the statement "x is greater than or equal to 3," we are implying that x could have the value of 3 or any number greater than 3. We write this as $x \geq 3$. We graph it as follows:

Case 3

Note that the closed circle at 3 indicates that we *do* include the point for the number 3.

Similarly, we can graph $x \leq -2$ as follows:

Case 4

Note: Be careful you do not confuse $\circ\!\!-\!\!\longrightarrow$ with $\bullet\!\!-\!\!\longrightarrow$. It is important to decide if you need an open circle or a closed one. Case 1 and Case 2 use open circles. Case 3 and Case 4 use closed circles.

EXAMPLE 2   State each mathematical relationship in words and then graph it.

(a) $x < -2$ **(b)** $-3 < x$ **(c)** $x \geq -2$ **(d)** $x \leq -6$

Solution

(a) We state that "x is less than -2."

(b) We can state that "-3 is less than x" or, equivalently, that "x is greater than -3." Be sure you see that $-3 < x$ is equivalent to $x > -3$. Although both statements are correct, we *usually write the variable first* in a simple inequality containing a variable and a numerical value.

(c) We state that "x is greater than or equal to -2."

(d) We state that "x is less than or equal to -6."

Practice Problem 2   State each mathematical relationship in words and then graph it on a number line in the margin.

(a) $x > 5$

(b) $x \leq -2$

(c) $3 > x$

(d) $x \geq -\dfrac{3}{2}$

Translating English Phrases into Algebraic Statements

We can translate many everyday situations into algebraic statements with an unknown value and an inequality symbol. This is the first step in solving word problems using inequalities.

EXAMPLE 3 Translate each English statement into an algebraic statement.

(a) The police on the scene said that the car was traveling more than 80 miles per hour. (Use the variable s for speed.)

(b) The owner of the trucking company said that the payload of a truck must never exceed 4500 pounds. (Use the variable p for payload.)

Solution

(a) Since the speed must be greater than 80, we have $s > 80$.

(b) If the payload of the truck can never exceed 4500 pounds, then the payload must be always less than or equal to 4500 pounds. Thus we write $p \leq 4500$.

NOTE TO STUDENT: Fully worked-out solutions to all of the Practice Problems can be found at the back of the text starting at page SP-1

Practice Problem 3 Translate each English statement into an inequality.

(a) During the drying cycle, the temperature inside the clothes dryer must never exceed 180 degrees Fahrenheit. (Use the variable t for temperature.)

(b) The bank loan officer said that the total consumer debt incurred by Wally and Mary must be less than \$15,000 if they want to qualify for a mortgage to buy their first home. (Use the variable d for debt.)

2.6 EXERCISES

Student Solutions Manual | CD/Video | PH Math Tutor Center | MathXL®Tutorials on CD | MathXL® | MyMathLab® | Interactmath.com

Verbal and Writing Skills

1. Is the statement $5 > -6$ equivalent to the statement $-6 < 5$? Why?

2. Is the statement $-8 < -3$ equivalent to the statement $-3 > -8$? Why?

Replace the ? by $<$ or $>$.

3. $9 \; ? \; -3$

4. $-2 \; ? \; 5$

5. $-4 \; ? \; -2$

6. $-3 \; ? \; -6$

7. $\dfrac{3}{5} \; ? \; \dfrac{4}{7}$

8. $\dfrac{4}{6} \; ? \; \dfrac{7}{9}$

9. $-4.2 \; ? \; 3.5$

10. $-2.6 \; ? \; 7.5$

11. $-\dfrac{13}{3} \; ? \; -4$

12. $-3 \; ? \; -\dfrac{15}{4}$

13. $-\dfrac{5}{8} \; ? \; -\dfrac{3}{5}$

14. $-\dfrac{2}{3} \; ? \; -\dfrac{3}{4}$

Which is greater?

 15. $\dfrac{123}{4986}$ or 0.0247?

 16. $\dfrac{997}{6384}$ or 0.15613?

Graph each inequality on the number line.

17. $x > 7$

18. $x < 1$

19. $x \geq -6$

20. $x \leq -2$

21. $x > \dfrac{3}{4}$

22. $x \geq -\dfrac{5}{2}$

23. $x \leq -5.3$

24. $x > -3.5$

25. $25 < x$

26. $35 \geq x$

Translate each graph to an inequality using the variable x.

27.

28.

29.

30.

31.

32.

Translate each English statement into an inequality.

33. *Weight Category* To box in the heavyweight category, your weight must be greater than 175 pounds. (Use the variable W for weight.)

34. *Cloud Formation* Cirrus clouds form more than 16,500 feet above the earth's surface. (Use the variable C for distance above the earth.)

35. *Full-Time Work* The number of hours for a full-time position at this company cannot be less than 37 in order to receive full-time benefits. (Use the variable h for hours.)

36. *Nurses on Duty* The number of nurses on duty on the floor can never exceed 6. (Use the variable n for the number of nurses.)

37. *Height Limit* In order for you to be allowed to ride the roller coaster at the theme park, your height must be at least 48 inches. (Use h for height.)

38. *College Credits* In order for you to avoid paying extra tuition for a semester, the number of credits you are taking must not exceed 18. (Use C for credits.)

To Think About

39. Suppose that the variable x must satisfy *all* of these conditions.

$$x \leq 2, \quad x > -3, \quad x < \frac{5}{2}, \quad x \geq -\frac{5}{2}$$

Graph on a number line the region that satisfies all of the conditions.

40. Suppose that the variable x must satisfy *all* of these conditions.

$$x < 4, \quad x > -4, \quad x \leq \frac{7}{2}, \quad x \geq -\frac{9}{2}$$

Graph on a number line the region that satisfies all of the conditions.

Cumulative Review

41. Find 16% of 38.

42. 18 is what percent of 120?

43. *Percent Accepted* For the most coveted graduate study positions, only 16 out of 800 students are accepted. What percent are accepted?

44. Write the fraction $\frac{3}{8}$ as a percent.

45. *Dry Cleaning* The Apex Dry Cleaners raised the cost of dry cleaning a sports coat from $8 to $10. If the same percent of increase applies to all other dry cleaning also, what will the new price for dry cleaning a men's suit be if the old price was $12?

46. *Radar Picture* A radar picture is taken of a portion of the Amazon rain forest. The radar detects that a cliff is 414 feet above sea level and that a giant tree trunk is directly below it in a body of water. The tree trunk is 81 feet below sea level. How far is the top of the cliff from the tree trunk?

2.7 SOLVING INEQUALITIES

 Solving and Graphing Inequalities on a Number Line

As stated in Section 2.6, the possible values that make an inequality true are called its **solutions.** Thus, when we **solve an inequality,** we are finding *all* the values that make it true. To solve an inequality, we simplify it to the point where we can clearly see all possible values for the variable. We've solved equations by adding, subtracting, multiplying by, and dividing by a particular value on both sides of the equation. Here we perform similar operations with inequalities with one important exception. We'll show some examples so that you can see how these operations can be used with inequalities just as with equations.

We will first examine the pattern that occurs when we perform these operations *with a positive value* on both sides of an inequality.

Student Learning Objective

After studying this section, you will be able to:

1 **Solve and graph inequalities on a number line.**

EXAMPLE 1

Original Inequality	Operations with a Positive Number	New Inequality
$4 < 6$	Add 2 to both sides.	$6 < 8$
	Subtract 2 from both sides.	$2 < 4$
	Multiply both sides by 2.	$8 < 12$
	Divide both sides by 2.	$2 < 3$

Notice that the inequality symbol remains the same when these operations are performed.

Practice Problem 1 Perform the given operation and write a new inequality.

(a) $9 > 6$ Add 4 to each side.
(b) $-2 < 5$ Subtract 3 from both sides.
(c) $1 > -3$ Multiply both sides by 2.
(d) $10 < 15$ Divide both sides by 5.

NOTE TO STUDENT: Fully worked-out solutions to all of the Practice Problems can be found at the back of the text starting at page SP-1

Now let us examine what happens when we perform these operations *with a negative value.*

Original Inequality	Operations with a Negative Number	New Inequality
$4 < 6$	Add -2 to both sides.	$2 < 4$
	Subtract -2 from both sides.	$6 < 8$
	Multiply both sides by -2.	$-8 \, ? -12$
	Divide both sides by -2.	$-2 \, ? -3$

What happens to the inequality sign when we multiply both sides by a negative number? Since -8 is to the right of -12 on the number line, we know that the new inequality should be $-8 > -12$ if we want the statement to remain true. Notice how we reverse the direction of the inequality from $<$ (less than) to $>$ (greater than). Thus we have the following.

$$4 < 6 \longrightarrow \text{Multiply both sides by } -2. \longrightarrow -8 > -12$$

The same thing happens when we divide by a negative number. The inequality is reversed from $<$ to $>$. We know this since -2 is to the right of -3 on the number line.

$$4 < 6 \longrightarrow \text{Divide both sides by } -2. \longrightarrow -2 > -3$$

EXAMPLE 2

Original Inequality			New Inequality
(a) $-2 < -1$	$\longrightarrow$	Multiply both sides by -3. $\longrightarrow$	$6 > 3$
(b) $0 > -4$	$\longrightarrow$	Divide both sides by -2. $\longrightarrow$	$0 < 2$
(c) $8 \geq 4$	$\longrightarrow$	Divide both sides by -4. $\longrightarrow$	$-2 \leq -1$

Notice that we perform the arithmetic with signed numbers just as we always do. But the new inequality signs reversed (from those of the original inequalities). *Whenever both sides of an inequality are multiplied or divided by a negative quantity, the direction of the inequality is reversed.*

Practice Problem 2

(a) $7 > 2$ Multiply each side by -2.

(b) $-3 < -1$ Multiply each side by -1.

(c) $-10 \geq -20$ Divide each side by -10.

(d) $-15 \leq -5$ Divide each side by -5.

NOTE TO STUDENT: Fully worked-out solutions to all of the Practice Problems can be found at the back of the text starting at page SP-1

PROCEDURE FOR SOLVING INEQUALITIES

You may use the same procedures to solve inequalities that you did to solve equations *except* that the direction of an inequality is *reversed* if you *multiply* or *divide* both sides *by a negative number.*

It may be helpful to think over quickly what we have discussed here. The inequalities remain the same when we add a number to both sides or subtract a number from both sides of the equation. The inequalities remain the same when we multiply both sides by a positive number or divide both sides by a positive number.

RULES WHEN INEQUALITIES REMAIN THE SAME

For all real numbers a, b, and c:

1. If $a > b$, then $a + c > b + c$.

2. If $a > b$, then $a - c > b - c$.

3. If $a > b$ and c is a **positive number** $(c > 0)$, then $ac > bc$.

4. If $a > b$ and c is a **positive number** $(c > 0)$, then $\dfrac{a}{c} > \dfrac{b}{c}$.

If we multiply both sides of an inequality by a negative number or if we divide both sides of an inequality by a negative number, then the inequality is reversed.

RULES WHEN INEQUALITIES ARE REVERSED

For all real numbers a, b, and c:

1. If $a > b$, and c is a **negative number** $(c < 0)$, then $ac < bc$.

2. If $a > b$, and c is a **negative number** $(c < 0)$, then $\dfrac{a}{c} < \dfrac{b}{c}$.

The pattern is fairly simple. We could also make similar rule boxes for the cases where $a \geq b$, $a < b$, and $a \leq b$, but they are really not necessary. We simply remember that the inequality is reversed when we multiply or divide by a negative number. Otherwise the inequality remains unchanged.

EXAMPLE 3   Solve and graph $3x + 7 \geq 13$.

Solution

$3x + 7 - 7 \geq 13 - 7$	Subtract 7 from both sides.
$3x \geq 6$	Simplify.
$\dfrac{3x}{3} \geq \dfrac{6}{3}$	Divide both sides by 3.
$x \geq 2$	Simplify. Note that the direction of the inequality is not changed, since we have divided by a positive number.

The graph is as follows:

Practice Problem 3   Solve and graph $8x - 2 < 3$.

EXAMPLE 4   Solve and graph $5 - 3x > 7$.

Solution

$5 - 5 - 3x > 7 - 5$	Subtract 5 from both sides.
$-3x > 2$	Simplify.
$\dfrac{-3x}{-3} < \dfrac{2}{-3}$	Divide by -3 and **reverse the inequality** since you are dividing by a negative number.
$x < -\dfrac{2}{3}$	Note the direction of the inequality.

The graph is as follows:

Practice Problem 4   Solve and graph $4 - 5x > 7$.

Just like equations, some inequalities contain parentheses and fractions. The initial steps to solve these inequalities will be the same as those used to solve equations with parentheses and fractions. When the variable appears on both sides of the inequality, it is advisable to collect the x-terms on the left side of the inequality symbol.

EXAMPLE 5   Solve and graph $-\dfrac{13x}{2} \leq \dfrac{x}{2} - \dfrac{15}{8}$.

Solution

$8\left(\dfrac{-13x}{2}\right) \leq 8\left(\dfrac{x}{2}\right) - 8\left(\dfrac{15}{8}\right)$	Multiply all terms by LCD $= 8$. We do **not** reverse the direction of the inequality symbol since we are multiplying by a positive number.
$-52x \leq 4x - 15$	Simplify.
$-52x - 4x \leq 4x - 15 - 4x$	Subtract $4x$ from both sides.
$-56x \leq -15$	Combine like terms.

$$\frac{-56x}{-56} \geq \frac{-15}{-56}$$

Divide both sides by -56. We **reverse** the direction of the inequality when we divide both sides by a negative number.

$$x \geq \frac{15}{56}$$

The graph is as follows:

NOTE TO STUDENT: *Fully worked-out solutions to all of the Practice Problems can be found at the back of the text starting at page SP-1*

Practice Problem 5 Solve and graph $\frac{1}{2}x + 3 < \frac{2}{3}x$.

EXAMPLE 6 Solve and graph $\frac{1}{3}(3 - 2x) \leq -4(x + 1)$.

Solution

$$1 - \frac{2x}{3} \leq -4x - 4 \qquad \text{Remove parentheses.}$$

$$3(1) - 3\left(\frac{2x}{3}\right) \leq 3(-4x) - 3(4) \qquad \text{Multiply all terms by LCD} = 3.$$

$$3 - 2x \leq -12x - 12 \qquad \text{Simplify.}$$

$$3 - 2x + 12x \leq -12x + 12x - 12 \qquad \text{Add } 12x \text{ to both sides.}$$

$$3 + 10x \leq -12 \qquad \text{Combine like terms.}$$

$$3 - 3 + 10x \leq -12 - 3 \qquad \text{Subtract 3 from both sides.}$$

$$10x \leq -15 \qquad \text{Simplify.}$$

$$\frac{10x}{10} \leq \frac{-15}{10} \qquad \begin{array}{l}\text{Divide both sides by 10. Since we are} \\ \text{dividing by a } \textbf{positive} \text{ number, the} \\ \text{inequality is } \textbf{not} \text{ reversed.}\end{array}$$

$$x \leq -\frac{3}{2}$$

The graph is as follows:

Practice Problem 6 Solve and graph $\frac{1}{2}(3 - x) \leq 2x + 5$.

CAUTION The most common error students make in solving inequalities is forgetting to reverse the direction of the inequality symbol when multiplying or dividing both sides of the inequality by a negative number.

Normally when solve inequalities you solve for x by putting the variables on the left side. If you solve by placing the variables on the right side, you will end up with statements like $3 > x$. This is equivalent to $x < 3$. It is wise to express your answer with the variable on the left side.

EXAMPLE 7   A hospital director has determined that the costs of operating one floor of the hospital for an eight-hour shift must never exceed $2370. An expression for the cost of operating one floor of the hospital is $130n + 1200$, where n is the number of nurses. This expression is based on an estimate of $1200 in fixed costs and a cost of $130 per nurse for an eight-hour shift. Solve the inequality $130n + 1200 \leq 2370$ to determine the number of nurses that may be on duty on this floor during an eight-hour shift if the director's cost control measure is to be followed.

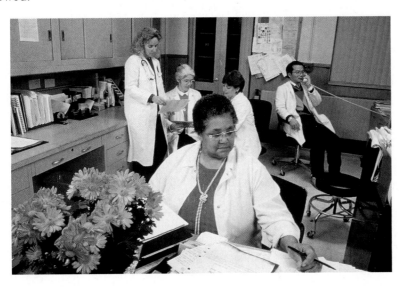

Solution

$$130n + 1200 \leq 2370 \qquad \text{The inequality we must solve.}$$

$$130n + 1200 - 1200 \leq 2370 - 1200 \qquad \text{Subtract 1200 from each side.}$$

$$130n \leq 1170 \qquad \text{Simplify.}$$

$$\frac{130n}{130} \leq \frac{1170}{130} \qquad \text{Divide each side by 130.}$$

$$n \leq 9$$

The number of nurses on duty on this floor during an eight-hour shift must always be less than or equal to nine.

Practice Problem 7   The company president of Staywell, Inc., wants the monthly profits never to be less than $2,500,000. He has determined that an expression for monthly profit for the company is $2000n - 700,000$. In the expression, n is the number of exercise machines manufactured each month. The profit on each machine is $2000, and the $-\$700,000$ in the expression represents the fixed costs of running the manufacturing division.

Solve the inequality $2000n - 700,000 \geq 2,500,000$ to find how many machines must be made and sold each month to satisfy these financial goals.

Solve and graph the result.

1. $x + 7 \le 4$

2. $x - 5 < -3$

3. $5x \le 25$

4. $7x \ge -35$

5. $-2x < 18$

6. $-6x > 24$

7. $\frac{1}{2}x \ge 4$

8. $\frac{1}{3}x \le 2$

9. $-\frac{1}{4}x > 3$

10. $-\frac{1}{5}x < 10$

11. $2x - 3 < 4$

12. $4x - 6 \ge -8$

13. $5 - 2x \le 9$

14. $3 - 3x > 12$

15. $-4 + 5x < -3x + 8$

16. $-6 - 4x < 1 - 6x$

17. $\dfrac{5x}{6} - 5 > \dfrac{x}{6} - 9$

18. $\dfrac{x}{4} - 2 < \dfrac{3x}{4} + 5$

19. $2(3x + 4) > 3(x + 3)$

20. $5(x - 3) \le 2(x - 3)$

Verbal and Writing Skills

21. Add -2 to both sides of the inequality $5 > 3$. What is the result? Why is the direction of the inequality not reversed?

22. Divide -3 into both sides of the inequality $-21 > -29$. What is the result? Why is the direction of the inequality reversed?

Mixed Practice

Solve. Collect the variable terms on the left side of the inequality.

23. $3x + 8 < 7x - 4$ **24.** $7x + 3 > 9x - 5$ **25.** $6x - 2 \ge 4x + 6$ **26.** $5x - 5 \le 2x + 10$

27. $0.3(x - 1) < 0.1x - 0.5$ **28.** $0.2(3 - x) + 0.1 > 0.1(x - 2)$ **29.** $3 + 5(2 - x) \ge -3(x + 5)$

30. $7 - 2(x - 4) \le 7(x - 3)$ **31.** $\dfrac{x + 6}{7} - \dfrac{6}{14} > \dfrac{x + 3}{2}$ **32.** $\dfrac{3x + 5}{4} + \dfrac{7}{12} > -\dfrac{x}{6}$

33. $\dfrac{1}{6} - \dfrac{1}{2}(3x + 2) < \dfrac{1}{3}\left(x - \dfrac{1}{2}\right)$ **34.** $\dfrac{2}{3}(2x - 5) + 3 \ge \dfrac{1}{4}(3x + 1) - 5$

Applications

35. **Course Average** To pass a course with a B grade, a student must have an average of 80 or greater. A student's grades on three tests are 75, 83, and 86. Solve the inequality $\dfrac{75 + 83 + 86 + x}{4} \ge 80$ to find what score the student must get on the next test to get a B average or better.

36. **Payment Options** Sharon sells very expensive European sports cars. She may choose to receive $10,000.00 or 8% of her sales as payment for her work. Solve the inequality $0.08x > 10,000$ to find how much she needs to sell to make the 8% offer a better deal.

37. *Elephant Weight* The average African elephant weighs 268 pounds at birth. During the first three weeks of life the baby elephant will usually gain about 4 pounds per day. Assuming that growth rate, solve the inequality $268 + 4x \geq 300$ to find how many days it will be until a baby elephant weighs at least 300 pounds.

38. *Car Loan* Rennie is buying a used car that costs $4500. The deal called for a $600 down payment, and payments of $260 monthly. He wants to know whether he can pay off the car within a year. Solve the inequality $600 + 260x \geq 4500$ to find out the minimum number of months it will take to pay off the car.

Cumulative Review

▲ **39.** *Tennis Court Fence* A rectangular tennis court measures 36 feet wide and 78 feet long. Robert is building a fence to surround the tennis court. He wants the fence to be 4 feet from each side of the court. How many feet of fence will he need?

▲ **40.** *Photo Enlargement* Jemma has a job with the school newspaper. She has taken a picture with the dimensions 3" high by 5" wide. The editor likes the picture, but the available space is 30% wider and 20% higher than the picture Jemma took. What dimensions must Jemma enlarge the picture to?

▲ **41.** *Basketball Poster* Melinda is making a poster for her college basketball team. On the poster she is placing a life-sized picture of a basketball. When inflated properly, a basketball has a diameter of 9 inches. What is the area of the basketball on her poster? Use $\pi \approx 3.14$. Round your answer to the nearest tenth.

▲ **42.** *Seating Area* The seats of an outdoor amphitheater are arranged in the shape of a trapezoid. The altitude of the trapezoid is 120 feet. The bases of the trapezoid are 90 feet and 170 feet. What is the area of this seating area?

Gasoline Consumption on a 27-Foot Outboard

Frank and his father have a 27-foot outboard sport fishing boat that they charter for part of the year. The boat can be equipped with either two 225-horsepower outboard motors or one 250-horsepower outboard motor. They tested the consumption of gasoline per hour at various speeds for each choice of motor and made the following chart.

Problems for Individual Study and Analysis

1. How many more gallons per hour does the boat use at 40 miles per hour if it is equipped with two 225-horsepower outboard motors rather than one 250-horsepower motor?

2. If the boat is used for 30 hours of fishing per week and the usual cruising speed is 40 miles per hour, how many more gallons per week will be consumed if the boat is equipped with two 225-horsepower outboard motors rather than one 250-horsepower motor?

Problems for Group Investigation and Cooperative Learning

For the boat equipped with two 225-horsepower outboard motors the rate of gasoline consumption in gallons per hour (y) can be approximated from knowing the speed in miles per hour (x) using the equation $y = 0.8x - 3.7$. Use this equation for problems 3 and 4.

3. How many gallons per hour are consumed if the boat travels at 45 miles per hour?

4. If the boat is consuming 40.3 gallons per hour, how fast is the boat traveling?

For the boat equipped with one 250-horsepower outboard motor the rate of gasoline consumption in gallons per hour (y) can be approximated from knowing the speed in miles per hour (x) using the equation $y = 0.5x - 0.1$. Use this equation for problems 5 and 6.

5. Solve the above equation for x.

6. Use the result of problem 5 to find how many miles per hour the boat is traveling if it is consuming 17.4 gallons per hour.

Topic	Procedure	Examples
Solving equations without parentheses or fractions, p. 146.	1. On each side of the equation, combine like terms if possible. 2. Add or subtract terms on both sides of the equation in order to get all terms with the variable on one side of the equation. 3. Add or subtract a value on both sides of the equation to get all terms not containing the variable on the other side of the equation. 4. Divide both sides of the equation by the coefficient of the variable. 5. If possible, simplify the solution. 6. Check your solution by substituting the obtained value into the original equation.	Solve for x. $$5x + 2 + 2x = -10 + 4x + 3$$ $$7x + 2 = -7 + 4x$$ $$7x - 4x + 2 = -7 + 4x - 4x$$ $$3x + 2 = -7$$ $$3x + 2 - 2 = -7 - 2$$ $$3x = -9$$ $$\frac{3x}{3} = \frac{-9}{3}$$ $$x = -3$$ *Check:* Is -3 the solution of $$5x + 2 + 2x = -10 + 4x + 3?$$ $$5(-3) + 2 + 2(-3) \stackrel{?}{=} -10 + 4(-3) + 3$$ $$-15 + 2 - 6 \stackrel{?}{=} -10 + (-12) + 3$$ $$-13 - 6 \stackrel{?}{=} -22 + 3$$ $$-19 = -19 \checkmark$$
Solving equations with parentheses and/or fractions, p. 148 and p. 153.	1. Remove any parentheses. 2. Simplify, if possible. 3. If fractions exist, multiply all terms on both sides by the least common denominator of all the fractions. 4. Now follow the remaining steps for solving an equation without parentheses or fractions.	Solve for y. $$5(3y - 4) = \frac{1}{4}(6y + 4) - 48$$ $$15y - 20 = \frac{3}{2}y + 1 - 48$$ $$15y - 20 = \frac{3}{2}y - 47$$ $$2(15y) - 2(20) = 2\left(\frac{3}{2}y\right) - 2(47)$$ $$30y - 40 = 3y - 94$$ $$30y - 3y - 40 = 3y - 3y - 94$$ $$27y - 40 = -94$$ $$27y - 40 + 40 = -94 + 40$$ $$27y = -54$$ $$\frac{27y}{27} = \frac{-54}{27}$$ $$y = -2$$
Solving formulas, p. 162	1. Remove any parentheses. 2. If fractions exist, multiply all terms on both sides by the LCD, which may be a variable. 3. Add or subtract terms on both sides of the equation in order to get all terms containing the *desired variable* on one side of the equation. 4. Add or subtract terms on both sides of the equation in order to get all other terms on the opposite side of the equation. 5. Divide both side of the equation by the coefficient of the desired variable. This division may involve other variables. 6. Simplify, if possible. 7. (Optional) Check your solution by substituting the obtained expression into the original equation.	Solve for z. $\quad B = \frac{1}{3}(hx + hz)$ First we remove parentheses. $$B = \frac{1}{3}hx + \frac{1}{3}hz$$ Now we multiply each term by 3. $$3(B) = 3\left(\frac{1}{3}hx\right) + 3\left(\frac{1}{3}hz\right)$$ $$3B = hx + hz$$ $$3B - hx = hx - hx + hz$$ $$3B - hx = hz$$ The coefficient of z is h, so we divide each side by h. $$\frac{3B - hx}{h} = z$$

Topic	Procedure	Examples
Solving inequalities, p. 176	**1.** Follow the steps for solving an equation up until the division step. **2.** If you divide both sides of the inequality by a *positive number,* the direction of the inequality is not reversed. **3.** If you divide both sides of the inequality by a *negative number,* the direction of the inequality is reversed.	Solve for *x* and graph your solution. $$\frac{1}{2}(3x - 2) \leq -5 + 5x - 3$$ First remove parentheses and simplify. $$\frac{3}{2}x - 1 \leq -8 + 5x$$ Now multiply each term by 2. $$2\left(\frac{3}{2}x\right) - 2(1) \leq 2(-8) + 2(5x)$$ $$3x - 2 \leq -16 + 10x$$ $$3x - 10x - 2 \leq -16 + 10x - 10x$$ $$-7x - 2 \leq -16$$ $$-7x - 2 + 2 \leq -16 + 2$$ $$-7x \leq -14$$ When we divide both sides by a negative number, the inequality is reversed. $$\frac{-7x}{-7} \geq \frac{-14}{-7}$$ $$x \geq 2$$ Graphical solution:

Chapter 2 Review Problems

Sections 2.1–2.3

Solve for the variable. Noninteger answers may be left in fractional form or decimal form.

1. $5x = -35$

2. $x - 19 = -22$

3. $6 - 18x = 4 - 17x$

4. $18 - 10x = 63 + 5x$

5. $6x - 2(x + 3) = 5$

6. $1 - 2(6 - x) = 3x + 2$

7. $x - (0.5x + 2.6) = 17.6$

8. $-0.2(x + 1) = 0.3(x + 11)$

9. $3(x - 2) = -4(5 + x)$

10. $\frac{2}{3}x = -18$

11. $\frac{3}{4}x = 15$

12. $4(2x + 3) = 5(x - 3)$

13. $3(x - 3) = 13x + 21$

14. $0.9x + 1.0 = 0.3x + 0.4$

15. $2.4 - 0.3x = 0.4(x - 1)$

16. $12 - x + 2 = 3x - 10 + 4x$

17. $36 = 9x - (3x - 18)$

18. $12 - 5x = -7x - 2$

19. $2(3 - x) = 1 - (x - 2)$

20. $4(x + 5) - 7 = 2(x + 3)$

21. $0.9y + 3 = 0.4y + 1.5$

22. $7y - 3.4 = 11.3$

23. $3 = 2x + 5 - 3(x - 1)$

24. $2(5x - 1) - 7 = 3(x - 1) + 5 - 4x$

Section 2.4

Solve for the variable. Noninteger answers may be left in fractional form or decimal form.

25. $\dfrac{3}{4}x - 3 = \dfrac{1}{2}x + 2$

26. $1 = \dfrac{5x}{6} + \dfrac{2x}{3}$

27. $\dfrac{7x}{5} = 5 + \dfrac{2x}{5}$

28. $\dfrac{7x - 3}{2} - 4 = \dfrac{5x + 1}{3}$

29. $\dfrac{3x - 2}{2} + \dfrac{x}{4} = 2 + x$

30. $\dfrac{-3}{2}(x + 5) = 1 - x$

31. $\dfrac{-4}{3}(2x + 1) = -x - 2$

32. $\dfrac{1}{3}(x - 2) = \dfrac{x}{4} + 2$

33. $\dfrac{1}{5}(x - 3) = 2 - \dfrac{x}{2}$

34. $\dfrac{4}{5} + \dfrac{1}{2}x = \dfrac{1}{5}x + \dfrac{1}{2}$

35. $2x - \dfrac{3}{4} + \dfrac{7}{2}x = \dfrac{1}{2}x + \dfrac{1}{4}$

36. $\dfrac{3}{2}x - \dfrac{5}{6} + x = \dfrac{1}{2}x + \dfrac{2}{3}$

37. $-\dfrac{8}{3}x - 8 + 2x - 5 = -\dfrac{5}{3}$

38. $3.5(2x + 3) = 2.4x - 1$

39. $\dfrac{7}{12}(x - 3) = \dfrac{1}{3}x + 4$

40. $\dfrac{1}{6} + \dfrac{1}{3}(x - 3) = \dfrac{1}{2}(x + 9)$

41. $\dfrac{1}{7}(x + 5) - \dfrac{6}{14} = \dfrac{1}{2}(x + 3)$

42. $\dfrac{1}{6}(8x + 3) = \dfrac{1}{2}(2x + 7)$

43. $-\dfrac{2}{5}(3x + 1) = \dfrac{1}{3}(2 - x)$

44. $\dfrac{7}{9}x + \dfrac{2}{3} = 5 + \dfrac{1}{3}x$

Section 2.5

Solve for the variable indicated.

45. Solve for y. $3x - y = 10$

46. Solve for y. $5x + 2y + 7 = 0$

47. Solve for r. $A = P(1 + rt)$

48. Solve for h. $A = 4\pi r^2 + 2\pi rh$

49. Solve for p. $H = \dfrac{1}{3}(a + 2p + 3)$

50. Solve for y. $ax + by = c$

51. **(a)** Solve for T. $C = \dfrac{WRT}{1000}$.

(b) Use your result to find T if $C = 0.36$, $W = 30$, and $R = 0.002$.

52. **(a)** Solve for y. $5x - 3y = 12$

(b) Use your result to find y if $x = 9$.

53. (a) Solve for h: $V = lwh$.

(b) Use your result to find h when $V = 48$, $l = 2$, and $w = 4$.

Sections 2.6–2.7

Solve each inequality and graph the result.

54. $7 - 2x \geq 4x - 5$

55. $2 - 3x \leq -5 + 4x$

56. $2x - 3 + x > 5(x + 1)$

57. $-x + 4 < 3x + 16$

58. $8 - \dfrac{1}{3}x \leq x$

59. $7 - \dfrac{3}{5}x > 4$

60. $-4x - 14 < 4 - 2(3x - 1)$

61. $3(x - 2) + 8 < 7x + 14$

62. $\dfrac{1}{2}(2x + 3) > 10$

63. $5(1 - x) < 3(x - 1) - 2(3 - x)$

Use an inequality to solve.

64. *Wages* Julian earns \$15 per hour as a plasterer's assistant. His employer determines that the current job allows him to pay \$480 in wages to Julian. What are the maximum number of hours that Julian can work on this job? (*Hint:* Use $15h \leq 480$.)

65. *Hiring a Temp* The cost of hiring a temporary secretary for a day is \$85. Let $n =$ the number of temporary secretaries. Set up an inequality to determine how many times a temporary secretary may be hired if the company budget for temporary secretaries is \$1445 per month. What is the maximum number of days a temporary secretary may be hired during the month? (*Hint:* Use $85n \leq 1445$.)

Mixed Practice

Solve for the variable. Noninteger answers may be left in fractional form or decimal form.

66. $8(3x + 5) - 10 = 9(x - 2) + 13$

67. $8 - 3x + 5 = 13 + 4x + 2$

68. $-2(x - 3) = -4x + 3(3x + 2)$

69. $\dfrac{1}{2} + \dfrac{5}{4}x = \dfrac{2}{5}x - \dfrac{1}{10} + 4$

70. $\dfrac{1}{6}x - \dfrac{2}{3} = \dfrac{1}{3}(x - 4)$

71. $\dfrac{1}{2}(x - 3) = \dfrac{1}{4}(3x - 1)$

Solve for the variable indicated.

72. Solve for d. $H + 2d = 6c - 3d$

73. Solve for b. $H = \dfrac{3c + 2b}{4}$

Solve each inequality and graph the result.

74. $5 - \dfrac{1}{2}x > 4$

75. $2(x - 1) \geq 3(2 + x)$

76. $\dfrac{1}{3}(x + 2) \leq \dfrac{1}{2}(3x - 5)$

77. $4(2 - x) - (-5x + 1) \geq -8$

Remember to use your Chapter Test Prep Video CD to see the worked-out solutions to the test problems you want to review.

Solve for the variable. Noninteger answers may be left in fractional form or decimal form.

1. $3x + 5.6 = 11.6$

2. $9x - 8 = -6x - 3$

3. $2(2y - 3) = 4(2y + 2)$

4. $\frac{1}{7}y + 3 = \frac{1}{2}y$

5. $4(7 - 4x) = 3(6 - 2x)$

6. $0.8x + 0.18 - 0.4x = 0.3(x + 0.2)$

7. $\frac{2y}{3} + \frac{1}{5} - \frac{3y}{5} + \frac{1}{3} = 1$

8. $3 - 2y = 2(3y - 2) - 5y$

9. $5(20 - x) + 10x = 165$

10. $5(x + 40) - 6x = 9x$

11. $-2(2 - 3x) = 76 - 2x$

12. $20 - (2x + 6) = 5(2 - x) + 2x$

In questions 13–17, solve for x.

13. $2x - 3 = 12 - 6x + 3(2x + 3)$

14. $\frac{1}{3}x - \frac{3}{4}x = \frac{1}{12}$

15. $\frac{3}{5}x + \frac{7}{10} = \frac{1}{3}x + \frac{3}{2}$

16. $\frac{15x - 2}{28} = \frac{5x - 3}{7}$

17. $\frac{2}{3}(x + 8) + \frac{3}{5} = \frac{1}{5}(11 - 6x)$

18. Solve for w. $A = 3w + 2P$

19. Solve for w. $\frac{2w}{3} = 4 - \frac{1}{2}(x + 6)$

20. Solve for a. $A = \frac{1}{2}h(a + b)$

21. Solve for y. $5ax(2 - y) = 3axy + 5$

22. Solve for B; $V = \frac{1}{3}Bh$

▲ **23.** Use your result from question 22 to find the area of the base (B) of a pyramid if the volume (V) is 140 cubic inches and the height (h) is 14 inches.

Solve and graph the inequality.

24. $3(x - 2) \geq 5x$

+----+----+----+----+----+----+ $\rightarrow x$
-5 -4 -3 -2 -1 0

25. $2 - 7(x + 1) - 5(x + 2) < 0$

+----+----+----+----+----+----+ $\rightarrow x$
$-\frac{3}{2}$ $-\frac{5}{4}$ -1 $-\frac{3}{4}$ $-\frac{1}{2}$ $-\frac{1}{4}$

26. $5 + 8x - 4 < 2x + 13$

+----+----+----+----+----+----+ $\rightarrow x$
0 1 2 3 4 5

27. $\frac{1}{4}x + \frac{1}{16} \leq \frac{1}{8}(7x - 2)$

+----+----+----+----+----+----+ $\rightarrow x$
-1 $-\frac{1}{2}$ 0 $\frac{1}{2}$ 1 $\frac{3}{2}$

1. _____

2. _____

3. _____

4. _____

5. _____

6. _____

7. _____

8. _____

9. _____

10. _____

11. _____

12. _____

13. _____

14. _____

15. _____

16. _____

17. _____

18. _____

19. _____

20. _____

21. _____

22. _____

23. _____

24. _____

25. _____

26. _____

27. _____

1. _____

2. _____

3. _____

4. _____

5. _____

6. _____

7. _____

8. _____

9. _____

10. _____

11. _____

12. _____

13. _____

14. _____

15. _____

16. _____

17. _____

18. _____

19. _____

20. _____

21. _____

22. _____

23. _____

Approximately one-half of this test covers the content of Chapters 0 and 1. The remainder covers the content of Chapter 2. In questions 1–4, simplify.

1. $\dfrac{6}{7} - \dfrac{2}{3}$

2. $1\dfrac{3}{4} + 2\dfrac{1}{5}$

3. $3\dfrac{1}{5} \div 1\dfrac{1}{2}$

4. $(1.23)(0.56)$

5. Divide. $0.144 \div 1.2$

6. What is 4.3% of 830?

7. Multiply. $(-3)(-5)(-1)(2)(-1)$

8. Collect like terms. $5ab - 7ab^2 - 3ab - 12ab^2 + 10ab - 9ab^2$

9. Simplify. $(5x)^2$

10. Simplify: $2\{3x - 4[5 - 3y(2 - x)]\}$

11. Solve for x. $4(7 - 2x) = 3x - 12$

12. Solve for x. $\dfrac{1}{3}(x + 5) = 2x - 5$

13. Solve for y. $\dfrac{2y}{3} - \dfrac{1}{4} = \dfrac{1}{6} + \dfrac{y}{4}$

14. Solve for a: $5a - 8b = 6 + 2a$

15. Solve for b. $H = \dfrac{2}{3}(b + 4a)$

16. Solve for t. $I = Prt$

17. Solve for a. $A = \dfrac{ha}{2} + \dfrac{hb}{2}$

In questions 18–22, solve and graph the inequality.

18. $-6x - 3 < 2x - 10x + 7$

19. $\dfrac{1}{2}(x - 5) \geq x - 4$

20. $4(2 - x) > 1 - 5x - 8$

21. $x + \dfrac{5}{9} \leq \dfrac{1}{3} + \dfrac{7}{9}x$

22. $7x - 13 \leq 3(4 - 5x) - 3$

23. The football team will not let Chuck play unless he passes biology with a C (70 or better) average. There are five tests in the semester, and he has failed (0) the first one. However, he found a tutor and received an 82, an 89, and an 87 on the next three tests. Solve the inequality $\dfrac{0 + 82 + 89 + 87 + x}{5} \geq 70$ to find what his minimum score must be on the last test in order to pass the course and play football.

3

Wild herds of caribou in Alaska are animals that are free to roam in one of the great wilderness areas of the world. Yet there is danger for these herds. Wildlife managers carefully measure the size of the herd and watch for population growth or losses. Predicting the size of the herd of caribou using a mathematical equation is highly desirable. Can you use your mathematical ability to create such an equation? Turn to the Putting Your Skills to Work problems on page 241 to find out.

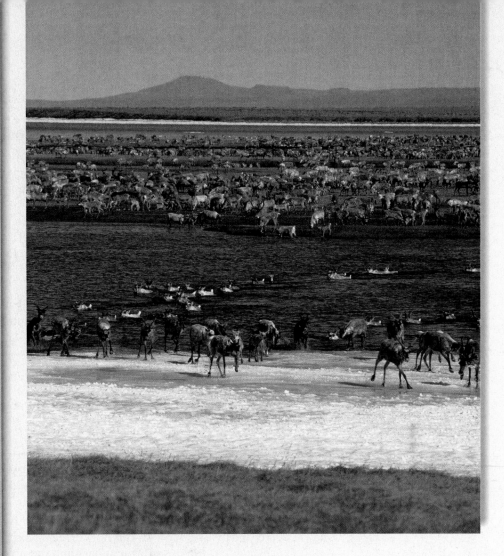

Solving Applied Problems

Student Learning Objectives

After studying this section, you will be able to:

1 Translate English phrases into algebraic expressions.

2 Write an algebraic expression to compare two or more quantities.

1 Translating English Phrases into Algebraic Expressions

One of the most useful applications of algebra is solving word problems. One of the first steps in solving word problems is translating the conditions of the problem into algebra. In this section we show you how to translate common English phrases into algebraic symbols. This process is similar to translating between languages like Spanish and French.

Several English phrases describe the operation of addition. If we represent an unknown number by the variable x, all of the following phrases can be translated into algebra as $x + 3$.

English Phrases Describing Addition	Algebraic Expression	Diagram
Three *more than* a number		
The *sum of* a number and three		
A number *increased by* three	$x + 3$	
Three is *added to* a number.		
Three *greater than* a number		
A number *plus* three		

In a similar way we can use algebra to express English phrases that describe the operations of subtraction, multiplication, and division.

CAUTION Since subtraction is not commutative, the order is essential. A number decreased by five is $x - 5$. It is not correct to say $5 - x$. Use extra care as you study each example. Make sure you understand the proper order.

English Phrases Describing Subtraction	Algebraic Expression	Diagram
A number *decreased by* four		
Four *less than* a number		
Four is *subtracted from* a number.		
Four *smaller than* a number	$x - 4$	
Four *fewer than* a number		
A number *diminished by* four		
A number *minus* 4		
The *difference between* a number and four		

English Phrases Describing Multiplication	Algebraic Expression	Diagram
Double a number		
Twice a number		
The *product* of two and a number	$2x$	
Two *of* a number		
Two *times* a number		

Since division is not commutative, the order is essential. A number divided by 3 is $\dfrac{x}{3}$. It is not correct to say $\dfrac{3}{x}$. Use extra care as you study each example.

English Phrases Describing Division	Algebraic Expression	Diagram

A number *divided by* five
One-*fifth* of a number
The quotient of a number and five

$\dfrac{x}{5}$

Often other words are used in English instead of the word *number*. We can use a variable, such as *x*, here also.

EXAMPLE 1

English Phrase	Algebraic Expression
(a) A *quantity* is increased by five.	$x + 5$
(b) Double the *value*	$2x$
(c) One-third of the *weight*	$\dfrac{x}{3}$ or $\dfrac{1}{3}x$
(d) Twelve *more than* a number	$x + 12$
(e) Seven *less than* a number	$x - 7$

Note that the algebraic expression for "seven less than a number" does not follow the order of the words in the English phrase. The variable or expression that follows the words *less than* always comes first.

seven less than x

$x - 7$

The variable or expression that follows the words *more than* technically comes before the plus sign. However, since addition is commutative, it also can be written after the plus sign.

NOTE TO STUDENT: Fully worked-out solutions to all of the Practice Problems can be found at the back of the text starting at page SP-1

Practice Problem 1 Write each English phrase as an algebraic expression.

(a) Four more than a number   **(b)** Triple a value

(c) Eight less than a number   **(d)** One-fourth of a height

More than one operation can be described in an English phrase. Sometimes parentheses must be used to make clear which operation is done first.

EXAMPLE 2

English Phrase	Algebraic Expression
(a) Seven more than double a number	$2x + 7$ *Note that these are **not** the same.*
(b) The value of the number is increased by seven and then doubled.	$2(x + 7)$ *Note that the word **then** tells us to add x and 7 before doubling.*
(c) One-half of the sum of a number and 3	$\dfrac{1}{2}(x + 3)$

Practice Problem 2   Write each English phrase as an algebraic expression.

(a) Eight more than triple a number

(b) A number is increased by eight and then it is tripled.

(c) One-third of the sum of a number and 4

② Writing an Algebraic Expression to Compare Two or More Quantities

Often in a word problem two or more quantities are described in terms of another. We will want to use a variable to represent one quantity and then write an algebraic expression using *the same variable* to represent the other quantity. Which quantity should we let the variable represent? We usually let the variable represent the quantity that is the basis of comparison: the quantity that the others are being *compared to*.

EXAMPLE 3   Use a variable and an algebraic expression to describe the two quantities in the English sentence "Mike's salary is $2000 more than Fred's salary."

Solution   The two quantities that are being compared are Mike's salary and Fred's salary. Since Mike's salary is being *compared to* Fred's salary, we let the variable represent Fred's salary. The choice of the letter *f* helps us to remember that the variable represents Fred's salary.

$$\text{Let } f = \text{Fred's salary.}$$

Then $f + \$2000 = $ Mike's salary.   *Since Mike's salary is $2000 more than Fred's.*

Practice Problem 3   Use a variable and an algebraic expression to describe the two quantities in the English sentence "Marie works 17 hours per week less than Ann."

NOTE TO STUDENT: Fully worked-out solutions to all of the Practice Problems can be found at the back of the text starting at page SP-1

EXAMPLE 4   The length of a rectangle is 3 meters shorter than twice the width. Use a variable and an algebraic expression to describe the length and the width. Draw a picture of the rectangle and label the length and width.

Solution   The length of the rectangle is being *compared to* the width. Use the letter w for width.

$$\text{Let } w = \text{the width.}$$

$$\underbrace{\text{3 meters shorter than twice the width}}$$

Then $2w - 3 = $ the length.

A picture of the rectangle is shown.

Practice Problem 4   The length of a rectangle is 5 meters longer than double the width. Use a variable and an algebraic expression to describe the length and the width. Draw a picture of the rectangle and label the length and width.

EXAMPLE 5   The first angle of a triangle is triple the second angle. The third angle of a triangle is 12° more than the second angle. Describe each angle algebraically. Draw a diagram of the triangle and label its parts.

Solution   Since the first and third angles are described in terms of the second angle, we let the variable represent the number of degrees in the second angle.

Let s = the number of degrees in
        the second angle.

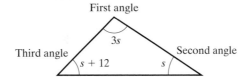

Then $3s$ = the number of degrees in
         the first angle.

And $s + 12$ = the number of degrees in
            the third angle.

Practice Problem 5   The first angle of a triangle is 16° less than the second angle. The third angle is double the second angle. Describe each angle algebraically. Draw a diagram of the triangle and label its parts.

Some comparisons will involve fractions.

EXAMPLE 6   A theater manager was examining the records of attendance for last year. The number of people attending the theater in January was one-half of the number of people attending the theater in February. The number of people attending the theater in March was three-fifths of the number of people attending the theater in February. Use algebra to describe the attendance each month.

Solution   What are we looking for? The *number of people* who attended the theater *each month*. The basis of comparison is February. That is where we begin.

Let f = the number of people who attended in February.

Then $\dfrac{1}{2}f$ = the number of people who attended in January.

And $\dfrac{3}{5}f$ = the number of people who attended in March.

Practice Problem 6   The college dean noticed that in the spring the number of students on campus was two-thirds of the number of students on campus in the fall. She also noticed that in the summer the number of students on campus was one-fifth the number of students on campus in the fall. Use algebra to describe the number of students on campus in each of these three time periods.

Verbal and Writing Skills

Write an algebraic expression for each quantity. Let x represent the unknown value.

1. a quantity increased by 5

2. nine greater than a number

3. six fewer than a quantity

4. a value decreased by seven

5. one-eighth of a quantity

6. one-half of a quantity

7. twice a quantity

8. triple a number

9. three more than half of a number

10. five more than one-third of a number

11. double a quantity increased by nine

12. six times a number increased by eight

13. one-third of the sum of a number and seven

14. one-fourth of the sum of a number and 5

15. one-third of a number reduced by twice the same number

16. one-fifth of a number reduced by double the same number

17. seven less than triple a number

18. four less than seven times a number

Write an algebraic expression for each of the quantities being compared.

19. *Stock Value* The value of a share of IBM stock on that day was $74.50 more than the value of a share of AT&T stock.

20. *Investments* The annual income from Dr. Smith's mutual fund was $833 less than the annual income from her retirement fund.

▲ **21.** *Geometry* The length of the rectangle is 7 inches more than double the width.

▲ **22.** *Geometry* The length of the rectangle is 3 meters more than triple the width.

23. *Cookie Sales* The number of boxes of cookies sold by Sarah was 43 fewer than the number of boxes of cookies sold by Keiko. The number of boxes of cookies sold by Imelda was 53 more than the number sold by Keiko.

24. *Fish Catch* The number of pounds of fish caught by Captain Jack was 813 pounds more than the amount of fish caught by Captain Sally. The amount of fish caught by Captain Ben was 623 pounds less than the amount of fish caught by Captain Sally.

▲ **25.** *Geometry* The first angle of a triangle is 16 degrees less than the second angle. The third angle of a triangle is double the second angle.

▲ **26.** *Geometry* The first angle of a triangle is 19 degrees more than the third angle. The second angle is triple the third angle.

27. *Exports* The value of the exports of Japan was twice the value of the exports of Canada.

28. *Olympic Medals* Mark Spitz won two more Olympic medals than Carl Lewis did.

▲ **29.** *Geometry* The first angle of a triangle is triple the second angle. The third angle of a triangle is 14 degrees less than the second angle.

30. *Book Cost* The cost of Hiro's biology book was $13 more than the cost of his history book. The cost of his English book was $27 less than the cost of his history book.

Applications

Use algebra to describe the situation.

31. *Land Area* Kentucky has about half the land area of Minnesota. The land area of Maine is approximately two-fifths the land area of Minnesota. Describe the land area of each of these three states.

32. *Middle School Population* A census of El Cerrito Middle School found that the number of seventh graders was fifty more than the number of eighth graders. The number of sixth graders was three-fourths the number of eighth graders. Describe the population of each grade.

33. *Archery* In an archery tournament, the number of points awarded for an arrow in the gold circle (bull's eye) is six less than triple the points awarded for an arrow in the blue ring. Write an expression for each of these scores in an archery tournament.

34. *Orbital Times* The orbital time of Pluto is 82 years less than double the orbital time of Neptune.

To Think About

Kayak Rentals *The following bar graph depicts the number of people renting sea kayaks at Essex Boat Rental during July 2004. Use the bar graph to answer exercises 35 and 36.*

35. Write an expression for the number of men who rent Sea Kayaks at Essex Boat Rental in each age category. Start by using x for the number of men aged 16 to 24 who rented kayaks.

36. Write an expression for the number of women who rent Sea Kayaks at Essex Boat Rental. Start by using y for the number of women aged 35 to 44 who rented kyacks.

Cumulative Review

Solve for the variable.

37. $x + \frac{1}{2}(x - 3) = 9$

38. $\frac{3}{5}x - 3(x - 1) = 9$

39. $5(x - 8) = 13 + x - 5$

40. $6(w - 1) - 3(2 + w) = 9$

3.2 USING EQUATIONS TO SOLVE WORD PROBLEMS

Student Learning Objectives

After studying this section, you will be able to:

1 Solve number problems.

2 Use the Mathematics Blueprint to solve applied word problems.

3 Use formulas to solve word problems.

In Section 0.7 we introduced a simple three-step procedure to solve applied problems. You have had an opportunity to use that approach to solve word problems in Exercises 0.7 and in the Cumulative Review sections in Chapters 1 and 2. Now we are going to focus our attention on solving applied problems that require the use of variables, translating English phrases into algebraic expressions, and setting up equations. The process is a little more involved. Some students find the following outline a helpful way to keep organized while solving such problems.

1. *Understand the problem.*
   (a) Read the word problem carefully to get an overview.
   (b) Determine what information you will need to solve the problem.
   (c) Draw a sketch. Label it with the known information. Determine what needs to be found.
   (d) Choose a variable to represent one unknown quantity.
   (e) If necessary, represent other unknown quantities in terms of that very same variable.

2. *Write an equation.*
   (a) Look for key words to help you to translate the words into algebraic symbols and expressions.
   (b) Use a given relationship in the problem or an appropriate formula to write an equation.

3. *Solve and state the answer.*

4. *Check.*
   (a) Check the solution in the original equation. Is the answer reasonable?
   (b) Be sure the solution to the equation answers the question in the word problem. You may need to do some additional calculations if it does not.

1 Solving Number Problems

EXAMPLE 1 Two-thirds of a number is eighty-four. What is the number?

Solution

1. **Understand the problem.** Draw a sketch.
   Let x = the unknown number.

2. **Write an equation.**

$$\underbrace{\text{Two-thirds of a number}}_{\dfrac{2}{3}x} \quad \underset{=}{\text{is}} \quad \underset{84}{\text{eighty-four.}}$$

3. Solve and state the answer.

$$\frac{2}{3}x = 84$$

$$3\left(\frac{2}{3}x\right) = 3(84) \quad \text{Multiply both sides of the equation by 3.}$$

$$2x = 252$$

$$\frac{2x}{2} = \frac{252}{2} \quad \text{Divide both sides by 2.}$$

$$x = 126$$

The number is 126.

4. Check. Is two-thirds of 126 eighty-four?

$$\frac{2}{3}(126) \stackrel{?}{=} 84$$

$$84 = 84 \quad ✓$$

Practice Problem 1   Three-fourths of a number is negative eighty-one. What is the number?

NOTE TO STUDENT: Fully worked-out solutions to all of the Practice Problems can be found at the back of the text starting at page SP-1

Learning to solve problems like Examples 1 and 2 is a very useful skill. You will find learning the material in Chapter 3 will be much easier if you can master the procedure used in these two examples.

EXAMPLE 2   Five more than six times a quantity is three hundred five. Find the number.

Solution

1. Understand the problem. Read the problem carefully. You may not need to draw a sketch.
Let x = the unknown quantity.

2. Write an equation.

Five more than	six times a number	is	three hundred five.
5 +	6x	=	305

3. Solve and state the answer. You may want to rewrite the equation to make it easier to solve.

$$6x + 5 = 305$$

$$6x + 5 - 5 = 305 - 5 \quad \text{Subtract 5 from both sides.}$$

$$6x = 300$$

$$\frac{6x}{6} = \frac{300}{6} \quad \text{Divide both sides by 6.}$$

$$x = 50$$

The number is 50.

4. Check. Is five more than six times 50 three hundred five?

$$6(50) + 5 \stackrel{?}{=} 305$$

$$300 + 5 \stackrel{?}{=} 305$$

$$305 = 305 \quad ✓$$

Practice Problem 2   Two less than triple a number is forty-nine. Find the number.

EXAMPLE 3 The larger of two numbers is three more than twice the smaller. The sum of the numbers is thirty-nine. Find each number.

Solution

1. *Understand the problem.* Read the problem carefully. The problem refers to *two* numbers. We must write an algebraic expression for *each number* before writing the equation. The larger number is being compared to the smaller number. We want to use *one variable* to describe each number.

$$\text{Let } s = \text{the smaller number.}$$
$$\text{Then } 2s + 3 = \text{the larger number.}$$

three more than twice the smaller number

2. *Write an equation.*   The sum of the numbers   is   thirty-nine.
$$s + (2s + 3) \qquad = \qquad 39$$

3. *Solve.*
$$s + (2s + 3) = 39$$
$$3s + 3 = 39 \quad \text{Collect like terms.}$$
$$3s = 36 \quad \text{Subtract 3 from each side.}$$
$$s = 12 \quad \text{Divide both sides by 3.}$$

4. *Check.*
$$12 + [2(12) + 3] \stackrel{?}{=} 39$$
$$39 = 39 \checkmark$$

The solution checks, but have we solved the word problem? We need to find *each* number. 12 is the smaller number. Substitute 12 into the expression $2s + 3$ to find the larger number.

$$2s + 3 = 2(12) + 3 = 27$$

The smaller number is 12. The larger number is 27.

Practice Problem 3 Consider two numbers. The second number is twelve less than triple the first number. The sum of the two numbers is twenty-four. Find each number.

NOTE TO STUDENT: Fully worked-out solutions to all of the Practice Problems can be found at the back of the text starting at page SP-1

② Using the Mathematics Blueprint to Solve Applied Word Problems

To facilitate understanding more involved word problems we will use a Mathematics Blueprint similar to the one we used in Section 0.7. This format is a simple way to organize facts, determine what to set variables equal to, and select a method or approach that will assist you in finding the desired quantity. You will find using this form helpful, particularly in those cases when you read through a word problem and mentally say to yourself, "Now where do I begin?" You begin by responding to the headings of the blueprint. Soon a procedure for solving the problem will emerge.

Mathematics Blueprint for Problem Solving

Gather the Facts	Assign the Variable	Basic Formula or Equation	Key Points to Remember

EXAMPLE 4   The mean annual snowfall in Juneau, Alaska, is 105.8 inches. This is 20.2 inches less than three times the annual snowfall in Boston. What is the annual snowfall in Boston?

Solution

Understand the problem and write an equation.

Mathematics Blueprint for Problem Solving

Gather the Facts	Assign the Variable	Basic Formula or Equation	Key Points to Remember
Snowfall in Juneau is 105.8 inches. This is 20.2 inches less than three times the snowfall in Boston.	We do not know the snowfall in Boston. Let b = annual snowfall in Boston. Then $3b - 20.2$ = annual snowfall in Juneau.	Set $3b - 20.2$ equal to 105.8, which is the snowfall in Juneau.	All measurements of snowfall are recorded in inches.

Juneau's snowfall is 20.2 less than three times Boston's snowfall.

$$105.8 = 3b - 20.2$$

Solve and state the answer. You may want to rewrite the equation to make it easier to solve.

$$3b - 20.2 = 105.8$$
$$3b = 126 \qquad \text{Add 20.2 to both sides.}$$
$$b = 42 \qquad \text{Divide both sides by 3.}$$

The annual snowfall in Boston is 42 inches.

Check. Reread the word problem. Work backward.

Three times 42 is 126.

126 less 20.2 is 105.8.

Is this the annual snowfall in Juneau? Yes. ✓

Practice Problem 4   The maximum recorded rainfall for a 24-hour period in the United States occurred in Alvin, Texas, on July 25–26, 1979. The maximum recorded rainfall for a 24-hour period in Canada occurred in Ucluelet Brynnor Mines, British Columbia, on October 6, 1977. Alvin, Texas received 43 inches of rain in that period. The amount recorded in Texas was 14 inches less than three times the amount recorded in Canada. How much rainfall was recorded for Ucluelet Brynnor Mines in Canada? (*Source:* National Oceanic and Atmospheric Administration.)

Some word problems require a simple translation of the facts. Others require a little more detective work. You will not always need to use the Mathematics Blueprint to solve every word problem. As you gain confidence in problem solving, you will no doubt leave out some of the steps. We suggest that you use the procedure when you find yourself on unfamiliar ground. It is a powerful organizational tool.

Using Formulas to Solve Word Problems

Sometimes the relationship between two quantities is so well understood that we have developed a formula to describe that relationship. We have already done some work with formulas in Section 2.5. The following examples show how you can use a formula to solve a word problem.

EXAMPLE 5 Two people travel in separate cars. They each travel a distance of 330 miles on an interstate highway. To maximize fuel economy, Fred travels at exactly 50 mph. Sam travels at exactly 55 mph. How much time did the trip take each person? (Use the formula distance = rate · time or $d = rt$.)

Sam's speed is 55 mph   Fred's speed is 50 mph   330 miles

Solution

Mathematics Blueprint for Problem Solving

Gather the Facts	Assign the Variable	Basic Formula or Equation	Key Points to Remember
Each person drives 330 miles. Fred drives at 50 mph. Sam drives at 55 mph.	Time is the unknown quantity for each driver. Use subscripts to denote different values of t. t_f = Fred's time t_s = Sam's time	distance = (rate)(time) or $d = rt$	The time is expressed in hours.

Substitute the known values into the formula and solve for t.

$$d = rt \qquad\qquad d = rt$$
$$330 = 50t_f \qquad 330 = 55t_s$$
$$6.6 = t_f \qquad\qquad 6 = t_s$$

It took Fred 6.6 hours to drive 330 miles.   It took Sam 6 hours to drive 330 miles.

Check. Is this reasonable? Yes, you would expect Fred to take longer to drive the same distance because Fred is driving at a lower rate of speed.

Note: You may wish to express 6.6 hours in hours and minutes. To change 0.6 hours to minutes, proceed as follows:

$$0.6 \text{ hour} \cdot \frac{60 \text{ minutes}}{1 \text{ hour}} = (0.6)(60) \text{ minutes} = 36 \text{ minutes}$$

Thus, Fred drove for 6 hours and 36 minutes.

NOTE TO STUDENT: Fully worked-out solutions to all of the Practice Problems can be found at the back of the text starting at page SP-1

Practice Problem 5 Sarah left the city to visit her aunt and uncle, who live in a rural area north of the city. She traveled the 220-mile trip in 4 hours. On her way home she took a slightly longer route, which measured 225 miles on the car odometer. The return trip took 4.5 hours.

(a) What was her average speed on the trip leaving the city?

(b) What was her average speed on the return trip?

(c) On which trip did she travel faster and by how much?

EXAMPLE 6   A teacher told Melinda that she had a course average of 78 based on her six math tests. When she got home, Melinda found five of her tests. She had scores of 87, 63, 79, 71, and 96 on the five tests. She could not find her sixth test. What score did she obtain on that test? (Use the formula that an average = the sum of scores ÷ the number of scores.)

Solution

Mathematics Blueprint for Problem Solving

Gather the Facts	Assign the Variable	Basic Formula or Equation	Key Points to Remember
Her five known test scores are 87, 63, 79, 71, and 96. Her course average is 78.	We do not know the score Melinda received on her sixth test. Let x = the score on the sixth test.	average = $\dfrac{\text{sum of scores}}{\text{number of scores}}$	Since there are six test scores, we will need to divide the sum by 6.

When you average anything, you total up the sum of all the values and then divide it by the number of values.

We now write the equation for the average of six items. This involves adding all the tests and dividing by 6.

$$\frac{87 + 63 + 79 + 71 + 96 + x}{6} = 78$$

$$\frac{396 + x}{6} = 78 \qquad \text{Add the numbers in the numerator.}$$

$$6\left(\frac{396 + x}{6}\right) = 6(78) \qquad \begin{array}{l}\text{Multiply both sides of the equation by} \\ \text{6 to remove the fraction.}\end{array}$$

$$396 + x = 468 \qquad \text{Simplify.}$$

$$x = 72 \qquad \text{Subtract 396 from both sides to find } x.$$

Melinda's score on the sixth test was 72.

Check. To verify that this is correct, we check that the average of the 6 tests is 78.

$$\frac{87 + 63 + 79 + 71 + 96 + 72}{6} \overset{?}{=} 78$$

$$\frac{468}{6} \overset{?}{=} 78$$

$$78 = 78 \quad \checkmark$$

The problem checks. We know that the score on the sixth test was 72.

Practice Problem 6   Barbara's math course has four tests and one final exam. The final exam counts as much as two tests. Barbara has test scores of 78, 80, 100, and 96 on her four tests. What grade does she need on the final exam if she wants to have a 90 average for the course?

Solve. Check your solution.

1. What number minus 543 gives 718?

2. What number added to 74 gives 265?

3. A number divided by eight is 296. What is the number?

4. Eighteen less than a number is 23. What is the number?

5. Seventeen greater than a number is 199. Find the number.

6. Three times a number is one. What is the number?

7. A number is doubled and then increased by seven. The result is ninety-three. What is the original number?

8. Four less than nine times a number is one hundred twenty-two. Find the original number.

9. When eighteen is reduced by two-thirds of a number, the result is 12. Find the number.

10. Twice a number is increased by one-third the same number. The result is 42. Find the number.

11. Five less than twice a number is the same as three times the number. Find the number.

12. Six less than five times a number is the same as seven times the number. Find the number.

13. A number, half of that number, and one-third of that number are added. The result is 22. What is the original number?

14. A number, twice that number, and one-third of that number are added. The result is 20. What is the original number?

Applications

Solve. Check to see if your answer is reasonable.

15. *Motorcycle Inventory* A Harley Davidson motorcycle shop maintains an inventory of four times as many new bikes as used bikes. If there are 60 new bikes, how many used bikes are now in stock?

16. *Shirt Sizes* The Enro Shirt Outlet store maintains an inventory of five times as many shirts in regular men's sizes (size XL and smaller) than it does in large men's sizes (size 2X and larger). There are 430 shirts in inventory in the store that are regular men's sizes. How many shirts in large men's sizes are in inventory?

17. *Wildfires* Between 1/1/02 and 7/27/02 there were 52,107 wildfires in the United States. This was 16,723 fewer than twice the number of wildfires between 1/1/03 and 7/27/03. Find the number of wildfires during the period 1/1/03 and 7/27/03. (*Source:* National Interagency Fire Center.)

18. *DVD Rentals* Harbor Video Rental charges $8 per day to rent a DVD player, and $2.25 per movie title. If Jeff rents the DVD player for 5 days, and he owes $53.50 to Harbor Video Rental, how many movies did he rent?

19. *CD Purchase* The sale price of a new Panasonic compact disc player is $218 at a local discount store. At the store where this sale is going on, each new CD is on sale for $11 each. If Kyle purchases a player and some CDs for $284, how many CDs did he purchase?

20. *Online Charges* Suellen subscribes to an online computer service that charges $9.50 per month for 30 hours online and $1.50 for each hour online in excess of 30 hours. Last month her bill was $20. How many extra hours was she charged for?

21. ***One-Day Sale*** Raquelle went to Weller's Department Store one-day sale. She bought two blouses for $38 each and a pair of shoes for $49. She also wanted to but some jewelry. Each item of jewelry was bargain priced at $11.50 each. If she brought $171 with her, how many pieces of jewelry could she buy?

22. ***Waiting Tables*** Brad is a waiter at the "Steaks-Are-Us" restaurant. He gets paid $5.75 per hour, and he can keep his tips. He knows that his tips average $8.80 per table. If he worked an eight-hour shift and took home $169.20, how many tables did he serve?

23. ***TGV*** On May 18, 1990 the fastest speed of any national railroad was achieved by the French high-speed train *Train à Grande Vitess* (TGV) as it traveled over a distance from Cortalain to Tours, France. A commentator said that this speed was so fast that if it continued at that rate, the train would travel 6404 miles in 20 hours. How fast did the train travel on that date? (*Source:* www.guinnessworld-records.com.)

24. ***Submarine Speed*** In 1958, the nuclear-powered submarine *Nautilus* took 6 days, 12 hours to travel submerged 5068 km across the Atlantic Ocean from Portsmouth, England, to New York City. What was its average speed, in kilometers per hour, for this trip? (Round to the nearest whole number.)

25. ***Gravity*** It has been shown that the force of gravity on a planet varies with the mass of the planet. The force of gravity on Jupiter, for example, is about two and a half times that of Earth. Using this information, approximately how much would a 220-lb astronaut weigh on Jupiter?

26. ***Sunday Comics*** Charles Schultz, the creator of *Peanuts*, once estimated that he had drawn close to 2600 Sunday comics over his career. At that time, about how many years had he been drawing *Peanuts?*

27. ***In-Line Skating*** Two in-line skaters, Nell and Kristin, start from the same point and skate in the same direction. Nell skates at 12 miles per hour and Kristin skates at 14 miles per hour. If they can keep up that pace for 2.5 hours, how far apart will they be at the end of that time?

28. ***Train Travel*** Two trains leave a train station at the same time. One train travels east at 50 miles per hour. The other train travels west at 55 miles per hour. In how many hours will the two trains be 315 miles apart?

29. ***Travel Routes*** Nella drove from Albuquerque, New Mexico, to the Garden of the Gods rock formation in Colorado Springs. It took her six hours to travel 312 miles over the mountain road. She came home on the highway. On the highway she took five hours to travel 320 miles. How fast did she travel using the mountain route? How much faster (in miles per hour) did she travel using the highway route?

30. ***Travel Speeds*** Allison drives 30 miles per hour through the city and 55 miles per hour on the New Jersey Turnpike. She drove 90 miles from Battery Park to the Jersey Shore. How much of the time was city driving if she spent 1.2 hours on the turnpike?

31. *High Temperatures* From July 13 to 21, 2003, Pueblo, Colorado had 9 consecutive days of temperatures above 100°F, with an average high temperature for that period of 104°F. Recently, Pueblo recorded eight days of high temperatures: 101°, 100°, 102.5°, 103°, 108°, 109°, 105.5°, and 102°. How high would the temperature need to be on the ninth day to tie that record from 2003? (*Source:* National Weather Service.)

32. *Grading System* Nyingwo's College Success Seminar teacher had a very complicated grading system. Each test counted once. Each project was worth two tests. The final exam was worth three tests. If Nyingwo received a 97, a 94, and an 89 on tests and a 91 and 93 on projects, what would she need to get on the final exam to receive an A for the course (A is 93 points)?

33. *Gas Mileage* The Ramirez family has three cars. The smallest car gets fuel mileage of 38 miles per gallon (mpg) in city driving. The second car gets 21 mpg in city driving. The third car is an ancient, heavy station wagon. Rita calculated that if all three cars were driven the same number of miles in city driving each year, the *average* miles-per-gallon rating of the three Ramirez cars would be $22\frac{2}{3}$ mpg in city driving. What is the miles-per-gallon rating of the old station wagon?

34. *Executive Salaries* A local hospital revealed that five officers of the hospital had an average salary of $125,000 per year. Three of the annual salaries were $50,000, $60,000, and $65,000. The annual salaries for the executive vice president and the president were not revealed. It was disclosed that the president makes double the salary of the executive vice president. Find the salaries of the president and the executive vice president.

To Think About

35. *Cricket Chirps* In warmer climates, approximate temperature predictions can be made by counting the number of chirps of a cricket during a minute. The Fahrenheit temperature decreased by forty is equivalent to one-fourth of the number of cricket chirps.
(a) Write an equation for this relationship.
(b) Approximately how many chirps per minute should be recorded if the temperature is 90°F?
(c) If a person recorded 148 cricket chirps in a minute, what would be the Fahrenheit temperature according to this formula?

Cumulative Review

Simplify.

36. $5x(2x^2 - 6x - 3)$

37. $-2a(ab - 3b + 5a)$

38. $7x - 3y - 12x - 8y + 5y$

39. $5x^2y - 7xy^2 - 8xy - 9x^2y$

40. *Apple Sales* The local Apple Factory produce market sells premium MacIntosh apples at the price of four apples for $3.60. They purchase the apples from the apple orchards for $5.40 per dozen. How many apples will they have to sell to make a profit of $1350?

41. *Anniversary Party* Jennifer is hosting a party for her grandparents' 50th wedding anniversary, to be held in the church fellowship hall. She has a budget of $1425.00 to pay a caterer. If the caterer gives her a price of $23.75 per person, how many people can she invite?

1 Solving Word Problems Involving Comparisons

Many real-life problems involve comparisons. We often compare quantities such as length, height, or income. Sometimes not all the information is known about the quantities that are being compared. You need to identify each quantity and write an algebraic expression that describes the situation in the word problem.

Student Learning Objective

After studying this section, you will be able to:

1 **Solve word problems involving comparisons.**

EXAMPLE 1   The Center City Animal Hospital treated a total of 18,360 dogs and cats last year. The hospital treated 1376 more dogs than cats. How many dogs were treated last year? How many cats were treated last year?

Solution

1. ***Understand the problem.***

   What information is given?   The combined number of dogs and cats is 18,360.
   What is being compared?   There were 1376 more dogs than cats.

   If you compare one quantity to another, usually the second quantity is represented by the variable. Since we are comparing the number of dogs to the number of cats, we start with the number of cats.

   Let c = the number of cats treated at the hospital.

   Then $c + 1376$ = the number of dogs treated at the hospital.

2. ***Write an equation.***

   The number of cats   plus   the number of dogs   is   18,360.
   $\qquad c \qquad\qquad + \qquad (c + 1376) \qquad = \qquad 18,360$

3. ***Solve and state the answer.***

$$c + c + 1376 = 18,360$$
$$2c + 1376 = 18,360 \qquad \text{Combine like terms.}$$
$$2c + 1376 - 1376 = 18,360 - 1376 \qquad \text{Subtract 1376 from both sides.}$$
$$2c = 16,984$$
$$\frac{2c}{2} = \frac{16,984}{2} \qquad \text{Divide both sides by 2.}$$
$$c = 8492 \qquad \text{The number of cats treated is 8492.}$$
$$c + 1376 = 8492 + 1376 = 9868 \qquad \text{The number of dogs treated is 9868.}$$

4. ***Check.*** The number of dogs treated plus the number of cats treated should total 18,360.

$$8492 + 9868 \stackrel{?}{=} 18,360$$
$$18,360 = 18,360 \quad ✓$$

Practice Problem 1   A deck hand on a fishing boat is working with a rope that measures 89 feet. He needs to cut it into two pieces. The long piece must be 17 feet longer than the short piece. Find the length of each piece of rope.

NOTE TO STUDENT: Fully worked-out solutions to all of the Practice Problems can be found at the back of the text starting at page SP-1

   If the word problem contains three unknown quantities, determine the basis of comparison for two of the quantities.

EXAMPLE 2   An airport filed a report showing the number of plane departures that took off from the airport during each month last year. The number of departures in March was 50 more than the number of departures in January. In July, the number of departures was 150 less than triple the number of departures in January. In those three months, the airport had 2250 departures. How many departures were recorded for each month?

Solution

1. *Understand the problem.* What is the basis of comparison?

   The number of departures in March is compared to the number in January.
   The number of departures in July is compared to the number in January.

   Express this algebraically. It may help to underline the key phrases.

   Let j = the departures in January.

   March was 50 more than January

   Then $j + 50$ = the departures in March.

   July was 150 less than triple January

   And $3j - 150$ = the departures in July.

2. *Write an equation.*

number of departures in January	number of departures in March	number of departures in July	three months' total departures
j	$+ (j + 50)$	$+ (3j - 150) =$	2250

3. *Solve and state the answer.*

   $j + (j + 50) + (3j - 150) = 2250$

   $5j - 100 = 2250$ Collect like terms.

   $5j = 2350$ Add 100 to each side.

   $j = 470$ Divide both sides by 5.

   Now, if $j = 470$, then

   $$j + 50 = 470 + 50 = 520$$

   and

   $$3j - 150 = 3(470) - 150 = 1410 - 150 = 1260.$$

   The number of departures in January was 470; the number of departures in March was 520; the number of departures in July was 1260.

4. *Check.* Do these answers seem reasonable? Yes. Do these answers agree with all the statements in the word problem?

   Is the number of departures in March 50 more than those in January?

   $$520 \overset{?}{=} 50 + 470$$
   $$520 = 520 \checkmark$$

   Is the number of departures in July 150 less than triple those in January?

   $$1260 \overset{?}{=} 3(470) - 150$$
   $$1260 \overset{?}{=} 1410 - 150$$
   $$1260 = 1260 \checkmark$$

   Is the total number of departures in the three months equal to 2250?

   $$470 + 520 + 1260 \overset{?}{=} 2250$$
   $$2250 = 2250 \checkmark$$

   Yes, all conditions are satisfied. The three answers are correct.

Practice Problem 2   A social services worker was comparing the cost in-curred by three families in heating their homes for the year. The first family had an annual heating bill that was $360 more than that of the second family. The third family had a heating bill that was $200 less than double the heating bill of the second family. The total annual heating bill for the three families was $3960. What was the annual heating bill for each family?

NOTE TO STUDENT: Fully worked-out solutions to all of the Practice Problems can be found at the back of the text starting at page SP-1

EXAMPLE 3   A small plot of land is in the shape of a rectangle. The length is 7 meters longer than the width. The perimeter of the rectangle is 86 meters. Find the dimensions of the rectangle.

Solution

1. ***Understand the problem.*** Read the problem: What information is given?

   The perimeter of a rectangle is 86 meters.
   What is being compared?
   The length is being compared to the width.

   Express this algebraically and draw a picture.

   Let w = the width.
   Then $w + 7$ = the length.

   Reread the problem: What are you being asked to do?

   Find the dimensions of the rectangle. The dimensions of the rectangle are the length and the width of the rectangle.

2. ***Write an equation.*** The perimeter is the total distance around the rectangle.

$$w + (w + 7) + w + (w + 7) = 86$$

3. ***Solve and state the answer.***

$$w + (w + 7) + w + (w + 7) = 86$$

   $4w + 14 = 86$ Combine like terms.

   $4w = 72$ Subtract 14 from both sides.

   $w = 18$ Divide both sides by 4.

   The width of the rectangle is 18 meters. What is the length?

   $w + 7$ = the length

   $18 + 7 = 25$

   The length of the rectangle is 25 meters.

4. ***Check.*** Put the actual dimensions in your drawing and add the lengths of the sides. Is the sum 86 meters? ✓

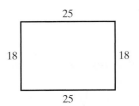

Practice Problem 3   A farmer purchased 720 meters of wire fencing to en-close a pasture. The pasture is in the shape of a triangle. The first side of the trian-gle is 30 meters less than the second side. The third side is one-half as long as the second side. Find the dimensions of the triangle.

Applications

Solve. Check to see if your answer is reasonable. Have you answered the question that was asked?

1. **Parade Banner** For their homecoming parade, the students of Wheaton College have created a colorful banner, 47 meters in length, that is made of two pieces of parachute material. The short piece is 17 meters shorter than the long piece. Find the length of each piece.

2. **Copper Wire** A copper conducting wire measures 84 centimeters in length. It is cut into two pieces. The shorter piece is 5 centimeters shorter than the long piece. Find the length of each piece.

3. **Salaries** Dave and Elsie moved to Hamilton after they got married and they both began new jobs. Elsie earned $2600 per year more than Dave. The total salary for the year for Dave and Elsie was $71,200. How much did each of them earn for the year?

4. **Auto Insurance** Dick and Barbara moved to Ipswich and purchased a home. They paid $ 1856 per year to insure their two cars. Dick said it cost $358 more per year to insure his car than it did to insure Barbara's car. How much did it cost to insure Dick's car? How much did it cost to insure Barbara's car?

5. **Volunteering** Three siblings—David, Kate, and Sarah—committed themselves to volunteer 100 hours for Habitat for Humanity to build new housing. Sarah worked 15 hours more than Dave. Kate worked 5 hours fewer than Dave. How many hours did each of these siblings volunteer?

6. **Fruit Basket** Jerome counted 24 pieces of fruit in a fruit basket. There were twice as many apples as pears. There were four fewer peaches than pears. How many of each fruit were there?

7. **Home States** Professor Fullerton was demonstrating the ease of making a bar graph using information of student's home states. He was amazed that his class of 32 represented only 3 states. There were 8 more students born in Texas than were born in Oklahoma. The number of students born in Arizona was 9 fewer than the number born in Oklahoma. How many were born in each state?

8. **Multiplex** An 800-seat multiplex is divided into three theaters. Theater One is the smallest. Theater Two has 180 more seats than Theater One. Theater Three has 60 seats fewer than twice the number in Theater One. How large is each theater?

▲ 9. **Bridge Support** The length of a rectangular piece of steel in a bridge is 3 meters less than double the width. The perimeter of the piece of steel is 42 meters. Find the length of the piece of steel. Find the width of the piece of steel.

▲ 10. **State Park** A state park in Colorado has a perimeter of 92 miles. The park is in the shape of a rectangle. The length of the park is 30 miles less than triple the width of the park. Find the length of the park. Find the width of the park.

▲ **11.** *Painting Frame* Rashau made a rectangular frame for her latest oil painting. The length is 20 centimeters more than triple the width. The perimeter of the frame is 96 centimeters. Find the length and the width of the frame.

▲ **12.** *Rug Dimensions* Theresa purchased a small rectangular rug for her dorm room. The perimeter of the rug is 1438 inches. The length of the rug is 23 inches more than double the width. Find the length and the width of the rug.

▲ **13.** *Jewelry Box* A solid-gold jewelry box was found in the underwater palace of Cleopatra just off the shore of Alexandria. The length of the rectangular box is 35 centimeters less than triple the width. The perimeter of the box is 190 centimeters. Find the length and width of Cleopatra's solid-gold jewelry box.

▲ **14.** *Giant Chocolate Bar* A giant rectangular chocolate bar was made for a special promotion. The length was 6 meters more than half the width. The perimeter of the chocolate bar was 24 meters. Find the length and width of the giant chocolate bar.

15. *Running Speed* The top running speed of a cheetah is double the top running speed of a jackal. The top running speed of an elk is 10 miles per hour faster than that of a jackal. If each of these three animals could run at top speed for an hour (which, of course, is not possible), they could run a combined distance of 150 miles. What is the top running speed of each of these three animals? (*Source:* American Museum of Natural History.)

16. *River Length* The Missouri River is 149 miles longer than double the length of the Snake River. The Potomac River is 796 miles shorter than the Snake River. The combined lengths of these three rivers is 3685 miles. What is the length of each of the three rivers? (*Source:* U.S. Department of the Interior.)

▲ **17.** *Kite Design* Mr. Shen wished to enter the Orange County Kite Competition with an innovative kite of his own design. He did tests that showed an irregular quadrilateral design was best. (See diagram.) The longest side (*A*) was twice the length of the shortest side (*C*). Side *B* was one and one-half times the shortest side (*C*). Side *D* needed to be 3" (3 inches) longer than side *C*. If the competition required that the perimeter of your kite must be 58", what were the dimensions of Mr. Shen's kite?

▲ **18.** *Pennant Design* The Tri-County schools are participating in a Renaissance Faire. The team from Appleton High School has designed a pennant to fly over its team tent. The pennant is in the shape of a triangle. The longest side is 5" (5 inches) less than twice the middle length side. The shortest side is 3" longer than half the middle length side. If the team captain bought 72" of satin edging for the banner, and there were 4" left over after the pennant was made, what are the dimensions of the pennant?

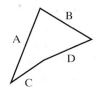

19. *Balloon Trip* A balloonist is trying to complete a nonstop trip around the world. During the course of his trip, he traveled 684 miles over land from Rockford, Illinois, to Washington, D.C. He traveled 1197 miles over water from Washington, D.C., to San Juan, Puerto Rico. The total trip was 39 hours. He traveled 3 more hours over water than he traveled over land.

(a) How long did he travel over land?

(b) How long did he travel over water?

(c) How much faster did he travel over water than over land?

20. *Waterline Breaks* In a typical year in New York City, 600 waterline breaks are reported along the 6181 miles of water pipes that twist and turn underground. Water breaks occur five times more often during the period from October to March than they do during the period from April to September.

(a) On average, how many waterline breaks would be expected during the period from October to March?

(b) On average, how many waterline breaks would be expected during the period from April to September?

(c) A water department official once complained to city hall that he expects a break in the waterline every five miles during the course of a year. Is his comment correct? Why?

To Think About

▲ **21.** *Geometry* A small square is constructed. Then a new square is made by increasing each side by 2 meters. The perimeter of the new square is 3 meters shorter than five times the length of one side of the original square. Find the dimensions of the original square.

▲ **22.** *Geometry* A rectangle is constructed. The length of this rectangle is double the width. Then a new rectangle is made by increasing each side by 3 meters. The perimeter of the new rectangle is 2 meters greater than four times the length of the old rectangle. Find the dimensions of the original rectangle.

Cumulative Review

Simplify.

23. $-4x(2x^2 - 3x + 8)$

24. $5a(ab + 6b - 2a)$

25. $-7x + 10y - 12x - 8y - 2$

26. $3x^2y - 6xy^2 + 7xy + 6x^2y$

27. *Hardware Shopping* Today there were 650 people shopping for hardware products in a home improvement store. Of the 650 shoppers, 117 were professional contractors and laborers. What percent of the shoppers were not professional contractors and laborers?

28. *Dance Lessons* A dance instructor is doing her best to teach engaged couples how to dance well before their wedding days. After giving lessons to 60 couples, she notes that only 48 couples are able to dance well without coaching. What percent of couples are not dancing very well?

How are you doing with your homework assignments in Sections 3.1 to 3.3? Do you feel you have mastered the material so far? Do you understand the concepts you have covered? Before you go further in the textbook, take some time to do each of the following problems.

3.1 *Write an algebraic expression. Use the variable x to represent the unknown number.*

1. Three times a quantity is then decreased by 40

2. One-half of a number is increased by 12.

3. First a number is multiplied by five. Then the result is divided by 12.

4. One-fourth of the sum of a number and 10.

3.2 *Translate the information into an equation. Then solve the equation in order to answer the question.*

5. A number is doubled and then increased by five. The result is −47. What is the original number?

6. A number is divided by two. Then nine is subtracted from the result. A value of 43 is obtained. What is the original number?

7. At Lucas Ford, the monthly rental on a Ford Focus is $310. The cost of a new car is $12,400. In how many months will the rental cost equal the cost of purchasing a new car?

8. Bill Tupper makes $24 per hour for a 40 hour week. He earns time and a half for every hour over 40 hours during that one week. If Bill earned $1140 last week, how many overtime hours did he work?

9. The average salary of five workers at Verizon is $50,200. The first person earns $40,000 per year. A second person earns $35,000 per year. A third person earns $13,500 more per year than the first person. The fourth person earns an unknown annual salary. The fifth person earns $2,500 per year more than the fourth person. How much is the annual salary of the fourth person? How much is the annual salary of the fifth person?

3.3

10. A rope measures 84 feet in length. It is cut into two pieces. The long piece is 18 feet longer than twice the length of the short piece. How long is each piece?

11. Andy is doing his reading for his History of World Civilizations class. He must read a total of 260 pages in the next three days. On Monday he reads 25 pages less than twice the number of pages he read on Sunday. On Tuesday he reads 15 pages less than three times the number of pages he read on Sunday. If he reached his reading goal, how many pages did he read on each day?

12. A farmer in southern New Jersey has three fields to plow. The first field takes three times as long to plow as the second field. The third field takes one fourth as long to plow as the second field. The farmer requires 510 hours to plow all three fields. How long does it take to plow each field?

Now turn to page SA-8 for the answer to each of these problems. Each answer also includes a reference to the objective in which the problem is first taught. If you missed any of these problems, you should stop and review the Examples and Practice Problems in the referenced objective. A little review now will help you master the material in the upcoming sections of the text.

1. _____

2. _____

3. _____

4. _____

5. _____

6. _____

7. _____

8. _____

9. _____

10. _____

11. _____

12. _____

3.4 SOLVING WORD PROBLEMS: THE VALUE OF MONEY AND PERCENTS

Student Learning Objectives

After studying this section, you will be able to:

1 Solve problems involving periodic rate charges.

2 Solve percent problems.

3 Solve investment problems involving simple interest.

4 Solve coin problems.

The problems we now present are frequently encountered in business. They deal with money: buying, selling, and renting items; earning and borrowing money; and the value of collections of stamps or coins. Many applications require an understanding of the use of percents and decimals. Review Sections 0.4 and 0.5 if you are weak in these skills.

1 Solving Problems Involving Periodic Rate Charges

EXAMPLE 1 A business executive rented a car. The Supreme Car Rental Agency charged $39 per day and $0.28 per mile. The executive rented the car for two days and the total rental cost was computed to be $176. How many miles did the executive drive the rented car?

Solution

1. **Understand the problem.** How do you calculate the cost of renting a car?

   total cost = per-day cost + mileage cost

   What is known?

   It cost $176 to rent the car for two days.

   What do you need to find?

   The number of miles the car was driven.

   Choose a variable:

   Let m = the number of miles driven in the rented car.

2. **Write an equation.** Use the relationship for calculating the total cost.

per-day cost	+	mileage cost	=	total cost
$(39)(2)$	+	$(0.28)m$	=	176

3. **Solve and state the answer.**

   $$(39)(2) + (0.28)(m) = 176$$

   $78 + 0.28m = 176$ Simplify the equation.

   $0.28m = 98$ Subtract 78 from both sides.

   $\dfrac{0.28m}{0.28} = \dfrac{98}{0.28}$ Divide both sides by 0.28.

   $m = 350$ Simplify.

   The executive drove 350 miles.

4. **Check.** Does this seem reasonable? If he drove the car 350 miles in two days, would it cost $176?

   (cost of $39 per day for 2 days) + (cost of $0.28 per mile for 350 miles) $\overset{?}{=}$ total cost of $176

   $(\$39)(2) + (350)(\$0.28) \overset{?}{=} \$176$

   $\$78 + \$98 \overset{?}{=} \$176$

   $\$176 = \$176$ ✓

Practice Problem 1   Alfredo wants to rent a truck to move to Florida. He has determined that the cheapest rental rates for a truck of the correct size are from a local company that will charge him $25 per day and $0.20 per mile. He has not yet completed an estimate of the mileage of the trip, but he knows that he will need the truck for three days. He has allowed $350 in his moving budget for the truck. How far can he travel for a rental cost of exactly $350?

NOTE TO STUDENT: Fully worked-out solutions to all of the Practice Problems can be found at the back of the text starting at page SP-1

② Solving Percent Problems

Many applied situations require finding a percent of an unknown number. If we want to find 23% of $400, we multiply 0.23 by 400: $0.23(400) = 92$. If we want to find 23% of an unknown number, we can express this using algebra by writing $0.23n$, where n represents the unknown number.

EXAMPLE 2   A sofa was marked with the following sign: "The price of this sofa has been reduced by 23%. You can save $138 if you buy now." What was the original price of the sofa?

Solution

1. **Understand the problem.**

    Let s = the original price of the sofa.

    Then $0.23s$ = the amount of the price reduction, which is $138.

2. **Write an equation and solve.**

    $$0.23s = 138 \quad \text{Write the equation.}$$
    $$\frac{0.23s}{0.23} = \frac{138}{0.23} \quad \text{Divide each side of the equation by 0.23.}$$
    $$s = 600 \quad \text{Simplify.}$$

    The original price of the sofa was $600.

3. **Check.** Is $600 a reasonable answer? ✓   Does 23% of $600 = $138? ✓

Practice Problem 2   John earns a commission of 38% of the cost of every set of encyclopedias that he sells. Last year he earned $4560 in commissions. What was the cost of the encyclopedias that he sold last year?

EXAMPLE 3   Hector received a pay raise this year. The raise was 6% of last year's salary. This year he will earn $15,900. What was his salary last year before the raise?

Solution

1. **Understand the problem.** What do we need to find?

    Hector's salary last year.

    What do we know?

    Hector received a 6% pay raise and now earns $15,900.

    What does this mean?

    Reword the problem: *This year's salary of $15,900 is 6% more than last year's salary.*

Choose a variable:

$$\text{Let } x = \text{Hector's salary last year.}$$
$$\text{Then } 0.06x = \text{the amount of the raise.}$$

2. Write an equation and solve.

Last year's salary	+	the amount of his raise	=	this year's salary	
x	+	$0.06x$	=	15,900	Write the equation.
$1.00x$	+	$0.06x$	=	15,900	Rewrite x as $1.00x$.
		$1.06x$	=	15,900	Combine like terms.
x			=	$\dfrac{15,900}{1.06}$	Divide by 1.06.
x			=	15,000	Simplify.

Thus Hector's salary was $15,000 last year before the raise.

3. Check. Does it seem reasonable that Hector's salary last year was $15,000? The check is up to you.

Practice Problem 3 The price of Betsy's new car is 7% more than the price of a similar model last year. She paid $13,910 for her car this year. What would a similar model have cost last year?

---

 3 ## Solving Investment Problems Involving Simple Interest

Interest is a charge for borrowing money or an income from investing money. Interest rates affect our lives. They affect the national economy and they affect a consumer's ability to borrow money for big purchases. For these reasons, a student of mathematics should be able to solve problems involving interest.

There are two basic types of interest: simple and compound. **Simple interest** is computed by multiplying the amount of money borrowed or invested (which is called the *principal*) times the rate of interest times the period of time over which it is borrowed or invested (usually measured in years unless otherwise stated).

$$\text{Interest} = \text{principal} \times \text{rate} \times \text{time}$$
$$I = prt$$

You often hear of banks offering a certain interest rate *compounded* quarterly, monthly, weekly, or daily. In **compound interest** the amount of interest is added to the amount of the original principal at the end of each time period, so future interest is based on the sum of both principal and previous interest. Most financial institutions use compound interest in their transactions.

Problems involving compound interest may be solved by:

1. Repeated calculations using the simple interest formula.
2. Using a compound interest table.
3. Using exponential functions, a topic that is usually covered in a higher-level college algebra course.

All examples and exercises in this chapter will involve **simple interest.**

Calculator

Interest

You can use a calculator to find simple interest. Find the interest on $450 invested at an annual rate of 6.5% for 15 months. Notice that the time is in months. Since the interest formula $I = prt$ is in years, you need to change 15 months to years by dividing 15 by 12.

Enter: $15 \div 12 =$

Display: 1.25

Leave this on the display and multiply as follows:

$1.25 \times 450 \times$

$6.5 \% =$

The display should read

36.5625

which would round to $36.56.

Try finding the simple interest in the following investments.

(a) $9516 invested at 12% for 30 months

(b) $593 borrowed at 8% for 5 months

Answers:

(a) $2854.80

(b) $19.97

EXAMPLE 4   Find the interest on $3000 borrowed at a simple interest rate of 18% for one year.

Solution

$$I = prt$$ The simple interest formula.

$$I = (3000)(0.18)(1)$$ Substitute the values of the variables: principal $= 3000$, the rate $= 18\% = 0.18$, the time $=$ one year.

$$I = 540$$

Thus the interest charge for borrowing $3000 for one year at a simple interest rate of 18% is $540.

Practice Problem 4   Find the interest on $7000 borrowed at a simple interest rate of 12% for one year.

NOTE TO STUDENT: *Fully worked-out solutions to all of the Practice Problems can be found at the back of the text starting at page SP-1*

Now we apply this concept to a word problem about investments.

EXAMPLE 5   A woman invested an amount of money in two accounts for one year. She invested some at 8% simple interest and the rest at 6% simple interest. Her total amount invested was $1250. At the end of the year she had earned $86 in interest. How much money had she invested in each account?

Solution

Mathematics Blueprint for Problem Solving

Gather the Facts	Assign the Variable	Basic Formula or Equation	Key Points to Remember
$1250 is invested: part at 8% interest, part at 6% interest. The total interest for the year is $86.	$x =$ amount invested at 8%. $1250 - x =$ amount invested at 6%. $0.08x =$ amount of interest for x dollars at 8%. $0.06(1250 - x) =$ amount of interest for $1250 - x$ dollars at 6%.	Interest earned at 8% $+$ interest earned at 6% $=$ total interest earned during the year, which is $86.	Be careful to write $1250 - x$ for the amount of money invested at 6%. The order of the amounts is $1250 - x$. Do not use $x - 1250$.

$$
\begin{array}{ccccc}
\text{interest} & & \text{interest} & & \text{total interest} \\
\text{earned} & + & \text{earned} & = & \text{earned during} \\
\text{at 8\%} & & \text{at 6\%} & & \text{the year} \\
\end{array}
$$

$$0.08x + 0.06(1250 - x) = 86$$

Note: Be sure you write $(1250 - x)$ for the amount of money invested at 6%. Students often write it backwards by mistake. It is *not* correct to use $(x - 1250)$ instead of $(1250 - x)$. The order of the terms is very important.

Solve and state the answer.

$$0.08x + 75 - 0.06x = 86 \qquad \text{Remove parentheses.}$$
$$0.02x + 75 = 86 \qquad \text{Combine like terms.}$$
$$0.02x = 11 \qquad \text{Subtract 75 from both sides.}$$
$$\frac{0.02x}{0.02} = \frac{11}{0.02} \qquad \text{Divide both sides by 0.02.}$$
$$x = 550 \qquad \text{The amount invested at 8\% interest is \$550.}$$
$$1250 - x = 1250 - 550 = 700 \qquad \text{The amount invested at 6\% interest is \$700.}$$

Check. Are these values reasonable? Yes. Do the amounts equal $1250?

$$\$550 + \$700 \overset{?}{=} \$1250$$
$$\$1250 = \$1250 \ \checkmark$$

Would these amounts earn $86 interest in one year invested at the specified rates?

$$0.08(\$550) + 0.06(\$700) \overset{?}{=} \$86$$
$$\$44 + \$42 \overset{?}{=} \$86$$
$$\$86 = \$86 \ \checkmark$$

Practice Problem 5 A woman invested her savings of $8000 in two accounts that each calculate interest only once per year. She placed one amount in a special notice account that yields 9% annual interest. The remainder she placed in a tax-free All-Savers account that yields 7% annual interest. At the end of the year, she had earned $630 in interest from the two accounts together. How much had she invested in each account?

NOTE TO STUDENT: Fully worked-out solutions to all of the Practice Problems can be found at the back of the text starting at page SP-1

 Solving Coin Problems

Coin problems provide an unmatched opportunity to use the concept of *value*. We must make a distinction between *how many coins* there are and the *value* of the coins.

Consider the next example. Here we know *the value* of some coins, but do not know *how many* we have.

EXAMPLE 6 When Bob got out of math class, he had to make a long-distance call. He had exactly enough dimes and quarters to make a phone call that would cost $2.55. He had one less quarter than he had dimes. How many coins of each type did he have?

Solution Let d = the number of dimes.

Then $d - 1$ = the number of quarters.

The total value of the coins was $2.55. How can we represent the value of the dimes and the value of the quarters? Think.

Each dime is worth $0.10.	Each quarter is worth $0.25.
5 dimes are worth $(5)(0.10) = 0.50$.	8 quarters are worth $(8)(0.25) = 2.00$.
d dimes are worth $(d)(0.10) = 0.10d$.	$(d - 1)$ quarters are worth $(d - 1)(0.25) = 0.25(d - 1)$.

Now we can write an equation for the total value.

$$(\text{value of dimes}) + (\text{value of quarters}) = \$2.55$$
$$0.10d \;+\; 0.25(d - 1) \;=\; 2.55$$

$0.10d + 0.25d - 0.25 = 2.55$	Remove parentheses.
$0.35d - 0.25 = 2.55$	Combine like terms.
$0.35d = 2.80$	Add 0.25 to both sides.
$\dfrac{0.35d}{0.35} = \dfrac{2.80}{0.35}$	Divide both sides by 0.35.
$d = 8$	Simplify.
$d - 1 = 7$	

Thus Bob had eight dimes and seven quarters.

Check. Is this answer reasonable? Yes. Does Bob have one less quarter than he has dimes?

$$8 - 7 \overset{?}{=} 1$$
$$1 = 1 \;\checkmark$$

Are eight dimes and seven quarters worth $2.55?

$$8(\$0.10) + 7(\$0.25) \overset{?}{=} \$2.55$$
$$\$0.80 + \$1.75 \overset{?}{=} \$2.55$$
$$\$2.55 = \$2.55 \;\checkmark$$

Practice Problem 6   Ginger has five more quarters than dimes. She has $5.10 in change. If she has only quarters and dimes, how many coins of each type does she have?

EXAMPLE 7   Michele and her two children returned from the grocery store with only $2.80 in change. She had twice as many quarters as nickels. She had two more dimes than nickels. How many nickels, dimes, and quarters did she have?

Solution

Mathematics Blueprint for Problem Solving

Gather the Facts	Assign the Variable	Basic Formula or Equation	Key Points to Remember
Michele had $2.80 in change. She had twice as many quarters as nickels. She had two more dimes than nickels.	x = number of nickels. $2x$ = number of quarters. $x + 2$ = number of dimes. $0.05x$ = value of the nickels. $0.25(2x)$ = value of the quarters. $0.10(x + 2)$ = value of the dimes.	The value of the nickels + the value of the dimes + the value of the quarters = $2.80.	Don't add the number of coins to get $2.80. You must add the value of the coins!

$$(\text{value of nickels}) + (\text{value of dimes}) + (\text{value of quarters}) = \$2.80$$
$$0.05x + 0.10(x + 2) + 0.25(2x) = 2.80$$

Solve.

$0.05x + 0.10x + 0.20 + 0.50x = 2.80$	Remove parentheses.
$0.65x + 0.20 = 2.80$	Combine like terms.
$0.65x = 2.60$	Subtract 0.20 from both sides.
$\dfrac{0.65x}{0.65} = \dfrac{2.60}{0.65}$	Divide both sides by 0.65.
$x = 4$	Simplify. Michele had four nickels.
$2x = 8$	She had eight quarters.
$x + 2 = 6$	She had six dimes.

When Michele left the grocery store she had four nickels, eight quarters, and six dimes.

Check. Is the answer reasonable? Yes. Did Michele have twice as many quarters as nickels?

$$(4)(2) \overset{?}{=} 8 \qquad 8 = 8 \ \checkmark$$

Did she have two more dimes than nickels?

$$4 + 2 \overset{?}{=} 6 \qquad 6 = 6 \ \checkmark$$

Do four nickels, eight quarters, and six dimes have a value of $2.80?

$$4(\$0.05) + 8(\$0.25) + 6(\$0.10) \overset{?}{=} \$2.80$$
$$\$0.20 + \$2.00 + \$0.60 \overset{?}{=} \$2.80$$
$$\$2.80 = \$2.80 \ \checkmark$$

Practice Problem 7 A young boy told his friend that he had twice as many nickels as dimes in his pocket. He also said that he had four more quarters than dimes. He said that he had $2.35 in change in his pocket. Can you determine how many nickels, dimes, and quarters he had?

NOTE TO STUDENT: *Fully worked-out solutions to all of the Practice Problems can be found at the back of the text starting at page SP-1*

Developing Your Study Skills

Applications or Word Problems

Applications or word problems are the very life of mathematics! They are the reason for doing mathematics because they teach you how to put into use the mathematical skills you have developed. Learning mathematics without ever doing word problems is similar to learning all the skills of a sport without ever playing a game or learning all the notes on an instrument without ever playing a song.

The key to success is practice. Make yourself do as many problems as you can. If you need help organizing your facts, use the Mathematics Blueprint. You may not be able to do all problems correctly at first, but keep trying. Do not give up whenever you reach a difficult one. If you cannot solve it, just try another one. Then come back and try it again later.

A misconception among students when they begin studying word problems is that each problem is different. At first the problems may seem this way, but as you practice more and more, you will begin to see the similarities, the different "types." You will see patterns in solving problems, which will enable you to solve them more easily.

Applications

Solve. All problems involving interest refer to simple interest.

1. **Raking Leaves** Paul has a job raking leaves for a neighbor. He makes $6.50 an hour, plus $0.75 for each bag he fills. Last Saturday he worked three hours and made a total of $27.75. How many bags of leaves did he fill?

2. **Waiting Tables** Marybelle is contemplating a job as a waitress. She would be paid $4 per hour plus tips. The other waitresses have told her that an average tip at that restaurant is $3 per table served. If she works 20 hours per week, how many tables would she have to serve in order to make $191 during the week?

3. **Overtime** Ramon has a summer job as a lifeguard to earn his college tuition. He gets paid $6.00 per hour for the first 40 hours and $9.00 per hour for each hour in the week worked more than 40 hours. This summer his goal is to earn at least $303.00 per week. How many hours of overtime per week will he need to achieve his goal?

4. **Overtime** Maria has a summer job as a tennis instructor to earn her college tuition. She gets paid $6.50 per hour for the first 40 hours and $9.75 for each hour in the week worked more than 40 hours. This summer her goal is to earn at least $338.00 per week. How many hours of overtime per week will she need to achieve her goal?

5. **Layaway** Mrs. Peterson's triplets all want to attend the local private academy, where they are required to wear uniforms. Mrs. Peterson knows that the only way she can afford these uniforms is to shop early, put the uniforms on layaway, and pay a little every week. The total uniform cost for the three girls is $1817.75. If she put down a deposit of $600 and she could afford $105 per week, how long would it take her to pay off the uniforms? (Round to nearest whole number.)

6. **Party Costs** The Swedish Chef Catering Company charges $50.00 for setup, $85.00 for cleanup, and the special menu that Judy Carter ordered cost $23.50 per person. If Judy Carter's bill came to $558.00, how many people came to the party?

7. **Camera Sale** The camera Melissa wanted for her birthday is on sale at 28% off the usual price. The amount of the discount is $100.80. What was the original price of the camera?

8. **Work Force** The number of women working full-time in Springfield has risen 12% this year. This means 216 more women have full-time jobs. What was the number of women working full-time last year?

9. **Salary** The cost of living last year went up 3%. Fortunately, Alice Swanson got a 3% raise in her salary from last year. This year she is earning $22,660. How much did she make last year?

10. **Stock Profit** A speculator bought stocks and later sold them for $5136, making a profit of 7%. How much did the stocks cost him?

11. **Investments** Don Williams invested some money at 9% simple interest. At the end of the year, the total amount of his original principal and the interest was $6540. How much did he originally invest?

12. **Investments** Robert Campbell invested some money at 8% simple interest. At the end of the year, the total amount of his original principal and the interest was $7560. How much did he originally invest?

13. ***Trust Fund*** Mr. and Mrs. Wright set up a trust fund for their children last year. Part of it is earning 7% simple interest per year while the rest of it is earning 5% simple interest per year. They placed $5000 in the trust fund. In one year the trust fund has earned $310. How much did they invest at each interest rate?

14. ***Investments*** Anne and Michael invested $8000 last year in tax-free bonds. Some of the bonds earned 8% simple interest while the rest earned 6% simple interest. At the end of the year, they had earned $580 in interest. How much did they invest at each interest rate?

15. ***Investments*** Plymouth Rock Bank invested $400,000 last year in mutual funds. The conservative fund earned 8% simple interest. The growth fund earned 12% simple interest. At the end of the year, the bank had earned $38,000 from these mutual funds. How much did it invest in each fund?

16. ***Investments*** Millennium Securities last year invested $600,000 in mutual funds. The international fund earned 11% simple interest. The high-tech fund earned 7% simple interest. At the end of the year, the company had earned $50,000 in interest. How much did it invest in each fund?

17. ***Investments*** Dave Horn invested half of his money at 5%, one-third of his money at 4%, and the rest of his money at 3.5%. If his total annual investment income was $530, how much had he invested?

18. ***Investments*** Last year Pete Pfeffer decided to invest half of his money in a credit union paying 4.5% interest, one-third of his money in a mutual fund paying 5% interest, and the rest of his money in a bank CD paying 4% interest. If his annual investment income was $357.50 last year, how much money had he invested?

19. ***Coin Bank*** Little Melinda has nickels and quarters in her bank. She has four fewer nickels than quarters. She has $3.70 in the bank. How many coins of each type does she have?

20. ***Coin Change*** Reggie's younger brother had several coins when he returned from his paper route. He had a total of $5.35 in dimes and quarters. He had six more quarters than he had dimes. How many of each coin did he have?

21. ***Coin Change*** A newspaper carrier has $3.75 in change. He has three more quarters than dimes but twice as many nickels as quarters. How many coins of each type does he have?

22. ***Coin Change*** Tim Whitman has $4.50 in change in his desk drawer to use on the vending machines downstairs. He has four more quarters than dimes. He has three times as many nickels as dimes. How many coins of each type does he have?

23. ***Stamps*** Ramon was office manager for a small law firm. He made the weekly postage purchase. He bought regular 37¢ stamps for most of the firm, but he also picked up some 40¢ breast cancer stamps at the request of some of the lawyers. He noticed that the number of breast cancer stamps was 8 fewer than the number of regular stamps and that he spent $19.90. How many of each stamp did he buy?

24. ***Ticket Prices*** Mario's Marionettes donated free tickets for their show to the local Boys & Girls Club. They claimed that the ticket value was $176.75. A child's ticket cost $5.50 and an adult ticket cost $8.75. If there were three times as many children's tickets as adult tickets, how many adults and children got to attend the show for free?

25. *Paper Currency* Charlie Saulnier cashed his paycheck and came home from the bank with $100 bills, $20 bills, and $10 bills. He has twice as many $20 bills as he has $10 bills. He has three more $100 bills than he has $10 bills. He is carrying $1500 in bills. How many of each denomination does he have?

26. *Paper Currency* Roberta Burgess came home with $325 in tips from two nights on her job as a waitress. She had $20 bills, $10 bills, and $5 bills. She discovered that she had three times as many $5 bills as she had $10 bills. She also found that she had 4 fewer $20 bills than she had $10 bills. How many of each denomination did she have?

27. *Salary* Madelyn Logan is an office furniture dealer who earns an $18,000 base salary. She also earns a 4% commission on sales. How much must she sell to earn a total of $55,000?

28. *Animal Shelter* The local animal shelter accepts abandoned cats and dogs. They usually receive twice as many cats as dogs. They estimate that 80% of the cats and 60% of the dogs that come in need some kind of medical treatment. If they treated 286 animals last year, how many cats and dogs did they take in?

To Think About

29. *Car Rentals* The West Suburban Car Rental Agency will rent a compact car for $35 per day and an additional charge of $0.24 per mile. The Golden Gate Car Rental Agency charges only $0.16 per mile but charges $41 per day. If a salesperson wanted to rent a car for three days, how many miles would that person have to drive to make the Golden Gate Car Rental Agency car a better bargain?

30. *Pumping Station* A Peabody pumping station pumps 2000 gallons of water per hour into an empty reservoir tank for a town's drinking supply. The station pumps for three hours. Then a leak in the reservoir tank is created by a large crack. Some water flows out of the reservoir tank at a constant rate. The pumping station continues pumping for six more hours while the leak is undetected. At the end of nine hours the reservoir contains 17,640 gallons of water. During the last six hours, how many gallons per hour are leaking from the reservoir?

Cumulative Review

Perform the operations in the proper order.

31. $5(3) + 6 \div (-2)$

32. $5(-3) - 2(12 - 15)^2 \div 9$

Evaluate for $x = -2$ and $y = 3$.

33. $2x^2 + 3xy - 2y^2$

34. $x^3 - 5x^2 + 3y - 6$

Satellite Parts   *Brian sells high-tech parts to satellite communications companies. In his negotiations he originally offers to sell one company 200 parts for a total of $22,400. However, after negotiations, he offers to sell that company the same parts at a 15% discount if the company agrees to sign a purchasing contract for 200 additional parts at some future date.*

35. What is the average cost per part if the parts are sold at the discounted price?

36. How much will the total bill be for the 200 parts at the discounted price?

1 Finding Area, Perimeter, and Missing Angles

In Section 1.8 we reviewed a number of area and perimeter formulas. We repeat this list of formulas here for convenience.

AREA AND PERIMETER FORMULAS

A **parallelogram** is a four-sided figure with opposite sides parallel. In a parallelogram, opposite sides are equal and opposite angles are equal.

Perimeter = the sum of all four sides

Area = ab

A **rectangle** is a parallelogram with all interior angles measuring 90°.

Perimeter = $2l + 2w$

Area = lw

A **square** is a rectangle with all four sides equal.

Perimeter = $4s$

Area = s^2

A **trapezoid** is a four-sided figure with two sides parallel. The parallel sides are called the bases of the trapezoid.

Perimeter = the sum of all four sides

Area = $\frac{1}{2}a(b_1 + b_2)$

A **triangle** is a closed plane figure with three sides.

Perimeter = the sum of the three sides

Area = $\frac{1}{2}ab$

A **circle** is a plane curve consisting of all points at an equal distance from a given point called the center.

Circumference is the distance around a circle. The **radius** is a line segment from the center of the circle to a point on the circle. The **diameter** is a line segment across the circle that passes through the center with endpoints on the circle.

Circumference = $2\pi r$

Area = πr^2

π is a constant associated with circles. It is an irrational number that is approximately 3.141592654. We usually use 3.14 as a sufficiently accurate approximation. Thus we write $\pi \approx 3.14$ for most of our calculations involving π.

We frequently encounter triangles in word problems. There are four important facts about triangles, which we list for convenient reference.

TRIANGLE FACTS

1. The sum of the interior angles of any triangle is 180°. That is,

measure of $\angle A$ + measure of $\angle B$ + measure of $\angle C = 180°$.

2. An **equilateral** triangle is a triangle with three sides equal in length and three angles that measure 60° each.

Equilateral triangle

3. An **isosceles** triangle is a triangle with two equal sides. The two angles opposite the equal sides are also equal.

Isosceles triangle

measure $\angle A$ = measure $\angle B$

4. A **right** triangle is a triangle with one angle that measures 90°.

Right triangle

▲ EXAMPLE 1 Find the area of a triangular window whose base is 16 inches and whose altitude is 19 inches.

Solution We can use the formula for the area of a triangle. Substitute the known values in the formula and solve for the unknown.

$A = \dfrac{1}{2}ab$ Write the formula for the area of a triangle.

$= \dfrac{1}{2}(16 \text{ in.})(19 \text{ in.})$ Substitute the known values in the formula.

$= \dfrac{1}{2}(16)(19)(\text{in.})(\text{in.})$ Simplify.

$A = 152 \text{ in.}^2$ or   152 square inches

The area of the triangle is 152 square inches.

19 in.

16 in.

▲ Practice Problem 1 Find the area of a triangle whose base is 14 inches and whose altitude is 20 inches.

NOTE TO STUDENT: *Fully worked-out solutions to all of the Practice Problems can be found at the back of the text starting at page SP-1*

▲ EXAMPLE 2 The area of an NBA basketball court is 4700 square feet. Find the width of the court if the length is 94 feet.

Solution Draw a diagram.

94 feet

4700 square feet

w

Write the formula for the area of a rectangle and solve for the unknown value.

$$A = lw$$

$$4700 \text{ (ft)}^2 = (94 \text{ ft})(w) \quad \text{Substitute the known values into the formula.}$$

$$\frac{4700 \text{ (ft)(ft)}}{94 \text{ ft}} = \frac{94 \text{ ft}}{94 \text{ ft}} w \quad \text{Divide both sides by 94 feet.}$$

$$50 \text{ feet} = w$$

The width of the basketball court is 50 feet.

▲ **Practice Problem 2** The area of a rectangular field is 120 square yards. If the width of the field is 8 yards, what is the length?

EXAMPLE 3 The area of a trapezoid is 400 square inches. The altitude is 20 inches and one of the bases is 15 inches. Find the length of the other base.

Solution

15 in.

20 in.

$$A = \frac{1}{2} a (b_1 + b_2) \quad \text{Write the formula for the area of a trapezoid.}$$

$$400(\text{in.})^2 = \frac{1}{2}(20 \text{ in.})(15 \text{ in.} + b_2) \quad \text{Substitute the known values.}$$

$$400(\text{in.})(\text{in.}) = 10(\text{in.})(15 \text{ in.} + b_2) \quad \text{Simplify.}$$

$$400(\text{in.})(\text{in.}) = 150(\text{in.})(\text{in.}) + (10 \text{ in.})b_2 \quad \text{Remove parentheses.}$$

$$250(\text{in.})(\text{in.}) = (10 \text{ in.})b_2 \quad \text{Subtract 150 in.}^2 \text{ from both sides.}$$

$$25 \text{ in.} = b_2 \quad \text{Divide both sides by 10 in.}$$

The other base is 25 inches long.

▲ **Practice Problem 3** The area of a trapezoid is 256 square feet. The bases are 12 feet and 20 feet. Find the altitude.

▲ **EXAMPLE 4** Find the area of a circular sign whose diameter is 14 inches. (Use 3.14 for π.) Round to the nearest square inch.

Solution $A = \pi r^2$ Write the formula for the area of a circle.

Note that the length of the diameter is given in the word problem. We need to know the length of the radius to use the formula. Since $d = 2r$, then 14 in. $= 2r$ and $r = 7$ in.

$$A = \pi(7 \text{ in.})(7 \text{ in.}) \quad \text{Substitute known values into the formula.}$$

$$= (3.14)(7 \text{ in.})(7 \text{ in.})$$

$$= (3.14)(49 \text{ in.}^2)$$

$$= 153.86 \text{ in.}^2$$

Rounded to the nearest square inch, the area of the circle is approximately 154 square inches.

In this example, you were asked to find the area of the circle. Do not confuse the formula for area with the formula for circumference. $A = \pi r^2$, while $C = 2\pi r$. Remember that area involves square units, so it is only natural that in the area formula you would square the radius.

▲ **Practice Problem 4** Find the circumference of a circle whose radius is 15 meters. (Use 3.14 for π.) Round your answer to the nearest meter.

▲ **EXAMPLE 5**   Find the perimeter of a parallelogram whose longer sides are 4 feet and whose shorter sides are 2.6 feet.

Solution   Draw a picture.

4 ft

2.6 ft

The perimeter is the distance around a figure. To find the perimeter, add the lengths of the sides. Since the opposite sides of a parallelogram are equal, we can write the following.

$$P = 2(4 \text{ feet}) + 2(2.6 \text{ feet})$$
$$= 8 \text{ feet} + 5.2 \text{ feet}$$
$$= 13.2 \text{ feet}$$

The perimeter of the parallelogram is 13.2 feet.

▲ **Practice Problem 5**   Find the perimeter of an equilateral triangle with a side that measures 15 centimeters.

▲ **EXAMPLE 6**   The smallest angle of an isosceles triangle measures 24°. The other two angles are larger. What are the measurements of the other two angles?

Solution   We know that in an isosceles triangle the measures of two angles are equal. We know that the sum of the measures of all three angles is 180°. Both of the larger angles must be different from 24°, therefore these two larger angles must be equal.

Let x = the measure in degrees of each of the larger angles.

Then we can write	$24° + x + x = 180°$	
	$24° + 2x = 180°$	Add like terms.
	$2x = 156°$	Subtract 24 from each side.
	$x = 78°$	Divide each side by 2.

24°

Thus the measures of each of the other two angles of the triangle must be 78°.

▲ **Practice Problem 6**   The largest angle on an isosceles triangle measures 132°. The other two angles are smaller. What are the measurements of the other two angles?

② Finding Volume and Surface Area

Now let's examine three-dimensional figures. **Surface area** is the total area of the faces of a figure. You can find surface area by calculating the area of each face and then finding the sum. **Volume** is the measure of the amount of space inside a figure. Some formulas for the surface area and the volume of regular figures can be found in the following table.

GEOMETRIC FORMULAS: THREE-DIMENSIONAL FIGURES

Rectangular prism   Note that all of the faces are rectangles.

h = height

w = width

l = length

Surface area $= 2lw + 2wh + 2lh$

Volume $= lwh$

GEOMETRIC FORMULAS: THREE-DIMENSIONAL FIGURES (continued)

Sphere

r = radius

$$\text{Surface area} = 4\pi r^2$$
$$\text{Volume} = \frac{4}{3}\pi r^3$$

Right circular cylinder

r = radius
h = height

$$\text{Surface area} = 2\pi rh + 2\pi r^2$$
$$\text{Volume} = \pi r^2 h$$

▲ **EXAMPLE 7**   Find the volume of a sphere with radius 4 centimeters. (Use 3.14 for π.) Round your answer to the nearest cubic centimeter.

Solution

$$V = \frac{4}{3}\pi r^3 \qquad \text{Write the formula for the volume of a sphere.}$$

$$= \frac{4}{3}(3.14)(4 \text{ cm})^3 \quad \text{Substitute the known values into the formula.}$$

$$= \frac{4}{3}(3.14)(64) \text{ cm}^3 \approx 267.946667 \text{ cm}^3$$

Rounded to the nearest whole number, the volume of the sphere is 268 cubic centimeters.

NOTE TO STUDENT: Fully worked-out solutions to all of the Practice Problems can be found at the back of the text starting at page SP-1

▲ **Practice Problem 7**   Find the surface area of a sphere with radius 5 meters. Round your answer to the nearest square meter.

▲ **EXAMPLE 8**   A can is made of aluminum. It has a flat top and a flat bottom. The height is 5 inches and the radius is 2 inches. How much aluminum is needed to make the can? How much aluminum is needed to make 10,000 cans?

Solution

Understand the problem. What do we need to find? Reword the problem.

We need to find the total surface area of a cylinder.

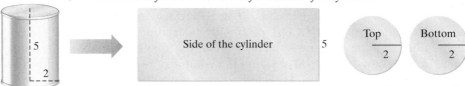

You may calculate the area of each piece and find the sum or you may use the formula.

We will use 3.14 to approximate π.

$$\text{Surface area} = 2\pi rh + 2\pi r^2$$
$$= 2(3.14)(2 \text{ in.})(5 \text{ in.}) + 2(3.14)(2 \text{ in.})^2$$
$$= 62.8 \text{ in.}^2 + 25.12 \text{ in.}^2 = 87.92 \text{ in.}^2$$

87.92 square inches of aluminum is needed to make each can.

Have we answered all the questions in the word problem? Reread the problem. How much aluminum is needed to make 10,000 cans?

(aluminum for 1 can)(10,000) = (87.92 in.²)(10,000) = 879,200 square inches

It would take 879,200 square inches of aluminum to make 10,000 cans.

▲ **Practice Problem 8**   Sand is stored in a cylindrical drum that is 4 feet high and has a radius of 3 feet. How much sand can be stored in the drum? ▪

③ Solving More-Involved Geometric Problems

▲ **EXAMPLE 9**   A quarter-circle (of radius 1.5 yards) is connected to two rectangles with dimensions as labeled on the sketch in the margin. You need to lay a strip of carpet in your house according to this sketch. How many square yards of carpeting will be needed on the floor? (Use $\pi \approx 3.14$. Round your final answer to the nearest tenth.)

Solution   The desired area is the sum of three areas, which we will call B, C, and D. Area B and area D are rectangular in shape. It is relatively easy to find the areas of these shapes.

$$A_B = (5 \text{ yd})(1.5 \text{ yd}) = 7.5 \text{ yd}^2 \qquad A_D = (3 \text{ yd})(1.5 \text{ yd}) = 4.5 \text{ yd}^2$$

Area C is one-fourth of a circle. The radius of the circle is 1.5 yd.

$$A_C = \frac{\pi r^2}{4} \approx \frac{(3.14)(1.5 \text{ yd})^2}{4} = \frac{(3.14)(2.25) \text{ yd}^2}{4} = 1.76625 \text{ yd}^2 \approx 1.8 \text{ yd}^2$$

The total area is $A_B + A_D + A_C \approx 7.5 \text{ yd}^2 + 4.5 \text{ yd}^2 + 1.8 \text{ yd}^2 \approx 13.8 \text{ yd}^2$.
    13.8 square yards of carpeting will be needed to cover the floor.

▲ **Practice Problem 9**   John has a swimming pool that measures 8 feet by 12 feet. He plans to make a concrete walkway around the pool that is 3 feet wide. What will be the total cost of the walkway at $12 a square foot? ▪

▲ **EXAMPLE 10**   Find the weight of the water in a full cylinder containing water if the height of the cylinder is 5 feet and the radius is 2 feet. 1 cubic foot of water weighs 62.4 pounds. Round your answer to the nearest ten pounds. Use 3.14 for π.

Solution
Understand the problem.
Draw a diagram.

Write the formula for the volume of a cylinder.

$$V = \pi r^2 h$$
$$= (3.14)(2 \text{ ft})^2(5 \text{ ft})$$
$$= (3.14)(4 \text{ ft}^2)(5 \text{ ft})$$
$$= 62.8 \text{ ft}^3$$

The volume of the cylinder is 62.8 cubic feet.
    Each cubic foot of water weighs 62.4 pounds. This can be written as 62.4 lb/ft³. Since the volume of the cylinder is 62.8 ft³, we multiply.

$$\text{Weight of the water} = (62.8 \text{ ft}^3)\left(\frac{62.4 \text{ lb}}{1 \text{ ft}^3}\right)$$
$$= (62.8)(62.4) \text{ lb} = 3918.72 \text{ lb}$$

The weight of the water in the cylinder is 3920 rounded to the nearest ten pounds.

▲ **Practice Problem 10**   Find the weight of the water in a full rectangular container that is 5 feet wide, 6 feet long, and 8 feet high. Remember, 1 cubic foot of water weighs 62.4 pounds. Round your answer to the nearest 100 pounds. ▪

Verbal and Writing Skills

Geometry   *Fill in the blank to complete each sentence.*

▲ **1.** Perimeter is the _____ a plane figure.

▲ **2.** _____ is the distance around a circle.

▲ **3.** Area is a measure of the amount of _____ in a region.

▲ **4.** A _____ is a four-sided figure with exactly two sides parallel.

▲ **5.** The sum of the interior angles of any triangle is _____.

▲ **6.** *Geometry* An isosceles triangle is a triangle with _____.

▲ **7.** *Geometry* Find the area of a triangle whose altitude is 24 inches and whose base is 13 inches.

▲ **8.** *Geometry* Find the area of a triangle whose altitude is 7 feet and whose base is 22 feet.

▲ **9.** *Geometry* Find the area of a parallelogram whose altitude is 14 inches and whose base is 7 inches.

▲ **10.** *Geometry* Find the area of a parallelogram whose altitude is 15 meters and whose base is 10 meters

▲ **11.** *Geometry* The area of a triangle is 80 square feet. Find the altitude if the base is 16 feet.

▲ **12.** *Geometry* The area of a triangle is 120 sq in. Find the altitude if the base is 24 in.

▲ **13.** *Geometry* The perimeter of a parallelogram is 46 inches. If the length of one side is 14 inches, what is the length of a side adjacent to it?

▲ **14.** *Geometry* If one side of a parallelogram measures 18.5 cm and an adjacent side measures 17.5 cm, what is the perimeter of the parallelogram?

▲ **15.** *Geometry* Find the area of a circular sign whose radius is 7.00 feet. (Use $\pi = 3.14$ as an approximate value.)

▲ **16.** *Flower Bed* Find the area of a circular flower bed whose diameter is 6 meters. (Use $\pi = 3.14$ as an approximate value.)

▲ **17.** *Mercury* The diameter of the planet Mercury is approximately 3032 miles. Find the distance around its equator. (Use 3.14 to approximate π.)

▲ **18.** *Frisbees* Find the area of an ultimate frisbee disc whose diameter is 27 cm. (Use 3.14 to approximate π.)

▲ **19.** *Geometry* The area of a trapezoid is 600 square inches and the bases are 20 inches and 30 inches. Find the altitude.

▲ **20.** *Geometry* The area of a trapezoid is 900 square inches. The bases are 40 inches and 50 inches. Find the altitude.

Applications

▲ **21.** *Streets of Manhattan* In midtown Manhattan, the street blocks have a uniform size of 80 meters north-south by 280 meters east-west. If a typical New York City neighborhood is two blocks east-west by three blocks north-south, how much area does it cover?

▲ **22.** *Computer Cases* A computer manufacturing plant makes its computer cases from plastic. Two sides of a triangular scrap of plastic measure 7 cm and 25 cm.
  (a) If the perimeter of the plastic piece is 56 cm, how long is the third side?

  (b) If each side of the triangular piece is decreased by 0.1 cm, what would the new perimeter be?

▲ **23.** *Geometry* The circumference of a circle is 31.4 centimeters. Find the radius. (Use $\pi = 3.14$ as an approximate value.)

▲ **24.** *Geometry* The perimeter of an equilateral triangle is 27 inches. Find the length of each side of the triangle.

▲ **25.** *Tires* An automobile tire has a diameter of 64 cm. What is the circumference of the tire? (Use $\pi = 3.14$ as an approximate value.)

▲ **26.** *Golden Ratio* The ancient Greeks discovered that rectangles where the length-to-width ratio was 8 to 5 (known as the golden ratio) were most pleasing to the eye. Kayla wants to create a painting that will be framed by some antique molding that she found. If she calculates that the molding can be used to make a frame of 104 inches in perimeter, what dimensions should her painting be so that it is visually appealing?

▲ **27.** *Geometry* Two angles of a triangle measure 47 degrees and 59 degrees. What is the measure of the third angle?

▲ **28.** *Geometry* A right triangle has one angle that measures 77 degrees. What does the other acute angle measure?

▲ **29.** *Geometry* Each of the equal angles of an isosceles triangle is 4 times as large as the third angle. What is the measure of each angle?

▲ **30.** *Geometry* In a triangle, the measure of the first angle is twice the measure of the second angle. The measure of the third angle is 20 degrees less than the second angle. What is the measure of each angle?

▲ **31.** *Boat Frame* The smallest angle of an isosceles triangle used in the wood frame of a boat measures 38°. The other two angles are larger. What are the measurements of the other two angles in this triangular part of the wood frame?

▲ **32.** *TV Antenna* The largest angle of an isosceles triangle used in holding a cable TV antenna measures 146°. The other two angles are smaller. What are the measurements of the other two angles in the triangular piece of the cable TV antenna?

▲ **33.** *Parking Lot* A triangular region of the community college parking lot was measured. The measure of the first angle of this triangle is double the measure of the second angle. The measure of the third angle is 19° greater than double the measure of the first angle. What are the measurements of all three angles?

▲ **34.** *Lobster Pot* A lobster pot used in Gloucester, Massachusetts, has a small triangular piece of wood at the entry point where the lobster enters. The measure of the first angle of this triangle is triple the measure of the second angle. The third angle of the triangle is 30° less than double the measure of the first angle. What are the measurements of all the three angles?

▲ **35.** *Geometry* A ladder makes a right triangle against a house. If the angle between the ladder and the house is 38°, what is the angle between the ladder and the ground?

▲ **36.** *Geometry* Two angles of a triangle measure 135° and 11°. What is the measure of the third angle?

▲ **37.** *Painting a Shed* The cost for paint to cover the exterior of the four sides of a shed was $40. A 5-gallon can of paint costs $20 and will cover 102 square feet. The shed consists of four sides, each shaped like a trapezoid. They each have one base of 8 feet and one base of 9 feet. What is the altitude of each trapezoid?

▲ **38.** *Painting a Sign* The cost for paint to cover both sides of a large road-side sign (front side and back side) was $150. A 5-gallon can of paint costs $25 and will cover 250 square feet. The sign is constructed in the shape of a trapezoid. It has one base measuring 35 feet and one base measuring 40 feet. What is the altitude of this trapezoid?

▲ **39.** *Geometry* What is the volume of a cylinder whose height is 8 inches and whose radius is 10 inches? (Use $\pi = 3.14$ as an approximate value.)

▲ **40.** *Gas Cylinder* A cylinder holds propane gas. The volume of the cylinder is 235.5 cubic feet. Find the height if the radius is 5 feet. (Use $\pi = 3.14$ as an approximate value.)

▲ **41.** *Weather Balloon* A spherical weather balloon needs to hold at least 175 cubic feet of helium to be buoyant enough to lift the meteorological instruments. Will a helium filled balloon with a diameter of 8 feet stay aloft?

▲ **42.** *Storage Cabinet* Mrs. Robinson has room for a storage cabinet that is 18 inches wide by 24 inches deep. If she calculates that she needs 15 cubic feet of storage space, how tall must the cabinet be?

▲ **43.** *Milk Container* A plastic cylinder made to hold milk is constructed with a solid top and bottom. The radius is 6 centimeters and the height is 4 centimeters. **(a)** Find the volume of the cylinder. **(b)** Find the total surface area of the cylinder. (Use $\pi = 3.14$ as an approximate value.)

▲ **44.** *Pyrex Sphere* A Pyrex glass sphere is made to hold liquids in a science lab. The radius of the sphere is 3 centimeters. **(a)** Find the volume of the sphere. **(b)** Find the total surface area of the sphere. (Use $\pi = 3.14$ as an approximate value.)

▲ **45.** *Goat Tether* Dan Perkin's goat is attached to the corner of the 8-foot-by-4-foot pump house by an 8-foot length of rope. The goat eats the grass around the pump house as shown in the diagram. This pattern consists of three-fourths of a circle of radius 8 feet and one-fourth of a circle of radius 4 feet. What area of grass can Dan Perkin's goat eat? (Use $\pi = 3.14$ as an approximate value.)

▲ **46.** *Seven-Layer Torte* For a party, Julian is making a seven-layer torte, which consists of seven 9-inch-diameter cake layers with a cream topping on each. He knows that one recipe of cream topping will cover 65 square inches of cake. How many batches of the topping must he make to complete his torte? (Use $\pi = 3.14$ as an approximate value.)

▲ **47.** *Walkway* A cement walkway is to be poured. It consists of the two rectangles with dimensions as shown in the diagram and a quarter of a circle with radius 1.5 yards. Use $\pi = 3.14$ as an approximate value. **(a)** How many square yards will the walkway be? **(b)** If a painter paints it for $2.50 per square yard, how much will the painting cost?

▲ **48.** *Aluminum Plate* An aluminum plate is made according to the following dimensions. The outer radius is 10 inches and the inner radius is 3 inches. (Use $\pi = 3.14$ as an approximate value.) **(a)** How many square inches in area is the aluminum plate (the shaded region in the sketch)? **(b)** If the cost to construct the plate is $7.50 per square inch, what is the construction cost?

49. *Toy Car Track* New Toy Company is producing a racing car set, complete with plastic track. The track is a solid surface and has a straight edge with circular ends. See the following sketch for dimensions. **(a)** What is the distance around the track? **(b)** If it is 0.5 inches thick, how much plastic does it take to make the track?

50. *Uniform Design* The Viceroy Soccer team has designed new uniforms. The front of their jerseys will have a blue rectangle, with a red V inside. (See the diagram.) The plain jersey shirt costs $19.00. The red color costs $0.05 per square inch and the blue color costs $0.02 per square inch.
(a) What is the area of the red V?

(b) What is the area of the blue color?

(c) What is the cost per jersey?

▲ **51. *Painting Frame*** It cost $120.93 to frame a painting with a flat aluminum strip. This was calculated at $1.39 per foot for materials and labor. Find the dimensions of the painting if its length is 3 feet less than four times its width.

▲ **52. *Driveway*** A concrete driveway measures 18 yards by 4 yards. The driveway will be 0.5 yards thick. **(a)** How many cubic yards of concrete will be needed? **(b)** How much will it cost if concrete is $3.50 per cubic yard?

To Think About

▲ **53. *Geometry*** An isosceles triangle has one side that measures 4 feet and another that measures 7.5 feet. The third side was not measured. Can you find one unique perimeter for this triangle? What are the two possible perimeters?

▲ **54. *Geometry*** Find the total surface area of **(a)** a sphere of radius 3 inches and **(b)** a right circular cylinder with radius 2 inches and height 0.5 inch. Which object has a greater surface area? Use $\pi = 3.14$ as an approximate value.

▲ **55. *Moon*** Assume that a very long rope supported by poles that are 3 feet tall is stretched around the moon. (Neglect gravitational pull and assume that the rope takes the shape of a large circle.) The radius of the moon is approximately 1080 miles. How much longer would you need to make the rope if you wanted the rope to be supported by poles that are 4 feet tall? Use $\pi = 3.14$ as an approximate value.

Cumulative Review

Simplify.

56. $2(x - 6) - 3[4 - x(x + 2)]$

57. $-5(x + 3) - 2[4 + 3(x - 1)]$

58. $-\{3 - 2[x - 3(x + 1)]\}$

59. $-2\{x + 3[2 - 4(x - 3)]\}$

1 Translating English Sentences into Inequalities

Real-life situations often involve inequalities. For example, if a student wishes to maintain a certain grade point average (GPA), you may hear him or her say, "I have to get a grade of at least an 87 on the final exam."

What does "at least an 87" mean? What grade does the student need to achieve? 87 would fit the condition. What about 88? The student would be happy with an 88. In fact the student would be happy with any grade that *is greater than or equal to* an 87. We can write this as

$$\text{grade} \geq 87.$$

Notice that the English phrase "at least an 87" translates mathematically as "greater than or equal to an 87."

Be careful when translating English phrases that involve inequalities into algebra. If you are not sure which symbol to use, try a few numbers to see if they fit the condition of the problem. Choose the mathematical symbol accordingly.

Here is a list that may help you.

English Phrases Describing Inequality	Algebraic Expression
Greater than	
Is more than	
Is greater than	$>$
Less than	
Is less than	
Is smaller than	$<$
Greater than or equal to	
Is greater than or equal to	
Is not less than	$\geq$
Is at least	
Cannot be less than	
Less than or equal to	
Is less than or equal to	
Is not greater than	$\leq$
Is at most	
Is not more than	

EXAMPLE 1 Translate each sentence into an inequality.

(a) The perimeter of the rectangle must be no more than 70 inches.

(b) The average number of errors must be less than two mistakes per page.

(c) His income must be at least $950 per week.

(d) The new car will cost at most $12,070.

Solution

(a) Perimeter must be *no more than* 70.

$$\text{Perimeter} \leq 70$$

Try 80. Does 80 satisfy this requirement? That is, is 80 more than 70? No. Do 60 and 50 satisfy this requirement? Yes. Use the "less than or equal to" symbol.

(b) Average number must be *less than* 2.

$$\text{Average} < 2$$

(c) Income must be *at least* $950.

$$\text{Income} \geq 950$$

"At least" means, "not less than." The income must be $950 or more. Use $\geq$.

(d) The new car will cost *at most* $12,070.

$$\text{Car} \quad \leq \quad 12{,}070$$

"At most $12,070" means it is the top price for the car. The cost of the car cannot be more than $12,070. Use $\leq$.

NOTE TO STUDENT: Fully worked-out solutions to all of the Practice Problems can be found at the back of the text starting at page SP-1

Practice Problem 1 Translate each sentence into an inequality.

(a) The height of the person must be no greater than 6 feet.

(b) The speed of the car must have been greater than 65 miles per hour.

(c) The area of the room must be greater than or equal to 560 square feet.

(d) The profit margin cannot be less than $50 per television set.

❷ Solving Applied Problems Using Inequalities

We can use the same problem-solving techniques we used previously to solve word problems involving inequalities.

EXAMPLE 2 A manufacturing company makes 60-watt light bulbs. For every 1000 bulbs manufactured, a random sample of five bulbs is selected and tested. To meet quality control standards, the average number of hours these five bulbs last must be at least 900 hours. The test engineer tested five bulbs. Four of them lasted 850 hours, 1050 hours, 1000 hours, and 950 hours, respectively. How many hours must the fifth bulb last in order for the batch to meet quality control standards?

Solution

1. *Understand the problem.*

Let n = the number of hours the fifth bulb must last

The average that the five bulbs last *must be at least* 900 hours. We could write

Average $\geq$ 900 hours.

2. *Write an inequality.*

Average of five bulbs $\geq$ 900 hours.

$$\frac{850 + 1050 + 1000 + 950 + n}{5} \geq 900$$

3. *Solve and state the answer.*

$$\frac{850 + 1050 + 1000 + 950 + n}{5} \geq 900$$

$$\frac{3850 + n}{5} \geq 900 \qquad \text{Simplify the numerator.}$$

$$3850 + n \geq 4500 \qquad \text{Multiply both sides by 5.}$$

$$n \geq 650 \qquad \text{Subtract 3850 from each side.}$$

If the fifth bulb burns for 650 hours or more, the quality control standards will be met.

4. *Check.* Any value of 650 or larger should satisfy the original inequality. Suppose we pick $n = 650$ and substitute it into the inequality.

$$\frac{850 + 1050 + 1000 + 950 + 650}{5} \overset{?}{\geq} 900$$

$$\frac{4500}{5} \overset{?}{\geq} 900$$

$$900 \geq 900 \quad \checkmark$$

If we try $n = 700$, the left side of the inequality will be 910. $910 \geq 900$. In fact, if we try any number greater than 650 for n, the left side of the inequality will be greater than 900. The inequality will be valid for $n = 650$ or any larger value.

Practice Problem 2   A manufacturing company makes coils of $\frac{1}{2}$-inch maritime rope. For every 300 coils of rope made, a random sample of four coils of rope is selected and tested. To meet quality control standards, the average number of pounds that each coil will hold must be at least 1100 pounds. The test engineer tested four coils of rope. The first three held 1050 pounds, 1250 pounds, and 950 pounds, respectively. How many pounds must the last coil hold if the sample is to meet quality control standards?

EXAMPLE 3   Juan is selling supplies for restaurants in the Southwest. He gets paid $700 per month plus 8% of all the supplies he sells to restaurants. If he wants to earn more than $3100 per month, what value of products will he need to sell?

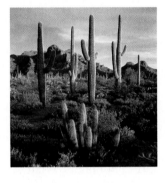

Solution

1. ***Understand the problem.*** Juan earns a fixed salary and a salary that depends on how much he sells (commission salary) each month.

   Let x = the value of the supplies he sells each month.

   Then $0.08x$ = the amount of the commission salary he earns each month.

   He wants his total income to be more than $3100 per month; thus

   $$\text{total income} > 3100.$$

2. ***Write the inequality.***

   The fixed income added to the commission income must be greater than 3100.

   $$700 \qquad + \qquad 0.08x \qquad\qquad > \qquad 3100$$

3. ***Solve and state the answer.***

   $$700 + 0.08x > 3100$$
   $$0.08x > 2400 \qquad \text{Subtract 700 from each side.}$$
   $$x > 30{,}000 \qquad \text{Divide each side by 0.08.}$$

   Juan must sell more than $30,000 worth of supplies each month.

4. ***Check.*** Any value greater than 30,000 should satisfy the original inequality.

   Suppose we pick $x = 30{,}001$ and substitute it into the inequality.

   $$700 + 0.08(30{,}001) \overset{?}{>} 3100$$
   $$700 + 2400.08 \overset{?}{>} 3100$$
   $$3100.08 > 3100 \quad \checkmark$$

Practice Problem 3   Rita is selling commercial fire and theft alarm systems to local companies. She gets paid $1400 per month plus 2% of the cost of all the systems she sells in one month. She wants to earn more than $2200 per month. What value of products will she need to sell each month?

Verbal and Writing Skills

Translate to an inequality.

1. The cost is greater than $67,000.

2. The clearance of the bridge is less than 12 feet.

3. The number of people is not more than 120.

4. The tax increase is more than $7890.

5. The height of stratocumulus clouds is not more than 1500 feet.

6. The volume of water in the aquarium must not drop below 35 gallons.

7. To earn an A in algebra, Ramon cannot get less than a 93 average.

8. The camp-ins at the Museum of Science in Boston cannot accommodate more than 600 people.

Applications

Solve using an inequality.

▲ **9.** ***Window Design*** A triangular window on Ross Camp's new Searay boat has two sides that measure 87 centimeters and 64 centimeters, respectively. The perimeter of the triangle must not exceed 291 centimeters. What are the possible values for the length of the third side of the window?

▲ **10.** ***Package Size*** The U.S. Postal Service provides guidelines for package size. One of the guidelines specifies the dimensions of the package: The length plus the girth (circumference around the widest part) cannot exceed 130 inches. If you are preparing a package that is 33 inches wide and 28 inches high, how long is the package permitted to be?

11. ***Flower Delivery*** Flora's Flowers shares the Speedy Delivery Service with Ruby's Rosebuds. Flora estimates that if she averages at least 24 deliveries per week, she will save money by hiring a student to deliver her flowers after school. Her contract with Speedy Delivery is up for renewal in 6 weeks. Flora tracks her deliveries for 5 weeks: 18, 40, 21, 7, and 36. At least how many deliveries must she average in week 6 to convince her to drop the delivery service?

12. ***Course Average*** Sophia has scored 85, 77, and 68 on three tests in her biology course. She has one final test to take and wants to obtain an average of 80 or more in the course. What possible values can she obtain on her final test and still achieve her goal?

▲ **13.** *Garden Design* Joel plans to plant a flower garden in the front of his house, where he can fit a garden that is 8 feet wide. He wishes to use the can of wildflower mix that he got for his birthday. The can of seeds claims that it will luxuriously carpet an area of no more than 60 square feet. What are the possible dimensions for the depth of Joel's garden, so that it will be covered with wildflowers?

14. *Class Size* The Lee County Elementary School District noted that there was an increase of 1785 pupils during the fall semester. How many new teachers must they add to keep their average class size under 28? (Assume one teacher for each class.)

15. *Truck Rental* Suzie and Freddie decided that to save money, they would move themselves from their apartment to their new house. They had to rent a truck to do this. Because their budget was very strict, they could afford to spend no more than $150. The truck costs $37.50 per day and 19.5 cents per mile. If they decided they would need the truck for two days, what was the limit on the mileage they could put on the truck?

16. *Salary* Robin just graduated from college and has taken her first job. She is working for a computer sales company, Compusell. While she hopes to eventually get into management, she is presently working as a salesperson. Her earnings consist of a monthly salary of $1300, plus 6% of all her sales. To meet her monthly obligations, Robin wants to earn at least $2800 per month. What is the amount of sales she must make monthly to meet her goal?

17. *Coolant* The coolant in Tom's car is effective for radiator temperatures of less than 110 degrees Celsius. What would be the corresponding temperature range in degrees Fahrenheit? (Use $C = \frac{5}{9}(F - 32)$ in your inequality.)

18. *IQ* A person's IQ is found by multiplying mental age (m), as indicated by standard tests, by 100 and dividing this result by the chronological age (c). This gives the formula $IQ = 100\frac{m}{c}$. If a group of 14-year-old children has an IQ of not less than 80, what is the mental age range of this group?

19. *Federal Aid* The amount of federal aid in billions of dollars given to state and local governments can be approximated by the equation $A = 9.2x + 77.1$, where x is the number of years since 1980. Use this equation to determine those years for which the amount of federal aid to state and local governments will exceed $307.1 billion. (*Source:* U.S. Office of Management and Budget)

20. *Social Security* The number of people in millions who are receiving Social Security benefits can be approximated by the equation $N = 0.57x + 34.8$, where x is the number of years since 1980. Use this equation to determine those years for which the number of people receiving Social Security benefits will exceed 50.19 million. (*Source:* U.S. Social Security Administration)

21. *Car Repair* Becky's car needed some major engine work in order to pass inspection. She decided that she would only have the work done if the total repair bill would be no more than than $1200. The dealer told her that the parts she needed would cost $840. He could not guarantee how many hours labor, at $45 per hour, the job would take. At most how many hours of labor could the repairs take to stay within her price?

22. *Apartment Furnishing* Rachel and Alissa are renting their first apartment. They need to buy lamps and throw rugs to complete the apartment. Their budget for these items is a maximum of $340. Rugs cost $48 each and lamps cost $35 each. They decide to buy 4 rugs. How many lamps can they buy?

To Think About

23. *CD Manufacturing* A compact disc recording company's weekly costs can be described by the equation cost = 5000 + 7*n*, where *n* is the number of compact discs manufactured in one week. The equation for the amount of income produced from selling these *n* discs is income = 18*n*. At least how many discs need to be manufactured and sold in one week for the income to be greater than the cost?

24. *VCR Manufacturing* A company that manufactures VCRs has weekly costs described by the equation cost = 12,000 + 100*n*, where *n* is the number of VCRs manufactured in one week. The equation for the amount of income produced from selling these *n* VCRs is income = 210*n*. At least how many VCRs need to be manufactured and sold in one week for the income to be greater than the cost?

Cumulative Review

Solve.

25. $10 - 3x > 14 - 2x$

26. $2x - 3 \geq 7x - 18$

27. $30 - 2(x + 1) \leq 4x$

28. $2(x + 3) - 22 < 4(x - 2)$

29. *Leaky Faucet* Frank has a leaky faucet in his bathtub. He closed the drain and measured that the faucet drips 8 gallons of water in one 24-hour period.
   (a) If the faucet isn't fixed, how many gallons of water will be wasted in one year?

   (b) If the faucet isn't fixed and water costs $8.50 per 1000 gallons, how much extra money will Frank have to pay on his water bill for the year?

Putting Your Skills to Work

Predicting the Population of Caribou in Alaska

In the Fortymile River area of Alaska, there is a huge herd of caribou. The herd seems to be slowly growing in number and is carefully monitored by the Alaska State Department of Fish and Game. Wildlife managers find it helpful to have an equation that predicts the number of caribou that will be in this Fortymile herd from year to year. The herd has fluctuated greatly in size. In 1920 the herd was estimated to be 568,000. In 1975 the herd was reduced to only 3982. Thanks to wildlife management techniques, the herd seems to be experiencing slow but positive growth. During the period 1986 to 1990 the following growth was observed:

Year	Estimate of Herd Size
1986	15,303
1988	19,975
1990	22,766

The desired equation that models the slow growth of the caribou in this herd is in the form

$$y = ax + b$$

    y is the size of the herd

    a is the yearly growth factor

    b is the size of the herd in 1986

    x is the number of years since 1986

Fortymile River

This type of equation will be discussed in more detail in Chapter 7. This type of linear equation can only be used to approximate change or growth that experiences approximately the same amount of growth or change each year.

Problems for Individual Investigation and Analysis

Use the preceding table to obtain the requested information.

1. Determine an estimate for the amount of increase in herd size per year by subtracting the 1986 figure from the 1990 figure and divide by 4. This will give you the value of a. Round your value of a to the nearest whole number. What is the value of b? Write the equation in the form $y = ax + b$.

2. Use your equation to predict what the herd size for 1995. Use the equation to predict the herd size for 1984.

Problems for Group Investigation and Cooperative Learning

3. Use the equation to predict the herd size for 2003. The actual measured population for 2003 was 43,375. How accurate was the equation? Does the equation predict a value that is too high or too low?

4. Using the population for 1986 in the table and the recorded population of 43,375 in the year 2003, calculate a new estimated value for the amount of increase in herd size per year. Round to the nearest whole number for a. Write a new equation that uses this value.

Procedure for Solving Applied Problems

1. **Understand the problem.**
   (a) Read the word problem carefully to get an overview.
   (b) Determine what information you will need to solve the problem.
   (c) Draw a sketch. Label it with the known information. Determine what needs to be found.
   (d) Choose a variable to represent one unknown quantity.
   (e) If necessary, represent other unknown quantities in terms of that same variable.

2. **Write an equation.**
   (a) Look for key words to help you to translate the words into algebraic symbols.
   (b) Use a given relationship in the problem or an appropriate formula in order to write an equation.

3. **Solve and state the answer.**

4. **Check.**
   (a) Check the solution in the original equation. Is the answer reasonable?
   (b) Be sure the solution to the equation answers the question in the word problem. You may need to do some additional calculations if it does not.

EXAMPLE A

The perimeter of a rectangle is 126 meters. The length of the rectangle is 6 meters less than double the width. Find the dimensions of the rectangle.

1. *Understand the problem.*
   We want to find the length and the width of a rectangle whose perimeter is 126 meters.

   The length is *compared to* the width, so we start with the width.

   The length is 6 meters less than double the width.

2. *Write an equation.*
   The perimeter of a rectangle is
   $P = 2w + 2l$.

3. *Solve.*

Length

Width   Width

Length

Let w = width.

Then $l = 2w - 6$.

$126 = 2w + 2(2w - 6)$

$126 = 2w + 4w - 12$
$126 = 6w - 12$
$138 = 6w$
$23 = w$ The width is 23 meters.

$2w - 6 = 2(23) - 6 =$
$46 - 6 = 40$

The length is 40 meters.

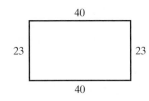

40

23   23

40

4. *Check:*
   Is this reasonable?
   Yes. A rectangle 23 meters wide and 40 meters long seems to be about right for the perimeter to be 126 meters.
   Is the perimeter exactly 126 meters?
   Is the length exactly 6 meters less than double the width?

$2(23) + 2(40) \overset{?}{=} 126$
$46 + 80 \overset{?}{=} 126$
$126 = 126$ ✓
$40 \overset{?}{=} 2(23) - 6$
$40 \overset{?}{=} 46 - 6$
$40 = 40$ ✓

EXAMPLE B

Gina saved some money for college. She invested $2400 for one year and earned $225 in simple interest. She invested part of it at 12% and the rest of it at 9%. How much did she invest at each rate?

1. *Understand the problem.*
   We want to find each amount that was invested.

   interest earned at 12% + interest earned at 9% = total interest of $225

   We let x represent one quantity of money.

   Let x = amount of money invested at 12%.

   We started with $2400. If we invest x at 12%, we still have $(2400 - x)$ left.

   Then $2400 - x =$ the amount of money invested at 9%.

   Interest $= prt$
   Interest at 12%
   Interest at 9%

   $I_1 = 0.12x$
   $I_2 = 0.09(2400 - x)$

2. *Write an equation.*
   $I_1 + I_2 = 225$.

   $0.12x + 0.09(2400 - x) = 225$

3. *Solve.*

   $0.12x + 216 - 0.09x = 225$
   $0.03x + 216 = 225$
   $0.03x = 9$
   $x = \dfrac{9}{0.03}$
   $x = 300$

   $300 was invested at 12%.
   $2400 - x = 2400 - 300 = 2100$
   $2100 was invested at 9%.

4. *Check:*
   Is this reasonable? Yes.
   Are the conditions of the problem satisfied? Does the total amount invested equal $2400?
   Will $2100 at 9% and $300 at 12% yield $225 in interest?

   $2100 + 300 \overset{?}{=} 2400$
   $2400 = 2400$ ✓

   $0.09(2100) + 0.12(300) \overset{?}{=} 225$
   $189 + 36 \overset{?}{=} 225$
   $225 = 225$ ✓

Topic	Procedure	Examples
Geometric formulas, *p. 224.*	Know the definitions of and formulas related to basic geometric shapes such as the triangle, rectangle, parallelogram, trapezoid, and circle. Review these formulas from the beginning of Section 3.5 as needed. Formulas for the volume and total surface area of three solids are also given there.	The volume of a right circular cylinder found in a four-cylinder car engine is determined by the formula $V = \pi r^2 h$. Find V if $r = 6$ centimeters and $h = 5$ centimeters. Use $\pi \approx 3.14$. $$V = 3.14(6 \text{ cm})^2(5 \text{ cm})$$ $$= 3.14(36)(5) \text{ cm}^3$$ $$= 565.2 \text{ cc (or cm}^3)$$ Four such cylinders would make this a 2260.8-cc engine.
Using inequalities to solve word problems, *p. 235.*	Be sure that the inequality you use fits the conditions of the word problem. **1.** *At least 87* means ≥ 87. **2.** *No less than $150* means $\geq \$150$. **3.** *At most $500* means $\leq \$500$. **4.** *No more than 70* means ≤ 70. **5.** *Less than 2* means < 2. **6.** *More than 10* means > 10.	Julie earns 2% on sales over $1000. How much must Julie make in sales to earn at least $300? Let x = amount of sales. $$(2\%)(x - \$1000) \geq \$300$$ $$0.02(x - 1000) \geq 300$$ $$x - 1000 \geq 15{,}000$$ $$x \geq 16{,}000$$ Julie must make sales of $16,000 or more (at least $16,000).

Chapter 3 Review Problems

Section 3.1

Write an algebraic expression. Use the variable x to represent the unknown value.

1. 19 more than a number

2. two-thirds of a number

3. half a number

4. 18 less than a number

5. triple the sum of a number and 4

6. twice a number decreased by three

Write an expression for each of the quantities compared. Use the letter specified.

7. **Workforce** The number of working people is four times the number of retired people. The number of unemployed people is one-half the number of retired people. (Use the letter r.)

8. **Geometry** The length of the rectangle is 5 meters more than triple the width. (Use the letter w.)

9. **Geometry** A triangle has three angles A, B, and C. The number of degrees in angle A is double the number of degrees in angle B. The number of degrees in angle C is 17 degrees less than the number in angle B. (Use the letter b.)

10. **Class Size** There are 29 more students in biology class than in algebra. There are one-half as many students in geology class as in algebra. (Use the letter a.)

Section 3.2

Solve

11. Eighteen is taken away from twice a number. The result is -8. What is the number?

12. When you reduce triple a number by 6, the result is 15. Find the number.

13. *Table Set* Six chairs and a table cost $450. If the table costs $210, how much does one chair cost?

14. *Age* Jon is twice as old as his cousin David. If Jon is 32, how old is David?

15. *Speed Rates* Two cars left Center City and drove to the shore 330 miles away. One car was driven at 50 miles an hour. The other car was driven at 55 miles an hour. How long did it take each car to make the trip?

16. *Math Score* A student took five math tests. She obtained scores of 85, 78, 65, 92, and 70. She needs to complete one more test. If she wants to get an average of 80 on all six tests, what score does she need on her last test?

Section 3.3

▲ **17.** *Geometry* An isosceles triangle has a perimeter of 85 feet. If one of the two equal sides measures 31 feet, what is the measure of the third side?

▲ **18.** *Geometry* The measure of the second angle of a triangle is three times the measure of the first angle. The measure of the third angle is 12 degrees less than twice the measure of the first. Find the measure of each of the three angles.

19. *Rope* A piece of rope 50 yards long is cut into two pieces. One piece is three-fifths as long as the other. Find the length of each piece.

20. *Income* Jon and Lauren have a combined income of $48,000. Lauren earns $3000 more than half of what Jon earns. How much does each earn?

Section 3.4

21. *Electric Bill* The electric bill at Jane's house this month was $71.50. The charge is based on a flat rate of $25 per month plus a charge of $0.15 per kilowatt-hour of electricity used. How many kilowatt-hours of electricity were used?

22. *Car Rental* Abel rented a car from Sunshine Car Rentals. He was charged $39 per day and $0.25 per mile. He rented the car for three days and paid a rental cost of $187. How many miles did he drive the car?

23. *Simple Interest* Boxell Associates has $7400 invested at 5.5% simple interest. After money was withdrawn, $242 was earned on the remaining funds. How much was withdrawn?

24. *CD Player* Jamie bought a new CD player for her car. When she bought it, she found that the price had been decreased by 18%. She was able to buy the CD player for $36 less than the original price. What was the original price?

25. *Investments* Peter and Shelly invested $9000 for one year. Part of it was invested at 12% and the remainder was invested at 8%. At the end of one year the couple had earned exactly $1000 in simple interest. How much did they invest at each rate?

26. *Investments* A man invests $5000 in a savings bank. He places part of it in a checking account that earns 4.5% and the rest in a regular savings account that earns 6%. His total annual income from this investment in simple interest is $270. How much was invested in each account?

27. *Coin Change* Mary has $3.75 in nickels, dimes, and quarters. She has three more quarters than dimes. She has twice as many nickels as quarters. How many of each coin does she have?

28. *Coin Change* Tim brought in $3.65 in coins. He had two more quarters than he had dimes. He had one fewer nickel than the number of dimes. How many coins of each type did he have?

Section 3.5

Solve.

▲ **29. *Geometry*** What is the surface area of a exercise ball with a diameter of 32 inches? (Use $\pi \approx 3.14$.)

▲ **30. *Trench Length*** Wilfredo and Tina bought a new screen house for camping. It was hexagonal (6 sides) in shape and 4 ft long on a side. If they wanted to dig a trench around their house so that rainwater would flow away, how long would the trench be? (You are looking for the perimeter of the screen house.)

▲ **31. *Geometry*** Find the third angle of a triangle if one angle measures 62 degrees and the second angle measures 47 degrees.

▲ **32. *Geometry*** Find the area of a triangle whose base is 8 miles and whose altitude is 10.5 miles.

▲ **33. *Geometry*** Find the volume of a rectangular solid whose width is 8 feet, whose length is 12 feet, and whose height is 20 feet.

▲ **34. *Geometry*** Find the volume of a sphere whose radius is 3 centimeters. (Use $\pi \approx 3.14$.)

▲ **35. *Geometry*** Find the area of a circle whose diameter is 18 centimeters. (Use $\pi \approx 3.14$.)

▲ **36. *Geometry*** The perimeter of an isosceles triangle is 46 inches. The two equal sides are each 17 inches long. How long is the third side?

▲ **37. *House Painting*** Jim wants to have the exterior siding on his house painted. Two sides of the house have a height of 18 feet and a length of 22 feet, and the other two sides of the house have a height of 18 feet and a length of 19 feet. The painter has quoted a price of $2.50 per square foot. The windows and doors of the house take up 100 square feet. How much will the painter charge to paint the siding?

▲ **38. *Aluminum Cylinder*** Find the cost to build a cylinder out of aluminum if the radius is 5 meters, the height is 14 meters, and the cylinder will have a flat top and flat bottom. The cost factor has been determined to be $40.00 per square meter. (*Hint:* First find the total surface area.) (Use $\pi \approx 3.14$.)

Section 3.6

Solve.

39. *Car Rental* Chelsea rents a car for $35 a day and 15 cents a mile. How far can she drive if she wants to rent the car for two days and does not want to spend more than $100?

40. *Income* Philip earns $12,000 per year in salary and 4% commission on his sales. How much were his sales if his annual income was more than $24,000?

41. *Transcontinental Railroad* In 1862, two companies were granted the rights to build the transcontinental railroad from Omaha, Nebraska to Sacramento, California. The Central Pacific railroad began in 1863 from Sacramento heading east. The Union Pacific began 24 months later, leaving from Omaha and heading west. The Central Pacific averaged 8.75 miles of track each month, while the Union Pacific averaged 20 miles of track per month. The two companies met at Promontory, Utah after completing 1590 miles of track. How much track did each build?

42. *Clothes Shopping* Michael plans to spend less than $70 on shirts and ties. He bought three shirts for $17.95 each. How much can he spend on ties?

Mixed Practice

Solve.

43. *TV Show* Wally timed a popular "30-minute" TV show. The show lasted 4 minutes longer than four times the amount of time that was devoted to commercials. How many minutes long was the actual show?

44. *Balanced Weight* Faye conducted an experiment. To verify the weight of four identical steel balls, she placed them on a balance. She placed three of the balls on one side of the balance. On the other side she placed an 8-ounce weight and one of the balls. She discovered that these two sides exactly balanced on the scale. How much does each steel ball weigh?

45. *Electricity Use* Fred and Nancy Sullivan are planning to purchase a new refrigerator. The most efficient model costs $870, but the electricity to operate it would cost only $41 per year. The less efficient model is the same size. It costs $744, but the electricity to operate it would cost $83 per year. How long will it be before the total cost is greater for the less efficient model?

46. *Buying vs. Renting* Custom Computer Works of Gloucester rents its office building and warehouse for $1800 per month. The owner is thinking of buying the building. He could purchase it with a down payment of $42,000. He would then make monthly mortgage payments of $1100. How many months would it take for his total costs to be less to purchase the building rather than rent it?

▲ 47. *Bookcase Construction* Judy Carter is planning to build a bookcase for her home office. She has only 38 feet of lumber to make the unit. The bookcase is constructed of shelves as shown in the diagram. She wants the width of the bookcase to be 2.5 times the height. Find the width and the height of the bookcase she needs to construct.

48. *Retirement Plan* Rick Ponticelli is investing some money in a retirement plan. He plans to invest $100 a month the first year, $200 a month the second year, $300 a month the third year, and continue with that pattern. How many years will it be before he has invested $66,000 in his retirement plan?

49. Basketball In a recent basketball game the Boston Celtics scored 99 points. They scored three times as many field goals as free throws. They also made 12 three-point baskets. How many field goals did they make? How many free throws did they make? (A field goal is worth two points and a free throw is worth one point.)

51. Deck Building Ryan and Carl spent 28 hours building a deck. Ryan worked 8 less than twice as many hours as Carl. How many hours did each man put in?

▲ **50. Room Dimensions** The perimeter of a rectangular room in Sally's house is 76 feet. The length of the room is 10 feet less than double the width of the room. Find the dimensions of the room.

52. Investments Sarah Tanguay has $2000 more invested at 8.5% than she does at 9.75%. Both investments are at simple interest rates. If the annual return from each investment is the same, how much is invested at each rate?

▲ **53. Geometry** The first angle of a triangle measures 20° larger than the second angle of a triangle. The third angle measures twice as large as the second angle. Find the measure of these three angles.

55. Final Exam Score Thelma has a math final exam that counts twice as much as a test. She had four tests with grades of 88, 77, 95, and 92. What score does she need to obtain on the final exam to get an average of 90 for the course?

57. Jet Travel A jet plane travels 32 miles in three minutes. How far can the plane travel in one hour if it continues at that speed? How many minutes would it take the jet to travel from Boston to Denver (approximately 1800 miles) at that same speed?

59. Field Trip A Ford Explorer can hold a driver and four passengers. A Dodge Caravan can hold a driver and six passengers. The biology department is taking a total of 95 people on a field trip. A total of 20 of the 95 people are available to drive. Not all potential drivers will be needed. The college wants to use twice as many Caravans as Explorers because the Caravans get better gas mileage. How many of each kind of vehicle are needed?

▲ **54. Sign Construction** A sign is made in the shape of a trapezoid. The altitude of the trapezoid is 12 feet. The two bases of the trapezoid are 15 feet and 19 feet. The sign is constructed of prime cedar that costs $6 per square foot. How much did the cedar that was used to make the sign cost?

56. Mileage Deduction Barbara drove her car 27,000 miles last year. She drove twice as much for pleasure as she did for business. She can deduct $0.33 per mile for all business miles. Last year she claimed a business mileage deduction of $2000. Was her deduction correct? Please explain your answer. If her deduction was not correct, please indicate the correct deduction.

58. Football A running back in a Wheaton College football game catches a pass and runs for the goal line 80 yards away. He can run a 50-yard dash in six seconds. If he runs at 80% of his speed in the 50-yard dash, how long will it take him teach to reach the goal line?

▲ **60. Geometry** The measure of the largest angle of a triangle is triple the measure of the smallest angle. The third angle is 12 degrees less than double the measure of the smallest angle. Find the measure of each angle.

61. Zorbing Zorbing is a sport, invented in New Zealand, where you are secured inside an inner capsule in a large transparent ball and rolled down a slope. The ball needs to be inflated with an inert gas before it can be used. An inflated zorb has a diameter of 3 meters. If the Zorb itself weighs 80 kg, the zorber weighs 75 kg, and the gas used to inflate the ball weighs 2 kg per cubic meter, what is the total weight of the zorb? (Use 3.14 to approximate π.)

Remember to use your Chapter Test Prep Video CD to see the worked-out solutions to the test problems you want to review.

Solve.

1. _____

2. _____

3. _____

4. _____

5. _____

6. _____

7. _____

8. _____

9. _____

10. _____

1. A number is doubled and then decreased by 11. The result is 59. What is the original number?

2. The sum of one-half of a number, one-ninth of the number, and one-twelfth of the number is twenty-five. Find the original number.

3. Double the sum of a number and 5 is the same as fourteen more than triple the same number. Find the number.

▲ 4. A triangular region has a perimeter of 66 meters. The first side is two-thirds of the second side. The third side is 14 meters shorter than the second side. What are the lengths of the three sides of the triangular region?

▲ 5. A rectangle has a length 7 meters longer than double the width. The perimeter is 134 meters. Find the dimensions of the rectangle.

6. Three harmful pollutants were measured by a consumer group in the city. The total sample contained 15 parts per million of three harmful pollutants. The amount of the first pollutant was double the second. The amount of the third pollutant was 75% of the second. How many parts per million of each pollutant were found?

7. Raymond has a budget of $1940 to rent a computer for his company office. The computer company he wants to rent from charges $200 for installation and service as a one-time fee. Then they charge $116 per month rental for the computer. How many months will Raymond be able to rent a computer with this budget?

8. Last year the yearly tuition at Westmont College went up 8%. This year's charge for tuition for the year is $22,680. What was it last year before the increase went into effect?

9. Franco invested $4000 in money market funds. Part was invested at 14% interest, the rest at 11% interest. At the end of each year the fund company pays interest. After one year he earned $482 in simple interest. How much was invested at each interest rate?

10. Mary has $3.50 in change. She has twice as many nickels as quarters. She has one less dime than she has quarters. How many of each coin does she have?

▲ **11.** Find the circumference of a circle with radius 34 inches. Use $\pi = 3.14$ as an approximation.

11. _____

▲ **12.** Find the area of a trapezoid if the two bases are 10 inches and 14 inches and the altitude is 16 inches.

12. _____

▲ **13.** Find the volume of a sphere with radius 10 inches. Use $\pi = 3.14$ as an approximation and round your answer to the nearest cubic inch.

13. _____

▲ **14.** Find the area of a parallelogram with a base of 12 centimeters and an altitude of 8 centimeters.

14. _____

▲ **15.** How much would it cost to carpet the area shown in the figure if carpeting costs $12 per square yard?

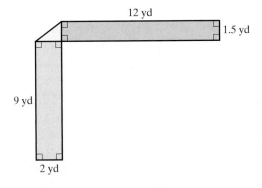

15. _____

16. _____

16. Cathy scored 76, 84, and 78 on three tests. How much must she score on the next test to have at least an 80 average?

17. _____

17. Carol earns $15,000 plus 5% commission on sales over $10,000. How much must she make in sales to earn more than $20,000?

▲ **18.** Andrew is putting down tile for a new gymnasium. He can lay 25 square feet of tile in an hour. How long will it take him to do a gymnasium 150 feet long and 100 feet wide?

18. _____

1. _____

2. _____

3. _____

4. _____

5. _____

6. _____

7. _____

8. _____

9. _____

10. _____

11. _____

12. _____

13. _____

14. _____

15. _____

Approximately one-half of this test covers the content of Chapters 0–2. The remainder covers the content of Chapter 3.

1. Divide. $2.4\overline{)8.856}$

2. Add. $\dfrac{3}{8} + \dfrac{5}{12} + \dfrac{1}{2}$

3. Simplify. $3(2x - 3y + 6) + 2(-3x - 7y)$

4. Evaluate. $12 - 3(2 - 4) + 12 \div 4$

5. Solve for b. $H = \dfrac{1}{2}(3a + 5b)$

6. Solve for x and graph your solution. $5x - 3 \le 2(4x + 1) + 4$

7. When a number is doubled and then increased by 15, the result is 1. Find the number.

8. The number of students in the psychology class is 34 fewer than the number of students in world history. The total enrollment for the two classes is 134 students. How many students are in each class?

▲ **9.** A rectangle has a perimeter of 78 centimeters. The length of the rectangle is 11 centimeters longer than triple the width. Find the dimensions of the rectangle.

10. This year the sales for Homegrown Video are 35% higher than last year. This year the sales are $182,250. What were the sales last year?

11. Hassan invested $7000 for one year. He invested part of it in a high-risk fund that pays 15% interest. He placed the rest in a safe, low-risk fund that pays 7% interest. At the end of one year he earned $730 in simple interest. How much did he invest in each of the two funds?

12. Linda has some dimes, nickels, and quarters in her purse. The total value of these coins is $2.55. She has three times as many dimes as quarters. She has three more nickels than quarters. How many coins of each type does she have?

▲ **13.** Find the area of a triangle with an altitude of 13 meters and a base of 25 meters. What is the cost to construct such a triangle out of sheet metal that costs $4.50 per square meter?

▲ **14.** Find the volume of a sphere that has a radius of 3.00 inches. How much will the contents of the sphere weigh if it is filled with a liquid that weighs 1.50 pounds per cubic inch? Use $\pi = 3.14$ as an approximation.

15. Melinda Tobey saw a glacier in Alaska that moves 65 feet per day. How long will it take the glacier to reach a cabin that is 1625 feet away?

Polynomials are an important tool of mathematics. They are used to create formulas that model many real-world phenomena. Let us take a simple example. Suppose you wanted to know how much more you might make over your lifetime if you obtain a bachelor's degree as opposed to just an associate's degree. Suppose you wanted to know what the expected annual income will be in the future if you obtain a bachelor's degree. Do you think you could use your mathematical skills to determine these values? Turn to the Putting Your Skills to Work problems on page 295 to find out.

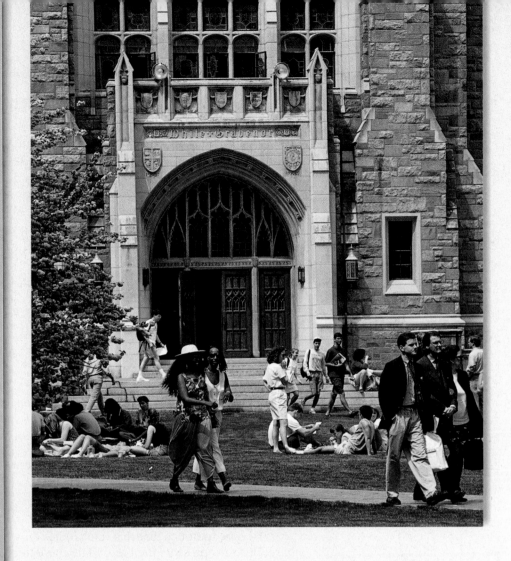

Exponents and Polynomials

Student Learning Objectives

After studying this section, you will be able to:

1 Use the product rule to multiply exponential expressions with like bases.

2 Use the quotient rule to divide exponential expressions with like bases.

3 Raise exponential expressions to a power.

1 Using the Product Rule to Multiply Exponential Expressions with Like Bases

Recall that x^2 means $x \cdot x$. That is, x appears as a factor two times. The 2 is called the **exponent.** The **base** is the variable x. The expression x^2 is called an **exponential expression.** What happens when we multiply $x^2 \cdot x^2$? Is there a pattern that will help us form a general rule?

$$(2^2)(2^3) = \overbrace{(2 \cdot 2)(2 \cdot 2 \cdot 2)}^{5 \text{ twos}} = 2^5$$

The exponent means 2 occurs 5 times as a factor.

$$(3^3)(3^4) = \overbrace{(3 \cdot 3 \cdot 3)(3 \cdot 3 \cdot 3 \cdot 3)}^{7 \text{ threes}} = 3^7$$

Notice that $3 + 4 = 7$.

$$(x^3)(x^5) = \overbrace{(x \cdot x \cdot x)(x \cdot x \cdot x \cdot x \cdot x)}^{8 \ x\text{'s}} = x^8$$

The sum of the exponents is $3 + 5 = 8$.

$$(y^4)(y^2) = \overbrace{(y \cdot y \cdot y \cdot y)(y \cdot y)}^{6 \ y\text{'s}} = y^6$$

The sum of the exponents is $4 + 2 = 6$.

We can state the pattern in words and then use variables.

> **THE PRODUCT RULE**
>
> To multiply two exponential expressions that have the same base, keep the base and *add the exponents.*
> $$x^a \cdot x^b = x^{a+b}$$

Be sure to notice that this rule applies only to expressions that have the *same base.* Here x represents the base, while the letters a and b represent the exponents that are added.

It is important that you apply this rule even when an exponent is 1. Every variable that does not have a written exponent is understood to have an exponent of 1. Thus $x^1 = x$, $y^1 = y$, and so on.

EXAMPLE 1 Multiply. **(a)** $x^3 \cdot x^6$ **(b)** $x \cdot x^5$

Solution

(a) $x^3 \cdot x^6 = x^{3+6} = x^9$

(b) $x \cdot x^5 = x^{1+5} = x^6$ Note that the exponent of the first x is 1.

Practice Problem 1 Multiply. **(a)** $a^7 \cdot a^5$ **(b)** $w^{10} \cdot w$

NOTE TO STUDENT: Fully worked-out solutions to all of the Practice Problems can be found at the back of the text starting at page SP-1

EXAMPLE 2 Simplify. **(a)** $y^5 \cdot y^{11}$ **(b)** $2^3 \cdot 2^5$ **(c)** $x^6 \cdot y^8$

Solution

(a) $y^5 \cdot y^{11} = y^{5+11} = y^{16}$

(b) $2^3 \cdot 2^5 = 2^{3+5} = 2^8$ Note that the base does not change! Only the exponent changes.

(c) $x^6 \cdot y^8$ The rule for multiplying exponential expressions does not apply since the bases are not the same. This cannot be simplified.

Practice Problem 2 Simplify, if possible.

(a) $x^3 \cdot x^9$ **(b)** $3^7 \cdot 3^4$ **(c)** $a^3 \cdot b^2$

We can now look at multiplying expressions such as $(2x^5)(3x^6)$.

The number 2 in $2x^5$ is called the **numerical coefficient.** Recall that a numerical co-efficient is a number that is multiplied by a variable. When we multiply two expressions such as $2x^5$ and $3x^6$, we first multiply the numerical coefficients; we multiply the variables with exponents separately.

EXAMPLE 3   Multiply.   **(a)** $(2x^5)(3x^6)$ **(b)** $(5x^3)(x^6)$ **(c)** $(-6x)(-4x^5)$

Solution

(a) $(2x^5)(3x^6) = (2 \cdot 3)(x^5 \cdot x^6)$ Multiply the numerical coefficients.

$\qquad\qquad\quad = 6(x^5 \cdot x^6)$ Use the rule for multiplying expressions with exponents. Add the exponents.

$\qquad\qquad\quad = 6x^{11}$

(b) Every variable that does not have a visible numerical coefficient is understood to have a numerical coefficient of 1. Thus x^6 has a numerical coefficient of 1.

$$(5x^3)(x^6) = (5 \cdot 1)(x^3 \cdot x^6) = 5x^9$$

(c) $(-6x)(-4x^5) = (-6)(-4)(x^1 \cdot x^5) = 24x^6$ Remember that x has an exponent of 1.

Practice Problem 3   Multiply.

(a) $(-a^8)(a^4)$ **(b)** $(3y^2)(-2y^3)$ **(c)** $(-4x^3)(-5x^2)$

Problems of this type may involve more than one variable or more than two factors.

EXAMPLE 4   Multiply. $(5ab)\left(-\frac{1}{3}a\right)(9b^2)$

Solution $(5ab)\left(-\frac{1}{3}a\right)(9b^2) = (5)\left(-\frac{1}{3}\right)(9)(a \cdot a)(b \cdot b^2)$

$\qquad\qquad\qquad\qquad\qquad\quad = -15a^2b^3$

As you do the following problems, keep in mind the rule for multiplying numbers with exponents and the rules for multiplying signed numbers.

Practice Problem 4   Multiply. $(2xy)\left(-\frac{1}{4}x^2y\right)(6xy^3)$

---

② Using the Quotient Rule to Divide Exponential Expressions with Like Bases

Frequently, we must divide exponential expressions. Since division by zero is unde-fined, in all problems in this chapter we assume that the denominator of any vari-able expression is not zero. We'll look at division in three separate parts.

Suppose that we want to simplify $x^5 \div x^2$. We could do the division the long way.

$$\frac{x^5}{x^2} = \frac{(x)(x)(x)\cancel{(x)}\cancel{(x)}}{\cancel{(x)}\cancel{(x)}} = x^3$$

Here we are using the arithmetical property of reducing fractions (see Section 0.1). When the same factor appears in both numerator and denominator, that factor can be removed.

A simpler way is to *subtract the exponents*. Notice that the base remains the same.

THE QUOTIENT RULE (PART 1)

$\dfrac{x^a}{x^b} = x^{a-b}$ Use this form if the larger exponent is in the numerator and $x \neq 0$.

EXAMPLE 5   Divide.   **(a)** $\dfrac{2^{16}}{2^{11}}$ **(b)** $\dfrac{x^5}{x^3}$ **(c)** $\dfrac{y^{16}}{y^7}$

Solution

(a) $\dfrac{2^{16}}{2^{11}} = 2^{16-11} = 2^5$ Note that the base does *not* change.

(b) $\dfrac{x^5}{x^3} = x^{5-3} = x^2$

(c) $\dfrac{y^{16}}{y^7} = y^{16-7} = y^9$

NOTE TO STUDENT: Fully worked-out solutions to all of the Practice Problems can be found at the back of the text starting at page SP-1

Practice Problem 5   Divide.   **(a)** $\dfrac{10^{13}}{10^7}$ **(b)** $\dfrac{x^{11}}{x}$ **(c)** $\dfrac{y^{18}}{y^8}$

Now we consider the situation where the larger exponent is in the denominator. Suppose that we want to simplify $x^2 \div x^5$.

$$\frac{x^2}{x^5} = \frac{\cancel{(x)}\cancel{(x)}}{\cancel{(x)}\cancel{(x)}(x)(x)(x)} = \frac{1}{x^3}$$

THE QUOTIENT RULE (PART 2)

$\dfrac{x^a}{x^b} = \dfrac{1}{x^{b-a}}$ Use this form if the larger exponent is in the denominator and $x \neq 0$.

EXAMPLE 6   Divide.   **(a)** $\dfrac{12^{17}}{12^{20}}$ **(b)** $\dfrac{b^7}{b^9}$ **(c)** $\dfrac{x^{20}}{x^{24}}$

Solution

(a) $\dfrac{12^{17}}{12^{20}} = \dfrac{1}{12^{20-17}} = \dfrac{1}{12^3}$ Note that the base does *not* change.

(b) $\dfrac{b^7}{b^9} = \dfrac{1}{b^{9-7}} = \dfrac{1}{b^2}$

(c) $\dfrac{x^{20}}{x^{24}} = \dfrac{1}{x^{24-20}} = \dfrac{1}{x^4}$

Practice Problem 6   Divide.   **(a)** $\dfrac{c^3}{c^4}$ **(b)** $\dfrac{10^{31}}{10^{56}}$ **(c)** $\dfrac{z^{15}}{z^{21}}$

When there are numerical coefficients, use the rules for dividing signed numbers to reduce fractions to lowest terms.

EXAMPLE 7   Divide.   (a) $\dfrac{5x^5}{25x^7}$ (b) $\dfrac{-12x^8}{4x^3}$ (c) $\dfrac{-16x^7}{-24x^8}$

Solution

(a) $\dfrac{5x^5}{25x^7} = \dfrac{1}{5x^{7-5}} = \dfrac{1}{5x^2}$ (b) $\dfrac{-12x^8}{4x^3} = -3x^{8-3} = -3x^5$

(c) $\dfrac{-16x^7}{-24x^8} = \dfrac{2}{3x^{8-7}} = \dfrac{2}{3x}$

Practice Problem 7   Divide.   (a) $\dfrac{-7x^7}{-21x^9}$ (b) $\dfrac{15x^{11}}{-3x^4}$ (c) $\dfrac{23x^8}{46x^9}$

You have to work very carefully if two or more variables are involved. Treat the coefficients and each variable separately.

EXAMPLE 8   Divide.   (a) $\dfrac{x^3y^2}{5xy^6}$ (b) $\dfrac{-3x^2y^5}{12x^6y^8}$

Solution

(a) $\dfrac{x^3y^2}{5xy^6} = \dfrac{x^2}{5y^4}$ (b) $\dfrac{-3x^2y^5}{12x^6y^8} = -\dfrac{1}{4x^4y^3}$

Practice Problem 8   Divide.   (a) $\dfrac{x^7y^9}{y^{10}}$ (b) $\dfrac{12x^5y^6}{-24x^3y^8}$

Suppose that a given base appears with the same exponent in the numerator and denominator of a fraction. In this case we can use the fact that *any nonzero number divided by itself is* 1.

EXAMPLE 9   Divide.   (a) $\dfrac{x^6}{x^6}$ (b) $\dfrac{3x^5}{x^5}$

Solution

(a) $\dfrac{x^6}{x^6} = 1$ (b) $\dfrac{3x^5}{x^5} = 3\left(\dfrac{x^5}{x^5}\right) = 3(1) = 3$

Practice Problem 9   Divide.   (a) $\dfrac{10^7}{10^7}$ (b) $\dfrac{12a^4}{15a^4}$

Do you see that if we had subtracted exponents when simplifying $\dfrac{x^6}{x^6}$ we would have obtained x^0 in Example 9? So we can surmise that any number (except 0) to the 0 power equals 1. We can write this fact as a separate rule.

THE QUOTIENT RULE (PART 3)

$$\dfrac{x^a}{x^a} = x^0 = 1 \quad \text{if } x \neq 0 \quad (0^0 \text{ remains undefined}).$$

TO THINK ABOUT: What Is 0 to the 0 Power?   What about 0^0? Why is it undefined? $0^0 = 0^{1-1}$. If we use the quotient rule, $0^{1-1} = \dfrac{0}{0}$. Since division by zero is undefined, we must agree that 0^0 is undefined.

EXAMPLE 10 Divide. **(a)** $\dfrac{4x^0 y^2}{8^0 y^5 z^3}$ **(b)** $\dfrac{5x^2 y}{10x^2 y^3}$

Solution

(a) $\dfrac{4x^0 y^2}{8^0 y^5 z^3} = \dfrac{4(1)y^2}{(1)y^5 z^3} = \dfrac{4y^2}{y^5 z^3} = \dfrac{4}{y^3 z^3}$ **(b)** $\dfrac{5x^2 y}{10x^2 y^3} = \dfrac{1x^0}{2y^2} = \dfrac{(1)(1)}{2y^2} = \dfrac{1}{2y^2}$

Practice Problem 10 Divide. **(a)** $\dfrac{-20a^3 b^8 c^4}{28a^3 b^7 c^5}$ **(b)** $\dfrac{5x^0 y^6}{10x^4 y^8}$

We can combine all three parts of the quotient rule we have developed.

THE QUOTIENT RULE

$\dfrac{x^a}{x^b} = x^{a-b}$ Use this form if the larger exponent is in the numerator and $x \neq 0$.

$\dfrac{x^a}{x^b} = \dfrac{1}{x^{b-a}}$ Use this form if the larger exponent is in the denominator and $x \neq 0$.

$\dfrac{x^a}{x^a} = x^0 = 1$ if $x \neq 0$.

We can combine the product rule and the quotient rule to simplify algebraic expressions that involve both multiplication and division.

EXAMPLE 11 Simplify. $\dfrac{(8x^2 y)(-3x^3 y^2)}{-6x^4 y^3}$

Solution $\dfrac{(8x^2 y)(-3x^3 y^2)}{-6x^4 y^3} = \dfrac{-24x^5 y^3}{-6x^4 y^3} = 4x$

Practice Problem 11 Simplify. $\dfrac{(-6ab^5)(3a^2 b^4)}{16a^5 b^7}$

NOTE TO STUDENT: Fully worked-out solutions to all of the Practice Problems can be found at the back of the text starting at page SP-1

3 **Raising Exponential Expressions to a Power**

How do we simplify an expression such as $(x^4)^3$? $(x^4)^3$ is x^4 raised to the third power. For this type of problem we say that we are raising a power to a power. A problem such as $(x^4)^3$ could be done by writing the following.

$$(x^4)^3 = x^4 \cdot x^4 \cdot x^4 \quad \text{By definition}$$
$$= x^{12} \quad \text{By adding exponents}$$

Notice that when we add the exponents we get $4 + 4 + 4 = 12$. This is the same as multiplying 4 by 3. That is, $4 \cdot 3 = 12$. This process can be summarized by the following rule.

RAISING A POWER TO A POWER

To raise a power to a power, keep the same base and multiply the exponents.

$$(x^a)^b = x^{ab}$$

Recall what happens when you raise a negative number to a power. $(-1)^2 = 1$. $(-1)^3 = -1$. In general,

$$(-1)^n = \begin{cases} +1 & \text{if } n \text{ is even} \\ -1 & \text{if } n \text{ is odd.} \end{cases}$$

EXAMPLE 12   Simplify.   **(a)** $\left(x^3\right)^5$ **(b)** $\left(2^7\right)^3$ **(c)** $(-1)^8$

Solution

(a) $\left(x^3\right)^5 = x^{3\cdot 5} = x^{15}$ **(b)** $\left(2^7\right)^3 = 2^{7\cdot 3} = 2^{21}$ **(c)** $(-1)^8 = +1$

Note that in both parts (a) and (b) the base does not change.

Practice Problem 12   Simplify.   **(a)** $\left(a^4\right)^3$ **(b)** $\left(10^5\right)^2$ **(c)** $(-1)^{15}$

Here are two rules involving products and quotients that are very useful: the product raised to a power rule and the quotient raised to a power rule. We'll illustrate each with an example.

If a product in parentheses is raised to a power, the parentheses indicate that *each factor* must be raised to that power.

$$(xy)^2 = x^2y^2 \qquad (xy)^3 = x^3y^3$$

PRODUCT RAISED TO A POWER

$$(xy)^a = x^a y^a$$

EXAMPLE 13   Simplify.   **(a)** $(ab)^8$ **(b)** $(3x)^4$ **(c)** $\left(-2x^2\right)^3$

Solution

(a) $(ab)^8 = a^8 b^8$ **(b)** $(3x)^4 = (3)^4 x^4 = 81x^4$

(c) $\left(-2x^2\right)^3 = (-2)^3 \cdot \left(x^2\right)^3 = -8x^6$

Practice Problem 13   Simplify.   **(a)** $(3xy)^3$ **(b)** $(yz)^{37}$ **(c)** $\left(-3x^3\right)^2$

If a fractional expression within parentheses is raised to a power, the parentheses indicate that both numerator and denominator must be raised to that power.

$$\left(\frac{x}{y}\right)^5 = \frac{x^5}{y^5} \qquad \left(\frac{x}{y}\right)^2 = \frac{x^2}{y^2} \qquad \text{if } y \neq 0$$

QUOTIENT RAISED TO A POWER

$$\left(\frac{x}{y}\right)^a = \frac{x^a}{y^a} \qquad \text{if } y \neq 0$$

EXAMPLE 14   Simplify.   **(a)** $\left(\frac{x}{y}\right)^5$ **(b)** $\left(\frac{7}{w}\right)^4$

Solution

(a) $\left(\frac{x}{y}\right)^5 = \frac{x^5}{y^5}$ **(b)** $\left(\frac{7}{w}\right)^4 = \frac{7^4}{w^4} = \frac{2401}{w^4}$

Practice Problem 14   Simplify.   **(a)** $\left(\frac{x}{5}\right)^3$ **(b)** $\left(\frac{4a}{b}\right)^6$

Many expressions can be simplified by using the previous rules involving exponents. Be sure to take particular care to determine the correct sign, especially if there is a negative numerical coefficient.

EXAMPLE 15 Simplify. $\left(\dfrac{-3x^2z^0}{y^3}\right)^4$

Solution

$$\left(\frac{-3x^2z^0}{y^3}\right)^4 = \left(\frac{-3x^2}{y^3}\right)^4$$ Simplify inside the parentheses first. Note that $z^0 = 1$.

$$= \frac{(-3)^4x^8}{y^{12}}$$ Apply the rules for raising a power to a power. Notice that we wrote $(-3)^4$ and not -3^4. We are raising -3 to the fourth power.

$$= \frac{81x^8}{y^{12}}$$ Simplify the coefficient: $(-3)^4 = +81$.

NOTE TO STUDENT: Fully worked-out solutions to all of the Practice Problems can be found at the back of the text starting at page SP-1

Practice Problem 15 Simplify. $\left(\dfrac{-2x^3y^0z}{4xz^2}\right)^5$

We list here the rules of exponents we have discussed in Section 4.1.

$$x^a \cdot x^b = x^{a+b}$$

$$\frac{x^a}{x^b} = \begin{cases} x^{a-b} & \text{if } a > b \\ \dfrac{1}{x^{b-a}} & \text{if } b > a \\ x^0 = 1 & \text{if } a = b \end{cases}$$

$$(x^a)^b = x^{ab}$$
$$(xy)^a = x^a y^a$$
$$\left(\frac{x}{y}\right)^a = \frac{x^a}{y^a} \quad y \neq 0$$

Developing Your Study Skills

Why Is Review Necessary?

You master a course in mathematics by learning the concepts one step at a time. Thus the study of mathematics is built step-by-step, with each step a supporting foundation for the next. The process is a carefully designed procedure, so no steps can be skipped. A student of mathematics needs to realize the importance of this building process to succeed.

Because new concepts depend on those previously learned, students often need to take time to review. The reviewing process will strengthen understanding and skills, which may be weak due to a lack of mastery or the passage of time. Review at the right time on the right concepts can strengthen previously learned skills and make progress possible.

Timely, periodic review of previously learned mathematical concepts is absolutely necessary for mastery of new concepts. You may have forgotten a concept or grown a bit rusty in applying it. Reviewing is the answer. Make use of the cumulative review problems in your textbook, whether they are assigned or not. Look back to previous chapters whenever you have forgotten how to do something. Review the chapter organizers from previous chapters. Study the examples and practice some exercises to refresh your understanding.

Be sure that you understand and can perform the computations of each new concept. This will enable you to move successfully on to the next one.

Remember, mathematics is a step-by-step building process. Learn each concept and reinforce and strengthen with review whenever necessary.

Verbal and Writing Skills

1. Write in your own words the product rule for exponents.

2. To be able to use the rules of exponents, what must be true of the bases?

3. If the larger exponent is in the denominator, the quotient rule states that $\dfrac{x^a}{x^b} = \dfrac{1}{x^{b-a}}$. Provide an example to show why this is true.

In exercises 4 and 5, identify the numerical coefficient, the base(s), and the exponent(s).

4. $-8x^5y^2$

5. $6x^{11}y$

6. Evaluate **(a)** $3x^0$ and **(b)** $(3x)^0$. **(c)** Why are the results different?

Write in simplest exponent form.

7. $2 \cdot 2 \cdot a \cdot a \cdot a \cdot b$

8. $5 \cdot x \cdot x \cdot x \cdot y \cdot y$

9. $(-3)(a)(a)(b)(c)(b)(c)(c)$

10. $(-7)(x)(y)(z)(y)(x)$

Multiply. Leave your answer in exponent form.

11. $(7^4)(7^6)$

12. $(4^3)(4^5)$

13. $(5^{10})(5^{16})$

14. $(6^5)(6^8)$

15. $x^4 \cdot x^8$

16. $x^9 \cdot x^6$

17. $w^{12} \cdot w^{20}$

18. $z^{16} \cdot z^{10}$

Multiply.

19. $-5x^4(4x^2)$

20. $6x^2(-9x^3)$

21. $(5x)(10x^2)$

22. $(-4x^2)(-3x)$

23. $(3x^2y)(8x^3y^3)$

24. $(6xy^3)(5x^4y)$

25. $\left(\dfrac{2}{5}xy^3\right)\left(\dfrac{1}{3}x^2y^2\right)$

26. $\left(\dfrac{4}{5}x^5y\right)\left(\dfrac{15}{16}x^2y^4\right)$

27. $(1.1x^2z)(-2.5xy)$

28. $(2.3x^4w)(-3.5xy^4)$

29. $(8a)(2a^3b)(0)$

30. $(5ab)(2a^2)(0)$

31. $(-16x^2y^4)(-5xy^3)$

32. $(-12x^4y)(-7x^5y^3)$

33. $(-8x^3y^2)(3xy^5)$

34. $(9x^2y^6)(-11x^3y^3)$

35. $(-2x^3y^2)(0)(-3x^4y)$

36. $(-4x^8y^2)(13y^3)(0)$

37. $(8a^4b^3)(-3x^2y^5)$

38. $(5x^3y)(-2w^4z)$

39. $(2x^2y)(-3y^3z^2)(5xz^4)$

40. $(3ab)(5a^2c)(-2b^2c^3)$

Divide. Leave your answer in exponent form. Assume that all variables in any denominator are nonzero.

41. $\dfrac{y^{12}}{y^5}$

42. $\dfrac{x^{13}}{x^3}$

43. $\dfrac{y^5}{y^8}$

44. $\dfrac{b^{20}}{b^{23}}$

45. $\dfrac{11^{18}}{11^{30}}$

46. $\dfrac{8^9}{8^{12}}$

47. $\dfrac{3^{18}}{3^{14}}$

48. $\dfrac{5^{16}}{5^{12}}$

49. $\dfrac{a^{13}}{4a^5}$

50. $\dfrac{b^{16}}{5b^{13}}$

51. $\dfrac{x^7}{y^9}$

52. $\dfrac{x^{20}}{y^3}$

53. $\dfrac{48x^5y^3}{24xy^3}$

54. $\dfrac{45a^4b^3}{15a^4b^2}$

55. $\dfrac{16x^5y}{-32x^2y^3}$

56. $\dfrac{-36x^3y^7}{72x^5y}$

57. $\dfrac{1.8f^4g^3}{54f^2g^8}$

58. $\dfrac{3.1s^5t^3}{62s^8t}$

59. $\dfrac{-51x^6y^8}{17x^3y^8}$

60. $\dfrac{-30x^5y^4}{5x^3y^4}$

61. $\dfrac{8^0x^2y^3}{16x^5y}$

62. $\dfrac{3^2x^3y^7}{3^0x^5y^2}$

63. $\dfrac{18a^6b^3c^0}{24a^5b^3}$

64. $\dfrac{12a^7b^8}{16a^3b^8c^0}$

65. $\dfrac{85a^2b}{45c^3}$

66. $\dfrac{28x}{63y^2z}$

To Think About

67. What expression can be multiplied by $(-3x^3yz)$ to obtain $81x^8y^2z^4$?

68. $63a^5b^6$ is divided by an expression and the result is $-9a^4b$. What is this expression?

Simplify.

69. $(x^2)^6$

70. $(w^5)^8$

71. $(xy^2)^7$

72. $(a^3b)^4$

73. $(rs^2)^6$

74. $(m^3n^2)^5$

75. $(3a^3b^2c)^3$

76. $(2x^4yz^3)^2$

77. $(-3a^4)^2$

78. $(-2a^5)^4$

79. $\left(\dfrac{x}{2m^4}\right)^7$

80. $\left(\dfrac{p^5}{6x}\right)^5$

81. $\left(\dfrac{5x}{7y^2}\right)^2$

82. $\left(\dfrac{2a^4}{3b^3}\right)^4$

83. $(-3a^2b^3c^0)^4$

84. $(-2a^5b^2c^0)^5$

85. $(-2x^3y^0z)^3$

86. $(-4xy^0z^4)^3$

87. $\dfrac{(3x)^5}{(3x^2)^3}$

88. $\dfrac{(6y^3)^4}{(6y)^2}$

Mixed Practice

89. $(3ab^2)^3(ab)$

90. $(-2a^2b^3)^3(ab^2)$

91. $\left(\dfrac{8}{y^5}\right)^2$

92. $\left(\dfrac{4}{x^6}\right)^3$

93. $\left(\dfrac{2x}{y^3}\right)^4$

94. $\left(\dfrac{3x^5}{y}\right)^3$

95. $\left(\dfrac{ab^2}{c^3d^4}\right)^4$

96. $\left(\dfrac{a^3b}{c^5d}\right)^5$

To Think About

97. What expression raised to the third power is $-27x^9y^{12}z^{21}$?

98. What expression raised to the fourth power is $16x^{20}y^{16}z^{28}$?

Cumulative Review

Simplify.

99. $-3 - 8$

100. $-17 + (-32) + (-24) + 27$

101. $\left(\dfrac{2}{3}\right)\left(-\dfrac{21}{8}\right)$

102. $\dfrac{-3}{4} \div \dfrac{-12}{80}$

Rain Forests *Brazil, the largest country in South America, boasts a land area of approximately 8,511,960 sq. km. Originally, 2,860,000 sq km in Brazil was covered by rain forest. As of 2002 the rain forest of that country only consists of 1,500,000 sq km. Round your answers to nearest tenth of a percent.*

103. What percentage of Brazil's land area was covered by rain forest 2002?

104. What was the percent decrease in the rain forest up to 2002?

105. If Brazil is said to lose 2.3% of its rain forest per year, how much rain forest was lost in the year after 2002?

(Information source: The Rain Forest Action Network)

4.2 NEGATIVE EXPONENTS AND SCIENTIFIC NOTATION

1 Using Negative Exponents

If n is an integer, and $x \neq 0$, then x^{-n} is defined as follows:

DEFINITION OF A NEGATIVE EXPONENT

$$x^{-n} = \frac{1}{x^n}, \quad x \neq 0$$

Student Learning Objectives

After studying this section, you will be able to:

1. Use negative exponents.

2. Write numbers in scientific notation.

EXAMPLE 1 Write with positive exponents.

(a) y^{-3} (b) z^{-6} (c) w^{-1}

Solution

(a) $y^{-3} = \dfrac{1}{y^3}$ (b) $z^{-6} = \dfrac{1}{z^6}$ (c) $w^{-1} = \dfrac{1}{w^1} = \dfrac{1}{w}$

Practice Problem 1 Write with positive exponents.

(a) x^{-12} (b) w^{-5} (c) z^{-2}

NOTE TO STUDENT: Fully worked-out solutions to all of the Practice Problems can be found at the back of the text starting at page SP-1

To evaluate a numerical expression with a negative exponent, first write the expression with a positive exponent. Then simplify.

EXAMPLE 2 Evaluate. (a) 3^{-2} (b) 2^{-5}

Solution

(a) $3^{-2} = \dfrac{1}{3^2} = \dfrac{1}{9}$ (b) $2^{-5} = \dfrac{1}{2^5} = \dfrac{1}{32}$

Practice Problem 2 Evaluate. (a) 4^{-3} (b) 2^{-4}

All the previously studied laws of exponents are true for any integer exponent. These laws are summarized in the following box. Assume that $x, y \neq 0$.

LAWS OF EXPONENTS

The Product Rule

$$x^a \cdot x^b = x^{a+b}$$

The Quotient Rule

$$\frac{x^a}{x^b} = x^{a-b} \quad \text{Use if } a > b, \qquad \frac{x^a}{x^b} = \frac{1}{x^{b-a}} \quad \text{Use if } a < b.$$

Power Rules

$$(xy)^a = x^a y^a, \qquad (x^a)^b = x^{ab}, \qquad \left(\frac{x}{y}\right)^a = \frac{x^a}{y^a}$$

By using the definition of a negative exponent and the properties of fractions, we can derive two more helpful properties of exponents. Assume that $x, y \neq 0$.

PROPERTIES OF NEGATIVE EXPONENTS

$$\frac{1}{x^{-n}} = x^n \qquad \frac{x^{-m}}{y^{-n}} = \frac{y^n}{x^m}$$

EXAMPLE 3 Simplify. Write the expression with no negative exponents.

(a) $\dfrac{1}{x^{-6}}$ **(b)** $\dfrac{x^{-3}y^{-2}}{z^{-4}}$ **(c)** $x^{-2}y^3$

Solution

(a) $\dfrac{1}{x^{-6}} = x^6$ **(b)** $\dfrac{x^{-3}y^{-2}}{z^{-4}} = \dfrac{z^4}{x^3y^2}$ **(c)** $x^{-2}y^3 = \dfrac{y^3}{x^2}$

NOTE TO STUDENT: *Fully worked-out solutions to all of the Practice Problems can be found at the back of the text starting at page SP-1*

Practice Problem 3 Simplify. Write the expression with no negative exponents.

(a) $\dfrac{3}{w^{-4}}$ **(b)** $\dfrac{x^{-6}y^4}{z^{-2}}$ **(c)** $x^{-6}y^{-5}$

EXAMPLE 4 Simplify. Write the expression with no negative exponents.

(a) $(3x^{-4}y^2)^{-3}$ **(b)** $\dfrac{x^2y^{-4}}{x^{-5}y^3}$

Solution

(a) $(3x^{-4}y^2)^{-3} = 3^{-3}x^{12}y^{-6} = \dfrac{x^{12}}{3^3y^6} = \dfrac{x^{12}}{27y^6}$

(b) $\dfrac{x^2y^{-4}}{x^{-5}y^3} = \dfrac{x^2x^5}{y^4y^3} = \dfrac{x^7}{y^7}$ First rewrite the expression so that only positive exponents appear. Then simplify using the product rule.

Practice Problem 4 Simplify. Write the expression with no negative exponents.

(a) $(2x^4y^{-5})^{-2}$ **(b)** $\dfrac{y^{-3}z^{-4}}{y^2z^{-6}}$

② Writing Numbers in Scientific Notation

One common use of negative exponents is in writing numbers in scientific notation. Scientific notation is most useful in expressing very large and very small numbers.

SCIENTIFIC NOTATION

A positive number is written in **scientific notation** if it is in the form $a \times 10^n$, where $1 \leq a < 10$ and n is an integer.

EXAMPLE 5   Write in scientific notation.   **(a)** 4567   **(b)** 157,000,000

Solution

(a) $4567 = 4.567 \times 1000$ To change 4567 to a number that is greater than 1 but less than 10, we move the decimal point three places to the left. We must then multiply the number by a power of 10 so that we do not change the value of the number. Use 1000.

$$= 4.567 \times 10^3$$

(b) $157{,}000{,}000 = \underset{\text{8 places}}{1.57000000} \times \underset{\text{8 zeros}}{100000000}$

$$= 1.57 \times 10^8$$

Practice Problem 5   Write in scientific notation.

(a) 78,200   **(b)** 4,786,000

Numbers that are smaller than 1 will have a negative power of 10 if they are written in scientific notation.

EXAMPLE 6   Write in scientific notation.   **(a)** 0.061   **(b)** 0.000052

Solution

(a) We need to write 0.061 as a number that is greater than 1 but less than 10. In which direction do we move the decimal point?

$0.061 = 6.1 \times 10^{-2}$ Move the decimal point 2 places to the right.

(b) $0.000052 = 5.2 \times 10^{-5}$ Why?

Practice Problem 6   Write in scientific notation.

(a) 0.98   **(b)** 0.000092

The reverse procedure transforms scientific notation into ordinary decimal notation.

EXAMPLE 7   Write in decimal notation.

(a) 1.568×10^2 **(b)** 7.432×10^{-3}

Solution

(a) $1.568 \times 10^2 = 1.568 \times 100$
$$= 156.8$$

Alternative Method

$1.568 \times 10^2 = 156.8$ The exponent 2 tells us to move the decimal point 2 places to the right.

(b) $7.432 \times 10^{-3} = 7.432 \times \dfrac{1}{1000}$

$$= 0.007432$$

Alternative Method

$7.432 \times 10^{-3} = 0.007432$ The exponent -3 tells us to move the decimal point 3 places to the left.

Practice Problem 7   Write in decimal notation.

(a) 1.93×10^6 **(b)** 8.562×10^{-5}

Calculator

Scientific Notation

Most calculators can display only eight digits at one time. Numbers with more than eight digits are usually shown in scientific notation. 1.12 E 08 or 1.12 8 means 1.12×10^8. You can use the calculator to compute with large numbers by entering the numbers using scientific notation. For example,

$$(7.48 \times 10^{24}) \times (3.5 \times 10^8)$$

is entered as follows.

7.48 [EXP]

24 [×] 3.5

[EXP] 8 [=]

Display: [2.618 E 33]

or [2.618   33]

Note: Some calculators have an [EE] key instead of [EXP].

Compute on a calculator.

1. 35,000,000,000 + 77,000,000,000

2. $(6.23 \times 10^{12}) \times (4.9 \times 10^5)$

3. $(2.5 \times 10^7)^5$

4. $3.3284 \times 10^{32} \div (6.28 \times 10^{24})$

5. How many seconds are there in 1000 years?

The distance light travels in one year is called a *light-year*. A light-year is a convenient unit of measure to use when investigating the distances between stars.

EXAMPLE 8 A light-year is a distance of 9,460,000,000,000,000 meters. Write this in scientific notation.

Solution $9,460,000,000,000,000 = 9.46 \times 10^{15}$ meters

Practice Problem 8 Astronomers measure distances to faraway galaxies in parsecs. A parsec is a distance of 30,900,000,000,000,000 meters. Write this in scientific notation.

To perform a calculation involving very large or very small numbers, it is usually helpful to write the numbers in scientific notation and then use the laws of exponents to do the calculation.

EXAMPLE 9 Use scientific notation and the laws of exponents to find the following. Leave your answer in scientific notation.

(a) $(32,000,000)(1,500,000,000,000)$ **(b)** $\dfrac{0.00063}{0.021}$

Solution

(a) $(32,000,000)(1,500,000,000,000)$

$= (3.2 \times 10^7)(1.5 \times 10^{12})$ Write each number in scientific notation.

$= 3.2 \times 1.5 \times 10^7 \times 10^{12}$ Rearrange the order. Remember that multiplication is commutative.

$= 4.8 \times 10^{19}$ Multiply 3.2×1.5. Multiply $10^7 \times 10^{12}$.

(b) $\dfrac{0.00063}{0.021} = \dfrac{6.3 \times 10^{-4}}{2.1 \times 10^{-2}}$ Write each number in scientific notation.

$= \dfrac{6.3}{2.1} \times \dfrac{10^{-4}}{10^{-2}}$ Rearrange the order. We are actually using the definition of multiplication of fractions.

$= \dfrac{6.3}{2.1} \times \dfrac{10^2}{10^4}$ Rewrite with positive exponents.

$= 3.0 \times 10^{-2}$

NOTE TO STUDENT: Fully worked-out solutions to all of the Practice Problems can be found at the back of the text starting at page SP-1

Practice Problem 9 Use scientific notation and the laws of exponents to find the following. Leave your answer in scientific notation.

(a) $(56,000)(1,400,000,000)$ **(b)** $\dfrac{0.000111}{0.00000037}$

When we use scientific notation, we are writing approximate numbers. We must include some zeros so that the decimal point can be properly located. However, all other digits except for these zeros are considered **significant digits.** The number 34.56 has four significant digits. The number 0.0049 has two significant digits. The zeros are considered placeholders. The number 634,000 has three significant digits (unless we have specific knowledge to the contrary). The zeros are considered placeholders. We sometimes round numbers to a specific number of significant digits. For example, 0.08746 rounded to two significant digits is 0.087. When we round 1,348,593 to three significant digits, we obtain 1,350,000.

| EXAMPLE 10 | The approximate distance from Earth to the star Polaris is 208 parsecs. A parsec is a distance of approximately 3.09×10^{13} kilometers. How long would it take a space probe traveling at 40,000 kilometers per hour to reach the star? Round to three significant digits.

Solution

1. **Understand the problem.** Recall that the distance formula is

$$\text{distance} = \text{rate} \times \text{time}.$$

We are given the distance and the rate. We need to find the time. Let's take a look at the distance. The distance is given in parsecs, but the rate is given in kilometers per hour. We need to change the distance to kilometers. We are told that a parsec is approximately 3.09×10^{13} kilometers. That is, there are 3.09×10^{13} kilometers per parsec. We use this information to change 208 parsecs to kilometers.

$$208 \text{ parsecs} = \frac{(208 \text{ parsecs})(3.09 \times 10^{13} \text{ kilometers})}{1 \text{ parsec}} = 642.72 \times 10^{13} \text{ kilometers}$$

2. **Write an equation.** Use the distance formula.

$$d = r \times t$$

3. **Solve the equation and state the answer.** Substitute the known values into the formula and solve for the unknown, time.

$$642.72 \times 10^{13} \text{ km} = \frac{40{,}000 \text{ km}}{1 \text{ hr}} \times t$$

$$6.4272 \times 10^{15} \text{ km} = \frac{4 \times 10^4 \text{ km}}{1 \text{ hr}} \times t \qquad \text{Change the numbers to scientific notation.}$$

$$\frac{6.4272 \times 10^{15} \text{ km}}{\dfrac{4 \times 10^4 \text{ km}}{1 \text{ hr}}} = t \qquad \text{Divide both sides by } \dfrac{4 \times 10^4 \text{ km}}{1 \text{ hr}}.$$

$$\frac{(6.4272 \times 10^{15} \text{ km})(1 \text{ hr})}{4 \times 10^4 \text{ km}} = t$$

$$1.6068 \times 10^{11} \text{ hr} = t$$

1.6068×10^{11} is 160.68×10^9 or 160.68 billion hours. The space probe will take approximately 160.68 billion hours to reach the star.

Reread the problem. Are we finished? What is left to do? We need to round the answer to three significant digits. Rounding to three significant digits, we have

$$160.68 \times 10^9 \approx 161 \times 10^9.$$

This is approximately 161 billion hours or a little more than 18 million years.

4. **Check.** Unless you have had a great deal of experience working in astronomy, it would be difficult to determine whether this is a reasonable answer. You may wish to reread your analysis and redo your calculations as a check.

Practice Problem 10 The average distance from Earth to the distant star Betelgeuse is 159 parsecs. How many hours would it take a space probe to travel from Earth to Betelgeuse at a speed of 50,000 kilometers per hour? Round to three significant digits.

Simplify. Express your answer with positive exponents. Assume that all variables are nonzero.

1. x^{-4}

2. y^{-5}

3. 3^{-4}

4. 5^{-3}

5. $\dfrac{1}{y^{-8}}$

6. $\dfrac{1}{z^{-10}}$

7. $\dfrac{x^{-4}y^{-5}}{z^{-6}}$

8. $\dfrac{x^{-6}y^{-2}}{z^{-5}}$

9. $x^{-5}y^6$

10. $x^{-4}y^7$

11. $(2x^{-3})^{-3}$

12. $(4x^{-4})^{-2}$

13. $3x^{-2}$

14. $4y^{-4}$

15. $(4x^2y)^{-2}$

16. $(2x^3y^5)^{-3}$

Mixed Practice

17. $\dfrac{3xy^{-2}}{z^{-3}}$

18. $\dfrac{4x^{-2}y^{-3}}{y^4}$

19. $\dfrac{(3x)^{-2}}{(3x)^{-3}}$

20. $\dfrac{(2ab^2)^{-3}}{(2ab^2)^{-4}}$

21. $wx^{-5}y^3z^{-2}$

22. $a^5b^{-3}c^{-4}$

23. $(8^{-2})(2^3)$

24. $(9^2)(3^{-3})$

25. $\left(\dfrac{3x^0y^2}{z^4}\right)^{-2}$

26. $\left(\dfrac{2a^3b^0}{c^2}\right)^{-3}$

27. $\dfrac{x^{-2}y^{-3}}{x^4y^{-2}}$

28. $\dfrac{a^{-6}b^3}{a^{-2}b^{-5}}$

Write in scientific notation.

29. 123,780

30. 5,786,100

31. 0.063

32. 0.000742

33. 889,610,000,000

34. 7,652,000,000

35. 0.00000001963

36. 0.000007618

In exercises 37–42, write in decimal notation.

37. 3.02×10^5 **38.** 8.137×10^7 **39.** 4.7×10^{-4} **40.** 6.53×10^{-3} **41.** 9.83×10^5 **42.** 3.5×10^{-8}

43. *Bamboo Growth* The growth rate of some species of bamboo is 0.0000237 miles per hour. Write this in scientific notation.

44. *Neptune* Neptune is 2.793×10^9 miles from the sun. Write this in decimal notation.

45. *Red Blood Cell* A single human red blood cell is about 7×10^{-6} meters in diameter. Write this in decimal notation.

46. *Gold Atom* The average volume of an atom of gold is 0.0000000000000000000001695 cubic centimeters. Write this in scientific notation.

Evaluate by using scientific notation and the laws of exponents. Leave your answer in scientific notation.

47. $(56{,}000{,}000{,}000)(780{,}000{,}000)$

48. $(35{,}000{,}000)(84{,}000{,}000{,}000)$

49. $\dfrac{(5{,}000{,}000)(16{,}000)}{8{,}000{,}000{,}000}$

50. $(0.0075)(0.0000002)(0.001)$

51. $(0.003)^4$

52. $(500{,}000)^4$

53. $(150{,}000{,}000)(0.00005)(0.002)(30{,}000)$

54. $\dfrac{(1{,}600{,}000)(0.00003)}{2400}$

Applications

National Debt For fiscal year 2003, the national debt was determined to be approximately 6.816×10^{12} dollars. *(Source: Treasury Department of Public Debt.)*

55. The census bureau estimates that in 2003, the entire population of the United States was 2.92×10^8 people. If the national debt were evenly divided among every person in the country, how much debt would be assigned to each individual? Round to three significant digits.

56. The census bureau estimates that in 2003, the number of people in the United States who were over age 18 was approximately 2.28×10^8 people. If the national debt were evenly divided among every person over age 18 in the country, how much debt would be assigned to each individual? Round to three significant digits.

Space Travel *A parsec is a distance of approximately* 3.09×10^{13} *kilometers.*

57. How long would it take a space probe to travel from Earth to the star Rigel, which is 276 parsecs from Earth? Assume that the space probe travels at 45,000 kilometers per hour. Round to three significant digits.

58. How long would it take a space probe to travel from Earth to the star Hadar, which is 150 parsecs from Earth? Assume that the space probe travels at 55,000 kilometers per hour. Round to three significant digits.

59. ***Watch Hand*** The tip of a $\frac{1}{3}$-inch-long hour hand on a watch travels at a speed of 0.00000275 miles per hour. How far has it traveled in a day?

60. ***Neutron Mass*** The mass of a neutron is approximately 1.675×10^{-27} kilogram. Find the mass of 180,000 neutrons.

61. ***007 Films*** The most profitable movie series made has been the James Bond series, which has grossed 3.2×10^9 over the 20 James Bond films. What is the average gross per film?

62. ***Moles per Molecule*** Avogadro's number says that there are approximately 6.02×10^{23} molecules/mole. How many molecules can one expect in 0.00483 mole?

63. ***Construction Cost*** In 1990 the cost for construction of new private buildings was estimated at 3.61×10^{11}. By 2010 the estimated cost for construction of new private buildings will be 9.36×10^{11}. What is the percent of increase from 1990 to 2010? Round to the nearest tenth of a percent. (*Source:* U.S. Census Bureau.)

64. ***Construction Cost*** In 1990 the cost for construction of new public buildings was estimated at 1.07×10^{11}. By 2010 the estimated cost for construction of new public buildings will be 3.06×10^{11}. What is the percent of increase from 1990 to 2010? Round to the nearest tenth of a percent. (*Source:* U.S. Census Bureau.)

Cumulative Review

Simplify.

65. $-2.7 - (-1.9)$

66. $(-1)^{33}$

67. $-\dfrac{3}{4} + \dfrac{5}{7}$

68. ***Hand Shakes*** A recent debate between two political candidates attracted 3540 people to a local gymnasium. At the conclusion of the debate, each person went to one side of the gymnasium to shake the hand of candidate #1 or to the other side of the room to shake the hand of candidate #2. Candidate #1 shook 524 less than triple the number of hands shaken by candidate #2. How many hands did each candidate shake?

69. ***Salaries*** Gina has a bachelor's degree. She earns $12,460 more a year than Mario, who holds an associate's degree. Alfonso, who has not yet been able to attend college, earns $8742 a year less than Mario. The combined annual salaries of the three people is $112,000. What is the annual salary of each person?

Student Learning Objectives

After studying this section, you will be able to:

1 Recognize polynomials and determine their degrees.

2 Add polynomials.

3 Subtract polynomials.

4 Evaluate polynomials to predict a value.

1 Recognizing Polynomials and Determining Their Degrees

A **polynomial** in x is the sum of a finite number of terms of the form ax^n, where a is any real number and n is a whole number. Usually these polynomials are written in descending powers of the variable, as in

$$5x^3 + 3x^2 - 2x - 5 \quad \text{and} \quad 3.2x^2 - 1.4x + 5.6.$$

A **multivariable polynomial** is a polynomial with more than one variable. The following are multivariable polynomials:

$$5xy + 8, \qquad 2x^2 - 7xy + 9y^2, \qquad 17x^3y^9$$

The **degree of a term** is the sum of the exponents of all of the variables in the term. For example, the degree of $7x^3$ is three. The degree of $4xy$ is two. The degree of $10x^4y^2$ is six.

The **degree of a polynomial** is the highest degree of all of the terms in the polynomial. For example, the degree of $5x^3 + 8x^2 - 20x - 2$ is three. The degree of $6xy - 4x^2y + 2xy^3$ is four.

The polynomial 0 is said to have **no degree.** Likewise, a polynomial consisting of a constant is said to have no degree.

There are special names for polynomials with one, two, or three terms.

A **monomial** has *one* term:

$$5a, \qquad 3x^3yz^4, \qquad 12xy$$

A **binomial** has *two* terms:

$$7x + 9y, \qquad -6x - 4, \qquad 5x^4 + 2xy^2$$

A **trinomial** has *three* terms:

$$8x^2 - 7x + 4, \qquad 2ab^3 - 6ab^2 - 15ab, \qquad 2 + 5y + y^4$$

EXAMPLE 1 State the degree of the polynomial, and whether it is a monomial, a binomial, or a trinomial.

(a) $5xy + 3x^3$ **(b)** $-7a^5b^2$ **(c)** $8x^4 - 9x - 15$

Solution

(a) This polynomial is of degree 3. It has two terms, so it is a binomial.

(b) The sum of the exponents is $5 + 2 = 7$. Therefore this polynomial is of degree 7. It has one term, so it is a monomial.

(c) This polynomial is of degree 4. It has three terms, so it is a trinomial.

Practice Problem 1 State the degree of the polynomial, and whether it is a monomial, a binomial, or a trinomial.

(a) $-7x^5 - 3xy$ **(b)** $22a^3b^4$ **(c)** $-3x^3 + 3x^2 - 6x$

NOTE TO STUDENT: Fully worked-out solutions to all of the Practice Problems can be found at the back of the text starting at page SP-1

2 Adding Polynomials

We usually write a polynomial in x so that the exponents on x decrease from left to right. For example, the polynomial

$$5x^2 - 6x + 2$$

is said to be written in **decreasing order** since each exponent is decreasing as we move from left to right. You can add, subtract, multiply, and divide polynomials. Let us take a look at addition. To add two polynomials, we add their like terms.

EXAMPLE 2   Add. $(5x^2 - 6x - 12) + (-3x^2 - 9x + 5)$

Solution

$$
\begin{aligned}
(5x^2 - 6x - 12) + (-3x^2 - 9x + 5) &= [5x^2 + (-3x^2)] + [-6x + (-9x)] + [-12 + 5] \\
&= [(5 - 3)x^2] + [(-6 - 9)x] + [-12 + 5] \\
&= 2x^2 + (-15x) + (-7) \\
&= 2x^2 - 15x - 7
\end{aligned}
$$

Practice Problem 2   Add. $(-8x^3 + 3x^2 + 6) + (2x^3 - 7x^2 - 3)$

The numerical coefficients of the polynomials may be any real number. Thus the polynomials may have numerical coefficients that are decimals or fractions.

EXAMPLE 3   Add. $\left(\frac{1}{2}x^2 - 6x + \frac{1}{3}\right) + \left(\frac{1}{5}x^2 - 2x - \frac{1}{2}\right)$

Solution

$$
\begin{aligned}
\left(\frac{1}{2}x^2 - 6x + \frac{1}{3}\right) + \left(\frac{1}{5}x^2 - 2x - \frac{1}{2}\right) &= \left[\frac{1}{2}x^2 + \frac{1}{5}x^2\right] + [-6x + (-2x)] + \left[\frac{1}{3} + \left(-\frac{1}{2}\right)\right] \\
&= \left[\left(\frac{1}{2} + \frac{1}{5}\right)x^2\right] + [(-6 - 2)x] + \left[\frac{1}{3} + \left(-\frac{1}{2}\right)\right] \\
&= \left[\left(\frac{5}{10} + \frac{2}{10}\right)x^2\right] + [-8x] + \left[\frac{2}{6} - \frac{3}{6}\right] \\
&= \frac{7}{10}x^2 - 8x - \frac{1}{6}
\end{aligned}
$$

Practice Problem 3   Add. $\left(-\frac{1}{3}x^2 - 6x - \frac{1}{12}\right) + \left(\frac{1}{4}x^2 + 5x - \frac{1}{3}\right)$

EXAMPLE 4   Add. $(1.2x^3 - 5.6x^2 + 5) + (-3.4x^3 - 1.2x^2 + 4.5x - 7)$

Solution   Group like terms.

$$
\begin{aligned}
(1.2x^3 - 5.6x^2 + 5) + (-3.4x^3 - 1.2x^2 + 4.5x - 7) &= (1.2 - 3.4)x^3 + (-5.6 - 1.2)x^2 + 4.5x + (5 - 7) \\
&= -2.2x^3 - 6.8x^2 + 4.5x - 2
\end{aligned}
$$

Practice Problem 4   Add.

$$(3.5x^3 - 0.02x^2 + 1.56x - 3.5) + (-0.08x^2 - 1.98x + 4)$$

As mentioned previously, polynomials may involve more than one variable.

3 Subtracting Polynomials

Recall that subtraction of real numbers can be defined as adding the opposite of the second number. Thus $a - b = a + (-b)$. That is, $3 - 5 = 3 + (-5)$. A similar method is used to subtract two polynomials.

To subtract two polynomials, change the sign of each term in the second polynomial and then add.

EXAMPLE 5 Subtract. $(7x^2 - 6x + 3) - (5x^2 - 8x - 12)$

Solution We change the sign of each term in the second polynomial and then add.

$$(7x^2 - 6x + 3) - (5x^2 - 8x - 12) = (7x^2 - 6x + 3) + (-5x^2 + 8x + 12)$$
$$= (7 - 5)x^2 + (-6 + 8)x + (3 + 12)$$
$$= 2x^2 + 2x + 15$$

NOTE TO STUDENT: Fully worked-out solutions to all of the Practice Problems can be found at the back of the text starting at page SP-1

Practice Problem 5 Subtract.

$$(5x^3 - 15x^2 + 6x - 3) - (-4x^3 - 10x^2 + 5x + 13)$$

When subtracting polynomials in two variables, you will need to use extra care in determining which terms are like terms. For example $6x^2y$ and $5x^2y$ are like terms. In a similar fashion, $3xy$ and $8xy$ are like terms. However $7xy^2$ and $15x^2y^2$ are not like terms. Every exponent of every variable in the two terms must be the same if the terms are to be like terms. You will use this concept in Example 6.

EXAMPLE 6 Subtract.

$$(-6x^2y - 3xy + 7xy^2) - (5x^2y - 8xy - 15x^2y^2)$$

Solution Change the sign of each term in the second polynomial and add. Look for like terms.

$$(-6x^2y - 3xy + 7xy^2) + (-5x^2y + 8xy + 15x^2y^2)$$
$$= (-6 - 5)x^2y + (-3 + 8)xy + 7xy^2 + 15x^2y^2$$
$$= -11x^2y + 5xy + 7xy^2 + 15x^2y^2$$

Nothing further can be done to combine these four terms.

Practice Problem 6 Subtract.

$$(x^3 - 7x^2y + 3xy^2 - 2y^3) - (2x^3 + 4xy - 6y^3)$$

Evaluating Polynomials to Predict a Value

Sometimes polynomials are used to predict a value. In such cases we need to **evaluate** the polynomial. We do this by substituting a known value for the variable and determining the value of the polynomial.

EXAMPLE 7   Automobiles sold in the United States have become more fuel efficient over the years due to regulations from Congress. The number of miles per gallon obtained by the average automobile in the United States can be described by the polynomial

$$0.3x + 12.9,$$

where x is the number of years since 1970. (*Source:* U.S. Federal Highway Administration.) Use this polynomial to estimate the number of miles per gallon obtained by the average automobile in

(a) 1972   **(b)** 2008

Solution

(a) The year 1972 is two years later than 1970, so $x = 2$.

Thus the number of miles per gallon obtained by the average automobile in 1972 can be estimated by evaluating $0.3x + 12.9$ when $x = 2$.

$$0.3(2) + 12.9 = 0.6 + 12.9$$
$$= 13.5$$

We estimate that the average car in 1972 obtained 13.5 miles per gallon.

(b) The year 2008 will be 38 years after 1970, so $x = 38$.

Thus the estimated number of miles per gallon obtained by the average automobile in 2008 can be predicted by evaluating $0.3x + 12.9$ when $x = 38$.

$$0.3(38) + 12.9 = 11.4 + 12.9$$
$$= 24.3$$

We therefore predict that the average car in 2008 will obtain 24.3 miles per gallon.

Practice Problem 7   The number of miles per gallon obtained by the average truck in the United States can be described by the polynomial $0.03x + 5.4$, where x is the number of years since 1970. (*Source:* U.S. Federal Highway Administration.) Use this polynomial to estimate the number of miles per gallon obtained by the average truck in

(a) 1974   **(b)** 2006

Verbal and Writing Skills

1. State in your own words a definition for a polynomial in x and give an example.

2. State in your own words a definition for a multivariable polynomial and give an example.

3. State in your own words how to determine the degree of a polynomial in x.

4. State in your own words how to determine the degree of a multivariable polynomial.

State the degree of the polynomial and whether it is a monomial, a binomial, or a trinomial.

5. $6x^3y$

6. $5xy^6$

7. $20x^5 + 6x^3 - 7x$

8. $13x^4 - 12x + 20$

9. $5xy^2 - 3x^2y^3$

10. $7x^3y + 5x^4y^4$

Add.

11. $(-3x + 15) + (8x - 43)$

12. $(5x - 11) + (-7x + 34)$

13. $(6x^2 + 5x - 6) + (-8x^2 - 3x + 5)$

14. $(3x^2 - 2x - 5) + (-7x^2 + 8x - 2)$

15. $\left(\frac{1}{2}x^2 + \frac{1}{3}x - 4\right) + \left(\frac{1}{3}x^2 + \frac{1}{6}x - 5\right)$

16. $\left(\frac{1}{4}x^2 - \frac{2}{3}x - 10\right) + \left(-\frac{1}{3}x^2 + \frac{1}{9}x + 2\right)$

17. $(3.4x^3 - 7.1x + 3.4) + (2.2x^2 - 6.1x - 8.8)$

18. $(-4.6x^3 + 5.6x - 0.3) + (9.8x^2 + 4.5x - 1.7)$

Subtract.

19. $(2x - 19) - (-3x + 5)$

20. $(5x - 5) - (6x - 3)$

21. $\left(\frac{2}{5}x^2 - \frac{1}{2}x + 5\right) - \left(\frac{1}{3}x^2 - \frac{3}{7}x - 6\right)$

22. $\left(\frac{3}{8}x^2 - \frac{2}{3}x - 7\right) - \left(\frac{2}{3}x^2 - \frac{1}{2}x + 2\right)$

23. $(-3x^2 + 5x) - (2x^3 - 3x^2 + 10)$

24. $(7x^3 - 3x^2 - 5x + 4) - (x^3 - 7x + 3)$

25. $(0.5x^4 - 0.7x^2 + 8.3) - (5.2x^4 + 1.6x + 7.9)$

26. $(1.3x^4 - 3.1x^3 + 6.3x) - (x^4 - 5.2x^2 + 6.5x)$

Perform the indicated operations.

27. $(4x - 3) - (7x - 2) + (-5x + 8)$

28. $(3 - 8x) - (9x + 7) - (-3x - 2)$

29. $(5x^2y - 6xy^2 + 2) + (-8x^2y + 12xy^2 - 6)$

30. $(7x^2y^2 - 6xy + 5) + (-15x^2y^2 - 6xy + 18)$

31. $(3x^4 - 4x^2 - 18) - (2x^4 + 3x^3 + 6)$

32. $(2b^3 + 3b - 5) - (-3b^3 + 5b^2 + 7b)$

To Think About

Gas Mileage   *Buses operated in this country have become more fuel efficient in recent years. The number of miles per gallon achieved by the average bus operated in the United States can be described by the polynomial* $0.04x + 5.2$, *where x is the number of years since 1970. (Source: U.S. Federal Highway Administration.)*

33. Estimate the number of miles per gallon obtained by the average bus in 1975.

34. Estimate the number of miles per gallon obtained by the average bus in 1995.

35. Estimate in what year the rating of the average bus will be 7.2 miles per gallon.

36. Estimate in what year the rating of the average bus will be 6.8 miles per gallon.

Prisons   *The average number of prisoners held in federal and state prisons increases each year. The number of prisoners measured in thousands can be described by the polynomial* $1.8x^2 + 22.2x + 325$, *where x is the number of years since 1980. (Source: U.S. Bureau of Justice Statistics.)*

37. Estimate the number of prisoners in 1990.

38. Estimate the number of prisoners in 1995.

39. According to the polynomial, by how much will the prison population increase from 2002 to 2006?

40. According to the polynomial, by how much will the prison population increase from 2004 to 2010?

Applications

▲ **41.** *Geometry* The lengths and the widths of the following three rectangles are labeled. Create a polynomial that describes the sum of the *area* of these three rectangles.

▲ **42.** *Geometry* The dimensions of the sides of the following figure are labeled. Create a polynomial that describes the *perimeter* of this figure.

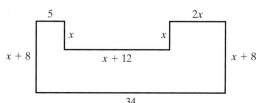

Cumulative Review

43. Solve for *x*. $3y - 8x = 2$

44. Solve for *B*. $3A = 2BCD$

45. Solve for *b*. $A = \frac{1}{2}h(b + c)$

46. Solve for *d*. $B = \frac{5xy}{d}$

47. *Health Care Costs* The total approximate expenditure for health care in the United States increased by 90% over the years 1990 to 2000. In 2000 a total of 1.324 trillion dollars was spent. How much was spent in 1990? (*Source:* U.S. Department of Health and Human Services.)

48. *World Population* In 1900, the population of the world was approximately 1.5×10^9. In 2000, the population had increased to 6×10^9. What was the percent of increase from 1900 to 2000?

1. _____

2. _____

How are you doing with your homework assignments in Sections 4.1 to 4.3? Do you feel you have mastered the material so far? Do you understand the concepts you have covered? Before you go further in the textbook, take some time to do each of the following problems.

Simplify all answers. Reduce all fractions.

4.1

Multiply

3. _____

1. $(-5xy)(3x^2y^5)$

Divide. Assume that all variables are nonzero. Leave your answer with only positive exponents.

4. _____

2. $-\dfrac{35xy^6}{25x^8y^3}$

3. $\dfrac{60x^7y^0}{15x^2y^9}$

5. _____

Simplify.

4. $(-3x^5y)^4$

4.2

6. _____

Simplify. Express your answer with positive exponents. Assume that all variables are nonzero.

5. $(4x^{-3}y^4)^{-2}$

6. $\dfrac{3x^{-2}y^6}{9xy^{-9}}$

7. _____

7. Write in scientific notation. 58,740

8. _____

8. Write in scientific notation. 0.00009362

4.3

Perform the indicated operations.

9. _____

9. $(2x^2 + 0.5x - 2) + (0.3x^2 - 0.9x - 3.4)$

10. $(5x^2 + 8x - 14) - (-2x^2 - 9x + 12)$

10. _____

11. $\left(\dfrac{1}{2}x^3 + \dfrac{1}{4}x^2 - 2x\right) - \left(\dfrac{1}{3}x^3 - \dfrac{1}{8}x^2 - 5x\right)$

11. _____

12. $\left(\dfrac{1}{16}x^2 + \dfrac{1}{8}\right) + \left(\dfrac{1}{4}x^2 - \dfrac{3}{10}x - \dfrac{1}{2}\right)$

Now turn to page SA-10 for the answer to each of these problems. Each answer also includes a reference to the objective in which the problem is first taught. If you missed any of these problems, you should stop and review the Examples and Practice Problems in the referenced objective. A little review now will help you master the material in the upcoming sections of the text.

12. _____

4.4 MULTIPLYING POLYNOMIALS

1 Multiplying a Monomial by a Polynomial

We use the distributive property to multiply a monomial by a polynomial. Remember, the distributive property states that for real numbers a, b, and c,

$$a(b + c) = ab + ac.$$

EXAMPLE 1 Multiply. $3x^2(5x - 2)$

Solution

$$3x^2(5x - 2) = 3x^2(5x) + 3x^2(-2) \qquad \text{Use the distributive property.}$$
$$= (3 \cdot 5)(x^2 \cdot x) + (3)(-2)x^2$$
$$= 15x^3 - 6x^2$$

Practice Problem 1 Multiply. $4x^3(-2x^2 + 3x)$

Try to do as much of the multiplication as you can mentally.

EXAMPLE 2 Multiply.

(a) $2x(x^2 + 3x - 1)$ **(b)** $-2xy^2(x^2 - 2xy - 3y^2)$

Solution

(a) $2x(x^2 + 3x - 1) = 2x^3 + 6x^2 - 2x$

(b) $-2xy^2(x^2 - 2xy - 3y^2) = -2x^3y^2 + 4x^2y^3 + 6xy^4$

Notice in part (b) that you are multiplying each term by the negative number $-2xy^2$. This will change the sign of each term in the product.

Practice Problem 2 Multiply.

(a) $-3x(x^2 + 2x - 4)$ **(b)** $6xy(x^3 + 2x^2y - y^2)$

When we multiply by a monomial, the monomial may be on the right side.

EXAMPLE 3 Multiply. $(x^2 - 2x + 6)(-2xy)$

Solution $(x^2 - 2x + 6)(-2xy) = -2x^3y + 4x^2y - 12xy$

Practice Problem 3 Multiply.

$$(-6x^3 + 4x^2 - 2x)(-3xy)$$

2 Multiplying Two Binomials

We can build on our knowledge of the distributive property and our experience with multiplying monomials to learn how to multiply two binomials. Let's suppose that we want to multiply $(x + 2)(3x + 1)$. We can use the distributive property. Since a can represent any number, let $a = x + 2$. Then let $b = 3x$ and $c = 1$. We now have the following.

$$a(b + c) = ab + ac$$
$$(x + 2)(3x + 1) = (x + 2)(3x) + (x + 2)(1)$$
$$= 3x^2 + 6x + x + 2$$
$$= 3x^2 + 7x + 2$$

NOTE TO STUDENT: *Fully worked-out solutions to all of the Practice Problems can be found at the back of the text starting at page SP-1*

Let's take another look at the original problem, $(x + 2)(3x + 1)$. This time we will assign a letter to each term in the binomials. That is, let $a = x$, $b = 2$, $c = 3x$, and $d = 1$. Using substitution, we have the following.

$$
\begin{aligned}
(x + 2)(3x + 1) &= (a + b)(c + d) \\
&= (a + b)c + (a + b)d \\
&= ac + bc + ad + bd \\
&= (x)(3x) + (2)(3x) + (x)(1) + (2)(1) \quad \text{By substitution} \\
&= 3x^2 + 6x + x + 2 \\
&= 3x^2 + 7x + 2
\end{aligned}
$$

How does this compare with the preceding result?

The distributive property shows us *how* the problem can be done and *why* it can be done. In actual practice there is a memory device to help students remember the steps involved. It is often referred to as FOIL. The letters FOIL stand for the following.

F	multiply the *First* terms
O	multiply the *Outer* terms
I	multiply the *Inner* terms
L	multiply the *Last* terms

The FOIL letters are simply a way to remember the four terms in the final product and how they are obtained. Let's return to our original problem.

$(x + 2)(3x + 1)$ F Multiply the *first* terms to obtain $3x^2$.

$(x + 2)(3x + 1)$ O Multiply the *outer* terms to obtain x.

$(x + 2)(3x + 1)$ I Multiply the *inner* terms to obtain $6x$.

$(x + 2)(3x + 1)$ L Multiply the *last* terms to obtain 2.

The result so far is $3x^2 + x + 6x + 2$. These four terms are the same four terms that we obtained when we multiplied using the distributive property. We can combine the like terms to obtain the final answer: $3x^2 + 7x + 2$. Now let's study the FOIL method in a few examples.

EXAMPLE 4

Multiply. $(2x - 1)(3x + 2)$

Solution

First   Last

$(2x - 1)(3x + 2)$

Inner

Outer

$$
\begin{aligned}
\text{First} &+ \text{Outer} + \text{Inner} + \text{Last} \\
F &\qquad O \qquad I \qquad L \\
= 6x^2 &+ \quad 4x \quad - \quad 3x \quad - \quad 2 \\
= 6x^2 &+ \quad x - 2 \quad \text{Collect like terms.}
\end{aligned}
$$

Practice Problem 4 Multiply. $(5x - 1)(x - 2)$

NOTE TO STUDENT: Fully worked-out solutions to all of the Practice Problems can be found at the back of the text starting at page SP-1

EXAMPLE 5 Multiply. $(3a - 2b)(4a - b)$

Solution

First   Last

$(3a - 2b)(4a - b)$ $= 12a^2 - 3ab - 8ab + 2b^2$

Inner $= 12a^2 - 11ab + 2b^2$

Outer

Practice Problem 5 Multiply. $(8a - 5b)(3a - b)$

After you have done several problems, you may be able to combine the outer and inner products mentally.

In some problems the inner and outer products cannot be combined.

EXAMPLE 6 Multiply. $(3x + 2y)(5x - 3z)$

Solution

First   Last

$(3x + 2y)(5x - 3z)$ $= 15x^2 - 9xz + 10xy - 6yz$

Inner

Outer   Since there are no like terms, we cannot combine any terms.

Practice Problem 6 Multiply. $(3a + 2b)(2a - 3c)$

EXAMPLE 7 Multiply. $(7x - 2y)^2$

Solution

$(7x - 2y)(7x - 2y)$ When we square a binomial, it is the same as multiplying the binomial by itself.

First   Last

$(7x - 2y)(7x - 2y)$ $= 49x^2 - 14xy - 14xy + 4y^2$

Inner $= 49x^2 - 28xy + 4y^2$

Outer

Practice Problem 7 Multiply. $(3x - 2y)^2$

We can multiply binomials containing exponents that are greater than 1. That is, we can multiply binomials containing x^2 or y^3, and so on.

EXAMPLE 8 Multiply. $(3x^2 + 4y^3)(2x^2 + 5y^3)$

Solution
$$(3x^2 + 4y^3)(2x^2 + 5y^3) = 6x^4 + 15x^2y^3 + 8x^2y^3 + 20y^6$$
$$= 6x^4 + 23x^2y^3 + 20y^6$$

Practice Problem 8 Multiply. $(2x^2 + 3y^2)(5x^2 + 6y^2)$

EXAMPLE 9 The width of a living room is $(x + 4)$ feet. The length of the room is $(3x + 5)$ feet. What is the area of the room in square feet?

3x + 5

x + 4

Solution
$$A = (\text{length})(\text{width}) = (3x + 5)(x + 4)$$
$$= 3x^2 + 12x + 5x + 20$$
$$= 3x^2 + 17x + 20$$

There are $(3x^2 + 17x + 20)$ square feet in the room.

NOTE TO STUDENT: Fully worked-out solutions to all of the Practice Problems can be found at the back of the text starting at page SP-1

Practice Problem 9 What is the area in square feet of a room that is $(2x - 1)$ feet wide and $(7x + 3)$ feet long?

Developing Your Study Skills

How to Review for an Exam

Reviewing adequately for an exam enables you to bring together the concepts you have learned over several sections. For your review, you will need to:

1. Reread your textbook. Make a list of any terms, rules, or formulas you need to know for the exam. Be sure you understand them all.

2. Reread your notes. Go over returned homework and quizzes. Redo the problems you missed.

3. Practice some of each type of problem covered in the chapter(s) you are to be tested on.

4. Use the end-of-chapter materials provided in your textbook. Read carefully through the chapter organizer. Do the review problems. Take the chapter test. When you are finished, check your answers. Redo any problems you missed.

5. Get help if any concepts give you difficulty.

4.4 EXERCISES

| Student Solutions Manual | CD/ Video | PH Math Tutor Center | MathXL®Tutorials on CD | MathXL® | MyMathLab® | Interactmath.com |

Multiply.

1. $-2x(6x^3 - x)$

2. $5x(-3x^4 + 4x)$

3. $3x^2(7x - 3)$

4. $2x^2(8x - 5)$

5. $2x^3(-2x^3 + 5x - 1)$

6. $5x^2(-3x^3 + 6x - 2)$

7. $\frac{1}{2}(2x + 3x^2 + 5x^3)$

8. $\frac{2}{3}(4x + 6x^2 - 2x^3)$

9. $(5x^3 - 2x^2 + 6x)(-3xy^2)$

10. $(2b^3 - 5b^2 + 3ab)(-3b^2)$

11. $(3x^3 + x^2 - 8x)(3xy)$

12. $(2x^3 + x^2 - 6x)(2xy)$

13. $(x^3 - 3x^2 + 5x - 2)(3x)$

14. $(-2x^3 + 4x^2 - 7x + 3)(2x)$

15. $(x^2y^2 - 6xy + 8)(-2xy)$

16. $(x^2y^2 + 5xy - 9)(-3xy)$

17. $(-7x^3 + 3x^2 + 2x - 1)(4x^2y)$

18. $(-5x^3 - 6x^2 + x - 1)(5xy^2)$

19. $(3d^4 - 4d^2 + 6)(-2c^2d)$

20. $(-4x^3 + 6x^2 - 5x)(-7xy^2)$

21. $6x^3(2x^4 - x^2 + 3x + 9)$

22. $8x^3(-2x^4 + 3x^2 - 5x - 14)$

23. $-7x^4(5x^3 - 8x^2 + 3x)$

24. $-5x^3(-2x^5 - 7x^2 + 4x)$

Multiply. Try to do most of the exercises mentally without writing down intermediate steps.

25. $(x + 10)(x + 3)$

26. $(x + 3)(x + 4)$

27. $(x + 6)(x + 2)$

28. $(x + 9)(x + 3)$

29. $(x - 8)(x + 2)$

30. $(x + 3)(x - 6)$

31. $(x - 5)(x - 4)$

32. $(x - 6)(x - 5)$

33. $(5x - 2)(-4x - 3)$

34. $(7x + 1)(-2x - 3)$

35. $(7x - 4)(x + 2y)$

36. $(x - 9)(3x + 4y)$

37. $(5x + 2)(3x - y)$

38. $(3x + 4)(5x - y)$

39. $(4y + 1)(5y - 3)$

40. $(5y + 1)(6y - 5)$

To Think About

41. What is wrong with this multiplication?
$(x - 2)(-3) = 3x - 6$

42. What is wrong with this answer? $-(3x - 7) = -3x - 7$

43. What is the missing term?
$(5x + 2)(5x + 2) = 25x^2 + $ _____ $+ 4$

44. Multiply the binomials and write a brief description of what is special about the result. $(5x - 1)(5x + 1)$

Mixed Practice

Multiply.

45. $(4x - 3y)(5x - 2y)$

46. $(3b - 5c)(2b - 7c)$

47. $(7x - 2)^2$

48. $(3x - 7)^2$

49. $(4a + 2b)^2$

50. $(5a + 3b)^2$

51. $(0.2x + 3)(4x - 0.3)$

52. $(0.5x - 2)(6x - 0.2)$

53. $\left(\frac{1}{2}x + \frac{1}{3}\right)\left(\frac{1}{2}x - \frac{1}{4}\right)$

54. $\left(\frac{1}{3}x + \frac{1}{5}\right)\left(\frac{1}{3}x - \frac{1}{2}\right)$

55. $(2a - 3b)(x - 5b)$

56. $(5b - 7c)(x - 2b)$

57. $(7x + 3y)(5x + 9y)$

58. $(4y - 3z)(5z - 3y)$

59. $(5y - 3z)(4y - 2z)$

Find the area of the rectangle.

▲ **60.**

5x + 2

2x − 3

▲ **61.**

7x + 3

4x − 6

Cumulative Review

62. Solve for x. $3(x - 6) = -2(x + 4) + 6x$

63. Solve for w. $3(w - 7) - (4 - w) = 11w$

64. Twice the sum of a number and six is five more than three times the same number. Find the number.

65. Jamie was buying decorative marbles to re-create a table centerpiece she saw in a decorating magazine. She bought twice as many cat's eye marbles as tiger's eye marbles. If the cat'e eye marbles cost $0.07 apiece and the tiger's eye marbles cost $0.11 apiece, and Jamie spent $6.00 for the marbles, how many of each kind did Jamie buy?

66. *Paper Currency* Heather returned from the bank with $375. She had one more $20 bill than she had $10 bills. The number of $5 bills she had was one less than triple the number of $10 bills. How many of each denomination did she have?

Video Game Playing *The polynomial* $1.2x + 16$ *can be used to describe the number of hours per week that young people age 12–18 spend playing video games, where x is the number of years since 1992. (Source: U.S. Department of Health and Human Services.)*

67. What was the average number of hours per week young people spent playing video games in 1994?

68. What was the average number of hours per week young people spent playing video games in 1996?

69. Predict the number of hours per week young people will spend playing video games in 2008.

70. Predict the number of hours per week young people will spend playing video games in 2009.

① Multiplying Binomials of the Type $(a + b)(a - b)$

The case when you multiply $(x + y)(x - y)$ is interesting and deserves special consideration. Using the FOIL method, we find

$$(x + y)(x - y) = x^2 - xy + xy - y^2 = x^2 - y^2.$$

Notice that the sum of the inner product and the outer product is zero. We see that

$$(x + y)(x - y) = x^2 - y^2.$$

This works in all cases when the binomials are the sum and difference of the same two terms. That is, in one factor the terms are added, while in the other factor the same two terms are subtracted.

$$(5a + 2b)(5a - 2b) = 25a^2 - 10ab + 10ab - 4b^2$$
$$= 25a^2 - 4b^2$$

The product is the difference of the squares of the terms. That is, $(5a)^2 - (2b)^2$ or $25a^2 - 4b^2$.

Many students find it helpful to memorize this equation.

> **MULTIPLYING BINOMIALS: A SUM AND A DIFFERENCE**
> $$(a + b)(a - b) = a^2 - b^2$$

You may use this relationship to find the product quickly in cases where it applies. The terms must be the same and there must be a sum and a difference.

EXAMPLE 1 Multiply. $(7x + 2)(7x - 2)$

Solution

$$(7x + 2)(7x - 2) = (7x)^2 - (2)^2 = 49x^2 - 4$$

Practice Problem 1 Multiply. $(6x + 7)(6x - 7)$

EXAMPLE 2 Multiply. $(5x - 8y)(5x + 8y)$

Solution

$$(5x - 8y)(5x + 8y) = (5x)^2 - (8y)^2 = 25x^2 - 64y^2$$

Practice Problem 2 Multiply. $(3x - 5y)(3x + 5y)$

Student Learning Objectives

After studying this section, you will be able to:

① Multiply binomials of the type $(a + b)(a - b)$.

② Multiply binomials of the type $(a + b)^2$ and $(a - b)^2$.

③ Multiply polynomials with more than two terms.

NOTE TO STUDENT: Fully worked-out solutions to all of the Practice Problems can be found at the back of the text starting at page SP-1

2 Multiplying Binomials of the Type $(a + b)^2$ and $(a - b)^2$

A second case that is worth special consideration is a binomial that is squared. Consider the following problem.

$$
\begin{aligned}
(3x + 2)^2 &= (3x + 2)(3x + 2) \\
&= 9x^2 + 6x + 6x + 4 \\
&= 9x^2 + 12x + 4
\end{aligned}
$$

If you complete enough problems of this type, you will notice a pattern. The answer always contains the square of the first term added to double the product of the first and last terms added to the square of the last term.

$3x$ is the first term		2 is the last term		Square the first term: $(3x)^2$		Double the product of the first and last terms: $2(3x)(2)$		Square the last term: $(2)^2$
↓		↓		↓		↓		↓
$(3x$	$+$	$2)^2$	$=$	$9x^2$	$+$	$12x$	$+$	4

We can show the same steps using variables instead of words.

$$(a + b)^2 = a^2 + 2ab + b^2$$

There is a similar formula for the square of a difference:

$$(a - b)^2 = a^2 - 2ab + b^2$$

We can use this formula to simplify $(2x - 3)^2$.

$$
\begin{aligned}
(2x - 3)^2 &= (2x)^2 - 2(2x)(3) + (3)^2 \\
&= 4x^2 - 12x + 9
\end{aligned}
$$

You may wish to multiply this product using FOIL to verify.

These two types of products, the square of a sum and the square of a difference, can be summarized as follows.

A BINOMIAL SQUARED

$$(a + b)^2 = a^2 + 2ab + b^2$$
$$(a - b)^2 = a^2 - 2ab + b^2$$

EXAMPLE 3 Multiply.

(a) $(5y - 2)^2$ **(b)** $(8x + 9y)^2$

Solution

(a) $(5y - 2)^2 = (5y)^2 - (2)(5y)(2) + (2)^2$
$$= 25y^2 - 20y + 4$$

(b) $(8x + 9y)^2 = (8x)^2 + (2)(8x)(9y) + (9y)^2$
$$= 64x^2 + 144xy + 81y^2$$

NOTE TO STUDENT: Fully worked-out solutions to all of the Practice Problems can be found at the back of the text starting at page SP-1

Practice Problem 3 Multiply.

(a) $(4a - 9b)^2$ **(b)** $(5x + 4)^2$

CAUTION: $(a + b)^2 \neq a^2 + b^2$! The two sides are not equal! Squaring the sum $(a + b)$ does not give $a^2 + b^2$! Beginning algebra students often make this error. Make sure you remember that when you square a binomial there is always a *middle term*.

$$(a + b)^2 = a^2 + 2ab + b^2$$

Sometimes a numerical example helps you to see this.

$$(3 + 4)^2 \neq 3^2 + 4^2$$
$$7^2 \neq 9 + 16$$
$$49 \neq 25$$

Notice that what is missing on the right is $2ab = 2 \cdot 3 \cdot 4 = 24$.

③ Multiplying Polynomials with More Than Two Terms

We used the distributive property to multiply two binomials $(a + b)(c + d)$, and we obtained $ac + ad + bc + bd$. We could also use the distributive property to multiply the polynomials $(a + b)$ and $(c + d + e)$, and we would obtain $ac + ad + ae + bc + bd + be$. Let us see if we can find a direct way to multiply products such as $(3x - 2)(x^2 - 2x + 3)$. It can be done quickly using an approach similar to that used in arithmetic for multiplying whole numbers. Consider the following arithmetic problem.

$$
\begin{array}{r}
128 \\
\times\ \ 43 \\
\hline
384 \\
512\ \ \ \\
\hline
5504
\end{array}
$$
  ← The product of 128 and 3
  ← The product of 128 and 4 moved one space to the left
  ← The sum of the two partial products

Let us follow a similar format to multiply the two polynomials. For example, multiply $(x^2 - 2x + 3)$ and $(3x - 2)$.

$$
\begin{array}{r}
x^2 - 2x + 3 \\
3x - 2 \\
\hline
-2x^2 + 4x - 6 \\
3x^3 - 6x^2 + 9x\ \ \ \ \ \\
\hline
3x^3 - 8x^2 + 13x - 6
\end{array}
$$
  ← The product $(x^2 - 2x + 3)(-2)$ This is often called **vertical multiplication.**
  ← The product $(x^2 - 2x + 3)(3x)$ moved one space to the left so that like terms are underneath each other
  ← The sum of the two partial products

EXAMPLE 4   Multiply vertically. $(3x^3 + 2x^2 + x)(x^2 - 2x - 4)$

Solution
$$
\begin{array}{r}
3x^3 + 2x^2 + x \\
x^2 - 2x - 4 \\
\hline
-12x^3 - 8x^2 - 4x \\
-6x^4 - 4x^3 - 2x^2\ \ \ \ \ \\
3x^5 + 2x^4 + x^3\ \ \ \ \ \ \ \ \ \ \\
\hline
3x^5 - 4x^4 - 15x^3 - 10x^2 - 4x
\end{array}
$$
We place one polynomial over the other.
  ← The product $(3x^3 + 2x^2 + x)(-4)$
  ← The product $(3x^3 + 2x^2 + x)(-2x)$
  ← The product $(3x^3 + 2x^2 + x)(x^2)$
  ← The sum of the three partial products

Note that the answers for each partial product are placed so that like terms are underneath each other.

Practice Problem 4 Multiply vertically. $(4x^3 - 2x^2 + x)(x^2 + 3x - 2)$

ALTERNATIVE METHOD: FOIL Horizontal Multiplication Some students prefer to do this type of multiplication using a horizontal format similar to the FOIL method. The following example illustrates this approach.

EXAMPLE 5 Multiply horizontally. $(x^2 + 3x + 5)(x^2 - 2x - 6)$

Solution We will use the distributive property repeatedly.

$$(x^2 + 3x + 5)(x^2 - 2x - 6) = x^2(x^2 - 2x - 6) + 3x(x^2 - 2x - 6) + 5(x^2 - 2x - 6)$$
$$= x^4 - 2x^3 - 6x^2 + 3x^3 - 6x^2 - 18x + 5x^2 - 10x - 30$$
$$= x^4 + x^3 - 7x^2 - 28x - 30$$

Practice Problem 5 Multiply horizontally. $(2x^2 + 5x + 3)(x^2 - 3x - 4)$

Some problems may need to be done in two or more separate steps.

EXAMPLE 6 Multiply. $(2x - 3)(x + 2)(x + 1)$

Solution We first need to multiply any two of the binomials. Let us select the first pair.

$$\underbrace{(2x - 3)(x + 2)}\,(x + 1)$$

Find this product first.

$$(2x - 3)(x + 2) = 2x^2 + 4x - 3x - 6$$
$$= 2x^2 + x - 6$$

Now we replace the first two factors with their resulting product.

$$\underbrace{(2x^2 + x - 6)}\,(x + 1)$$

Result of first product

We then multiply again.

$$(2x^2 + x - 6)(x + 1) = (2x^2 + x - 6)x + (2x^2 + x - 6)1$$
$$= 2x^3 + x^2 - 6x + 2x^2 + x - 6$$
$$= 2x^3 + 3x^2 - 5x - 6$$

The vertical format of Example 4 is an alternative method for this type of problem.

$$
\begin{array}{r}
2x^2 + x - 6 \\
x + 1 \\
\hline
2x^2 + x - 6 \\
2x^3 + x^2 - 6x \\
\hline
2x^3 + 3x^2 - 5x - 6
\end{array}
$$

Be sure to use special care in writing the exponents correctly in problems with more than one variable.

Thus we have

$$(2x - 3)(x + 2)(x + 1) = 2x^3 + 3x^2 - 5x - 6.$$

Note that it does not matter which two binomials are multiplied first. For example, you could first multiply $(2x - 3)(x + 1)$ to obtain $2x^2 - x - 3$ and then multiply that product by $(x + 2)$ to obtain the same result.

NOTE TO STUDENT: Fully worked-out solutions to all of the Practice Problems can be found at the back of the text starting at page SP-1

Practice Problem 6 Multiply. $(3x - 2)(2x + 3)(3x + 2)$
(*Hint:* Rearrange the factors.)

Sometimes we encounter a binomial raised to a third power. In such cases we would write out the binomial three times as a product and then multiply. So to evaluate $(3x + 4)^3$ we would first write $(3x + 4)(3x + 4)(3x + 4)$ and then follow the method of Example 6.

Verbal and Writing Skills

1. In the special case of $(a + b)(a - b)$, a binomial times a binomial is a _____.

2. Identify which of the following could be the answer to a problem using the formula for $(a + b)(a - b)$. Why?
(a) $9x^2 - 16$
(b) $4x^2 + 25$
(c) $9x^2 + 12x + 4$
(d) $x^4 - 1$

3. A student evaluated $(4x - 7)^2$ as $16x^2 + 49$. What is missing? State the correct answer.

4. The square of a binomial, $(a - b)^2$, always produces which of the following?
(a) binomial
(b) trinomial
(c) four-term polynomial

Use the formula $(a + b)(a - b) = a^2 - b^2$ to multiply.

5. $(y - 7)(y + 7)$

6. $(x + 5)(x - 5)$

7. $(x - 9)(x + 9)$

8. $(x + 6)(x - 6)$

9. $(7x - 4)(7x + 4)$

10. $(4x - 9)(4x + 9)$

11. $(2x - 7)(2x + 7)$

12. $(3x - 10)(3x + 10)$

13. $(5x - 3y)(5x + 3y)$

14. $(8a - 3b)(8a + 3b)$

15. $(0.6x + 3)(0.6x - 3)$

16. $(5x - 0.2)(5x + 0.2)$

Use the formula for a binomial squared to multiply.

17. $(3y + 1)^2$

18. $(4x - 1)^2$

19. $(5x - 4)^2$

20. $(6x + 5)^2$

21. $(7x + 3)^2$

22. $(8x - 3)^2$

23. $(3x - 7)^2$

24. $(2x + 3y)^2$

25. $\left(\dfrac{2}{3}x + \dfrac{1}{4}\right)^2$

26. $\left(\dfrac{3}{4}x + \dfrac{1}{2}\right)^2$

27. $(9xy + 4z)^2$

28. $(7y - 3xz)^2$

Mixed Practice

Multiply. Use the special formula that applies.

29. $(7x + 3y)(7x - 3y)$

30. $(12x - 5y)(12x + 5y)$

31. $(7c - 6d)^2$

32. $(4c - 5d)^2$

Multiply.

33. $(9a - 10b)(9a + 10b)$

34. $(11a + 6b)(11a - 6b)$

35. $(5x + 9y)^2$

36. $(4x + 8y)^2$

37. $(x^2 + 5x - 3)(x - 2)$

38. $(x^2 - 4x + 5)(x - 3)$

39. $(4x + 1)(x^3 - 2x^2 + x - 1)$

40. $(3x - 1)(x^3 + x^2 - 4x - 2)$

41. $(a^2 - 3a + 2)(a^2 + 4a - 3)$

42. $(x^2 + 4x - 5)(x^2 - 3x + 4)$

43. $(x + 3)(x - 1)(3x - 8)$

44. $(x - 7)(x + 4)(2x - 5)$

45. $(3x + 5)(x - 2)(x - 4)$

46. $(2x - 7)(x + 1)(x - 2)$

47. $(a - 5)(2a + 3)(a + 5)$

48. $(b + 7)(3b - 2)(b - 7)$

To Think About

▲ **49.** Find the volume of this object.

$(2x + 1)$

$(3x - 2)$

$(4x + 3)$

▲ **50.** *Geometry* The volume of a pyramid is $V = \dfrac{1}{3}Bh$, where B is the area of the base and H is the height. Find the volume of the following pyramid.

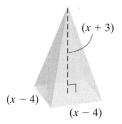

$(x + 3)$

$(x - 4)$

$(x - 4)$

Cumulative Review

51. *Investments* An executive invested $18,000 in two accounts, one a cash-on-reserve account that yielded 7% simple interest, and the other a long-term certificate paying 11% simple interest. At the end of one year, the two accounts had earned her a total of $1540. How much had she invested in each account?

▲ **52.** *Room Dimensions* The perimeter of a rectangular room measures 34 meters. The width is 2 meters more than half the length. Find the dimensions of the room.

53. *Boiling Point of Ammonia* Ammonia has a boiling point of $-33.4°C$. What is the boiling point of ammonia measured in degrees Fahrenheit? (Use $F = 1.8C + 32$.)

54. *Red Blood Cells* A man with 5.5 liters of blood has approximately 2.75×10^{10} red blood cells. Each red blood cell is a small disk that measures 7.5×10^{-6} meters in diameter. If all of the red blood cells were arranged in a long straight line, how long would that line be?

Student Learning Objectives

After studying this section, you will be able to:

 Divide a polynomial by a monomial.

 Divide a polynomial by a binomial.

1 Dividing a Polynomial by a Monomial

To divide a polynomial by a monomial, divide each term of the numerator by the denominator; then write the sum of the results. We are using the property of fractions that states that

$$\frac{a + b}{c} = \frac{a}{c} + \frac{b}{c}.$$

DIVIDING A POLYNOMIAL BY A MONOMIAL

1. Divide each term of the polynomial by the monomial.
2. When dividing variables, use the property $\frac{x^a}{x^b} = x^{a-b}$.

EXAMPLE 1 Divide. $\dfrac{8y^6 - 8y^4 + 24y^2}{8y^2}$

Solution $\dfrac{8y^6 - 8y^4 + 24y^2}{8y^2} = \dfrac{8y^6}{8y^2} - \dfrac{8y^4}{8y^2} + \dfrac{24y^2}{8y^2} = y^4 - y^2 + 3$

Practice Problem 1 Divide. $\dfrac{15y^4 - 27y^3 - 21y^2}{3y^2}$

NOTE TO STUDENT: Fully worked-out solutions to all of the Practice Problems can be found at the back of the text starting at page SP-1

2 Dividing a Polynomial by a Binomial

Division of a polynomial by a binomial is similar to long division in arithmetic. Notice the similarity in the following division problems.

Division of a three-digit number by a two-digit number

$$\begin{array}{r} 32 \\ 21\overline{)672} \\ \underline{63} \\ 42 \\ \underline{42} \\ 0 \end{array}$$

Division of a polynomial by a binomial

$$\begin{array}{r} 3x + 2 \\ 2x + 1\overline{)6x^2 + 7x + 2} \\ \underline{6x^2 + 3x} \\ 4x + 2 \\ \underline{4x + 2} \\ 0 \end{array}$$

DIVIDING A POLYNOMIAL BY A BINOMIAL

1. Place the terms of the polynomial and binomial in descending order. Insert a 0 for any missing term.
2. Divide the first term of the polynomial by the first term of the binomial. The result is the first term of the answer.
3. Multiply the first term of the answer by the binomial and subtract the result from the first two terms of the polynomial. Bring down the next term to obtain a new polynomial.
4. Divide the new polynomial by the binomial using the process described in step 2.
5. Continue dividing, multiplying, and subtracting until the degree of the remainder is less than the degree of the binomial divisor.
6. Write the remainder as the numerator of a fraction that has the binomial divisor as its denominator.

EXAMPLE 2   Divide. $(x^3 + 5x^2 + 11x + 14) \div (x + 2)$

Solution

Step 1   The terms are arranged in descending order. No terms are missing.

Step 2   Divide the first term of the polynomial by the first term of the binomial. In this case, divide x^3 by x to get x^2.

$$\begin{array}{r} x^2 \\ x + 2 \overline{\smash{)}x^3 + 5x^2 + 11x + 14} \end{array}$$

Step 3   Multiply x^2 by $x + 2$ and subtract the result from the first two terms of the polynomial, $x^3 + 5x^2$ in this case.

$$\begin{array}{r} x^2 \\ x + 2 \overline{\smash{)}x^3 + 5x^2 + 11x + 14} \\ \underline{x^3 + 2x^2} \downarrow \\ 3x^2 + 11x \end{array}$$

Bring down the next term.

Step 4   Continue to use the step 2 process. Divide $3x^2$ by x. Write the resulting $3x$ as the next term of the answer.

$$\begin{array}{r} x^2 + 3x \\ x + 2 \overline{\smash{)}x^3 + 5x^2 + 11x + 14} \\ \underline{x^3 + 2x^2} \\ 3x^2 + 11x \end{array}$$

Step 5   Continue multiplying, dividing, and subtracting until the degree of the remainder is less than the degree of the divisor. In this case, we stop when the remainder does not have an x.

$$\begin{array}{r} x^2 + 3x + 5 \\ x + 2 \overline{\smash{)}x^3 + 5x^2 + 11x + 14} \\ \underline{x^3 + 2x^2} \\ 3x^2 + 11x \\ \underline{3x^2 + 6x} \\ 5x + 14 \\ \underline{5x + 10} \\ 4 \end{array}$$

$\longleftarrow$ The remainder is 4.

Step 6   The answer is $x^2 + 3x + 5 + \dfrac{4}{x + 2}$.

To check the answer, we multiply $(x + 2)(x^2 + 3x + 5)$ and add the remainder 4.

$$(x + 2)(x^2 + 3x + 5) + 4 = x^3 + 5x^2 + 11x + 10 + 4 = x^3 + 5x^2 + 11x + 14$$

This is the original polynomial. It checks.

Practice Problem 2   Divide. $(x^3 + 10x^2 + 31x + 35) \div (x + 4)$

Take great care with the subtraction step when negative numbers are involved.

EXAMPLE 3 Divide. $(5x^3 - 24x^2 + 9) \div (5x + 1)$

Solution We must first insert $0x$ to represent the missing x-term. Then we divide $5x^3$ by $5x$.

$$
\begin{array}{r}
x^2 \\
5x + 1 \overline{)5x^3 - 24x^2 + 0x + 9} \\
\underline{5x^3 + x^2} \\
- 25x^2
\end{array}
$$

Note that we are subtracting:
$$-24x^2 - (+1x^2) = -24x^2 - 1x^2$$
$$= -25x^2$$

Next we divide $-25x^2$ by $5x$.

$$
\begin{array}{r}
x^2 - 5x \\
5x + 1 \overline{)5x^3 - 24x^2 + 0x + 9} \\
\underline{5x^3 + x^2} \\
- 25x^2 + 0x \\
\underline{- 25x^2 - 5x} \\
5x
\end{array}
$$

Note that we are subtracting:
$$0x - (-5x) = 0x + 5x = 5x$$

Finally, we divide $5x$ by $5x$.

$$
\begin{array}{r}
x^2 - 5x + 1 \\
5x + 1 \overline{)5x^3 - 24x^2 + 0x + 9} \\
\underline{5x^3 + x^2} \\
- 25x^2 + 0x \\
\underline{- 25x^2 - 5x} \\
5x + 9 \\
\underline{5x + 1} \\
8
\end{array}
$$

$\longleftarrow$ The remainder is 8.

The answer is $x^2 - 5x + 1 + \dfrac{8}{5x + 1}$.

To check, multiply $(5x + 1)(x^2 - 5x + 1)$ and add the remainder 8.

$$(5x + 1)(x^2 - 5x + 1) + 8 = 5x^3 - 24x^2 + 1 + 8 = 5x^3 - 24x^2 + 9$$

This is the original polynomial. Our answer is correct.

NOTE TO STUDENT: *Fully worked-out solutions to all of the Practice Problems can be found at the back of the text starting at page SP-1*

Practice Problem 3 Divide. $(2x^3 - x^2 + 1) \div (x - 1)$

Now we will perform the division by writing a minimum of steps. See if you can follow each step.

EXAMPLE 4   Divide and check. $(12x^3 - 11x^2 + 8x - 4) \div (3x - 2)$

Solution

$$
\begin{array}{r}
4x^2 - x + 2 \\
3x - 2\overline{)12x^3 - 11x^2 + 8x - 4} \\
\underline{12x^3 - 8x^2} \\
-3x^2 + 8x \\
\underline{-3x^2 + 2x} \\
6x - 4 \\
\underline{6x - 4} \\
0
\end{array}
$$

Check. $(3x - 2)(4x^2 - x + 2) = 12x^3 - 3x^2 + 6x - 8x^2 + 2x - 4$

$$= 12x^3 - 11x^2 + 8x - 4 \quad \text{Our answer is correct.}$$

Practice Problem 4   Divide and check.

$$(20x^3 - 11x^2 - 11x + 6) \div (4x - 3)$$

Developing Your Study Skills

Taking the Exam

Allow yourself plenty of time to get to your exam. You may even find it helpful to arrive a little early in order to collect your thoughts and ready yourself. This will help you feel more relaxed.

After you get your exam, you will find it helpful to do the following.

1. Take two or three moderately deep breaths. Inhale, then exhale slowly. You will feel your entire body begin to relax.

2. Write down on the back of the exam any formulas or ideas that you need to remember.

3. Look over the entire test quickly in order to pace yourself and use your time wisely. Notice how many points each question is worth. Spend more time on items of greater worth.

4. Read directions carefully and be sure to answer all questions clearly. Keep your work neat and easy to read.

5. Ask your instructor about anything that is not clear to you.

6. Answer the questions that are easiest for you first. Then come back to the more difficult ones.

7. Do not get bogged down on one question for too long because it may jeopardize your chances of finishing other problems. Leave the tough question and come back to it when you have time later.

8. Check your work. This will help you to catch minor errors.

9. Stay calm if others leave before you do. You are entitled to use the full amount of allotted time. You will do better on the exam if you take your time and work carefully.

Divide.

1. $\dfrac{25x^4 - 15x^2 + 20x}{5x}$

2. $\dfrac{18b^5 - 12b^3 + 6b^2}{3b}$

3. $\dfrac{8y^4 - 12y^3 - 4y^2}{4y^2}$

4. $\dfrac{10y^4 - 35y^3 + 5y^2}{5y^2}$

5. $\dfrac{81x^7 - 36x^5 - 63x^3}{9x^3}$

6. $\dfrac{49x^8 - 35x^6 - 56x^3}{7x^3}$

7. $(48x^7 - 54x^4 + 36x^3) \div 6x^3$

8. $(72x^8 - 56x^5 - 40x^3) \div 8x^3$

Divide. Check your answers for exercises 9–16 by multiplication.

9. $\dfrac{6x^2 + 13x + 5}{2x + 1}$

10. $\dfrac{12x^2 + 19x + 5}{3x + 1}$

11. $\dfrac{x^2 - 8x - 17}{x - 5}$

12. $\dfrac{x^2 - 9x - 5}{x - 3}$

13. $\dfrac{3x^3 - x^2 + 4x - 2}{x + 1}$

14. $\dfrac{2x^3 - 3x^2 - 3x + 6}{x - 1}$

15. $\dfrac{4x^3 + 4x^2 - 19x - 15}{2x + 5}$

16. $\dfrac{6x^3 + 11x^2 - 8x + 5}{2x + 5}$

17. $\dfrac{10x^3 + 11x^2 - 11x + 2}{5x - 2}$

18. $\dfrac{5x^3 - 28x^2 - 20x + 21}{5x - 3}$

19. $\dfrac{12y^3 - 12y^2 - 25y + 31}{2y - 3}$

20. $\dfrac{9y^3 - 30y^2 + 31y - 4}{3y - 5}$

21. $(y^3 - y^2 - 13y - 12) \div (y + 3)$

22. $(y^3 - 2y^2 - 26y - 4) \div (y + 4)$

23. $(y^4 - 9y^2 - 5) \div (y - 2)$

24. $(2y^4 + 3y^2 - 5) \div (y - 2)$

To Think About

25. $(8y^3 + 3y - 7) \div (4y - 1)$

    (*Hint:* The answer contains fractions.)

26. $(6y^3 - 3y^2 + 4) \div (2y + 1)$

    (*Hint:* The answer contains fractions.)

Cumulative Review

27. ***Water Conservation*** The average home in the United States uses 110,000 gallons of water per year. Most environmental groups feel that with improved efficiency of water-using devices (washers, toilets, faucets, shower heads), this could be reduced by 30%. (*Source:* U.S. Energy Information Administration.) How many gallons of water per year would be used in the average home if this reduction were accomplished?

28. ***Flour Price*** In 1965, the price for 5 kg of flour was $0.75. In 1970, the price increased by 15%. In 1971, the 1970 price decreased by 5%. What was the cost of 5 kg of flour in 1971? (Be sure to round each amount to the nearest cent.)

29. ***Cat Health*** In a veterinary study of a group of cats over age 5 who experienced regular exercise and playtime, 184 of the cats, or 92%, had no health problems. How many cats were in the study?

30. ***Internet Research*** Marlena's job requires doing research on the Internet for a marketing company. Today, she has downloaded a huge file. She has been reading all morning and has just completed two consecutive pages whose page numbers add to 1039. What are the page numbers?

31. ***Hurricane Pattern*** The National Hurricane Center has noticed an interesting pattern in the number of Atlantic Hurricanes that begin during the period August 1 to October 31 each year. Examine the bar graph. Then answer the following questions. Round all answers to the nearest tenth.

    (a) What was the mean number of hurricanes per year during the years 1993 to 1997?

    (b) What was the mean number of hurricanes per year during the years 1998 to 2002?

    (c) What was the percent of increase from the five year period 1993 to 1997 to the five year period 1998 to 2002?

    (d) If the same percent of increase for a five-year period continued to the years 2003 to 2007, what would be the mean number hurricanes during the period 2003 to 2007?

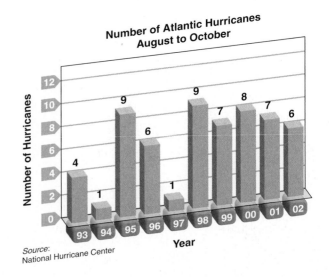

Source: National Hurricane Center

The Value of an Education

Is it worth your time, effort, and money to stay in school and earn a higher degree? Based on data from the Census Bureau's current population survey, the answer is yes! Below is a bar graph demonstrating the relationship between education and earnings, for people aged 18 and over, during the 1999 calendar year. You can see that more education means greater earnings over a year's time as well as greater earnings over a lifetime.

Problems for Individual Investigation and Analysis

1. You are trying to decide whether to continue on to a four-year college, or end your education with an associate's degree. How much more would you make per year if you continued on to earn a bachelor's degree?

2. You expect to have a working career of 42 years. How much more money will you expect to earn over your working career if you go on to complete your bachelor's degree?

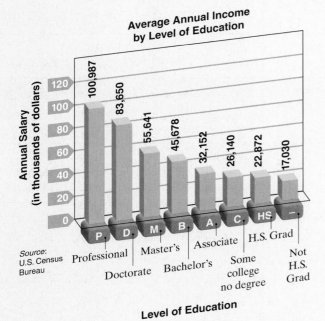

Average Annual Income by Level of Education

Annual Salary (in thousands of dollars)

P — 100,987 — Professional
D — 83,650 — Doctorate
M — 55,641 — Master's
B — 45,678 — Bachelor's
A — 32,152 — Associate
C — 26,140 — Some college no degree
HS — 22,872 — H.S. Grad
— — 17,030 — Not H.S. Grad

Source: U.S. Census Bureau

Level of Education

Problems for Group Investigation and Cooperative Learning

"Professional" denotes a person such as a doctor or lawyer who must obtain a doctorate and a professional license, as well.

3. You can calculate your lifetime earnings based on this chart of 1999 dollars, but it will not be completely accurate. The reality is that the value of the dollar continually changes.

   In 1992, professionals were estimated to earn \$74,560 per year. What percentage of growth in average annual income did professionals show from 1992 to 1999?

4. If this rate of growth remains constant for professionals over the next 7 years, what would a profession expect to earn in 2006?

5. Some economists use the income polynomial to predict the average annual income of people in the United States who have a bachelor's degree. It is given by the equation $I = 35x^3 - 620x^2 + 2820x + 43,112$, where x is the number of years since 1998 and I is the annual income in dollars. Using the income polynomial as just defined, predict the average annual income of a person with a bachelor's degree in the year 2004.

6. Using the income polynomial in exercise 5, predict the average annual income of a person with a bachelor's degree in the year 2006.

Chapter 4 Organizer

Topic	Procedure	Examples
Multiplying monomials, p. 252.	$x^a \cdot x^b = x^{a+b}$ 1. Multiply the numerical coefficients. 2. Add the exponents of a given base.	$3^{12} \cdot 3^{15} = 3^{27}$ $(-3x^2)(6x^3) = -18x^5$ $(2ab)(4a^2b^3) = 8a^3b^4$
Dividing monomials, p. 254.	$\dfrac{x^a}{x^b} = \begin{cases} x^{a-b} & \text{Use if } a \text{ is greater than } b. \\ \dfrac{1}{x^{b-a}} & \text{Use if } b \text{ is greater than } a. \end{cases}$ 1. Divide or reduce the fraction created by the quotient of the numerical coefficients. 2. Subtract the exponents of a given base.	$\dfrac{16x^7}{8x^3} = 2x^4$ $\dfrac{5x^3}{25x^5} = \dfrac{1}{5x^2}$ $\dfrac{-12x^5y^7}{18x^3y^{10}} = -\dfrac{2x^2}{3y^3}$
Exponent of zero, p. 255.	$x^0 = 1 \quad \text{if } x \neq 0$	$5^0 = 1 \qquad \dfrac{x^6}{x^6} = 1$ $w^0 = 1 \qquad 3x^0y = 3y$
Raising a power to a power, p. 256.	$(x^a)^b = x^{ab}$ $(xy)^a = x^a y^a$ $\left(\dfrac{x}{y}\right)^a = \dfrac{x^a}{y^a} \quad (y \neq 0)$ 1. Raise the numerical coefficient to the power outside the parentheses. 2. Multiply the exponent outside the parentheses times the exponent inside the parentheses.	$(x^9)^3 = x^{27}$ $(3x^2)^3 = 27x^6$ $\left(\dfrac{2x^2}{y^3}\right)^3 = \dfrac{8x^6}{y^9}$ $(-3x^4y^5)^4 = 81x^{16}y^{20}$ $(-5ab)^3 = -125a^3b^3$
Negative exponents, p. 261.	If $x \neq 0$ and $y \neq 0$, then $x^{-n} = \dfrac{1}{x^n}, \quad \dfrac{1}{x^{-n}} = x^n, \quad \dfrac{x^{-m}}{y^{-n}} = \dfrac{y^n}{x^m}$	Write with positive exponents. $x^{-6} = \dfrac{1}{x^6}, \quad \dfrac{1}{w^{-3}} = w^3, \quad \dfrac{w^{-12}}{z^{-5}} = \dfrac{z^5}{w^{12}}$
Scientific notation, p. 262.	A positive number is written in scientific notation if it is in the form $a \times 10^n$, where $1 \leq a < 10$ and n is an integer.	$2{,}568{,}000 = 2.568 \times 10^6$ $0.0000034 = 3.4 \times 10^{-6}$
Add polynomials, p. 268.	To add two polynomials, we add their like terms.	$(-7x^3 + 2x^2 + 5) + (x^3 + 3x^2 + x)$ $\qquad = -6x^3 + 5x^2 + x + 5$
Subtracting polynomials, p. 270.	To subtract the polynomials, change all signs of the second polynomial and add the result to the first polynomial. $(a) - (b) = (a) + (-b)$	$(5x^2 - 6) - (-3x^2 + 2) = (5x^2 - 6) + (+3x^2 - 2)$ $\qquad = 8x^2 - 8$
Multiplying a monomial by a polynomial, p. 275.	Use the distributive property. $a(b + c) = ab + ac$ $(b + c)a = ba + ca$	Multiply. $-5x(2x^2 + 3x - 4) = -10x^3 - 15x^2 + 20x$ $(6x^3 - 5xy - 2y^2)(3xy) = 18x^4y - 15x^2y^2 - 6xy^3$
Multiplying two binomials, p. 275.	1. The product of the sum and difference of the same two terms: $(a + b)(a - b) = a^2 - b^2$ 2. The square of a binomial: $(a + b)^2 = a^2 + 2ab + b^2$ $(a - b)^2 = a^2 - 2ab + b^2$ 3. Use FOIL for other binomial multiplication. The middle terms can often be combined, giving a trinomial.	$(3x + 7y)(3x - 7y) = 9x^2 - 49y^2$ $(3x + 7y)^2 = 9x^2 + 42xy + 49y^2$ $(3x - 7y)^2 = 9x^2 - 42xy + 49y^2$ $(3x - 5)(2x + 7) = 6x^2 + 21x - 10x - 35$ $\qquad = 6x^2 + 11x - 35$

Topic	Procedure	Examples
Multiplying two polynomials, p. 283.	To multiply two polynomials, multiply each term of one by each term of the other. This method is similar to the multiplication of many-digit numbers.	Vertical method: $$\begin{array}{r} 3x^2 - 7x + 4 \\ \times\quad 3x - 1 \\ \hline -3x^2 + 7x - 4 \\ 9x^3 - 21x^2 + 12x \\ \hline 9x^3 - 24x^2 + 19x - 4 \end{array}$$ Horizontal method: $(5x + 2)(2x^2 - x + 3)$ $= 10x^3 - 5x^2 + 15x + 4x^2 - 2x + 6$ $= 10x^3 - x^2 + 13x + 6$
Multiplying three or more polynomials, p. 284.	1. Multiply any two polynomials. 2. Multiply the result by any remaining polynomials.	$(2x + 1)(x - 3)(x + 4) = (2x^2 - 5x - 3)(x + 4)$ $= 2x^3 + 3x^2 - 23x - 12$
Dividing a polynomial by a monomial, p. 288.	1. Divide each term of the polynomial by the monomial. 2. When dividing variables, use the property $$\frac{x^a}{x^b} = x^{a-b}.$$	Divide. $(15x^3 + 20x^2 - 30x) \div (5x)$ $= \dfrac{15x^3}{5x} + \dfrac{20x^2}{5x} + \dfrac{-30x}{5x}$ $= 3x^2 + 4x - 6$
Dividing a polynomial by a binomial, p. 288.	1. Place the terms of the polynomial and binomial in descending order. Insert a 0 for any missing term. 2. Divide the first term of the polynomial by the first term of the binomial. 3. Multiply the partial answer by the binomial, and subtract the result from the first two terms of the polynomial. Bring down the next term to obtain a new polynomial. 4. Divide the new polynomial by the binomial using the process described in step 2. 5. Continue dividing, multiplying, and subtracting until the degree of the remainder is less than the degree of the binomial divisor. 6. Write the remainder as the numerator of a fraction that has the binomial divisor as its denominator.	Divide. $(8x^3 - 13x + 2x^2 + 7) \div (4x - 1)$ We rearrange the terms. $$\begin{array}{r} 2x^2 + \;\; x - 3 \\ 4x - 1 \overline{)8x^3 + 2x^2 - 13x + 7} \\ \underline{8x^3 - 2x^2} \\ 4x^2 - 13x \\ \underline{4x^2 - \;\; x} \\ -12x + 7 \\ \underline{-12x + 3} \\ 4 \end{array}$$ The answer is $$2x^2 + x - 3 + \frac{4}{4x - 1}.$$

Chapter 4 Review Problems

Section 4.1

Simplify. Leave your answer in exponent form.

1. $(-6a^2)(3a^5)$

2. $(5^{10})(5^{13})$

3. $(3xy^2)(2x^3y^4)$

4. $(2x^3y^4)(-7xy^5)$

5. $\dfrac{7^{15}}{7^{27}}$

6. $\dfrac{x^{12}}{x^{17}}$

7. $\dfrac{y^{30}}{y^{16}}$

8. $\dfrac{9^{13}}{9^{24}}$

9. $\dfrac{-15xy^2}{25x^6y^6}$

10. $\dfrac{-12a^3b^6}{18a^2b^{12}}$

11. $(x^3)^8$

12. $(b^5)^6$

13. $(-3a^3b^2)^2$

14. $(3x^3y)^4$

15. $\left(\dfrac{5ab^2}{c^3}\right)^2$

16. $\left(\dfrac{x^0y^3}{4w^5z^2}\right)^3$

Section 4.2

Simplify. Write with positive exponents.

17. $a^{-3}b^5$

18. $m^8 p^{-5}$

19. $\dfrac{2x^{-6}}{y^{-3}}$

20. $(3x^{-4}y^3)^{-2}$

21. $(5x^2y^{-4})^{-2}$

22. $\dfrac{3x^{-3}}{y^{-2}}$

23. $\dfrac{4x^{-5}y^{-6}}{w^{-2}z^8}$

24. $\dfrac{3^{-3}a^{-2}b^5}{c^{-3}d^{-4}}$

Write in scientific notation.

25. 156,340,200,000

26. 179,632

27. 0.0078

28. 0.00006173

Write in decimal notation.

29. 1.2×10^5

30. 6.034×10^6

31. 3×10^6

32. 2.5×10^{-1}

33. 4.32×10^{-5}

34. 6×10^{-9}

Perform the indicated calculation. Leave your answer in scientific notation.

35. $\dfrac{(28,000,000)(5,000,000,000)}{7000}$

36. $(3.12 \times 10^5)(2.0 \times 10^6)(1.5 \times 10^8)$

37. $(1.6 \times 10^{-3})(3.0 \times 10^{-5})(2.0 \times 10^{-2})$

38. $\dfrac{(0.00078)(0.000005)(0.00004)}{0.002}$

39. *Space Travel* If a space probe travels at 40,000 kilometers per hour for 1 year, how far will it travel? (Assume that 1 year = 365 days.)

40. *Atomic Clock* An atomic clock is based on the fact that cesium emits 9,192,631,770 cycles of radiation in one second. How many of these cycles occur in one day? Round to three significant digits.

41. *Computer Speed* Today's fastest modern computers can perform one operation in 1×10^{-8} second. How many operations can such a computer perform in 1 minute?

Section 4.3

Combine.

42. $(4.3x^2 - 2.6x - 3.1) + (7.1x^2 - 4.7x + 8.4)$

43. $(1.2x^2 - 3.4x + 6) + (5.5x^2 - 7.6x - 3)$

44. $(x^3 + x^2 - 6x + 2) - (2x^3 - x^2 - 5x - 6)$

45. $(4x^3 - x^2 - x + 3) - (-3x^3 + 2x^2 + 5x - 1)$

46. $\left(\dfrac{3}{5}x^2y - \dfrac{1}{3}x + \dfrac{3}{4}\right) - \left(\dfrac{1}{2}x^2y + \dfrac{2}{7}x + \dfrac{1}{3}\right)$

47. $\dfrac{1}{2}x^2 - \dfrac{3}{4}x + \dfrac{1}{5} - \left(\dfrac{1}{4}x^2 - \dfrac{1}{2}x + \dfrac{1}{10}\right)$

48. $(5x^2 + 3x) + (-6x^2 + 2) - (5x - 8)$

49. $(2x^2 - 7) - (3x^2 - 4) + (-5x^2 - 6x)$

Section 4.4

Multiply.

50. $(3x + 1)(5x - 1)$

51. $(7x - 2)(4x - 3)$

52. $(2x + 3)(10x + 9)$

53. $5x(2x^2 - 6x + 3)$

54. $(6x^3 - 2x^2 + x - 4)(-3x)$

55. $(2xy^2 - 3xy - 4y)(-3x^2y)$

56. $(5a + 7b)(a - 3b)$

57. $(2x^2 - 3)(4x^2 - 5y)$

58. $-3x^2y(5x^4y + 3x^2 - 2)$

Section 4.5

Multiply.

59. $(3x - 2)^2$

60. $(5x + 3)(5x - 3)$

61. $(7x + 6y)(7x - 6y)$

62. $(5a - 2b)^2$

63. $(8x + 9y)^2$

64. $(x^2 + 7x + 3)(4x - 1)$

65. $(x - 6)(2x - 3)(x + 4)$

Section 4.6

Divide.

66. $(12y^3 + 18y^2 + 24y) \div (6y)$

67. $(30x^5 + 35x^4 - 90x^3) \div (5x^2)$

68. $(16x^3y^2 - 24x^2y + 32xy^2) \div (4xy)$

69. $(106x^6 - 24x^5 + 38x^4 + 26x^3) \div (2x^3)$

70. $(15x^2 + 11x - 14) \div (5x + 7)$

71. $(12x^2 - x - 63) \div (4x + 9)$

72. $(6x^3 + x^2 + 6x + 5) \div (2x - 1)$

73. $(2x^3 - x^2 + 3x - 1) \div (x + 2)$

74. $(12x^2 + 11x + 2) \div (3x + 2)$

75. $(8x^2 - 6x + 6) \div (2x + 1)$

76. $(x^3 - x - 24) \div (x - 3)$

77. $(2x^3 - 3x + 1) \div (x - 2)$

Mixed Practice

Simplify. Leave your answer in exponent form.

78. $(4^6)(4^{23})$

79. $\dfrac{6^{50}}{6^{35}}$

Simplify. Write your answer with only positive exponents.

80. $(-5x^3y^4)(-25x^5y^{10})$

81. $\dfrac{-36x^6y^5}{18xy^8}$

82. $(-4x^5y^6)^3$

83. Write in Scientific notation. 0.0000786

Remove parentheses and simplify.

84. $(-6x^3 - 7x + 8) - (-2x^3 + 5x - 9)$

85. $-6x^2(2x^3 + 5x^2 - 2)$

86. Multiply. $(7x - 3)(2x - 1)(x + 2)$

87. Divide. $(6x^3 + 11x^2 - x - 2) \div (3x - 2)$

Applications

Solve. Express your answer in scientific notation.

88. *Foreign Aid* In 2002, the United States donated $\$12.9 \times 10^9$ in foreign aid to poor countries. If the population of the United States in 2002 was estimated at 2.8×10^8 people, how much foreign aid per person did the United States donate? Leave your answer in dollars and cents.

89. *Population* In 2002, the estimated population of India was 1.132×10^9 people, while the estimated population of Bangladesh was 1.31×10^8 people. (*Source:* United Nations Statistical Bureau.) What was the total population in those two countries?

90. *Electron Mass* The mass of an electron is approximately 9.11×10^{-28} gram. Find the mass of 30,000 electrons.

91. *Sun Energy* The sun radiates energy into space at the rate of 3.9×10^{26} joules per second. How many joules are emitted in a day?

To Think About

Find a polynomial that describes the shaded area.

▲ **92.**

▲ **93.**

Remember to use your Chapter Test Prep Video CD to see the worked-out solutions to the test problems you want to review.

Simplify. Leave your answer in exponent form.

1. $(3^{10})(3^{24})$

2. $\dfrac{25^{18}}{25^{34}}$

3. $(8^4)^6$

In questions 4–8, simplify.

4. $(-3xy^4)(-4x^3y^6)$

5. $\dfrac{-35x^8y^{10}}{25x^5y^{10}}$

6. $(-5xy^6)^3$

7. $\left(\dfrac{7a^7b^2}{3c^0}\right)^2$

8. $\dfrac{(3x^2)^3}{(6x)^2}$

9. Evaluate. 4^{-3}

In questions 10 and 11, simplify and write with only positive exponents.

10. $6a^{-4}b^{-3}c^5$

11. $\dfrac{3x^{-3}y^2}{x^{-4}y^{-5}}$

12. Write in scientific notation. 0.0005482

13. Write in decimal notation. 5.82×10^8

14. Multiply. Leave your answer in scientific notation.
$(4.0 \times 10^{-3})(3.0 \times 10^{-8})(2.0 \times 10^4)$

Combine.

15. $(2x^2 - 3x - 6) + (-4x^2 + 8x + 6)$

16. $(3x^3 - 4x^2 + 3) - (14x^3 - 7x + 11)$

Multiply.

17. $-7x^2(3x^3 - 4x^2 + 6x - 2)$

18. $(5x^2y^2 - 6xy + 2)(3x^2y)$

19. $(5a - 4b)(2a + 3b)$

20. $(3x + 2)(2x + 1)(x - 3)$

21. $(7x^2 + 2y^2)^2$

22. $(5s - 11t)(5s + 11t)$

23. $(3x - 2)(4x^3 - 2x^2 + 7x - 5)$

24. $(3x^2 - 5xy)(x^2 + 3xy)$

Divide.

25. $15x^6 - 5x^4 + 25x^3 \div 5x^3$

26. $(8x^3 - 22x^2 - 5x + 12) \div (4x + 3)$

27. $(2x^3 - 6x - 36) \div (x - 3)$

Solve. Express your answer in scientific notation. Round to the nearest hundredth.

28. At the end of 2002, Saudi Arabia was estimated to have 2.618×10^{11} barrels of oil reserves. By one estimate, they have 86 years of reserves remaining. If they disbursed the oil by an equal amount each year, how many barrels of oil would they pump each year? (*Source:* United Nations Review of World Energy, 2003)

29. A space probe is traveling from Earth to the planet Pluto at a speed of 2.49×10^4 miles per hour. How far would this space probe travel in one week?

1. _____

2. _____

3. _____

4. _____

5. _____

6. _____

7. _____

8. _____

9. _____

10. _____

11. _____

12. _____

13. _____

14. _____

15. _____

16. _____

17. _____

18. _____

19. _____

20. _____

21. _____

22. _____

23. _____

24. _____

25. _____

26. _____

27. _____

28. _____

29. _____

1. _____

2. _____

3. _____

4. _____

5. _____

6. _____

7. _____

8. _____

9. _____

10. _____

11. _____

12. _____

13. _____

14. _____

15. _____

16. _____

17. _____

18. _____

19. _____

20. _____

21. _____

22. _____

23. _____

24. _____

25. _____

26. _____

Approximately one-half of this test covers the content of Chapters 0 through 3. The remainder covers the content of Chapter 4.

In questions 1–5, simplify.

1. $\dfrac{5}{12} - \dfrac{7}{8}$

2. $(-3.7) \times (0.2)$

3. $\left(-4\dfrac{1}{2}\right) \div \left(5\dfrac{1}{4}\right)$

4. Find 6% of 1842.5

5. $7x(3x - 4) - 5x(2x - 3) - (3x)^2$

6. Evaluate $2x^2 - 3xy + y^2$ when $x = -2$ and $y = 3$.

In questions 7–9, solve.

7. $7x - 3(4 - 2x) = 14x - (3 - x)$

8. $\dfrac{2}{3}x + 6 = 4(x - 11)$

9. $4 - 7x < 11$

10. Solve for d: $8(x + d) = 3(y - d)$

11. A national walkout of 11,904 employees of the VBM Corp. occurred last month. This was 96% of the total number of employees. How many employees does VBM have?

12. How much interest would $3320 earn in one year if invested at 6% simple interest?

13. Multiply. $(5x + 6)(4x - 3)$

14. Multiply. $(3x - 5)^2$

15. Multiply. $(3x + 2)(2x + 1)(x - 4)$

In questions 16–18, simplify.

16. $(-4x^4y^5)(5xy^3)$

17. $\dfrac{14x^8y^3}{-21x^5y^{12}}$

18. $(-2x^3y^2z^0)^4$

19. Write with only positive exponents. $\dfrac{9x^{-3}y^{-4}}{w^2z^{-8}}$

20. Write in scientific notation. 1,360,000,000,000,000

21. Write in scientific notation. 0.00056

22. Calculate. Leave your answer in scientific notation. $\dfrac{(2.0 \times 10^{-12})(8.0 \times 10^{-20})}{4.0 \times 10^3}$

23. Subtract. $(x^3 - 3x^2 - 5x + 20) - (-4x^3 - 10x^2 + x - 30)$

Multiply.

24. $-6xy^2(6x^2 - 3xy + 8y^2)$

25. $(2x^2 - 3x + 1)(3x - 5)$

26. Divide. $(x^2 + 2x - 12) \div (x - 3)$

5

A touchdown. An exciting time for any NFL game. But was there a penalty on the play? Is the call of the official correct? Should there be an instant replay to determine whether the call is correct? If current trends continue, more replay reviews will take place each year. Can you determine the mathematical pattern? Turn to the Putting Your Skills to Work problems on page 347 to find out.

Factoring

1 ## Factoring Polynomials Whose Terms Contain a Common Factor

Recall that when two or more numbers, variables, or algebraic expressions are multiplied, each is called a **factor**.

$$\underbrace{3}_{\text{factor}} \cdot \underbrace{2}_{\text{factor}} \qquad \underbrace{3x^2}_{\text{factor}} \cdot \underbrace{5x^3}_{\text{factor}} \qquad \underbrace{(2x-3)}_{\text{factor}}\underbrace{(x+4)}_{\text{factor}}$$

When you are asked **to factor** a number or an algebraic expression, you are being asked, "What factors, when multiplied, will give that number or expression?"

For example, you can factor 6 as $3 \cdot 2$ since $3 \cdot 2 = 6$. You can factor $15x^5$ as $3x^2 \cdot 5x^3$ since $3x^2 \cdot 5x^3 = 15x^5$. Factoring is simply the reverse of multiplying. 6 and $15x^5$ are simple expressions to factor and can be factored in different ways.

The factors of the polynomial $2x^2 + x - 12$ are not so easy to recognize. Factoring a polynomial changes an addition and/or subtraction problem into a multiplication problem. In this chapter we will be learning techniques for finding the factors of a polynomial. We will begin with **common factors.**

EXAMPLE 1 Factor. **(a)** $3x - 6y$ **(b)** $9x + 2xy$

Solution Begin by looking for a common factor, a factor that both terms have in common. Then rewrite the expression as a product.

(a) $3x - 6y = 3(x - 2y)$ This is true because $3(x - 2y) = 3x - 6y$.
(b) $9x + 2xy = x(9 + 2y)$ This is true because $x(9 + 2y) = 9x + 2xy$.

Some people find it helpful to think of factoring as the distributive property in reverse. When we write $3x - 6y = 3(x - 2y)$, we are doing the reverse of distributing the 3. This is the main point of this section. We are doing problems of the form $ca + cb = c(a + b)$. The common factor c becomes the first factor of our answer.

Practice Problem 1 Factor. **(a)** $21a - 7b$ **(b)** $5xy + 8x$

When we factor, we begin by looking for the **greatest common factor.** For example, in the polynomial $48x - 16y$, a common factor is 2. We could factor $48x - 16y$ as $2(24x - 8y)$. However, this is not complete. To factor $48x - 16y$ completely, we look for the greatest common factor of 48 and of 16.

$$48x - 16y = 16(3x - y)$$

EXAMPLE 2 Factor. $24xy + 12x^2 + 36x^3$. Remember to remove the greatest common factor.

Solution Find the greatest common factor of 24, 12, and 36. You may want to factor each number, or you may notice that 12 is a common factor. 12 is the greatest numerical common factor.

Notice also that x is a factor of each term. Thus, $12x$ is the greatest common factor.

$$24xy + 12x^2 + 36x^3 = 12x(2y + x + 3x^2)$$

Practice Problem 2 Factor. $12a^2 + 16ab^2 - 12a^2b$. Be careful to remove the greatest common factor.

FACTORING A POLYNOMIAL WITH COMMON FACTORS

1. Determine the greatest common numerical factor by asking, "What is the largest integer that will divide into the coefficient of all the terms?"

2. Determine the greatest common variable factor by asking, "What variables are common to all the terms, and what is the smallest exponent on each of those variables?"

3. The common factors are the first part of the answer.

4. After removing the common factors what remains is placed in a parentheses as the second factor.

EXAMPLE 3   Factor.   **(a)** $12x^2 - 18y^2$ **(b)** $x^2y^2 + 3xy^2 + y^3$

Solution

(a) Note that the largest integer that is common to both terms is 6 (not 3 or 2).

$$12x^2 - 18y^2 = 6(2x^2 - 3y^2)$$

(b) Although y is common to all of the terms, we factor out y^2 since 2 is the largest exponent of y that is common to all terms. We do not factor out x, since x is not common to all of the terms.

$$x^2y^2 + 3xy^2 + y^3 = y^2(x^2 + 3x + y)$$

Practice Problem 3   Factor.   **(a)** $16a^3 - 24b^3$ **(b)** $r^3s^2 - 4r^4s + 7r^5$ ■

Checking Is Very Important!   You can check any factoring problem by multiplying the factors you obtain. The result should be the same as the original polynomial.

EXAMPLE 4   Factor. $8x^3y + 16x^2y^2 - 24x^3y^3$

Solution   We see that 8 is the largest integer that will divide evenly into the three numerical coefficients. We can factor an x^2 out of each term. We can also factor y out of each term.

$$8x^3y + 16x^2y^2 - 24x^3y^3 = 8x^2y(x + 2y - 3xy^2)$$

Check.

$$8x^2y(x + 2y - 3xy^2) = 8x^3y + 16x^2y^2 - 24x^3y^3 \checkmark$$

Practice Problem 4   Factor. $18a^3b^2c - 27ab^3c^2 - 45a^2b^2c^2$ ■

EXAMPLE 5   Factor. $9a^3b^2 + 9a^2b^2$

Solution   We observe that both terms contain a common factor of 9. We can factor a^2 and b^2 out from each term. Thus the common factor is $9a^2b^2$.

$$9a^3b^2 + 9a^2b^2 = 9a^2b^2(a + 1)$$

CAUTION   Don't forget to include the 1 inside the parentheses in Example 5. The solution is wrong without it. You will see why if you try to check a result written without the 1.

Practice Problem 5   Factor and check. $30x^3y^2 - 24x^2y^2 + 6xy^2$ ■

EXAMPLE 6 Factor. $3x(x - 4y) + 2(x - 4y)$

Solution Be sure you understand what are terms and what are factors of the polynomial in this example. There are two terms. The expression $3x(x - 4y)$ is one term. The expression $2(x - 4y)$ is the second term. Each term is made up of two factors. Observe that the binomial $(x - 4y)$ is a common factor of the terms. A common factor may be any type of polynomial. Thus we can factor out the common factor $(x - 4y)$.

$$3x(x - 4y) + 2(x - 4y) = (x - 4y)(3x + 2)$$

Practice Problem 6 Factor. $3(a + 5b) + x(a + 5b)$

EXAMPLE 7 Factor. $7x^2(2x - 3y) - (2x - 3y)$

Solution The common factor of the terms is $(2x - 3y)$. What happens when we factor out $(2x - 3y)$? What are we left with in the second term?
Recall that $(2x - 3y) = 1(2x - 3y)$. Thus

$$7x^2(2x - 3y) - (2x - 3y) = 7x^2(2x - 3y) - 1(2x - 3y) \quad \text{Rewrite the original expression.}$$

$$= (2x - 3y)(7x^2 - 1) \quad \text{Factor out } (2x - 3y).$$

Practice Problem 7 Factor. $8y(9y^2 - 2) - (9y^2 - 2)$

NOTE TO STUDENT: Fully worked-out solutions to all of the Practice Problems can be found at the back of the text starting at page SP-1

EXAMPLE 8 A computer programmer is writing a program to find the area of 4 circles. She uses the formula $A = \pi r^2$. The radii of the circles are a, b, c, and d, respectively. She wants the final answer to be in factored form with the value of π occurring only once, in order to minimize round-off error. Write the total area with a formula that has π occurring only once.

Solution

For each circle, $A = \pi r^2$, where $r = a, b, c$, or d.
The total area is $\pi a^2 + \pi b^2 + \pi c^2 + \pi d^2$.
In factored form the total area $= \pi(a^2 + b^2 + c^2 + d^2)$.

Practice Problem 8 Use $A = \pi r^2$ to find the shaded area. The radius of the larger circle is b. The radius of the smaller circle is a. Write the total area formula in factored form so that π appears only once.

Verbal and Writing Skills

In exercises 1 and 2, write a word or words to complete each sentence.

1. In the expression $3x^2 \cdot 5x^3$, $3x^2$ and $5x^3$ are called _____.

2. In the expression $3x^2 + 5x^3$, $3x^2$ and $5x^3$ are called _____.

3. We can factor $30a^4 + 15a^3 - 45a^2$ as $5a(6a^3 + 3a^2 - 9a)$. Is the factoring complete? Why or why not?

4. We can factor $4x^3 - 8x^2 + 20x$ as $4(x^3 - 2x^2 + 10x)$. Is the factoring complete?

Remove the largest possible common factor. Check your answers for exercises 5–28 by multiplication.

5. $3a^2 + 3a$ **6.** $2c^2 + 2c$ **7.** $21ab - 14ab^2$ **8.** $18wz - 27w^2z$

9. $2\pi rh + 2\pi r^2$ **10.** $9a^2b^2 - 36ab$ **11.** $5x^3 + 25x^2 - 15x$ **12.** $8x^3 - 10x^2 - 14x$

13. $12ab - 28bc + 20ac$ **14.** $12xy - 18yz - 36xz$ **15.** $16x^5 + 24x^3 - 32x^2$ **16.** $36x^6 + 45x^4 - 18x^2$

17. $14x^2y - 35xy - 63x$ **18.** $40a^2 - 16ab - 24a$ **19.** $54x^2 - 45xy + 18x$ **20.** $48xy - 24y^2 + 40y$

21. $3xy^2 - 2ay + 5xy - 2y$ **22.** $2ab^3 + 3xb^2 - 5b^4 + 2b^2$ **23.** $24x^2y - 40xy^2$

24. $35abc^2 - 49ab^2c$ **25.** $7x^3y^2 + 21x^2y^2$ **26.** $8x^3y^2 + 32xy^2$

27. $9x^2y - 18xy^2 - 27xy$ **28.** $10x^2y - 15xy^2 + 20xy$

Hint: In exercises 29–42, refer to Examples 6 and 7.

29. $7a(x + 2y) - b(x + 2y)$ **30.** $6(3a + b) - z(3a + b)$ **31.** $3x(x - 4) - 2(x - 4)$

32. $5x(x - 7) + 3(x - 7)$ **33.** $6b(2a - 3c) - 5d(2a - 3c)$ **34.** $7x(3y + 5z) - 6t(3y + 5z)$

35. $7c(b - a^2) - 5d(b - a^2) + 2f(b - a^2)$ **36.** $3b(y^2 - x) - 4a(y^2 - x) + 6c(y^2 - x)$

37. $2a(xy - 3) - 4(xy - 3) - z(xy - 3)$ **38.** $3c(bc - 3a) - 2(bc - 3a) - 6b(bc - 3a)$

39. $4a^3(a - 3b) + (a - 3b)$ **40.** $(c + 4d) + 7c^2(c + 4d)$

41. $(2a + 3) - 7x(2a + 3)$ **42.** $d(5x - 3) - (5x - 3)$

To Think About

▲ **43.** *Geometry* Find a formula for the area of four rectangles of width 2.786 inches. The lengths of the rectangles are *a*, *b*, *c*, and *d* inches. Write the formula in factored form.

44. *Wholesale Purchases* Find a formula for the total cost of all purchases by four people. Each person went to the local wholesale warehouse and spent $29.95 per item. Harry bought *a* items, Richard bought *b* items, Lyle bought *c* items, and Selena bought *d* items. Write the formula in factored form.

Cumulative Review

45. *Integers* The sum of three consecutive odd integers is 40 more than the smallest of these integers. Find each of these three consecutive odd integers. (*Hint:* Consecutive odd integers are numbers included in the following pattern: $-3, -1, 1, 3, 5, 7, 9, \ldots, x, x + 2, \ldots$).

46. *College Costs* Tuition, room, and board at a college in Colorado cost Lisa $27,040. This was a 4% increase over the cost of these items last year. What did Lisa pay last year for tuition, room, and board?

47. *Videotape Speed* A videotape plays 6.72 feet per minute for 2 hours at standard speed. It plays for 6 hours at extended long play speed. What is the rate of the tape at the slower speed?

Fruit Production *In 2001 the top six fruit producing countries of the world produced 258,000,000 tons of fruit. The percent of this 258 million tons produced by each of these countries is shown in the following graph. Use the graph to answer the following questions.*

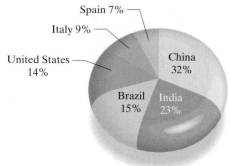

Percent of Fruit Produced by the Top Six Fruit Producing Countries of the World

Spain 7%
Italy 9%
United States 14%
China 32%
Brazil 15%
India 23%

Source: Food and Agriculture Organization of the United Nations

48. How many tons of fruit were produced in Italy in 2001?

49. How many tons of fruit were produced in Brazil in 2001?

50. The population of Italy in 2001 was approximately 57,000,000 people. How many pounds of fruit was produced in the country for each person? (Round your answer to nearest whole number. Remember than 1 ton = 2000 pounds.)

51. The population of Brazil in 2001 was approximately 174,000,000 people. How many pounds of fruit was produced in the country for each person? (Round your answer to nearest whole number.)

5.2 FACTORING BY GROUPING

1 Factoring Expressions with Four Terms by Grouping

A common factor of a polynomial can be a number, a variable, or an algebraic expression. Sometimes the polynomial is written so that it is easy to recognize the common factor. This is especially true when the common factor is enclosed by parentheses.

EXAMPLE 1 Factor. $x(x - 3) + 2(x - 3)$

Solution Observe each term:

$$\underbrace{x(x - 3)}_{\substack{\text{first} \\ \text{term}}} + \underbrace{2(x - 3)}_{\substack{\text{second} \\ \text{term}}}$$

The common factor of the first and second terms is the quantity $(x - 3)$, so we have

$$x(\,x - 3\,) + 2(\,x - 3\,) = (\,x - 3\,)(x + 2).$$

Practice Problem 1 Factor. $3y(2x - 7) - 8(2x - 7)$

Now let us face a new challenge. Think carefully. Try to follow this new idea. Suppose the polynomial in Example 1, $x(x - 3) + 2(x - 3)$, were written in the form $x^2 - 3x + 2x - 6$. (Note that this form is obtained by multiplying the factors of the first and second terms.) How would we factor a four-term polynomial like this?

In such cases we remove a common factor from the first two terms and a different common factor from the second two terms. That is, we would factor x from $x^2 - 3x$ and 2 from $2x - 6$.

$$x^2 - 3x + 2x - 6 = x(x - 3) + 2(x - 3)$$

Because the resulting terms have a common factor (the binomial enclosed by the parentheses), we would then proceed as we did in Example 1. This procedure for factoring is often called **factoring by grouping.**

EXAMPLE 2 Factor. $2x^2 + 3x + 6x + 9$

Solution

$$\underset{\substack{\text{Factor out a common} \\ \text{factor of } x \text{ from} \\ \text{the first two terms.}}}{2x^2 + 3x} \qquad + \qquad \underset{\substack{\text{Factor out a common} \\ \text{factor of 3 from} \\ \text{the second two terms.}}}{6x + 9}$$

$$\underbrace{x(2x + 3)}_{} \qquad\qquad \underbrace{3(2x + 3)}_{}$$

Note that the sets of parentheses in the two terms contain the same expression at this step.

The expression in parentheses is now a common factor of the terms. Now we finish the factoring.

$$2x^2 + 3x + 6x + 9 = x(\,2x + 3\,) + 3(\,2x + 3\,)$$
$$= (\,2x + 3\,)(x + 3)$$

Practice Problem 2 Factor. $6x^2 - 15x + 4x - 10$

Student Learning Objective

After studying this section, you will be able to:

1 Factor expressions with four terms by grouping.

NOTE TO STUDENT: Fully worked-out solutions to all of the Practice Problems can be found at the back of the text starting at page SP-1

EXAMPLE 3 Factor. $4x + 8y + ax + 2ay$

Solution

Factor out a common factor of 4 from the first two terms.

$$\overbrace{4x + 8y} + \underbrace{ax + 2ay} = \overbrace{4(x + 2y)} + \underbrace{a(x + 2y)}$$

Factor out a common factor of a from the second two terms.

$4(x + 2y) + a(x + 2y) = (x + 2y)(4 + a)$ The common factor of the terms is the expression in parentheses, $x + 2y$.

NOTE TO STUDENT: *Fully worked-out solutions to all of the Practice Problems can be found at the back of the text starting at page SP-1*

Practice Problem 3 Factor by grouping. $ax + 2a + 4bx + 8b$

In some problems the terms are out of order. We have to rearrange the order of the terms first so that the first two terms have a common factor.

EXAMPLE 4 Factor. $bx + 4y + 4b + xy$

Solution

$$
\begin{aligned}
bx + 4y + 4b + xy &= bx + 4b + xy + 4y & &\text{Rearrange the terms so that the first terms have a common factor.}\\
&= b(x + 4) + y(x + 4) & &\text{Factor out the common factor of } b \text{ from the first two terms.}\\
&= (x + 4)(b + y) & &\text{Factor out the common factor of } y \text{ from the second two terms.}
\end{aligned}
$$

Practice Problem 4 Factor. $6a^2 + 5bc + 10ab + 3ac$

Sometimes you will need to factor out a negative common factor from the second two terms to obtain two terms that contain the same parenthetical expression.

EXAMPLE 5 Factor. $2x^2 + 5x - 4x - 10$

Solution

$$
\begin{aligned}
2x^2 + 5x - 4x - 10 &= x(2x + 5) - 2(2x + 5) & &\text{Factor out the common factor of } x \text{ from the first two terms and the common factor of } -2 \text{ from the second two terms.}\\
&= (2x + 5)(x - 2)
\end{aligned}
$$

Notice that if you factored out a common factor of $+2$ in the first step, the two resulting terms would not contain the same parenthetical expression. If the expressions inside the two sets of parentheses are not exactly the same, you cannot express the polynomial as a product of two factors!

Practice Problem 5 Factor. $6xy + 14x - 15y - 35$

EXAMPLE 6   Factor. $2ax - a - 2bx + b$

Solution

$$2ax - a - 2bx + b = a(2x - 1) - b(2x - 1)$$ Factor out the common factor of a from the first two terms. Factor out the common factor of $-b$ from the second two terms.

$$= (2x - 1)(a - b)$$ Since the two resulting terms contain the same parenthetical expression, we can complete the factoring.

Practice Problem 6   Factor. $3x - 10ay + 6y - 5ax$

CAUTION   Many students find that they make a factoring error in the first step of problems like Example 6. Watch that step very carefully. If you take your answer and multiply it by FOIL, you should obtain the original problem.

EXAMPLE 7   Factor and check your answer. $8ad + 21bc - 6bd - 28ac$

Solution   We observe that the first two terms do not have a common factor.

$$8ad + 21bc - 6bd - 28ac = 8ad - 6bd - 28ac + 21bc$$ Rearrange the order using the commutative property of addition.

$$= 2d(4a - 3b) - 7c(4a - 3b)$$ Factor out the common factor of $2d$ from the first two terms and the common factor of $-7c$ from the last two terms.

$$= (4a - 3b)(2d - 7c)$$ Factor out the common factor of $(4a - 3b)$.

To check, we multiply the two binomials using the FOIL procedure.

$$(4a - 3b)(2d - 7c) = 8ad - 28ac - 6bd + 21bc$$

$$= 8ad + 21bc - 6bd - 28ac \quad \checkmark$$ Rearrange the order of the terms. This is the original problem. Thus it checks.

Practice Problem 7   Factor and check your answer.

$$10ad + 27bc - 6bd - 45ac$$

Verbal and Writing Skills

1. To factor $3x^2 - 6xy + 5x - 10y$, we must first remove a common factor of $3x$ from the first two terms. What do we do with the last two terms? What should we get for the answer?

2. To factor $5x^2 + 15xy - 2x - 6y$, we must first remove a common factor of $4x$ from the first two terms. What do we do with the last two terms? What should we get for the answer?

Factor by grouping. Check your answers for exercises 3–26.

3. $ab - 3a + 4b - 12$

4. $xy - x + 4y - 4$

5. $x^3 - 4x^2 + 3x - 12$

6. $x^3 - 6x^2 + 2x - 12$

7. $2ax + 6bx - ay - 3by$

8. $4x + 8y - 3wx - 6wy$

9. $3ax + bx - 6a - 2b$

10. $ad + 3a - d^2 - 3d$

11. $5a + 12bc + 10b + 6ac$

12. $4u^2 + v + 4uv + u$

13. $5a - 5b - 2ax + 2xb$

14. $xy - 4x - 3y + 12$

15. $y^2 - 2y - 3y + 6$

16. $x^2 - 4x + 3x - 12$

17. $14 - 7y + 2y - y^2$

18. $xa + 2bx - a - 2b$

19. $6ax - y + 2ay - 3x$

20. $6tx + r - 3t - 2rx$

21. $2x^2 + 8x - 3x - 12$

22. $3y^2 - y + 9y - 3$

23. $t^3 - t^2 + t - 1$

24. $x^2 - 2x - xy + 2y$

25. $28x^2 + 8xy^2 + 21xw + 6y^2w$

26. $8xw + 10x^2 + 35xy^2 + 28y^2w$

To Think About

27. Although $6a^2 - 12bd - 8ad + 9ab = 6(a^2 - 2bd) - a(8d - 9b)$ is true, it is not the correct solution to the problem "Factor $6a^2 - 12bd - 8ad + 9ab$." Explain. Can this expression be factored?

28. Tim was trying to factor $5x^2 - 3xy - 10x + 6y$. In his first step he wrote down $x(5x - 3y) + 2(-5x + 3y)$. Was he doing the problem correctly? What is the answer?

Cumulative Review

29. *Braking* A train was traveling at 73 miles per hour when the engineer spotted a stalled truck on the railroad crossing ahead. He jammed on the brakes but was unable to stop the train before it collided with the stalled truck. For every second the engineer applied the brakes, the train slowed down by 4 miles per hour. The accident reconstruction team found that the train was still traveling at 41 miles per hour at the time of impact. For how many seconds were the brakes applied?

30. *Car Sales* Tim Brown made an average profit of $320 per car for each car he sold at Donahue Motors over the first 5 months of the year. In January and February, he averaged only $200 per car. In March, he averaged $300 per car, but in April he averaged $480 per car. Assuming that he sells around the same number of cars each month, what was his average profit per car for the month of May?

31. *Music Sales* In 1998, the Recording Industry Association of America reported a 6.8% increase from the previous year in U.S. unit sales of all musical recordings. This translates to an 11.9% increase in dollar value, or a 15.1 million increase. (*Source:* Bureau of Economic Analysis.) If the dollar value increase is equal to $15.1 million dollars, what was the total dollar value of musical recordings in the United States in 1997? Round to the nearest tenth of a million.

32. *Ticket Sales* A community theater sold 900 tickets to a play. Floor seats cost $12 each and balcony seats cost $10. The total receipts were $9780. How many of each type of ticket were sold?

33. *Coffee Production* In 1995, 5.5 million metric tons of coffee were produced in the world. By the year 2000, total world production had increased by 33%. How many metric tons were produced in the world in 2000? It is estimated that from 2000 to 2005 the total world production of coffee will increase by 28%. If that is true, how many metric tons will be produced in 2005? (*Source:* Statistical Division of the United Nations)

1 Factoring Polynomials of the Form $x^2 + bx + c$

Suppose that you wanted to factor $x^2 + 5x + 6$. After some trial and error you *might* obtain $(x + 2)(x + 3)$, or you might get discouraged and not get an answer. If you did get these factors, you could check this answer by the FOIL method.

$$(x + 2)(x + 3) = x^2 + 3x + 2x + 6$$
$$= x^2 + 5x + 6$$

But trial and error can be a long process. There is another way. Let's look at the preceding equation again.

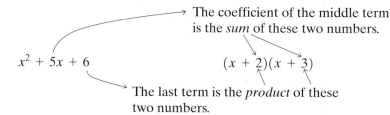

The first thing to notice is that the product of the first terms in the factors gives the first term of the polynomial. That is, $x \cdot x = x^2$.

> The first term is the product of these terms.
> $x^2 + 5x + 6 \qquad = \qquad (x + 2)(x + 3)$

The next thing to notice is that the sum of the products of the outer and inner terms in the factors produces the middle term of the polynomial. That is, $(x \cdot 3) + (2 \cdot x) = 3x + 2x = 5x$. Thus we see that the sum of the second terms in the factors, $2 + 3$, gives the coefficient of the middle term, 5.

Finally, note that the product of the last terms of the factors gives the last term of the polynomial. That is, $2 \cdot 3 = 6$.

> The coefficient of the middle term is the *sum* of these two numbers.
> $x^2 + 5x + 6 \qquad\qquad (x + 2)(x + 3)$
> The last term is the *product* of these two numbers.

Let's summarize our observations in general terms and then try a few examples.

FACTORING TRINOMIALS OF THE FORM $x^2 + bx + c$

1. The answer will be of the form $(x + m)(x + n)$.

2. m and n are numbers such that:
   (a) When you multiply them, you get the last term, which is c.
   (b) When you add them, you get the coefficient of the middle term, which is b.

EXAMPLE 1   Factor. $x^2 + 7x + 12$

Solution   The answer is of the form $(x + m)(x + n)$. We want to find the two numbers, m and n, that you can multiply to get 12 and add to get 7. The numbers are 3 and 4.

$$x^2 + 7x + 12 = (x + 3)(x + 4)$$

Practice Problem 1   Factor. $x^2 + 8x + 12$

EXAMPLE 2 Factor. $x^2 + 12x + 20$

Solution We want two numbers that have a product of 20 and a sum of 12. The numbers are 10 and 2.

$$x^2 + 12x + 20 = (x + \underline{10})(x + \underline{2})$$

Note: If you cannot think of the numbers in your head, write down the possible factors whose product is 20.

Product	**Sum**
$1 \cdot 20 = 20$	$1 + 20 = 21$
$2 \cdot 10 = 20$	$2 + 10 = 12 \leftarrow$
$4 \cdot 5 = 20$	$4 + 5 = 9$

Then select the pair whose sum is 12.   Select this pair.

Practice Problem 2 Factor. $x^2 + 17x + 30$
You may find that it is helpful to list all the factors whose product is 30 first.

So far we have factored only trinomials of the form $x^2 + bx + c$, where b and c are positive numbers. The same procedure applies if b is a negative number and c is positive. Because m and n have a positive product and a negative sum, they must both be negative.

EXAMPLE 3 Factor. $x^2 - 8x + 15$

Solution We want two numbers that have a product of $+15$ and a sum of -8. They must be negative numbers since the sign of the middle term is negative and the sign of the last term is positive.

$$\text{the sum } -5 + (-3)$$
$$x^2 - 8x + 15 = (x - 5)(x - 3)$$
$$\text{the product } (-5)(-3)$$

Think: $(-5)(-3) = +15$
and $-5 + (-3) = -8$.

Multiply using FOIL to check.

Practice Problem 3 Factor. $x^2 - 11x + 18$

EXAMPLE 4 Factor. $x^2 - 9x + 14$

Solution We want two numbers whose product is 14 and whose sum is -9. The numbers are -7 and -2. So

$$x^2 - 9x + 14 = (x - 7)(x - 2) \text{ or } (x - 2)(x - 7).$$

Practice Problem 4 Factor. $x^2 - 11x + 24$

All the examples so far have had a positive last term. What happens when the last term is negative? If the last term is negative, one of the numbers m or n must be a positive and the other must be a negative. Why? The product of a positive number and a negative number is negative.

NOTE TO STUDENT: *Fully worked-out solutions to all of the Practice Problems can be found at the back of the text starting at page SP-1*

EXAMPLE 5   Factor. $x^2 - 3x - 10$

Solution   We want two numbers whose product is -10 and whose sum is -3. The two numbers are -5 and $+2$.

$$x^2 - 3x - 10 = (x - 5)(x + 2)$$

Practice Problem 5   Factor. $x^2 - 5x - 24$

What if we made a sign error and *incorrectly* factored the trinomial $x^2 - 3x - 10$ as $(x + 5)(x - 2)$? We could detect the error immediately since the sum of $+5$ and -2 is 3. We need a sum of -3!

EXAMPLE 6   Factor. $y^2 + 10y - 24$. Check your answer.

Solution   The two numbers whose product is -24 and whose sum is $+10$ are the numbers $+12$ and -2.

$$y^2 + 10y - 24 = (y + 12)(y - 2)$$

CAUTION   It is very easy to make a sign error in these problems. Make sure that you mentally multiply your answer back by FOIL to obtain the original expression. Check each sign carefully.

Check. $(y + 12)(y - 2) = y^2 - 2y + 12y - 24 = y^2 + 10y - 24$ ✓

Practice Problem 6   Factor. $y^2 + 17y - 60$
Multiply your answer to check.

EXAMPLE 7   Factor. $x^2 - 16x - 36$

Solution   We want two numbers whose product is -36 and whose sum is -16.
List all the possible factors of 36 (without regard to sign). Find the pair that has a difference of 16. We are looking for a difference because the signs of the factors are different.

Factors of 36	The Difference Between the Factors
36 and 1	35
18 and 2	16 ← This is the value we want.
12 and 3	9
9 and 4	5
6 and 6	0

Once we have picked the pair of numbers (18 and 2), it is not difficult to find the signs. For the coefficient of the middle term to be -16, we will have to add the numbers -18 and $+2$.

$$x^2 - 16x - 36 = (x - 18)(x + 2)$$

Practice Problem 7   Factor. $x^2 - 7x - 60$
You may find it helpful to list the pairs of numbers whose product is 60.

At this point you should work several problems to develop your factoring skill. This is one section where you really need to drill by doing many problems.

Feel a little confused about the signs? If you do, you may find these facts helpful.

FACTS ABOUT FACTORING TRINOMIALS OF THE FORM $x^2 + bx + c$

The *two numbers m and n* will have the *same sign* if the last term of the polynomial is *positive*.

1. They will both be *positive* if the *coefficient* of the *middle* term is *positive*.

2. They will both be *negative* if the *coefficient* of the *middle* term is *negative*.

$$x^2 + bx + c = (x \quad m)(x \quad n)$$

$$x^2 + 5x + 6 = (x + 2)(x + 3)$$

$$x^2 - 5x + 6 = (x - 2)(x - 3)$$

The two numbers *m* and *n* will have *opposite signs* if the last term is *negative*.

1. The *larger* of the absolute values of the two numbers will be given a plus sign if the coefficient of the *middle term* is *positive*.

$$x^2 + 6x - 7 = (x + 7)(x - 1)$$

2. The larger of the absolute values of the two numbers will be given a negative sign if the coefficient of the *middle term* is *negative*.

$$x^2 - 6x - 7 = (x - 7)(x + 1)$$

Do not memorize these facts; rather, try to understand the pattern.

Sometimes the exponent of the first term of the polynomial will be greater than 2. If the exponent is an even power, it is a square. For example, $x^4 = (x^2)(x^2)$. Likewise, $x^6 = (x^3)(x^3)$.

EXAMPLE 8   Factor. $y^4 - 2y^2 - 35$

Solution   Think: $y^4 = (y^2)(y^2)$ This will be the first term of each parentheses.

$$(y^2 \quad)(y^2 \quad)$$
$$(y^2 + \quad)(y^2 - \quad) \quad \text{The last term of the polynomial is negative.}$$
$$(y^2 + 5)(y^2 - 7) \quad \text{Thus the signs of } m \text{ and } n \text{ will be different.}$$
$$\text{Now think of factors of 35 whose difference is 2.}$$

Practice Problem 8   Factor. $a^4 + a^2 - 42$

---

② Factoring Polynomials That Have a Common Factor and a Factor of the Form $x^2 + bx + c$

Some factoring problems require two steps. Often we must first factor out a common factor from each term of the polynomial. Once this is done, we may find that the other factor is a trinomial that can be factored using the methods previously discussed in this section.

EXAMPLE 9   Factor. $2x^2 + 36x + 160$

Solution

$$2x^2 + 36x + 160 = 2(x^2 + 18x + 80) \quad \text{First factor out the common factor of 2 from each term of the polynomial.}$$
$$= 2(x + 8)(x + 10) \quad \text{Then factor the remaining polynomial.}$$

The final answer is $2(x + 8)(x + 10)$. *Be sure to list all parts of the answer.*

Check. $2(x + 8)(x + 10) = 2(x^2 + 18x + 80) = 2x^2 + 36x + 160$ ✓

Thus we are sure that the answer is $2(x + 8)(x + 10)$.

Practice Problem 9   Factor. $3x^2 + 45x + 150$

EXAMPLE 10 Factor. $3x^2 + 9x - 162$

Solution

$3x^2 + 9x - 162 = 3(x^2 + 3x - 54)$ First factor out the common factor of 3 from each term of the polynomial.

$= 3(x - 6)(x + 9)$ Then factor the remaining polynomial.

The final answer is $3(x - 6)(x + 9)$. *Be sure you included the 3.*

Check. $3(x - 6)(x + 9) = 3(x^2 + 3x - 54) = 3x^2 + 9x - 162$ ✓

Thus we are sure that the answer is $3(x - 6)(x + 9)$.

Practice Problem 10   Factor. $4x^2 - 8x - 140$

NOTE TO STUDENT: Fully worked-out solutions to all of the Practice Problems can be found at the back of the text starting at page SP-1

CAUTION.   Don't forget the common factor!

It is quite easy to forget to look for a greatest common factor as the first step of factoring a trinomial. Therefore, it is a good idea to examine your final answer in any factoring problem and ask yourself, "Can I factor out a common factor from any binomial contained inside a set of parentheses?" Often you will be able to see a common factor at that point if you missed it in the first step of the problem.

▲ **EXAMPLE 11**   Find a polynomial in factored form for the shaded area in the figure.

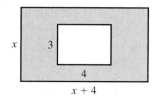

Solution   To obtain the shaded area, we find the area of the larger rectangle and subtract from it the area of the smaller rectangle. Thus we have the following:

$$\text{shaded area} = x(x + 4) - (4)(3)$$
$$= x^2 + 4x - 12$$

Now we factor this polynomial to obtain the shaded area $= (x + 6)(x - 2)$.

▲ **Practice Problem 11**   Find a polynomial in factored form for the shaded area in the figure.

5.3 EXERCISES

Student Solutions Manual | CD/Video | PH Math Tutor Center | MathXL®Tutorials on CD | MathXL® | MyMathLab® | Interactmath.com

Verbal and Writing Skills

Fill in the blanks.

1. To factor $x^2 + 5x + 6$, find two numbers whose _____ is 6 and whose _____ is 5.

2. To factor $x^2 + 5x - 6$, find two numbers whose _____ is -6 and whose _____ is 5.

Factor.

3. $x^2 + 2x + 1$

4. $x^2 + 11x + 30$

5. $x^2 + 12x + 20$

6. $x^2 + 10x + 21$

7. $x^2 - 4x + 3$

8. $x^2 - 6x + 8$

9. $x^2 - 11x + 28$

10. $x^2 - 13x + 12$

11. $x^2 + x - 20$

12. $x^2 + x - 42$

13. $x^2 - 13x - 14$

14. $x^2 - 6x - 16$

15. $x^2 + 2x - 35$

16. $x^2 + 4x - 12$

17. $x^2 - 2x - 24$

18. $x^2 - 11x - 26$

19. $x^2 + 12x + 32$

20. $x^2 + 18x + 72$

21. $x^2 - 10x + 24$

22. $x^2 - 13x + 42$

23. $x^2 + 13x + 30$

24. $x^2 + 9x + 20$

25. $x^2 - 6x + 5$

26. $y^2 - 8y + 7$

Mixed Practice

Look over your answers to exercises 3–26 carefully. Be sure that you are clear on your sign rules. Exercises 27–42 contain a mixture of all the types of problems in this section. Make sure you can do them all. Check your answers by multiplication.

27. $a^2 + 6a - 16$

28. $a^2 - 13a + 30$

29. $x^2 - 12x + 32$

30. $x^2 - 6x - 27$

31. $x^2 + 4x - 21$

32. $x^2 - 9x + 18$

33. $x^2 + 15x + 56$

34. $x^2 + 20x + 99$

35. $x^2 - 21x - 46$

36. $x^2 + 12x - 45$

37. $x^2 + 9x - 36$

38. $x^2 - 13x + 36$

39. $x^2 - 2xy - 15y^2$

40. $x^2 - 2xy - 35y^2$

41. $x^2 - 16xy + 63y^2$

42. $x^2 + 19xy + 48y^2$

In exercises 43–54, first factor out the greatest common factor from each term. Then factor the remaining polynomial. Refer to Examples 9 and 10.

43. $4x^2 + 24x + 20$

44. $4x^2 + 28x + 40$

45. $6x^2 + 18x + 12$

46. $6x^2 + 24x + 18$

47. $2x^2 - 20x + 42$

48. $2x^2 - 22x + 56$

49. $3x^2 - 6x - 72$

50. $3x^2 - 12x - 63$

51. $7x^2 + 21x - 70$

52. $7x^2 + 7x - 84$

53. $3x^2 - 18x + 15$

54. $3x^2 - 33x + 54$

▲ 55. **Geometry** The circle just touches the middle of each side of the square in this sketch. Find a formula for the shaded area. Place the formula in factored form. Evaluate it when $x = 8$. Use $\pi \approx 3.14$.

▲ 56. **Geometry** Find a polynomial that describes how much greater the perimeter of the square is than the circumference of the circle in the sketch for exercise 55. (*Hint:* Find the perimeter of the square. Find the circumference of the circle. Subtract the circumference from the perimeter. Use $\pi \approx 3.14$.)

▲ 57. **Geometry** Find a polynomial in factored form for the shaded area. Both figures are rectangles with dimensions as labeled.

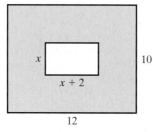

▲ 58. **Geometry** How much larger is the perimeter of the large rectangle than the perimeter of the small rectangle?

Cumulative Review

59. Solve for t. $A = P + Prt$

60. Solve $2 - 3x \le 7$ for x. Graph the solution on the number line.

61. **Travel Speed** A new car that maintains a constant speed travels from Watch Hill, Rhode Island, to Greenwich, Connecticut, in 2 hours. A train, traveling 20 mph faster, makes the trip in $1\frac{1}{2}$ hours. How far is it from Watch Hill to Greenwich? (*Hint:* Let c = the car's speed. First find the speed of the car and the speed of the train. Then be sure to answer the question.)

62. **Salary** Kerri works as a radio advertising sales rep. She earns a guaranteed minimum salary of $500 per month plus 3% commission on her sales. She wants to earn $4400 or more this month. At least how much must she generate in sales?

DJ Charges *The Golden Sounds DJ Service charges $65 per hour plus $85 for each 30-minute period after midnight.*

63. If the bill for Marcia's wedding reception was $515 and the reception started at 8 P.M., when did the reception end?

64. If the bill for Melissa's wedding reception was $535 and the reception ended at 2 A.M., when did the reception start?

Average Temperature *The equation $T = 19 + 2M$ has been used by some meteorologists to predict the monthly average temperature for the small island of Menorca off the coast of Spain during the first 6 months of the year. The variable T represents the average monthly temperature measured in degrees Celsius. The variable M represents the number of months since January.*

65. What is the average temperature of Menorca during the month of April?

66. During what month will the average temperature be 29°C?

5.4 FACTORING TRINOMIALS OF THE FORM $ax^2 + bx + c$

1 Using the Trial-and-Error Method

When the coefficient of the x^2-term in a trinomial of the form $ax^2 + bx + c$ is not 1, the trinomial is more difficult to factor. Several possibilities must be considered.

Student Learning Objectives

After studying this section, you will be able to:

1 Factor a trinomial of the form $ax^2 + bx + c$ by the trial-and-error method.

2 Factor a trinomial of the form $ax^2 + bx + c$ by the grouping method.

3 Factor a trinomial of the form $ax^2 + bx + c$ after a common factor has been factored out of each term.

EXAMPLE 1 Factor. $2x^2 + 5x + 3$

Solution In order for the coefficient of the x^2-term of the polynomial to be 2, the coefficients of the x-terms in the factors must be 2 and 1. Thus $2x^2 + 5x + 3 = (2x \quad)(x \quad)$.

In order for the last term of the polynomial to be 3, the constants in the factors must be 3 and 1.

Since all signs in the polynomial are positive, we know that each factor in parentheses will contain only positive signs. However, we still have two possibilities. They are as follows:

$$(2x + 3)(x + 1)$$
$$(2x + 1)(x + 3)$$

We check them by multiplying by the FOIL method.

$$(2x + 1)(x + 3) = 2x^2 + 7x + 3 \quad \text{Wrong middle term}$$

$$(2x + 3)(x + 1) = 2x^2 + 5x + 3 \quad \text{Correct middle term}$$

Thus the correct answer is

$$(2x + 3)(x + 1) \text{ or } (x + 1)(2x + 3).$$

Practice Problem 1 Factor. $2x^2 + 7x + 5$

Some problems have many more possibilities.

EXAMPLE 2 Factor. $4x^2 - 13x + 3$

Solution

The Different Factors of 4 Are:	The Factors of 3 Are:
2 and 2	1 and 3
1 and 4	

Let us list the possible factoring combinations and compute the middle term by the FOIL method. Note that the signs of the constants in both factors will be negative. Why?

Possible Factors	Middle Term	Correct?
$(2x - 3)(2x - 1)$	$-8x$	No
$(4x - 3)(x - 1)$	$-7x$	No
$(4x - 1)(x - 3)$	$-13x$	Yes

The correct answer is $(4x - 1)(x - 3)$ or $(x - 3)(4x - 1)$.
This method is called the **trial-and-error method.**

Practice Problem 2 Factor. $9x^2 - 64x + 7$

NOTE TO STUDENT: Fully worked-out solutions to all of the Practice Problems can be found at the back of the text starting at page SP-1

EXAMPLE 3 Factor. $3x^2 - 2x - 8$

Solution

Factors of 3	Factors of 8
3 and 1	8 and 1
	4 and 2

Let us list only one-half of the possibilities. We'll let the constant in the first factor of each product be positive.

Possible Factors	Middle Term	Correct Factors?
$(x + 8)(3x - 1)$	$+23x$	No
$(x + 1)(3x - 8)$	$-5x$	No
$(x + 4)(3x - 2)$	$+10x$	No
$(x + 2)(3x - 4)$	$+2x$	No (but only because the sign is wrong)

So we just *reverse* the signs of the constants in the factors.

	Middle Term	Correct Factor?
$(x - 2)(3x + 4)$	$-2x$	Yes

The correct answer is

$$(x - 2)(3x + 4) \text{ or } (3x + 4)(x - 2).$$

Practice Problem 3 Factor. $3x^2 - x - 14$

NOTE TO STUDENT: Fully worked-out solutions to all of the Practice Problems can be found at the back of the text starting at page SP-1

It takes a good deal of practice to readily factor problems of this type. The more problems you do, the more proficient you will become. The following method will help you factor more quickly.

2 Using the Grouping Method

One way to factor a trinomial of the form $ax^2 + bx + c$ is to write it with four terms and factor by grouping, as we did in Section 5.2. For example, the trinomial $2x^2 + 13x + 20$ can be written as $2x^2 + 5x + 8x + 20$. Using the methods of Section 5.2, we factor it as follows.

$$2x^2 + 5x + 8x + 20 = x(2x + 5) + 4(2x + 5)$$
$$= (2x + 5)(x + 4)$$

We can factor all factorable trinomials of the form $ax^2 + bx + c$ in this way. We will use the following procedure.

GROUPING NUMBER METHOD FOR FACTORING TRINOMIALS OF THE FORM $ax^2 + bx + c$

1. Obtain the grouping number ac.
2. Find the two numbers whose product is the grouping number and whose sum is b.
3. Use those numbers to write bx as the sum of two terms.
4. Factor by grouping.
5. Multiply to check.

Let's try the problem from Example 1.

EXAMPLE 4

Factor by grouping. $2x^2 + 5x + 3$

Solution

1. The grouping number is $(2)(3) = 6$.
2. The factors of 6 are $6 \cdot 1$ and $3 \cdot 2$. We choose the numbers 3 and 2 because their product is 6 and their sum is 5.
3. We write $5x$ as the sum $3x + 2x$.
4. Factor by grouping.

$$
\begin{aligned}
2x^2 + 5x + 3 &= 2x^2 + 2x + 3x + 3 \\
&= 2x(x + 1) + 3(x + 1) \\
&= (x + 1)(2x + 3)
\end{aligned}
$$

5. Multiply to check.

$$
\begin{aligned}
(x + 1)(2x + 3) &= 2x^2 + 3x + 2x + 3 \\
&= 2x^2 + 5x + 3 \quad \checkmark
\end{aligned}
$$

Practice Problem 4   Factor by grouping. $2x^2 + 7x + 5$

EXAMPLE 5

Factor by grouping. $4x^2 - 13x + 3$

Solution

1. The grouping number is $(4)(3) = 12$.
2. The factors of 12 are $(12)(1)$ or $(4)(3)$ or $(6)(2)$. Note that the middle term of the polynomial is negative. Thus we choose the numbers -12 and -1 because their product is still 12 and their sum is -13.
3. We write $-13x$ as the sum $-12x + (-1x)$.
4. Factor by grouping.

$$
\begin{aligned}
4x^2 - 13x + 3 &= 4x^2 - 12x - 1x + 3 \\
&= 4x(x - 3) - 1(x - 3) \\
&= (x - 3)(4x - 1)
\end{aligned}
$$

Remember to factor out a -1 from the last two terms so that both sets of parentheses contain the same expression.

Practice Problem 5   Factor by grouping. $9x^2 - 64x + 7$

EXAMPLE 6

Factor by grouping. $3x^2 - 2x - 8$

Solution

1. The grouping number is $(3)(-8) = -24$.
2. We want two numbers whose product is -24 and whose sum is -2. They are -6 and 4.
3. We write $-2x$ as a sum $-6x + 4x$.
4. Factor by grouping.

$$
\begin{aligned}
3x^2 - 6x + 4x - 8 &= 3x(x - 2) + 4(x - 2) \\
&= (x - 2)(3x + 4)
\end{aligned}
$$

Practice Problem 6   Factor by grouping. $3x^2 + 4x - 4$

To factor polynomials of the form $ax^2 + bx + c$, use the method, either trial-and-error or grouping, that works best for you.

3 Using the Common Factor Method

Some problems require first factoring out the greatest common factor and then factoring the trinomial by one of the two methods of this section.

EXAMPLE 7 Factor. $9x^2 + 3x - 30$

Solution

$$9x^2 + 3x - 30 = 3(3x^2 + 1x - 10)$$ We first factor out the common factor of 3 from each term of the trinomial.

$$= 3(3x - 5)(x + 2)$$ We then factor the trinomial by the grouping method or by the trial-and-error method.

NOTE TO STUDENT: *Fully worked-out solutions to all of the Practice Problems can be found at the back of the text starting at page SP-1*

Practice Problem 7 Factor. $8x^2 + 8x - 6$

Be sure to remove the greatest common factor in the very first step.

EXAMPLE 8 Factor. $32x^2 - 40x + 12$

Solution

$$32x^2 - 40x + 12 = 4(8x^2 - 10x + 3)$$ We first factor out the greatest common factor of 4 from each term of the trinomial.

$$= 4(2x - 1)(4x - 3)$$ We then factor the trinomial by the grouping method or by the trial-and-error method.

Practice Problem 8 Factor. $24x^2 - 38x + 10$

Factor by the trial-and-error method. Check your answers using FOIL.

1. $4x^2 + 13x + 3$ **2.** $3x^2 + 11x + 10$ **3.** $5x^2 + 7x + 2$ **4.** $3x^2 + 4x + 1$

5. $3x^2 - 4x - 7$ **6.** $5x^2 + 7x - 6$ **7.** $2x^2 - 5x - 3$ **8.** $2x^2 - x - 6$

Factor by the grouping number method. Check your answers by using FOIL.

9. $9x^2 + 9x + 2$ **10.** $4x^2 + 11x + 6$ **11.** $15x^2 - 34x + 15$ **12.** $10x^2 - 29x + 10$

13. $2x^2 + 3x - 20$ **14.** $6x^2 + 11x - 10$ **15.** $8x^2 + 10x - 3$ **16.** $5x^2 + 2x - 7$

Factor by any method.

17. $6x^2 - 5x - 6$ **18.** $3x^2 - 13x - 10$ **19.** $6x^2 + x - 15$ **20.** $12x^2 + 5x - 3$

21. $7x^2 - 5x - 18$ **22.** $9x^2 - 22x - 15$ **23.** $9y^2 - 13y + 4$ **24.** $5y^2 - 11y + 2$

25. $5a^2 - 13a - 6$ **26.** $3a^2 - 10a - 8$ **27.** $14x^2 - 3x - 5$ **28.** $40x^2 - x - 6$

29. $15x^2 + 4x - 4$ **30.** $8x^2 - 11x + 3$ **31.** $12x^2 + 28x + 15$ **32.** $24x^2 + 17x + 3$

33. $12x^2 - 16x - 3$ **34.** $12x^2 + x - 6$ **35.** $2x^4 + 15x^2 - 8$ **36.** $4x^4 - 11x^2 - 3$

37. $2x^2 + 11xy + 15y^2$ **38.** $15x^2 + 14xy + 3y^2$ **39.** $5x^2 + 16xy - 16y^2$ **40.** $12x^2 + 11xy - 5y^2$

Factor by first factoring out the greatest common factor. See Examples 7 and 8.

41. $4x^2 + 34x + 42$　　**42.** $10x^2 + 22x + 12$　　**43.** $8x^2 - 26x + 6$　　**44.** $12x^2 - 24x + 9$

45. $10x^2 - 25x - 15$　　**46.** $20x^2 - 25x - 30$　　**47.** $6x^3 + 9x^2 - 60x$　　**48.** $6x^3 - 16x^2 - 6x$

Mixed Practice

Factor.

49. $5x^2 + 3x - 2$　　**50.** $6x^2 + x - 2$　　**51.** $12x^2 - 38x + 20$　　**52.** $20x^2 - 38x + 12$

53. $12x^2 - 20x + 3$　　**54.** $6x^2 - 23x + 20$　　**55.** $8x^2 + 16x - 10$　　**56.** $16x^2 + 36x - 10$

Cumulative Review

57. Solve. $7x - 3(4 - 2x) = 2(x - 3) - (5 - x)$　　**58.** Solve for x. $7x - 3y = 5$

Carbon Dioxide Emissions　*Answer each of the following questions using the carbon dioxide emissions graph.*

59. From 1995 to the estimated value for 2010, what is the percent of increase for carbon dioxide emissions for the United States? (Round to the nearest tenth.)

60. From 1995 to the estimated value for 2010, what is the percent of increase for carbon dioxide emissions for China? (Round to the nearest tenth.)

61. Which country has the greatest percent of increase for the period 1995 to the estimated value for 2010?

62. During what five-year period is the increase in carbon dioxide emissions expected to be the greatest in the United States?

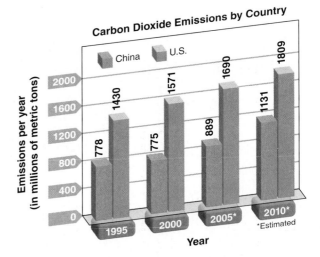

How are you doing with your homework assignments in Sections 5.1 to 5.4? Do you feel you have mastered the material so far? Do you understand the concepts you have covered? Before you go further in the textbook, take some time to do each of the following problems.

Factor each of the following problems. Be sure to remove the greatest common factor in each case.

5.1

1. $6xy - 15z + 21$

2. $20x^2 - 32xy + 12x$

3. $7(4x - 5) - b(4x - 5)$

4. $2x(8y + 3z) - 5y(8y + 3z)$

5.2

5. $18 + 3x - 6y - xy$

6. $15x - 9xb + 20w - 12bw$

7. $x^3 - 5x^2 - 3x + 15$

8. $7a + 21b + 2ab + 6b^2$

5.3

9. $x^2 - 15x + 56$

10. $x^2 + 12x - 64$

11. $x^2 + 13xy + 40y^2$

12. $7x^2 - 14x - 245$
(*Hint:* First factor out the greatest common factor.)

5.4

13. $10x^2 + x - 2$

14. $3x^2 - 23x + 14$

15. $6x^2 + 17xy + 12y^2$

16. $14x^3 - 20x^2 - 16x$
(*Hint:* First factor out the greatest common factor.)

Now turn to page SA-12 for the answer to each of these problems. Each answer also includes a reference to the objective in which the problem is first taught. If you missed any of these problems, you should stop and review the Examples and Practice Problems in the referenced objective. A little review now will help you master the material in the upcoming sections of the text.

1. _____

2. _____

3. _____

4. _____

5. _____

6. _____

7. _____

8. _____

9. _____

10. _____

11. _____

12. _____

13. _____

14. _____

15. _____

16. _____

5.5 SPECIAL CASES OF FACTORING

Student Learning Objectives

After studying this section, you will be able to:

1 Recognize and factor expressions of the type $a^2 - b^2$ (difference of two squares).

2 Recognize and factor expressions of the type $a^2 + 2ab + b^2$ (perfect-square trinomial).

3 Recognize and factor expressions that require factoring out a common factor and then using a special-case formula.

NOTE TO STUDENT: Fully worked-out solutions to all of the Practice Problems can be found at the back of the text starting at page SP-1

As we proceed in this section you will be able to reduce the time it takes you to factor a polynomial by quickly recognizing and factoring two special types of polynomials: the difference of two squares and perfect-square trinomials.

1 Factoring the Difference of Two Squares

Recall the formula from Section 4.5:

$$(a + b)(a - b) = a^2 - b^2.$$

In reverse form we can use it for factoring.

DIFFERENCE OF TWO SQUARES
$$a^2 - b^2 = (a + b)(a - b)$$

We can state it in words in this way: "The difference of two squares can be factored into the sum and difference of those values that were squared."

EXAMPLE 1 Factor. $9x^2 - 1$

Solution We see that the problem is in the form of the difference of two squares. $9x^2$ is a square and 1 is a square. So using the formula we can write the following.

$$9x^2 - 1 = (3x + 1)(3x - 1) \quad \text{Because } 9x^2 = (3x)^2 \text{ and } 1 = (1)^2$$

Practice Problem 1 Factor. $64x^2 - 1$

EXAMPLE 2 Factor. $25x^2 - 16$

Solution Again we use the formula for the difference of squares.

$$25x^2 - 16 = (5x + 4)(5x - 4) \quad \text{Because } 25x^2 = (5x)^2 \text{ and } 16 = (4)^2$$

Practice Problem 2 Factor. $36x^2 - 49$

Sometimes the polynomial contains two variables.

EXAMPLE 3 Factor. $4x^2 - 49y^2$

Solution We see that

$$4x^2 - 49y^2 = (2x + 7y)(2x - 7y).$$

Practice Problem 3 Factor. $100x^2 - 81y^2$

CAUTION Please note that the Difference of Two Squares formula only works if the last term is negative. So if Example 3 had been to factor $4x^2 + 49y^2$ we would **not** have been able to factor the problem. We will examine this in more detail in Section 5.6.

Some problems may involve more than one step.

EXAMPLE 4   Factor. $81x^4 - 1$

Solution   We see that

$$81x^4 - 1 = (9x^2 + 1)(9x^2 - 1) \qquad \text{Because } 81x^4 = (9x^2)^2 \text{ and } 1 = (1)^2$$

Is the factoring complete? We can factor $9x^2 - 1$.

$$81x^4 - 1 = (9x^2 + 1)(3x - 1)(3x + 1) \quad \text{Because } (9x^2 - 1) = (3x - 1)(3x + 1)$$

Practice Problem 4   Factor. $x^8 - 1$

---

② Factoring Perfect-Square Trinomials

There is a formula that will help us to factor very quickly certain trinomials, called **perfect-square trinomials.** Recall from Section 4.5 the formulas for a binomial squared.

$$(a + b)^2 = a^2 + 2ab + b^2$$
$$(a - b)^2 = a^2 - 2ab + b^2$$

We can use these two equations in reverse form for factoring.

> **PERFECT-SQUARE TRINOMIALS**
>
> $$a^2 + 2ab + b^2 = (a + b)^2$$
> $$a^2 - 2ab + b^2 = (a - b)^2$$

A perfect-square trinomial is a trinomial that is the result of squaring a binomial. How can we recognize a perfect-square trinomial?

1. The first and last terms are *perfect squares*.
2. The middle term is twice the product of the values whose squares are the first and last terms.

EXAMPLE 5   Factor. $x^2 + 6x + 9$

Solution   This is a perfect-square trinomial.

1. The first and last terms are perfect squares because $x^2 = (x)^2$ and $9 = (3)^2$.
2. The middle term, $6x$, is twice the product of x and 3.

Since $x^2 + 6x + 9$ is a perfect-square trinomial, we can use the formula

$$a^2 + 2ab + b^2 = (a + b)^2$$

with $a = x$ and $b = 3$. So we have

$$x^2 + 6x + 9 = (x + 3)^2.$$

Practice Problem 5   Factor. $x^2 + 10x + 25$

EXAMPLE 6 Factor. $4x^2 - 20x + 25$

Solution This is a perfect-square trinomial. Note that $20x = 2(2x \cdot 5)$. Also note the negative sign. Thus we have the following.

$$4x^2 - 20x + 25 = (2x - 5)^2 \quad \text{Since } a^2 - 2ab + b^2 = (a - b)^2$$

Practice Problem 6 Factor. $25x^2 - 30x + 9$

NOTE TO STUDENT: *Fully worked-out solutions to all of the Practice Problems can be found at the back of the text starting at page SP-1*

A polynomial may have more than one variable and its exponents may be higher than 2. The same principles apply.

EXAMPLE 7 Factor.

(a) $49x^2 + 42xy + 9y^2$ **(b)** $36x^4 - 12x^2 + 1$

Solution

(a) This is a perfect-square trinomial. Why?

$$49x^2 + 42xy + 9y^2 = (7x + 3y)^2 \quad \text{Because } 49x^2 = (7x)^2, 9y^2 = (3y)^2, \text{ and } 42xy = 2(7x \cdot 3y)$$

(b) This is a perfect-square trinomial. Why?

$$36x^4 - 12x^2 + 1 = (6x^2 - 1)^2 \quad \text{Because } 36x^4 = (6x^2)^2, 1 = (1)^2, \text{ and } 12x^2 = 2(6x^2 \cdot 1)$$

Practice Problem 7 Factor.

(a) $25x^2 + 60xy + 36y^2$ **(b)** $64x^6 - 48x^3 + 9$

Some polynomials appear to be perfect-square trinomials but are not. They were factored in other ways in Section 5.4.

EXAMPLE 8 Factor. $49x^2 + 35x + 4$

Solution This is *not* a perfect-square trinomial! Although the first and last terms are perfect squares since $(7x)^2 = 49x^2$ and $(2)^2 = 4$, the middle term, $35x$, is not double the product of 2 and $7x$! $35x \neq 28x$! So we must factor by trial and error or by grouping to obtain

$$49x^2 + 35x + 4 = (7x + 4)(7x + 1).$$

Practice Problem 8 Factor. $9x^2 + 15x + 4$

3   Factoring Out a Common Factor Before Using a Special-Case Formula

For some polynomials, we will first factor out the greatest common factor. Then we will find an opportunity to use the difference-of-two-squares formula or one of the perfect-square trinomial formulas.

     Look carefully at example 9. Do you know what is the greatest common factor?

EXAMPLE 9   Factor. $12x^2 - 48$

Solution

$$12x^2 - 48 = 12(x^2 - 4)$$ First we factor out the greatest common factor, 12.

$$= 12(x + 2)(x - 2)$$ Then we use the difference-of-two-squares formula, $a^2 - b^2 = (a + b)(a - b)$.

Practice Problem 9   Factor. $20x^2 - 45$

     Look carefully at example 10. Do you know what is the greatest common factor?

EXAMPLE 10   Factor. $24x^2 - 72x + 54$

Solution

$$24x^2 - 72x + 54 = 6(4x^2 - 12x + 9)$$ First we factor out the greatest common factor, 6.

$$= 6(2x - 3)^2$$ Then we use the perfect-square trinomial formula, $a^2 - 2ab + b^2 = (a - b)^2$.

Practice Problem 10   Factor. $75x^2 - 60x + 12$

Factor by using the difference-of-two-squares formula.

1. $9x^2 - 1$

2. $25x^2 - 1$

3. $81x^2 - 16$

4. $100x^2 - 49$

5. $x^2 - 49$

6. $x^2 - 100$

7. $4x^2 - 25$

8. $16x^2 - 25$

9. $x^2 - 25$

10. $x^2 - 36$

11. $1 - 16x^2$

12. $1 - 64x^2$

13. $16x^2 - 49y^2$

14. $25x^2 - 81y^2$

15. $36x^2 - 169y^2$

16. $64x^2 - 81y^2$

17. $81x^2 - 100$

18. $25a^2 - 16$

19. $25a^2 - 81b^2$

20. $9x^2 - 49y^2$

Factor by using the perfect-square trinomial formula.

21. $9x^2 + 6x + 1$

22. $25x^2 + 10x + 1$

23. $y^2 - 10y + 25$

24. $y^2 - 12y + 36$

25. $36x^2 - 60x + 25$

26. $16x^2 - 72x + 81$

27. $49x^2 + 28x + 4$

28. $25x^2 + 30x + 9$

29. $x^2 + 14x + 49$

30. $x^2 + 8x + 16$

31. $25x^2 - 40x + 16$

32. $49x^2 - 42x + 9$

33. $81x^2 + 36xy + 4y^2$

34. $36x^2 + 60xy + 25y^2$

35. $25x^2 - 30xy + 9y^2$

36. $4x^2 - 28xy + 49y^2$

Mixed Practice

Factor by using either the difference-of-two-squares or the perfect-square trinomial formulas.

37. $16a^2 + 72ab + 81b^2$

38. $169a^2 + 26ab + b^2$

39. $49x^2 - 42xy + 9y^2$

40. $9x^2 - 30xy + 25y^2$

41. $49x^2 + 42x + 9$

42. $25x^2 + 40x + 16$

43. $49x^2 - 9$

44. $9x^2 - 25$

45. $x^4 - 36$

46. $x^4 - 49$

47. $9x^4 - 12x^2 + 4$

48. $36x^4 - 60x^2 + 25$

To Think About

49. In Example 4, first we factored $81x^4 - 1$ as $(9x^2 + 1)(9x^2 - 1)$, and then we factored $9x^2 - 1$ as $(3x + 1)(3x - 1)$. Show why you cannot factor $9x^2 + 1$.

50. What two numbers could replace the b in $25x^2 + bx + 16$ so that the resulting trinomial would be a perfect square? (*Hint*: One number is negative.)

51. What value could you give to c so that $16y^2 - 56y + c$ would become a perfect-square trinomial? Is there only one answer or more than one?

52. Jerome says that he can find two values of b so that $100x^2 + bx - 9$ will be a perfect square. Kesha says there is only one that fits, and Larry says there are none. Who is correct and why?

Factor by first looking for a greatest common factor. See Examples 9 and 10.

53. $16x^2 - 36$ **54.** $27x^2 - 75$ **55.** $147x^2 - 3y^2$ **56.** $16y^2 - 100x^2$

57. $12x^2 - 36x + 27$ **58.** $125x^2 - 100x + 20$ **59.** $98x^2 + 84x + 18$ **60.** $128x^2 + 96x + 18$

Mixed Practice

Factor. Be sure to look for common factors first.

61. $x^2 - 15x + 44$ **62.** $x^2 - 16x + 63$ **63.** $2x^2 + 5x - 3$ **64.** $15x^2 - 11x + 2$

65. $16x^2 - 121$ **66.** $9x^2 - 100y^2$ **67.** $9x^2 + 42x + 49$ **68.** $9x^2 + 30x + 25$

69. $3x^2 + 6x - 45$ **70.** $4x^2 + 24x + 32$ **71.** $5x^2 - 80$ **72.** $13x^2 - 13$

73. $7x^2 + 42x + 63$ **74.** $6x^2 + 60x + 150$ **75.** $2x^2 - 32x + 126$ **76.** $2x^2 - 32x + 110$

Cumulative Review

77. Divide. $(x^3 + x^2 - 2x - 11) \div (x - 2)$ **78.** Divide. $(6x^3 + 11x^2 - 11x - 20) \div (3x + 4)$

__Iguana Diet__ The green iguana can reach a length of 6 feet and weigh up to 18 pounds. Of the basic diet of the iguana, 40% should consist of greens such as lettuce, spinach, and parsley; 35% should consist of bulk vegetables such as broccoli, zucchini, and carrots; and 25% should consist of fruit.

79. If a certain iguana weighing 150 ounces has a daily diet equal to 2% of its body weight, compose a diet for it in ounces that will meet the iguana's one-day requirement for nutrition.

80. If another iguana weighing 120 ounces has a daily diet equal to 3% of its body weight, compose a diet for it in ounces that will meet the iguana's one-day requirement for nutrition.

__Altitude__ The peak of Mount Washington is at an altitude of 6288 feet above sea level. The altitude A in feet of a car driving down the mountain road from the peak a distance M measured in miles is given by $A = 6288 - 700M$.

81. What is the altitude of a car that has driven from the mountain peak down a distance of 3.5 miles?

82. A car drives from the peak down the mountain road to a point where the altitude is 2788 feet above sea level. How many miles down the road has the car driven?

5.6 A BRIEF REVIEW OF FACTORING

Student Learning Objectives

After studying this section, you will be able to:

 Identify and factor any polynomial that can be factored.

2 **Determine whether a polynomial is prime.**

1 Identifying and Factoring Polynomials

Often the various types of factoring problems are all mixed together. We need to be able to identify each type of polynomial quickly. The following table summarizes the information we have learned about factoring.

Many polynomials require more than one factoring method. When you are asked to factor a polynomial, it is expected that you will factor it completely. Usually, the first step is factoring out a common factor; then the next step will become apparent.

Carefully go through each example in the following Factoring Organizer. Be sure you understand each step that is involved.

Factoring Organizer

Number of Terms in the Polynomial	Identifying Name and/or Formula	Example
A. Any number of terms	**Common factor** The terms have a common factor consisting of a number, a variable, or both.	$2x^2 - 16x = 2x(x - 8)$ $3x^2 + 9y - 12 = 3(x^2 + 3y - 4)$ $4x^2y + 2xy^2 - wxy + xyz = xy(4x + 2y - w + z)$
B. Two terms	**Difference of two squares** First and last terms are perfect squares. $a^2 - b^2 = (a + b)(a - b)$	$16x^2 - 1 = (4x + 1)(4x - 1)$ $25y^2 - 9x^2 = (5y + 3x)(5y - 3x)$
C. Three terms	**Perfect-square trinomial** First and last terms are perfect squares. $a^2 + 2ab + b^2 = (a + b)^2$ $a^2 - 2ab + b^2 = (a - b)^2$	$25x^2 - 10x + 1 = (5x - 1)^2$ $16x^2 + 24x + 9 = (4x + 3)^2$
D. Three terms	**Trinomial of the form $x^2 + bx + c$** It starts with x^2. The constants of the two factors are numbers whose product is c and whose sum is b.	$x^2 - 7x + 12 = (x - 3)(x - 4)$ $x^2 + 11x - 26 = (x + 13)(x - 2)$ $x^2 - 8x - 20 = (x - 10)(x + 2)$
E. Three terms	**Trinomial of the form $ax^2 + bx + c$** It starts with ax^2, where a is any number but 1.	Use trial-and-error or the grouping number method to factor $12x^2 - 5x - 2$. 1. The grouping number is -24. 2. The two numbers whose product is -24 and whose sum is -5 are -8 and 3. 3. $12x^2 - 5x - 2 = 12x^2 + 3x - 8x - 2$ $\qquad = 3x(4x + 1) - 2(4x + 1)$ $\qquad = (4x + 1)(3x - 2)$
F. Four terms	**Factor by grouping** Rearrange the order if the first two terms do not have a common factor.	$wx - 6yz + 2wy - 3xz = wx + 2wy - 3xz - 6yz$ $\qquad = w(x + 2y) - 3z(x + 2y)$ $\qquad = (x + 2y)(w - 3z)$

EXAMPLE 1 Factor.

(a) $25x^3 - 10x^2 + x$ **(b)** $20x^2y^2 - 45y^2$

(c) $2ax + 4ay + 4x + 8y$ **(d)** $15x^2 - 3x^3 + 18x$

Solution

(a) $25x^3 - 10x^2 + x = x(25x^2 - 10x + 1)$ Factor out the common factor of x. The other factor is a perfect-square trinomial.

$\qquad\qquad\qquad\qquad = x(5x - 1)^2$

(b) $20x^2y^2 - 45y^2 = 5y^2(4x^2 - 9)$

Factor out the common factor of $5y^2$. The other factor is a difference of squares.

$$= 5y^2(2x - 3)(2x + 3)$$

(c) $2ax + 4ay + 4x + 8y = 2[ax + 2ay + 2x + 4y]$

Factor out the common factor of 2.

$$= 2[a(x + 2y) + 2(x + 2y)]$$

Factor the terms inside the bracket by the grouping method.

$$= 2[(x + 2y)(a + 2)]$$

Factor out the common factor of $(x + 2y)$.

(d) $15x^2 - 3x^3 + 18x = -3x^3 + 15x^2 + 18x$

Rearrange the terms in descending order of powers of x.

$$= -3x(x^2 - 5x - 6)$$

Factor out the common factor of $-3x$.

$$= -3x(x - 6)(x + 1)$$

Factor the trinomial.

Practice Problem 1   Factor. Be careful. These practice problems are mixed.

(a) $3x^2 - 36x + 108$

(b) $9x^4y^2 - 9y^2$

(c) $5x^3 - 15x^2y + 10x^2 - 30xy$

(d) $12x - 9 - 4x^2$

NOTE TO STUDENT: Fully worked-out solutions to all of the Practice Problems can be found at the back of the text starting at page SP-1

2 Determining Whether a Polynomial Is Prime

Not all polynomials can be factored using the methods in this chapter. If we cannot factor a polynomial by elementary methods, we will identify it as a **prime** polynomial. If, after you have mastered the factoring techniques in this chapter, you encounter a polynomial that you cannot factor with these methods, you should feel comfortable enough to say, "The polynomial cannot be factored with the methods in this chapter, so it is prime," rather than "I can't do it—I give up!"

EXAMPLE 2   Factor, if possible. $x^2 + 6x + 12$

Solution   The factors of 12 are

$$(1)(12) \text{ or } (2)(6) \text{ or } (3)(4).$$

None of these pairs add up to 6, the coefficient of the middle term. Thus the problem cannot be factored by the methods of this chapter. It is prime.

Practice Problem 2   Factor. $x^2 - 9x - 8$

EXAMPLE 3   Factor, if possible. $25x^2 + 4$

Solution   We have a formula to factor the difference of two squares. There is no way to factor the sum of two squares. That is, $a^2 + b^2$ cannot be factored. Thus

$$25x^2 + 4 \text{ is prime.}$$

Practice Problem 3   Factor, if possible. $25x^2 + 82x + 4$

Review the six basic types of factoring in the Factoring Organizer on page 334. Each of the six types is included in exercises 1–12. Be sure you can find two of each type.

Factor. Check your answer by multiplying.

1. $6a^2 + 2ab - 3a$

2. $6x^2 - 3xy + 5x$

3. $16x^2 - 25y^2$

4. $64x^2 - 9y^2$

5. $9x^2 - 12xy + 4y^2$

6. $16x^2 + 24xy + 9y^2$

7. $x^2 + 8x + 15$

8. $x^2 + 15x + 54$

9. $15x^2 + 7x - 2$

10. $6x^2 + 13x - 5$

11. $ax - 3cx + 3ay - 9cy$

12. $bx - 2dx + 5by - 10dy$

Mixed Practice

Factor, if possible. Be sure to factor completely. Always factor out the greatest common factor first, if one exists.

13. $y^2 + 14y + 49$

14. $y^2 + 16y + 64$

15. $4x^2 - 12x + 9$

16. $16x^2 - 40x + 25$

17. $2x^2 - 11x + 12$

18. $3x^2 - 10x + 8$

19. $x^2 - 3xy - 70y^2$

20. $x^2 - 6xy - 16y^2$

21. $ax - 5a + 3x - 15$

22. $by + 7b - 6y - 42$

23. $45x - 5x^3$

24. $18y^2 + 3y - 6$

25. $5x^3y^3 - 10x^2y^3 + 5xy^3$

26. $2x^4y - 12x^3y + 18x^2y$

27. $27xyz^2 - 12xy$

28. $12x^2 - 2x - 18x^3$

29. $3x^2 + 6x - 105$

30. $4x^2 - 28x - 72$

31. $3x^3 + 3x^2 - 36x$

32. $7x^2 + 3x - 2$

33. $7x^2 - 2x^4 + 4$

34. $14x^2 - x^3 + 32x$

35. $6x^2 - 3x + 2$

36. $4x^3 + 8x^2 - 60x$

Remove the greatest common factor first. Then continue to factor, if possible.

37. $5x^2 + 10xy - 30y$

38. $7a^2 + 21b - 42$

39. $30x^3 + 3x^2y - 6xy^2$

40. $30x^3 - 25x^2y - 30xy^2$

41. $24x^2 - 58x + 30$

42. $12x^2 - 30x + 12$

To Think About

43. A polynomial that cannot be factored by the methods of this chapter is called _____ .

44. A binomial of the form $x^2 - d$ can be quickly factored or identified as prime. If it can be factored, what is true of the number d?

Cumulative Review

45. *DVD Player Sale* A DVD player was reduced in price by $19.56. This was 20% of the original price. Find the sale price.

46. *Antiviral Drug* A major pharmaceutical company is testing a new, powerful antiviral drug. It kills 13 strains of virus every hour. If there are presently 294 live strains of virus in the test container, how many live strains were there 6 hours ago?

Compensation Costs   *If you were to compare the relative costs to produce items in different countries, you would need a basis of comparison. Let us assume that it costs $100 in 1980 in the United States to pay production workers to manufacture some product. In the same year it would cost $38 to pay production workers in Israel and $117 to pay production workers in Norway to produce the same object. Use the chart below to compare these hourly compensation costs and answer the following questions:*

47. Suppose the same percent of increase in Israel that was observed from 1990 to 2000 continued from 2000 to 2010. If it cost $100 for a production worker to produce an item in the United States in 2010, what would it cost for a production worker in Israel in 2010? Round to the nearest dollar.

48. Suppose the same percent of decrease in Norway that was observed from 1990 to 2000 continued from 2000 to 2010. If it cost $100 for a production worker to produce an item in the United States in 2010, what would it cost for a production worker in Norway in 2010? Round to the nearest dollar.

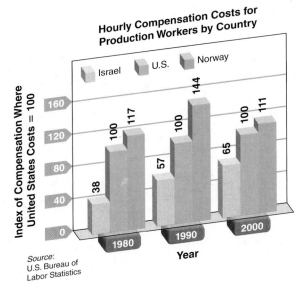

49. Write an equation of the form $y = ax + b$, where x is the number of years since 1990 that will predict the hourly compensation costs (y) for a production worker in Israel. Assume that the change over the period from 1990 to 2000 will continue at the same rate.

50. Write an equation of the form $y = ax + b$, where x is the number of years since 1990 that will predict the hourly compensation costs (y) for a production worker in Norway. Assume that the change over the period from 1990 to 2000 will continue at the same rate.

Student Learning Objectives

After studying this section, you will be able to:

1 Solve quadratic equations by factoring.

2 Use quadratic equations to solve applied problems.

1 Solving Quadratic Equations by Factoring

In Chapter 2, we learned how to solve linear equations such as $3x + 5 = 0$ by finding the root (or value of x) that satisfied the equation. Now we turn to the question of how to solve equations like $3x^2 + 5x + 2 = 0$. Such equations are called **quadratic equations.** A quadratic equation is a polynomial equation in one variable that contains a variable term of degree 2 and no terms of higher degree.

> The *standard form* of a quadratic equation is $ax^2 + bx + c = 0$, where a, b, and c are real numbers and $a \neq 0$.

In this section, we will study quadratic equations in standard form, where a, b, and c are integers.

Many quadratic equations have two real number solutions (also called real **roots**). But how can we find them? The most direct approach is the factoring method. This method depends on a very powerful property.

> **ZERO FACTOR PROPERTY**
> If $a \cdot b = 0$, then $a = 0$ or $b = 0$.

Notice the word *or* in the zero factor property. When we make a statement in mathematics using this word, we intend it to mean *one or the other or both*. Therefore, the zero factor property states that if the product $a \cdot b$ is zero, then a can equal zero or b can equal zero or *both a and b can equal zero*. We can use this principle to solve a quadratic equation. Before you start, make sure that the equation is in standard form.

> 1. Make sure the equation is set equal to zero.
> 2. Factor, if possible, the quadratic expression that equals 0.
> 3. Set each factor containing a variable equal to 0.
> 4. Solve the resulting equations to find each root.
> 5. Check each root.

EXAMPLE 1 Solve the equation to find the two roots. $3x^2 + 5x + 2 = 0$

Solution

$$3x^2 + 5x + 2 = 0 \qquad \text{The equation is in standard form.}$$
$$(3x + 2)(x + 1) = 0 \qquad \text{Factor the quadratic expression.}$$
$$3x + 2 = 0 \qquad x + 1 = 0 \qquad \text{Set each factor equal to 0.}$$
$$3x = -2 \qquad x = -1 \qquad \text{Solve the equations to find the two roots.}$$
$$x = -\frac{2}{3}$$

The two roots (that is, solutions) are $-\frac{2}{3}$ and -1.

Check. We can determine if the two numbers $-\frac{2}{3}$ and -1 are solutions to the equation. Substitute $-\frac{2}{3}$ for x in the *original equation*. If an identity results, $-\frac{2}{3}$ is a solution. Do the same for -1.

$$3x^2 + 5x + 2 = 0$$

$$3\left(-\frac{2}{3}\right)^2 + 5\left(-\frac{2}{3}\right) + 2 \stackrel{?}{=} 0$$

$$3\left(\frac{4}{9}\right) + 5\left(-\frac{2}{3}\right) + 2 \stackrel{?}{=} 0$$

$$\frac{4}{3} - \frac{10}{3} + 2 \stackrel{?}{=} 0$$

$$\frac{4}{3} - \frac{10}{3} + \frac{6}{3} \stackrel{?}{=} 0$$

$$0 = 0 \checkmark$$

$$3x^2 + 5x + 2 = 0$$

$$3(-1)^2 + 5(-1) + 2 \stackrel{?}{=} 0$$

$$3(1) + 5(-1) + 2 \stackrel{?}{=} 0$$

$$3 - 5 + 2 \stackrel{?}{=} 0$$

$$-2 + 2 \stackrel{?}{=} 0$$

$$0 = 0 \checkmark$$

Thus $-\frac{2}{3}$ and -1 are both roots of the equation $3x^2 + 5x + 2 = 0$.

Practice Problem 1 Solve the equation by factoring to find the two roots and check. $10x^2 - x - 2 = 0$

NOTE TO STUDENT: *Fully worked-out solutions to all of the Practice Problems can be found at the back of the text starting at page SP-1*

EXAMPLE 2 Solve the equation to find the two roots. $2x^2 + 13x - 7 = 0$

Solution

$$2x^2 + 13x - 7 = 0 \qquad \text{The equation is in standard form.}$$

$$(2x - 1)(x + 7) = 0 \qquad \text{Factor.}$$

$$2x - 1 = 0 \qquad x + 7 = 0 \qquad \text{Set each factor equal to 0.}$$

$$2x = 1 \qquad\qquad x = -7 \qquad \text{Solve the equations to find the two roots.}$$

$$x = \frac{1}{2}$$

The two roots are $\frac{1}{2}$ and -7.

Check. If $x = \frac{1}{2}$, then we have the following.

$$2\left(\frac{1}{2}\right)^2 + 13\left(\frac{1}{2}\right) - 7 = 2\left(\frac{1}{4}\right) + 13\left(\frac{1}{2}\right) - 7$$

$$= \frac{1}{2} + \frac{13}{2} - \frac{14}{2} = 0 \checkmark$$

If $x = -7$, then we have the following.

$$2(-7)^2 + 13(-7) - 7 = 2(49) + 13(-7) - 7$$

$$= 98 - 91 - 7 = 0 \checkmark$$

Thus $\frac{1}{2}$ and -7 are both roots of the equation $2x^2 + 13x - 7 = 0$.

Practice Problem 2 Solve the equation to find the two roots. $3x^2 + 11x - 4 = 0$

If the quadratic equation $ax^2 + bx + c = 0$ has no visible constant term, then $c = 0$. All such quadratic equations can be solved by factoring out a common factor and then using the zero factor property to obtain two solutions that are real numbers.

EXAMPLE 3 Solve the equation to find the two roots. $7x^2 - 3x = 0$

Solution

$$7x^2 - 3x = 0 \quad \text{The equation is in standard form. Here } c = 0.$$

$$x(7x - 3) = 0 \quad \text{Factor out the common factor.}$$

$$x = 0 \qquad 7x - 3 = 0 \quad \text{Set each factor equal to 0 by the zero factor property.}$$

$$7x = 3 \quad \text{Solve the equations to find the two roots.}$$

$$x = \frac{3}{7}$$

The two roots are 0 and $\frac{3}{7}$.

Check. Verify that 0 and $\frac{3}{7}$ are the roots of $7x^2 - 3x = 0$.

NOTE TO STUDENT: *Fully worked-out solutions to all of the Practice Problems can be found at the back of the text starting at page SP-1*

Practice Problem 3 Solve the equation to find the two roots.
$7x^2 + 11x = 0$

If the quadratic equation is not in standard form, we use the same basic algebraic methods we studied in Sections 2.1–2.3 to place the terms on one side and zero on the other so that we can use the zero factor property.

EXAMPLE 4 Solve. $x^2 = 12 - x$

Solution

$$x^2 = 12 - x \quad \text{The equation is not in standard form.}$$

$$x^2 + x - 12 = 0 \quad \text{Add } x \text{ and } -12 \text{ to both sides of the equation so that the left side is equal to zero; we can now factor.}$$

$$(x - 3)(x + 4) = 0 \quad \text{Factor.}$$

$$x - 3 = 0 \qquad x + 4 = 0 \quad \text{Set each factor equal to 0 by the zero factor property.}$$

$$x = 3 \qquad x = -4 \quad \text{Solve the equations for } x.$$

Check. If $x = 3$: $(3)^2 \overset{?}{=} 12 - 3$ If $x = -4$: $(-4)^2 \overset{?}{=} 12 - (-4)$

$$9 \overset{?}{=} 12 - 3 \qquad\qquad\qquad 16 \overset{?}{=} 12 + 4$$

$$9 = 9 \ \checkmark \qquad\qquad\qquad\qquad 16 = 16 \ \checkmark$$

Both roots check.

Practice Problem 4 Solve. $x^2 - 6x + 4 = -8 + x$

EXAMPLE 5 Solve. $\dfrac{x^2 - x}{2} = 6$

Solution We must first clear the fractions from the equation.

$$2\left(\frac{x^2 - x}{2}\right) = 2(6) \quad \text{Multiply each side by 2.}$$

$$x^2 - x = 12 \quad \text{Simplify.}$$

$$x^2 - x - 12 = 0 \quad \text{Place in standard form.}$$

$$(x - 4)(x + 3) = 0 \quad \text{Factor.}$$

$$x - 4 = 0 \qquad x + 3 = 0 \quad \text{Set each factor equal to zero.}$$

$$x = 4 \qquad x = -3 \quad \text{Solve the equations for } x.$$

The check is left to the student.

Practice Problem 5 Solve. $\dfrac{2x^2 - 7x}{3} = 5$

2 Using Quadratic Equations to Solve Applied Problems

Certain types of word problems—for example, some geometry applications—lead to quadratic equations. We'll show how to solve such word problems in this section.

It is particularly important to check the apparent solutions to the quadratic equation with conditions stated in the word problem. Often a particular solution to the quadratic equation will be eliminated by the conditions of the word problem.

▲ **EXAMPLE 6**   Carlos lives in Mexico City. He has a rectangular brick walkway in front of his house. The length of the walkway is 3 meters longer than twice the width. The area of the walkway is 44 square meters. Find the length and width of the rectangular walkway.

Solution

1. *Understand the problem.*
   Draw a picture.

   Let w = the width in meters

   Then $2w + 3$ = the length in meters

2. *Write an equation.*

   $$\text{area} = (\text{width})(\text{length})$$
   $$44 = w(2w + 3)$$

3. *Solve and state the answer.*

$44 = w(2w + 3)$	
$44 = 2w^2 + 3w$	Remove parentheses.
$0 = 2w^2 + 3w - 44$	Subtract 44 from both sides.
$0 = (2w + 11)(w - 4)$	Factor.
$2w + 11 = 0 \qquad w - 4 = 0$	Set each factor equal to 0.
$2w = -11 \qquad w = 4$	Simplify and solve.
$w = -5\dfrac{1}{2}$	Although $-5\dfrac{1}{2}$ is a solution to the quadratic equation, it is not a valid solution to the word problem. It would not make sense to have a rectangle with a negative number as a width.

   Since $w = 4$, the width of the walkway is 4 meters. The length is $2w + 3$, so we have $2(4) + 3 = 8 + 3 = 11$. Thus the length of the walkway is 11 meters.

4. *Check.*   Is the length 3 meters more than twice the width?

   $$11 \overset{?}{=} 3 + 2(4) \qquad 11 = 3 + 8 \;\checkmark$$

   Is the area of the rectangle 44 square meters?

   $$4 \times 11 \overset{?}{=} 44 \qquad 44 = 44 \;\checkmark$$

▲ **Practice Problem 6**   The length of a rectangle is 2 meters longer than triple the width. The area of the rectangle is 85 square meters. Find the length and width of the rectangle.

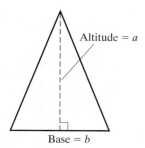

Altitude = a

Base = b

▲ **EXAMPLE 7** The top of a local cable television tower has several small triangular reflectors. The area of each triangle is 49 square centimeters. The altitude of each triangle is 7 centimeters longer than the base. Find the altitude and the base of one of the triangles.

Solution

Let b = the length of the base in centimeters

$b + 7$ = the length of the altitude in centimeters

To find the area of a triangle, we use

$$\text{area} = \frac{1}{2}(\text{altitude})(\text{base}) = \frac{1}{2}ab = \frac{ab}{2}.$$

$\dfrac{ab}{2} = 49$	Write an equation.
$\dfrac{(b+7)(b)}{2} = 49$	Substitute the expressions for altitude and base.
$\dfrac{b^2 + 7b}{2} = 49$	Simplify.
$b^2 + 7b = 98$	Multiply each side of the equation by 2.
$b^2 + 7b - 98 = 0$	Place the quadratic equation in standard form.
$(b - 7)(b + 14) = 0$	Factor.
$b - 7 = 0 \qquad b + 14 = 0$	Set each factor equal to zero.
$b = 7 \qquad\qquad b = -14$	Solve the equations for b.

We cannot have a base of -14 centimeters, so we reject the negative answer. The only possible solution is 7. So the base is 7 centimeters. The altitude is $b + 7 = 7 + 7 = 14$. The altitude is 14 centimeters. The triangular reflector has a base of 7 centimeters and an altitude of 14 centimeters.

Check. When you do the check, answer the following two questions.

1. Is the altitude 7 centimeters longer than the base?

2. Is the area of a triangle with a base of 7 centimeters and an altitude of 14 centimeters actually 49 square centimeters?

NOTE TO STUDENT: Fully worked-out solutions to all of the Practice Problems can be found at the back of the text starting at page SP-1

▲ **Practice Problem 7** A triangle has an area of 35 square centimeters. The altitude of the triangle is 3 centimeters shorter than the base. Find the altitude and the base of the triangle.

Many problems in the sciences require the use of quadratic equations. You will study these in more detail if you take a course in physics or calculus in college. Often a quadratic equation is given as part of the problem.

When an object is thrown upward, its height (S) in meters is given, approximately, by the quadratic equation

$$S = -5t^2 + vt + h.$$

The letter h represents the initial height in meters. The letter v represents the initial velocity of the object thrown. The letter t represents the time in seconds starting from the time the object is thrown.

EXAMPLE 8   A tennis ball is thrown upward with an initial velocity of 8 meters/second. Suppose that the initial height above the ground is 4 meters. At what time *t* will the ball hit the ground?

4 m

Solution   In this case $S = 0$ since the ball will hit the ground. The initial upward velocity is $v = 8$ meters/second. The initial height is 4 meters, so $h = 4$.

$$S = -5t^2 + vt + h \quad \text{Write an equation.}$$
$$0 = -5t^2 + 8t + 4 \quad \text{Substitute all values into the equation.}$$
$$5t^2 - 8t - 4 = 0 \qquad \text{Isolate the terms on the left side. (Most}$$
$$\text{students can factor more readily if the}$$
$$\text{squared variable is positive.)}$$
$$(5t + 2)(t - 2) = 0 \qquad \text{Factor.}$$
$$5t + 2 = 0 \qquad t - 2 = 0 \qquad \text{Set each factor} = 0.$$
$$5t = -2 \qquad t = 2 \qquad \text{Solve the equations for } t.$$
$$t = -\frac{2}{5}$$

We want a positive time *t* in seconds; thus we do not use $t = -\frac{2}{5}$. Therefore, the ball will strike the ground 2 seconds after it is thrown.

Check.   Verify the solution.

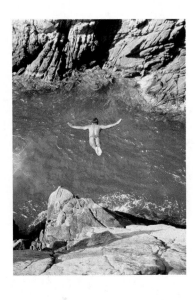

Practice Problem 8   A Mexican cliff diver does a dive from a cliff 45 meters above the ocean. This constitutes free fall, so the initial velocity is $v = 0$, and if there is no upward spring, then $h = 45$ meters. How long will it be until he breaks the water's surface?

Using the factoring method, solve for the roots of each quadratic equation. Be sure to place the equation in standard form before factoring. Check your answers.

1. $x^2 - 4x - 21 = 0$ **2.** $x^2 - x - 20 = 0$ **3.** $x^2 + 16x + 39 = 0$ **4.** $x^2 + 11x + 18 = 0$

5. $2x^2 - 7x + 6 = 0$ **6.** $2x^2 - 11x + 12 = 0$ **7.** $6x^2 - 13x = -6$ **8.** $10x^2 + 19x = 15$

9. $x^2 + 13x = 0$ **10.** $8x^2 - x = 0$ **11.** $8x^2 = 72$ **12.** $9x^2 = 81$

13. $5x^2 + 3x = 8x$ **14.** $6x^2 - 4x = 3x$ **15.** $6x^2 = 16x - 8$ **16.** $24x^2 = -10x + 4$

17. $(x - 5)(x + 2) = -4(x + 1)$ **18.** $(x - 5)(x + 4) = 2(x - 5)$

19. $4x^2 - 3x + 1 = -7x$ **20.** $9x^2 - 2x + 4 = 10x$ **21.** $\dfrac{x^2}{2} - 8 + x = -8$ **22.** $4 + \dfrac{x^2}{3} = 2x + 4$

23. $\dfrac{x^2 + 7x}{4} = -3$ **24.** $\dfrac{x^2 + 5x}{6} = 4$ **25.** $\dfrac{10x^2 - 25x}{12} = 5$ **26.** $\dfrac{12x^2 - 4x}{5} = 8$

To Think About

27. Why can an equation in standard form with $c = 0$ (that is, an equation of the form $ax^2 + bx = 0$) always be solved?

28. Martha solved $(x + 3)(x - 2) = 14$ as follows:

$$x + 3 = 14 \quad \text{or} \quad x - 2 = 14$$
$$x = 11 \quad \text{or} \quad x = 16$$

Josette said this had to be wrong because these values do not check. Explain what is wrong with Martha's method.

Applications

▲ **29.** ***Geometry*** The area of a rectangular garden is 140 square meters. The width is 3 meters longer than one-half of the length. Find the length and the width of the garden.

▲ **30.** ***Geometry*** The area of a triangular sign is 33 square meters. The base of the triangle is 1 meter less than double the altitude. Find the altitude and the base of the sign.

Forming Groups *Suppose the number of students in a mathematics class is x. The teacher insists that each student participate in group work each class. The number of possible groups is:*

$$G = \frac{x^2 - 3x + 2}{2}$$

31. The class has 13 students. How many possible groups are there?

32. Three students withdraw from the class. How many fewer groups can be formed?

33. A teacher claims each student could be in 36 different groups. How many students are there?

34. The teacher wants each student to participate in a different group each day. There are 45 class days in the semester. How many students must be in the class?

Falling Object   *Use the following information for exercises 35 and 36. When an object is thrown upward, its height (S), in meters, is given (approximately) by the quadratic equation*

$$S = -5t^2 + vt + h,$$

where $v =$ the upward initial velocity in meters/second,

$t =$ the time of flight in seconds, and

$h =$ the height above level ground from which the object is thrown.

35. Johnny is standing on a platform 6 meters high and throws a ball straight up as high as he can at a velocity of 13 meters/second. At what time t will the ball hit the ground? How far from the ground is the ball after 2 seconds have elapsed from the time of the throw? (Assume that the ball is 6 meters from the ground when it leaves Johnny's hand.)

36. You are standing on the edge of a cliff near Acapulco, overlooking the ocean. The place where you stand is 180 meters from the ocean. You drop a pebble into the water. ("Dropping" the pebble implies that there is no initial velocity, so $v = 0$.) How many seconds will it take to hit the water? How far has the pebble dropped after 3 seconds?

37. *Overtime Costs* Paying overtime to employees can be very expensive, but if profits are realized, it is worth putting a shift on overtime. In a high-tech helicopter company, the extra hourly cost in dollars for producing x additional helicopters is given by the cost equation $C = 2x^2 - 7x$. If the extra hourly cost is $15, how many additional helicopters are produced?

38. *Producing Power* A boat generator on a Gloucester fishing boat is required to produce 64 watts of power. The amount of current I measured in amperes needed to produce the power for this generator is given by the equation $P = 40I - 4I^2$. What is the *minimum* number of amperes required to produce the necessary power?

Internal Phone Calls *The technology and communication office of a local company has set up a new telephone system so that each employee has a separate telephone and extension number. They are studying the possible number of telephone calls that can be made from people in the office to other people in the office. They have discovered that the total number of possible telephone calls T is described by the equation $T = 0.5(x^2 - x)$, where x is the number of people in the office. Use this information to answer exercises 39–42.*

39. If 70 people are presently employed at the office, how many possible telephone calls can be made between these 70 people?

40. If the company hires 10 new employees next year, how many possible telephone calls can be made between the 80 people that will be employed next year?

41. One Saturday, only a small number of employees were working at the office. It has been determined that on that day, a total of 153 different phone calls could have been made from people working in the office to other people working in the office. How many people worked on that Saturday?

42. On the day after Thanksgiving, only a small number of employees were working at the office. It has been determined that on that day, a total of 105 different phone calls could have been made from people working in the office to other people working in the office. How many people worked on the day after Thanksgiving?

Cumulative Review

Simplify.

43. $(2x^2y^3)(-5x^3y)$

44. $(3a^4b^5)(4a^6b^8)$

45. $\dfrac{21a^5b^{10}}{-14ab^{12}}$

46. $\dfrac{18x^3y^6}{54x^8y^{10}}$

Putting Your Skills to Work

The Mathematics of Instant Replay

In the National Football League the outcome of the call of game officials can be challenged and reviewed by looking at a video replay of the play that is being considered. Over the years since 1999 the number of replay reviews during the regular season has been increasing steadily. The number of reversals (where the original decision of game officials has been overturned after careful review) has been increasing also.

Consider the following chart which displays the data for a five-year period.

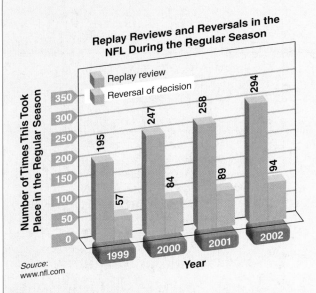

Replay Reviews and Reversals in the NFL During the Regular Season

Source: www.nfl.com

4. Find the percentage of reversals in 2001. Find the percentage of reversals in 2002. By how much did the percentage change? (Round your answer to nearest whole percent.)

Problems for Individual Investigation and Analysis

1. If the same pattern of yearly increase in the number of replay reviews continues from 2002 to 2005 that occurred from 2001 to 2002, how many replay reviews would you expect during the 2005 playing season?

2. If the same pattern of yearly increase in the number of reversals continues from 2002 to 2006 that occurred from 2001 to 2002, how many reversals of an official game decision would you expect during the 2006 playing season?

3. Find the percentage of reversals in 1999. Find the percentage of reversals in 2000. By how much did the percentage change? (Round your answer to nearest whole percent.)

Problems for Group Investigation and Cooperative Learning

5. A possible predictor of the number of replays (y) for a given NFL season is given by the equation $y = 30x + 294$, where x is the number of years after 2002. Use that equation to predict the number of replays in the 2004 season.

6. A possible predictor of the number of reversals (y) for a given NFL season is given by the equation $y = 11x + 99$, where x is the number of years after 2002. Use that equation to predict the number of reversals in the 2005 season.

7. Write the equation in Exercise 5 in factored form.

8. Write the equation in Exercise 6 in factored form.

Topic	Procedure	Examples
A. Common factor, p. 304.	Factor out the largest common factor from each term.	$2x^2 - 2x = 2x(x - 1)$ $3a^2 + 3ab - 12a = 3a(a + b - 4)$ $8x^4y - 24x^3 = 8x^3(xy - 3)$
Special cases **B. Difference of squares,** p. 328. **C. Perfect-square trinomials,** p. 329.	If you recognize the special cases, you will be able to factor quickly. $a^2 - b^2 = (a + b)(a - b)$ $a^2 + 2ab + b^2 = (a + b)^2$ $a^2 - 2ab + b^2 = (a - b)^2$	$25x^2 - 36y^2 = (5x + 6y)(5x - 6y)$ $16x^4 - 1 = (4x^2 + 1)(2x + 1)(2x - 1)$ $25x^2 + 10x + 1 = (5x + 1)^2$ $49x^2 - 42xy + 9y^2 = (7x - 3y)^2$
D. Trinomials of the form $x^2 + bx + c$, p. 314.	Factor trinomials of the form $x^2 + bx + c$ by asking what two numbers have a product of c and a sum of b. If each term of the trinomial has a common factor, factor it out as the first step.	$x^2 - 18x + 77 = (x - 7)(x - 11)$ $x^2 + 7x - 18 = (x + 9)(x - 2)$ $5x^2 - 10x - 40 = 5(x^2 - 2x - 8)$ $\qquad = 5(x - 4)(x + 2)$
E. Trinomials of the form $ax^2 + bx + c$, where $a \neq 1$, p. 321.	Factor trinomials of the form $ax^2 + bx + c$ by the grouping number method or by the trial-and-error method.	$6x^2 + 11x - 10$ Grouping number $= -60$ Two numbers whose product is -60 and whose sum is $+11$ are $+15$ and -4. $6x^2 + 15x - 4x - 10 = 3x(2x + 5) - 2(2x + 5)$ $\qquad = (2x + 5)(3x - 2)$
F. Four terms. Factor by grouping, p. 309.	Rearrange the terms if necessary so that the first two terms have a common factor. Then factor out the common factors. $ax + ay - bx - by = a(x + y) - b(x + y)$ $\qquad = (a - b)(x + y)$	$2ax^2 + 21 + 3a + 14x^2$ $= 2ax^2 + 14x^2 + 3a + 21$ $= 2x^2(a + 7) + 3(a + 7)$ $= (a + 7)(2x^2 + 3)$
Prime polynomials, p. 335.	A polynomial that is not factorable is called prime.	$x^2 + y^2$ is prime. $x^2 + 5x + 7$ is prime.
Multistep factoring, p. 334.	Many problems require two or three steps of factoring. Always try to factor out the greatest common factor as the first step.	$3x^2 - 21x + 36 = 3(x^2 - 7x + 12)$ $\qquad = 3(x - 4)(x - 3)$ $2x^3 - x^2 - 6x = x(2x^2 - x - 6)$ $\qquad = x(2x + 3)(x - 2)$ $25x^3 - 49x = x(25x^2 - 49)$ $\qquad = x(5x + 7)(5x - 7)$ $8x^2 - 24x + 18 = 2(4x^2 - 12x + 9)$ $\qquad = 2(2x - 3)^2$
Solving quadratic equations by factoring, p. 338.	**1.** Write as $ax^2 + bx + c = 0$. **2.** Factor. **3.** Set each factor equal to 0. **4.** Solve the resulting equations.	Solve: $3x^2 + 5x = 2$ $3x^2 + 5x - 2 = 0$ $(3x - 1)(x + 2) = 0$ $3x - 1 = 0 \quad$ or $\quad x + 2 = 0$ $x = \dfrac{1}{3} \quad$ or $\quad x = -2$

Topic	Procedure	Examples
Using quadratic equations to solve applied problems, p. 341.	Some word problems, like those involving the product of two numbers, area, and formulas with a squared variable, can be solved using the factoring methods we have shown.	The length of a rectangle is 4 less than three times the width. Find the length and width if the area is 55 square inches. Let w = width. Then $3w - 4$ = length. $$55 = w(3w - 4)$$ $$55 = 3w^2 - 4w$$ $$0 = 3w^2 - 4w - 55$$ $$0 = (3w + 11)(w - 5)$$ $$w = -\tfrac{11}{3} \quad \text{or} \quad w = 5$$ $-\tfrac{11}{3}$ is not a valid solution. Thus width = 5 inches and length = 11 inches.

Chapter 5 Review Problems

Section 5.1

Factor out the greatest common factor.

1. $12x^3 - 20x^2y$

2. $10x^3 - 35x^3y$

3. $7x^2y - 14xy^2 - 21x^3y^3$

4. $50a^4b^5 - 25a^4b^4 + 75a^5b^5$

5. $3a^3 + 6a^2 - 9ab + 12a$

6. $2x - 4y + 6z + 12$

7. $2a(a + 3b) - 5(a + 3b)$

8. $15x^3y + 6xy^2 + 3xy$

Section 5.2

Factor by grouping.

9. $3ax - 7a - 6x + 14$

10. $a^2 + 5ab - 4a - 20b$

11. $x^2y + 3y - 2x^2 - 6$

12. $30ax - 15ay + 42x - 21y$

13. $15x^2 - 3x + 10x - 2$

14. $30w^2 - 18w + 5wz - 3z$

Section 5.3

Factor completely. Be sure to factor out any common factors as your first step.

15. $x^2 - 2x - 35$

16. $x^2 - 10x + 24$

17. $x^2 + 14x + 48$

18. $x^2 + 8xy + 15y^2$

19. $x^4 + 13x^2 + 42$

20. $x^4 - 2x^2 - 35$

21. $6x^2 + 30x + 36$

22. $3x^2 + 39x + 36$

23. $2x^2 - 28x + 96$

24. $4x^2 - 44x + 120$

Section 5.4

Factor completely. Be sure to factor out any common factors as your first step.

25. $4x^2 + 7x - 15$

26. $12x^2 + 11x - 5$

27. $15x^2 + 7x - 4$

28. $6x^2 - 13x + 6$

29. $2x^2 - x - 3$

30. $3x^2 + 2x - 8$

31. $20x^2 + 48x - 5$

32. $20x^2 + 21x - 5$

33. $6a^2 + 11a - 10$

34. $6a^2 - 19a + 10$

35. $6x^2 + 4x - 10$

36. $6x^2 - 4x - 10$

37. $4x^2 - 26x + 30$

38. $4x^2 - 20x - 144$

39. $12x^2 - 22x - 20$

40. $18x^2 + 51x - 42$

41. $6x^2 - 19xy + 10y^2$

42. $6x^2 - 32xy + 10y^2$

Section 5.5

Factor these special cases. Be sure to factor out any common factors.

43. $49x^2 - y^2$

44. $16x^2 - 36y^2$

45. $25x^2 + 30x + 9$

46. $49x^2 - 28x + 4$

47. $25x^2 - 36$

48. $100x^2 - 9$

49. $y^2 - 36x^2$

50. $9y^2 - 25x^2$

51. $36x^2 + 12x + 1$

52. $25x^2 - 20x + 4$

53. $16x^2 - 24xy + 9y^2$

54. $49x^2 - 28xy + 4y^2$

55. $2x^2 - 18$

56. $3x^2 - 75$

57. $28x^2 + 140x + 175$

58. $72x^2 - 192x + 128$

Section 5.6

If possible, factor each polynomial completely. If a polynomial cannot be factored, state that it is prime.

59. $4x^2 - 9y^2$

60. $x^2 + 6x + 9$

61. $x^2 - 9x + 18$

62. $x^2 + 13x - 30$

63. $6x^2 + x - 7$

64. $10x^2 + x - 2$

65. $24x - 60$

66. $8x^2y^2 - 4xy$

67. $50x^3y^2 + 30x^2y^2 - 10x^2y^2$

68. $26a^3b - 13ab^3 + 52a^2b^4$

69. $x^3 - 16x^2 + 64x$

70. $2x^2 + 40x + 200$

71. $3x^2 - 18x + 27$

72. $25x^3 - 60x^2 + 36x$

73. $7x^2 - 9x - 10$

74. $4x^2 - 13x - 12$

75. $9x^3y - 4xy^3$

76. $3x^3a^3 - 11x^4a^2 - 20x^5a$

77. $12a^2 + 14ab - 10b^2$

78. $121a^2 + 66ab + 9b^2$

79. $7a - 7 - ab + b$

Mixed Practice

If possible, factor each polynomial completely. If a polynomial cannot be factored, state that it is prime.

80. $8b - 10 + 28bc - 35c$

81. $18b - 21 + 14x - 12bx$

82. $5xb - 35x + 4by - 28y$

83. $2a^2x - 15ax + 7x$

84. $x^5 - 17x^3 + 16x$

85. $x^4 - 81y^{12}$

86. $6x^4 - x^2 - 15$

87. $28yz - 16xyz + x^2yz$

88. $12x^3 + 17x^2 + 6x$

89. $16w^2 - 2w - 5$

90. $12w^2 - 12w + 3$

91. $4y^3 + 10y^2 - 6y$

92. $10y^2 + 33y - 7$

93. $8y^{10} - 16y^8$

94. $9x^4 - 144$

95. $x^2 - 6x + 12$

96. $8x^2 - 19x - 6$

97. $8y^5 + 4y^3 - 60y$

98. $9xy^2 + 3xy - 42x$

99. $16x^4y^2 - 56x^2y + 49$

100. $128x^3y - 2xy$

101. $2ax + 5a - 10b - 4bx$

102. $2x^3 - 9 + x^2 - 18x$

Section 5.7

Solve the following equations by factoring.

103. $x^2 - 3x - 18 = 0$

104. $2x^2 + 5x - 12 = 0$

105. $5x^2 = 2x - 7x^2$

106. $5x^2 - x = 4x^2 + 12$

107. $2x^2 + 9x - 5 = 0$

108. $x^2 + 11x + 24 = 0$

109. $x^2 + 14x + 45 = 0$

110. $5x^2 = 7x + 6$

111. $3x^2 + 6x = 2x^2 - 9$

112. $4x^2 + 9x - 9 = 0$

113. $5x^2 - 11x + 2 = 0$

Solve.

▲ **114.** *Geometry* The area of a triangle is 35 square centimeters. The base is 3 centimeters longer than the altitude of the triangle. Find the length of the base and the altitude.

▲ **115.** *Geometry* The area of a rectangle is 105 square feet. The length of the rectangle is 1 foot longer than double the width. Find the length and width of the rectangle.

116. *Rocket Height* The height in feet that a model rocket attains is given by $h = -16t^2 + 80t + 96$, where *t* is the time measured in seconds. How many seconds will it take until the rocket finally reaches the ground? (*Hint:* At ground level $h = 0$.)

117. *Output Power* An electronic technician is working with a 100-volt electric generator. The output power of the generator is given by the equation $p = -5x^2 + 100x$, where *x* is the amount of current measured in amperes and *p* is measured in watts. The technician wants to find the value for *x* when the power is 480 watts. Can you find the two answers?

Remember to use your Chapter Test Prep Video CD to see the worked-out solutions to the test problems you want to review.

If possible, factor each polynomial completely. If a polynomial cannot be factored, state that it is prime.

1. $x^2 + 12x - 28$

2. $16x^2 - 81$

3. $10x^2 + 27x + 5$

4. $9a^2 - 30a + 25$

5. $7x - 9x^2 + 14xy$

6. $10xy + 15by - 8x - 12b$

7. $6x^3 - 20x^2 + 16x$

8. $5a^2c - 11ac + 2c$

9. $81x^2 - 100$

10. $9x^2 - 15x + 4$

11. $20x^2 - 45$

12. $36x^2 + 1$

13. $3x^3 + 11x^2 + 10x$

14. $60xy^2 - 20x^2y - 45y^3$

15. $81x^2 - 1$

16. $81y^4 - 1$

17. $2ax + 6a - 5x - 15$

18. $aw^2 - 8b + 2bw^2 - 4a$

19. $3x^2 - 3x - 90$

20. $2x^3 - x^2 - 15x$

Solve.

21. $x^2 + 14x + 45 = 0$

22. $14 + 3x(x + 2) = -7x$

23. $2x^2 + x - 10 = 0$

24. $x^2 - 3x - 28 = 0$

Solve using a quadratic equation.

▲ **25.** The park service is studying a rectangular piece of land that has an area of 91 square miles. The length of this piece of land is 1 mile shorter than double the width. Find the length and width of this rectangular piece of land.

1. _____

2. _____

3. _____

4. _____

5. _____

6. _____

7. _____

8. _____

9. _____

10. _____

11. _____

12. _____

13. _____

14. _____

15. _____

16. _____

17. _____

18. _____

19. _____

20. _____

21. _____

22. _____

23. _____

24. _____

25. _____

Cumulative Test for Chapters 0–5

1.
2.
3.
4.
5.
6.
7.
8.
9.
10.
11.
12.
13.
14.
15.
16.
17.
18.
19.
20.
21.
22.
23.
24.

Approximately one-half of this test covers the content of Chapters 0–4. The remainder covers the content of Chapter 5.

1. What % of 480 is 72?

2. What is 13% of 3.8?

Simplify.

3. $4.3 - 2(3.4 - 7.2) - 6.1$

4. $(-2x^3y^4)(-4xy^6)$

5. $(-3)^4$

6. $(9x - 4)(3x + 2)$

7. $(2x^2 - 6x + 1)(x - 3)$

Solve.

8. $3x - 4 \geq 6x + 5$

9. $3x - (7 - 5x) = 3(4x - 5)$

10. $\dfrac{1}{2}x - 3 = \dfrac{1}{4}(3x + 3)$

11. Solve for q: $3(q + 1) = 8q - p$

Factor each polynomial completely. If a polynomial cannot be factored, state that it is prime.

12. $6x^2 - 5x + 1$

13. $6x^2 + 5x - 4$

14. $9x^2 + 3x - 2$

15. $121x^2 - 64y^2$

16. $4x + 120 - 80x^2$

17. $x^2 + 5x + 9$

18. $16x^3 + 40x^2 + 25x$

19. $81x^4 - 16b^4$

20. $2ax - 4bx + 3a - 6b$

21. $15x^2 + 6x - 9$

Solve.

22. $x^2 + 5x - 24 = 0$

23. $3x^2 - 11x + 10 = 0$

Solve using a quadratic equation.

▲ **24.** Mr. Jerome's garden covered 300 square feet. If the width was 5 feet less than the length, find the measurements of the garden.

Imagine you and a friend have begun a business venture. You have decided to open a calculator store in a shopping mall. The profits are good, but they seem to vary greatly depending on how much advertising you do for the business. Do you think you could use your mathematics to help predict your monthly profits? Turn to the Putting Your Skills to Work problems on page 395 to find out.

Rational Expressions and Equations

6.1 SIMPLIFYING RATIONAL EXPRESSIONS

Student Learning Objective

After studying this section, you will be able to:

 Simplify rational expressions by factoring.

Recall that a rational number is a number that can be written as one integer divided by another integer, such as $3 \div 4$ or $\frac{3}{4}$. We usually use the word *fraction* to mean $\frac{3}{4}$. We can extend this idea to algebraic expressions. A **rational expression** is a polynomial divided by another polynomial, such as

$$(3x + 2) \div (x + 4) \quad \text{or} \quad \frac{3x + 2}{x + 4}.$$

The last fraction is sometimes also called a **fractional algebraic expression.** There is a special restriction for all fractions, including fractional algebraic expressions. The denominator of the fraction cannot be 0. For example, in the expression

$$\frac{3x + 2}{x + 4},$$

the denominator cannot be 0. Therefore, the value of x cannot be -4. The following important restriction will apply throughout this chapter. We state it here to avoid having to mention it repeatedly throughout this chapter.

RESTRICTION

The denominator of a rational expression cannot be zero. Any value of the variable that would make the denominator zero is not allowed.

We have discovered that fractions can be simplified (or reduced) in the following way.

$$\frac{15}{25} = \frac{3 \cdot \cancel{5}}{5 \cdot \cancel{5}} = \frac{3}{5}$$

This is sometimes referred to as the **basic rule of fractions** and can be stated as follows.

BASIC RULE OF FRACTIONS

For any rational expression $\frac{a}{b}$ and any polynomial a, b, and c (where $b \neq 0$ and $c \neq 0$),

$$\frac{ac}{bc} = \frac{a}{b}.$$

We will examine several examples where a, b, and c are real numbers, as well as more involved examples where a, b, and c are polynomials. In either case we shall make extensive use of our factoring skills in this section.

One essential property is revealed by the basic rule of fractions: If the numerator and denominator of a given fraction are multiplied by the same nonzero quantity, an equivalent fraction is obtained. The rule can be used two ways. You can start with $\frac{ac}{bc}$ and end with the equivalent fraction $\frac{a}{b}$. Or, you can start with $\frac{a}{b}$ and end with the equivalent fraction $\frac{ac}{bc}$. In this section we focus on the first method.

EXAMPLE 1 Reduce $\frac{21}{39}$.

Solution $\frac{21}{39} = \frac{7 \cdot \cancel{3}}{13 \cdot \cancel{3}} = \frac{7}{13}$ Use the rule $\frac{ac}{bc} = \frac{a}{b}$. Let $c = 3$ because 3 is the greatest common factor of 21 and 39.

Practice Problem 1 Reduce $\frac{28}{63}$.

NOTE TO STUDENT: Fully worked-out solutions to all of the Practice Problems can be found at the back of the text starting at page SP-1

Simplifying Rational Expressions by Factoring

The process of reducing the fraction shown previously is sometimes called *dividing out* common factors. Remember, only **factors** of both the numerator and the denominator can be divided out. To apply the basic rule of fractions, it is usually necessary that the numerator and denominator of the fraction be completely factored. You will need to use your factoring skills from Chapter 5 to accomplish this step. When you apply this rule, you are **simplifying the fraction.**

EXAMPLE 2 Simplify. $\dfrac{4x + 12}{5x + 15}$

Solution $\dfrac{4x + 12}{5x + 15} = \dfrac{4(x + 3)}{5(x + 3)}$ Factor 4 from the numerator.

$\qquad\qquad\qquad\qquad$ Factor 5 from the denominator.

$\qquad\qquad = \dfrac{4\,\cancel{(x + 3)}}{5\,\cancel{(x + 3)}}$ Apply the basic rule of fractions.

$\qquad\qquad = \dfrac{4}{5}$

Practice Problem 2 Simplify. $\dfrac{12x - 6}{14x - 7}$

EXAMPLE 3 Simplify. $\dfrac{x^2 + 9x + 14}{x^2 - 4}$

Solution $= \dfrac{(x + 7)(x + 2)}{(x - 2)(x + 2)}$ Factor the numerator.

$\qquad\qquad\qquad\qquad$ Factor the denominator.

$\qquad = \dfrac{(x + 7)\,\cancel{(x + 2)}}{(x - 2)\,\cancel{(x + 2)}}$ Apply the basic rule of fractions.

$\qquad = \dfrac{x + 7}{x - 2}$

CAUTION Do not try to remove terms that are added. In Example 3, do not try to remove the x^2 in the top and the x^2 in the bottom of the fraction. The basic rule of fractions only applies to quantities that are factors of both numerator and denominator.

Practice Problem 3 Simplify. $\dfrac{4x^2 - 9}{2x^2 - x - 3}$

Some problems may involve more than one step of factoring. Always remember to factor out any common factors as the first step, if it is possible to do so.

EXAMPLE 4 Simplify. $\dfrac{x^3 - 9x}{x^3 + x^2 - 6x}$

Solution $= \dfrac{x(x^2 - 9)}{x(x^2 + x - 6)}$ Factor out a common factor from the polynomials in the numerator and the denominator.

$\qquad = \dfrac{\cancel{x}\,\cancel{(x + 3)}\,(x - 3)}{\cancel{x}\,\cancel{(3 + x)}\,(x - 2)}$ Factor each polynomial and apply the basic rule of fractions.

$\qquad = \dfrac{x - 3}{x - 2}$

Practice Problem 4 Simplify. $\dfrac{x^3 - 16x}{x^3 - 2x^2 - 8x}$

When you are simplifying, be on the lookout for the special situation where *a factor in the denominator is the opposite of a factor in the numerator*. In such a case you should factor a negative number from one of the factors so that it becomes equivalent to the other factor and it can be divided out. Look carefully at the following two examples.

EXAMPLE 5 Simplify. $\dfrac{5x - 15}{6 - 2x}$

Solution Notice that the variable term in the numerator, $5x$, and the variable term in the denominator, $-2x$, *are opposite in sign*. Likewise, the numerical terms -15 and 6 *are opposite in sign*. Factor out a negative number from the denominator.

$$\frac{5x - 15}{6 - 2x} = \frac{5(x - 3)}{-2(-3 + x)}$$

Factor 5 from the numerator.

Factor -2 from the denominator.

Note that $(x - 3)$ and $(-3 + x)$ are equivalent since $(+x - 3) = (-3 + x)$.

$$= \frac{5\cancel{(x - 3)}}{-2\cancel{(-3 + x)}}$$

Apply the basic rule of fractions.

$$= -\frac{5}{2}$$

Note that $\frac{5}{-2}$ is not considered to be in simple form. We usually avoid leaving a negative number in the denominator. Therefore, to simplify, give the result as $-\frac{5}{2}$ or $\frac{-5}{2}$.

NOTE TO STUDENT: *Fully worked-out solutions to all of the Practice Problems can be found at the back of the text starting at page SP-1*

Practice Problem 5 Simplify. $\dfrac{2x - 5}{5 - 2x}$

EXAMPLE 6 Simplify. $\dfrac{2x^2 - 11x + 12}{16 - x^2}$

Solution

$$= \frac{(x - 4)(2x - 3)}{(4 - x)(4 + x)}$$

Factor the numerator and the denominator.

Observe that $(x - 4)$ and $(4 - x)$ are opposites.

$$= \frac{(x - 4)(2x - 3)}{-1(-4 + x)(4 + x)}$$

Factor -1 out of $(+4 - x)$ to obtain $-1(-4 + x)$.

$$= \frac{\cancel{(x - 4)}(2x - 3)}{-1\cancel{(-4 + x)}(4 + x)}$$

Apply the basic rule of fractions since $(x - 4)$ and $(-4 + x)$ are equivalent.

$$= \frac{2x - 3}{-1(4 + x)}$$

$$= -\frac{2x - 3}{4 + x}$$

Practice Problem 6 Simplify. $\dfrac{4x^2 + 3x - 10}{25 - 16x^2}$

After doing Examples 5 and 6, you will notice a pattern. Whenever the factor in the numerator and the factor in the denominator are opposites, the value -1 results. We could actually make this a definition of that property.

For all monomials A and B where $A \neq B$, it is true that

$$\frac{A - B}{B - A} = -1.$$

You may use this definition in reducing fractions if it is helpful to you. Otherwise, you may use the factoring method discussed in Examples 5 and 6.

Some problems will involve two or more variables. In such cases, you will need to factor carefully and make sure that each set of parentheses contains the correct letters.

EXAMPLE 7 Simplify. $\dfrac{x^2 - 7xy + 12y^2}{2x^2 - 7xy - 4y^2}$

Solution

$$= \frac{(x - 4y)(x - 3y)}{(2x + y)(x - 4y)} \qquad \text{Factor the numerator.}$$
$$\text{Factor the denominator.}$$

$$= \frac{\cancel{(x - 4y)}(x - 3y)}{(2x + y)\cancel{(x - 4y)}} \qquad \text{Apply the basic rule of fractions.}$$

$$= \frac{x - 3y}{2x + y}$$

Practice Problem 7 Simplify. $\dfrac{4x^2 - 9y^2}{4x^2 + 12xy + 9y^2}$

EXAMPLE 8 Simplify. $\dfrac{6a^2 + ab - 7b^2}{36a^2 - 49b^2}$

Solution

$$= \frac{(6a + 7b)(a - b)}{(6a + 7b)(6a - 7b)} \qquad \text{Factor the numerator.}$$
$$\text{Factor the denominator.}$$

$$= \frac{\cancel{(6a + 7b)}(a - b)}{\cancel{(6a + 7b)}(6a - 7b)} \qquad \text{Apply the basic rule of fractions.}$$

$$= \frac{a - b}{6a - 7b}$$

Practice Problem 8 Simplify. $\dfrac{25a^2 - 16b^2}{10a^2 + 3ab - 4b^2}$

Simplify.

1. $\dfrac{3a - 9b}{a - 3b}$

2. $\dfrac{5x + 2y}{35x + 14y}$

3. $\dfrac{6x + 18}{x^2 + 3x}$

4. $\dfrac{8x - 12}{2x^3 - 3x^2}$

5. $\dfrac{9x^2 + 6x + 1}{1 - 9x^2}$

6. $\dfrac{4x^2 + 4x + 1}{1 - 4x^2}$

7. $\dfrac{3a^2b(a - 2b)}{6ab^2}$

8. $\dfrac{9ab^2}{6a^2b^2(b + 3a)}$

9. $\dfrac{x^2 + x - 2}{x^2 - x}$

10. $\dfrac{x^2 + x - 12}{x^2 - 3x}$

11. $\dfrac{x^2 - 3x - 10}{3x^2 + 5x - 2}$

12. $\dfrac{4x^2 - 10x + 6}{2x^2 + x - 3}$

13. $\dfrac{x^2 + 4x - 21}{x^3 - 49x}$

14. $\dfrac{x^3 - 3x^2 - 40x}{x^2 + 10x + 25}$

15. $\dfrac{3x^2 + 7x - 6}{x^2 + 7x + 12}$

16. $\dfrac{x^2 - 5x - 14}{2x^2 - x - 10}$

17. $\dfrac{3x^2 - 8x + 5}{4x^2 - 5x + 1}$

18. $\dfrac{3y^2 + 10y + 3}{3y^2 - 14y - 5}$

19. $\dfrac{5x^2 - 27x + 10}{5x^2 + 3x - 2}$

20. $\dfrac{2x^2 + 2x - 12}{x^2 + 3x - 4}$

Mixed Practice

Take some time to review exercises 1–20 before you proceed with exercises 21–34.

21. $\dfrac{10 - 2x}{4x^2 - 20x}$

22. $\dfrac{6 - 2ab}{ab^2 - 3b}$

23. $\dfrac{2x^2 - 7x - 15}{25 - x^2}$

24. $\dfrac{49 - x^2}{2x^2 - 9x - 35}$

25. $\dfrac{(4x + 5)^2}{8x^2 + 6x - 5}$

26. $\dfrac{6x^2 - 13x - 8}{(3x - 8)^2}$

27. $\dfrac{2x^2 + 9x - 18}{30 - x - x^2}$

28. $\dfrac{4y^2 - y - 3}{8 - 7y - y^2}$

29. $\dfrac{a^2 + 2ab - 3b^2}{2a^2 + 5ab - 3b^2}$

30. $\dfrac{a^2 + 3ab - 10b^2}{3a^2 - 7ab + 2b^2}$

31. $\dfrac{9x^2 - 4y^2}{9x^2 + 12xy + 4y^2}$

32. $\dfrac{16x^2 - 24xy + 9y^2}{16x^2 - 9y^2}$

To Think About

Be sure to remove a common factor first when solving exercises 33 and 34.

33. $\dfrac{x^3 - bx^2 - abx + ax^2}{bx^2 - b^2x - b^2c + bcx}$

34. $\dfrac{bxy + bx^2 - axy - ax^2}{ay^2 + axy + 2by^2 + 2bxy}$

Cumulative Review

Multiply.

35. Multiply. $(3x - 7)^2$

36. Multiply: $(7x + 6y)(7x - 6y)$

37. $(2x + 3)(x - 4)(x - 2)$

38. $(x^2 - 3x - 5)(2x^2 + x + 3)$

39. *Dividing Acreage* Walter and Ann Perkins wish to divide $4\frac{7}{8}$ acres of farmland into three equal-sized house lots. What will be the acreage of each lot?

40. *Roasting a Turkey* Ron and Mary Larson are planning to cook a $17\frac{1}{2}$-pound turkey. The directions suggest a cooking time of 22 minutes per pound for turkeys that weigh between 16 and 20 pounds. How many hours and minutes should they allow for an approximate cooking time?

Languages *The number of people on our planet who speak the Chinese Mandarin language is 158,000,000 more than twice the number of people who speak Spanish. The number of people on Earth who speak English is 17,000,000 less than the number of those who speak Spanish. There are 341,000,000 people who speak English on our planet. (Source: United Nations Statistical Division)*

41. How many people speak Chinese Mandarin?

42. How many people speak Spanish?

6.2 MULTIPLYING AND DIVIDING RATIONAL EXPRESSIONS

Student Learning Objectives

After studying this section, you will be able to:

 1 Multiply rational expressions.

 2 Divide rational expressions.

1 Multiplying Rational Expressions

To multiply two rational expressions, we multiply the numerators and multiply the denominators. As before, the denominators cannot equal zero.

> For any two rational expressions $\frac{a}{b}$ and $\frac{c}{d}$ where $b \neq 0$ and $d \neq 0$,
> $$\frac{a}{b} \cdot \frac{c}{d} = \frac{ac}{bd}.$$

Simplifying or reducing fractions *prior to multiplying them* usually makes the computations easier to do. Leaving the reducing step until the end makes the simplifying process longer and increases the chance for error. This long approach should be avoided.

As an example, let's do the same problem two ways to see which one is easier. Let's simplify the following problem by multiplying first and then reducing the result.

$$\frac{5}{7} \times \frac{49}{125}$$

$$\frac{5}{7} \times \frac{49}{125} = \frac{245}{875} \quad \text{Multiply the numerators and multiply the denominators.}$$

$$= \frac{7}{25} \quad \text{Reduce the fraction. (}\textit{Note:}\text{ It takes a bit of trial and error to discover how to reduce it.)}$$

Compare this with the following method, where we reduce the fractions prior to multiplying them.

$$\frac{5}{7} \times \frac{49}{125}$$

$$\frac{5}{7} \times \frac{7 \cdot 7}{5 \cdot 5 \cdot 5} \quad \text{\textbf{Step 1.} It is easier to factor first. We factor the numerator and denominator of the second fraction.}$$

$$\frac{5 \cdot 7 \cdot 7}{7 \cdot 5 \cdot 5 \cdot 5} \quad \text{\textbf{Step 2.} We express the product as one fraction (by the definition of multiplication of fractions).}$$

$$\frac{\cancel{5} \cdot \cancel{7} \cdot 7}{\cancel{7} \cdot \cancel{5} \cdot 5 \cdot 5} = \frac{7}{25} \quad \text{\textbf{Step 3.} Then we apply the basic rule of fractions to divide the common factors of 5 and 7 that appear in the numerator and in the denominator.}$$

A similar approach can be used with the multiplication of rational expressions. We first factor the numerator and denominator of each fraction. Then we divide out any factor that is common to a numerator and a denominator. Finally, we multiply the remaining numerators and the remaining denominators.

EXAMPLE 1 Multiply. $\dfrac{x^2 - x - 12}{x^2 - 16} \cdot \dfrac{2x^2 + 7x - 4}{x^2 - 4x - 21}$

Solution

$$\frac{(x - 4)(x + 3)}{(x - 4)(x + 4)} \cdot \frac{(x + 4)(2x - 1)}{(x + 3)(x - 7)} \qquad \text{Factoring is always the first step.}$$

$$= \frac{\cancel{(x - 4)}\cancel{(x + 3)}}{\cancel{(x - 4)}\cancel{(x + 4)}} \cdot \frac{\cancel{(x + 4)}(2x - 1)}{\cancel{(x + 3)}(x - 7)} \qquad \begin{array}{l}\text{Apply the basic rule of fractions.}\\ \text{(Three pairs of factors divide out.)}\end{array}$$

$$= \frac{2x - 1}{x - 7} \qquad \text{The final answer.}$$

NOTE TO STUDENT: Fully worked-out solutions to all of the Practice Problems can be found at the back of the text starting at page SP-1

Practice Problem 1 Multiply. $\dfrac{6x^2 + 7x + 2}{x^2 - 7x + 10} \cdot \dfrac{x^2 + 3x - 10}{2x^2 + 11x + 5}$

In some cases, a given numerator can be factored more than once. You should always check for a *common factor* as your first step.

EXAMPLE 2   Multiply. $\dfrac{x^4 - 16}{x^3 + 4x} \cdot \dfrac{2x^2 - 8x}{4x^2 + 2x - 12}$

Solution

$= \dfrac{(x^2 + 4)(x^2 - 4)}{x(x^2 + 4)} \cdot \dfrac{2x(x - 4)}{2(2x^2 + x - 6)}$
Factor each numerator and denominator. Factoring out the common factor first is very important.

$= \dfrac{(x^2 + 4)(x + 2)(x - 2)}{x(x^2 + 4)} \cdot \dfrac{2x(x - 4)}{2(x + 2)(2x - 3)}$
Factor again where possible.

$= \dfrac{\cancel{(x^2 + 4)}\cancel{(x + 2)}(x - 2)}{\cancel{x(x^2 + 4)}} \cdot \dfrac{\cancel{2x}(x - 4)}{\cancel{2}\cancel{(x + 2)}(2x - 3)}$
Divide out factors that appear in both the numerator and the denominator. (There are four such pairs of factors.)

$= \dfrac{(x - 2)(x - 4)}{(2x - 3)}$ or $\dfrac{x^2 - 6x + 8}{2x - 3}$
Write the answer as one fraction. (Usually, if there is more than one factor in a numerator, the answer is left in factored form.)

Practice Problem 2   Multiply. $\dfrac{2y^2 - 6y - 8}{y^2 - y - 2} \cdot \dfrac{y^2 - 5y + 6}{2y^2 - 32}.$

Dividing Rational Expressions

For any two fractions $\frac{a}{b}$ and $\frac{c}{d}$, the operation of division can be performed by inverting the second fraction and multiplying it by the first fraction. When we invert a fraction, we are finding its *reciprocal*. Two numbers are **reciprocals** of each other if their product is 1. The reciprocal of $\frac{3}{5}$ is $\frac{5}{3}$. The reciprocal of 7 is $\frac{1}{7}$. The reciprocal of $\frac{a}{b}$ is $\frac{b}{a}$. Sometimes people state the rule for dividing fractions this way: "To divide two fractions, keep the first fraction unchanged and multiply by the reciprocal of the second fraction."

The definition for division of fractions is

$$\frac{a}{b} \div \frac{c}{d} = \frac{a}{b} \cdot \frac{d}{c}.$$

This property holds whether a, b, c, and d are polynomials or numerical values. (It is assumed, of course, that no denominator is zero.)

In the first step for dividing two rational expressions, invert the second fraction and rewrite the quotient as a product. Then follow the procedure for multiplying rational expressions.

EXAMPLE 3   Divide. $\dfrac{6x + 12y}{2x - 6y} \div \dfrac{9x^2 - 36y^2}{4x^2 - 36y^2}$

Solution

$= \dfrac{6x + 12y}{2x - 6y} \cdot \dfrac{4x^2 - 36y^2}{9x^2 - 36y^2}$
Invert the second fraction and write the problem as the product of two fractions.

$= \dfrac{6(x + 2y)}{2(x - 3y)} \cdot \dfrac{4(x^2 - 9y^2)}{9(x^2 - 4y^2)}$
Factor each numerator and denominator.

$$= \frac{(3)(2)(x + 2y)}{2(x - 3y)} \cdot \frac{(2)(2)(x + 3y)(x - 3y)}{(3)(3)(x + 2y)(x - 2y)}$$

Factor again where possible.

$$= \frac{\cancel{(3)}\cancel{(2)}\cancel{(x + 2y)}}{\cancel{2}\cancel{(x - 3y)}} \cdot \frac{(2)(2)(x + 3y)\cancel{(x - 3y)}}{\cancel{(3)}(3)\cancel{(x + 2y)}(x - 2y)}$$

Divide out factors that appear in both numerator and denominator.

$$= \frac{(2)(2)(x + 3y)}{3(x - 2y)}$$

Write the result as one fraction.

$$= \frac{4(x + 3y)}{3(x - 2y)}$$

Simplify. (Usually, answers are left in this form.)

Although it is correct to write this answer as $\dfrac{4x + 12y}{3x - 6y}$, it is customary to leave the answer in factored form to ensure that the final answer is simplified.

NOTE TO STUDENT: Fully worked-out solutions to all of the Practice Problems can be found at the back of the text starting at page SP-1

Practice Problem 3   Divide. $\dfrac{x^2 + 5x + 6}{x^2 + 8x} \div \dfrac{2x^2 + 5x + 2}{2x^2 + x}$

A polynomial that is not in fraction form can be written as a fraction if you give it a denominator of 1.

EXAMPLE 4   Divide. $\dfrac{15 - 3x}{x + 6} \div (x^2 - 9x + 20)$

Solution   Note that $x^2 - 9x + 20$ can be written as $\dfrac{x^2 - 9x + 20}{1}$.

$$= \frac{15 - 3x}{x + 6} \cdot \frac{1}{x^2 - 9x + 20}$$

Invert and multiply.

$$= \frac{-3(-5 + x)}{x + 6} \cdot \frac{1}{(x - 5)(x - 4)}$$

Factor where possible. Note that we had to factor -3 from the first numerator so that it would contain a factor in common with the second denominator.

$$= \frac{-3\cancel{(-5 + x)}}{x + 6} \cdot \frac{1}{\cancel{(x - 5)}(x - 4)}$$

Divide out the common factor. $(-5 + x)$ is equivalent to $(x - 5)$.

$$= \frac{-3}{(x + 6)(x - 4)}$$

The final answer. Note that the answer can be written in several equivalent forms.

or $-\dfrac{3}{(x + 6)(x - 4)}$ or $\dfrac{3}{(x + 6)(4 - x)}$

Practice Problem 4   Divide. $\dfrac{x + 3}{x - 3} \div (9 - x^2)$

CAUTION: It is logical to assume that the problems included in Section 6.2 have at least one common factor that can be divided out. Therefore if, after factoring, you do not observe any common factors, you should be somewhat suspicious. In such cases, it would be wise to double check your factoring steps to see if an error has been made.

Verbal and Writing Skills

1. Before multiplying rational expressions, we should always first try to

_____

_____ .

2. Division of two rational expressions is done by keeping the first fraction unchanged and then

_____ .

Multiply.

3. $\dfrac{2x - 10}{x - 4} \cdot \dfrac{x^2 + 5x + 4}{x^2 - 4x - 5}$

4. $\dfrac{7x + 7}{x + 4} \cdot \dfrac{x^2 - x - 20}{7x^2 - 42x - 49}$

5. $\dfrac{24x^3}{4x^2 - 36} \cdot \dfrac{2x^2 + 6x}{16x^2}$

6. $\dfrac{14x^2}{10x + 50} \cdot \dfrac{5x^3 - 125x}{35x^4}$

7. $\dfrac{x^2 + 3x - 10}{x^2 + x - 20} \cdot \dfrac{x^2 - 3x - 4}{x^2 + 4x + 3}$

8. $\dfrac{x^2 - x - 20}{x^2 - 3x - 10} \cdot \dfrac{x^2 + 7x + 10}{x^2 + 4x - 5}$

Divide.

9. $\dfrac{x + 6}{x - 8} \div \dfrac{x + 5}{x^2 - 6x - 16}$

10. $\dfrac{x - 4}{x + 9} \div \dfrac{x - 7}{x^2 + 13x + 36}$

11. $(6x - 5) \div \dfrac{36x^2 - 25}{6x^2 + 17x + 10}$

12. $\dfrac{4x^2 - 9}{4x^2 + 12x + 9} \div (6x - 9)$

13. $\dfrac{3x^2 + 12xy + 12y^2}{x^2 + 4xy + 3y^2} \div \dfrac{4x + 8y}{x + y}$

14. $\dfrac{5x^2 + 10xy + 5y^2}{x^2 + 5xy + 6y^2} \div \dfrac{3x + 3y}{x + 2y}$

Mixed Practice

Perform the operation indicated.

15. $\dfrac{(x + 5)^2}{3x^2 - 7x + 2} \cdot \dfrac{x^2 - 4x + 4}{x + 5}$

16. $\dfrac{3x^2 - 10x - 8}{(4x + 5)^2} \cdot \dfrac{4x + 5}{(x - 4)^2}$

17. $\dfrac{x^2 + x - 30}{10 - 2x} \div \dfrac{x^2 + 4x - 12}{5x + 15}$

18. $\dfrac{x^2 + 3x - 28}{x^2 + 14x + 49} \div \dfrac{12 - 3x^2}{x^2 + 5x - 14}$

19. $\dfrac{y^2 + 4y - 12}{y^2 + 2y - 24} \cdot \dfrac{y^2 - 16}{y^2 + 2y - 8}$

20. $\dfrac{5y^2 + 17y + 6}{10y^2 + 9y + 2} \cdot \dfrac{4y^2 - 1}{2y^2 + 5y - 3}$

21. $\dfrac{x^2 + 7x + 12}{2x^2 + 9x + 4} \div \dfrac{x^2 + 6x + 9}{2x^2 - x - 1}$

22. $\dfrac{x^2 + 8x + 15}{2x^2 + 11x + 5} \div \dfrac{x^2 + 6x + 9}{2x^2 - 7x - 4}$

To Think About

23. Consider the problem $\dfrac{x + 5}{x - 2} \div \dfrac{x + 7}{x - 6}$. Explain why 2, −7, and 6 are not allowable replacements for the variable x.

24. Consider the problem $\dfrac{x - 8}{x + 5} \div \dfrac{x - 9}{x + 4}$. Explain why −5, 9, and −4 are not allowable replacements for the variable x.

Cumulative Review

25. Solve. $6x^2 + 3x - 18 = 5x - 2 + 6x^2$

26. Multiply. $(7x^2 - x - 1)(x - 3)$

27. *Correct Dosage* Suzanne Hartling is a nurse at Meadowbrook Rehabilitation Center. She must give an 80-kilogram patient a medication based on his weight. The dosage is $\frac{1}{16}$ milligram of medication for each kilogram of patient weight. Find the amount of medication she should give this patient.

▲ **28.** *Golden Gate Bridge* The Golden Gate Bridge has a total length (including approaches) of 8981 feet and a road width of 90 feet. The width of the sidewalk is 10.5 feet. (The sidewalk spans the entire length of the bridge.) Assume it would cost $x per square foot to resurface the road or the sidewalk. Write an expression for how much more it would cost to resurface the road than the sidewalk.

▲ **29.** *Garden Design* Harold Rafton planted a square garden bed. George Avis also planted a garden that was 2 feet less in width, but 3 feet longer in length than Harold's garden. If the area of George's garden was 36 square feet, find the dimensions of each garden.

6.3 ADDING AND SUBTRACTING RATIONAL EXPRESSIONS

1 Adding and Subtracting Rational Expressions with a Common Denominator

If rational expressions have the same denominator, they can be combined in a fashion similar to that used for fractions in arithmetic. The numerators are added or subtracted and the denominator remains the same.

Student Learning Objectives

After studying this section, you will be able to:

1 Add and subtract rational expressions with a common denominator.

2 Determine the LCD for rational expressions with different denominators.

3 Add and subtract rational expressions with different denominators.

ADDING RATIONAL EXPRESSIONS

For any rational expressions $\dfrac{a}{b}$ and $\dfrac{c}{b}$,

$$\frac{a}{b} + \frac{c}{b} = \frac{a + c}{b} \qquad \text{where } b \neq 0.$$

EXAMPLE 1 Add. $\dfrac{5a}{a + 2b} + \dfrac{6a}{a + 2b}$

Solution

$$\frac{5a}{a + 2b} + \frac{6a}{a + 2b} = \frac{5a + 6a}{a + 2b} = \frac{11a}{a + 2b}$$

Note that the denominators are the same. Only add the numerators. Keep the same denominator.
Do not change the denominator.

Practice Problem 1 Add. $\dfrac{2s + t}{2s - t} + \dfrac{s - t}{2s - t}$

SUBTRACTING RATIONAL EXPRESSIONS

For any rational expressions $\dfrac{a}{b}$ and $\dfrac{c}{b}$,

$$\frac{a}{b} - \frac{c}{b} = \frac{a - c}{b} \qquad \text{where } b \neq 0.$$

EXAMPLE 2 Subtract. $\dfrac{3x}{(x + y)(x - 2y)} - \dfrac{8x}{(x + y)(x - 2y)}$

Solution

$$\frac{3x}{(x + y)(x - 2y)} - \frac{8x}{(x + y)(x - 2y)} = \frac{3x - 8x}{(x + y)(x - 2y)} \qquad \text{Write as one fraction.}$$

$$= \frac{-5x}{(x + y)(x - 2y)} \qquad \text{Simplify.}$$

Practice Problem 2 Subtract. $\dfrac{b}{(a - 2b)(a + b)} - \dfrac{2b}{(a - 2b)(a + b)}$

NOTE TO STUDENT: Fully worked-out solutions to all of the Practice Problems can be found at the back of the text starting at page SP-1

2 Determining the LCD for Rational Expressions with Different Denominators

How do we add or subtract rational expressions when the denominators are not the same? First we must find the **least common denominator** (LCD). You need to be clear on how to find a least common denominator and how to add and subtract fractions from arithmetic before you attempt this section. Review Sections 0.1 and 0.2 if you have any questions about this topic.

HOW TO FIND THE LCD OF TWO OR MORE RATIONAL EXPRESSIONS

1. Factor each denominator completely.
2. The LCD is a product containing each *different factor*.
3. If a factor occurs more than once in any one denominator, the LCD will contain that factor repeated the greatest number of times that it occurs in any one denominator.

EXAMPLE 3 Find the LCD. $\dfrac{5}{2x-4}, \dfrac{6}{3x-6}$

Solution Factor each denominator.

$$2x - 4 = 2(x - 2) \qquad 3x - 6 = 3(x - 2)$$

The different factors are 2, 3, and $(x - 2)$. Since no factor appears more than once in any one denominator, the LCD is the product of these three factors.

$$\text{LCD} = (2)(3)(x - 2) = 6(x - 2)$$

NOTE TO STUDENT: Fully worked-out solutions to all of the Practice Problems can be found at the back of the text starting at page SP-1

Practice Problem 3 Find the LCD. $\dfrac{7}{6x+21}, \dfrac{13}{10x+35}$

EXAMPLE 4 Find the LCD.

(a) $\dfrac{5}{12ab^2c}, \dfrac{13}{18a^3bc^4}$ (b) $\dfrac{8}{x^2-5x+4}, \dfrac{12}{x^2+2x-3}$

Solution If a factor occurs more than once in any one denominator, the LCD will contain that factor repeated the greatest number of times that it occurs in any one denominator.

(a) $12ab^2c = 2 \cdot 2 \cdot 3 \cdot \quad a \cdot \quad\quad b \cdot b \cdot c$

$18a^3bc^4 = \quad 2 \cdot 3 \cdot 3 \cdot a \cdot a \cdot a \cdot b \cdot \quad c \cdot c \cdot c \cdot c$

$\quad\quad\quad\quad 2 \cdot 2 \cdot 3 \cdot 3 \cdot a \cdot a \cdot a \cdot b \cdot b \cdot c \cdot c \cdot c \cdot c$

$$\text{LCD} = 2^2 \cdot 3^2 \cdot a^3 \cdot b^2 \cdot c^4 = 36a^3b^2c^4$$

(b) $x^2 - 5x + 4 = (x - 4)(x - 1)$

$x^2 + 2x - 3 = \quad (x - 1)(x + 3)$

$$\text{LCD} = (x - 4)(x - 1)(x + 3)$$

Practice Problem 4 Find the LCD.

(a) $\dfrac{3}{50xy^2z}, \dfrac{19}{40x^3yz}$ (b) $\dfrac{2}{x^2+5x+6}, \dfrac{6}{3x^2+5x-2}$

Adding and Subtracting Rational Expressions with Different Denominators

If two rational expressions have different denominators, we first change them to equivalent rational expressions with the least common denominator. Then we add or subtract the numerators and keep the common denominator.

EXAMPLE 5   Add. $\dfrac{5}{xy} + \dfrac{2}{y}$

Solution   The denominators are different. We must find the LCD. The two factors are x and y. We observe that the LCD is xy.

$$\frac{5}{xy} + \frac{2}{y} = \frac{5}{xy} + \frac{2}{y} \cdot \frac{x}{x} \qquad \text{Multiply the second fraction by } \frac{x}{x}.$$

$$= \frac{5}{xy} + \frac{2x}{xy} \qquad \text{Now each fraction has a common denominator of } xy.$$

$$= \frac{5 + 2x}{xy} \qquad \text{Write the sum as one fraction.}$$

Practice Problem 5   Add. $\dfrac{7}{a} + \dfrac{3}{abc}$

EXAMPLE 6   Add. $\dfrac{3x}{x^2 - y^2} + \dfrac{5}{x + y}$

Solution   We factor the first denominator so that $x^2 - y^2 = (x + y)(x - y)$. Thus, the factors of the denominators are $(x + y)$ and $(x - y)$. We observe that the LCD $= (x + y)(x - y)$.

$$\frac{3x}{(x + y)(x - y)} + \frac{5}{(x + y)} \cdot \frac{x - y}{x - y} \qquad \text{Multiply the second fraction by } \frac{x - y}{x - y}.$$

$$= \frac{3x}{(x + y)(x - y)} + \frac{5x - 5y}{(x + y)(x - y)} \qquad \text{Now each fraction has a common denominator of } (x + y)(x - y).$$

$$= \frac{3x + 5x - 5y}{(x + y)(x - y)} \qquad \text{Write the sum of the numerators over one common denominator.}$$

$$= \frac{8x - 5y}{(x + y)(x - y)} \qquad \text{Collect like terms.}$$

Practice Problem 6   Add. $\dfrac{2a - b}{a^2 - 4b^2} + \dfrac{2}{a + 2b}$

It is important to remember that the LCD is the smallest algebraic expression into which each denominator can be divided. For rational expressions the LCD must contain *each factor* that appears in any denominator. If the factor is repeated, the LCD must contain that factor the greatest number of times that it appears in any one denominator.

In many cases, the denominators in an addition or subtraction problem are not in factored form. You must factor each denominator to determine the LCD. Collect like terms in the numerator; then determine whether that final numerator can be factored. If so, you may be able to simplify the fraction.

EXAMPLE 7 Add. $\dfrac{5}{x^2 - y^2} + \dfrac{3x}{x^3 + x^2y}$

Solution

$$\frac{5}{x^2 - y^2} + \frac{3x}{x^3 + x^2y}$$

Factor the two denominators. Observe that the LCD is $x^2(x + y)(x - y)$.

$$= \frac{5}{(x + y)(x - y)} + \frac{3x}{x^2(x + y)}$$

$$= \frac{5}{(x + y)(x - y)} \cdot \frac{x^2}{x^2} + \frac{3x}{x^2(x + y)} \cdot \frac{x - y}{x - y}$$

Multiply each fraction by the appropriate value to obtain a common denominator of $x^2(x + y)(x - y)$.

$$= \frac{5x^2}{x^2(x + y)(x - y)} + \frac{3x^2 - 3xy}{x^2(x + y)(x - y)}$$

$$= \frac{5x^2 + 3x^2 - 3xy}{x^2(x + y)(x - y)}$$

Write the sum of the numerators over one common denominator.

$$= \frac{8x^2 - 3xy}{x^2(x + y)(x - y)}$$

Collect like terms.

$$= \frac{x(8x - 3y)}{x^2(x + y)(x - y)}$$

Divide out the common factor x in the numerator and denominator and simplify.

$$= \frac{8x - 3y}{x(x + y)(x - y)}$$

NOTE TO STUDENT: Fully worked-out solutions to all of the Practice Problems can be found at the back of the text starting at page SP-1

Practice Problem 7 Add. $\dfrac{7a}{a^2 + 2ab + b^2} + \dfrac{4}{a^2 + ab}$

It is very easy to make a sign mistake when subtracting two fractions. You will find it helpful to place parentheses around the numerator of the second fraction so that you will not forget to subtract the entire numerator.

EXAMPLE 8 Subtract. $\dfrac{3x + 4}{x - 2} - \dfrac{x - 3}{2x - 4}$

Solution Factor the second denominator.

$$= \frac{3x + 4}{x - 2} - \frac{x - 3}{2(x - 2)}$$

Observe that the LCD is $2(x - 2)$.

$$= \frac{2}{2} \cdot \frac{(3x + 4)}{x - 2} - \frac{x - 3}{2(x - 2)}$$

Multiply the first fraction by $\frac{2}{2}$ so that the resulting fraction will have the common denominator.

$$= \frac{2(3x + 4) - (x - 3)}{2(x - 2)}$$

Write the indicated subtraction as one fraction. Note the parentheses around $x - 3$.

$$= \frac{6x + 8 - x + 3}{2(x - 2)}$$

Remove the parentheses in the numerator.

$$= \frac{5x + 11}{2(x - 2)}$$

Collect like terms.

Practice Problem 8 Subtract. $\dfrac{x + 7}{3x - 9} - \dfrac{x - 6}{x - 3}$

To avoid making errors when subtracting two fractions, some students find it helpful to change subtraction to addition of the opposite of the second fraction. In other words, we use the property that $\dfrac{a}{b} - \dfrac{c}{b} = \dfrac{a}{b} + \dfrac{-c}{b}$.

EXAMPLE 9   Subtract. $\dfrac{8x}{x^2 - 16} - \dfrac{4}{x - 4}$

Solution

$\dfrac{8x}{x^2 - 16} - \dfrac{4}{x - 4}$

$= \dfrac{8x}{(x + 4)(x - 4)} + \dfrac{-4}{x - 4}$ Factor the first denominator. Use the property that $\dfrac{a}{b} - \dfrac{c}{b} = \dfrac{a}{b} + \dfrac{-c}{b}$.

$= \dfrac{8x}{(x + 4)(x - 4)} + \dfrac{-4}{x - 4} \cdot \dfrac{x + 4}{x + 4}$ Multiply the second fraction by $\dfrac{x + 4}{x + 4}$.

$= \dfrac{8x + (-4)(x + 4)}{(x + 4)(x - 4)}$ Write the sum of the numerators over one common denominator.

$= \dfrac{8x - 4x - 16}{(x + 4)(x - 4)}$ Remove parentheses.

$= \dfrac{4x - 16}{(x + 4)(x - 4)}$ Collect like terms. Note that the numerator can be factored.

$= \dfrac{4\cancel{(x - 4)}}{(x + 4)\cancel{(x - 4)}}$ Since $(x - 4)$ is a *factor* of the numerator *and* the denominator, we may divide out the common factor.

$= \dfrac{4}{x + 4}$

Practice Problem 9   Subtract and simplify. $\dfrac{x - 2}{x^2 - 4} - \dfrac{x + 1}{2x^2 + 4x}$

Verbal and Writing Skills

1. Suppose two rational expressions have denominators of $(x + 3)(x + 5)$ and $(x + 3)^2$. Explain how you would determine the LCD.

2. Suppose two rational expressions have denominators of $(x - 4)^2(x + 7)$ and $(x - 4)^3$. Explain how you would determine the LCD.

Perform the operation indicated. Be sure to simplify.

3. $\dfrac{x}{x + 5} + \dfrac{2x + 1}{5 + x}$

4. $\dfrac{8}{7 + 2x} + \dfrac{x + 3}{2x + 7}$

5. $\dfrac{3x}{x + 3} - \dfrac{x + 5}{x + 3}$

6. $\dfrac{2x + 7}{x - 8} - \dfrac{x + 1}{x - 8}$

7. $\dfrac{8x + 3}{5x + 7} - \dfrac{6x + 10}{5x + 7}$

8. $\dfrac{5x - 9}{3x + 4} - \dfrac{4x + 14}{3x + 4}$

Find the LCD. Do not combine fractions.

9. $\dfrac{13}{3ab}, \dfrac{7}{a^2b^2}$

10. $\dfrac{12}{5a^2}, \dfrac{9}{a^3}$

11. $\dfrac{5}{18x^2y^5}, \dfrac{7}{30x^3y^3}$

12. $\dfrac{11}{16x^2y^3}, \dfrac{17}{56xy^4}$

13. $\dfrac{8}{5x - 15}, \dfrac{7}{3x - 9}$

14. $\dfrac{9}{2x - 8}, \dfrac{11}{5x - 20}$

15. $\dfrac{8}{x + 3}, \dfrac{15}{x^2 - 9}$

16. $\dfrac{13}{x^2 - 16}, \dfrac{7}{x - 4}$

17. $\dfrac{7}{3x^2 + 14x - 5}, \dfrac{4}{9x^2 - 6x + 1}$

18. $\dfrac{4}{2x^2 - 9x - 35}, \dfrac{3}{4x^2 + 20x + 25}$

Add.

19. $\dfrac{7}{ab} + \dfrac{3}{b}$

20. $\dfrac{8}{cd} + \dfrac{9}{d}$

21. $\dfrac{3}{x + 7} + \dfrac{8}{x^2 - 49}$

22. $\dfrac{5}{x^2 - 2x + 1} + \dfrac{3}{x - 1}$

23. $\dfrac{3y}{y + 2} + \dfrac{y}{y - 2}$

24. $\dfrac{2}{y - 1} + \dfrac{2}{y + 1}$

25. $\dfrac{6}{5a} + \dfrac{5}{3a + 2}$

26. $\dfrac{3}{2a + 5} + \dfrac{4}{7a}$

27. $\dfrac{2}{3xy} + \dfrac{1}{6yz}$

28. $\dfrac{5}{4xy} + \dfrac{5}{12yz}$

Subtract.

29. $\dfrac{5x + 6}{x - 3} - \dfrac{x - 2}{2x - 6}$

30. $\dfrac{7x + 3}{x - 4} - \dfrac{x - 3}{2x - 8}$

31. $\dfrac{3x}{x^2 - 25} - \dfrac{2}{x + 5}$

32. $\dfrac{7x}{x^2 - 9} - \dfrac{6}{x + 3}$

33. $\dfrac{a + 3b}{2} - \dfrac{a - b}{5}$

34. $\dfrac{3b - a}{6} - \dfrac{a + b}{5}$

35. $\dfrac{8}{2x - 3} - \dfrac{6}{x + 2}$

36. $\dfrac{6}{3x - 4} - \dfrac{5}{4x - 3}$

37. $\dfrac{x}{x^2 + 2x - 3} - \dfrac{x}{x^2 - 5x + 4}$

38. $\dfrac{1}{x^2 - 2x} - \dfrac{5}{x^2 - 4x + 4}$

39. $\dfrac{2}{x^2 + 5x + 6} + \dfrac{3}{x^2 + 7x + 10}$

40. $\dfrac{3}{x^2 - x - 12} + \dfrac{2}{x^2 + x - 20}$

41. $\dfrac{3x - 8}{x^2 - 5x + 6} + \dfrac{x + 2}{x^2 - 6x + 8}$

42. $\dfrac{3x + 5}{x^2 + 4x + 3} + \dfrac{-x + 5}{x^2 + 2x - 3}$

Mixed Practice

Add or subtract the following problems.

43. $\dfrac{6x}{y - 2x} - \dfrac{5x}{2x - y}$

44. $\dfrac{8b}{2b - a} - \dfrac{3b}{a - 2b}$

45. $\dfrac{3y}{8y^2 + 2y - 1} - \dfrac{5y}{2y^2 - 9y - 5}$

46. $\dfrac{2x}{x^2 + 5x + 6} - \dfrac{x + 1}{x^2 + 2x - 3}$

47. $\dfrac{4y}{y^2 + 4y + 3} + \dfrac{2}{y + 1}$

48. $\dfrac{y - 23}{y^2 - y - 20} + \dfrac{2}{y - 5}$

To Think About

Add the following.

49. $\dfrac{2x}{x-3} + \dfrac{3x}{x+2} + \dfrac{7}{x^2 - x - 6}$

50. $\dfrac{x}{x-4} + \dfrac{5}{x+4} + \dfrac{10}{x^2 - 16}$

Cumulative Review

51. Solve. $\dfrac{1}{3}(x-2) + \dfrac{1}{2}(x+3) = \dfrac{1}{4}(3x+1)$

52. Solve for y. $5ax = 2(ay - 3bc)$

53. Solve for x. $\dfrac{1}{2}x < \dfrac{1}{3}x + \dfrac{1}{4}$

54. Simplify. $(3x^3y^4)^4$

55. *Commuting Costs* A subway token costs $1.50. A monthly unlimited ride subway pass costs $50. How many days per month would you have to use the subway to go to work (assume one subway token to get to work and one subway token to get back home) in order for it to be cheaper to buy a monthly subway pass?

56. *Unemployment in Finland* In Finland the unemployment rate went from 3.5% in 1990 to 14.6% in 1999. (*Source:* United Nations Statistics Division.) If the working age population of Finland was 5,100,000 during this period, how many more people were unemployed in 1999 than in 1990?

57. *Anchovy Population* During the 1972–1973 El Niño weather pattern, the waters around the country of Peru became very warm. The cool-water-loving anchovy fish population dropped from 20 million to 2 million. This particular El Niño lasted 20 months. (*Source:* National Oceanic and Atmospheric Administration.) **(a)** If during this time the population level dropped at a constant rate, how much did it drop per month? **(b)** Fishermen worried that the entire anchovy population would be destroyed. If the weather pattern had continued beyond the 20 months and if the anchovy population had continued to decline at the same rate, how many more months would it have been before the entire anchovy population was destroyed?

58. *DVD Rentals* Renting a DVD from the local video store costs $3.50. If you sign up for a rental service on the Web, you pay $21.00 each month, plus a $0.50 mailing charge per DVD. How many DVDs must you rent each month so that the Web service is cheaper than renting locally?

How are you doing with your homework assignments in Sections 6.1 to 6.3? Do you feel you have mastered the material so far? Do you understand the concepts you have covered? Before you go further in the textbook, take some time to do each of the following problems.

Simplify.

6.1

1. $\dfrac{8x - 48}{x^2 - 6x}$

2. $\dfrac{2x^2 - 7x - 15}{x^2 - 12x + 35}$

3. $\dfrac{y^2 + 6y + 9}{27x^2 - 3x^2y^2}$

4. $\dfrac{5x^2 - 23x + 12}{5x^2 + 7x - 6}$

6.2

5. $\dfrac{8x^3}{3x + 12} \cdot \dfrac{9x + 36}{16x^2}$

6. $\dfrac{x - 5}{x^2 + 5x - 14} \cdot \dfrac{x^2 + 12x + 35}{15 - 3x}$

7. $\dfrac{x^2 - 9}{2x + 6} \div \dfrac{2x^2 - 5x - 3}{4x^2 - 1}$

8. $\dfrac{3a^2 + 7a + 2}{4a^2 + 11a + 6} \div \dfrac{6a^2 - 13a - 5}{16a^2 - 9}$

6.3

9. $\dfrac{x - 3y}{xy} - \dfrac{4a - y}{ay}$

10. $\dfrac{7}{2x - 4} + \dfrac{-14}{x^2 - 4}$

11. $\dfrac{2x}{x^2 + 10x + 21} + \dfrac{x - 3}{x + 7}$

12. $\dfrac{2}{x^2 + 3x - 10} - \dfrac{5x}{x^2 - 6x + 8}$

Now turn to page SA-14 for the answer to each of these problems. Each answer also includes a reference to the objective in which the problem is first taught. If you missed any of these problems, you should stop and review the Examples and Practice Problems in the referenced objective. A little review now will help you master the material in the upcoming sections of the text.

1. _____

2. _____

3. _____

4. _____

5. _____

6. _____

7. _____

8. _____

9. _____

10. _____

11. _____

12. _____

6.4 Simplifying Complex Rational Expressions

Student Learning Objectives

After studying this section, you will be able to:

1 Simplify complex rational expressions by adding or subtracting in the numerator and denominator.

2 Simplify complex rational expressions using the LCD.

1 Simplifying Complex Rational Expressions by Adding or Subtracting in the Numerator and Denominator

A **complex rational expression** (also called a **complex fraction**) has a fraction in the numerator or in the denominator, or both.

$$\frac{3 + \dfrac{2}{x}}{\dfrac{x}{7} + 2} \qquad \frac{\dfrac{x}{y} + 1}{2} \qquad \frac{\dfrac{a + b}{3}}{\dfrac{x - 2y}{4}}$$

The bar in a complex rational expression is both a grouping symbol and a symbol for division.

$$\frac{\dfrac{a + b}{3}}{\dfrac{x - 2y}{4}} \quad \text{is equivalent to} \quad \left(\frac{a + b}{3}\right) \div \left(\frac{x - 2y}{4}\right)$$

We need a procedure for simplifying complex rational expressions.

PROCEDURE TO SIMPLIFY A COMPLEX RATIONAL EXPRESSION ADDING AND SUBTRACTING

1. Add or subtract so that you have a single fraction in the numerator and in the denominator.
2. Divide the fraction in the numerator by the fraction in the denominator. This is done by inverting the fraction in the denominator and multiplying it by the numerator.

EXAMPLE 1 Simplify. $\dfrac{\dfrac{1}{x}}{\dfrac{2}{y^2} + \dfrac{1}{y}}$

Solution

Step 1 Add the two fractions in the denominator.

$$\frac{\dfrac{1}{x}}{\dfrac{2}{y^2} + \dfrac{1}{y} \cdot \dfrac{y}{y}} = \frac{\dfrac{1}{x}}{\dfrac{2 + y}{y^2}}$$

Step 2 Divide the fraction in the numerator by the fraction in the denominator.

$$\frac{1}{x} \div \frac{2 + y}{y^2} = \frac{1}{x} \cdot \frac{y^2}{2 + y} = \frac{y^2}{x(2 + y)}$$

Practice Problem 1 Simplify. $\dfrac{\dfrac{1}{a} + \dfrac{1}{b}}{\dfrac{2}{ab^2}}$

NOTE TO STUDENT: Fully worked-out solutions to all of the Practice Problems can be found at the back of the text starting at page SP-1

A complex rational expression may contain two or more fractions in the numerator and the denominator.

EXAMPLE 2 Simplify. $\dfrac{\dfrac{1}{x} + \dfrac{1}{y}}{\dfrac{3}{a} - \dfrac{2}{b}}$

Solution We observe that the LCD of the fractions in the numerator is xy. The LCD of the fractions in the denominator is ab.

$= \dfrac{\dfrac{1}{x} \cdot \dfrac{y}{y} + \dfrac{1}{y} \cdot \dfrac{x}{x}}{\dfrac{3}{a} \cdot \dfrac{b}{b} - \dfrac{2}{b} \cdot \dfrac{a}{a}}$ Multiply each fraction by the appropriate value to obtain common denominators.

$= \dfrac{\dfrac{y + x}{xy}}{\dfrac{3b - 2a}{ab}}$ Add the two fractions in the numerator.

Subtract the two fractions in the denominator.

$= \dfrac{y + x}{xy} \cdot \dfrac{ab}{3b - 2a}$ Invert the fraction in the denominator and multiply it by the numerator.

$= \dfrac{ab(y + x)}{xy(3b - 2a)}$ Write the answer as one fraction.

Practice Problem 2 Simplify. $\dfrac{\dfrac{1}{a} + \dfrac{1}{b}}{\dfrac{1}{a} - \dfrac{1}{b}}$

For some complex rational expressions, factoring may be necessary to determine the LCD and to combine fractions.

EXAMPLE 3 Simplify. $\dfrac{\dfrac{1}{x^2 - 1} + \dfrac{2}{x + 1}}{x}$

Solution We need to factor $x^2 - 1$.

$= \dfrac{\dfrac{1}{(x + 1)(x - 1)} + \dfrac{2}{(x + 1)} \cdot \dfrac{x - 1}{x - 1}}{x}$ The LCD for the fractions in the numerator is $(x + 1)(x - 1)$.

$= \dfrac{\dfrac{1 + 2x - 2}{(x + 1)(x - 1)}}{x}$ Add the two fractions in the numerator.

$= \dfrac{2x - 1}{(x + 1)(x - 1)} \cdot \dfrac{1}{x}$ Simplify the numerator. Invert the fraction in the denominator and multiply.

$= \dfrac{2x - 1}{x(x + 1)(x - 1)}$ Write the answer as one fraction.

Practice Problem 3 Simplify. $\dfrac{\dfrac{x}{x^2 + 4x + 3} + \dfrac{2}{x + 1}}{x + 1}$

When simplifying complex rational expressions, always check to see if the final fraction can be reduced or simplified.

EXAMPLE 4 Simplify. $\dfrac{\dfrac{3}{a+b} - \dfrac{3}{a-b}}{\dfrac{5}{a^2 - b^2}}$

Solution The LCD of the two fractions in the numerator is $(a+b)(a-b)$.

$$\dfrac{\dfrac{3}{a+b} \cdot \dfrac{a-b}{a-b} - \dfrac{3}{a-b} \cdot \dfrac{a+b}{a+b}}{\dfrac{5}{a^2 - b^2}}$$

$$= \dfrac{\dfrac{3a - 3b}{(a+b)(a-b)} - \dfrac{3a + 3b}{(a+b)(a-b)}}{\dfrac{5}{a^2 - b^2}}$$

Study carefully how we combine the two fractions in the numerator. Do you see how we obtain $-6b$?

$$= \dfrac{\dfrac{-6b}{(a+b)(a-b)}}{\dfrac{5}{(a+b)(a-b)}} \qquad \text{Factor } a^2 - b^2 \text{ as } (a+b)(a-b).$$

$$= \dfrac{-6b}{(a+b)(a-b)} \cdot \dfrac{(a+b)(a-b)}{5} \qquad \begin{array}{l}\text{Since } (a+b)(a-b) \text{ are factors in both}\\ \text{numerator and denominator, they may be}\\ \text{divided out.}\end{array}$$

$$= \dfrac{-6b}{5} \quad \text{or} \quad -\dfrac{6b}{5}$$

NOTE TO STUDENT: *Fully worked-out solutions to all of the Practice Problems can be found at the back of the text starting at page SP-1*

Practice Problem 4 Simplify. $\dfrac{\dfrac{6}{x^2 - y^2}}{\dfrac{1}{x-y} + \dfrac{3}{x+y}}$

② Simplifying Complex Rational Expressions Using the LCD

There is another way to simplify complex rational expressions: Multiply the numerator and denominator of the complex fraction by the least common denominator of all the denominators appearing in the complex fraction.

PROCEDURE TO SIMPLIFY A COMPLEX RATIONAL EXPRESSION: MULTIPLYING BY THE LCD

1. Determine the LCD of all individual denominators occurring in the numerator and denominator of the complex rational expression.

2. Multiply both the numerator and the denominator of the complex rational expression by the LCD.

3. Simplify, if possible.

EXAMPLE 5   Simplify by multiplying by the LCD. $\dfrac{\dfrac{5}{ab^2} - \dfrac{2}{ab}}{3 - \dfrac{5}{2a^2b}}$

Solution   The LCD of all the denominators in the complex rational expression is $2a^2b^2$.

$$\dfrac{2a^2b^2\left(\dfrac{5}{ab^2} - \dfrac{2}{ab}\right)}{2a^2b^2\left(3 - \dfrac{5}{2a^2b}\right)}$$

$$= \dfrac{2a^2b^2\left(\dfrac{5}{ab^2}\right) - 2a^2b^2\left(\dfrac{2}{ab}\right)}{2a^2b^2(3) - 2a^2b^2\left(\dfrac{5}{2a^2b}\right)} \qquad \text{Multiply each term by } 2a^2b^2.$$

$$= \dfrac{10a - 4ab}{6a^2b^2 - 5b} \qquad \text{Simplify.}$$

Practice Problem 5   Simplify by multiplying by the LCD. $\dfrac{\dfrac{2}{3x^2} - \dfrac{3}{y}}{\dfrac{5}{xy} - 4}$

So that you can compare the two methods, we will redo Example 4 by multiplying by the LCD.

EXAMPLE 6   Simplify by multiplying by the LCD. $\dfrac{\dfrac{3}{a + b} - \dfrac{3}{a - b}}{\dfrac{5}{a^2 - b^2}}$

Solution   The LCD of all individual fractions contained in the complex fraction is $(a + b)(a - b)$.

$$\dfrac{(a + b)(a - b)\left(\dfrac{3}{a + b}\right) - (a + b)(a - b)\left(\dfrac{3}{a - b}\right)}{(a + b)(a - b)\left(\dfrac{5}{(a + b)(a - b)}\right)} \qquad \text{Multiply each term by the LCD.}$$

$$= \dfrac{3(a - b) - 3(a + b)}{5} \qquad \text{Simplify.}$$

$$= \dfrac{3a - 3b - 3a - 3b}{5} \qquad \text{Remove parentheses.}$$

$$= -\dfrac{6b}{5} \qquad \text{Simplify.}$$

Practice Problem 6   Simplify by multiplying by the LCD. $\dfrac{\dfrac{6}{x^2 - y^2}}{\dfrac{7}{x - y} + \dfrac{3}{x + y}}$

Simplify.

1. $\dfrac{\dfrac{3}{x}}{\dfrac{2}{x^2} + \dfrac{5}{x}}$

2. $\dfrac{\dfrac{4}{x}}{\dfrac{7}{x} + \dfrac{2}{x^2}}$

3. $\dfrac{\dfrac{1}{x} + \dfrac{1}{y}}{\dfrac{1}{xy}}$

4. $\dfrac{\dfrac{3}{x} + \dfrac{4}{y}}{\dfrac{6}{xy}}$

5. $\dfrac{\dfrac{x}{6} - \dfrac{1}{3}}{\dfrac{2}{3x} + \dfrac{5}{6}}$

6. $\dfrac{\dfrac{x}{8} - \dfrac{1}{2}}{\dfrac{3}{8x} + \dfrac{3}{4}}$

7. $\dfrac{\dfrac{7}{5x} - \dfrac{1}{x}}{\dfrac{3}{5} + \dfrac{2}{x}}$

8. $\dfrac{\dfrac{8}{x} - \dfrac{2}{3x}}{\dfrac{2}{3} + \dfrac{5}{x}}$

9. $\dfrac{\dfrac{5}{x} + \dfrac{3}{y}}{3x + 5y}$

10. $\dfrac{\dfrac{1}{x} + \dfrac{1}{y}}{x + y}$

11. $\dfrac{4 - \dfrac{1}{x^2}}{2 + \dfrac{1}{x}}$

12. $\dfrac{1 - \dfrac{36}{x^2}}{1 - \dfrac{6}{x}}$

13. $\dfrac{\dfrac{10}{x^2 - 25}}{\dfrac{3}{x + 5} + \dfrac{2}{x - 5}}$

14. $\dfrac{\dfrac{8}{x^2 - 4}}{\dfrac{7}{x + 2} - \dfrac{3}{x - 2}}$

15. $\dfrac{a + \dfrac{3}{a}}{\dfrac{a^2 + 2}{3a}}$

16. $\dfrac{x + \dfrac{4}{x}}{\dfrac{x^2 + 3}{4x}}$

17. $\dfrac{\dfrac{3}{x - 3}}{\dfrac{1}{x^2 - 9} + \dfrac{2}{x + 3}}$

18. $\dfrac{\dfrac{5}{x + 4}}{\dfrac{1}{x - 4} - \dfrac{2}{x^2 - 16}}$

19. $\dfrac{\dfrac{2}{y - 1} + 2}{\dfrac{2}{y + 1} - 2}$

20. $\dfrac{\dfrac{y}{y + 1} + 1}{\dfrac{2y + 1}{y - 1}}$

To Think About

21. Consider the complex fraction $\dfrac{\dfrac{4}{x + 3}}{\dfrac{5}{x} - 1}$. What values are not allowable replacements for the variable x?

22. Consider the complex fraction $\dfrac{\dfrac{5}{x - 2}}{\dfrac{6}{x} + 1}$. What values are not allowable replacements for the variable x?

Cumulative Review

23. Solve for w. $P = 2(l + w)$

24. Solve and graph. $7 + x < 11 + 5x$

$$\xleftarrow{\hspace{0.5cm}}\underset{-5}{+}\ \underset{-4}{+}\ \underset{-3}{+}\ \underset{-2}{+}\ \underset{-1}{+}\ \underset{0}{+}\ \underset{1}{+}\ \underset{2}{+}\ \underset{3}{+}\ \underset{4}{+}\ \underset{5}{+}\xrightarrow{\hspace{0.3cm}} x$$

25. *Loan Repayment* Manuela is paying back a $4000 loan with no interest from her favorite sister. If Manuela has already paid $125 per month for 17 months, how much of the loan does she have yet to pay?

26. *Urban Sprawl* A neighborhood in Hopewell, New Jersey, is upset because a big office complex is being built in their neighborhood. Many of the citizens of Hopewell moved to their town to get away from urban sprawl. An investment firm will be building a complex planned to cover 3.5 million square feet. This initial estimate could grow to 5.5 million square feet. What percent of increase would this be?

① Solving Equations Involving Rational Expressions That Have Solutions

In Section 2.4 we developed procedures to solve linear equations containing fractions whose denominators were numerical values. In this section we use a similar approach to solve equations containing fractions whose denominators are polynomials. It would be wise for you to review Section 2.4 briefly *before you begin this section*. It will be especially helpful to carefully study Examples 1 and 2.

TO SOLVE AN EQUATION CONTAINING RATIONAL EXPRESSIONS

1. Determine the LCD of all the denominators.
2. Multiply each term of the equation by the LCD.
3. Solve the resulting equation.
4. Check your solution. Exclude from your solution any value that would make the LCD equal to zero. If such a value is obtained, there is *no solution*.

EXAMPLE 1 Solve for x and check your solution. $\dfrac{5}{x} + \dfrac{2}{3} = -\dfrac{3}{x}$

Solution

$$3x\left(\frac{5}{x}\right) + 3x\left(\frac{2}{3}\right) = 3x\left(-\frac{3}{x}\right) \qquad \text{Observe that the LCD is } 3x. \text{ Multiply each term by } 3x.$$

$$15 + 2x = -9$$
$$2x = -9 - 15 \qquad \text{Subtract 15 from both sides.}$$
$$2x = -24$$
$$x = -12 \qquad \text{Divide both sides by 2.}$$

Check. $\qquad \dfrac{5}{-12} + \dfrac{2}{3} \overset{?}{=} -\dfrac{3}{-12} \qquad$ Replace each x by -12.

$$-\frac{5}{12} + \frac{8}{12} \overset{?}{=} \frac{3}{12}$$

$$\frac{3}{12} = \frac{3}{12} \qquad \checkmark \quad \text{It checks.}$$

Practice Problem 1 Solve for x and check your solution. $\dfrac{3}{x} + \dfrac{4}{5} = -\dfrac{2}{x}$

NOTE TO STUDENT: Fully worked-out solutions to all of the Practice Problems can be found at the back of the text starting at page SP-1

EXAMPLE 2 Solve and check. $\dfrac{6}{x+3} = \dfrac{3}{x}$

Solution Observe that the LCD $= x(x+3)$.

$$x(x+3)\left(\frac{6}{x+3}\right) = x(x+3)\left(\frac{3}{x}\right) \qquad \text{Multiply both sides by } x(x+3).$$

$$6x = 3(x+3) \qquad \text{Simplify. Do you see how this is done?}$$
$$6x = 3x + 9 \qquad \text{Remove parentheses.}$$
$$3x = 9 \qquad \text{Subtract } 3x \text{ from both sides.}$$
$$x = 3 \qquad \text{Divide both sides by 3.}$$

Check.
$$\frac{6}{3 + 3} \overset{?}{=} \frac{3}{3} \qquad \text{Replace each } x \text{ by 3.}$$
$$\frac{6}{6} = \frac{3}{3} \quad \checkmark \qquad \text{It checks.}$$

Practice Problem 2 Solve and check. $\dfrac{4}{2x + 1} = \dfrac{6}{2x - 1}$

It is sometimes necessary to factor denominators before the correct LCD can be determined.

EXAMPLE 3 Solve and check. $\dfrac{3}{x + 5} - 1 = \dfrac{4 - x}{2x + 10}$

Solution

$$\frac{3}{x + 5} - 1 = \frac{4 - x}{2(x + 5)} \qquad \begin{array}{l} \text{Factor } 2x + 10. \text{ We} \\ \text{determine that the} \\ \text{LCD is } 2(x + 5). \end{array}$$

$$2(x + 5)\left(\frac{3}{x + 5}\right) - 2(x + 5)(1) = 2(x + 5)\left[\frac{4 - x}{2(x + 5)}\right] \qquad \begin{array}{l} \text{Multiply each term by} \\ \text{the LCD.} \end{array}$$

$$2(3) - 2(x + 5) = 4 - x \qquad \text{Simplify.}$$
$$6 - 2x - 10 = 4 - x \qquad \text{Remove parentheses.}$$
$$-2x - 4 = 4 - x \qquad \text{Collect like terms.}$$
$$-4 = 4 + x \qquad \text{Add } 2x \text{ to both sides.}$$
$$-8 = x \qquad \begin{array}{l} \text{Subtract 4 from both} \\ \text{sides.} \end{array}$$

Check. $\dfrac{3}{-8 + 5} - 1 \overset{?}{=} \dfrac{4 - (-8)}{2(-8) + 10}$ Replace each x in the original equation by -8.

$$\frac{3}{-3} - 1 \overset{?}{=} \frac{4 + 8}{-16 + 10}$$

$$-1 - 1 \overset{?}{=} \frac{12}{-6}$$

$$-2 = -2 \quad \checkmark \qquad \text{It checks. The solution is } -8.$$

NOTE TO STUDENT: Fully worked-out solutions to all of the Practice Problems can be found at the back of the text starting at page SP-1

Practice Problem 3 Solve and check. $\dfrac{x - 1}{x^2 - 4} = \dfrac{2}{x + 2} + \dfrac{4}{x - 2}$

② Determining Whether an Equation Involving Rational Expressions Has No Solution

Equations containing rational expressions sometimes appear to have solutions when in fact they do not. By this we mean that the "solutions" we get by using completely correct methods are, in actuality, not solutions.

In the case where a value makes a denominator in the equation equal to zero, we say it is not a solution to the equation. Such a value is called an **extraneous solution.** An extraneous solution is an apparent solution that does *not* satisfy the original equation. If all of the apparent solutions of an equation are extraneous solutions, we say that the equation has **no solution.** It is important that you check all apparent solutions in the original equation.

EXAMPLE 4   Solve and check. $\dfrac{y}{y-2} - 4 = \dfrac{2}{y-2}$

Solution   Observe that the LCD is $y - 2$.

$$(y-2)\left(\dfrac{y}{y-2}\right) - (y-2)(4) = (y-2)\left(\dfrac{2}{y-2}\right)$$ Multiply each term by $(y-2)$.

$$y - 4(y-2) = 2$$ Simplify. Do you see how this is done?

$$y - 4y + 8 = 2$$ Remove parentheses.

$$-3y + 8 = 2$$ Collect like terms.

$$-3y = -6$$ Subtract 8 from both sides.

$$\dfrac{-3y}{-3} = \dfrac{-6}{-3}$$ Divide both sides by -3.

$$y = 2$$ 2 is only an apparent solution.

This equation has no solution.

Why? We can see immediately that $y = 2$ is not a solution for the original equation. When we substitute 2 for y in a denominator, the denominator is equal to zero and the expression is undefined.

Check. $\dfrac{y}{y-2} - 4 = \dfrac{2}{y-2}$ Suppose that you try to check the apparent solution by substituting 2 for y.

$$\dfrac{2}{2-2} - 4 \overset{?}{=} \dfrac{2}{2-2}$$

$$\dfrac{2}{0} - 4 = \dfrac{2}{0}$$ This does not check since you do not obtain a real number when you divide by zero.

       ↑   ↑

   These expressions are not defined.

There is no such number as $2 \div 0$. We see that 2 does *not* check. This equation has **no solution.**

Practice Problem 4   Solve and check. $\dfrac{2x}{x+1} = \dfrac{-2}{x+1} + 1$

6.5 EXERCISES

Student Solutions Manual | CD/Video | PH Math Tutor Center | MathXL®Tutorials on CD | MathXL® | MyMathLab® | Interactmath.com

Solve and check problems 1–16.

1. $\dfrac{7}{x} + \dfrac{3}{4} = \dfrac{-2}{x}$

2. $\dfrac{8}{x} + \dfrac{2}{5} = \dfrac{-2}{x}$

3. $\dfrac{-1}{4x} + \dfrac{3}{2} = \dfrac{5}{x}$

4. $\dfrac{1}{3x} + \dfrac{5}{6} = \dfrac{2}{x}$

5. $\dfrac{5x + 3}{3x} = \dfrac{7}{3} - \dfrac{9}{x}$

6. $\dfrac{3x - 7}{4x} = \dfrac{2}{x} - \dfrac{9}{4}$

7. $\dfrac{x + 1}{2x} = \dfrac{2}{3}$

8. $\dfrac{x - 2}{4x} = \dfrac{1}{6}$

9. $\dfrac{6}{3x - 5} = \dfrac{3}{2x}$

10. $\dfrac{1}{x + 6} = \dfrac{4}{x}$

11. $\dfrac{2}{2x + 5} = \dfrac{4}{x - 4}$

12. $\dfrac{3}{x + 5} = \dfrac{3}{3x - 2}$

13. $\dfrac{2}{x} + \dfrac{x}{x + 1} = 1$

14. $\dfrac{5}{2} = 3 + \dfrac{2x + 7}{x + 6}$

15. $\dfrac{85 - 4x}{x} = 7 - \dfrac{3}{x}$

16. $\dfrac{63 - 2x}{x} = 2 - \dfrac{5}{x}$

Mixed Practice

Solve and check. If there is no solution, so indicate.

17. $\dfrac{3}{x + 2} - 4 = \dfrac{4x - 1}{x + 2}$

18. $\dfrac{4}{x + 3} - 2 = \dfrac{5x - 9}{x + 3}$

19. $\dfrac{2}{x - 6} - 5 = \dfrac{2(x - 5)}{x - 6}$

20. $7 - \dfrac{x}{x + 5} = \dfrac{5}{5 + x}$

21. $\dfrac{2}{x + 1} - \dfrac{1}{x - 1} = \dfrac{2x}{x^2 - 1}$

22. $\dfrac{8x}{4x^2 - 1} = \dfrac{3}{2x + 1} + \dfrac{3}{2x - 1}$

23. $\dfrac{y + 1}{y^2 + 2y - 3} = \dfrac{1}{y + 3} - \dfrac{1}{y - 1}$

24. $\dfrac{6}{x - 5} + \dfrac{3x + 1}{x^2 - 2x - 15} = \dfrac{5}{x + 3}$

25. $\dfrac{2x}{x + 4} - \dfrac{8}{x - 4} = \dfrac{2x^2 + 32}{x^2 - 16}$

26. $\dfrac{4x}{x + 3} - \dfrac{12}{x - 3} = \dfrac{4x^2 + 36}{x^2 - 9}$

27. $\dfrac{4}{x^2 - 1} + \dfrac{7}{x + 1} = \dfrac{5}{x - 1}$

28. $\dfrac{7}{4x^2 - 1} + \dfrac{2}{2x + 1} = \dfrac{3}{2x - 1}$

29. $\dfrac{x + 11}{x^2 - 5x + 4} + \dfrac{3}{x - 1} = \dfrac{5}{x - 4}$

30. $\dfrac{6}{x - 3} = \dfrac{-5}{x - 2} + \dfrac{-5}{x^2 - 5x + 6}$

To Think About

In each of the following equations, what values are not allowable replacements for the variable x? Do not solve the equation.

31. $\dfrac{3x}{x - 2} - \dfrac{4x}{x - 4} = \dfrac{3}{x^2 - 6x + 8}$

32. $\dfrac{3x}{x - 1} - \dfrac{5x}{2x + 3} = \dfrac{7}{2x^2 + 5x + 3}$

Cumulative Review

33. Factor. $6x^2 - x - 12$

34. Solve. $4 - (3 - x) = 7(x + 3)$

▲ **35.** *Geometry* The perimeter of a rectangular sign is 54 meters. The length is 1 meter less than triple the width. Find the dimensions of the sign.

36. *Mortgage Payments* Wally and Adele Panzas plan to purchase a new home and borrow $115,000. They plan to take out a 25-year mortgage at an annual interest rate of 8.75%. The bank will charge them $8.23 per month for each $1000 of mortgage. What will their monthly payments be?

Immigrants   *The accompanying bar graph depicts the number of immigrants to the United States from Europe and Latin America for each decade from 1960 to 2010. Use the graph to answer exercises 37–40.*

37. What was the percent of decrease in immigrants from Europe from the decade 1960–1969 to the decade 1970–1979? Round to the nearest tenth.

38. What was the percent of increase in immigrants from Latin America from the decade 1980–1989 to the decade 1990–1999? Round to the nearest tenth.

39. If the rate of increase in immigrants from Latin America from the decade starting in 1990 to the decade starting in 2000 continues for another two decades, what will be the expected number of immigrants from Latin America to the United States in 2010? Round to the nearest tenth.

40. If the rate of decrease in immigrants from Europe from the decade starting in 1990 to the decade starting in 2000 continues for another two decades, what will be the expected number of immigrants from Europe to the United States in the decade starting in 2010?

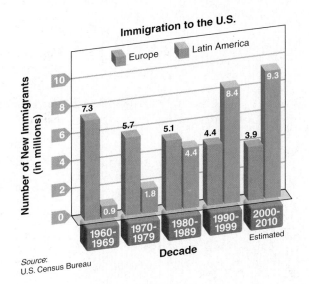

1 Solving Problems Involving Ratio and Proportion

A **ratio** is a comparison of two quantities. You may be familiar with ratios that compare miles to hours or miles to gallons. A ratio is often written as a quotient in the form of a fraction. For example, the ratio of 7 to 9 can be written as $\frac{7}{9}$.

A **proportion** is an equation that states that two ratios are equal. For example,

$$\frac{7}{9} = \frac{21}{27}, \quad \frac{2}{3} = \frac{10}{15}, \quad \text{and} \quad \frac{a}{b} = \frac{c}{d} \quad \text{are proportions.}$$

Let's take a closer look at the last proportion. We can see that the LCD of the fractional equation is bd.

$$(b\,d)\frac{a}{b} = (b\,d)\frac{c}{d} \qquad \text{Multiply each side by the LCD.}$$
$$da = bc$$
$$ad = bc \qquad \text{Since multiplication is commutative, } da = ad.$$

Thus we have proved the following.

THE PROPORTION EQUATION

If

$$\frac{a}{b} = \frac{c}{d}, \quad \text{then} \quad ad = bc$$

for all real numbers a, b, c, and d, where $b \neq 0$ and $d \neq 0$.

This is sometimes called **cross multiplying.** It can be applied only if you have *one* fraction and nothing else on each side of the equation.

EXAMPLE 1 Michael took 5 hours to drive 245 miles on the turnpike. At the same rate, how many hours will it take him to drive a distance of 392 miles?

Solution

1. *Understand the problem.* Let x = the number of hours it will take to drive 392 miles. If 5 hours are needed to drive 245 miles, then x hours are needed to drive 392 miles.

2. *Write an equation.* We can write this as a proportion. Compare time to distance in each ratio.

$$\begin{array}{c} \text{Time} \\ \text{Distance} \end{array} \xrightarrow{} \frac{5 \text{ hours}}{245 \text{ miles}} = \frac{x \text{ hours}}{392 \text{ miles}} \xleftarrow{} \begin{array}{c} \text{Time} \\ \text{Distance} \end{array}$$

3. *Solve and state the answer.*

$$5(392) = 245x \qquad \text{Cross multiply.}$$
$$\frac{1960}{245} = x \qquad \text{Divide both sides by 245.}$$
$$8 = x$$

It will take Michael 8 hours to drive 392 miles.

4. *Check.* Is $\frac{5}{245} = \frac{8}{392}$? Do the computation and see.

Practice Problem 1 It took Brenda 8 hours to drive 420 miles. At the same rate, how long would it take her to drive 315 miles?

EXAMPLE 2 If $\frac{3}{4}$ inch on a map represents an actual distance of 20 miles, how long is the distance represented by $4\frac{1}{8}$ inches on the same map?

Solution Let x = the distance represented by $4\frac{1}{8}$ inches.

Initial measurement on map $\longrightarrow \dfrac{3}{4}$ $4\frac{1}{8}$ $\longleftarrow$ Second measurement on the map

Initial distance $\longrightarrow \dfrac{3}{4}{20} = \dfrac{4\frac{1}{8}}{x}$ $\longleftarrow$ Second distance

$$\left(\frac{3}{4}\right)(x) = (20)\left(4\frac{1}{8}\right) \quad \text{Cross multiply.}$$

$$\left(\frac{3}{4}\right)(x) = (\overset{5}{\cancel{20}})\left(\frac{33}{\underset{2}{\cancel{8}}}\right) \quad \text{Write } 4\frac{1}{8} \text{ as } \frac{33}{8} \text{ and simplify.}$$

$$\frac{3x}{4} = \frac{165}{2} \quad \text{Multiplication of fractions.}$$

$$4\left(\frac{3x}{4}\right) = \overset{2}{\cancel{4}}\left(\frac{165}{\cancel{2}}\right) \quad \text{Multiply each side by 4.}$$

$$3x = 330 \quad \text{Simplify.}$$

$$x = 110 \quad \text{Divide both sides by 3.}$$

$4\frac{1}{8}$ inches on the map represents an actual distance of 110 miles.

Practice Problem 2 If $\frac{5}{8}$ inch on a map represents an actual distance of 30 miles, how long is the distance represented by $2\frac{1}{2}$ inches on the same map?

② Solving Problems Involving Similar Triangles

Similar triangles are triangles that have the same shape, but may be different sizes. For example, if you draw a triangle on a sheet of paper, place the paper in a photocopy machine, and make a copy that is reduced by 25%, you would create a triangle that is similar to the original triangle. The two triangles will have the same shape. The corresponding sides of the triangles will be proportional. The corresponding angles of the triangles will also be equal.

Original triangle   25% reduction

You can use the proportion equation to show that the corresponding sides of the preceding triangles are proportional. In fact, you can use the proportion equation to find an unknown length of a side of one of the two similar triangles.

EXAMPLE 3 A ramp is 32 meters long and rises up 15 meters. A ramp at the same angle is 9 meters long. How high is the second ramp?

Solution To answer this question, we find the length of side x in the following two similar triangles.

Ramp A, longest side $\longrightarrow$ $\dfrac{32}{9} = \dfrac{15}{x}$ $\longleftarrow$ Shortest side, ramp A
Ramp B, longest side $\longrightarrow$ $\longleftarrow$ Shortest side, ramp B

$$32x = (9)(15) \qquad \text{Cross multiply.}$$
$$32x = 135 \qquad \text{Multiply.}$$
$$x = \frac{135}{32} \qquad \text{Divide both sides by 32.}$$

$$\text{or} \quad x = 4\frac{7}{32} \text{ meters}$$

NOTE TO STUDENT: *Fully worked-out solutions to all of the Practice Problems can be found at the back of the text starting at page SP-1*

▲ **Practice Problem 3** Triangle C is similar to triangle D. Find the length of side x. Leave your answer as a fraction.

13 centimeters   16 centimeters x 18 centimeters

Triangle C   Triangle D

We can also use similar triangles for indirect measurement—for instance, to find the measure of an object that is too tall to measure using standard measuring devices. When the sun shines on two vertical objects at the same time, the shadows and the objects form similar triangles.

▲ **EXAMPLE 4** A woman who is 5 feet tall casts a shadow that is 8 feet long. At the same time of day, a building casts a shadow that is 72 feet long. How tall is the building?

Solution

1. *Understand the problem.* First we draw a sketch. We do not know the height of the building, so we call it x.

Building:
x feet

Woman:
5 feet

8 foot shadow   72 foot shadow

2. *Write an equation and solve.*

Height of woman $\longrightarrow$ $\dfrac{5}{8} = \dfrac{x}{72}$ $\longleftarrow$ Height of building
Length of woman's shadow $\longrightarrow$ $\longleftarrow$ Length of building's shadow

$$(5)(72) = 8x \qquad \text{Cross multiply.}$$
$$360 = 8x$$
$$45 = x$$

The height of the building is 45 feet.

▲ **Practice Problem 4** A man who is 6 feet tall casts a shadow that is 7 feet long. At the same time of day, a large flagpole casts a shadow that is 38.5 feet long. How tall is the flagpole?

In problems such as Example 4, we are assuming that the building and the person are standing exactly perpendicular to the ground. In other words, each triangle is assumed to be a right triangle. In other similar triangle problems, if the triangles are not right triangles you must be careful that the angles between the objects and the ground are the same.

③ Solving Distance Problems Involving Rational Expressions

Some distance problems are solved using equations with rational expressions. We will need the formula Distance = Rate × Time, $D = RT$, which we can write in the form $T = \dfrac{D}{R}$. In the United States, distances are usually measured in miles. In European countries, distances are usually measured in kilometers.

Les Ailes
Parce que vous
n'êtes pas un oiseau

EXAMPLE 5   A French commuter airline flies from Paris to Avignon. Plane A flies at a speed that is 50 kilometers per hour faster than plane B. Plane A flies 500 kilometers in the amount of time that plane B flies 400 kilometers. Find the speed of each plane.

Solution

1. **Understand the problem.** Let s = the speed of plane B in kilometers per hour.
   Let $s + 50$ = the speed of plane A in kilometers per hour.
   Make a simple table for D, R, and T.

	D	R	$T = \dfrac{D}{R}$
Plane A	500	$s + 50$	?
Plane B	400	s	?

   Since $T = \dfrac{D}{R}$, for each plane we divide the expression for D by the expression for R and write it in the table in the column for time.

	D	R	$T = \dfrac{D}{R}$
Plane A	500	$s + 50$	$\dfrac{500}{s + 50}$
Plane B	400	s	$\dfrac{400}{s}$

2. **Write an equation and solve.** Each plane flies for the same amount of time. That is, the time for plane A equals the time for plane B.

$$\frac{500}{s + 50} = \frac{400}{s}$$

   You can solve this equation using the methods in Section 6.5 or you may cross multiply. Here we will cross multiply.

$500s = (s + 50)(400)$	Cross multiply.
$500s = 400s + 20,000$	Remove parentheses.
$100s = 20,000$	Subtract $400s$ from each side.
$s = 200$	Divide each side by 100.

Plane B travels 200 kilometers per hour. Since

$$s + 50 = 200 + 50 = 250,$$

plane A travels 250 kilometers per hour.

Practice Problem 5   Two European freight trains traveled toward Paris for the same amount of time. Train A traveled 180 kilometers, while train B traveled 150 kilometers. Train A traveled 10 kilometers per hour faster than train B. What was the speed of each train?

4 Solving Work Problems

Some applied problems involve the length of time needed to do a job. These problems are often referred to as work problems.

EXAMPLE 6 Reynaldo can sort a huge stack of mail on an old sorting machine in 9 hours. His brother Carlos can sort the same amount of mail using a newer sorting machine in 8 hours. How long would it take them to do the job working together? Express your answer in hours and minutes. Round to the nearest minute.

Solution

1. **Understand the problem.** Let's do a little reasoning.

   If Reynaldo can do the job in 9 hours, then in *1 hour* he could do $\frac{1}{9}$ of the job.

   If Carlos can do the job in 8 hours, then in *1 hour* he could do $\frac{1}{8}$ of the job.

   Let x = the number of hours it takes Reynaldo and Carlos to do the job together. In *1 hour* together they could do $\frac{1}{x}$ of the job.

2. **Write an equation and solve.** The amount of work Reynaldo can do in 1 hour plus the amount of work Carlos can do in 1 hour must be equal to the amount of work they could do together in 1 hour.

Amount of work done by Reynaldo	+	Amount of work done by Carlos	=	Amount of work done together
$\dfrac{1}{9}$	+	$\dfrac{1}{8}$	=	$\dfrac{1}{x}$

Let us solve for x. We observe that the LCD is $72x$.

$$72x\left(\frac{1}{9}\right) + 72x\left(\frac{1}{8}\right) = 72x\left(\frac{1}{x}\right) \quad \text{Multiply each term by the LCD.}$$

$$8x + 9x = 72 \qquad \text{Simplify.}$$

$$17x = 72 \qquad \text{Collect like terms.}$$

$$x = \frac{72}{17} \qquad \text{Divide each side by 17.}$$

$$x = 4\frac{4}{17}$$

To change $\frac{4}{17}$ of an hour to minutes we multiply.

$$\frac{4}{17}\ \text{hour} \times \frac{60\ \text{minutes}}{1\ \text{hour}} = \frac{240}{17}\ \text{minutes, which is approximately 14.118 minutes}$$

To the nearest minute this is 14 minutes. Thus doing the job together will take 4 hours and 14 minutes.

Calculator

Reciprocals

You can find $\frac{1}{x}$ for any value of x on a scientific calculator by using the key labeled $\boxed{x^{-1}}$ or the key labeled $\boxed{1/x}$. For example, to find $\frac{1}{9}$, we use $9\ \boxed{x^{-1}}$ or $9\ \boxed{1/x}$. The display will read 0.11111111. Therefore we can solve Example 6 as follows:

$$9\ \boxed{x^{-1}}\ \boxed{+}\ 8\ \boxed{x^{-1}}\ \boxed{=}$$

The display will read 0.2361111.
Thus we have obtained the equation $0.2361111 = \frac{1}{x}$.
Now this is equivalent to $x = \frac{1}{0.2361111}$.
(Do you see why?)
Thus we enter 0.2361111 $\boxed{x^{-1}}$, and the display reads 4.235294118.
If we round to the nearest hundredth, we have $x \approx 4.24$ hours, which is approximately equal to our answer of $4\frac{4}{17}$ hours.

NOTE TO STUDENT: Fully worked-out solutions to all of the Practice Problems can be found at the back of the text starting at page SP-1

Practice Problem 6 John Tobey and Dave Wells obtained night custodian jobs at a local factory while going to college part time. Using the buffer machine, John can buff all the floors in the building in 6 hours. Dave takes a little longer and can do all the floors in the building in 7 hours. Their supervisor bought another buffer machine. How long will it take John and Dave to do all the floors in the building working together, each with his own machine? Express your answer in hours and minutes. Round to the nearest minute.

6.6 EXERCISES

| Student Solutions Manual | CD/ Video | PH Math Tutor Center | MathXL®Tutorials on CD | MathXL® | MyMathLab® | Interactmath.com |

Solve.

1. $\dfrac{4}{9} = \dfrac{8}{x}$

2. $\dfrac{5}{12} = \dfrac{x}{8}$

3. $\dfrac{x}{17} = \dfrac{12}{5}$

4. $\dfrac{16}{x} = \dfrac{3}{4}$

5. $\dfrac{8}{5} = \dfrac{x}{7}$

6. $\dfrac{6}{x} = \dfrac{45}{12}$

7. $\dfrac{7}{x} = \dfrac{40}{130}$

8. $\dfrac{x}{18} = \dfrac{13}{2}$

Applications

Use a proportion to answer exercises 9–16.

9. Map Scale The scale on the AAA map of Colorado is approximately $\frac{3}{4}$ inch to 15 miles. If the distance from Denver to Pueblo measures 5.5 inches on the map, how far apart are the two cities?

10. Recipe Ratios Nella Coates's recipe for Shoofly Pie contains $\frac{3}{4}$ cup of unsulfured molasses. This recipe makes a small pie that serves 8 people. If she makes the larger pie that serves 12 people and the ratio of molasses to people remains the same, how much molasses will she need for the larger pie?

11. Exchange Rates Jocelyn Robinson spent a year studying in Australia. The day she arrived, the exchange rate was 1.49 Australian dollars per U.S. dollar. When she arrived she exchanged $700 U.S.
   (a) How many Australian dollars were her $700 worth?
   (b) Three days later, the value of the Australian dollar rose to 1.68 Australian dollars per U.S. dollar. How much more money would she have received had she waited the three days to exchange her money?

12. Exchange Rates Sarah Tanguay spent last summer touring thoughout Europe. On the day she landed in Paris, the exchange rate for the Euro was 0.88 Euros per U.S. dollar. Jen converted $650 to euros that day.
   (a) How many euros did Jen receive for her $650?
   (b) On her way back home, Jen decided to stop in England for a week. She had €250 (250 Euros) that she wanted to change into British pounds for her stay in London. If the exchange rate on the day she landed was 0.69 British pounds per Euro, how many pounds did she receive?

13. Speed Units Alfonse and Melinda are taking a drive in Mexico. They know that a speed of 100 kilometers per hour is approximately equal to 62 miles per hour. They are now driving on a Mexican road that has a speed limit of 90 kilometers per hour. How many miles per hour is the speed limit? Round to the nearest mile per hour.

14. Baggage Weight Dick and Anne took a trip to France. Their suitcases were weighed at the airport and the weight recorded was 39 kilograms. If 50 kilograms is equivalent to 110 pounds, how many pounds did their suitcases weigh? Round to the nearest pound.

15. Map Scale On a map the distance between two mountains is $3\frac{1}{2}$ inches. The actual distance between the mountains is 136 miles. Russ is camped at a location that on the map is $\frac{3}{4}$ inch from the base of the mountain. How many miles is he from the base of the mountain? Round to the nearest mile.

16. Construction Scales Maria is adding a porch to her house that is 18 feet long. On the drawing done by the carpenter the length is shown as 11 inches. The drawing shows that the width of the porch is 8 inches. How many feet wide will the porch be? Round to the nearest foot.

Geometry   *Triangles A and B are similar. Use them to answer exercises 17–20. Leave your answers as fractions.*

Triangle A

Triangle B

▲ **17.** If $x = 20$ in., $y = 29$ in., and $m = 13$ in., find the length of side n.

▲ **18.** If $p = 14$ in., $m = 17$ in., and $z = 23$ in., find the length of side x.

▲ **19.** If $x = 175$ meters, $n = 40$ meters, and $m = 35$ meters find the length of side y.

▲ **20.** If $z = 18$ cm, $y = 25$ cm, and $n = 9$ cm, find the length of side p.

Geometry   *Just as we have discussed similar triangles, other geometric shapes can be similar. Similar geometric shapes will have sides that are proportional. Quadrilaterals abcd and ghjk are similar. Use them to answer exercises 21–24. Leave your answers as fractions.*

▲ **21.** If $a = 7$ in., $g = 9$ in., and $k = 12$ in., find the length of side d.

▲ **22.** If $b = 8$ ft, $c = 7$ ft, and $j = 11$ ft, find the length of side h.

▲ **23.** If $b = 20$ m, $h = 24$ m, and $d = 32$ m, find the length of side k.

▲ **24.** If $a = 16$ cm, $d = 19$ cm, and $k = 23$ cm, find the length of side g.

Use a proportion to solve.

▲ **25.** **Geometry** A rectangle whose width-to-length ratio is approximately 5 to 8 is called a **golden rectangle** and is said to be pleasing to the eye. Using this ratio, what should the length of a rectangular picture be if its width is to be 30 inches?

▲ **26.** **Shadows** Jake is 6 feet tall and notices that he casts a shadow of 8 feet. At the same time, the new public sculpture in the park casts a shadow of 23 feet. How tall is the sculpture? (Round your answer to the nearest foot.)

▲ **27.** **Floral Displays** Floral designers often create arrangements where the flower height to container ratio is 5 to 3. The FIU Art Museum wishes to create a floral display for the opening of a new show. They know they want to use an antique Chinese vase from their collection that is 13 inches high. How tall will the entire flower arrangement be if they use this standard ratio? (Round your answer to the nearest inch.)

▲ **28.** *Securing Wires* A wire line helps to secure a radio transmission tower. The wire measures 23 meters from the tower to the ground anchor pin. The wire is secured 14 meters up on the tower. If a second wire is secured 130 meters up on the tower and is extended from the tower at the same angle as the first wire, how long would the second wire need to be to reach an anchor pin on the ground? Round to the nearest meter.

130 meters

14 meters   23 meters

29. *Acceleration* Ben Hale is driving his new Toyota Camry on Interstate 90 at 45 miles per hour. He accelerates at the rate of 3 miles per hour every 2 seconds. How fast will he be traveling after accelerating for 11 seconds?

30. *Braking* Tim Newitt is driving a U-Haul truck to Chicago. He is driving at 55 miles per hour and has to hit the brakes because of heavy traffic. His truck slows at the rate of 2 miles per hour for every 3 seconds. How fast will he be traveling 10 seconds after he hits the brakes?

31. *Flight Speeds* A Montreal commuter airliner travels 40 kilometers per hour faster than the television news helicopter over the city. The commuter airliner travels 1250 kilometers during the same time that the television news helicopter travels only 1050 kilometers. How fast does the commuter airliner fly? How fast does the television news helicopter fly?

32. *Driving Speeds* Melissa drove to Dallas while Marcia drove to Houston in the same amount of time. Melissa drove 360 kilometers, while Marcia drove 280 kilometers. Melissa traveled 20 kilometers per hour faster than Marcia on her trip. What was the average speed in kilometers per hour for each woman?

33. *Fluff Containers* Marshmallow fluff comes in only two sizes, a $7\frac{1}{2}$ - oz glass jar and a 16-oz plastic tub. At a local market, the 16-oz tub costs $1.49 and the $7\frac{1}{2}$ - oz jar costs $0.79.
   (a) How much does marshmallow fluff in the glass jar cost per oz? (Round your answer to the nearest cent.)
   (b) How much does marshmallow fluff in the plastic tub cost per oz? (Round your answer to the nearest cent.)
   (c) If the marshmallow fluff company decided to add a third size, a 40-oz bucket, how much would the price be if it was at the same unit price as the 16-oz plastic tub? (*Hint:* Set up a proportion. Do *not* use your answer from part (b). Round your answer to the nearest cent.)

34. *Green Tea* Won Ling is a Chinese tea importer in Boston's Chinatown. He charges $12.25 for four sample packs of his famous green tea. The packs are in the following sizes: 25 grams, 40 grams, 50 grams, and 60 grams.
   (a) How much is Won Ling charging per gram for his green tea?
   (b) How much would you pay for a 60-gram pack if he were willing to sell that one by itself?
   (c) How much would you pay for an 800-gram package of green tea if it cost the same amount per gram?

35. *Lawn Mowing* It takes a person using a large rotary mower 4 hours to mow all the lawns at the town park. It takes a person using a small rotary mower 5 hours to mow these same lawns. How long should it take two people using these mowers to mow these lawns together? Round to the nearest minute.

36. *Tank Cleaning* Juan Reyes works at Sea World. Every fish tank needs a thorough cleaning once a week. It takes Juan 5 hours to clean all the tanks in the store. When Juan is not there, Chet James does the cleaning. Since he is not as experienced as Juan, it takes Chet 7 hours to do the same job. At the start of the July sales promotion, the boss wanted all the tanks cleaned as quickly as possible so that Juan could help with the customers. He had both Chet and Juan cleaning the tanks. How long did it take them to do the job? (Round your answer to the nearest minute.)

37. *Raking Leaves* When all the leaves have fallen at Fred and Suzie's house in Concord, New Hampshire, Suzie can rake the entire yard in 6 hours. When Fred does it alone, it takes him 8 hours. How long would it take them to rake the yard together? Round to the nearest minute.

38. *Typing Speed* Professor Matthews can type his course syllabus in 12 hours. The departmental secretary can type it in 7 hours. How long would it take the two people to type it together? Round to the nearest minute.

Cumulative Review

39. Write in scientific notation. 0.000892465

40. Write in decimal notation. 5.82×10^8

41. Write with positive exponents. $\dfrac{x^{-3}y^{-2}}{z^4 w^{-8}}$

42. Evaluate. $\left(\dfrac{2}{3}\right)^{-3}$

Determining Monthly Profit of a Business

Rick and Wally discovered a great need for a calculator store in their city. They opened a store called Calculator Corner in a major shopping mall. As long as they advertised in local newspapers, they discovered that they made a good profit. When they spent $1000 a month in advertising, the Calculator Corner store made a monthly profit of $6000. When they spent $3000 a month in advertising, the store made a monthly profit of $12,000. The following chart shows the observed relationship between monthly advertising and the monthly profit of the store in thousands of dollars. They measured the profit after subtracting all expenses, including advertising, salaries and benefits of employees, taxes, rent, utilities, and shipping expenses.

Monthly Profit of Calculator Corner

Monthly advertising expenses (in dollars)

Problems for Individual Investigation and Analysis

1. When they increased the monthly advertising budget from $3000 to $5000, how much did the monthly profit of the store increase?

2. When they increased the monthly advertising budget from $7000 to $9000, how much did the monthly profit of the store increase?

Problems for Group Investigation and Cooperative Learning

Rich and Wally discovered that the predicted monthly profit of the business (measured in thousands of dollars) could be obtained from the equation

$$P = \frac{24x^2 + 120x}{x^2 + 8x + 15}$$

where x is the number of thousands of dollars in the monthly advertising budget.

3. Using your skills learned in this chapter, factor the numerator and denominator and then simplify the equation.

4. Use the simplified equation obtained in exercise 3 to predict the monthly profit if the monthly advertising budget is $11,000. (Round to the nearest dollar.)

5. Use the simplified equation obtained in exercise 3 to predict how large a monthly advertising budget would be needed to obtain a monthly profit of $20,000.

6. Use the simplified equation obtained in exercise 3 to predict how large a monthly advertising budget would be needed to obtain a monthly profit of $21,000.

Topic	Procedure	Examples
Simplifying rational expressions, p. 356.	1. Factor the numerator and denominator. 2. Divide out any factor common to both the numerator and denominator.	$$\frac{36x^2 - 16y^2}{18x^2 + 24xy + 8y^2} = \frac{4(3x - 2y)(3x + 2y)}{2(3x + 2y)(3x + 2y)}$$ $$= \frac{2(3x - 2y)}{3x - 2y}$$
Multiplying rational expressions, p. 362.	1. Factor all numerators and denominators and rewrite the product as one fraction. 2. Simplify the resulting rational expression as described above.	$$\frac{x^2 - y^2}{x^2 + 2xy + y^2} \cdot \frac{x^2 + 4xy + 3y^2}{x^2 - 4xy + 3y^2}$$ $$= \frac{(x + y)(x - y)(x + y)(x + 3y)}{(x + y)(x + y)(x - y)(x - 3y)}$$ $$= \frac{x + 3y}{x - 3y}$$
Dividing rational expressions, p. 363.	1. Invert the second fraction and rewrite the problem as a product. 2. Multiply the rational expressions.	$$\frac{14x^2 + 17x - 6}{x^2 - 25} \div \frac{4x^2 - 8x - 21}{x^2 + 10x + 25}$$ $$= \frac{(2x + 3)(7x - 2)}{(x + 5)(x - 5)} \cdot \frac{(x + 5)(x + 5)}{(2x - 7)(2x + 3)}$$ $$= \frac{(7x - 2)(x + 5)}{(x - 5)(2x - 7)}$$
Adding rational expressions, p. 367.	1. If the denominators differ, factor them and determine the least common denominator (LCD). 2. Change each fraction by multiplication into an equivalent one with the LCD. 3. Add numerators; put the answer over the LCD. 4. Simplify as needed.	$$\frac{x - 1}{x^2 - 4} + \frac{x - 1}{3x + 6} = \frac{x - 1}{(x + 2)(x - 2)} + \frac{x - 1}{3(x + 2)}$$ $$\text{LCD} = 3(x + 2)(x - 2)$$ $$\frac{x - 1}{(x + 2)(x - 2)} = \frac{?}{3(x + 2)(x - 2)}$$ Need to multiply by $\frac{3}{3}$. $$\frac{(x - 1)}{3(x + 2)} = \frac{?}{3(x + 2)(x - 2)}$$ Need to multiply by $\frac{x - 2}{x - 2}$. $$\frac{x - 1}{(x + 2)(x - 2)} + \frac{x - 1}{3(x + 2)}$$ $$= \frac{(x - 1) \cdot 3}{3(x + 2)(x - 2)} + \frac{(x - 1)(x - 2)}{3(x + 2)(x - 2)}$$ $$= \frac{3x - 3 + x^2 - 3x + 2}{3(x + 2)(x - 2)}$$ $$= \frac{x^2 - 1}{3(x + 2)(x - 2)}$$
Subtracting rational expressions, p. 367.	Move a subtraction sign to the numerator of the second fraction. Add. Simplify if possible. $$\frac{a}{b} - \frac{c}{b} = \frac{a}{b} + \frac{-c}{b}$$	$$\frac{5x}{x - 2} - \frac{3x + 4}{x - 2} = \frac{5x}{x - 2} + \frac{-(3x + 4)}{x - 2}$$ $$= \frac{5x - 3x - 4}{x - 2}$$ $$= \frac{2x - 4}{x - 2}$$ $$= \frac{2(x - 2)}{x - 2} = 2$$

Topic	Procedure	Examples
Simplifying complex rational expressions, p. 376.	1. Add the two fractions in the numerator. 2. Add the two fractions in the denominator. 3. Divide the fraction in the numerator by the fraction in the denominator. This is done by inverting the fraction in the denominator and multiplying by the numerator. 4. Simplify.	$$\dfrac{\dfrac{x}{x^2-4}+\dfrac{1}{x+2}}{\dfrac{3}{x+2}-\dfrac{4}{x-2}}$$ $$=\dfrac{\dfrac{x}{(x+2)(x-2)}+\dfrac{1(x-2)}{(x+2)(x-2)}}{\dfrac{3(x-2)}{(x+2)(x-2)}+\dfrac{-4(x+2)}{(x+2)(x-2)}}$$ $$=\dfrac{\dfrac{x+x-2}{(x+2)(x-2)}}{\dfrac{3x-6-4x-8}{(x+2)(x-2)}}$$ $$=\dfrac{2x-2}{(x+2)(x-2)}\div\dfrac{-x-14}{(x+2)(x-2)}$$ $$=\dfrac{2(x-1)}{(x+2)(x-2)}\cdot\dfrac{(x+2)(x-2)}{-x-14}$$ $$=\dfrac{2x-2}{-x-14}\ \text{or}\ \dfrac{-2x+2}{x+14}$$
Solving equations involving rational expressions, p. 381.	1. Determine the LCD of all denominators. 2. Note what values will make the LCD equal to 0. These are excluded from your solutions. 3. Multiply each side by the LCD, distributing as needed. 4. Solve the resulting polynomial equation. 5. Check. Be sure to exclude those values found in step 2.	$$\dfrac{3}{x-2}=\dfrac{4}{x+2}$$ LCD $=(x-2)(x+2)$. Since LCD $\neq 0$, then $x\neq 2,-2$. $$(x-2)(x+2)\dfrac{3}{x-2}=\dfrac{4}{x+2}(x-2)(x+2)$$ $$3(x+2)=4(x-2)$$ $$3x+6=4x-8$$ $$-x=-14$$ $$x=14$$ (Since $x\neq 2,-2$, this solution should check unless an error has been made.) $Check:\quad \dfrac{3}{14-2}\overset{?}{=}\dfrac{4}{14+2}$ $$\dfrac{3}{12}\overset{?}{=}\dfrac{4}{16}$$ $$\dfrac{1}{4}=\dfrac{1}{4}\ \checkmark$$
Solving applied problems with proportions, p. 386.	1. Organize the data. 2. Write a proportion equating the respective parts. Let x represent the value that is not known. 3. Solve the proportion.	Renee can make five cherry pies with 3 cups of flour. How many cups of flour does she need to make eight cherry pies? $$\dfrac{5\ \text{cherry pies}}{3\ \text{cups flour}}=\dfrac{8\ \text{cherry pies}}{x\ \text{cups flour}}$$ $$\dfrac{5}{3}=\dfrac{8}{x}$$ $$5x=24$$ $$x=\dfrac{24}{5}$$ $$x=4\dfrac{4}{5}$$ $4\dfrac{4}{5}$ cups of flour are needed for eight cherry pies.

Chapter 6 Review Problems

Section 6.1

Simplify.

1. $\dfrac{bx}{bx - by}$

2. $\dfrac{4x - 4y}{5y - 5x}$

3. $\dfrac{2x^2 + 5x - 3}{2x^2 - 9x + 4}$

4. $\dfrac{3x^2 + 7x + 2}{3x^2 + 13x + 4}$

5. $\dfrac{x^2 - 9}{x^2 - 10x + 21}$

6. $\dfrac{2x^2 + 18x + 40}{3x + 15}$

7. $\dfrac{x^3 + 3x^2}{x^3 - 2x^2 - 15x}$

8. $\dfrac{4x^2 + 4x - 3}{4x^2 - 2x}$

9. $\dfrac{2x^2 - 2xy - 24y^2}{2x^2 + 5xy - 3y^2}$

10. $\dfrac{4 - y^2}{3y^2 + 5y - 2}$

11. $\dfrac{5x^3 - 10x^2}{25x^4 + 5x^3 - 30x^2}$

12. $\dfrac{16x^2 - 4y^2}{4x - 2y}$

Section 6.2

Multiply or divide.

13. $\dfrac{3x^2 - 13x - 10}{3x^2 + 2x} \cdot \dfrac{x^2 - 25x}{x^2 - 25}$

14. $\dfrac{2y^2 - 18}{3y^2 + 3y} \div \dfrac{y^2 + 6y + 9}{y^2 + 4y + 3}$

15. $\dfrac{2y^2 + 3y - 2}{2y^2 + y - 1} \div \dfrac{2y^2 + y - 1}{2y^2 - 3y - 2}$

16. $\dfrac{6y^2 + 13y - 5}{9y^2 + 3y} \div \dfrac{4y^2 + 20y + 25}{12y^2}$

17. $\dfrac{3xy^2 + 12y^2}{2x^2 - 11x + 5} \div \dfrac{2xy + 8y}{8x^2 + 2x - 3}$

18. $\dfrac{11}{x - 2} \cdot \dfrac{2x^2 - 8}{44}$

19. $\dfrac{x^2 - 5xy - 24y^2}{2x^2 - 2xy - 24y^2} \cdot \dfrac{4x^2 + 4xy - 24y^2}{x^2 - 10xy + 16y^2}$

20. $\dfrac{2x^2 + 10x + 2}{8x - 8} \cdot \dfrac{3x - 3}{4x^2 + 20x + 4}$

Section 6.3

Add or subtract.

21. $\dfrac{7}{x + 1} + \dfrac{4}{2x}$

22. $5 + \dfrac{1}{x} + \dfrac{1}{x + 1}$

23. $\dfrac{7}{x + 2} + \dfrac{3}{x - 4}$

24. $\dfrac{2}{x^2 - 9} + \dfrac{x}{x + 3}$

25. $\dfrac{x}{y} + \dfrac{3}{2y} + \dfrac{1}{y + 2}$

26. $\dfrac{4}{a} + \dfrac{2}{b} + \dfrac{3}{a + b}$

27. $\dfrac{3x + 1}{3x} - \dfrac{1}{x}$

28. $\dfrac{x + 4}{x + 2} - \dfrac{1}{2x}$

29. $\dfrac{27}{x^2 - 81} + \dfrac{3}{2(x + 9)}$

30. $\dfrac{1}{x^2 + 7x + 10} - \dfrac{x}{x + 5}$

Section 6.4

Simplify.

31. $\dfrac{\dfrac{3}{2y} - \dfrac{1}{y}}{\dfrac{4}{y} + \dfrac{3}{2y}}$

32. $\dfrac{\dfrac{2}{x} + \dfrac{1}{2x}}{x + \dfrac{x}{2}}$

33. $\dfrac{w - \dfrac{4}{w}}{1 + \dfrac{2}{w}}$

34. $\dfrac{1 - \dfrac{w}{w - 1}}{1 + \dfrac{w}{1 - w}}$

35. $\dfrac{1 + \dfrac{1}{y^2 - 1}}{\dfrac{1}{y + 1} - \dfrac{1}{y - 1}}$

36. $\dfrac{\dfrac{1}{y} + \dfrac{1}{x + y}}{1 + \dfrac{2}{x + y}}$

37. $\dfrac{\dfrac{1}{a+b} - \dfrac{1}{a}}{b}$

38. $\dfrac{\dfrac{2}{a+b} - \dfrac{3}{b}}{\dfrac{1}{a+b}}$

39. $\dfrac{x+5y}{x-6y} \div \left(\dfrac{1}{5y} - \dfrac{1}{x+5y} \right)$

40. $\left(\dfrac{1}{x+2y} - \dfrac{1}{x-y} \right) \div \dfrac{2x-4y}{x^2 - 3xy + 2y^2}$

Section 6.5

Solve for the variable. If there is no solution, say so.

41. $\dfrac{8a-1}{6a+8} = \dfrac{3}{4}$

42. $\dfrac{8}{a-3} = \dfrac{12}{a+3}$

43. $\dfrac{2x-1}{x} - \dfrac{1}{2} = -2$

44. $\dfrac{5-x}{x} - \dfrac{7}{x} = -\dfrac{3}{4}$

45. $\dfrac{5}{2} - \dfrac{2y+7}{y+6} = 3$

46. $\dfrac{5}{4} - \dfrac{1}{2x} = \dfrac{1}{x} + 2$

47. $\dfrac{7}{8x} - \dfrac{3}{4} = \dfrac{1}{4x} + \dfrac{1}{2}$

48. $\dfrac{1}{3x} + 2 = \dfrac{5}{6x} - \dfrac{1}{2}$

49. $\dfrac{3}{y-3} = \dfrac{3}{2} + \dfrac{y}{y-3}$

50. $\dfrac{x-8}{x-2} = \dfrac{2x}{x+2} - 2$

51. $\dfrac{9}{2} - \dfrac{7y-4}{y+2} = -\dfrac{1}{4}$

52. $\dfrac{3y-1}{3y} - \dfrac{6}{5y} = \dfrac{1}{y} - \dfrac{4}{15}$

53. $\dfrac{y+18}{y^2 - 16} = \dfrac{y}{y+4} - \dfrac{y}{y-4}$

54. $\dfrac{4}{x^2 - 1} = \dfrac{2}{x-1} + \dfrac{2}{x+1}$

55. $\dfrac{9y-3}{y^2 + 2y} - \dfrac{5}{y+2} = \dfrac{3}{y}$

56. $\dfrac{2}{3-3y} + \dfrac{2}{2y-1} = \dfrac{4}{3y-3}$

Section 6.6

Solve. Leave your answer in decimal form. Round to the nearest tenth.

57. $\dfrac{x}{4} = \dfrac{12}{17}$

58. $\dfrac{8}{5} = \dfrac{2}{x}$

59. $\dfrac{33}{10} = \dfrac{x}{8}$

60. $\dfrac{5}{x} = \dfrac{22}{9}$

61. $\dfrac{13.5}{0.6} = \dfrac{360}{x}$

62. $\dfrac{2\frac{1}{2}}{3\frac{1}{4}} = \dfrac{7}{x}$

Use a proportion to answer each question.

63. *Paint Needs* A 5-gallon can of paint will cover 240 square feet. How many gallons of paint will be needed to cover 400 square feet? Round to the nearest tenth of a gallon.

64. *Recipe Ratios* Aunt Lexie uses 3 pounds of sugar to make 100 cookies. How many cookies can she make with 5 pounds of sugar? Round to the nearest whole cookie.

65. *Gas Usage* Ron found that his car used 7 gallons of gas to travel 200 miles. He plans to drive 1300 miles from his home to Denver, Colorado. How many gallons of gas will he use if his car continues to consume gas at the same rate? Round to the nearest gallon.

66. *Map Scale* The distance on a map between two cities is 4 inches. The actual distance between these cities is 122 miles. Two lakes are 3 inches apart on the same map. How many miles apart are these two lakes?

67. *Travel Speeds* A train travels 180 miles in the same time that a car travels 120 miles. The speed of the train is 20 miles per hour faster than the speed of the car. Find the speed of the train and the speed of the car.

68. *Interior Painting* A professional painter can paint the interior of the Jacksons' house in 5 hours. John Jackson can do the same job in 8 hours. How long would it take these two people working together on the painting job? Round to the nearest minute.

▲ **69.** *Shadows* Mary takes a walk across a canyon in New Mexico. She stands 5.75 feet tall and her shadow is 3 feet long. At the same time, the shadow from the peak of the canyon wall casts a shadow that is 95 feet long. How tall is the peak of the canyon? Round to the nearest foot.

▲ **70.** *Shadows* A flagpole that is 8 feet tall casts a shadow of 3 feet. At the same time of day, a tall office building in the city casts a shadow of 450 feet. How tall is the office building?

71. *House Painting* Fred is an experienced painter. He can paint the sides of an average house in 5 hours. His new assistant is still being trained. It takes the assistant 10 hours to paint the sides of an average house. How long would it take Fred and his assistant to paint the sides of an average house if they worked together?

72. *Plowing Fields* Sally runs the family farm in Boone, Iowa. She can plow the fields of the farm in 20 hours. Her daughter Brenda can plow the fields of the farm in 30 hours. If they have two identical tractors, how long would it take Brenda and Sally to plow the fields of the farm if they worked together?

Mixed Practice

Perform the indicated operation. Simplify.

73. $\dfrac{a^2 + a - 12}{48a^2 - 16a^3}$

74. $\dfrac{6b^3 - 24b^2}{b^2 + b - 20}$

75. $\dfrac{x^2 - y^2}{x^2 + 4xy + 3y^2} \cdot \dfrac{x^2 + xy - 6y^2}{x^2 + xy - 2y^2}$

76. $\dfrac{x^2 - 6xy - 16y^2}{x^2 + 4xy - 21y^2} \cdot \dfrac{x^2 - 8xy + 15y^2}{x^2 - 3xy - 10y^2}$

77. $\dfrac{x}{x + 3} + \dfrac{9x + 18}{x^2 + 3x}$

78. $\dfrac{x - 30}{x^2 - 5x} + \dfrac{x}{x - 5}$

79. $\dfrac{a + b}{ax + ay} - \dfrac{a + b}{bx + by}$

80. $\dfrac{\dfrac{5}{3x} + \dfrac{2}{9x}}{\dfrac{3}{x} + \dfrac{8}{3x}}$

81. $\dfrac{\dfrac{4}{5y} - \dfrac{8}{y}}{y + \dfrac{y}{5}}$

82. $\dfrac{x - 3y}{x + 2y} \div \left(\dfrac{2}{y} - \dfrac{12}{x + 3y} \right)$

83. $\dfrac{10}{x - 3} = \dfrac{9}{x - 5}$

84. $\dfrac{7x - 4}{6x - 5} = \dfrac{8}{7}$

85. $2 + \dfrac{4}{b - 1} = \dfrac{4}{b^2 - b}$

86. ***Wire Weight*** If 30 meters of wire weighs 8 kilograms, what will 40 meters of the same kind of wire weigh? Round your answer to the nearest tenth.

87. ***Driving Rates*** Ming drove 270 miles in $4\dfrac{1}{2}$ hours. At that same rate, how far would she drive in 7 hours?

Remember to use your Chapter Test Prep Video CD to see the worked-out solutions to the test problems you want to review.

Perform the operation indicated. Simplify.

1. $\dfrac{2ac + 2ad}{3a^2c + 3a^2d}$

2. $\dfrac{8x^2 - 2x^2y^2}{y^2 + 4y + 4}$

3. $\dfrac{x^2 + 2x}{2x - 1} \cdot \dfrac{10x^2 - 5x}{12x^3 + 24x^2}$

4. $\dfrac{x + 2y}{12y^2} \cdot \dfrac{4y}{x^2 + xy - 2y^2}$

5. $\dfrac{2a^2 - 3a - 2}{a^2 + 5a + 6} \div \dfrac{a^2 - 5a + 6}{a^2 - 9}$

6. $\dfrac{1}{a^2 - a - 2} + \dfrac{3}{a - 2}$

7. $\dfrac{x - y}{xy} - \dfrac{a - y}{ay}$

8. $\dfrac{3x}{x^2 - 3x - 18} - \dfrac{x - 4}{x - 6}$

9. $\dfrac{\dfrac{x}{3y} - \dfrac{1}{2}}{\dfrac{4}{3y} - \dfrac{2}{x}}$

10. $\dfrac{\dfrac{6}{b} - 4}{\dfrac{5}{bx} - \dfrac{10}{3x}}$

11. $\dfrac{2x^2 + 3xy - 9y^2}{4x^2 + 13xy + 3y^2}$

12. $\dfrac{1}{x + 4} - \dfrac{2}{x^2 + 6x + 8}$

In questions 13–18, solve for x. Check your answers. If there is no solution, say so.

13. $\dfrac{4}{3x} - \dfrac{5}{2x} = 5 - \dfrac{1}{6x}$

14. $\dfrac{x - 3}{x - 2} = \dfrac{2x^2 - 15}{x^2 + x - 6} - \dfrac{x + 1}{x + 3}$

15. $3 - \dfrac{7}{x + 3} = \dfrac{x - 4}{x + 3}$

16. $\dfrac{3}{3x - 5} = \dfrac{7}{5x + 4}$

17. $\dfrac{9}{x} = \dfrac{13}{5}$

18. $\dfrac{9.3}{2.5} = \dfrac{x}{10}$

19. A random check of America West air flights last month showed that 113 of the 150 flights checked arrived on time. If the inspectors check 200 flights next month, how many can be expected to be on time? (Round to the nearest whole number.)

20. In northern Michigan the Gunderson family heats their home with firewood. They used $100 worth of wood in 25 days. Mr. Gunderson estimates that he needs to burn wood at that rate for about 92 days during the winter. If that is so, how much will the 92-day supply of wood cost?

▲ **21.** A hiking club is trying to construct a rope bridge across a canyon. A 6-foot construction pole held upright casts a 7-foot shadow. At the same time of day, a tree at the edge of the canyon casts a shadow that exactly covers the distance that is needed for the rope bridge. The tree is exactly 87 feet tall. How long should the rope bridge be? Round to the nearest foot.

1. _____

2. _____

3. _____

4. _____

5. _____

6. _____

7. _____

8. _____

9. _____

10. _____

11. _____

12. _____

13. _____

14. _____

15. _____

16. _____

17. _____

18. _____

19. _____

20. _____

21. _____

1. _____

2. _____

3. _____

4. _____

5. _____

6. _____

7. _____

8. _____

9. _____

10. _____

11. _____

12. _____

13. _____

14. _____

15. _____

16. _____

17. _____

18. _____

19. _____

20. _____

21. _____

22. _____

Approximately one-half of this test covers the content of Chapters 0–5. The remainder covers the content of Chapter 6.

1. To the nearest thousandth, how much error can be tolerated in the length of a wire that is supposed to be 2.57 centimeters long if specifications allow an error of no more than 0.25%?

2. There are 252 students in the drama club at Western Community College. If there are 720 students at the Community College, what percent are in the drama club?

3. Carla is thinking of buying a new car for $18,500. Her state presently has a sales tax of 5.5%. In a few weeks the state will raise the sales tax to 6%. How much will she save in sales tax if she purchases the car before the sales tax rate is raised?

4. Solve. $5(x - 3) - 2(4 - 2x) = 7(x - 1) - (x - 2)$

5. Solve for h. $A = \pi r^2 h$

6. Solve and graph on a number line. $4(2 - x) < 3$

$$\xleftarrow{\quad}\underset{0}{+}\ +\ \underset{1}{+}\ +\ \underset{2}{+}\ +\ \xrightarrow{\quad} x$$

7. Solve for x. $\dfrac{3}{5}x - 2 \le \dfrac{1}{2}x + \dfrac{3}{5}$

8. Factor. $3ax + 3bx - 2ay - 2by$ 9. Factor. $8a^3 - 38a^2b - 10ab^2$

10. Simplify. Leave the answer in positive exponents. $\dfrac{-12x^{-2}y^4}{3x^{-5}y^{-8}}$

11. Simplify. $\dfrac{4x^2 - 25}{2x^2 + 9x - 35}$

Perform the indicated operations.

12. $\dfrac{x^2 - 4}{x^2 - 25} \cdot \dfrac{3x^2 - 14x - 5}{3x^2 + 6x}$ 13. $\dfrac{2x^2 - 9x + 9}{8x - 12} \div \dfrac{x^2 - 3x}{2x}$

14. $\dfrac{5}{2x + 4} + \dfrac{3}{x - 3}$ 15. $\dfrac{8}{c^2 - 4} - \dfrac{2}{c^2 - 5c + 6}$

Solve for x.

16. $\dfrac{3x - 2}{3x + 2} = 2$ 17. $\dfrac{x - 3}{x} = \dfrac{x + 2}{x + 3}$

In questions 18 and 19, simplify.

18. $\dfrac{\dfrac{1}{x - 3} + \dfrac{5}{x^2 - 9}}{\dfrac{6x}{x + 3}}$ 19. $\dfrac{\dfrac{3}{a} + \dfrac{2}{b}}{\dfrac{5}{a^2} - \dfrac{2}{b^2}}$ 20. Solve for x. $\dfrac{2x + 1}{3} = \dfrac{4}{5}$

21. Jane is looking at a road map. The distance between two cities is 130 miles. This distance is represented by $2\frac{1}{2}$ inches on the map. She sees that the distance she has to drive today is a totally straight interstate highway that is 4 inches long on the map. How many miles does she have to drive?

22. Roberto is working as a telemarketing salesperson for a large corporation. In the last 22 telephone calls he has made, he was able to make a sale five times. His goal is to make 110 sales this month. If his rate for making a sale continues as in the past, how many phone calls must he make?

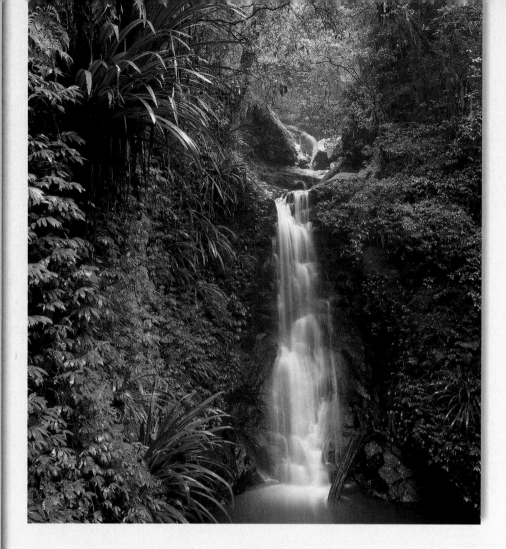

CHAPTER 7

A rainforest is one of the most amazing ecosystems in the entire world. It is a huge region with its own climate, an amazing variety of animals and plant life, and home to many species that do not exist anywhere else in the world. However, all is not well. The amount of land covered by rainforests is decreasing. Can you predict how rapidly this decrease is taking place? Turn to the Putting Your Skills to Work problems on page 459 to find out.

Graphing and Functions

Student Learning Objectives

After studying this section, you will be able to:

1 Plot a point, given the coordinates.

2 Determine the coordinates of a plotted point.

3 Find ordered pairs for a given linear equation.

1 Plotting a Point, Given the Coordinates

Oftentimes we can better understand an idea if we see a picture. This is the case with many mathematical concepts, including those relating to algebra. We can illustrate algebraic relationships with drawings called **graphs.** Before we can draw a graph, however, we need a frame of reference.

In Chapter 1 we showed that any real number can be represented on a number line. Look at the following number line. The arrow indicates the positive direction.

To form a **rectangular coordinate system,** we draw a second number line vertically. We construct it so that the 0 point on each number line is exactly at the same place. We refer to this location as the **origin.** The horizontal number line is often called the **x-axis.** The vertical number line is often called the **y-axis.** Arrows show the positive direction for each axis.

We can represent a point in this rectangular coordinate system by using an **ordered pair** of numbers. For example, (5, 2) is an ordered pair that represents a point in the rectangular coordinate system. The numbers in an ordered pair are often referred to as the **coordinates** of the point. The first number is called the **x-coordinate** and it represents the distance from the origin measured along the horizontal or x-axis. If the x-coordinate is positive, we count the proper number of squares to the right (that is, in the positive direction). If the x-coordinate is negative, we count to the left. The second number in the pair is called the **y-coordinate** and it represents the distance from the origin measured along the y-axis. If the y-coordinate is positive, we count the proper number of squares upward (that is, in the positive direction). If the y-coordinate is negative, we count downward.

$$(5, 2)$$
x-coordinate ⌐↑ ↑⌐ y-coordinate

Suppose the directory for the map on the left indicated that you would find a certain street in the region B5. To find the street you would first scan across the horizontal scale until you found section B; from there you would scan up the map until you hit section 5 along the vertical scale. As we will see in the next example, plotting a point in the rectangular coordinate system is much like finding a street on a map with grids.

> **EXAMPLE 1** Plot the point (5, 2) on a rectangular coordinate system. Label this as point A.
>
> **Solution** Since the x-coordinate is 5, we first count 5 units to the right on the x-axis. Then, because the y-coordinate is 2, we count 2 units up from the point

where we stopped on the *x*-axis. This locates the point corresponding to (5, 2). We mark this point with a dot and label it *A*.

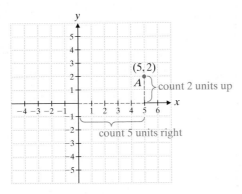

Practice Problem 1   Plot the point (3, 4) on the preceding coordinate system. Label this as point *B*.

NOTE TO STUDENT: Fully worked-out solutions to all of the Practice Problems can be found at the back of the text starting at page SP-1

It is important to remember that the first number in an ordered pair is the *x*-coordinate and the second number is the *y*-coordinate. The ordered pairs (5, 2) and (2, 5) represent different points.

EXAMPLE 2   Plot each point on the following coordinate system. Label the points *F*, *G*, and *H*, respectively.

(a) $(-5, 3)$ **(b)** $(2, -6)$ **(c)** $(-4, -5)$

Solution

(a) Notice that the *x*-coordinate, -5, is negative. On the coordinate grid, negative *x*-values appear to the left of the origin. Thus, we will begin by counting 5 squares to the left starting at the origin. Since the *y*-coordinate, 3, is positive, we will count 3 units up from the point where we stopped on the *x*-axis.

(b) The *x*-coordinate is positive. Begin by counting 2 squares to the right of the origin. Then count down because the *y*-coordinate is negative.

(c) The *x*-coordinate is negative. Begin by counting 4 squares to the left of the origin. Then count down because the *y*-coordinate is negative.

PRACTICE PROBLEM 2

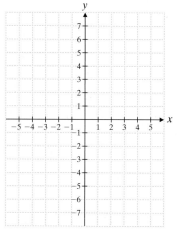

Practice Problem 2 Use the coordinate system in the margin to plot each point. Label the points *I*, *J*, and *K*, respectively.

(a) $(-2, -4)$ **(b)** $(-4, 5)$ **(c)** $(4, -2)$

EXAMPLE 3 Plot the following points.

$F: (0, 5)$ $G: \left(3, \frac{3}{2}\right)$ $H: (-6, 4)$ $I: (-3, -4)$

$J: (-4, 0)$ $K: (2, -3)$ $L: (6.5, -7.2)$

Solution These points are plotted in the figure.

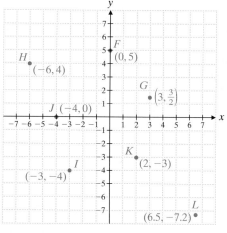

Note: When you are plotting decimal values like $(6.5, -7.2)$, plot the point halfway between 6 and 7 in the *x*-direction (for the 6.5) and at your best approximation of -7.2 in the *y*-direction.

PRACTICE PROBLEM 3

NOTE TO STUDENT: Fully worked-out solutions to all of the Practice Problems can be found at the back of the text starting at page SP-1

Practice Problem 3 Plot the following points. Label each point with both the letter and the ordered pair. Use the coordinate system provided in the margin.

$A: (3, 7)$ $B: (0, -6)$ $C: (3, -4.5)$ $D: \left(-\frac{7}{2}, 2\right)$

---

② Determining the Coordinates of a Plotted Point

Sometimes we need to find the coordinates of a point that has been plotted. First, we count the units we need on the *x*-axis to get as close as possible to the point. Next we count the units up or down we need to go from the *x*-axis to reach the point.

EXAMPLE 4   What ordered pair of numbers represents point A in the graph below?

Solution   If we move along the x-axis until we get as close as possible to A, we end up at the number 5. Thus we obtain 5 as the first number of the ordered pair. Then we count 4 units upward on a line parallel to the y-axis to reach A. So we obtain 4 as the second number of the ordered pair. Thus point A is represented by the ordered pair $(5, 4)$.

Practice Problem 4   What ordered pair of numbers represents point B in the graph below?

EXAMPLE 5   Write the coordinates of each point plotted in the following graph.

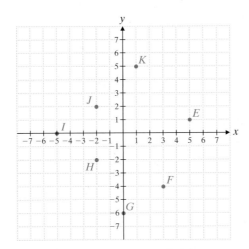

Solution   The coordinates of each point are as follows.

$E = (5, 1)$ $I = (-5, 0)$
$F = (3, -4)$ $J = (-2, 2)$ Be very careful that you put the x-coordinate
$G = (0, -6)$ $K = (1, 5)$ first and the y-coordinate second. Be careful
$H = (-2, -2)$ that each sign is correct.

Practice Problem 5   Give the coordinates of each point plotted in the graph in the margin.

PRACTICE PROBLEM 5

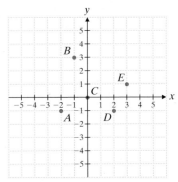

③ Finding Ordered Pairs for a Given Linear Equation

Equations such as $3x + 2y = 5$ and $6x + y = 3$ are called linear equations in two variables.

> A **linear equation in two variables** is an equation that can be written in the form $Ax + By = C$ where A, B, and C are real numbers but A and B are not *both* zero.

Replacement values for x and y that make *true mathematical statements* of the equation are called *truth values,* and an ordered pair of these truth values is called a **solution.**

Consider the equation $3x + 2y = 5$. The ordered pair $(1, 1)$ is a solution to the equation because when we replace x by 1 and y by 1 in the equation, we obtain a true statement.

$$3(1) + 2(1) = 5 \quad \text{or} \quad 3 + 2 = 5$$

Likewise $(-1, 4)$, $(3, -2)$, $(5, -5)$, and $(7, -8)$ are also solutions to the equation. In fact, there are an infinite number of solutions for any given linear equation in two variables.

The linear equations that we work with are not always written in the form $Ax + By = C$, but are sometimes solved for y, as in $y = -6x + 3$. Consider the equation $y = -6x + 3$. The ordered pair $(2, -9)$ is a solution to the equation. When we replace x by 2 and y by -9 we obtain a true mathematical statement:

$$(-9) = -6(2) + 3 \quad \text{or} \quad -9 = -12 + 3.$$

If one value of an ordered-pair solution to a linear equation is known, the other can be quickly obtained. To do so, we replace the proper variable in the equation by the known value. Then using the methods learned in Chapter 2, we solve the resulting equation for the other variable.

EXAMPLE 6 Find the missing coordinate to complete the following ordered-pair solutions for the equation $2x + 3y = 15$.

(a) $(0, ?\,)$ **(b)** $(\,?, 1)$

Solution

(a) For the ordered pair $(0, ?\,)$ we know that $x = 0$. Replace x by 0 in the equation.

$$2x + 3y = 15$$
$$2(0) + 3y = 15$$
$$0 + 3y = 15$$
$$y = 5$$

Thus we have the ordered pair $(0, 5)$.

(b) For the ordered pair $(\,?, \ 1)$, we *do not know* the value of x. However, we do know that $y = 1$. So we start by replacing the variable y by 1. We will end up with an equation with one variable, x. We can then solve for x.

$$2x + 3y = 15$$
$$2x + 3(1) = 15$$
$$2x + 3 = 15$$
$$2x = 12$$
$$x = 6$$

Thus we have the ordered pair $(6, 1)$.

NOTE TO STUDENT: Fully worked-out solutions to all of the Practice Problems can be found at the back of the text starting at page SP-1

Practice Problem 6 Find the missing coordinate to complete the following ordered-pair solutions for the equation $3x - 4y = 12$.

(a) $(0, ?)$ **(b)** $(?, 3)$ **(c)** $(?, -6)$

In examining data from real-world situations, we often find that plotting data points shows useful trends. In such cases, it is often necessary to use a different scale, one that displays only positive values.

EXAMPLE 7   The number of motor vehicle accidents in millions is recorded in the following table for the years 1980 to 2000.

(a) Plot points that represent this data on the given coordinate system.

(b) What trends are apparent from the plotted data?

Number of Years Since 1980	Number of Motor Vehicle Accidents (in Millions)
0	18
5	19
10	12
15	11
20	15

Source: U.S. National Highway Traffic Safety Administration

Solution

(a)

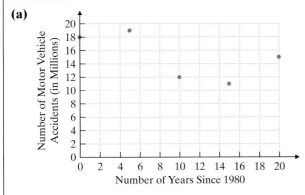

(b) From 1980 to 1985, there was a slight increase in the number of accidents. From 1985 to 1995, there was a significant decrease in the number of accidents. From 1995 to 2000, there was a moderate increase in the number of accidents.

Practice Problem 7   The number of motor vehicle deaths in thousands is recorded in the following table for the years 1980 to 2000.

(a) Plot points that represent this data on the given coordinate system.

(b) What trends are apparent from the plotted data?

Number of Years Since 1980	Number of Motor Vehicle Deaths (in Thousands)
0	51
5	44
10	45
15	42
20	43

Source: U.S. National Highway Traffic Safety Administration

Verbal and Writing Skills

1. What is the *x*-coordinate of the origin?

2. What is the *y*-coordinate of the origin?

3. Explain why (5, 1) is referred to as an *ordered* pair of numbers.

4. Explain how you would locate a point (4, 3) on graph paper.

5. Plot the following points.
 J: (−4, 3.5) *K*: (6, 0)
 L: (5, −6) *M*: (0, −4)

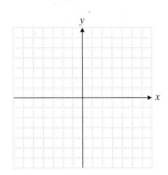

6. Plot the following points.
 R: (−3, 0) *S*: (3.5, 4)
 T: (−2, −2.5) *V*: (0, 5)

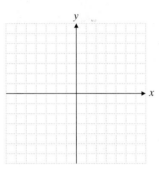

Consider the points plotted in the graph at right.

7. Give the coordinates for points *R*, *S*, *X*, and *Y*.

8. Give the coordinates for points *T*, *V*, *W*, and *Z*.

In exercises 9 and 10, 6 points are plotted in the figure. List all the ordered pairs needed to represent the points.

9.

10.

Find the missing coordinate to complete the ordered-pair solution to the given linear equation.

11. $y = 4x + 7$
 (a) $(0, \quad)$
 (b) $(2, \quad)$

12. $y = 6x + 5$
 (a) $(0, \quad)$
 (b) $(3, \quad)$

13. $y = -4x + 2$
 (a) $(-5, \quad)$
 (b) $(4, \quad)$

14. $y = -2x + 3$
 (a) $(-6, \quad)$
 (b) $(3, \quad)$

15. $3x - 4y = 11$
 (a) $(-3, \quad)$
 (b) $(\quad, 1)$

16. $5x - 2y = 9$
 (a) $(7, \quad)$
 (b) $(\quad, -7)$

17. $2y + 3x = -6$
 (a) $(-2, \quad)$
 (b) $(\quad, 3)$

18. $-4x + 5y = -20$
 (a) $(10, \quad)$
 (b) $(\quad, -8)$

19. $2x + \dfrac{1}{5}y = 6$
 (a) $(\quad, 20)$
 (b) $\left(\dfrac{9}{5}, \quad\right)$

20. $3x + \dfrac{1}{4}y = 11$ **(a)** $(\quad, 8)$ **(b)** $\left(\dfrac{13}{4}, \quad\right)$

Using Road Maps *The preceding map shows a portion of New York, Connecticut, and Massachusetts. Like many maps used in driving or flying, it has horizontal and vertical grid markers for ease of use. For example, Newburgh, New York, is located in grid B3. Use the grid labels to indicate the locations of the following cities.*

21. Lynbrook, New York

22. Hampton Bays, New York

23. Athol, Massachusetts

24. Pittsfield, Massachusetts

25. Hartford, Connecticut

26. Waterbury, Connecticut

27. *Number of Cellular Phones* The number of cellular phones per thousand people in the United States from 1997 to 2002 is recorded in the following table. For example, 315 in the second column means that in that year, 315 out of every 1000 Americans owned cellular phones.

(a) Plot points that represent this data on the given rectangular coordinate system.

(b) What trends are obvious from the plotted data?

Number of Years Since 1997	Number of Cellular Phones (per 1000 People in U.S.)
0	200
1	250
2	315
3	400
4	450
5	485

Source: U.S. Federal Communications Commission

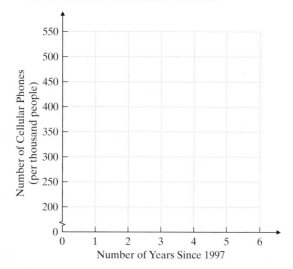

28. *Oil Production* The number of barrels of oil produced in the U.S. for selected years starting in 1940 is recorded in the following table. The number of barrels of oil is measured in hundred thousands. For example, 14 in the second column means 14 hundred thousand, or 1,400,000.

(a) Plot points on the given rectangular coordinate system that represent this data.

(b) What trends are apparent from the plotted data?

Number of Years Since 1940	Number of Barrels of Crude Oil (in Hundred Thousands)
0	14
10	20
20	26
30	35
40	31
50	27
60	21

Source: U.S. Department of Energy

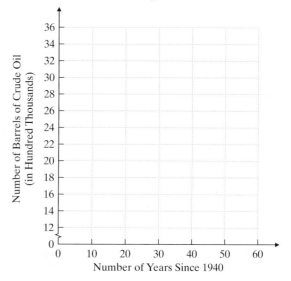

29. *Buying Books Online* The number of people buying books online continues to increase at an amazing rate. The following chart records the number of people for each of the years from 2000 to 2003.

Year	Number of People (in millions) Buying Books Online
2000	23
2001	32
2002	41
2003	49

Source: U.S. Department of Labor

(a) Plot points on the given rectangular coordinate system that represent this data.

(b) Based on your graph, what would you estimate would be the number of people buying books online for the year 2004?

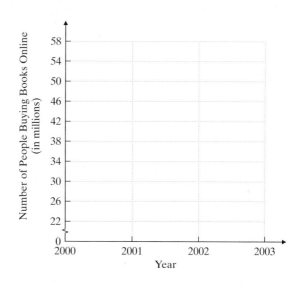

30. *Number of DVD Players* The number of households in the United States that have a DVD player continues to increase at a significant rate. The following chart records the number of households that have this device for each of the years from 2000 to 2003.

Year	Number of Households in Millions in U.S. That Have a DVD Player
2000	7
2001	19
2002	37
2003	48

Source: Federal Communications Commission

(a) Plot points on the given rectangular coordinate system that represents this data.

(b) Based on your graph, what would you estimate would be the number of households that have a DVD player in the year 2004?

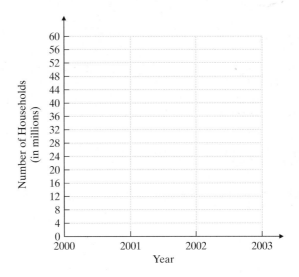

Cumulative Review

31. Simplify. $\dfrac{8x}{x + y} + \dfrac{8y}{x + y}$

32. Simplify. $\dfrac{x^2 - y^2}{x^2 + 4xy + 3y^2} \cdot \dfrac{x^2 - 9y^2}{x^2 - 6xy + 9y^2}$

▲ **33.** *Circular Swimming Pool* The circular pool at the hotel where Bob and Linda stayed in Orlando, Florida, has a radius of 19 yards. What is the area of the pool? (Use $\pi \approx 3.14$.)

34. *Membership at a Gym* The membership at David's gym has increased by 16% over the last 2 years. It used to have 795 members. How many members does it have now?

35. *Circulation of a Newspaper* A major Russian newspaper called *Izvestia* had a circulation of 6,109,005 before the dissipation of the Soviet Union. Currently the circulation is 358,205. What is the percent of decrease of readership of *Izvestia?* Round to the nearest tenth.

▲ **36.** *Prices of Persian Rugs* An expensive Persian rug that measures 30 feet by 22 feet is priced at $44,020.00. A customer negotiates with the owner of the rug, and they agree upon a price of $36,300.

(a) What is the cost per square foot of the rug at the discounted price?

(b) The rug dealer has a smaller matching Persian rug measuring 14 feet by 8 feet. He says he will sell it to the customer at the same cost per square foot as the larger one. How much will the small rug cost?

Student Learning Objectives

After studying this section, you will be able to:

1 Graph a linear equation by plotting three ordered pairs.

2 Graph a straight line by plotting its intercepts.

3 Graph horizontal and vertical lines.

1 Graphing a Linear Equation by Plotting Three Ordered Pairs

We have seen that a solution to a linear equation in two variables is an ordered pair. The graph of an ordered pair is a point. Thus we can graph an equation by graphing the points corresponding to its ordered-pair solutions.

A linear equation in two variables has an infinite number of ordered-pair solutions. We can see that this is true by noting that we can substitute any number for x in the equation and solve it to obtain a y-value. For example, if we substitute $x = 0, 1, 2, 3, \ldots$ into the equation $y = -x + 3$ and solve for y, we obtain the ordered-pair solutions $(0, 3), (1, 2), (2, 1), (3, 0), \ldots$. (If desired, substitute these values into the equation to convince yourself.) If we plot these points on a rectangular coordinate system, we notice that they form a straight line.

 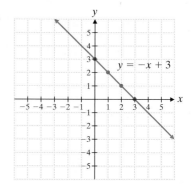

It turns out that all of the points corresponding to the ordered-pair solutions of $y = -x + 3$ lie on this line, and the line extends forever in both directions. A similar statement can be made about any linear equation in two variables.

> The graph of any linear equation in two variables is a straight line.

From geometry, we know that two points determine a line. Thus to graph a linear equation in two variables, we need only graph two ordered-pair solutions of the equation and then draw the line that passes through them. Having said this, we recommend that you use three points to graph a line. Two points will determine where the line is. The third point verifies that you have drawn the line correctly. For ease in plotting, it is better if the ordered pairs contain integers.

> **TO GRAPH A LINEAR EQUATION**
>
> **1.** Look for three ordered pairs that are solutions to the equation.
> **2.** Plot the points.
> **3.** Draw a line through the points.

EXAMPLE 1 Find three ordered pairs that satisfy $2x + y = 4$. Then graph the resulting straight line.

Solution Since we can choose any value for x, choose numbers that are convenient. To organize the results, we will make a table of values. We will let $x = 0$, $x = 1$, and $x = 3$, respectively. We write these numbers under x in our table of values. For each of these x-values, we find the corresponding y-value in the equation $2x + y = 4$.

Table of Values

x	y
0	4
1	2
3	-2

$2x + y = 4$	$2x + y = 4$	$2x + y = 4$
$2(0) + y = 4$	$2(1) + y = 4$	$2(3) + y = 4$

$$0 + y = 4 \qquad\qquad 2 + y = 4 \qquad\qquad 6 + y = 4$$
$$y = 4 \qquad\qquad\qquad y = 2 \qquad\qquad\qquad y = -2$$

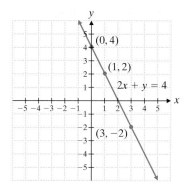

We record these results by placing each y-value in the table next to its corresponding x-value. Keep in mind that these values represent ordered pairs, each of which is a solution to the equation. To make calculating and graphing easier, we choose integer values whenever possible. If we plot these ordered pairs and connect the three points, we get a straight line that is the graph of the equation $2x + y = 4$. The graph of the equation is shown in the figure at the right.

Practice Problem 1   Find 3 ordered pairs that satisfy $x + y = 10$. Then graph the resulting straight line. Use the given coordinate system.

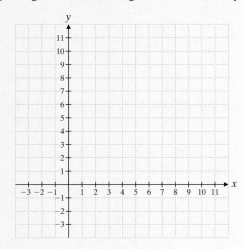

NOTE TO STUDENT: Fully worked-out solutions to all of the Practice Problems can be found at the back of the text starting at page SP-1

EXAMPLE 2   Graph $5x - 4y + 2 = 2$.

Solution   First, we simplify the equation $5x - 4y + 2 = 2$ by adding -2 to each side.

$$5x - 4y + 2 - 2 = 2 - 2$$
$$5x - 4y = 0$$

Since we are free to choose any value of x, $x = 0$ is a natural choice. Calculate the value of y when $x = 0$.

$$5(0) - 4y = 0$$
$$-4y = 0$$
$$y = 0 \qquad \text{Remember: Any number times 0 is 0.}$$
$$\qquad\qquad \text{Since } -4y = 0, y \text{ must equal 0.}$$

Now let's see what happens when $x = 1$.

$$5(1) - 4y = 0$$
$$5 - 4y = 0$$
$$-4y = -5$$
$$y = \frac{-5}{-4} \quad \text{or} \quad \frac{5}{4} \qquad \text{This is not an easy number to graph.}$$

A better choice for a replacement of x is a number that is divisible by 4. Let's see why. Let $x = 4$ and let $x = -4$.

$$5(4) - 4y = 0 \qquad\qquad 5(-4) - 4y = 0$$
$$20 - 4y = 0 \qquad\qquad -20 - 4y = 0$$
$$-4y = -20 \qquad\qquad -4y = 20$$
$$y = \frac{-20}{-4} \quad \text{or} \quad 5 \qquad\qquad y = \frac{20}{-4} \quad \text{or} \quad -5$$

PRACTICE PROBLEM 2

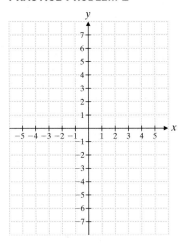

NOTE TO STUDENT: Fully worked-out solutions to all of the Practice Problems can be found at the back of the text starting at page SP-1

Now we can put these numbers into our table of values and graph the line.

Table of Values	
x	*y*
0	0
4	5
−4	−5

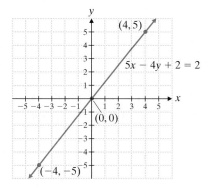

Practice Problem 2   Graph $7x + 3 = -2y + 3$ on the coordinate system in the margin.

TO THINK ABOUT: An Alternative Approach   In Example 2, we picked values of *x* and found the corresponding values for *y*. An alternative approach is to solve the equation for the variable *y* *first*.

$$5x - 4y = 0$$
$$-4y = -5x \qquad \text{Add } -5x \text{ to each side.}$$
$$\frac{-4y}{-4} = \frac{-5x}{-4} \qquad \text{Divide each side by } -4.$$
$$y = \frac{5}{4}x$$

Now let $x = -4$, $x = 0$, and $x = 4$, and find the corresponding values of *y*. Explain why you would choose multiples of 4 as replacements of *x* in this equation. Graph the equation and compare it to the graph in Example 2.

In the previous two examples we began by picking values for *x*. We could just as easily have chosen values for *y*.

② Graphing a Straight Line by Plotting Its Intercepts

What values should we pick for *x* and *y*? Which points should we use for plotting? For many straight lines it is easiest to pick the two *intercepts*. Some lines have only one intercept. We will discuss these separately.

> The **x-intercept** of a line is the point where the line crosses the *x*-axis; it has the form $(a, 0)$. The **y-intercept** of a line is the point where the line crosses the *y*-axis; it has the form $(0, b)$.

INTERCEPT METHOD OF GRAPHING

To graph an equation using intercepts, we:

1. Find the *x*-intercept by letting $y = 0$ and solving for *x*.
2. Find the *y*-intercept by letting $x = 0$ and solving for *y*.
3. Find one additional ordered pair so that we have three points with which to plot the line.

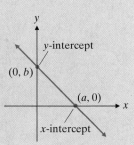

EXAMPLE 3   Use the intercept method to graph $5y - 3x = 15$.

Solution

Let $y = 0$. $5(0) - 3x = 15$ Replace y by 0.

$-3x = 15$ Divide both sides by -3.

$x = -5$

The ordered pair $(-5, 0)$ is the x-intercept.

Let $x = 0$. $5y - 3x = 15$

$5y - 3(0) = 15$ Replace x by 0.

$5y = 15$ Divide both sides by 5.

$y = 3$

The ordered pair $(0, 3)$ is the y-intercept.
    We find another ordered pair to have a third point.

Let $y = 6$. $5(6) - 3x = 15$ Replace y by 6.

$30 - 3x = 15$ Simplify.

$-3x = -15$ Subtract 30 from both sides.

$x = \dfrac{-15}{-3}$ or   5

The ordered pair is $(5, 6)$.
    Our table of values is

x	y
-5	0
0	3
5	6

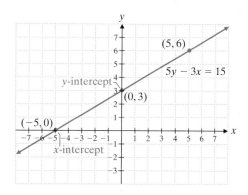

Practice Problem 3   Use the intercept method to graph $2y - x = 6$. Use the given coordinate system.

TO THINK ABOUT: Lines That Go Through the Origin   Can you draw all straight lines by the intercept method? Not really. Some straight lines may go through the origin and have only one intercept. If a line goes through the origin, it will have an equation of the form $Ax + By = 0$, where $A \neq 0$ or $B \neq 0$ or both. Examples are $3x + 4y = 0$ and $5x - 2y = 0$. In such cases you should plot two additional points besides the origin. Be sure to simplify each equation before attempting to graph it.

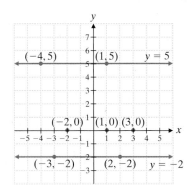

3 Graphing Horizontal and Vertical Lines

You will notice that the x-axis is a horizontal line. It is the line $y = 0$, since for any value of x, the value of y is 0. Try a few points. The points $(1, 0)$, $(3, 0)$, and $(-2, 0)$ all lie on the x-axis. Any horizontal line will be parallel to the x-axis. Lines such as $y = 5$ and $y = -2$ are horizontal lines. What does $y = 5$ mean? It means that for any value of x, y is 5. Likewise $y = -2$ means that for any value of x, $y = -2$.

How can we recognize the equation of a line that is horizontal, that is, parallel to the x-axis?

> If the graph of an equation is a straight line that is parallel to the x-axis (that is, a horizontal line), the equation will be of the form $y = b$, where b is some real number.

EXAMPLE 4 Graph $y = -3$.

Solution You could write the equation as $0x + y = -3$. Then it is clear that for any value of x that you substitute, you will always obtain $y = -3$. Thus, as shown in the figure, $(4, -3)$, $(0, -3)$, and $(-3, -3)$ are all ordered pairs that satisfy the equation $y = -3$. Since the y-coordinate of every point on this line is -3, it is easy to see that the horizontal line will be 3 units below the x-axis.

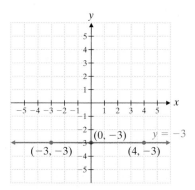

NOTE TO STUDENT: *Fully worked-out solutions to all of the Practice Problems can be found at the back of the text starting at page SP-1*

Practice Problem 4 Graph $2y - 3 = 0$ on the given coordinate system.

Notice that the y-axis is a vertical line. This is the line $x = 0$, since for any y, x is 0. Try a few points. The points $(0, 2)$, $(0, -3)$, and $\left(0, \frac{1}{2}\right)$ all lie on the y-axis. Any vertical line will be parallel to the y-axs. Lines such as $x = 2$ and $x = -3$ are vertical lines. Think of what $x = 2$ means. It means that for any value of y, x is 2. The graph of $x = 2$ is a vertical line two units to the right of the y-axis.

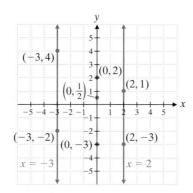

How can we recognize the equation of a line that is vertical, that is, parallel to the y-axis?

> If the graph of an equation is a straight line that is parallel to the y-axis (that is, a vertical line), the equation will be of the form $x = a$, where a is some real number.

EXAMPLE 5   Graph $x = 5$.

Solution   This can be done immediately by drawing a vertical line 5 units to the right of the origin. The x-coordinate of every point on this line is 5.

     The equation $x - 5 = 0$ can be rewritten as $x = 5$ and graphed as shown.

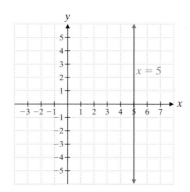

Practice Problem 5   Graph $x + 3 = 0$ on the following coordinate system.

7.2 EXERCISES

Student Solutions Manual | CD/Video | PH Math Tutor Center | MathXL®Tutorials on CD | MathXL® | MyMathLab® | Interactmath.com

Verbal and Writing Skills

1. Is the point $(-2, 5)$ a solution to the equation $2x + 5y = 0$? Why or why not?

2. The graph of a linear equation in two variables is a _____ _____.

3. The x-intercept of a line is the point where the line crosses the _____.

4. The graph of the equation $y = b$ is a _____ line.

Complete the ordered pairs so that each is a solution of the given linear equation. Then plot each solution and graph the equation by connecting the points by a straight line.

5. $y = -2x + 1$
(0,)
(-2,)
(1,)

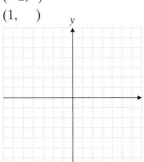

6. $y = -3x - 4$
(-2,)
(-1,)
(0,)

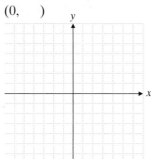

7. $y = x - 4$
(0,)
(2,)
(4,)

8. $y = 3x - 1$
(0,)
(2,)
(-2,)

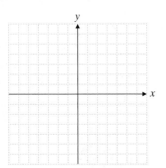

9. $y = -2x + 3$
(0,)
(2,)
(4,)

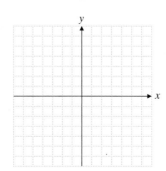

10. $y = 2x - 5$
(0,)
(2,)
(4,)

11. $y = 3x + 2$
(-1,)
(0,)
(1,)

Graph each equation by plotting three points and connecting them.

12. $y = -3x + 2$

13. $3x - 2y = 0$

14. $2y - 5x = 0$

15. $y = -\dfrac{3}{4}x + 3$

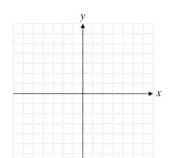

16. $y = -\dfrac{2}{5}x - 2$

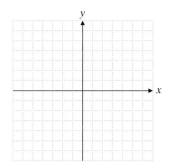

17. $4x + 3y = 12$

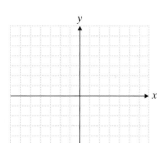

18. $3x + 2y = 6$

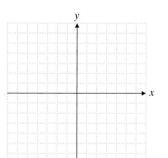

Graph each equation by plotting the intercepts and one other point.

19. $y = 6 - 2x$

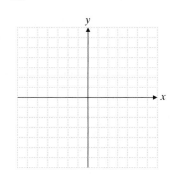

20. $y = 4 - 2x$

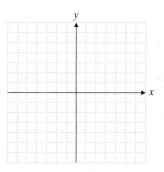

21. $x + 3 = 6y$

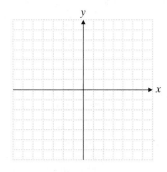

22. $x - 6 = 2y$

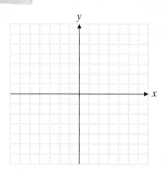

Mixed Practice

Graph the equation. Be sure to simplify the equation before graphing it.

23. $y - 2 = 3y$

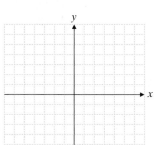

24. $3y + 1 = 7$

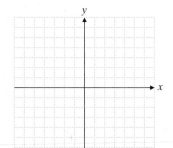

25. $2x + 9 = 5x$

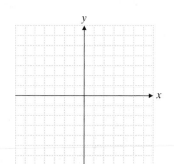

26. $3x - 4 = -13$

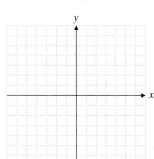

27. $2x + 5y - 2 = -12$

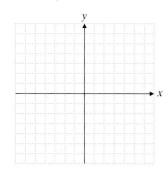

28. $3x - 4y - 5 = -17$

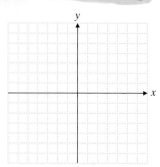

Applications

29. ***Cross-Country Skiing*** The number of calories burned by an average person while cross-country skiing is given by the equation $C = 8m$, where m is the number of minutes. (*Source:* National Center for Health Statistics.) Graph the equation for $m = 0, 15, 30, 45, 60,$ and $75.$

Calories Burned While Cross-Country Skiing

Minutes spent skiing

30. ***Calories Burned While Jogging*** The number of calories burned by an average person while jogging is given by the equation $C = \dfrac{28}{3}m$, where m is the number of minutes. (*Source:* National Center for Health Statistics.) Graph the equation for $m = 0,$ $15, 30, 45, 60,$ and $75.$

Calories Burned While Jogging

Minutes spent jogging

31. ***Foreign Students in the United States*** The number of foreign students enrolled in college in the United States is approximated by the equation $S = 12t + 280$, where t stands for the number of years since 1980, and S is the number of foreign students (in thousands). (*Source:* Statistical Abstract of the United States, 2002) Graph the equation for $t = 0, 6, 15, 21.$

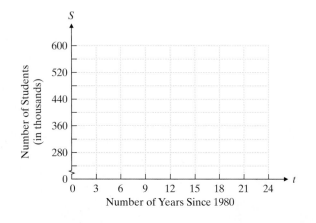

32. ***Population of the United States*** The approximate population of the United States is given by the equation $P = 2.3t + 179$ where P is the population in millions and t is the number of years since 1960. (*Source:* Bureau of the Census, U.S. Department of Commerce.) Graph the equation for $t = 0,$ $10, 30, 40.$

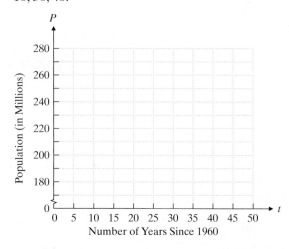

Cumulative Review

33. Solve. $\dfrac{x}{x+4} + \dfrac{x-1}{x-4} = 2$

34. Solve and graph on a number line. $4 - 3x \le 18$

▲ 35. *Geometry* A rectangle's width is 1 meter more than half of its length. Find the dimensions if the perimeter measures 53 meters.

36. *Education* Last semester there were 29 students for every 2 faculty members at Skyline University. At that time the school had 3074 students. How many faculty members were there last semester?

37. *Foreign Mail Costs* Hank is working for six months in Japan while his wife is back in California with their two children. Each of them sends the other a letter each week. The cost of mailing a letter from Japan and the cost of mailing a letter from the United States totals $1.84. The cost of mailing a letter from the United States is two cents less than half the cost of mailing a letter from Japan. What is the cost of mailing a letter from Japan?

38. *Caffeine in Coffee* Each day Melinda has a cup of instant coffee at home in the morning and a cup of percolated coffee at the office in the afternoon. She consumes 194 milligrams of caffeine each day with these two cups. If the instant coffee has seventeen more milligrams than half the number of milligrams of caffeine in the percolated coffee, how many milligrams of caffeine does a cup of the instant coffee contain?

▲ 39. *Geometry* One base of a trapezoid is 3 less than twice the length of the other base. The altitude of the trapezoid is 14 inches. If the area of the trapezoid is 63 square inches, find the length of each base.

7.3 The Slope of a Line

Student Learning Objectives

After studying this section, you will be able to:

1. Find the slope of a line given two points on the line.

2. Find the slope and *y*-intercept of a line given its equation.

3. Write the equation of a line given the slope and *y*-intercept.

4. Graph a line using the slope and *y*-intercept.

5. Find the slopes of parallel and perpendicular lines.

1 Finding the Slope of a Line Given Two Points on the Line

We often use the word *slope* to describe the incline (the steepness) of a hill. A carpenter or a builder will refer to the *pitch* or *slope* of a roof. The slope is the change in the vertical distance compared to the change in the horizontal distance as you go from one point to another point along the roof. If the change in the vertical distance is greater than the change in the horizontal distance, the slope will be steep. If the change in the horizontal distance is greater than the change in the vertical distance, the slope will be gentle.

In a coordinate plane, the **slope** of a straight line is defined by the change in *y* divided by the change in *x*.

$$\text{Slope} = \frac{\text{change in } y}{\text{change in } x}$$

Consider the line drawn through points *A* and *B* in the figure. If we measure the change from point *A* to point *B* in the *x*-direction and the *y*-direction, we will have an idea of the steepness (or the slope) of the line. From point *A* to point *B* the change in *y* values is from 2 to 4, a *change of* 2. From point *A* to point *B* the change in *x* values is from 1 to 5, a *change of* 4. Thus

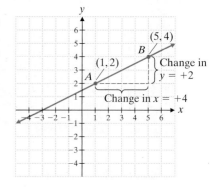

$$\text{slope} = \frac{\text{change in } y}{\text{change in } x} = \frac{2}{4} = \frac{1}{2}$$

Informally, we can describe this move as the rise over the run: $\text{slope} = \frac{\text{rise}}{\text{run}}$.

We now state a more formal (and more frequently used) definition.

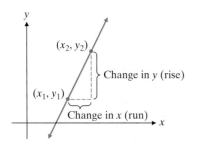

DEFINITION OF SLOPE OF A LINE

The **slope** of any *nonvertical* straight line that contains the points with coordinates (x_1, y_1) and (x_2, y_2) has a slope defined by the difference ratio

$$\text{slope} = m = \frac{y_2 - y_1}{x_2 - x_1} \quad \text{where } x_2 \neq x_1.$$

The use of subscripted terms such as x_1, x_2, and so on, is just a way of indicating that the first *x*-value is x_1 and the second *x*-value is x_2. Thus (x_1, y_1) are the coordinates of the first point and (x_2, y_2) are the coordinates of the second point. The letter *m* is commonly used for the slope.

EXAMPLE 1   Find the slope of the line that passes through $(2, 0)$ and $(4, 2)$.

Solution   Let $(2, 0)$ be the first point (x_1, y_1) and $(4, 2)$ be the second point (x_2, y_2).

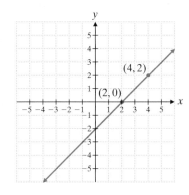

$$(2, 0) \quad\quad (4, 2) \qquad m = \frac{2}{2} = 1$$

By the formula, $\text{slope} = m = \dfrac{y_2 - y_1}{x_2 - x_1} = \dfrac{2 - 0}{4 - 2} = \dfrac{2}{2} = 1.$

The sketch of the line is shown in the figure at the right.

Note that the slope of the line will be the same if we let $(4, 2)$ be the first point (x_1, y_1) and $(2, 0)$ be the second point (x_2, y_2).

$$(4, 2) \quad\quad (2, 0)$$

$$m = \frac{y_2 - y_1}{x_2 - x_1} = \frac{0 - 2}{2 - 4} = \frac{-2}{-2} = 1$$

Thus, given two points, it does not matter which you call (x_1, y_1) and which you call (x_2, y_2).

CAUTION   Be careful, however, not to put the x's in one order and the y's in another order when finding the slope from two points on a line.

Practice Problem 1   Find the slope of the line that passes through $(6, 1)$ and $(-4, -1)$.

NOTE TO STUDENT: Fully worked-out solutions to all of the Practice Problems can be found at the back of the text starting at page SP-1

It is a good idea to have some concept of the values of slopes. In downhill skiing, a very gentle slope used for teaching beginning skiers might drop one foot vertically for each 10 feet horizontally. The slope would be $\frac{1}{10}$. The speed of a skier on a hill with such a gentle slope would be only about 6 miles per hour.

A triple diamond slope for experts might drop 11 feet vertically for each 10 feet horizontally. The slope would be $\frac{11}{10}$. The speed of a skier on such an expert trail would be in the range of 60 miles per hour.

It is important to see how positive and negative slopes affect the graphs of lines.

Slope $= \frac{1}{10}$

1 foot

10 feet

Slope $= \frac{11}{10}$

11 feet

10 feet

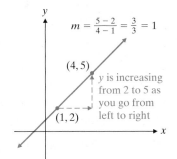

$m = \dfrac{5 - 2}{4 - 1} = \dfrac{3}{3} = 1$

$(4, 5)$

y is increasing from 2 to 5 as you go from left to right

$(1, 2)$

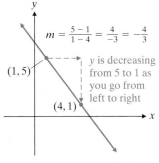

$m = \dfrac{5 - 1}{1 - 4} = \dfrac{4}{-3} = -\dfrac{4}{3}$

$(1, 5)$

y is decreasing from 5 to 1 as you go from left to right

$(4, 1)$

1. If the y-values increase as you go from left to right, the slope of the line is positive.

2. If the y-values decrease as you go from left to right, the slope of the line is negative.

EXAMPLE 2 Find the slope of the line that passes through $(-3, 2)$ and $(2, -4)$.

Solution Let $(-3, 2)$ be (x_1, y_1) and $(2, -4)$ be (x_2, y_2).

$$m = \frac{y_2 - y_1}{x_2 - x_1} = \frac{-4 - 2}{2 - (-3)} = \frac{-4 - 2}{2 + 3} = \frac{-6}{5} = -\frac{6}{5}$$

The slope of this line is negative. We would expect this, since the y-value decreased from 2 to -4 as the x-value increased. What does the graph of this line look like? Plot the points and draw the line to verify.

Practice Problem 2 Find the slope of the line that passes through $(2, 0)$ and $(-1, 1)$.

TO THINK ABOUT: Using the Slope to Describe a Line Describe the line in Practice Problem 2 by looking at its slope. Then verify by drawing the graph.

EXAMPLE 3 Find the slope of the line that passes through the given points.

(a) $(0, 2)$ and $(5, 2)$ **(b)** $(-4, 0)$ and $(-4, -4)$

Solution

(a) Take a moment to look at the y-values. What do you notice? What does this tell you about the line? Now calculate the slope.

$$m = \frac{2 - 2}{5 - 0} = \frac{0}{5} = 0$$

Since any two points on a horizontal line will have the same y-value, the slope of a horizontal line is 0.

(b) Take a moment to look at the x-values. What do you notice? What does this tell you about the line? Now calculate the slope.

$$m = \frac{-4 - 0}{-4 - (-4)} = \frac{-4}{0}$$

Recall that division by 0 is undefined.

Example 3(a)

Example 3(b)

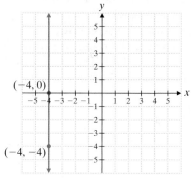

The slope of a vertical line is undefined. We say that a vertical line has **no slope.**

Notice in our definition of slope that $x_2 \neq x_1$. Thus it is not appropriate to use the formula for slope for the points in (b). We did so to illustrate what would happen if $x_2 = x_1$. We get an impossible situation, $\frac{y_2 - y_1}{0}$. Now you can see why we include the restriction $x_2 \neq x_1$ in our definition.

Practice Problem 3 Find the slope of the line that passes through the given points.

(a) $(-5, 6)$ and $(-5, 3)$ **(b)** $(-7, -11)$ and $(3, -11)$

SLOPE OF A STRAIGHT LINE

Positive slope
Line goes upward to the right

1. *Lines with positive slopes go upward as x increases* (that is, as you go from left to right).

Negative slope
Line goes downward to the right

2. *Lines with negative slopes go downward as x increases* (that is, as you go from left to right).

Zero slope
Horizontal line

3. Horizontal lines have a slope of 0.

Undefined slope
Vertical line

4. A vertical line is said to have undefined slope. The slope of a vertical line is not defined. In other words, a vertical line has no slope.

② Finding the Slope and *y*-Intercept of a Line Given Its Equation

Recall that the equation of a line is a linear equation in two variables. This equation can be written in several different ways. A very useful form of the equation of a straight line is the slope–intercept form. This form can be derived in the following way. Suppose that a straight line with slope m crosses the y-axis at a point $(0, b)$. Consider any other point on the line and label the point (x, y). Then we have the following.

$$\frac{y_2 - y_1}{x_2 - x_1} = m \qquad \text{Definition of slope.}$$

$$\frac{y - b}{x - 0} = m \qquad \text{Substitute } (0, b) \text{ for } (x_1, y_1) \text{ and } (x, y) \text{ for } (x_2, y_2).$$

$$\frac{y - b}{x} = m \qquad \text{Simplify.}$$

$$y - b = mx \qquad \text{Multiply both sides by } x.$$

$$y = mx + b \qquad \text{Add } b \text{ to both sides.}$$

This form of a linear equation immediately reveals the slope of the line, m, and the y-coordinate of the point where the line intercepts (crosses) the y-axis, b.

SLOPE–INTERCEPT FORM OF A LINE

The slope–intercept form of the equation of the line that has slope m and the y-intercept $(0, b)$ is given by

$$y = mx + b.$$

By using algebraic operations, we can write any linear equation in slope–intercept form and use this form to identify the slope and the y-intercept of the line.

EXAMPLE 4 What is the slope and the y-intercept of the line $5x + 3y = 2$?

Solution We want to solve for y and get the equation in the form $y = mx + b$. We need to isolate the y-variable.

$$5x + 3y = 2$$
$$3y = -5x + 2 \qquad \text{Subtract } 5x \text{ from both sides.}$$
$$y = \frac{-5x + 2}{3} \qquad \text{Divide both sides by 3.}$$
$$y = -\frac{5}{3}x + \frac{2}{3} \qquad \text{Using the property } \frac{a + b}{c} = \frac{a}{c} + \frac{b}{c}, \text{ write the right-hand side as two fractions.}$$

The *slope* is $-\dfrac{5}{3}$. The y-intercept is $\left(0, \dfrac{2}{3}\right)$.

NOTE TO STUDENT: *Fully worked-out solutions to all of the Practice Problems can be found at the back of the text starting at page SP-1*

Practice Problem 4 What is the slope and the y-intercept of the line $4x - 2y = -5$?

③ Writing the Equation of a Line Given the Slope and y-Intercept

If we know the slope of a line and the y-intercept, we can write the equation of the line, $y = mx + b$.

EXAMPLE 5 Find the equation of the line with slope $\frac{2}{5}$ and y-intercept $(0, -3)$.

(a) Write the equation in slope–intercept form, $y = mx + b$.

(b) Write the equation in the form $Ax + By = C$.

Solution

(a) We are given that $m = \frac{2}{5}$ and $b = -3$. Thus we have the following.

$$y = mx + b$$
$$y = \frac{2}{5}x + (-3)$$
$$y = \frac{2}{5}x - 3$$

(b) Recall, for the form $Ax + By = C$, that A, B, and C are integers. We first clear the equation of fractions. Then we move the x-term to the left side.

$$5y = 5\left(\frac{2x}{5}\right) - 5(3) \qquad \text{Multiply each term by 5.}$$
$$5y = 2x - 15 \qquad \text{Simplify.}$$
$$-2x + 5y = -15 \qquad \text{Subtract } 2x \text{ from each side.}$$
$$2x - 5y = 15 \qquad \text{Multiply each term by } -1. \text{ The form } Ax + By = C \text{ is usually written with } A \text{ as a positive integer.}$$

Practice Problem 5   Find the equation of the line with slope $-\frac{3}{7}$ and y-intercept $\left(0, \frac{2}{7}\right)$.

(a) Write the equation in slope–intercept form.

(b) Write the equation in the form $Ax + By = C$.

4 Graphing a Line Using the Slope and y-Intercept

If we know the slope of a line and the y-intercept, we can draw the graph of the line.

EXAMPLE 6   Graph the line with slope $m = \frac{2}{3}$ and y-intercept $(0, -3)$. Use the given coordinate system.

Solution   Recall that the y-intercept is the point where the line crosses the y-axis. We need a point to start the graph with. So we plot the point $(0, -3)$ on the y-axis.

Recall that slope $= \dfrac{\text{rise}}{\text{run}}$. Since the slope for this line is $\frac{2}{3}$, we will go up (rise) 2 units and go over (run) to the right 3 units from the point $(0, -3)$. Look at the figure below. This is the point $(3, -1)$. Plot the point. Draw a line that connects the two points $(0, -3)$ and $(3, -1)$.

This is the graph of the line with slope $\frac{2}{3}$ and y-intercept $(0, -3)$.

PRACTICE PROBLEM 6

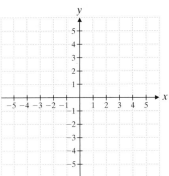

Practice Problem 6   Graph the line with slope $= \frac{3}{4}$ and y-intercept $(0, -1)$. Use the coordinate system in the margin.

EXAMPLE 7   Graph the equation $y = -\frac{1}{2}x + 4$. Use the following coordinate system.

Solution   Begin with the y-intercept. Since $b = 4$, plot the point $(0, 4)$. Now look at the slope $-\frac{1}{2}$. This can be written as $\frac{-1}{2}$. Begin at $(0, 4)$ and go *down* 1 unit and to the right 2 units. This is the point $(2, 3)$. Plot the point. Draw the line that connects the points $(0, 4)$ and $(2, 3)$.

This is the graph of the equation $y = -\frac{1}{2}x + 4$.

PRACTICE PROBLEM 7

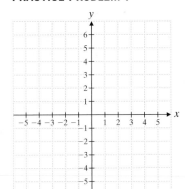

Practice Problem 7   Graph the equation $y = -\frac{2}{3}x + 5$. Use the coordinate system in the margin.

TO THINK ABOUT: Verifying Your Answer Explain why you count down 1 unit and move to the right 2 units to represent the slope $-\frac{1}{2}$. Could you have done this in another way? Try it. Verify that this is the same line.

5 Finding the Slopes of Parallel and Perpendicular Lines

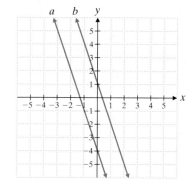

Parallel lines are two straight lines that never touch. Look at the parallel lines in the figure. Notice that the slope of line a is -3 and the slope of line b is also -3. Why do you think the slopes must be equal? What would happen if the slope of line b were -1? Graph it and see.

PARALLEL LINES

Parallel lines are two straight lines that never touch.
Parallel lines have the same slope but different y-intercepts.

$$m_1 = m_2$$

Perpendicular lines are two lines that meet at a 90° angle. Look at the perpendicular lines in the figure at left. The slope of line c is -3. The slope of line d is $\frac{1}{3}$. Notice that

$$(-3)\left(\frac{1}{3}\right) = \left(-\frac{3}{1}\right)\left(\frac{1}{3}\right) = -1.$$

You may wish to draw several pairs of perpendicular lines to determine whether the product of their slopes is always -1.

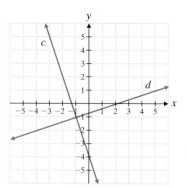

PERPENDICULAR LINES

Perpendicular lines are two lines that meet at a 90° angle.
Perpendicular lines have slopes whose product is -1. If m_1 and m_2 are slopes of perpendicular lines, then

$$m_1 m_2 = -1 \quad \text{or} \quad m_1 = -\frac{1}{m_2}.$$

EXAMPLE 8 Line h has a slope of $-\frac{2}{3}$.

(a) If line f is parallel to line h, what is its slope?

(b) If line g is perpendicular to line h, what is its slope?

Solution

(a) Parallel lines have the same slope. Line f has a slope of $-\frac{2}{3}$.

(b) Perpendicular lines have slopes whose product is -1.

$$m_1 m_2 = -1$$

$$-\frac{2}{3} m_2 = -1 \qquad \text{Substitute } -\frac{2}{3} \text{ for } m_1.$$

$$\left(-\frac{3}{2}\right)\left(-\frac{2}{3}\right) m_2 = -1\left(-\frac{3}{2}\right) \qquad \text{Multiply both sides by } -\frac{3}{2}.$$

$$m_2 = \frac{3}{2}$$

Thus line g has a slope of $\frac{3}{2}$.

Practice Problem 8 Line h has a slope of $\frac{1}{4}$.

(a) If line j is parallel to line h, what is its slope?

(b) If line k is perpendicular to line h, what is its slope?

NOTE TO STUDENT: Fully worked-out solutions to all of the Practice Problems can be found at the back of the text starting at page SP-1

Graphing Calculator

Graphing Parallel Lines

If two equations are in the form $y = mx + b$, then it will be obvious that they are parallel because the slope will be the same. On a graphing calculator graph both of these equations:

$$y = -2x + 6$$
$$y = -2x - 4$$

Use the window of -10 to 10 for both x and y.
Display:

EXAMPLE 9 The equation of line l is $y = -2x + 3$.

(a) What is the slope of a line that is parallel to line l?

(b) What is the slope of a line that is perpendicular to line l?

Solution

(a) Looking at the equation, we can see that the slope of line l is -2. The slope of a line that is parallel to line l is -2.

(b) Perpendicular lines have slopes whose product is -1.

$$m_1 m_2 = -1$$

$$(-2) m_2 = -1 \qquad \text{Substitute } -2 \text{ for } m_1.$$

$$m_2 = \frac{1}{2} \qquad \text{Because } (-2)\left(\frac{1}{2}\right) = -1.$$

The slope of a line that is perpendicular to line l is $\frac{1}{2}$.

Practice Problem 9 The equation of line n is $y = \frac{1}{4}x - 1$.

(a) What is the slope of a line that is parallel to line n?

(b) What is the slope of a line that is perpendicular to line n?

7.3 EXERCISES

Student Solutions Manual | CD/ Video | PH Math Tutor Center | MathXL®Tutorials on CD | MathXL® | MyMathLab® | Interactmath.com

Verbal and Writing Skills

1. Can you find the slope of the line passing through $(5, -12)$ and $(5, -6)$? Why or why not?

2. Can you find the slope of the line passing through $(6, -2)$ and $(-8, -2)$? Why or why not?

Find the slope of a straight line that passes through the given pair of points.

3. $(4, 1)$ and $(6, 7)$ **4.** $(7, 5)$ and $(12, 10)$ **5.** $(6, 6)$ and $(9, 3)$

6. $(11, 2)$ and $(5, 14)$ **7.** $(-2, 1)$ and $(3, 4)$ **8.** $(5, 6)$ and $(-3, 1)$

9. $(-6, -5)$ and $(2, -7)$ **10.** $(-8, -3)$ and $(4, -9)$ **11.** $(-3, 0)$ and $(0, -4)$

12. $(0, 5)$ and $(5, 3)$ **13.** $\left(\dfrac{3}{4}, -4\right)$ and $(2, -8)$ **14.** $\left(\dfrac{5}{3}, -2\right)$ and $(3, 6)$

Find the slope and the y-intercept.

15. $y = 8x + 9$ **16.** $y = 2x + 10$ **17.** $y = -3x + 4$ **18.** $y = -8x - 7$

19. $y = -\dfrac{8}{7}x + \dfrac{3}{4}$ **20.** $y = \dfrac{5}{3}x - \dfrac{4}{5}$ **21.** $y = -6x$ **22.** $y = -2$

23. $6x + y = \dfrac{4}{5}$ **24.** $2x + y = -\dfrac{3}{4}$ **25.** $5x + 2y = 3$ **26.** $7x + 3y = 4$

27. $7x - 3y = 4$ **28.** $9x - 4y = 18$

Write the equation of the line in slope–intercept form.

29. $m = \dfrac{3}{5}$, y-intercept $(0, 3)$ **30.** $m = \dfrac{2}{3}$, y-intercept $(0, 5)$

31. $m = 6$, y-intercept $(0, -3)$ **32.** $m = 5$, y-intercept $(0, -6)$

33. $m = -\dfrac{5}{4}$, y-intercept $\left(0, -\dfrac{3}{4}\right)$ **34.** $m = -4$, y-intercept $\left(0, \dfrac{1}{2}\right)$

Graph the line $y = mx + b$ for the given values.

35. $m = \dfrac{3}{4}$, $b = -4$

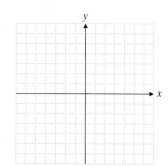

36. $m = \dfrac{1}{3}$, $b = -2$

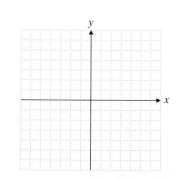

37. $m = -\dfrac{5}{3}, b = 2$

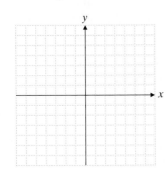

38. $m = -\dfrac{3}{2}, b = 4$

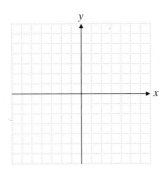

In exercises 39–44, graph the line.

39. $y = \dfrac{2}{3}x + 2$

40. $y = \dfrac{3}{4}x + 1$

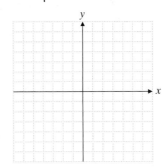

41. $y + 2x = 3$

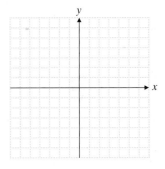

42. $y + 4x = 5$

43. $y = 2x$

44. $y = 3x$

45. A line has a slope of $\dfrac{5}{6}$.
   (a) What is the slope of the line parallel to it?
   (b) What is the slope of the line perpendicular to it?

46. A line has a slope of $\dfrac{11}{4}$.
   (a) What is the slope of the line parallel to it?
   (b) What is the slope of the line perpendicular to it?

47. A line has a slope of 6.
   (a) What is the slope of the line parallel to it?
   (b) What is the slope of the line perpendicular to it?

48. A line has a slope of $-\dfrac{5}{8}$.
   (a) What is the slope of the line parallel to it?
   (b) What is the slope of the line perpendicular to it?

49. The equation of a line is $y = \dfrac{1}{3}x + 2$.
   (a) What is the slope of a line parallel to it?
   (b) What is the slope of a line perpendicular to it?

50. The equation of a line is $y = \dfrac{3}{5}x - 5$.
   (a) What is the slope of a line parallel to it?
   (b) What is the slope of a line perpendicular to it?

To Think About

51. Do the points $(3, -4)$, $(18, 6)$, and $(9, 0)$ all lie on the same line? If so, what is the equation of the line?

52. Do the points $(2, 1)$, $(-3, -2)$ and $(7, 4)$ lie on the same line? If so, what is the equation of the line?

53. *Federal Budget* During the years from 1980 to 2005, the total income for the U.S. federal budget can be approximated by the equation $y = 14(4x + 35)$, where x is the number of years since 1980 and y is the amount of money in billions of dollars. (*Source:* U.S. Office of Management and Budget)
(a) Write the equation in slope–intercept form.

(b) Find the slope and the y-intercept.

(c) In this specific equation, what is the meaning of the slope? What does it indicate?

54. *Civilian Employment* During the years from 1970 to 2005, the approximate number of civilians employed in the United States could be predicted by the equation $y = \frac{1}{10}(22x + 830)$, where x is the number of years since 1970 and y is the number of civilians employed, measured in millions. (*Source:* U.S. Bureau of Labor Statistics)
(a) Write the equation in slope–intercept form.

(b) Find the slope and the y-intercept.

(c) In this specific equation, what is the meaning of the slope? What does it indicate?

Cumulative Review

Solve for x and graph the solution.

55. $3x + 8 > 2x + 12$

56. $\frac{1}{4}x + 3 > \frac{2}{3}x + 2$

57. $\frac{1}{2}(x + 2) \le \frac{1}{3}x + 5$

58. $7x - 2(x + 3) \le 4(x - 7)$

59. *Fishing Charters* Captain Charles Allen has a 35-foot charter offshore fishing boat in Falmouth, Massachusetts. His boat has a cruising range of 390 miles when traveling at 25 miles per hour. However, the cruising range is decreased by 62% at the boat's maximum speed of 45 miles per hour. What is the cruising range at maximum speed? (Round to the nearest mile.)

60. *Camp Counselors* New England Camp Cherith tries to maintain a ratio of 3 counselors for every 25 campers. One week last summer there were 189 campers. What is the minimum number of counselors who should have been there that week?

61. *Subway Ridership* New York City subway ridership rose from 206 million a year in 1990 to 240 million in 1999. (*Source:* Bureau of Transportation Statistics.) The mayor predicts it will rise to 288 million by 2008. How much larger is the projected percent of increase from 1999 to 2008 than the observed percent of increase from 1990 to 1999?

How are you doing with your homework assignments in Sections 7.1 to 7.3? Do you feel you have mastered the material so far? Do you understand the concepts you have covered? Before you go further in the textbook, take some time to do each of the following problems.

1. _____

7.1

1. Plot the following points.

E $(-6, 3)$
F $(5, -4)$
G $(-2, -8)$
H $(6.5, 3.5)$

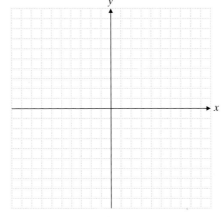

2. _____

2. Give the coordinates for points A, B, C, D.

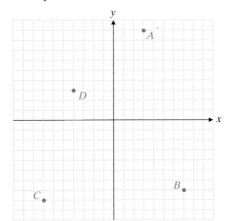

3. _____

3. Complete the ordered pairs for the equation $y = -7x + 3$.

$(4, \quad)$
$(0, \quad)$
$(-2, \quad)$

4. Complete the table for the equation $6x - 5y = -30$.

x	y
0	
	0
	−6

4. _____

7.2

5. _____

5. Graph $y = -4x - 3$.

6. Graph $y = \dfrac{3}{4}x - 1$.

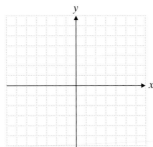

6. _____

7. _____

7. Graph $3x - 5y = 0$.

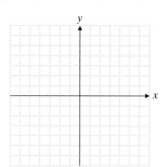

8. _____

8. Graph $-5x + 2 = y + 3$.

9. _____

7.3

9. Find the slope of a straight line that passes through $(-6, -8)$ and $(2, -3)$.

10. _____

10. Find the slope and the y-intercept of $9x - 3y - 5 = 0$.

11. A line has a slope of $\dfrac{3}{4}$. What is the slope of a line perpendicular to it?

11. _____

12. A equation of a line is $3y + 11x + 7 = 0$. What is the slope of a line parallel to that line?

12. _____

Now turn to page SA-18 for the answer to each of these problems. Each answer also includes a reference to the objective in which the problem is first taught. If you missed any of these problems, you should stop and review the Examples and Practice Problems in the referenced objective. A little review now will help you master the material in the upcoming sections of the text.

Writing the Equation of a Line Given a Point and a Slope

If we know the slope of a line and the y-intercept, we can write the equation of the line in slope–intercept form. Sometimes we are given the slope and a point on the line. We use the information to find the y-intercept. Then we can write the equation of the line.

It may be helpful to summarize our approach.

Student Learning Objectives

After studying this section, you will be able to:

1 Write the equation of a line given a point and a slope.

2 Write the equation of a line given two points.

3 Write the equation of a line given a graph of the line.

TO FIND THE EQUATION OF A LINE GIVEN A POINT AND A SLOPE

1. Substitute the given values of x, y, and m into the equation $y = mx + b$.
2. Solve for b.
3. Use the values of b and m to write the equation in the form $y = mx + b$.

EXAMPLE 1 Find an equation of the line that passes through $(-3, 6)$ with slope $-\frac{2}{3}$.

Solution We are given the values $m = -\frac{2}{3}$, $x = -3$, and $y = 6$.

$$y = mx + b$$

$$6 = \left(-\frac{2}{3}\right)(-3) + b \quad \text{Substitute known values.}$$

$$6 = 2 + b$$

$$4 = b$$

The equation of the line is $y = -\frac{2}{3}x + 4$.

Practice Problem 1 Find an equation of the line that passes through $(-8, 12)$ with slope $-\frac{3}{4}$.

NOTE TO STUDENT: Fully worked-out solutions to all of the Practice Problems can be found at the back of the text starting at page SP-1

---

2 **Writing the Equation of a Line Given Two Points**

Our procedure can be extended to the case for which two points are given.

EXAMPLE 2 Find an equation of the line that passes through $(2, 5)$ and $(6, 3)$.

Solution We first find the slope of the line. Then we proceed as in Example 1.

$$m = \frac{y_2 - y_1}{x_2 - x_1}$$

$$m = \frac{3 - 5}{6 - 2} \quad \text{Substitute } (x_1, y_1) = (2, 5) \text{ and } (x_2, y_2) = (6, 3) \text{ into the formula.}$$

$$= \frac{-2}{4} = -\frac{1}{2}$$

Choose either point, say $(2, 5)$, to substitute into $y = mx + b$ as in Example 1.

$$5 = -\frac{1}{2}(2) + b$$

$$5 = -1 + b$$

$$6 = b$$

The equation of the line is $y = -\frac{1}{2}x + 6$.

Note: We could have substituted the slope and the other point, $(6, 3)$, into the slope–intercept form and arrived at the same answer. Try it.

Practice Problem 2 Find an equation of the line that passes through $(3, 5)$ and $(-1, 1)$.

NOTE TO STUDENT: *Fully worked-out solutions to all of the Practice Problems can be found at the back of the text starting at page SP-1*

3 Writing the Equation of a Line Given a Graph of the Line

EXAMPLE 3 What is the equation of the line in the figure at right?

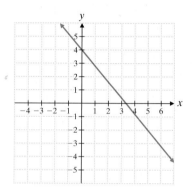

Solution First, look for the y-intercept. The line crosses the y-axis at $(0, 4)$. Thus $b = 4$.
Second, find the slope.

$$m = \frac{\text{change in } y}{\text{change in } x}$$

Look for another point on the line. We choose $(5, -2)$. Count the number of vertical units from 4 to -2 (rise). Count the number of horizontal units from 0 to 5 (run).

$$m = \frac{-6}{5}$$

Now, using $m = -\frac{6}{5}$ and $b = 4$, we can write the equation of the line.

$$y = mx + b$$

$$y = -\frac{6}{5}x + 4$$

PRACTICE PROBLEM 3

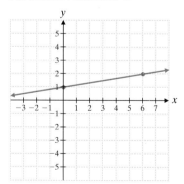

Practice Problem 3 What is the equation of the line in the figure at the left?

7.4 EXERCISES

Student Solutions Manual | CD/Video | PH Math Tutor Center | MathXL®Tutorials on CD | MathXL® | MyMathLab® | Interactmath.com

Find an equation of the line that has the given slope and passes through the given point.

1. $m = 4, (-3, 0)$ **2.** $m = 3, (2, -2)$ **3.** $m = -3, (2, 5)$ **4.** $m = -2, (3, 2)$

5. $m = -3, \left(\dfrac{1}{2}, 2\right)$ **6.** $m = -2, \left(3, \dfrac{1}{3}\right)$ **7.** $m = -\dfrac{2}{5}, (5, -3)$ **8.** $m = \dfrac{2}{3}, (3, -2)$

Write an equation of the line passing through the given points.

9. $(3, -12)$ and $(-4, 2)$ **10.** $(-3, 9)$ and $(2, -11)$ **11.** $(2, -6)$ and $(-1, 6)$ **12.** $(3, -11)$ and $(-2, 14)$

13. $(3, 5)$ and $(-1, -15)$ **14.** $(-1, -19)$ and $(2, 2)$ **15.** $\left(1, \dfrac{5}{6}\right)$ and $\left(3, \dfrac{3}{2}\right)$ **16.** $(2, 0)$ and $\left(\dfrac{3}{2}, \dfrac{1}{2}\right)$

Mixed Practice

17. Find the equation of a line with a slope of -2 that passes through the point $(4, 3)$.

18. Find the equation of a line with a slope of -4 that passes through the point $(5, 7)$.

19. Find the equation of a line that passes through $(2, -3)$ and $(-1, 6)$.

20. Find the equation of a line that passes through $(1, -8)$ and $(2, -14)$.

Write an equation of each line.

21.

22.

23.

24.

25.

26.

27.

28.

To Think About

Find an equation of the line that fits each description.

29. Passes through $(7, -2)$ and has zero slope

30. Passes through $(4, 6)$ and has undefined slope

31. Passes through $(4, -6)$ and is perpendicular to the *x*-axis.

32. Passes through $(-3, 5)$ and is perpendicular to the *y*-axis.

33. Passes through $(0, -4)$ and is parallel to $y = \dfrac{3}{4}x + 2$

34. Passes through $(0, -4)$ and is perpendicular to $y = \dfrac{3}{4}x + 2$

35. Passes through $(2, 3)$ and is perpendicular to $y = 2x - 9$

36. Passes through $(2, 9)$ and is parallel to $y = 5x - 3$

37. *Population Growth* The growth of the population of the United States during the period from 1980 to 2005 can be approximated by an equation of the form $y = mx + b$, where *x* is the number of years since 1980 and *y* is the population measured in millions. (*Source:* U.S. Census Bureau.) Find the equation if two ordered pairs that satisfy it are $(0, 227)$ and $(10, 251)$.

38. *Home Equity Loans* The amount of debt outstanding on home equity loans in the United States during the period from 1993 to 2005 can be approximated by an equation of the form $y = mx + b$, where *x* is the number of years since 1993 and *y* is the debt measured in billions of dollars. (*Source:* Board of Governors of the Federal Reserve System.) Find the equation if two ordered pairs that satisfy it are $(1, 280)$ and $(6, 500)$.

Cumulative Review

39. $\dfrac{3}{t} - \dfrac{2}{t-1} = \dfrac{4}{t}$

40. Simplify. $\dfrac{4x^2 - 25}{4x^2 - 20x + 25} \cdot \dfrac{x^2 - 9}{2x^2 + x - 15}$

41. *Basketball Sneakers* A pair of basketball sneakers sells for $80. The next week the sneakers go on sale for 15% off. The third week there is a coupon in the newspaper offering a 10% discount off the second week's price. How much would you have paid for the sneakers during the third week if you had used the coupon?

42. *Cell Phone Costs* Dave and Jane Wells have a cell phone. The plan they subscribe to costs $50 per month. It includes 200 free minutes of calling time. Each minute after the 200 is charged at the rate of $0.21 per minute. Last month their cell phone bill was $68.90. How many total minutes did they use their cell phone during the month?

43. *Defective Furniture* A Scandinavian furniture wholesaler had a shipment of oak entertainment centers shipped by boat across the Atlantic. There were 35 cargo containers, each containing 60 entertainment centers. The entire shipment was found to contain 104 defective entertainment centers. What percent of the shipment was defective? Round to the nearest hundredth of a percent.

44. *Investment Income* Jackie earned $2122.40 in interest after investing $24,000 for one year. She placed part of the money in an account earning 7% simple interest. She took a risk on some stock options with the rest of the savings, which yielded her 10.5%. How much was invested in the stock options?

▲ **45.** *Archeology* Archeologists are searching a rectangular region in Mexico for evidence of a primitive civilization over 3000 years old. The perimeter of the region is 3440 feet. The length of the region is 70 feet longer than twice the width. Find the dimensions of the region.

7.5 GRAPHING LINEAR INEQUALITIES

In Section 2.7 we discussed inequalities in one variable. Look at the inequality $x < -2$ (x is less than -2). Some of the solutions to the inequality are -3, -5, and $-5\frac{1}{2}$. In fact all numbers to the left of -2 on the number line are solutions. The graph of the inequality is given in the following figure. Notice that the open circle at -2 indicates that -2 is *not* a solution.

$x < -2$

Student Learning Objective

After studying this section, you will be able to:

 Graph linear inequalities in two variables.

Now we will extend our discussion to consider linear inequalities in two variables.

1 Graphing Linear Inequalities in Two Variables

Consider the inequality $y \geq x$. The solution of the inequality is the set of all possible ordered pairs that when substituted into the inequality will yield a true statement. Which ordered pairs will make the statement $y \geq x$ true? Let's try some.

$(0, 6)$	$(-2, 1)$	$(1, -2)$	$(3, 5)$	$(4, 4)$
$6 \geq 0$, true	$1 \geq -2$, true	$-2 \geq 1$, false	$5 \geq 3$, true	$4 \geq 4$, true

$(0, 6)$, $(-2, 1)$, $(3, 5)$, and $(4, 4)$ are solutions to the inequality $y \geq x$. In fact, every point at which the y-coordinate is greater than or equal to the x-coordinate is a solution to the inequality. This is shown by the solid line and the shaded region in the graph at the right.

Is there an easier way to graph a linear inequality in two variables? It turns out that we can graph such an inequality by first graphing the associated linear equation and then testing one point that is not on that line. That is, we can change the inequality symbol to an equal sign and graph the equation. If the inequality symbol is $\geq$ or $\leq$, we use a solid line to indicate that the points on the line are included in the solution of the inequality. If the inequality symbol is $>$ or $<$, we use a dashed line to indicate that the points on the line are not included in the solution of the inequality. Then we test one point that is not on the line. If the point is a solution to the inequality, we shade the region on the side of the line that includes the point. If the point is not a solution, we shade the region on the other side of the line.

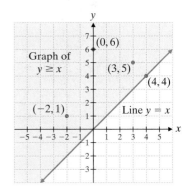

Graph of $y \geq x$

Line $y = x$

EXAMPLE 1   Graph $5x + 3y > 15$. Use the given coordinate system.

Solution   We begin by graphing the line $5x + 3y = 15$. You may use any method discussed previously to graph the line. Since there is no equal sign in the inequality, we will draw a dashed line to indicate that the line is *not* part of the solution set.

Look for a test point. The easiest point to test is $(0, 0)$. Substitute $(0, 0)$ for (x, y) in the inequality.

$$5x + 3y > 15$$
$$5(0) + 3(0) > 15$$
$$0 > 15 \quad \text{false}$$

$(0, 0)$ is not a solution. Shade the side of the line that does *not* include $(0, 0)$.

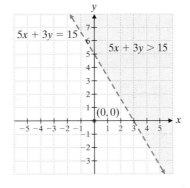

$5x + 3y = 15$

$5x + 3y > 15$

$(0, 0)$

Practice Problem 1   Graph $x - y \geq -10$. Use the coordinate system in the margin.

PRACTICE PROBLEM 1

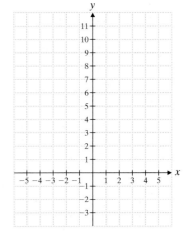

NOTE TO STUDENT: Fully worked-out solutions to all of the Practice Problems can be found at the back of the text starting at page SP-1

GRAPHING LINEAR INEQUALITIES

1. Replace the inequality symbol by an equality symbol. Graph the line.
   (a) The line will be solid if the inequality is $\geq$ or $\leq$.
   (b) The line will be dashed if the inequality is $>$ or $<$.

2. Test the point $(0, 0)$ in the inequality if $(0, 0)$ does not lie on the graphed line in step 1.
   (a) If the inequality is true, shade the side of the line that includes $(0, 0)$.
   (b) If the inequality is false, shade the side of the line that does not include $(0, 0)$.

3. If the point $(0, 0)$ is a point on the line, choose another test point and proceed accordingly.

PRACTICE PROBLEM 2

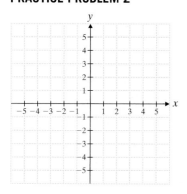

NOTE TO STUDENT: Fully worked-out solutions to all of the Practice Problems can be found at the back of the text starting at page SP-1

EXAMPLE 2   Graph $2y \leq -3x$.

Solution

Step 1 Graph $2y = -3x$. Since $\leq$ is used, the line should be a solid line.

Step 2 We see that the line passes through $(0, 0)$.

Step 3 Choose another test point. We will choose $(-3, -3)$.

$$2y \leq -3x$$
$$2(-3) \leq -3(-3)$$
$$-6 \leq 9 \quad \text{true}$$

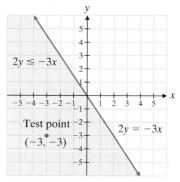

Shade the region that includes $(-3, -3)$, that is, the region below the line.

Practice Problem 2   Graph $2y > x$ on the coordinate system in the margin.

If we are graphing the inequality $x < -2$ on the coordinate plane, the solution will be a region. Notice that this is very different from the solution $x < -2$ on the number line discussed earlier.

EXAMPLE 3   Graph $x < -2$.

Solution

Step 1 Graph $x = -2$. Since $<$ is used, the line should be dashed.

PRACTICE PROBLEM 3

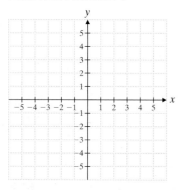

Step 2 Test $(0, 0)$ in the inequality.

$$x < -2$$
$$0 < -2 \quad \text{false}$$

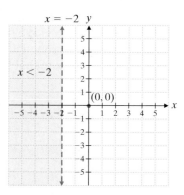

Shade the region that does not include $(0, 0)$, that is, the region to the left of the line $x = -2$. Observe that every point in the shaded region has an x-value that is less than -2.

Practice Problem 3   Graph $y \geq -3$ on the coordinate system in the margin.

Verbal and Writing Skills

1. Does it matter what point you use as your test point? Justify your response.

2. Explain when to use a solid line or a dashed line when graphing linear inequalities in two variables.

Graph the region described by the inequality.

3. $y > 2 - 3x$

4. $y > 3x - 1$

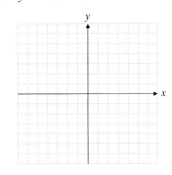

5. $2x - 3y < 6$

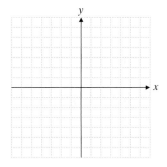

6. $3x + 2y < -6$

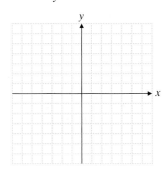

7. $2x - y \geq 3$

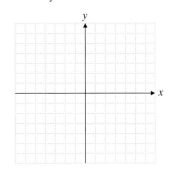

8. $3x - y \geq 4$

9. $y \geq 4x$

10. $y \leq -2x$

11. $y < -\dfrac{1}{2}x$

12. $y > \dfrac{1}{5}x$

13. $x \geq 2$

14. $y \leq -2$

15. $2x - 3y + 6 \geq 0$

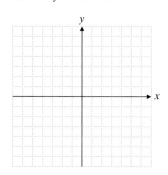

16. $3x + 4y - 8 \leq 0$

17. $2x > -3y$

18. $3x \leq -2y$

19. $2x > 3 - y$

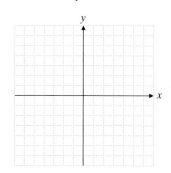

20. $3x > 1 + y$

21. $x > -2y$

22. $x < -3y$

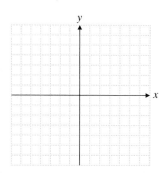

To Think About

23. Graph the inequality $3x + 3 > 5x - 3$ by graphing the lines $y = 3x + 3$ and $y = 5x - 3$ on the same coordinate plane.

   (a) For which values of x does the graph of $y = 3x + 3$ lie above the graph of $y = 5x - 3$?

   (b) For which values of x is $3x + 3 > 5x - 3$ true?

   (c) Explain how the answer to (a) can help you to find the answer to (b).

Cumulative Review

24. Suppose you tried to use the method of problem 23 to solve the inequality $2x + 4 < 2x - 3$.

   (a) If you graphed the lines $y = 2x + 4$ and $y = 2x - 3$, what kind of graph would you obtain?

   (b) Why is it not possible to find a solution for the inequality $2x + 4 < 2x - 3$?

25. Solve for x: $5x^2 + 12x - 9 = 0$

26. What is the slope–intercept form of the equation for the line through $(3, 0)$ and $(6, -7)$?

27. *Leasing a Car* Angela has a new sales job that allows her to lease a car. Her lease budget is $8400 for a 3-year period. She wants to lease a new PT Cruiser at $195 per month for the next 36 months. However, all mileage in excess of 36,000 miles driven over this 3-year period will be charged at $0.15 per mile. How many miles above the 36,000-mile limit can Angela drive and still not exceed the leasing budget?

28. *Telemarketing Sales* Samantha is a great telemarketer. She has great success in selling subscriptions to a travel club. She has an average of 0.492 after having made 1000 telephone calls. (She sells a subscription 49.2% of the time.) Assuming that she makes a sale with each of her next telephone calls, what is the minimum number of calls she needs to make in order to raise her average to 0.500?

29. *Metal Composition of an Antenna* A new communications satellite has an antenna support made of a special brass alloy. This brass alloy is composed of copper and zinc only. It is formulated in the ratio of 4 parts copper to 3 parts zinc. If a total of 210 ounces of brass is made, how much zinc is used?

1 Understanding the Meanings of a Relation and a Function

Thus far you have studied linear equations in two variables. You have seen that such an equation can be represented by a table of values, by the algebraic equation itself, and by a graph.

x	0	1	3
y	4	1	−5

$$y = -3x + 4$$

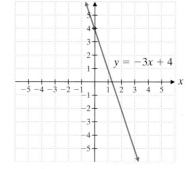

The solutions to the linear equation are all the ordered pairs that satisfy the equation (make the equation true). They are all the points that lie on the graph of the line. These ordered pairs can be represented in a table of values. Notice the relationship between the ordered pairs. We can choose any value for *x*. But once we have chosen a value for *x*, the value of *y* is determined. For the preceding equation, if *x* is 0, then *y* must be 4. We say that *x* is the **independent variable** and that *y* is the **dependent variable.**

Mathematicians call such a pairing of two values a *relation*.

DEFINITION OF A RELATION

A **relation** is any set of ordered pairs.

All the first coordinates in all of the ordered pairs of the relation make up the **domain** of the relation. All the second coordinates in all of the ordered pairs make up the **range** of the relation. Notice that the definition of a relation is very broad. Some relations cannot be described by an equation. These relations may simply be a set of discrete ordered pairs.

EXAMPLE 1   State the domain and range of the relation.

$$\{(5, 7), (9, 11), (10, 7), (12, 14)\}$$

Solution   The domain consists of all the first coordinates in the ordered pairs.

Domain

$$\{(5, 7), (9, 11), (10, 7), (12, 14)\}$$

Range

The range consists of all of the second coordinates in the ordered pairs.

The domain is $\{5, 9, 10, 12\}$.

The range is $\{7, 11, 14\}$.   We list 7 only once.

Practice Problem 1   State the domain and range of the relation.

$$\{(-3, -5), (3, 5), (0, -5), (20, 5)\}$$

Some relations have the special property that no two different ordered pairs have the same first coordinate. Such relations are called **functions.** The relation $y = -3x + 4$ is a function. If we substitute a value for x, we get just one value for y. Thus no two ordered pairs will have the same x-coordinate and different y-coordinates.

DEFINITION OF A FUNCTION

A **function** is a relation in which no two different ordered pairs have the same first coordinate.

EXAMPLE 2   Determine whether the relation is a function.

(a) $\{(3, 9), (4, 16), (5, 9), (6, 36)\}$

(b) $\{(7, 8), (9, 10), (12, 13), (7, 14)\}$

Solution

(a) Look at the ordered pairs. No two ordered pairs have the same first coordinate. Thus this set of ordered pairs defines a function. Note that the ordered pairs $(3, 9)$ and $(5, 9)$ have the same second coordinate, but the relation is still a function. It is the first coordinates that cannot be the same.

(b) Look at the ordered pairs. Two different ordered pairs, $(7, 8)$ and $(7, 14)$, have the same first coordinate. Thus this relation is *not* a function.

Practice Problem 2   Determine whether the relation is a function.

(a) $\{(-5, -6), (9, 30), (-3, -3), (8, 30)\}$

(b) $\{(60, 30), (40, 20), (20, 10), (60, 120)\}$

A functional relationship is often what we find when we analyze two sets of data. Look at the following table of values, which compares Celsius temperature with Fahrenheit temperature. Is there a relationship between degrees Fahrenheit and degrees Celsius? Is the relation a function?

Temperature

°F	23	32	41	50
°C	-5	0	5	10

Since every Fahrenheit temperature produces a unique Celsius temperature, we would expect this to be a function. We can verify our assumption by looking at the formula $C = \frac{5}{9}(F - 32)$ and its graph. The formula is a linear equation, and its graph is a line with slope $\frac{5}{9}$ and y-intercept at about -17.8. The relation is a function. In the equation given here, notice that the *dependent variable* is C, since the value of C *depends* on the value of F. We say that F is the *independent variable*. The *domain* can be described as the set of possible values of the independent variable. The *range* is the set of corresponding values of the dependent variable. Scientists believe that the coldest temperature possible is approximately $-273°C$. They call this temperature **absolute zero.** Thus,

Domain = {all possible Fahrenheit temperatures from absolute zero to infinity}

Range = {all corresponding Celsius temperatures from $-273°C$ to infinity}.

EXAMPLE 3 Each of the following tables contains some data pertaining to a relation. Determine whether the relation suggested by the table is a function. If it is a function, identify the domain and range.

(a)
Circle

Radius	1	2	3	4	5
Area	3.14	12.56	28.26	50.24	78.5

(b) $4000 Loan at 8% for a Minimum of One Year

Time (yr)	1	2	3	4	5
Interest	$320	$665.60	$1038.85	$1441.96	$1877.31

Solution

(a) Looking at the table, we see that no two different ordered pairs have the same first coordinate. The area of a circle is a function of the length of the radius.

Next we need to identify the independent variable to determine the domain. Sometimes it is easier to identify the dependent variable. Here we notice that the area of the circle depends on the length of the radius. Thus radius is the independent variable. Since a negative length does not make sense, the radius cannot be a negative number. Although only integer radius values are listed in the table, the radius of a circle can be any nonnegative real number.

Domain = {all nonnegative real numbers}

Range = {all nonnegative real numbers}

(b) No two different ordered pairs have the same first coordinate. Interest is a function of time.

Since the amount of interest paid on a loan depends on the number of years (term of the loan), interest is the dependent variable and time is the independent variable. Negative numbers do not apply in this situation. Although the table includes only integer values for the time, the length of a loan in years can be any real number that is greater than or equal to 1.

Domain = {all real numbers greater than or equal to 1}

Range = {all positive real numbers greater than or equal to $320}

NOTE TO STUDENT: Fully worked-out solutions to all of the Practice Problems can be found at the back of the text starting at page SP-1

Practice Problem 3 Determine whether the relation suggested by the table is a function. If it is a function, identify the domain and the range.

(a)
28 Mpg at $2.05 per Gallon

Distance	0	28	42	56	70
Cost	$0	$2.05	$3.08	$4.10	$5.13

(b)
Store's Inventory of Shirts

Number of Shirts	5	10	5	2	8
Price of Shirt	$20	$25	$30	$45	$50

TO THINK ABOUT Is It a Function? Look at the following bus schedule. Determine whether the relation is a function. Which is the independent variable? Explain your choice.

Bus Schedule

Bus Stop	Main St.	8th Ave.	42nd St.	Sunset Blvd.	Cedar Lane
Time	7:00	7:10	7:15	7:30	7:39

② Graphing Simple Nonlinear Equations

Thus far in this chapter we have graphed linear equations in two variables. We now turn to graphing a few nonlinear equations. We will need to plot more than three points to get a good idea of what the graph of a nonlinear equation will look like.

EXAMPLE 4   Graph $y = x^2$.

Solution   Begin by constructing a table of values. We select values for x and then determine by the equation the corresponding values of y. We will include negative values for x as well as positive values. We then plot the ordered pairs and connect the points with a smooth curve.

x	$y = x^2$	y
-2	$y = (-2)^2 = 4$	4
-1	$y = (-1)^2 = 1$	1
0	$y = (0)^2 = 0$	0
1	$y = (1)^2 = 1$	1
2	$y = (2)^2 = 4$	4

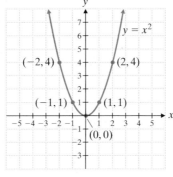

This type of curve is called a *parabola*. We will study this type of curve more extensively in Chapter 10.

Practice Problem 4   Graph $y = x^2 - 2$ on the coordinate system in the margin.

Some equations are solved for x. Usually, in those cases we pick values of y and then obtain the corresponding values of x from the equation.

EXAMPLE 5   Graph $x = y^2 + 2$.

Solution   Since the equation is solved for x, we start by picking a value of y. We will find the value of x from the equation in each case. We will select $y = -2$ first. Then we substitute it into the equation to obtain x. For convenience in graphing, we will repeat the y column at the end so that it is easy to write the ordered pairs (x, y).

y	$x = y^2 + 2$	x	y
-2	$x = (-2)^2 + 2 = 4 + 2 = 6$	6	-2
-1	$x = (-1)^2 + 2 = 1 + 2 = 3$	3	-1
0	$x = (0)^2 + 2 = 0 + 2 = 2$	2	0
1	$x = (1)^2 + 2 = 1 + 2 = 3$	3	1
2	$x = (2)^2 + 2 = 4 + 2 = 6$	6	2

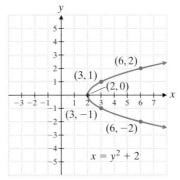

Practice Problem 5   Graph $x = y^2 - 1$ on the coordinate system in the margin.

If the equation involves fractions with variables in the denominator, we must use extra caution. Remember that you may never divide by zero.

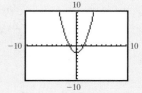

Graphing Calculator

Graphing Nonlinear Equations

You can graph nonlinear equations solved for y using a graphing calculator. For example, graph $y = x^2 - 2$ on a graphing calculator using an appropriate window. Display:

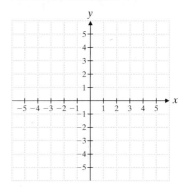

Try graphing $y = \dfrac{5}{x}$.

PRACTICE PROBLEM 4

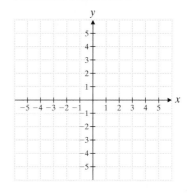

PRACTICE PROBLEM 5

EXAMPLE 6   Graph $y = \dfrac{4}{x}$.

Solution   It is important to note that x cannot be zero because division by zero is not defined. $y = \frac{4}{0}$ is not allowed! Observe that when we draw the graph we get two separate branches that do not touch.

PRACTICE PROBLEM 6

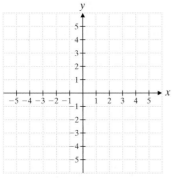

x	$y = \dfrac{4}{x}$	y
-4	$y = \dfrac{4}{-4} = -1$	-1
-2	$y = \dfrac{4}{-2} = -2$	-2
-1	$y = \dfrac{4}{-1} = -4$	-4
0	We cannot divide by zero.	There is no value.
1	$y = \dfrac{4}{1} = 4$	4
2	$y = \dfrac{4}{2} = 2$	2
4	$y = \dfrac{4}{4} = 1$	1

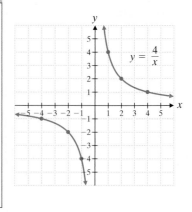

NOTE TO STUDENT: Fully worked-out solutions to all of the Practice Problems can be found at the back of the text starting at page SP-1

Practice Problem 6   Graph $y = \dfrac{6}{x}$ on the coordinate system in the margin.

③ Determining Whether a Graph Represents a Function

Can we tell whether a graph represents a function? Recall that a function cannot have two different ordered pairs with the same first coordinate. That is, each value of x must have a separate unique value of y. Look at the graph of the function $y = x^2$ in Example 4. Each x-value has a unique y-value. Look at the graph of $x = y^2 + 2$ in Example 5. At $x = 3$ there are two y-values, 1 and -1. In fact, for every x-value greater than 2 there are two y-values. $x = y^2 + 2$ is not a function.

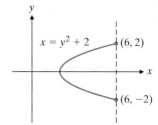

Observe that we can draw a vertical line through $(6, 2)$ and $(6, -2)$. Any graph that is not a function will have at least one region in which a vertical line will cross the graph more than once.

> **VERTICAL LINE TEST**
>
> If a vertical line can intersect the graph of a relation more than once, the relation is not a function. If no such line can be drawn, then the relation is a function.

EXAMPLE 7   Determine whether each of the following is the graph of a function.

(a)

(b)

(c)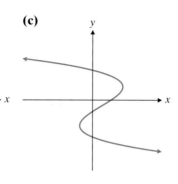

Solution

(a) The graph of the straight line is a function. Any vertical line will cross this straight line in only one location.

(b) and (c) Each of these graphs is not the graph of a function. In each case there exists a vertical line that will cross the curve in more than one place.

(a)

(b)

(c)

Practice Problem 7   Determine whether each of the following is the graph of a function.

(a)

(b)

(c)

④ Using Function Notation

We have seen that an equation like $y = 2x + 7$ is a function. For each value of x, the equation assigns a unique value to y. We could say "y is a function of x." If we name the function f, this statement can be symbolized by using the **function notation** $y = f(x)$. Many times we avoid using the y-variable completely and write the function as $f(x) = 2x + 7$.

CAUTION   Be careful. The notation $f(x)$ does not mean f multiplied by x.

EXAMPLE 8 If $f(x) = 3x^2 - 4x + 5$, find each of the following.

(a) $f(-2)$ **(b)** $f(4)$ **(c)** $f(0)$

Solution

(a) $f(-2) = 3(-2)^2 - 4(-2) + 5 = 3(4) - 4(-2) + 5 = 12 + 8 + 5 = 25$

(b) $f(4) = 3(4)^2 - 4(4) + 5 = 3(16) - 4(4) + 5 = 48 - 16 + 5 = 37$

(c) $f(0) = 3(0)^2 - 4(0) + 5 = 3(0) - 4(0) + 5 = 0 - 0 + 5 = 5$

NOTE TO STUDENT: Fully worked-out solutions to all of the Practice Problems can be found at the back of the text starting at page SP-1

Practice Problem 8   If $f(x) = -2x^2 + 3x - 8$, find each of the following.

(a) $f(2)$ **(b)** $f(-3)$ **(c)** $f(0)$

When evaluating a function, it is helpful to place parentheses around the value that is being substituted for x. Taking the time to do this will minimize sign errors in your work.

Some functions are useful in medicine. For example, the approximate width of a man's elbow is given by the function $e(x) = 0.03x + 0.6$, where x is the height of the man in inches. If a man is 68 inches tall, $e(68) = 0.03(68) + 0.6 = 2.64$. A man 68 inches tall would have an elbow width of approximately 2.64 inches.

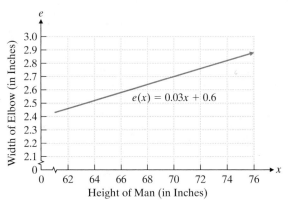

Source: U.S. National Center for Health Statistics

Verbal and Writing Skills

1. What are the three ways you can describe a function?

2. What is the difference between a function and a relation?

3. The domain of a function is the set of _____ _____ of the _____ variable.

4. The range of a function is the set of _____ _____ of the _____ variable.

5. How can we tell whether a graph is the graph of a function?

6. Without drawing a graph, how could you tell if the equation $x = y^2$ is a function or not?

(a) *Find the domain and range of the relation.* **(b)** *Determine whether the relation is a function.*

7. $\left\{ \left(\frac{2}{5}, 4 \right), \left(3, \frac{2}{5} \right), \left(-3, \frac{2}{5} \right), \left(\frac{2}{5}, -1 \right) \right\}$

8. $\left\{ \left(\frac{3}{8}, -6 \right), (-6, 5), \left(\frac{3}{8}, 2 \right), (5, -6) \right\}$

9. $\{ (7, 3.1), (5, 0), (7, 2.3) \}$

10. $\{ (7.3, 1), (0, 8), (2, 1) \}$

11. $\{ (12, 1), (14, 3), (1, 12), (9, 12) \}$

12. $\{ (5.6, 8), (5.8, 6), (6, 5.8), (5, 6) \}$

13. $\{ (3, 75), (5, 95), (3, 85), (7, 100) \}$

14. $\{ (85, 3), (95, 11), (110, 15), (110, 20) \}$

Graph the equation.

15. $y = x^2 + 3$

16. $y = x^2 - 1$

17. $y = 2x^2$

18. $y = \dfrac{1}{2}x^2$

19. $x = -2y^2$

20. $x = \dfrac{1}{2}y^2$

21. $x = y^2 - 4$

22. $x = 2y^2$

23. $y = \dfrac{2}{x}$

24. $y = -\dfrac{2}{x}$

25. $y = \dfrac{4}{x^2}$

26. $y = -\dfrac{6}{x^2}$

27. $x = (y + 1)^2$

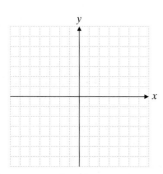

28. $y = (x - 3)^2$

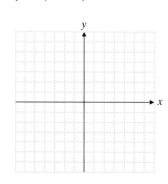

29. $y = \dfrac{4}{x - 2}$

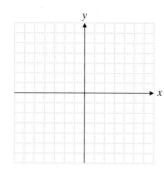

30. $x = \dfrac{2}{y + 1}$

Determine whether each relation is a function.

31.

32.

33.

34.

35.

36.

37.

38.

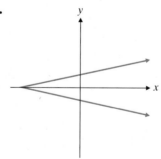

Given the following functions, find the indicated values.

39. $f(x) = 2 - 3x$
(a) $f(-8)$ (b) $f(0)$ (c) $f(2)$

40. $f(x) = 3 - 5x$
(a) $f(-1)$ (b) $f(1)$ (c) $f(2)$

41. $f(x) = 2x^2 - x + 3$
(a) $f(0)$ (b) $f(-3)$ (c) $f(2)$

42. $f(x) = -x^2 + 3x - 2$
(a) $f(0)$ (b) $f(-2)$ (c) $f(2)$

43. $f(x) = \dfrac{3}{x + 2}$

    (a) $f(3)$ **(b)** $f(0)$ **(c)** $f(-5)$

44. $f(x) = \dfrac{5}{6 - x}$

    (a) $f(0)$ **(b)** $f(-2)$ **(c)** $f(4)$

Applications

Smoking Rates *The accompanying graph represents the approximate percent of people in the United States who smoked during selected years from 1980 to 2005. Source: National Center for Health Statistics.*

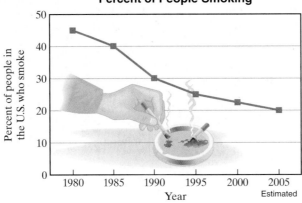

45. Does the graph represent a function?

46. Find $f(1990)$.

47. If $f(x) = 45\%$, what year is represented by x?

48. Between what two years is the difference in function values equal to 10%?

49. ***Nevada Population*** During a recent population growth period in Nevada from 1980 to 1990, the approximate population of the state measured in thousands could be predicted by the function $f(x) = 2x^2 + 20x + 800$, where x is the number of years since 1980. (*Source:* U.S. Census Bureau.) Find $f(0)$, $f(3)$, $f(6)$, and $f(10)$. Graph the function. What pattern do you observe?

50. ***Alaska Population*** During a recent population growth period in Alaska from 1980 to 2000, the approximate population of the state measured in thousands could be predicted by the function $f(x) = -0.6x^2 + 22x + 411$, where x is the number of years since 1980. (*Source:* U.S. Census Bureau.) Find $f(0)$, $f(4)$, $f(10)$, and $f(20)$. Graph the function. What pattern do you observe?

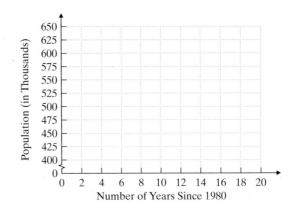

Cumulative Review

Simplify.

51. $(5x^2 - 6x + 2) + (-7x^2 - 8x + 4)$

52. $(-2x^2 - 3x + 8) - (x^2 + 4x - 3)$

53. $(4x^2 - 2x + 3)(x + 2)$

54. $(6x^2 + 10x - 5) \div (3x - 1)$

Graphing the Size of Rainforests

Most of us are aware that the world's rainforests are disappearing, but do you realize how quickly? Rainforests cover only 2% of the Earth's surface, but are home to almost half of all life forms (plants and animals) on the planet. As deforestation continues, many species are becoming extinct and the world's climate is being negatively affected (a phenomenon known as the greenhouse effect).

In 1990, there were 1756 million hectares of rainforest on the planet. Since that time, rainforest destruction has occurred at 2.47 acres per second. (This area is about as large as two football fields.)

Problems for Individual Investigation and Analysis

1. How many acres of rainforest are destroyed each day? Each year? (Round to the nearest million.)

2. Approximately how many acres of rainforest were there in 1990? Round to the nearest million. (1 hectare = 2.47 acres)

3. Use your answers from questions 1 and 2 to determine how many acres of rainforest remained in 1992. How many acres of rainforest were there in 1995?

Problems for Group Investigation and Cooperative Learning

Suppose that the current rate of rainforest destruction continues without intervention. How many acres of rainforest will there be 20 years from now? 35 years from now? In what year will there be no rainforests left in the world? Answer the following questions to find out.

4. Write a function that gives the number of acres of rainforest r (in millions of acres) that remain t years after 1990.

5. Complete the following table.

Number of Years after 1990	Amount of Rainforest Remaining on Earth (in Millions of Acres)
0	
5	
10	
15	
20	

6. Graph the information from question 5. What is the dependent variable? What is the independent variable? Be sure to use appropriate scales on the axes.

7. Using the graph, determine how many acres of rainforest will remain in the year 2035.

8. Using the graph, determine when there will be no rainforests left on Earth.

Chapter 7 Organizer

Topic	Procedure	Examples
Graphing straight lines, p. 416.	An equation of the form $$Ax + By = C$$ has a graph that is a straight line. To graph such an equation, plot any three points; two give the line and the third checks it. (Where possible, use the *x*- and *y*-intercepts.)	Graph $3x + 2y = 6$. $\begin{array}{\|c\|c\|} \hline x & y \\ \hline 0 & 3 \\ \hline 4 & -3 \\ \hline 2 & 0 \\ \hline \end{array}$
Finding the slope given two points, p. 426.	Nonvertical lines passing through distinct points (x_1, y_1) and (x_2, y_2) have slope $$m = \frac{y_2 - y_1}{x_2 - x_1}.$$ The slope of a horizontal line is 0. The slope of a vertical line is undefined.	What is the slope of the line through $(2, 8)$ and $(5, 1)$? $$m = \frac{1 - 8}{5 - 2} = -\frac{7}{3}$$
Finding the slope and y-intercept of a line given the equation, p. 430.	**1.** Rewrite the equation in the form $y = mx + b$. **2.** The slope is m. **3.** The *y*-intercept is $(0, b)$.	Find the slope and *y*-intercept. $$3x - 4y = 8$$ $$-4y = -3x + 8$$ $$y = \frac{3}{4}x - 2$$ The slope is $\frac{3}{4}$. The *y*-intercept is $(0, -2)$.
Finding the equation of a line given the slope and y-intercept, p. 430.	The slope–intercept form of the equation of a line is $$y = mx + b$$ The slope is m and the *y*-intercept is $(0, b)$.	Find the equation of the line with *y*-intercept $(0, 7)$ and with slope $m = 3$. $$y = 3x + 7$$
Graphing a line using slope and y-intercept, p. 431.	**1.** Plot the *y*-intercept. **2.** Starting from $(0, b)$, plot a second point using the slope. $$\text{slope} = \frac{\text{rise}}{\text{run}}$$ **3.** Draw a line that connects the two points.	Graph $y = -4x + 1$. First plot the *y*-intercept at $(0, 1)$. Slope $= -4$ or $\frac{-4}{1}$
Finding the slope of parallel and perpendicular lines, p. 432.	Parallel lines have the same slope. Perpendicular lines have slopes whose product is -1.	Line q has a slope of 2. The slope of a line parallel to q is 2. The slope of a line perpendicular to q is $-\frac{1}{2}$.
Finding the equation of a line through a point with a given slope, p. 439.	**1.** Substitute the known values in the equation $y = mx + b$. **2.** Solve for b. **3.** Use the values of m and b to write the general equation.	Find the equation of the line through $(3, 2)$ with slope $m = \frac{4}{5}$. $$y = mx + b \qquad 2 = \frac{4}{5}(3) + b$$ $$2 = \frac{12}{5} + b$$ $$-\frac{2}{5} = b$$ The equation is $y = \frac{4}{5}x - \frac{2}{5}$.

Topic	Procedure	Examples
Finding the equation of a line through two points, p. 439.	**1.** Find the slope. **2.** Use the procedure when given a point and the slope.	Find the equation of the line through $(3, 2)$ and $(13, 10)$. $$m = \frac{y_2 - y_1}{x_2 - x_1} = \frac{10 - 2}{13 - 3} = \frac{8}{10} = \frac{4}{5}$$ We choose the point $(3, 2)$. $$y = mx + b \qquad 2 = \frac{4}{5}(3) + b$$ $$2 = \frac{12}{5} + b$$ $$-\frac{2}{5} = b$$ The equation is $y = \frac{4}{5}x - \frac{2}{5}$.
Graphing linear inequalities, p. 443.	**1.** Graph as if it were an equation. If the inequality symbol is $>$ or $<$, use a dashed line. If the inequality symbol is $\geq$ or $\leq$, use a solid line. **2.** Look for a test point. The easiest test point is $(0, 0)$, unless the line passes through $(0, 0)$. In that case, choose another test point. **3.** Substitute the coordinates of the test point into the inequality. **4.** If it is a true statement, shade the side of the line containing the test point. If it is a false statement, shade the side of the line that does *not* contain the test point.	Graph $y \geq 3x + 2$. Graph the line $y = 3x + 2$. Use a solid line. $$\text{Test } (0, 0). \quad 0 \geq 3(0) + 2$$ $$0 \geq 2 \quad \text{false}$$ Shade the side of the line that does not contain $(0, 0)$.
Determining whether a relation is a function, p. 449.	A function is a relation in which no two different ordered pairs have the same first coordinate.	Is this relation a function? $$\{(5, 7), (3, 8), (5, 10)\}$$ It is *not* a function since $(5, 7)$ and $(5, 10)$ are two different ordered pairs with the same x-coordinate, 5.
Determining whether a graph represents a function, p. 452.	If a vertical line can intersect the graph of a relation more than once, the relation is not a function. If no such line exists, the relation is a function.	Is this graph a function? Yes. Any vertical line will intersect it at most once.

Section 7.1

1. Plot and label the following points.
A: $(2, -3)$ B: $(-1, 0)$ C: $(3, 2)$ D: $(-2, -3)$

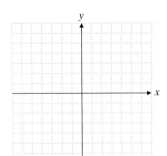

2. Give the coordinates of each point.

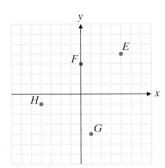

Complete the ordered pairs so that each is a solution to the given equation.

3. $y = 7 - 3x$
(a) $(0, \ )$ **(b)** $(\ , 10)$

4. $2x + 5y = 12$
(a) $(1, \ )$ **(b)** $(\ , 4)$

5. $x = 6$
(a) $(\ , -1)$ **(b)** $(\ , 3)$

Section 7.2

6. Graph $3y = 2x + 6$.

7. Graph $5y + x = -15$.

8. Graph $2y + 4x = -8 + 2y$.

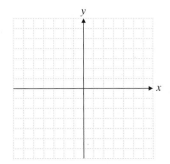

Section 7.3

9. Find the slope of the line passing through $(5, -3)$ and $\left(2, -\frac{1}{2}\right)$.

10. Find the slope and y-intercept of the line $9x - 11y + 15 = 0$.

11. Write an equation of the line with slope $-\frac{1}{2}$ and y-intercept $(0, 3)$.

12. The equation of a line is $y = \frac{3}{5}x - 2$. What is the slope of a line perpendicular to that line?

13. Graph $y = -\dfrac{1}{2}x + 3$.

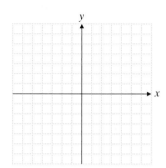

14. Graph $2x - 3y = -12$.

15. Graph $6x - 2y = 6 + 6x$.

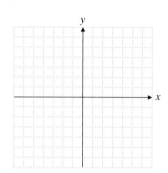

Section 7.4

16. Write an equation of the line passing through $(5, 6)$ having a slope of 2.

17. Write an equation of the line passing through $(3, -4)$ having a slope of -6.

18. Write an equation of the line passing through $(-4, 3)$ having a slope of $-\frac{3}{2}$.

19. Write an equation of the line passing through $(3, 7)$ and $(-6, 7)$.

20. What is the slope of a line parallel to a line whose equation is $y = -\frac{2}{3}x + 4$?

21. What is the slope of a line perpendicular to a line whose equation is $-3x + 4y = 8$?

Write an equation of the graph.

22.

23.

24.

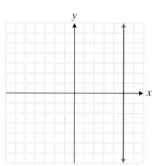

Section 7.5

25. Graph $y < \frac{1}{3}x + 2$.

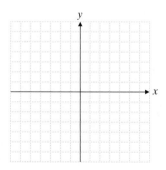

26. Graph $3y + 2x \geq 12$.

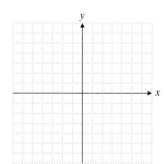

27. Graph $x \leq 2$.

Section 7.6

Determine the domain and range of the relation. Determine whether the relation is a function.

28. $\{(5, -6), (-6, 5), (-5, 5), (-6, -6)\}$

29. $\{(2, -3)(5, -3)(6, 4)(-2, 4)\}$

In exercises 30–32, determine whether the graphs represent a function.

30.

31.

32.

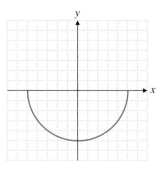

33. Graph $y = x^2 - 5$.

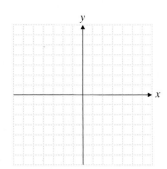

34. Graph $x = y^2 + 3$.

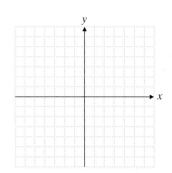

35. Graph $y = (x - 3)^2$.

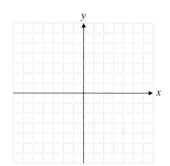

Given the following functions, find the indicated values.

36. $f(x) = 7 - 6x$ **(a)** $f(0)$ **(b)** $f(-4)$

37. $g(x) = -2x^2 + 3x + 4$ **(a)** $g(-1)$ **(b)** $g(3)$

38. $f(x) = 8x - 3$ **(a)** $f(-1)$ **(b)** $f\left(\dfrac{1}{2}\right)$

39. $f(x) = \dfrac{2}{x + 4}$ **(a)** $f(-2)$ **(b)** $f(6)$

40. $f(x) = x^2 - 2x + \dfrac{3}{x}$ **(a)** $f(-1)$ **(b)** $f(3)$

Applications

Monthly Electric Bill   *Russ and Norma Camp found that their monthly electric bill could be calculated by the equation $y = 30 + 0.09x$. In this equation y represents the amount of the monthly bill in dollars and x represents the number of kilowatt-hours used during the month.*

41. What would be their monthly bill if they used 2000 kilowatt-hours of electricity?

42. What would be their monthly bill if they used 1600 kilowatt-hours of electricity?

43. Write the equation in the form $y = mx + b$, and determine the numerical value of the y-intercept. What is the significance of this y-intercept? What does it tell us?

44. If the equation is placed in the form $y = mx + b$, what is the numerical value of the slope? What is the significance of this slope? What does it tell us?

45. If Russ and Norma have a monthly bill of $147, how many kilowatt-hours of electricity did they use?

46. If Russ and Norma have a monthly bill of $246, how many kilowatt-hours of electricity did they use?

Petroleum Consumption   *Scientists have found that the amount of petroleum consumed in the United States for a given year can be predicted by the equation $y = 0.4x + 23.6$, where x is the number of years since 1960 and y is the amount of energy consumed measured by quadrillions of British thermal units (Btu). Use this equation to answer the following questions. (Source: U.S. Energy Information Administration)*

47. How many British thermal units will be consumed from petroleum use in 2008?

48. How many British thermal units will be consumed from petroleum use in 2010?

49. If the equation is in the form $y = mx + b$, determine the numerical value of the slope. What is the significance of this slope? What does it tell us?

50. If the equation is in the form $y = mx + b$, determine the numerical value of the y-intercept. What is the significance of this y-intercept? What does it tell us?

51. In what year would the amount of petroleum consumed in the United States be 47.6 quadrillion Btu?

52. In what year would the amount of petroleum consumed in the United States be 51.6 quadrillion Btu?

Mixed Practice

53. Graph $5x + 3y = -15$.

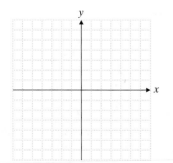

54. Graph $y = \frac{3}{4}x - 3$.

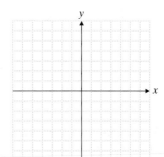

55. Find the slope of a line through $(2, -7)$ and $(-3, -5)$.

56. What is the slope and the y-intercept of the line $7x + 6y - 10 = 0$?

57. Write an equation of a line that passes through $(3, -5)$ and has a slope of $\frac{2}{3}$.

58. Graph $y < -2x + 1$.

Job Losses in Manufacturing *Economists in the Labor Department are concerned about the continued job loss in the manufacturing industry. The number of people employed in the manufacturing industry in the United States can be predicted by the equation $y = -130x + 22,000$, where x is the number of years since 1980 and y is the number of people **in thousands** employed in the manufacturing industry. Use this data to answer the following questions. (Source: U.S. Bureau of Labor Statistics)*

59. How many people were employed in manufacturing in 1980? in 1990? in 2000?

60. Use your answers for 59 to draw a graph of the equation $y = -130x + 22,000$.

61. What is the slope of this equation? What is its significance?

62. What is the y-intercept of this equation? What is its significance?

63. Use the equation to predict in what year the number of manufacturing jobs will be 18,100,000.

64. Use the equation to predict in what year the number of manufacturing jobs will be 16,800,000.

Remember to use your Chapter Test Prep Video CD to see the worked-out solutions to the test problems you want to review.

1. Plot and label the following points.

$B: (6, 1)$ $C: (-4, -3)$
$D: (-3, 0)$ $E: (5, -2)$

2. Graph the line $6x - 3 = 5x - 2y$.

1. _____

2. _____

3. _____

3. Graph the line $8x + 2y = -4$.

4. Graph $y = \frac{2}{3}x - 4$.

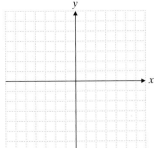

4. _____

5. _____

5. What is the slope and the y-intercept of the line $3x + 2y - 5 = 0$?

6. _____

6. Find the slope of the line that passes through $(8, 6)$ and $(-3, -5)$.

7. _____

7. Write an equation for the line that passes through $(4, -2)$ and has a slope of $\frac{1}{2}$.

8. _____

8. Find the slope of the line through $(-3, 11)$ and $(6, 11)$.

9. _____

9. Find an equation for the line passing through $(5, -4)$ and $(-3, 8)$.

10. Write an equation for the line through $(2, 7)$ and $(2, -2)$. What is the slope of this line?

10. _____

11. _____

12. _____

13. _____

14. _____

15. _____

16.   (a) _____

       (b) _____

17.   (a) _____

       (b) _____

11. Graph the region described by $4y \le 3x$.

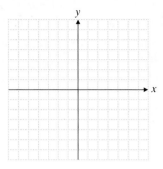

12. Graph the region described by $-3x - 2y > 10$.

13. Is this relation a function? $\{(2, -8)(3, -7)(2, 5)\}$ Why?

14. Look at the relation graphed below. Is this relation a function? Why?

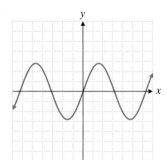

15. Graph $y = 2x^2 - 3$.

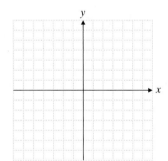

x	y
-2	
-1	
0	
1	
2	

16. For $f(x) = -x^2 - 2x - 3$:

   (a) Find $f(0)$.

   (b) Find $f(-2)$.

17. For $g(x) = \dfrac{3}{x - 4}$:

   (a) Find $g(3)$.

   (b) Find $g(-11)$.

Approximately one-half of this test covers the content of Chapters 0–6. The remainder covers the content of Chapter 7.

1. Simplify. $(-3x^3y^4z)^3$

2. Simplify; leave the answer in positive exponents only. $\dfrac{-16x^{-3}y^4}{24x^5y^{-6}}$

Solve.

3. $2 - 3(4 - x) = x - (3 - x)$

4. $\dfrac{600}{R} = \dfrac{600}{R + 50} + \dfrac{500(R + 20)}{R^2 + 50R}$

5. $x = 2^3 - 6 \div 3 - 1$

6. Solve $A = lw + lh$ for w.

7. Factor. $50a^2 - 98b^2$

8. Factor. $4x^2 + 4x - 15$

9. Simplify. $\dfrac{x - \dfrac{1}{x}}{\dfrac{1}{2} + \dfrac{1}{2x}}$

10. Find an equation of the line through $(6, 8)$ and $(7, 11)$.

In questions 11 and 12, give the equation of the line that fits each description.

11. Passes through $(7, -4)$ and is vertical.

12. Passes through $(3, -5)$ with slope $-\dfrac{2}{3}$.

13. Find the slope of the line through $(-8, -3)$ and $(11, -3)$.

14. What is the slope of the line $3x - 7y = -2$?

1. _____

2. _____

3. _____

4. _____

5. _____

6. _____

7. _____

8. _____

9. _____

10. _____

11. _____

12. _____

13. _____

14. _____

15. _____

16. _____

17. _____

18. _____

19. _____

20. **(a)** _____

(b) _____

15. Graph $y = \frac{2}{3}x - 4$.

16. Graph $3x + 8 = 5x$.

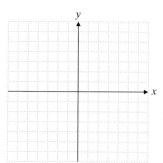

17. Graph the region $2x + 5y \leq -10$.

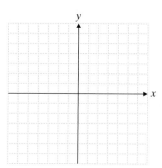

18. Graph $x = y^2 + 3$. Is this a function?

x	y
	−2
	−1
	0
	1
	2

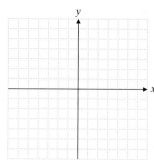

19. Is this relation a function? $\{(3, 10), (10, -3), (0, -3), (5, 10)\}$

20. For $f(x) = 3x^2 - 2x + \dfrac{4}{x}$:

(a) Find $f(-1)$.   **(b)** Find $f(2)$.

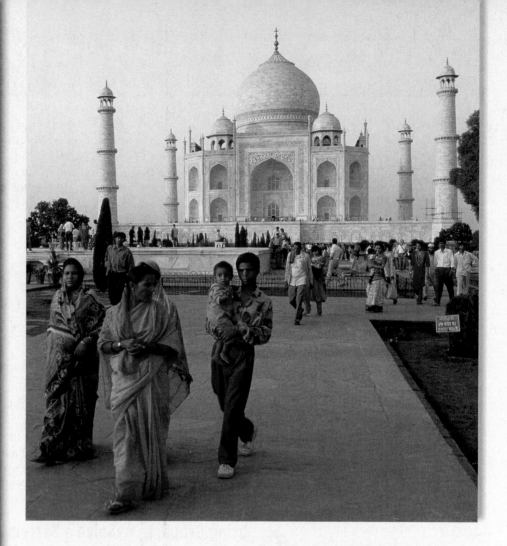

The population of different regions of the world is increasing at different rates. By comparing these differences we can make predictions about the population and about the changes this will cause. Do you think you can use your mathematical skills to investigate these kinds of issues? Turn to the Putting Your Skills to Work on page 508.

Systems of Equations

Student Learning Objectives

After studying this section, you will be able to:

1 Use graphing to solve a system of linear equations with a unique solution.

2 Graph a system of linear equations to determine the type of solution.

3 Solve applied problems by graphing.

In Chapter 7 we examined linear equations and inequalities in two variables. In this chapter we will examine systems of linear equations and inequalities. Two or more equations or inequalities in several variables that are considered simultaneously are called a **system of equations** or a **system of inequalities.** Recall that the graph of a linear equation is a straight line. If we have the graphs of two linear equations on one coordinate plane, how might they relate to one another?

The lines may intersect.

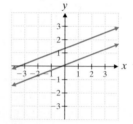
The lines may be parallel.

The lines may coincide.

The solutions to a system of linear equations are those points that both lines have in common. Note that parallel lines have no points in common. Thus the second system has no solution. Intersecting lines intersect at one point. Thus the first system has one solution. Lines that coincide have an infinite number of points in common. Thus the third system has an infinite number of solutions. You can determine the solution to a system of equations by graphing.

Graphing Calculator

 Graphing a System of Linear Equations

If the two equations in a system are not in the form $y = mx + b$, you must write the equations in this form before graphing.

To solve the system

$$3x + 4y = -28$$
$$x - 4y = 12,$$

we first need to solve each equation for y. This will give us the equivalent system

$$y = -0.75x - 7$$
$$y = 0.25x - 3.$$

We graph this system with a scale of -10 to 10 for both x and y.
Display.

If we use the Zoom function, we will locate the intersection at $x = -4$, $y = -4$. Try it to see if you can obtain those values.

1 ## Using Graphing to Solve a System of Linear Equations with a Unique Solution

To solve a system of equations by graphing, graph both equations on the same coordinate system. If the lines intersect, the system has a unique solution. The ordered pair of coordinates of the point of intersection is the solution to the system of equations.

EXAMPLE 1 Solve by graphing.

$$-2x + 3y = 6$$
$$2x + 3y = 18$$

Solution Graph both equations on the same coordinate plane and determine the point of intersection.

We will graph the system by first obtaining a table of values for each equation. First we solve the first equation for y.

$$-2x + 3y = 6$$
$$3y = 2x + 6$$

$$y = \frac{2}{3}x + 2$$

Let $x = 0$. $y = \frac{2}{3}(0) + 2 = 0 + 2 = 2$

Let $x = -3$. $y = \frac{2}{3}(-3) + 2 = -2 + 2 = 0$

Let $x = 6$. $y = \frac{2}{3}(6) + 2 = 4 + 2 = 6$

x	y
0	2
-3	0
6	6

This allows us to use the three sets of ordered pairs to graph $-2x + 3y = 6$. Next we solve the second equation for y.

$$2x + 3y = 18$$
$$3y = -2x + 18$$
$$\boxed{y = -\frac{2}{3}x + 6}$$

Let $x = 0$. $y = -\frac{2}{3}(0) + 6 = 0 + 6 = 6$

Let $x = -3$. $y = -\frac{2}{3}(-3) + 6 = 2 + 6 = 8$

Let $x = 6$. $y = -\frac{2}{3}(6) + 6 = -4 + 6 = 2$

x	y
0	6
-3	8
6	2

This allows us to use these three ordered pairs to graph $2x + 3y = 18$. (Of course, each equation could be graphed by any method we have discussed in the text. It is not necessary to use this method.) The graph of the system is shown in the figure at the right. The lines intersect at the point for which $x = 3$ and $y = 4$. Thus the unique solution to the system of equations is $(3, 4)$. Always check your answer.

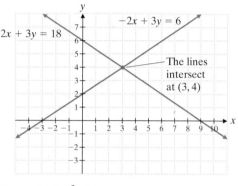

$$-2(3) + 3(4) \overset{?}{=} 6 \qquad 2(3) + 3(4) \overset{?}{=} 18$$
$$6 = 6 \qquad\qquad 18 = 18 \;\checkmark$$

A linear system of equations that has one solution is said to be **consistent**.

Practice Problem 1   Solve by graphing. $x + y = 12$
$$-x + y = \;\; 4$$

PRACTICE PROBLEM 1

NOTE TO STUDENT: Fully worked-out solutions to all of the Practice Problems can be found at the back of the text starting at page SP-1

② Graphing a System of Linear Equations to Determine the Type of Solution

Two lines in a plane may intersect, may never intersect (be parallel lines), or may be the same line. The corresponding system of equations will have one solution, no solution, or an infinite number of solutions, respectively. Thus far we have focused on systems of linear equations that have a unique solution. We now draw your attention to the other two cases.

EXAMPLE 2   Solve by graphing.
$$3x - y = \;\; 1$$
$$3x - y = -7$$

Solution   We graph both equations on the same coordinate plane. The graph is at the right. Notice that the lines are parallel. They do not intersect. Hence there is no solution to this system of equations. A system of linear equations that has no solution is called **inconsistent**.

PRACTICE PROBLEM 2

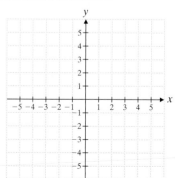

Practice Problem 2   Solve by graphing. $4x + 2y = 8$
$$-6x - 3y = 6$$

PRACTICE PROBLEM 3

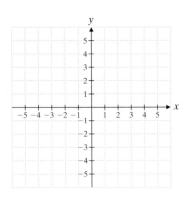

NOTE TO STUDENT: Fully worked-out solutions to all of the Practice Problems can be found at the back of the text starting at page SP-1

EXAMPLE 3   Solve by graphing.

$$x + y = 4$$
$$3x + 3y = 12$$

Solution   We graph both equations on the same coordinate plane at the right. Notice that the equations represent the same line (coincide). The coordinates of every point on the line will satisfy both equations. That is, every point on the line represents a solution to $x + y = 4$ and to $3x + 3y = 12$. Thus there is *an infinite number of solutions* to this system. Such equations are said to be **dependent.**

Practice Problem 3   Solve by graphing.

$$3x - 9y = 18$$
$$-4x + 12y = -24$$

TO THINK ABOUT: Determining the Type of Solution Without Graphing

Without graphing, how could you tell that the system in Example 2 represented parallel lines? Without graphing, how could you tell that the equations in Example 3 are dependent? Without graphing, determine whether each of the following systems is inconsistent (has no solution) or consistent (has a solution). Are any systems of equations dependent?

(a) $2x - 3y = 6$
     $6y = 4x + 1$

(b) $3x + y = 2$
     $2x + y = 3$

(c) $x - 2y = 1$
     $3x - 6y = 3$

③ Solving Applied Problems by Graphing

EXAMPLE 4   Walter and Barbara need some plumbing repairs done at their house. They called two plumbing companies for estimates of the work that needed to be done. Roberts Plumbing and Heating charges $40 for a house call and then $35 per hour for labor. Instant Plumbing Repairs charges $70 for a house call and then $25 per hour for labor.

(a) Create a cost equation for each company where y is the total cost of plumbing repairs and x is the number of hours of labor. Write a system of equations.

(b) Graph the two equations using the values $x = 0, 3,$ and 6.

(c) Determine from your graph how many hours of plumbing repairs would be required for the two companies to charge the same.

(d) Determine from your graph which company charges less if the estimated amount of time to complete the plumbing repairs is 4 hours.

Solution

(a) For each company we will obtain an equation of the form

$$y = (\text{cost of house call}) + (\text{cost per hour})x.$$

Roberts Plumbing and Heating charges $40 for a house call and $35 per hour. Thus we obtain the first equation, $y = 40 + 35x$.

Instant Plumbing Repairs charges $70 for a house call and $25 per hour. Thus we obtain the second equation, $y = 70 + 25x$.

This yields the following system of equations:

$$y = 40 + 35x$$
$$y = 70 + 25x.$$

(b) We will graph the system by first obtaining a table of values for each equation.

Roberts Plumbing and Heating $y = 40 + 35x$

Let $x = 0$. $y = 40 + 35(0) = 40$

Let $x = 3$. $y = 40 + 35(3) = 145$

Let $x = 6$. $y = 40 + 35(6) = 250$

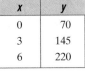

x	y
0	40
3	145
6	250

Instant Plumbing Repairs $y = 70 + 25x$

Let $x = 0$. $y = 70 + 25(0) = 70$

Let $x = 3$. $y = 70 + 25(3) = 145$

Let $x = 6$. $y = 70 + 25(6) = 220$

x	y
0	70
3	145
6	220

The graph of the system follows.

(c) We see that the graphs of the two lines intersect at $(3, 145)$. Thus the two companies will charge the same if 3 hours of plumbing repairs are required.

(d) We draw a dashed line at $x = 4$. We see that the blue line representing Roberts Plumbing and Heating is higher than the red line representing Instant Plumbing Repairs after 3 hours. Thus the cost would be less if we use Instant Plumbing Repairs for 4 hours of work.

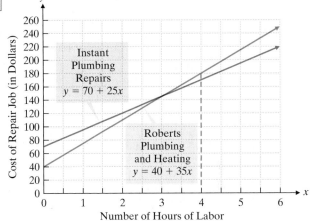

Practice Problem 4   Jonathan and Joanne Wells need some electrical work done on their new house. They obtained estimates from two companies. Bill Tupper's Electrical Service charges $100 for a house call and $30 per hour for labor. Wire for Hire charges $50 for a house call and $40 per hour for labor.

(a) Create a cost equation for each company where y is the total cost of the electrical work and x is the number of hours of labor. Write a system of equations.

(b) Graph the two equations using the values $x = 0, 4,$ and 8.

(c) Determine from your graph how many hours of electrical repairs would be required for the two companies to charge the same.

(d) Determine from your graph which company charges less if the estimated amount of time to complete the electrical repairs to the house is 6 hours.

Verbal and Writing Skills

1. In the system $y = 2x - 5$ and $y = 2x + 6$, the lines have the same slope but different y-intercepts. What does this tell you about the graph of this system?

2. In the system $y = -3x + 4$ and $y = -3x + 4$, the lines have the same slope and the same y-intercepts. What does this tell you about the graph of the system?

3. Before you graph a system of equations, if you notice that the lines have different slopes, you can conclude that they _____ and that the system has _____ solution.

4. If lines have the same slope, you cannot conclude that the lines are parallel. Why? How else might the lines relate to each other?

5. Two lines have different slopes but the same y-intercept. What does this tell you about the graphs of the lines and about the solution of the system?

6. In the system $y = -3x + 5$ and $y = -3x - 2$, what can you say about the number of solutions?

Solve by graphing. If there isn't a unique solution to the system, state the reason.

7. $x - y = 3$
$x + y = 5$

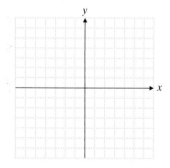

8. $x - y = 4$
$-x - y = 6$

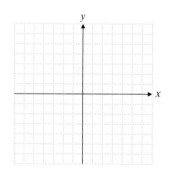

9. $2x + y = 0$
$-3x + y = 5$

10. $x + 2y = 5$
$-3x + 2y = 17$

11. $4x + y = 5$
$3x - 2y = 12$

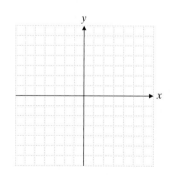

12. $x + 2y = 10$
$3x - 2y = 6$

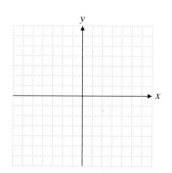

13. $-2x + y - 3 = 0$
$4x + y + 3 = 0$

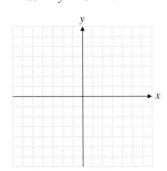

14. $x - 2y - 10 = 0$
$2x + 3y - 6 = 0$

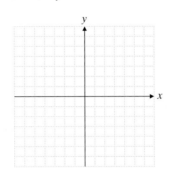

15. $3x - 2y = -18$
$2x + 3y = 14$

Mixed Practice

Solve by graphing. If there isn't a unique solution to the system, state the reason.

16. $3x + 2y = -10$
$-2x + 3y = 24$

17. $y = \dfrac{3}{4}x + 7$

$y = -\dfrac{1}{2}x + 2$

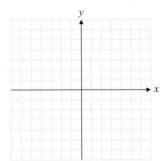

18. $y = \dfrac{5}{7}x - 2$

$y = \dfrac{1}{3}x + \dfrac{2}{3}$

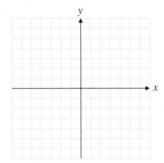

19. $3x - 2y = -4$
$-9x + 6y = -9$

20. $4x - 6y = 8$
$-2x + 3y = -4$

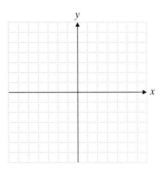

21. $y - 2x - 6 = 0$

$\dfrac{1}{2}y - 3 = x$

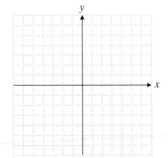

22. $2y + x - 6 = 0$

$y + \dfrac{1}{2}x = 4$

23. $y = \dfrac{1}{2}x - 2$

$y = \dfrac{2}{3}x - 1$

24. $2y + 6x - 8 = 0$

$y = \dfrac{3}{2}x + 4$

25. *Telephone Charges* The Newtown Telephone Company advertises that if you use their 10-10 service, you will always pay only 7¢ per minute, no matter where you call within the continental U.S. Its major rival, the Tryus Telephone Company, charges only 3¢ per minute, but you must pay a 36¢ "connection fee" per call.

(a) Create a cost equation for each telephone company, where y is the total cost for each phone call and x is the number of minutes the phone call lasts. Write a system of equations.

(b) Graph the two equations using the values $x = 1, 4, 8, 12$. Fill in a table for those values of x and y.

(c) Determine from your graph how many minutes long your phone call must be for the cost of using each telephone company to be the same.

(d) Determine from your graph which company would cost you less if you typically make phone calls of 10 minutes or more.

26. *Landscaping Costs* Fred and Amy want to have shrubs planted in front of their new house. They have obtained estimates from two landscaping companies. Camp Property Care charges $100 for an initial consultation and $60 an hour for labor. Manchester Landscape Designs charges $200 for an initial consultation and $40 an hour for labor.

(a) Create a cost equation for each company where y is the total cost of the landscaping work and x is the number of hours of labor. Write a system of equations.

(b) Graph the two equations using the values $x = 0, 5,$ and 10. Fill in a table for those values of x and y.

(c) Determine from your graph how many hours of labor would be required for the cost of landscaping from each of the two companies to be the same.

(d) Determine from your graph which company would charge less if 7 hours of landscaping are required.

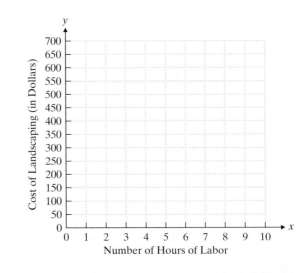

If you have a graphing calculator, solve each system by graphing the equations and estimating the point of intersection to the nearest hundredth. You will need to adjust the window for the appropriate x- and y-values in order to see the region where the straight lines intersect.

27. $y = 56x + 1808$
$ y = -62x - 2086$

28. $y = 47x - 960$
$ y = -36x + 783$

29. $88x + 57y = 683.10$
$ 95x - 48y = 7460.64$

30. $64x + 99y = -8975.52$
$ 58x - 73y = 3503.58$

Cumulative Review

31. Solve. $7 - 3(4 - x) \le 11x - 6$

32. Solve for d. $3dx - 5y = 2(dx + 8)$

33. Multiply. $(3x - 7)^2$

34. Factor. $25x^2 - 60x + 36$

Tourist Visits to Machu Picchu *In 1999, Machu Picchu, the 15th-century Incan citadel perched between the Andes and the Amazon jungle, drew a record 360,000 visitors. This was 20% more than the year before and four times higher than in 1991. According to the government, the ideal number of tourists would be just under two million a year. Critics state that the ruins can sustain only 1000 to 1500 visitors a day without suffering further deterioration. The general manager of a company that plans to install and operate a tram there believes that 400,000 to 500,000 visitors annually would be a better target than two million. (Source: Peru Department of Tourism)*

35. How many people visited Machu Picchu in 1991?

36. How many people visited Machu Picchu in 1998?

37. In the view of the critics, what is the maximum number of people per year that should be allowed to visit Machu Picchu?

38. If the target of 500,000 visitors offered by the general manager of the tram company is achieved, what will be the percent of increase in the number of visitors since 1999?

8.2 SOLVING A SYSTEM OF EQUATIONS IN TWO VARIABLES BY THE SUBSTITUTION METHOD

1 Solving a System of Two Linear Equations with Integer Coefficients by the Substitution Method

Since the solution to a system of two linear equations may be a point where the two lines intersect, we would expect the solution to be an ordered pair. Since the point must lie on both lines, the ordered pair (x, y) must satisfy both equations. For example, the solution to the system $3x + 2y = 6$ and $x + y = 3$ is $(0, 3)$. Let's check.

$$3x + 2y = 6 \qquad\qquad x + y = 3$$
$$3(0) + 2(3) \overset{?}{=} 6 \qquad\qquad 0 + 3 \overset{?}{=} 3$$
$$6 = 6 \;\checkmark \qquad\qquad\qquad 3 = 3 \;\checkmark$$

Given a system of linear equations, we can develop algebraic methods for finding a solution. These methods reduce the system to one equation in one variable, which we already know how to solve. Once we have solved for this variable, we can substitute this known value into any one of the original equations and solve for the second variable.

The first method we will discuss is the **substitution method.**

PROCEDURE FOR SOLVING A SYSTEM OF EQUATIONS BY THE SUBSTITUTION METHOD

1. Solve one of the two equations for one variable. If possible, solve for a variable with a coefficient of 1 or -1.

2. Substitute the expression you obtain in step 1 for this variable in the *other* equation.

3. You now have one equation with one variable. Solve this equation to find the value for that one variable.

4. Substitute this value for the variable into one of the original equations to obtain a value for the other variable.

5. Check the solution in each equation to verify your results.

EXAMPLE 1   Find the solution.

$$x - 2y = 7$$
$$-5x + 4y = -5$$

Solution

Step 1   Solve one equation for one variable.

$\quad x - 2y = 7$ The first equation is the easiest one in which to isolate a variable.

$\qquad x = 7 + 2y$ Add $2y$ to both sides to solve for x.

Step 2   Substitute the expression into the other equation.

$\qquad\qquad -5x + 4y = -5$ This is the original second equation.

$\quad -5(\,7 + 2y\,) + 4y = -5$ Substitute the value $7 + 2y$ for x in this equation.

Step 3   Solve this equation.

$\qquad\qquad -35 - 10y + 4y = -5$ Remove the parentheses.

$\qquad\qquad\qquad -35 - 6y = -5$ Collect like terms.

$\qquad\qquad\qquad\qquad -6y = 30$ Add 35 to both sides.

$\qquad\qquad\qquad\qquad\quad y = -5$ Divide both sides by -6.

Step 4 Obtain the value of the other variable. We will now use the value $y = -5$ in one of the equations that contains both variables to find the value for x.

$x = 7 + 2y$ The easiest equation to use is the one that we solved for x.

$= 7 + 2(-5)$ Replace y by -5.

$= 7 - 10$

$= -3$

The solution is $(-3, -5)$.

Step 5 Check. To be sure that we have the correct solution, we will need to check that the values obtained for x and y can be substituted into *both* *original* equations to obtain true mathematical statements.

$$x - 2y = 7 \qquad\qquad -5x + 4y = -5$$
$$-3 - 2(-5) \overset{?}{=} 7 \qquad -5(-3) + 4(-5) \overset{?}{=} -5$$
$$-3 + 10 \overset{?}{=} 7 \qquad\qquad 15 - 20 \overset{?}{=} -5$$
$$7 = 7 \ \checkmark \qquad\qquad\qquad -5 = -5 \ \checkmark$$

Practice Problem 1   Find the solution.

$$5x + 3y = 19$$
$$2x - \ y = 12$$

NOTE TO STUDENT: Fully worked-out solutions to all of the Practice Problems can be found at the back of the text starting at page SP-1

② Solving a System of Two Linear Equations with Fractional Coefficients by the Substitution Method

If a system of equations contains fractions, clear the equations of fractions *before* performing any other steps.

EXAMPLE 2   Find the solution.

$$\frac{3}{2}x + \ y = \ \frac{5}{2}$$
$$-y + 2x = -1$$

Solution   We want to clear the first equation of fractions. We observe that the LCD of the fractions is 2.

$$2\left(\frac{3}{2}x\right) + 2\,(y) = 2\left(\frac{5}{2}\right) \quad \text{Multiply each term of the equation by the LCD.}$$
$$3x + 2y = 5 \qquad \text{This is equivalent to } \frac{3}{2}x + y = \frac{5}{2}.$$

Now follow the five-step procedure.

Step 1 Solve for one variable. We will use the second equation.

$$-y + 2x = -1$$
$$-y = -1 - 2x \quad \text{Add } -2x \text{ to both sides.}$$
$$y = 1 + 2x \quad \text{Multiply each term by } -1.$$

Step 2 Substitute the resulting expression into the other equation.

$$3x + 2(1 + 2x) = 5$$

Finding an Approximate Solution to a System of Equations

We can solve systems of equations graphically by using a graphing calculator. For example, to solve the system of equations in Example 2, first rewrite each equation in slope–intercept form.

$$y = -\frac{3}{2}x + \frac{5}{2}$$
$$y = 2x + 1$$

Then graph $y_1 = -\frac{3}{2}x + \frac{5}{2}$ and $y_2 = 2x + 1$ on the same screen.
Display:

Next you can use the Trace and Zoom features to find the intersection of the two lines. Some graphing calculators have a command to find and calculate the intersection.
Display:

Rounded to four decimal places, the solution is (0.4286, 1.8571). Observe that

$$\frac{3}{7} = 0.4285714\ldots$$

and $\frac{13}{7} = 1.8571428\ldots$,

so the answer agrees with the answer found in Example 2.

Step 3 Solve this equation for the variable.

$$3x + 2 + 4x = 5 \qquad \text{Remove parentheses.}$$
$$7x + 2 = 5$$
$$7x = 3$$
$$x = \frac{3}{7}$$

Step 4 Find the value of the second variable. We will use the second equation.

$$-y + 2x = -1$$
$$-y + 2\left(\frac{3}{7}\right) = -1$$
$$-y + \frac{6}{7} = -1$$
$$y - \frac{6}{7} = 1$$
$$y = 1 + \frac{6}{7} = \frac{13}{7}$$

The solution to the system is $\left(\frac{3}{7}, \frac{13}{7}\right)$.

Step 5 Check.

$$\frac{3}{2}x + y = \frac{5}{2} \qquad\qquad\qquad -y + 2x = -1$$

$$\frac{3}{2}\left(\frac{3}{7}\right) + \left(\frac{13}{7}\right) \overset{?}{=} \frac{5}{2} \qquad\qquad -\frac{13}{7} + 2\left(\frac{3}{7}\right) \overset{?}{=} -1$$

$$\frac{9}{14} + \frac{13}{7} \overset{?}{=} \frac{5}{2} \qquad\qquad -\frac{13}{7} + \frac{6}{7} \overset{?}{=} -\frac{7}{7}$$

$$\frac{9}{14} + \frac{26}{14} \overset{?}{=} \frac{35}{14} \qquad\qquad -\frac{7}{7} = -\frac{7}{7} \ \checkmark$$

$$\frac{35}{14} = \frac{35}{14} \ \checkmark$$

Practice Problem 2 Find the solution. Be sure to check your answer.

$$\frac{1}{3}x - \frac{1}{2}y = 1$$
$$x + 4y = -8$$

Find the solution to each system of equations by the substitution method. Check your answers. Place your solution in the form $(x, y), (a, b), (p, q),$ *or* (s, t).

1. $4x + 3y = 9$
  $x = 3y + 6$

2. $x = 4 - 4y$
  $-x + 2y = 2$

3. $2x + y = 4$
  $2x - y = 0$

4. $7x - 3y = -10$
  $x + 3y = 2$

5. $5x + 2y = 5$
  $3x + y = 4$

6. $4x + 2y = 4$
  $3x + y = 4$

7. $3x - 2y = -8$
  $x + y = 4$

8. $-x + 3y = -6$
  $2x - y = 2$

9. $3a - 5b = 2$
  $3a + b = 32$

10. $a - 5b = 21$
  $3a + 4b = -13$

11. $3x - y = 3$
  $x + 3y = 11$

12. $4x - y = 9$
  $3x - 5y = 11$

13. $p + 2q - 4 = 0$
  $7p - q - 3 = 0$

14. $7s + 2t - 10 = 0$
  $5s - t + 5 = 0$

15. $3x - y - 9 = 0$
  $8x + 5y - 1 = 0$

16. $8x + 2y - 7 = 0$
  $-2x - y + 2 = 0$

Mixed Practice

Find the solution to each system of equations. Place your answer in the form (x, y).

17. $\frac{5}{3}x + \frac{1}{3}y = -3$
  $-2x + 3y = 24$

18. $-x + y = -4$
  $\frac{3}{7}x + \frac{2}{3}y = 5$

19. $4x + 5y = 2$
  $\frac{1}{5}x + y = -\frac{7}{5}$

20. $x + 2y = -6$
  $\frac{2}{3}x + \frac{1}{3}y = -3$

21. $\frac{4}{7}x + \frac{2}{7}y = 2$
  $3x + y = 13$

22. $x - 4y = 11$
  $\frac{2}{3}x + \frac{1}{2}y = 1$

23. $3x + y = 6$
  $3x - 2y = -3$

24. $4x + y = 5$
  $3x - 2y = 12$

25. $3x - 4y = 36$
  $y + 3 = 3(x + 1)$

26. $2x + 3y = -15$
  $y + 15 = 2(x + 1)$

27. $2x = 4(2y + 2)$
  $3(x - 3y) + 2 = 17$

28. $2(2x - y - 11) = 1 - y$
  $3(3x + y) - 17 = 2(4x + y)$

To Think About

29. How many equations do you think you would need to solve a system with three unknowns? With seven unknowns? Explain your reasoning.

30. The point where the graphs of two linear equations intersect is the _____ to the system of linear equations.

31. The solution to a system of two equations must satisfy _____ equations.

32. How many solutions will a system of equations have if the graphs of the lines intersect? Justify your answer.

33. How many solution(s) will a system of equations have if the graphs of the equations are parallel lines? Why?

Applications

34. *Construction Costs* Tim Martinez is involved in a large construction project in the city. He needs to rent a heavy construction crane for several weeks. He is considering renting from one of two companies. Boston Construction will rent a crane for an initial delivery charge of $1500 and a rental fee of $900 per week. North End Contractors will rent a crane for an initial delivery charge of $500 and a rental fee of $1000 per week.

    (a) Create a cost equation for each company where y is the total cost of renting a crane and x is the number of weeks the crane is rented. Write a system of equations.

    (b) Solve the system of equations by the substitution method to find out how many weeks would be required for the two companies to charge the same. What would the cost be for each company?

    (c) Tim remembers that last year he considered renting from the same two companies. For the number of weeks he needed the crane, the cost of renting from one company was $4,000 more than the other. How many weeks was the rental last year? Which company was less expensive for that period last year?

35. *Snow Removal Costs* Fred Driscoll has just moved from the sunny Southwest to the wintry Northeast, outside Buffalo, New York. His farmhouse has a very long driveway and Fred has little experience with snow. He decides to hire a snow removal company for the winter. Lake Erie Plowing requires a $50 nonrefundable reservation fee at the start of the winter and charges $90 every time they come out to plow. Adirondack Plowing requires a reservation fee of $100 but only charges $80 each time they come out to plow.

    (a) Create a cost equation for each plowing company where y is the total cost of plowing and x is the number of times Fred needs to be plowed out. Write a system of equations.

    (b) Solve the system of equations by the substitution method to find out how many times Fred would have to be plowed out for the cost of the plowing companies to be the same. How many times would this be? What would be the cost for each company?

    (c) Fred calculated that if he chose one company over the other, based on the average number of times that his neighbors claimed he would need to be plowed, he would save himself $80. How many times do his neighbors expect Fred to need plowing? Which company should he hire to save that money?

Cumulative Review

36. Find the slope and y-intercept of $7x + 11y = 19$.

37. Graph $6x + 3y \geq 9$.

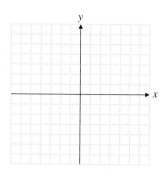

38. *Athletic Training* Find the running speed r of a 85-kilogram runner who is using up energy at the rate of 35 kilocalories per minute. Use the equation $85(dr - g) = 35$, where d is 0.0011, g is 0.026, and r is measured in meters per minute. Round to the nearest whole number.

39. *Middle School Construction* West Coast Scholastic Construction has been given a contract to build the new wing of the Benson Memorial Middle School for $3.8 million. The construction company has been offered a 3% bonus if the work is completed 3 months ahead of schedule. What will be the total cost of the construction if the work is completed 3 months ahead of schedule?

40. *Travel Agency Operations* The average American leisure traveler usually vacations with three other people for three days and three nights. If each week the average travel agency books lodgings for 14 leisure travelers (who are making arrangements for their traveling companions) and all of the travelers request single rooms, how many nights in rooms will the travel agency have to book in one year?

41. *Automobile Travel Expenses* The average American family that owns a car will take five major family trips by car per year. The average car trip amounts to 360 miles. The average American car owned by a family gets about 20 miles per gallon. If gas costs $1.80 per gallon, what is the total cost of gasoline for family trips for the average American family in one year?

Student Learning Objectives

After studying this section, you will be able to:

1 Use the addition method to solve a system of two linear equations with integer coefficients.

2 Use the addition method to solve a system of two linear equations with fractional coefficients.

3 Use the addition method to solve a system of two linear equations with decimal coefficients.

1 Using the Addition Method to Solve a System of Two Linear Equations with Integer Coefficients

The substitution method is useful for solving a system of equations when the coefficient of one variable is 1. In the following system, the variable x in the first equation has a coefficient of 1, and we can easily solve for x.

$$x - 2y = 7 \Rightarrow x = 7 + 2y$$
$$-5x + 4y = -5$$

This makes the substitution method a natural choice for solving this system. But we may not always have 1 as a coefficient of a variable. We need a method of solving systems of equations that will work for integer coefficients, fractional coefficients, and decimal coefficients. One such method is the **addition method.**

EXAMPLE 1 Solve by addition.

$$5x + 2y = 7 \qquad \textbf{(1)}$$
$$3x - y = 13 \qquad \textbf{(2)}$$

Solution We would like the coefficients of the y-terms to be opposites. One should be $+2y$ and the other $-2y$. Therefore, we multiply each term of equation **(2)** by 2.

$2(3x) - 2(y) = 2(13)$	Multiply each term of equation **(2)** by 2.
$5x + 2y = 7$	We now have an equivalent system of equations.
$\underline{6x - 2y = 26}$	Add the two equations. This will eliminate the y-variable.
$11x = 33$	
$x = 3$	Divide both sides by 11.
$5(3) + 2y = 7$	Substitute $x = 3$ into one of the original equations. Arbitrarily, we pick equation **(1)**.
$15 + 2y = 7$	Remove parentheses.
$2y = -8$	Subtract 15 from both sides.
$y = -4$	Divide both sides by 2.

The solution is $(3, -4)$.

Check: Replace x by 3 and y by -4 in *both* original equations.

$$5x + 2y = 7 \qquad\qquad 3x - y = 13$$
$$5(3) + 2(-4) \stackrel{?}{=} 7 \qquad\qquad 3(3) - (-4) \stackrel{?}{=} 13$$
$$15 - 8 \stackrel{?}{=} 7 \qquad\qquad 9 + 4 \stackrel{?}{=} 13$$
$$7 = 7 \checkmark \qquad\qquad 13 = 13 \checkmark$$

Practice Problem 1 Solve by addition.

$$3x + y = 7$$
$$5x - 2y = 8$$

NOTE TO STUDENT: Fully worked-out solutions to all of the Practice Problems can be found at the back of the text starting at page SP-1

Notice that when we added the two equations together, one variable was eliminated. The addition method is therefore often called the **elimination method.**

For convenience, we will make a list of the steps we use to solve a system of equations by the addition method.

> **PROCEDURE FOR SOLVING A SYSTEM OF EQUATIONS BY THE ADDITION METHOD**
>
> **1.** Multiply each term of one or both equations by some nonzero integer so that the coefficients of one of the variables are opposites.
>
> **2.** Add the equations of this new system so that one variable is eliminated.
>
> **3.** Solve the resulting equation for the remaining variable.
>
> **4.** Substitute the value found in step 3 into either of the original equations to find the value of the other variable.
>
> **5.** Check your solution in both of the original equations.

Care should be used when the solution of a system contains fractions.

EXAMPLE 2 Solve by addition.

$$3x + 4y = 17 \quad \textbf{(1)}$$
$$2x + 7y = 19 \quad \textbf{(2)}$$

Solution Suppose we want the x-terms to have opposite coefficients. One equation could have $+6x$ and the other $-6x$. This would happen if we multiply equation **(1)** by $+2$ and equation **(2)** by -3.

$$(2)(3x) + (2)(4y) = (2)(17) \qquad \text{Multiply each term of equation (1) by 2.}$$
$$(-3)(2x) + (-3)(7y) = (-3)(19) \qquad \text{Multiply each term of equation (2) by } -3.$$

(Often this multiplication step can be done mentally.)

$$
\begin{aligned}
6x + 8y &= 34 &&\text{We now have an equivalent system of equations.}\\
-6x - 21y &= -57 &&\text{The coefficients of the } x\text{-terms are opposites.}\\
\hline
-13y &= -23 &&\text{Add the two equations to eliminate the variable } x.\\
y &= \frac{-23}{-13} = \frac{23}{13} &&\text{Divide both sides by } -13.
\end{aligned}
$$

Substitute $y = \dfrac{23}{13}$ into one of the original equations.

$$
\begin{aligned}
3x + 4\left(\frac{23}{13}\right) &= 17 \\
3x + \frac{92}{13} &= 17 &&\text{Solve for } x.\\
13(3x) + 13\left(\frac{92}{13}\right) &= 13(17) &&\text{Multiply both sides by 13 to clear the fraction.}\\
39x + 92 &= 221 &&\text{Remove the parentheses.}\\
39x &= 129 &&\text{Subtract 92 from both sides.}\\
x &= \frac{43}{13} &&\text{Divide both sides by 39. The solution is } \left(\frac{43}{13}, \frac{23}{13}\right).
\end{aligned}
$$

Alternative solution: You could also have eliminated the y-variable. For example, if you multiply equation **(1)** by 7 and equation **(2)** by -4, you obtain the equivalent system shown here.

$$21x + 28y = 119$$
$$-8x - 28y = -76$$

You can add these equations to eliminate the y-variable. Since the numbers involved in this approach are somewhat larger, it is probably wiser to eliminate the x-variable in this example.

NOTE TO STUDENT: *Fully worked-out solutions to all of the Practice Problems can be found at the back of the text starting at page SP-1*

CAUTION: A common error occurs when students forget that we want to obtain two terms with *opposite signs* that when added will equal zero. If all the coefficients in the equations are positive, such as in Example 2, it will be necessary to multiply *one* of the equations by a negative number.

Practice Problem 2 Solve by addition.

$$4x + 5y = 17$$
$$3x + 7y = 12$$

2 Using the Addition Method to Solve a System of Two Linear Equations with Fractional Coefficients

If the system of equations has fractional coefficients, you should first clear each equation of the fractions. To do so, you will need to multiply each term in the equation by the LCD of the fractions.

EXAMPLE 3 Solve.

$$x - \frac{5}{2}y = \frac{5}{2} \quad \textbf{(1)}$$

$$\frac{4}{3}x + y = \frac{23}{3} \quad \textbf{(2)}$$

Solution

$$2(x) - 2\left(\frac{5}{2}y\right) = 2\left(\frac{5}{2}\right) \quad \text{Multiply each term of equation \textbf{(1)} by 2.}$$
$$2x - 5y = 5$$
$$3\left(\frac{4}{3}x\right) + 3(y) = 3\left(\frac{23}{3}\right) \quad \text{Multiply each term of equation \textbf{(2)} by 3.}$$
$$4x + 3y = 23$$

We now have an equivalent system of equations that does not contain fractions.

$$2x - 5y = 5 \quad \textbf{(3)}$$
$$4x + 3y = 23 \quad \textbf{(4)}$$

Let us eliminate the *x*-variable. We want the coefficients of *x* to be opposites.

$$(-2)(2x) - (-2)5y = (-2)(5) \quad \text{Multiply each term of equation \textbf{(3)} by } -2.$$
$$-4x + 10y = -10 \quad \text{We now have an equivalent system of equations.}$$
$$\underline{4x + 3y = 23} \quad \text{The coefficients of the } x\text{-terms are opposites.}$$
$$13y = 13 \quad \text{Add the two equations to eliminate the variable } x.$$
$$y = 1 \quad \text{Divide both sides by 13.}$$
$$4x + 3(1) = 23 \quad \text{Substitute } y = 1 \text{ into equation \textbf{(4)}.}$$
$$4x + 3 = 23$$
$$4x = 20$$
$$x = 5$$

The solution is (5, 1).

Check: Check the solution in both of the *original* equations.

Practice Problem 3 Solve.

$$\frac{2}{3}x - \frac{3}{4}y = 3$$

$$-2x + y = 6$$

Using the Addition Method to Solve a System of Two Linear Equations with Decimal Coefficients

Some linear equations will have decimal coefficients. It will be easier to work with the equations if we change the decimal coefficients to integer coefficients. To do so, we will multiply each term of the equation by a power of 10.

EXAMPLE 4 Solve.

$$0.12x + 0.05y = -0.02 \quad \textbf{(1)}$$
$$0.08x - 0.03y = -0.14 \quad \textbf{(2)}$$

Solution Since the decimals are hundredths, we will multiply each term of both equations by 100.

$$100(0.12x) + 100(0.05y) = 100(-0.02)$$
$$100(0.08x) + 100(-0.03y) = 100(-0.14)$$

$$12x + 5y = -2 \quad \textbf{(3)} \qquad \text{We now have an equivalent system of equations that has integer coefficients.}$$

$$8x - 3y = -14 \quad \textbf{(4)}$$

We will eliminate the variable y. We want the coefficients of the y-terms to be opposites.

$$\begin{aligned} 36x + 15y &= -6 \\ \underline{40x - 15y} &= \underline{-70} \end{aligned} \qquad \text{Multiply equation } \textbf{(3)} \text{ by 3 and equation } \textbf{(4)} \text{ by 5.}$$

$$76x \qquad = -76 \qquad \text{Add the equations.}$$
$$x = -1 \qquad \text{Solve for } x.$$
$$12(-1) + 5y = -2 \qquad \text{Replace } x \text{ by } -1 \text{ in equation } \textbf{(3)}, \text{ and solve for } y.$$
$$-12 + 5y = -2$$
$$5y = -2 + 12$$
$$5y = 10$$
$$y = 2$$

The solution is $(-1, 2)$.

Check: We substitute $x = -1$ and $y = 2$ into the original equations. Most students probably would rather not use equation **(1)** and equation **(2)**! However, *we must check the solutions in the original equations* if we want to be sure of our answers. It is possible to make an error going from equations **(1)** and **(2)** to equations **(3)** and **(4)**. The solutions could satisfy equations **(3)** and **(4)**, but might not satisfy equations **(1)** and **(2)**.

$$0.12x + 0.05y = -0.02 \qquad\qquad 0.08x - 0.03y = -0.14$$
$$0.12(-1) + 0.05(2) \overset{?}{=} -0.02 \qquad 0.08(-1) - 0.03(2) \overset{?}{=} -0.14$$
$$-0.12 + 0.10 \overset{?}{=} -0.02 \qquad\qquad -0.08 - 0.06 \overset{?}{=} -0.14$$
$$-0.02 = -0.02 \ \checkmark \qquad\qquad\qquad -0.14 = -0.14 \ \checkmark$$

Practice Problem 4 Solve. $0.2x + 0.3y = -0.1$
$$0.5x - 0.1y = -1.1$$

CAUTION: A common error when solving a system is to find the value for x and then stop. You have not solved a system in x and y until you find both the x- and y-values that make both equations true.

Student Solutions Manual CD/Video PH Math Tutor Center MathXL®Tutorials on CD MathXL® MyMathLab® Interactmath.com

Verbal and Writing Skills

Look at the following systems of equations. Decide which variable to eliminate in each system, and explain how you would eliminate that variable.

1. $3x + 2y = 5$
$5x - y = 3$

2. $2x - 9y = 1$
$2x + 3y = 2$

3. $4x - 3y = 10$
$5x + 4y = 0$

4. $7x + 6y = 13$
$-2x + 5y = 3$

Find the solution by the addition method. Check your answers.

5. $-x + y = -3$
$-2x - y = 6$

6. $-x + y = 2$
$x + y = 4$

7. $2x + 3y = 1$
$x - 2y = 4$

8. $2x + y = 4$
$3x - 2y = -1$

9. $4x - 2y = -2$
$4x + 3y = -12$

10. $2x + y = 0$
$2x - 3y = -8$

11. $5x - 15y = 9$
$-x + 10y = 1$

12. $-4x + 5y = -16$
$8x + y = -1$

13. $2x + 5y = 2$
$3x + y = 3$

14. $5x - 3y = 14$
$2x - y = 6$

15. $8x + 6y = -2$
$10x - 9y = -8$

16. $4x + 9y = 0$
$8x - 5y = -23$

17. $2x + 3y = -8$
$5x + 4y = -34$

18. $3x - 5y = 11$
$4x + 3y = 5$

19. $2x - 7y = 8$
$3x - 4y = -1$

20. $-5x + 4y = -13$
$11x + 6y = -1$

21. $12x - 6y = -2$
$-9x - 7y = -10$

22. $2x - 4y = -22$
$-6x + 3y = 3$

Verbal and Writing Skills

Before using the addition method to solve each system, you will need to simplify the equation(s). Explain how you would change the fractional coefficients to integer coefficients in each system.

23. $\frac{1}{4}x - \frac{3}{4}y = 2$

$2x + 3y = 1$

24. $5x + 2y = 9$

$\frac{4}{9}x - \frac{2}{3}y = 4$

25. $\frac{1}{2}x + \frac{2}{3}y = \frac{1}{3}$

$\frac{3}{4}x - \frac{4}{5}y = 2$

26. $\frac{1}{4}x + \frac{1}{3}y = 15$

$\frac{1}{5}x + \frac{3}{10}y = 13$

Find the solution. Check your answers.

27. $x + \frac{5}{4}y = \frac{9}{4}$

$\frac{2}{5}x - y = \frac{3}{5}$

28. $\frac{2}{3}x + y = 2$

$x + \frac{1}{2}y = 7$

29. $\frac{x}{6} + \frac{y}{2} = -\frac{1}{2}$

$x - 9y = 21$

30. $\frac{3}{2}x - \frac{y}{8} = -1$

$16x + 3y = -28$

31. $\frac{5}{6}x + y = -\frac{1}{3}$

$-8x + 9y = 28$

32. $\frac{2}{9}x - \frac{1}{3}y = 1$

$4x + 9y = -2$

33. $\frac{2}{3}x + \frac{3}{5}y = -\frac{1}{5}$

$\frac{1}{4}x + \frac{1}{3}y = \frac{1}{4}$

34. $\frac{x}{5} + \frac{1}{2}y = \frac{9}{10}$

$\frac{x}{3} + \frac{3}{4}y = \frac{5}{4}$

Verbal and Writing Skills

Explain how you would change each system to an equivalent system with integer coefficients. Write the equivalent system.

35. $0.5x - 0.3y = 0.1$
$5x + 3y = 6$

36. $0.08x + y = 0.05$
$2x - 0.1y = 3$

37. $4x + 0.5y = 9$
$0.2x - 0.05y = 1$

38. $7x + 0.3y = 5$
$0.6x - 1.14y = 2$

Solve for x and y using the addition method.

39. $0.5x - 0.2y = 0.4$
$-0.3x + 0.4y = 0.6$

40. $0.2x + 0.3y = 0.4$
$0.5x + 0.4y = 0.3$

41. $0.02x - 0.04y = 0.26$
$0.07x - 0.09y = 0.66$

42. $0.04x - 0.03y = 0.05$
$0.05x + 0.08y = -0.76$

43. $0.4x - 5y = -1.2$
$-0.03x + 0.5y = 0.14$

44. $-0.6x - 0.08y = -4$
$3x + 2y = 4$

Mixed Practice

Solve for x and y using the addition method.

45. $2x + y = 11$
$x + 3y = 18$

46. $4x - 3y = 6$
$2x + 4y = 14$

47. $\dfrac{5}{4}x + \phantom{\dfrac{5}{3}}y = 16$
$\phantom{\dfrac{5}{4}}x - \dfrac{5}{3}y = -2$

48. $\dfrac{1}{3}x - \dfrac{1}{4}y = 0$
$\dfrac{1}{6}x - \dfrac{1}{2}y = -6$

49. $0.2x + 0.1y = 1.1$
$0.01x + 0.03y = 0.18$

50. $0.05x - 0.04y = -0.08$
$0.2x + 0.3y = 0.6$

To Think About

Find the solution.

51. $5(3x - y) = 2 - (y - x)$
$9(2x - 2) = y - 3x$

52. $4(x - y) = 3(1 - 2x) - y$
$1 = -3(5x + 4y)$

Cumulative Review

53. *Air Traffic* At any given moment of the day, between 5000 and 7000 commercial aircraft are flying in U.S. airspace. If 89% of the air traffic is flying over the contiguous states, how many commercial airplanes are flying over Alaska and Hawaii?

54. *Used Car Purchase* A used car is priced at $5800, if you pay cash. An installment plan requires $1000 down, plus $230 a month for 24 months. What percent more does one pay for the car under the installment plan? Round to the nearest tenth of a percent.

55. *Race Car Speeds* Carlos and Bobby are professional race car drivers. They are in a 100-lap race on an oval track that is 4.75 miles long. Carlos makes 4 complete laps during the time that Bobby is out of the race for repairs to his car. If Carlos maintains an average speed of 95 miles per hour for the entire race, what average speed will Bobby have to achieve in order to beat Carlos? (Round to the nearest whole number.)

56. *Baseball* At one point during a recent season, the Boston Red Sox had won 45 baseball games and lost 55 games. Tricia and Mindy were discussing the potential for the Red Sox to win 50 percent of all their games for the season. If the team had 62 games remaining at that point, how many of the future games would they need to win in order to achieve that percent?

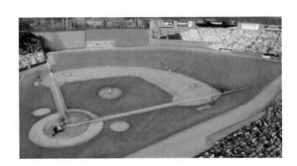

57. Factor $75x^2 - 3$ completely.

58. Factor. $6x^2 - 25x + 25$

1. _____

2. _____

3. _____

4. _____

5. _____

6. _____

7. _____

8. _____

9. _____

10. _____

How are you doing with your homework assignments in Sections 8.1 to 8.3? Do you feel you have mastered the material so far? Do you understand the concepts you have covered? Before you go further in the textbook, take some time to do each of the following problems.

8.1   *Solve by graphing.*

1. $2x - y = -8$
$-x + 2y = 10$

2. $-2x + y = 4$
$x - 3y = 3$

    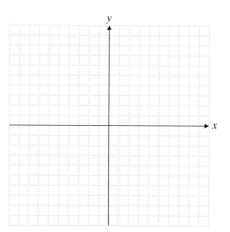

8.2   *Solve by the substitution method.*

3. $5x + y = 26$
$3x + 2y = 10$

4. $4x - 3y = 18$
$5x - y = 17$

5. $2x + 3y = 5$
$\dfrac{1}{2}x + \dfrac{1}{6}y = -\dfrac{1}{2}$

6. $7x - 5y - 17 = 0$
$x + 8y + 15 = 0$

8.3   *Solve by the addition method.*

7. $3x - 2y = 19$
$x + y = 8$

8. $9x - 8y = 3$
$6x + 3y = 27$

9. $\dfrac{2}{3}x - \dfrac{3}{4}y = 3$
$\dfrac{1}{9}x + \dfrac{1}{12}y = -2$

10. $0.6x - 0.5y = 2.1$
$0.4x - 0.3y = 1.3$

Now turn to page SA-25 for the answer to each of these problems. Each answer also includes a reference to the objective in which the problem is first taught. If you missed any of these problems, you should stop and review the Examples and Practice Problems in the referenced objective. A little review now will help you master the material in the upcoming sections of the text.

1 ## Choosing an Appropriate Method to Solve a System of Equations Algebraically

At this point we will review the algebraic methods for solving systems of linear equations and discuss the advantages and disadvantages of each method.

Method	Advantage	Disadvantage
Substitution	Works well if one or more variable has a coefficient of 1 or -1.	Often becomes difficult to use if no variable has a coefficient of 1 or -1.
Addition	Works well if equations have fractional or decimal coefficients. Works well if no variable has a coefficient of 1 or -1.	None

Student Learning Objectives

After studying this section, you will be able to:

1 Choose an appropriate method to solve a system of equations algebraically.

2 Using algebraic methods to identify inconsistent and dependent systems.

EXAMPLE 1 Select a method and solve each system of equations.

(a) $\quad x + y = 3080$
$\quad 2x + 3y = 8740$

(b) $\quad 5x - 2y = 19$
$\quad -3x + 7y = 35$

Solution

(a) Since there are x- and y-values that have coefficients of 1, we will select the substitution method.

$$y = 3080 - x \qquad \text{Solve the first equation for } y.$$
$$2x + 3(3080 - x) = 8740 \qquad \text{Substitute the expression into the second equation.}$$
$$2x + 9240 - 3x = 8740 \qquad \text{Remove parentheses.}$$
$$-1x = -500 \qquad \text{Simplify.}$$
$$x = 500 \qquad \text{Divide each side by } -1.$$
$$y = 3080 - x$$
$$= 3080 - 500 \qquad \text{Substitute 500 for } x.$$
$$= 2580 \qquad \text{Simplify.}$$

The solution is $(500, 2580)$.

(b) Because none of the x- and y-variables have a coefficient of 1 or -1, we select the addition method. We choose to eliminate the y-variable. Thus, we would like the coefficients of y to be -14 and 14.

$$7(5x) - 7(2y) = 7(19) \qquad \text{Multiply each term of the first equation by 7.}$$
$$2(-3x) + 2(7y) = 2(35) \qquad \text{Multiply each term of the second equation by 2.}$$
$$35x - 14y = 133 \qquad \text{We now have an equivalent system of equations.}$$
$$\underline{-6x + 14y = 70}$$
$$29x = 203 \qquad \text{Add the two equations.}$$
$$x = 7 \qquad \text{Divide each side by 29.}$$

Substitute $x = 7$ into one of the original equations.

$$5(7) - 2y = 19$$
$$35 - 2y = 19 \qquad \text{Solve for } y.$$
$$-2y = -16$$
$$y = 8$$

The solution is $(7, 8)$.

NOTE TO STUDENT: Fully worked-out solutions to all of the Practice Problems can be found at the back of the text starting at page SP-1

Practice Problem 1   Select a method and solve each system of equations.

(a) $3x + 5y = 1485$

  $x + 2y = 564$

(b) $7x + 6y = 45$

  $6x - 5y = -2$

② Using Algebraic Methods to Identify Inconsistent and Dependent Systems

Recall that an inconsistent system of linear equations is a system of parallel lines. Since parallel lines never intersect, the system has no solution. Can we determine this algebraically?

EXAMPLE 2   Solve algebraically.

$$3x - y = -1$$
$$3x - y = -7$$

Solution   Clearly, the addition method would be very convenient in this case.

$\begin{array}{ll} 3x - y = -1 & \text{Keep the first equation unchanged.} \\ \underline{-3x + y = +7} & \text{Multiply each term in the second equation by } -1. \\ 0 = 6 & \text{Add the two equations.} \end{array}$

Notice we have $0 = 6$, which we know is not true. The statement $0 = 6$ is inconsistent with known mathematical facts. No possible x- and y-values can make this equation true. Thus there is **no solution to this system of equations.**

This example shows us that, if we obtain a mathematical statement that is not true (inconsistent with known facts), we can identify the entire system as inconsistent. If graphed, these lines would be parallel.

Practice Problem 2   Solve algebraically.

$$4x + 2y = 2$$
$$-6x - 3y = 6$$

What happens if we try to solve a dependent system of equations algebraically? Recall that a dependent system consists of two lines that coincide (are the same line).

EXAMPLE 3   Solve algebraically.

$$x + y = 4$$
$$3x + 3y = 12$$

Solution   Let us use the substitution method.

$\begin{array}{ll} y = 4 - x & \text{Solve the first equation for } y. \\ 3x + 3(4 - x) = 12 & \text{Substitute } 4 - x \text{ for } y \text{ in the second equation.} \\ 3x + 12 - 3x = 12 & \text{Remove parentheses.} \\ 12 = 12 \end{array}$

Notice we have $12 = 12$, which is always true. Thus we obtain an equation that is true for any value of x. An equation that is always true is called an **identity.** All the solutions of one equation of the system are also solutions of the other equation. Thus the lines coincide. There are an **infinite number of solutions to this system of equations**.

This example shows us that, if we obtain a mathematical statement that is always true (an identity), we can identify the equations as dependent. There are an unlimited number of solutions to a system that has dependent equations.

Practice Problem 3   Solve algebraically.

$$3x - 9y = 18$$
$$-4x + 12y = -24$$

TO THINK ABOUT: Two Linear Equations with Two Variables   Now is a good time to look back over what we have learned. When you graph a system of two linear equations, what possible kinds of graphs will you obtain?

What will happen when you try to solve a system of two linear equations using algebraic methods? How many solutions are possible in each case? The following chart may help you to organize your answers to these questions.

Graph	Number of Solutions	Algebraic Interpretation
Two lines intersect at one point $(6, -3)$	**One unique solution**	You obtain one value for x and one value for y. For example, $$x = 6, \quad y = -3.$$
Parallel lines	**No solution**	You obtain an equation that is inconsistent with known facts. For example, $$0 = 6.$$ The system of equations is inconsistent.
Lines coincide	**Infinite number of solutions**	You obtain an equation that is always true. For example, $$8 = 8.$$ The equations are dependent.

Think about each of the three possibilities in this chart. You need to understand each one. As you do each of the exercises 5–28 you will need to identify which of the three possibilities is involved.

Verbal and Writing Skills

1. If there is no solution to a system of linear equations, the graphs of the equations are _____. Solving the system algebraically, you will obtain an equation that is _____ with known facts.

2. If an algebraic attempt at solving a system of linear equations results in an identity, the system is said to be _____. There are an _____ number of solutions, and the graphs of the lines _____.

3. If there is exactly one solution, the graphs of the equations _____. This system is said to be _____ and _____.

4. A student solves a system of equations and obtains the equation $-36x = 0$. He is unsure of what to do next. What should he do next? What should he conclude about the system of equations?

If possible, solve by an algebraic method (substitution or addition method, without the use of graphing). Otherwise, state that the problem has no solution or an infinite number of solutions.

5. $-2x - 3y = 15$
$5x + 2y = 1$

6. $3x + 4y = 8$
$-5x - 3y = 5$

7. $4x - 5y = 1$
$-3x + 4y = 4$

8. $2x + 3y = 13$
$5x + 2y = -6$

9. $2x - 4y = 5$
$-4x + 8y = 9$

10. $3x + 6y = 12$
$x + 2y = 7$

11. $-5x + 2y = 2$
$15x - 6y = -6$

12. $6x - 4y = 8$
$-9x + 6y = -12$

13. $5x - 3y = 13$
$7x + 2y = 43$

14. $-4x + 5y = 10$
$6x - 7y = 8$

15. $3x - 2y = 70$
$0.6x + 0.5y = 50$

16. $5x - 3y = 10$
$0.9x + 0.4y = 30$

17. $0.3x - 0.5y = 0.4$
$0.6x + 1.0y = 2.8$

18. $-0.05x + 0.02y = -0.01$
$0.10x + 0.04y = -0.22$

19. $\frac{4}{3}x + \frac{1}{2}y = 1$
$\frac{1}{3}x - y = -\frac{1}{2}$

20. $\frac{3}{5}x - \frac{3}{4}y = \frac{1}{5}$
$2x + \frac{3}{2}y = 2$

21. $\frac{2}{3}x + \frac{1}{6}y = 2$
$\frac{1}{4}x - \frac{1}{2}y = -\frac{3}{4}$

22. $\frac{1}{2}x + \frac{2}{5}y = \frac{9}{10}$
$\frac{2}{3}x + \frac{2}{15}y = \frac{1}{2}$

Mixed Practice

Solve by any algebraic method. If there is not one solution to a system of equations, state why this is the case.

23. $4x + 3y = -2$
$5x - 2y = 9$

24. $3x = 4(y + 9)$
$y + 3 = 3(x + 1)$

25. $\frac{2}{3}x + y = 1$
$\frac{3}{4}x + \frac{1}{2}y = \frac{3}{4}$

26. $0.8x - 0.3y = 0.7$
$1.2x + 0.6y = 4.2$

27. $x - 2y + 2 = 0$
$3(x - 2y + 1) = 15$

28. $3(x - 2y + 2) = -3$
$-\dfrac{1}{3}x + \dfrac{2}{3}y = 1$

To Think About

Remove parentheses and solve the system for (a, b).

29. $2(a + 3) = b + 1$
$3(a - b) = a + 1$

30. $3(a + b) - 2 = b + 2$
$a - 2(b - 1) = 3a - 4$

Cumulative Review

Solve for x.

31. $20x + 12\left(\dfrac{9}{2} - x\right) = 70$

32. $0.2(x + 0.3) + x = 0.3(x + 0.5)$

In 2003, a federal income tax rate of 15% applied to single people who had a taxable income of $23,350 or less and to married people filing jointly who had a combined taxable income of $39,000 or less.

33. *Income Tax* What was Robert's taxable income if he paid an income tax of $3225 for 2003 and he filed as a single person?

34. *Income Tax* What was the combined taxable income of Alberto and Maria Lazano if they paid an income tax of $5775 for 2003 and they filed as a married couple?

35. *Traffic Tickets* On a given day, approximately 92,500 auto drivers in the United States are stopped by police officers and given tickets. (*Source:* National Highway Traffic Safety Administration.) If New York, New Jersey, Maryland, California, Texas, and Illinois as a group account for 40% of the tickets, what is the average number of tickets written per day in each of the remaining states? (Round to the nearest whole number.)

36. *Automobile Expenses* The cost of owning and operating a midsize car in 1960 was about 10 cents per mile. Today, it is 45.9 cents per mile. If fuel prices increase by 47% next year and if 30% of the cost of operating a midsize car is based on fuel, how much will it cost to run a car next year if other car maintenance costs remain the same?

1 Using a System of Equations to Solve Word Problems

Word problems can be solved in a variety of ways. In Chapter 3 and throughout the text, you have used one equation with one unknown to describe situations found in real-life applications. Some of these same problems can be solved using two equations in two unknowns. The method you use will depend on what you find works best for you. Become familiar with both methods before you decide.

EXAMPLE 1 A worker in a large post office is trying to verify the rate at which two electronic card-sorting machines operate. Yesterday the first machine sorted for 3 minutes and the second machine sorted for 4 minutes. The total workload both machines processed during that time period was 10,300 cards. Two days ago the first machine sorted for 2 minutes and the second machine for 3 minutes. The total workload both machines processed during that time period was 7400 cards. Can you determine the number of cards per minute sorted by each machine?

Solution

1. **Understand the problem.** The number of cards processed by two different machines on two different days provides the basis for two linear equations with two unknowns.

   The two equations will represent what occurred on two different days:

   (1) yesterday and

   (2) two days ago.

   The number of cards sorted by each machine is what we need to find.
   Let x = the number of cards per minute sorted by the first machine and
   y = the number of cards per minute sorted by the second machine.

2. **Write the equation(s).** Yesterday the first machine sorted for 3 minutes and the second machine sorted for 4 minutes, processing a total of 10,300 cards.

first machine		second machine	total no. of cards
3 (no. of cards per minute)		4 (no. of cards per minute)	
$3x$	$+$	$4y$	$=$ 10,300 **(1)**

   Two days ago the first machine sorted for 2 minutes and the second machine sorted for 3 minutes, processing a total of 7400 cards.

first machine		second machine	total no. of cards
2 (no. of cards per minute)		3 (no. of cards per minute)	
$2x$	$+$	$3y$	$=$ 7400 **(2)**

3. **Solve and state the answer.** We will use the addition method to solve this system.

$$3x + 4y = 10{,}300 \quad \textbf{(1)}$$
$$2x + 3y = 7400 \quad \textbf{(2)}$$
$$6x + 8y = 20{,}600 \quad \text{Multiply equation (1) by 2.}$$
$$\underline{-6x - 9y = -22{,}200} \quad \text{Multiply equation (2) by } -3.$$
$$-1y = -1600 \quad \text{Add the two equations to eliminate } x \text{ and solve for } y.$$
$$y = 1600$$
$$2x + 3(1600) = 7400 \quad \text{Substitute the value of } y \text{ into equation (2).}$$
$$2x + 4800 = 7400 \quad \text{Simplify and solve for } x.$$
$$2x = 2600$$
$$x = 1300$$

Thus, the first machine sorts 1300 cards per minute and the second machine sorts 1600 cards per minute.

4. Check. We can verify each statement in the problem.

Yesterday: Were 10,300 cards sorted?

$$3(1300) + 4(1600) \overset{?}{=} 10{,}300$$
$$3900 + 6400 \overset{?}{=} 10{,}300$$
$$10{,}300 = 10{,}300 \ \checkmark$$

Two days ago: Were 7400 cards sorted?

$$2(1300) + 3(1600) \overset{?}{=} 7400$$
$$2600 + 4800 \overset{?}{=} 7400$$
$$7400 = 7400 \ \checkmark$$

Practice Problem 1   After a recent disaster, the Red Cross brought in a pump to evacuate water from a basement. After 3 hours a larger pump was brought in and the two ran for an additional 5 hours, removing 49,000 gallons of water. The next day, after running both pumps for 3 hours, the larger one was taken to another site. The smaller pump finished the job after another 2 hours. If on the second day 30,000 gallons were pumped, how much water does each pump remove per hour?

NOTE TO STUDENT Fully worked-out solutions to all of the Practice Problems can be found at the back of the text starting at page SP-1

EXAMPLE 2   Fred is considering working for a company that sells encyclopedias. Fred found out that all starting representatives receive the same annual base salary and a standard commission of a certain percentage of the sales they make during the first year. He has been told that one representative sold $50,000 worth of encyclopedias her first year and that she earned $14,000. He was able to find out that another representative sold $80,000 worth of encyclopedias and that he earned $17,600. Determine the base salary and the commission rate of a beginning sales representative.

Solution   The earnings of the two different sales representatives will represent the two equations, each of the form

$$\text{base salary} + \text{commission} = \text{total earnings.}$$

What is unknown? The base salary and the commission rate are what we need to find.

Let b = the amount of base salary and
c = the commission rate.

$b + 50{,}000c = 14{,}000$ **(1)**   Earnings for the first sales representative.
$b + 80{,}000c = 17{,}600$ **(2)**   Earnings for the second sales representative.

We will use the addition method to solve these equations.

$\quad -b - 50{,}000c = -14{,}000$ Multiply equation **(1)** by -1.
$\underline{\quad b + 80{,}000c = 17{,}600}$ Leave equation **(2)** unchanged.
$\quad\quad\quad\quad 30{,}000c = 3600$ Add the two equations and solve for c.

$$c = \frac{3600}{30{,}000}$$

$\quad\quad\quad\quad\quad c = 0.12$ The commission rate is 12%.
$b + (50{,}000)(0.12) = 14{,}000$ Substitute this value into equation **(1)**.
$\quad\quad\quad b + 6000 = 14{,}000$ Simplify and solve for b.
$\quad\quad\quad\quad\quad\quad b = 8000$ The base salary is $8000 per year.

Thus the base salary is $8000 per year, and the commission rate is 12%.

NOTE TO STUDENT: Fully worked-out solutions to all of the Practice Problems can be found at the back of the text starting at page SP-1

Practice Problem 2 Last week Rick bought 7 gallons of unleaded premium and 8 gallons of unleaded regular gasoline and paid $27.22. This week he purchased 8 gallons of unleaded premium and 4 gallons of unleaded regular and paid $22.16. He forgot to record how much each type of fuel cost per gallon. Can you determine these values?

EXAMPLE 3 A lab technician is required to prepare 200 liters of a solution. The prepared solution must contain 42% fungicide. The technician wishes to combine a solution that contains 30% fungicide with a solution that contains 50% fungicide. How much of each solution should he use?

Solution *Understand the problem.* The technician needs to make one solution from two different solutions. We need to find the amount (number of liters) of each solution that will give us 200 liters of a 42% solution. The two unknowns are easy to identify.

Let x = the number of liters of the 30% solution needed and
y = the number of liters of the 50% solution needed.

Now, what are the two equations? We know that the 200 liters of the final solution will be made by combining x and y. $x + y = 200$ **(1)**

The second piece of information concerns the percent of fungicide in each solution. See the picture at left. $0.3x + 0.5y = 0.42(200)$ **(2)**

Thus our system of equations is

$$x + y = 200 \quad \textbf{(1)}$$
$$0.3x + 0.5y = 84. \quad \textbf{(2)}$$

We will solve this system by the addition method.

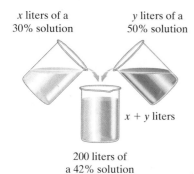

x liters of a 30% solution   *y* liters of a 50% solution

$x + y$ liters

200 liters of a 42% solution

$$
\begin{array}{ll}
1.0x + 1.0y = 200 & \text{Rewrite equation \textbf{(1)} in an equivalent form.} \\
\underline{-0.6x - 1.0y = -168} & \text{Multiply equation \textbf{(2)} by } -2. \\
0.4x = 32 & \text{Solve for } x. \\
\end{array}
$$

$$\frac{0.4x}{0.4} = \frac{32}{0.4}$$

$$x = 80$$

$$80 + y = 200$$

$$y = 120 \qquad \text{Substitute the value } x = 80 \text{ into one of the}$$
original equations and solve for y.

The technician should use 80 liters of the 30% fungicide and 120 liters of the 50% fungicide to obtain the required solution.

Check. Do the amounts total 200 liters?

$$80 + 120 = 200 \quad ✓$$

Do the amounts yield a 42% strength mixture?

$$0.30x + 0.50y = 0.42(200)$$
$$0.30(80) + 0.50(120) \overset{?}{=} 84$$
$$24 + 60 = 84 \quad ✓$$

Practice Problem 3 Dwayne Laboratories needs to ship 4000 liters of H_2SO_4 (sulfuric acid). Dwayne keeps in stock solutions that are 20% H_2SO_4 and 80% H_2SO_4. If the strength of the solution shipped is to be 65%, how should Dwayne's stock be mixed?

EXAMPLE 4   Mike recently rode his boat on Lazy River. He took a 48-mile trip up the river traveling against the current in exactly 3 hours. He refueled and made the return trip in exactly 2 hours. What was the speed of his boat in still water and the speed of the current on the river?

Solution   Let x = the speed of the boat in still water in miles/hour and
y = the speed of the river current in miles/hour.

You may want to draw a picture to show the boat traveling in each direction.

Distance is 48 miles

Against the current boat takes 3 hours

With the current boat takes 2 hours

Traveling *against* the current, the boat took 3 hours to travel 48 miles. We can calculate the rate (speed) the boat traveled upstream. Use distance = rate · time or $d = r \cdot t$.

$$48 = 3 \cdot r \qquad r = 16 \text{ miles/hour}$$

The boat was traveling *against* the current. Thus the speed of the boat in still water minus the speed of the current is 16.

$$x - y = 16 \quad \textbf{(1)}$$

Traveling with the current, the boat took 2 hours to travel 48 miles. We can calculate the rate (speed) the boat traveled downstream. Use $d = r \cdot t$.

$$48 = 2 \cdot r \qquad r = 24 \text{ miles/hour}$$

The boat was traveling *with* the current. Thus the speed of the boat in still water plus the speed of the current is 24.

$$x + y = 24 \quad \textbf{(2)}$$

This system is solved most easily by the addition method.

$$
\begin{array}{ll}
x - y = 16 & \textbf{(1)} \\
\underline{x + y = 24} & \textbf{(2)} \\
2x = 40 & \text{Add the two equations.} \\
x = 20 & \text{Solve for } x \text{ by dividing both sides by 2.} \\
x + y = 24 \quad \textbf{(2)} & \text{Substitute the value of } x = 20 \text{ into one of the original} \\
& \text{equations and solve for } y. \text{ We'll use equation } \textbf{(2)}. \\
20 + y = 24 & \\
y = 4 &
\end{array}
$$

Thus the speed of Mike's boat in still water was 20 miles/hour and the speed of the current in Lazy River was 4 miles/hour.

Check.   Can you verify these answers?

Practice Problem 4   Mr. Caminetti traveled downstream in his boat a distance of 72 miles in a time of 3 hours. His return trip up the river against the current took 4 hours. Find the speed of the boat in still water. Find the speed of the current in the river.

8.5 EXERCISES

Student Solutions Manual | CD/Video | PH Math Tutor Center | MathXL®Tutorials on CD | MathXL® | MyMathLab® | Interactmath.com

Applications

Solve using two equations with two variables.

1. **Airline Travel** Sarah took the advertising department from her company on a round trip to Chicago to meet with a potential client. Including Sarah a total of 14 people took the trip. She was able to purchase coach tickets for $250 and first class tickets for $1150. She used her total budget for airfare for the trip, which was $10,700. How many first class tickets did she buy? How many coach tickets did she buy?

2. **Baseball** Walter had to purchase tickets for 11 people in his office to go to a Red Sox game. He paid $449 to buy the 11 tickets. Infield Grandstand seats cost $44. Right Field Roof seats cost $37. How many of each kind of ticket did he purchase?

3. **Art Dealer** An art dealer sold 12 etchings for $455. He sold some of them at $35 each and the rest at $40 each. How many etchings did he sell at each price?

4. **High School Musical** El Segundo High School put on their annual musical. The students sold 650 tickets for a value of $4375. If orchestra seats cost $7.50 and balcony seats cost $3.50, how many of each kind of seat were sold?

5. **Income of a Nurse** Sharyn worked as a Registered Nurse in London last summer. The first week she was there she worked a total of 23 hours at a city hospital. She was paid $13.50 an hour for day shift and $16.50 an hour for night shift work. She earned $352.50 for her first week of work. How many hours did she spend on the day shift? How many hours did she spend on the night shift?

6. **Income of a Substitute Teacher** Tony worked as a substitute teacher last month. He substituted a total of 18 days. The first part of the month he worked in Tucson, Arizona, where he was paid $75 a day. Then he moved to Oregon, where he was paid $135 a day to substitute. During the month he earned $2010. How many days did he substitute in Tucson? How many days did he substitute in Oregon?

 7. **Home Office Construction** The basement of your house is only partially finished. Currently, there is a room that you use for storage. You want to convert this space to a home office, but you want more space than is available in the room now. You decide to knock down two walls and enlarge the room. The existing room has a perimeter of 38 feet. If you extend the width of the room by 5 feet and make the new length 6 feet less than double the original length, the new perimeter will be 56 feet. What are the dimensions of the existing storage room? What are the dimensions of the new room?

existing room   enlarged room

8. *Engineering Club* The engineering club at MCTC ran a coffee and donut sale to raise money for the club's activities. The first week, they made a profit of $200. The second week, they sold three times as many donuts, but half the number of coffees, and made a profit of $225. If the profit for each donut was $0.50 and the profit for each coffee was $0.75, how many donuts and coffees did they sell the first week?

9. *Automobile Radiators* During her beginning mechanics course, Rose learned that her leaking radiator holds 16 quarts of water. She temporarily patched the leak when the radiator was partially full. At that time it was 50% antifreeze. After she filled the remaining space in the radiator with an 80% antifreeze solution and waited a little while, she found that it was 65% antifreeze. How many quarts of solution were in the radiator just before Rose added antifreeze? How many quarts of 80% antifreeze solution did she add?

10. *Manufacturing Chocolate* A candy company wants to produce 20 kilograms of a special 45% fat content chocolate to conform with customer dietary demands. To obtain this, a 50% fat content Hawaiian chocolate is combined with a 30% fat content domestic chocolate in a special melting process. How many kilograms of the 50% fat content chocolate and how many kilograms of the 30% fat content chocolate are used to make the required 20 kilograms?

11. *Hiking Club* The hiking club wants to sell 50 pounds of an energy snack food that will cost them only $1.80 per pound. The mixture will be made up of nuts that cost $2.00 per pound and raisins that cost $1.50 per pound. How much of each should they buy to obtain 50 pounds of the desired mixture?

12. *Flower Costs* How many bunches of local daisies that cost $2.50 per bunch should be mixed with imported daisies that cost $3.50 per bunch to obtain 100 bunches of daisies that will cost $2.90 per bunch?

13. *Airline Travel* An airplane traveled between two cities that are 3000 kilometers apart. The trip against the wind took 6 hours. The return trip with the benefit of the wind was 5 hours. What is the wind speed in kilometers per hour? What is the speed of the plane in still air?

14. *Airline Travel* An airplane flies between Boston and Cleveland. During a round-trip flight, the plane flew a distance of 630 miles each way. The trip with a tailwind (the wind traveling the same direction as the plane) took 3 hours. The return trip traveling against the wind took 3.5 hours. Find the speed of the wind. Find the speed of the airplane in still air.

Mixed Practice

Solve the following applied problems using a system of equations.

15. Department Store Operations Soledad Rivera's job was to ready the cash registers for the day's business at Foley's Department Store, in Houston. Each cash register began the day with the same amount of cash, $320. Each cash register received $25 in $1 bills and $55 in $5 bills. But from here each department had its own requirements. The Cosmetics department wanted two less than double the number of $10 bills and four fewer $20 bills than Jewelry wanted. How many $10 bills and $20 bills did these departments want?

16. Canoe Trip The outing club of the College of the Canyons sponsored a canoe trip. The participants paddled 8 miles upstream from the canoe rental headquarters to a popular swimming and picnicking spot. They lunched, swam, rested, and headed back down to where they started. If the trip upstream took 2 hours and the trip back took only 1 hour, how fast were they paddling on average, and how fast was the current?

17. Fruit Purchase Hank purchased 7 pounds of oranges and 5 pounds of apples and it cost him $8.78. His wife came in the next day and purchased for the same prices per pound a total of 3 pounds of oranges and 8 pounds of apples for a total of $8.39. What is the price per pound of oranges? What is the price per pound of apples?

18. Work Rates Tom and Dave worked sorting discarded bottles for a recycle center. Tom worked for 3 hours and Dave worked for 2 hours on Saturday. They were able to sort through 2250 bottles. On Monday Tom worked for 2 hours and Dave worked for 3 hours. On Monday they sorted through 2650 bottles. How many bottles per hour can Tom sort? How many bottles per hour can Dave sort?

19. Talent Show The North Shore Community College annual talent show sold tickets for the event. General admission tickets were $4.50 and student admission tickets were $3. On the first night the show was attended by 290 people. The receipts for the night were $1110. How many general admission tickets were sold? How many student admission tickets were sold?

20. Airplane Flight Walter and Mary Jones flew from Boston to Denver to visit their daughter in college. The flight of the plane was a distance of 2000 miles. It took 5 hours to fly west against the jet stream. Coming back home with the jet stream the trip only took 4 hours. What was the speed of the jet stream? What was the speed of the plane in still air?

21. Population Growth In 2003 the estimated population of Maryland was 5,275,000 and is projected to grow moderately. In the same year, the population of Missouri was 5,540,000 and is projected to grow at a slower rate. The relationship between the population of Maryland in thousands (y) and the number of years (x) since 2003 can be approximated by $40x - y = -5275$. A similar relationship for Missouri can be approximated by $28x - 2y = -11,078$. (*Source:* U.S. Census Bureau.) Solve the system of equations to determine the approximate year that the two states will have the same population. What will be the population of each state during that year?

22. Population Growth In 2003 the estimated population of Iowa was 2,900,000 and was projected to grow at a slow rate. In the same year, the population of Kansas was 2,668,000 and was projected to grow more rapidly. The relationship between the population of Iowa in thousands (y) and the number of years (x) since 2003 can be approximated by $56x - 10y = -29,500$. A similar relationship for Kansas can by approximated by $176x - 10y = -27,100$. (*Source:* U.S. Census Bureau.) Solve the system of equations to determine the approximate year that the two states will have the same population. What will be the population of each state during that year?

23. Commission Sales Sam's commission is a given percent of his total sales of surgical transplant equipment. He receives an 8% commission for selling heart transplant kits and a 6% commission for selling kidney transplant kits. Last month he earned $4480 in commissions. This month he sold the same quantity of heart and kidney kits; however, his company gave him a raise, and he now earns a 10% commission for selling heart kits and an 8% commission for selling kidney kits. This month he earned $5780. What is the total value of the heart kits that he sold last month? What is the total value of the kidney kits?

24. Population Growth The present population of the town of Springfield is 52,500. The town is growing at the rate of 500 people per year. The population of the city of Essex is decreasing at the rate of 600 people per year. There are presently 80,000 people in Essex. How many years will it be until the populations of Springfield and Essex are the same? What will each population be?

Cumulative Review

25. Add. $\dfrac{1}{x-3} - \dfrac{1}{x+3} + 2$

26. Evaluate when $x = -2$ and $y = 3$.
$2x^2 - 3y + 4y^2$

27. Simplify. $\left(-4x^3y^2\right)^3$

28. Food Quality Control Jillian inspected a 32-ounce bottle of cranberry-raspberry drink. She discovered that it contained only 11.2 ounces of pure fruit juice. What percent of the contents of the bottle is pure fruit juice?

29. Shelf Space Rafaela ordered a beautiful 3-volume set of leather bound books with her birthday money. Before she rearranged all her display shelves, she needed to know how much space she needed for the books. According to the catalogue, the front and back cover of each book was $\frac{1}{8}$ inch thick and the page sections were each $1\frac{3}{4}$ inches thick. How much space did she need to house these books?

Predicting Future Population

The population of North America is growing rapidly. The equation that predicts this growth is $-5.4x + y = 372$, where x is the number of years since 1980 and y is the population in millions of people. The population of South America is also growing rapidly. The equation that predicts this growth is $-5.25x + y = 242$, where x is the number of years since 1980 and y is the population in millions of people.

Problems for Individual Investigation and Analysis

1. According to these equations, what was the population of North America in 1980? In 2000? What will the estimated population be in 2010?

2. According to these equations, what was the population of South America in 1980? In 2000? What will the estimated population be in 2010?

3. On which continent was the increase in population greater? How great is this difference in the period of time from 1980 to 2000?

4. What will be the estimated difference between the population of North America and the population of South America in the year 2020? Will the difference become larger in future years after 2020?

Problems for Group Investigation and Cooperative Learning

5. Using the information obtained in problems 1 and 2, graph the population of North America and South America using the graph grid below.

6. Do the two straight lines intersect on your graph paper? What does this mean?

7. Can you use algebraic methods to solve the system of equations? What is the solution to the system?

8. Does the solution of problem 7 have a real-world meaning? Do you think this meaning is correct?

Topic	Procedure	Examples
Solving a system of equations by graphing, p. 472.	Graph both equations on the same coordinate system. One of the following will be true. 1. The lines intersect. The point of intersection is the solution to the system. Such a system is consistent.	**1.**
	2. The lines are parallel. There is no point of intersection. The system has no solution. Such a system is inconsistent.	**2.**
	3. The lines coincide. Every point on the line represents a solution. There are an infinite number of solutions. Such a system has dependent equations.	**3.**
Solving a system of equations: substitution method, p. 480.	1. Solve for one variable in terms of the other variable in one of the two equations. 2. Substitute the expression you obtain for this variable into the other equation. 3. Solve this equation to find the value of the variable. 4. Substitute this value for the variable into one of the equations to obtain a value for the other variable. 5. Check the solution in both equations.	Solve: $$2x + 3y = -6$$ $$3x - y = 13$$ Solve the second equation for y. $-y = -3x + 13$ $$y = 3x - 13$$ Substitute this expression into the first equation. $$2x + 3(3x - 13) = -6$$ $$2x + 9x - 39 = -6$$ $$11x - 39 = -6$$ $$x = 3$$ Substitute $x = 3$ into $y = 3x - 13$. $$y = 3(3) - 13 = 9 - 13 = -4$$ *Check:* $2(3) + 3(-4) \stackrel{?}{=} -6 \qquad 3(3) - (-4) \stackrel{?}{=} 13$ $6 - 12 \stackrel{?}{=} -6 \qquad\qquad 9 - (-4) \stackrel{?}{=} 13$ $-6 = -6 \;\checkmark \qquad\qquad 9 + 4 = 13 \;\checkmark$ The solution is $(3, -4)$.

Topic	Procedure	Examples
Solving a system of equations: addition method, p. 486.	1. Multiply each term of one or both equations by some nonzero integer so that the coefficients of one of the variables are opposites. 2. Add the equations of this new system so that one variable is eliminated. 3. Solve the resulting equation for the remaining variable. 4. Substitute the value found into either of the original equations to find the value of the other variable. 5. Check your solution in both of the original equations.	Solve: $3x - 2y = 0$ **(1)** $\qquad -4x + 3y = 1$ **(2)** Eliminate the x's: a common multiple of 3 and 4 is 12. Multiply equation **(1)** by 4 and equation **(2)** by 3; then add. $\qquad\quad 12x - 8y = 0$ $\qquad\underline{-12x + 9y = 3}$ $\qquad\qquad\qquad\ y = 3$ Substitute $y = 3$ into either equation and solve for x. Let us pick $3x - 2y = 0$. $\quad 3x - 2(3) = 0 \qquad 3x - 6 = 0, \qquad x = 2$ *Check:* $3(2) - 2(3) \stackrel{?}{=} 0 \qquad\qquad -4(2) + 3(3) \stackrel{?}{=} 1$ $\qquad 6 - 6 \stackrel{?}{=} 0 \qquad\qquad\qquad -8 + 9 \stackrel{?}{=} 1$ $\qquad\quad\ 0 = 0\ \checkmark \qquad\qquad\qquad\quad\ 1 = 1\ \checkmark$ The solution is $(2, 3)$.
Choosing a method to solve a system of equations, p. 495.	1. Substitution works well if at least one variable has a coefficient of 1 or -1. 2. Addition works well for integer, fractional, or decimal coefficients. 3. Graphing gives you a picture of the system. It works well if the system has integer solutions or if the system is inconsistent or dependent.	Use substitution. $\qquad\quad x + y = 8$ $\qquad\ 2x - 3y = 9$ Use addition. $\qquad \dfrac{3}{5}x - \dfrac{1}{4}y = 4$ $\qquad \dfrac{1}{5}x + \dfrac{3}{4}y = 8$ Graph to get a picture of the system. If the system is consistent and if it is difficult to determine the solutions because they are fractions, use an algebraic method.
To identify inconsistent and dependent systems algebraically, p. 496.	1. If you obtain an equation that is inconsistent with known facts, such as $0 = 2$, the system is inconsistent. There is no solution. 2. If you obtain an equation that is always true, such as $0 = 0$, the system is dependent. We often say that in this case the equations are dependent. There are an infinite number of solutions.	Solve: $\quad x + 2y = 1$ $\qquad\qquad 3x + 6y = 12$ Multiply the first equation by -3. $\qquad\qquad -3x - 6y = -3$ $\qquad\qquad\underline{3x + 6y = 12}$ $\qquad\qquad\qquad\qquad\ 0 = 9$ 0 is not equal to 9, so there is *no solution*. The system is inconsistent. Solve: $2x + y = 1$ $\qquad\quad 6x + 3y = 3$ Multiply the first equation by -3. $\qquad\qquad -6x - 3y = -3$ $\qquad\qquad\ \underline{6x + 3y = 3}$ $\qquad\qquad\qquad\qquad 0 = 0$ Since 0 is always equal to 0, the system is dependent. There are an infinite number of solutions.

Topic	Procedure	Examples
Solving word problems with a system of equations, p. 500.	1. Understand the problem. Choose a variable to represent each unknown quantity. 2. Write a system of equations in two variables. 3. Solve the system of equations and state the answer. 4. Check the answer.	Apples sell for $0.35 per pound. Oranges sell for $0.40 per pound. Nancy bought 12 pounds of apples and oranges for $4.45. How many pounds of each fruit did she buy? Let x = number of pounds of apples and y = number of pounds of oranges. 12 pounds in all were purchased. $x + y = 12$ The purchase cost $4.45. $0.35x + 0.40y = 4.45$ Multiply the second equation by 100 to obtain the following system. $$x + y = 12$$ $$35x + 40y = 445$$ Multiply the top equation by -35 and add the equations. $$-35x - 35y = -420$$ $$\underline{35x + 40y = \quad 445}$$ $$5y = \quad 25$$ $$y = 5$$ Substitute $y = 5$ into $x + y = 12$. $$x + 5 = 12$$ $$x = 7$$ Nancy purchased 7 pounds of apples and 5 pounds of oranges. *Check:* Are there 12 pounds of apples and oranges? $$7 + 5 = 12 \;\checkmark$$ Would this purchase cost $4.45? $$0.35(7) + 0.40(5) \overset{?}{=} 4.45$$ $$2.45 + 2.00 \overset{?}{=} 4.45$$ $$4.45 = 4.45 \;\checkmark$$

Chapter 8 Review Problems

Section 8.1

Solve by graphing the lines and finding the point of intersection.

1. $2x + 3y = 0$
   $-x + 3y = 9$

2. $-3x + y = -2$
   $-2x - y = -8$

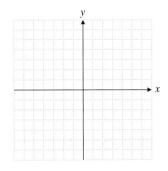

3. $2x - y = 6$
$6x + 3y = 6$

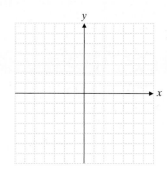

4. $2x - y = 1$
$3x + y = -6$

Section 8.2

Solve by the substitution method. Check your solution.

5. $x + y = 6$
$-2x + y = -3$

6. $x + 3y = 18$
$2x + y = 11$

7. $3x - 2y = 3$
$x - \dfrac{1}{3}y = 8$

8. $0.5x + y = 16$
$4x - 2y = 8$

Section 8.3

Solve by the addition method. Check your solution.

9. $6x - 2y = 10$
$2x + 3y = 7$

10. $4x - 5y = -4$
$-3x + 2y = 3$

11. $5x - 4y = 12$
$-2x + 3y = -9$

12. $7x + 2y = 4$
$2x - 3y = 19$

Section 8.4

Solve by any appropriate method. If it is not possible to obtain one solution, state the reason. Place the solution in the form (x, y), (a, b), (s, t) or (m, n).

13. $5x - 3y = 32$
$5x + 7y = -8$

14. $4x + 3y = 46$
$-2x + 3y = -14$

15. $7x + 3y = 2$
$-8x - 7y = 2$

16. $7x + 5y = 7$
$-3x - 4y = 1$

17. $4x + 5y = 4$
$-5x - 8y = 2$

18. $7x - 2y = 1$
$-5x + 3y = -7$

19. $2x - 4y = 6$
$-3x + 6y = 7$

20. $4x - 7y = 8$
$5x + 9y = 81$

21. $2x - 9y = 0$
$3x + 5 = 6y$

22. $2x + 10y = 1$
$-4x - 20y = -2$

23. $5x - 8y = 3x + 12$
$7x + y = 6y - 4$

24. $1 + x - y = y + 4$
$4(x - y) = 3 - x$

25. $3x + y = 9$
$x - 2y = 10$

26. $x + 12y = 0$
$3x - 5y = -2$

27. $2(x + 3) = y + 4$
$4x - 2y = -4$

28. $x + y = 3000$
$x - 2y = -120$

Mixed Practice

Solve by any appropriate method. If it is not possible to obtain the solution, state the reason.

29. $5x + 4y + 3 = 23$
$8x - 3y - 4 = 75$

30. $4x - 3y + 1 = 6$
$5x + 8y + 2 = -74$

31. $\dfrac{2x}{3} - \dfrac{3y}{4} = \dfrac{7}{12}$
$8x + 5y = 9$

32. $\dfrac{x}{2} + \dfrac{y}{5} = 4$
$\dfrac{x}{3} + \dfrac{y}{5} = \dfrac{10}{3}$

33. $\dfrac{1}{5}a + \dfrac{1}{2}b = 6$
$\dfrac{3}{5}a - \dfrac{1}{2}b = 2$

34. $\dfrac{2}{3}a + \dfrac{3}{5}b = -17$
$\dfrac{1}{2}a - \dfrac{1}{3}b = -1$

35. $4.8 + 0.6m = 0.9n$
$0.2m - 0.3n = 1.6$

36. $8.4 - 0.8m = 0.4n$
$0.2m + 0.1n = 2.1$

37. $3s - 8t = 5$
$4s + 3t = 2$

38. $5s + 6t = 2$
$3s - 5t = 4$

39. $3(x + 2) = -2 - (x + 3y)$
$3(x + y) = 3 - 2(y - 1)$

40. $13 - x = 3(x + y) + 1$
$14 + 2x = 5(x + y) + 3x$

41. $0.2b = 1.4 - 0.3a$
$0.1b + 0.6 = 0.5a$

42. $0.3a = 1.1 - 0.2b$
$0.3b = 0.4a - 0.9$

43. $\dfrac{b}{5} = \dfrac{2}{5} - \dfrac{a - 3}{2}$
$4(a - b) = 3b - 2(a - 2)$

44. $9(b + 4) = 2(2a + 5b)$
$\dfrac{b}{5} + \dfrac{a}{2} = \dfrac{18}{5}$

Section 8.5

Solve.

45. *Ticket Sales* During a concert, $3280 worth of tickets were sold. A total of 760 students attended the concert. Reserved seats were $6 and general admission seats were $4. How many of each kind of ticket were sold?

46. *Fruit Packaging* A farmer near Napa has 252 pounds of apples to take to the farmer's market. He packages them into 2-lb and 5-lb bags to sell. If the number of 5-lb bags he used was double the number of 2-lb bags, how many of each type of bags did he prepare to bring to market?

47. *Airplane Travel* A plane travels 1500 miles in 5 hours with the benefit of a tailwind. On the return trip it requires 6 hours to fly against the wind. Can you find the speed of the plane in still air and the wind speed?

48. *Chemical Solutions* A chemist needs to combine two solutions of a certain acid. How many liters of a 30% acid solution should be added to 40 liters of a 12% acid solution to obtain a final mixture that is 20% acid?

49. *Scout Car Wash* The girls and boys sponsoring the Scouts Car Wash had 86 customers on Saturday. They charged $4 to wash regular-sized cars and $3.50 to wash compact cars. The gross receipts for the day were $323. How many cars of each type were washed?

50. *Sanding Roads in Maine* Carl manages a highway department garage in Maine. A mixture of salt and sand is stored in the garage for use during the winter. Carl needs 24 tons of a salt/sand mixture that is 25% salt. He will combine shipments of 15% salt and shipments of 30% salt to achieve the desired 24 tons. How much of each type should he use?

51. *Water Ski Boat* A water ski boat traveled 23 kilometers/hour going with the current. It went back in the opposite direction against the current and traveled only 15 kilometers/hour. What was the speed of the boat? What was the speed of the current?

52. *Manufacturing Costs* Fred is analyzing the cost of producing two different items at an electronics company. An electrical sensing device uses 5 grams of copper and requires 3 hours to assemble. A smaller sensing device made by the same company uses 4 grams of copper but requires 5 hours to assemble. The first device has a production cost of $27. The second device has a production cost of $32. How much does it cost the company for a gram of this type of copper? What is the hourly labor cost at this company? (Assume that production cost is obtained by adding the copper cost and the labor cost.)

Use the following information for exercises 53 and 54.

Toronto Blue Jays 2004 Season Ticket Prices			
In the Action Seats	$165	Field Level Bases	$34
Premium Dugout Level	$45	Lower Skydeck	$22
Field Level Infield	$42	Upper Skydeck	$6.50

53. Connie and Dan Lacorazza took the youth group from their church to a Toronto Blue Jays game. They purchased 23 tickets. Some of the tickets were Lower Skydeck seats, while the rest were Upper Skydeck seats. The cost of all the tickets was $320. How many of each type of ticket did they buy?

54. Russ and Norma Camp took a group of senior citizens to a Toronto Blue Jays game. They purchased 27 tickets. Some of the tickets were Field Level Infield seats, while the rest were Field Level Bases seats. The cost of all the tickets was $1070. How many of each type of ticket did they buy?

Remember to use your Chapter Test Prep Video CD to see the worked-out solutions to the test problems you want to review.

Solve by the method specified.

1. Substitution method
$$3x - y = -5$$
$$-2x + 5y = -14$$

2. Addition method
$$3x + 4y = 7$$
$$2x + 3y = 6$$

3. Graphing method
$$2x - y = 4$$
$$4x + y = 2$$

4. Any method
$$x + 3y = 12$$
$$2x - 4y = 4$$

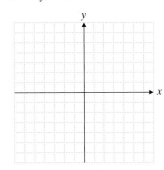

Solve by any method. If there is not one solution to a system, state why.

5. $2x - y = 5$
$-x + 3y = 5$

6. $2x + 3y = 13$
$3x - 5y = 10$

7. $\dfrac{2}{3}x - \dfrac{1}{5}y = 2$

$\dfrac{4}{3}x + 4y = 4$

8. $3x - 6y = 5$

$-\dfrac{1}{2}x + y = \dfrac{7}{2}$

9. $5x - 2 = y$
$\quad 10x = 4 + 2y$

10. $0.3x + 0.2y = 0$
$1.0x + 0.5y = -0.5$

11. $2(x + y) = 2(1 - y)$
$5(-x + y) = 2(23 - y)$

12. $3(x - y) = 12 + 2y$
$8(y + 1) = 6x - 7$

Solve.

13. Twice one number plus three times a second is one. Twice the second plus three times the first is nine. Find the numbers.

14. Both $8 and $12 tickets were sold for a basketball game. In all, 30,500 people paid for admission, resulting in a "gate" of $308,000. How many of each type of ticket were sold?

15. Five shirts and three pairs of slacks cost $172. Three of the same shirts and four pairs of the same slacks cost $156. How much does each shirt cost? How much is a pair of slacks?

16. You are producing a commemorative booklet for the 25th anniversary of your church. Ace Printers charges a $200 typesetting fee and $0.25 per booklet. Wong Printers charges a $250 typesetting fee but only $0.20 per booklet. How many booklets must you order for the cost to be the same? What is the cost?

17. A jet plane flew 2000 kilometers against the wind in a time of 5 hours. It refueled and flew back the same distance with the wind in 4 hours. Find the wind speed in kilometers per hour. Find the speed of the jet in still air in kilometers per hour.

1. _____

2. _____

3. _____

4. _____

5. _____

6. _____

7. _____

8. _____

9. _____

10. _____

11. _____

12. _____

13. _____

14. _____

15. _____

16. _____

17. _____

1. _____

2. _____

3. _____

4. _____

5. _____

6. _____

7. _____

8. _____

9. _____

10. _____

11. _____

12. _____

13. _____

14. _____

15. _____

16. _____

17. _____

18. _____

Approximately one-half of this test covers the content of Chapters 0–7. The remainder covers the content of Chapter 8.

Simplify.

1. 11% of $37.20
(Round to the nearest cent.)

2. $\dfrac{-2^4 \cdot (2^2)}{2^3}$

3. $(3x^2 - 2x + 1)(5x - 3)$

4. $\dfrac{x^2 - 7x + 12}{x^2 - 9} \div (x^2 - 8x + 16)$

In questions 5 and 6, solve.

5. $2(x + 3) < 3(x - 5)$

6. $6x^2 + x - 2 = 0$

7. 12% of what number is 360?

8. Solve for x. $\dfrac{1}{10}x + \dfrac{1}{3} = \dfrac{2}{5}x - \dfrac{1}{6}$

9. Factor $10t^2 - 25t^3 + 15t^4$.

10. Solve for x. $4x = 2(12 - 2x)$

Solve by any method. If there is not one solution to a system, state why.

11. $2x + y = 8$
$3x + 4y = -8$

12. $\dfrac{1}{3}x - \dfrac{1}{2}y = 4$
$\dfrac{3}{8}x - \dfrac{1}{4}y = 7$

13. $1.3x - 0.7y = 0.4$
$-3.9x + 2.1y = -1.2$

14. $10x - 5y = 45$
$3x - 8y = 7$

Solve.

15. The equations in a system are graphed and parallel lines are obtained. What is this system called? What does that mean?

16. How much water must be added to 60 kilograms of an 80% acid solution to produce a 50% solution?

17. A boat trip 6 miles upstream takes 3 hours. The return takes 90 minutes. How fast is the stream?

18. Two printers were used for a 5-hour job of printing 15,000 labels. The next day, a second run of the same labels took 7 hours because one printer broke down after 2 hours. How many labels can each printer process per hour?

Have you ever thought how the International Space Station gets its electric power? Orbiting 250 miles above earth, the only power source is the sun. A highly efficient array of solar panels is directed toward the sun to collect the necessary power. Could you use your mathematical skill to make some calculations about the amount of electrical power produced by one of these solar panels? Turn to the Putting Your Skills to Work problems on page 561 to find out.

Radicals

9.1 SQUARE ROOTS

Student Learning Objectives

After studying this section, you will be able to:

1 Evaluate the square root of a perfect square.

2 Approximate the square root of a number that is not a perfect square.

1 Evaluating the Square Root of a Perfect Square

How long is the side of a square whose area is 4? Recall the formula for the area of a square.

area of a square $= s^2$ Our question then becomes, what number times itself is 4?

$$s^2 = 4$$
$$s = 2 \quad \text{because } (2)(2) = 4$$
$$s = -2 \quad \text{because } (-2)(-2) = 4$$

We say that 2 is a **square root** of 4 because $(2)(2) = 4$. We can also say that -2 is a square root of 4 because $(-2)(-2) = 4$. Note that 4 is a **perfect square.** A square root of a perfect square is an integer.

The symbol $\sqrt{\ }$ is used in mathematics for the **principal square root** of a number, which is the nonnegative square root. The symbol itself $\sqrt{\ }$ is called the **radical sign.** If we want to find the negative square root of a number, we use the symbol $-\sqrt{\ }$.

DEFINITION OF PRINCIPAL SQUARE ROOT

For all nonnegative numbers N, the principal square root of N (written $\sqrt{N}$) is defined to be the nonnegative number a if and only if $a^2 = N$.

Notice in the definition that we did not use the words "for all *positive* numbers N" because we also want to include the square root of 0. $\sqrt{0} = 0$ since $0^2 = 0$.

Let us now examine the use of this definition with a few examples.

EXAMPLE 1   Find.   **(a)** $\sqrt{144}$ **(b)** $-\sqrt{9}$

Solution

(a) $\sqrt{144} = 12$ since $(12)^2 = (12)(12) = 144$ and $12 \geq 0$.

(b) The symbol $-\sqrt{9}$ is read "the opposite of the square root of 9."

Because $\sqrt{9} = 3$, we have $-\sqrt{9} = -3$.

Practice Problem 1   Find.   **(a)** $\sqrt{64}$ **(b)** $-\sqrt{121}$

NOTE TO STUDENT: Fully worked-out solutions to all of the Practice Problems can be found at the back of the text starting at page SP-1

The number beneath the radical sign is called the **radicand.** The radicand will not always be an integer.

EXAMPLE 2   Find.   **(a)** $\sqrt{\dfrac{1}{4}}$ **(b)** $-\sqrt{\dfrac{4}{9}}$

Solution

(a) $\sqrt{\dfrac{1}{4}} = \dfrac{1}{2}$ since $\left(\dfrac{1}{2}\right)^2 = \left(\dfrac{1}{2}\right)\left(\dfrac{1}{2}\right) = \dfrac{1}{4}$.

(b) $-\sqrt{\dfrac{4}{9}} = -\dfrac{2}{3}$ since $\sqrt{\dfrac{4}{9}} = \dfrac{2}{3}$.

Practice Problem 2   Find.   **(a)** $\sqrt{\dfrac{9}{25}}$ **(b)** $-\sqrt{\dfrac{121}{169}}$

EXAMPLE 3   Find.   **(a)** $\sqrt{0.09}$ **(b)** $\sqrt{1600}$ **(c)** $\sqrt{225}$

Solution

(a) $\sqrt{0.09} = 0.3$ since $(0.3)(0.3) = 0.09$.

Notice that $\sqrt{0.09}$ is *not* 0.03 because $(0.03)(0.03) = 0.0009$. Remember to count the decimal places when you multiply decimals. When finding the square root of a decimal, you should multiply to check your answer.

(b) $\sqrt{1600} = 40$ since $(40)(40) = 1600$.

(c) $\sqrt{225} = 15$

Practice Problem 3   Find.

(a) $-\sqrt{0.0036}$ **(b)** $\sqrt{2500}$ **(c)** $\sqrt{196}$

2 Approximating the Square Root of a Number That Is Not a Perfect Square

Not all numbers are perfect squares. How can we find the square root of 2? That is, what number times itself equals 2? $\sqrt{2}$ is an irrational number. It cannot be written as $\frac{p}{q}$, where p and q are integers and $q \neq 0$. $\sqrt{2}$ is a nonterminating, nonrepeating decimal, and the best we can do is approximate its value. To do so, we can use the square root key on a calculator.

To use a calculator, enter 2 and push the $\boxed{\sqrt{x}}$ or $\boxed{\sqrt{}}$ key.

$$\boxed{2} \quad \boxed{\sqrt{x}} \quad \boxed{1.4142136}$$

Remember, this is only an approximation. Most calculators are limited to an eight-digit display.

To use the table of square roots, look under the x column until you find the number 2. Then look across the row to the $\sqrt{x}$ column. The value there is rounded to the nearest thousandth.

$$\sqrt{2} \approx 1.414$$

x	$\sqrt{x}$
1	1.000
2	1.414
3	1.732
4	2.000
5	2.236
6	2.449
7	2.646

EXAMPLE 4   Use a calculator or a table to approximate. Round to the nearest thousandth.

(a) $\sqrt{28}$ **(b)** $\sqrt{191}$

Solution

(a) Using a calculator, we have 28 $\boxed{\sqrt{x}}$ 5.291502622. Thus $\sqrt{28} \approx 5.292$.

(b) Using a calculator, we have 191 $\boxed{\sqrt{x}}$ 13.82027496. Thus $\sqrt{191} \approx 13.820$.

Practice Problem 4   Approximate.

(a) $\sqrt{3}$ **(b)** $\sqrt{35}$ **(c)** $\sqrt{127}$

On a professional softball diamond, the distance from home plate to second base is exactly $\sqrt{7200}$ feet. Approximate this value using a calculator.

SIDELIGHT: A Topic for Further Thought

Let's look at the radicand carefully. Does the radicand need to be a nonnegative number? What if we write $\sqrt{-4}$? Is there a number that you can square to get -4?

Obviously, there is *no real number* that you can square to get -4. We know that $(2)^2 = 4$ and $(-2)^2 = 4$. Any real number that is squared will be nonnegative. We therefore conclude that $\sqrt{-4}$ does not represent a real number. Because we want to work with real numbers, our definition of $\sqrt{n}$ requires that n be nonnegative. Thus $\sqrt{-4}$ is *not a real number*.

You may encounter a more sophisticated number system called **complex numbers** in a higher-level mathematics course. In the complex number system, negative numbers have square roots.

Verbal and Writing Skills

1. Define a principal square root in your own words.

2. Is $\sqrt{576} = 24$? Why or why not?

3. Is $\sqrt{0.9} = 0.3$? Why or why not?

4. How would you find $\sqrt{90}$?

Find the two square roots of each number.

5. 9

6. 4

7. 49

8. 36

Find the square root. Do not use a calculator or a table of square roots.

9. $\sqrt{16}$

10. $\sqrt{100}$

11. $\sqrt{81}$

12. $\sqrt{25}$

13. $-\sqrt{36}$

14. $-\sqrt{25}$

15. $\sqrt{0.81}$

16. $\sqrt{0.64}$

17. $\sqrt{\dfrac{36}{121}}$

18. $\sqrt{\dfrac{25}{81}}$

19. $\sqrt{\dfrac{49}{64}}$

20. $\sqrt{\dfrac{49}{100}}$

21. $\sqrt{400}$

22. $\sqrt{1225}$

23. $-\sqrt{10,000}$

24. $-\sqrt{40,000}$

Mixed Practice

Find the square root. Do not use a calculator or a table of square roots.

25. $\sqrt{169}$

26. $\sqrt{256}$

27. $-\sqrt{\dfrac{1}{64}}$

28. $\sqrt{\dfrac{81}{100}}$

29. $\sqrt{\dfrac{9}{16}}$

30. $-\sqrt{\dfrac{16}{49}}$

31. $\sqrt{0.0036}$

32. $\sqrt{0.0009}$

33. $\sqrt{16,900}$

34. $\sqrt{22,500}$

35. $-\sqrt{289}$

36. $-\sqrt{324}$

Use a calculator or the square root table to approximate to the nearest thousandth.

37. $\sqrt{27}$

38. $\sqrt{33}$

39. $\sqrt{74}$

40. $\sqrt{79}$

41. $-\sqrt{133}$

42. $-\sqrt{146}$

43. $-\sqrt{195}$

44. $-\sqrt{182}$

Applications

▲ **45.** *Tree Care* A tree is 12 feet tall. A wire that stretches from a point that is 5 feet from the base of the tree to the top of the tree is $\sqrt{169}$ feet long. Exactly how many feet long is the wire?

▲ **46.** *Extension Ladder* A building is 15 ft tall. A ladder that stretches from a point that is 8 feet from the base of the building to the top of the building is $\sqrt{289}$ feet long. Exactly how many feet long is the ladder?

Time for an Object to Fall If we ignore wind resistance, the time in seconds it takes for an object to fall s feet is given by $t = \frac{1}{4}\sqrt{s}$. Thus an object will fall 100 feet in $t = \frac{1}{4}\sqrt{100} = \frac{1}{4}(10) = 2.5$ seconds.

47. How long would it take an object dropped off a building to fall 900 feet?

48. How long would it take an object dropped out of a tower to fall 1600 feet?

49. The Sears Tower in Chicago is one of the world's tallest buildings. It is 1454 feet tall. If a metal frame fell off the building from a height of 1444 feet, how long would it take the metal frame to hit the ground?

50. In the movie *The Gods Must Be Crazy*, a Coke bottle is dropped out of a plane to land near an aboriginal tribe's village. If the plane was flying at a height of 1225 feet, how long did it take the Coke bottle to hit the ground?

To Think About: Finding Higher Roots

In addition to finding the square root of a number, we can also find the cube root of a number. To find the cube root of N (which is written as $\sqrt[3]{N}$), we find a number that, when cubed, will equal N.

For example, $\sqrt[3]{8} = 2$ since $2^3 = 8$. We can also find the cube root of a negative number—for example, $\sqrt[3]{-8} = -2$ since $(-2)^3 = -8$. The cube root of a positive number is positive. The cube root of a negative number is negative. Therefore, each real number has only one real cube root.

The concept of roots can be extended to fourth and fifth roots. For example, $\sqrt[4]{16} = 2$ since $2^4 = 16$. Also, $\sqrt[5]{32} = 2$ since $2^5 = 32$.

Find each higher root.

51. $\sqrt[3]{27}$ **52.** $\sqrt[3]{125}$ **53.** $\sqrt[3]{-216}$ **54.** $\sqrt[3]{-27}$

55. $\sqrt[4]{81}$ **56.** $\sqrt[4]{625}$ **57.** $\sqrt[5]{243}$ **58.** $\sqrt[5]{1024}$

59. Is there a real number that equals $\sqrt[4]{-16}$? Why?

60. Is there a real number that equals $\sqrt[4]{-81}$? Why?

Cumulative Review

61. Solve. $3x + 2y = 8$
$7x - 3y = 11$

62. Solve if possible. $2x - 3y = 1$
$-8x + 12y = 4$

63. *Snowboarding* A snowboard racer bought three new snowboards and two pairs of racing goggles last month for $850. This month, she bought four snowboards and three pairs of goggles for $1150. How much did each snowboard cost? How much did each pair of goggles cost?

64. *Flying with a Wind* With a tailwind, the Qantas flight from Los Angeles to Sydney, Australia, covered 7280 miles in 14 hours, nonstop. The return flight of 7140 miles took 17 hours because of a headwind. If the airspeed of the plane remained constant, and if the return flight encountered a constant headwind equal to the earlier tailwind, what was the plane's airspeed in still air?

65. *Hospital Care in France* The average length of stay for inpatient hospital care in France in 1990 was 13.3 days. In 2000 the average length of stay was 10.6 days. In the year 2010 the average length of stay is projected to be 9.4 days. What will be the percent of decrease from 2000 to 2010? If the same percent of decrease continues from 2010 to 2020, what will be the projected length of stay in 2020? Round all answers to the nearest tenth. (*Source:* Organization for Economic Cooperation and Development, Paris, France.)

66. *Hospital Care in the U.S.* The average length of stay for inpatient hospital care in the United States in 1990 was 9.1 days. In 2000 the average length of stay was 6.9 days. In the year 2010 the average length of stay is projected to be 5.8 days. What will be the percent of decrease from 2000 to 2010? Using the data obtained in problem 65, determine which country (France or the United States) will be experiencing the larger percent of decrease from 2000 to 2010. How great is this difference? Round all answers to the nearest tenth. (*Source:* Organization for Economic Cooperation and Development, Paris, France.)

9.2 SIMPLIFYING RADICAL EXPRESSIONS

1 Simplifying a Radical Expression with a Radicand That Is a Perfect Square

We know that $\sqrt{25} = \sqrt{5^2} = 5$. $\sqrt{9} = \sqrt{3^2} = 3$ and $\sqrt{36} = \sqrt{6^2} = 6$. If we write the radicand as a square of a nonnegative number, the square root is this nonnegative base. That is, $\sqrt{x^2} = x$ when x is nonnegative. We can use this idea to simplify radicals.

Student Learning Objectives

After studying this section, you will be able to:

1. Simplify a radical expression with a radicand that is a perfect square.

2. Simplify a radical expression with a radicand that is not a perfect square.

EXAMPLE 1 Find. **(a)** $\sqrt{9^4}$ **(b)** $\sqrt{126^2}$ **(c)** $\sqrt{17^6}$

Solution

(a) Using the law of exponents, we rewrite 9^4 as a square.

$\sqrt{9^4} = \sqrt{(9^2)^2}$ Raise a power to a power: $(9^2)^2 = 9^4$.

$\quad\quad = 9^2$ To check, multiply: $(9^2)(9^2) = 9^4$.

(b) $\sqrt{126^2} = 126$

(c) $\sqrt{17^6} = \sqrt{(17^3)^2}$ Raise a power to a power: $(17^3)^2 = 17^6$.

$\quad\quad = 17^3$ To check, multiply: $(17^3)(17^3) = 17^6$.

Practice Problem 1 Find. **(a)** $\sqrt{6^2}$ **(b)** $\sqrt{13^4}$ **(c)** $\sqrt{18^{12}}$

NOTE TO STUDENT: Fully worked-out solutions to all of the Practice Problems can be found at the back of the text starting at page SP-1

This same concept can be used with variable expressions. We will restrict the values of each variable to be nonnegative if the variable is under the radical sign. Thus *all variable radicands in this chapter* are assumed to *represent positive numbers or zero*.

EXAMPLE 2 Find. **(a)** $\sqrt{x^6}$ **(b)** $\sqrt{y^{24}}$

Solution

(a) $\sqrt{x^6} = \sqrt{(x^3)^2}$ By the law of exponents, $(x^3)^2 = x^6$.

$\quad\quad = x^3$

(b) $\sqrt{y^{24}} = y^{12}$

Practice Problem 2 Find. **(a)** $\sqrt{y^{18}}$ **(b)** $\sqrt{x^{30}}$

In each of these examples we found the square root of a perfect square, and thus the radical sign disappeared. When you find the square root of a perfect square, do *not* leave a radical sign in your answer.

Often the coefficient of a variable radical will not be written in exponent form. You will want to be able to recognize the squares of numbers from 1 to 15, 20, and 25. If you do, you will be able to find many square roots faster mentally than by using a calculator or a table of square roots.

Before you begin, you need to know the multiplication rule for square roots.

MULTIPLICATION RULE FOR SQUARE ROOTS

For all nonnegative numbers a and b,

$$\sqrt{a} \cdot \sqrt{b} = \sqrt{ab} \quad \text{and} \quad \sqrt{ab} = \sqrt{a} \cdot \sqrt{b}.$$

EXAMPLE 3 Find.

(a) $\sqrt{225x^2}$ (b) $\sqrt{x^6 y^{14}}$ (c) $\sqrt{169x^8 y^{10}}$

Solution

(a) $\sqrt{225x^2} = 15x$ Using the multiplication rule for square roots,
$\sqrt{225x^2} = \sqrt{225}\sqrt{x^2} = 15x$.

(b) $\sqrt{x^6 y^{14}} = x^3 y^7$

(c) $\sqrt{169x^8 y^{10}} = 13x^4 y^5$

    To check each answer, square it. The result should be the expression under the radical sign.

NOTE TO STUDENT: Fully worked-out solutions to all of the Practice Problems can be found at the back of the text starting at page SP-1

Practice Problem 3 Find.

(a) $\sqrt{625y^4}$ (b) $\sqrt{x^{16} y^{22}}$ (c) $\sqrt{121x^{12} y^6}$

② Simplifying a Radical Expression with a Radicand That Is Not a Perfect Square

Most of the time when we encounter a square root, the radicand is not a perfect square. Thus, when we simplify the square root algebraically, we must retain the radical sign. Still, being able to simplify such radical expressions is a useful skill because it allows us to combine them, as we shall see in Section 9.3.

EXAMPLE 4 Simplify.

(a) $\sqrt{20}$ (b) $\sqrt{50}$ (c) $\sqrt{48}$

Solution   To begin, when you look at a radicand, look for perfect squares.

(a) $\sqrt{20} = \sqrt{4 \cdot 5}$ Note that 4 is a perfect square, and we can write 20 as $4 \cdot 5$.

$\qquad = \sqrt{4}\sqrt{5}$ Recall that $\sqrt{4} = 2$.

$\qquad = 2\sqrt{5}$

(b) $\sqrt{50} = \sqrt{25 \cdot 2}$

$\qquad = \sqrt{25}\sqrt{2}$

$\qquad = 5\sqrt{2}$

(c) $\sqrt{48} = \sqrt{16 \cdot 3}$

$\qquad = \sqrt{16}\sqrt{3}$

$\qquad = 4\sqrt{3}$

Practice Problem 4 Simplify.

(a) $\sqrt{98}$ (b) $\sqrt{12}$ (c) $\sqrt{75}$

    The same procedure can be used with square root radical expressions containing variables. The key is to think of squares. That is, think, "How can I rewrite the expression so that it contains a perfect square?"

EXAMPLE 5   Simplify.   **(a)** $\sqrt{x^3}$ **(b)** $\sqrt{x^7 y^9}$

Solution

(a) Recall that the law of exponents tells us to add the exponents when multiplying two expressions. Because we want one of these expressions to be a perfect square, we think of the exponent 3 as a sum of an even number and 1: $3 = 2 + 1$.

$$\sqrt{x^3} = \sqrt{x^2}\sqrt{x} \quad \text{Because } (x^2)(x) = x^3 \text{ by the law of exponents.}$$
$$= x\sqrt{x}$$

In general, to simplify a square root that has a variable with an exponent in the radicand, first write the square root as a product of the form $\sqrt{x^n}\sqrt{x}$, where n is the largest possible even exponent.

(b) $\sqrt{x^7 y^9} = \sqrt{x^6}\sqrt{x}\sqrt{y^8}\sqrt{y}$

$$= x^3\sqrt{x}\,y^4\sqrt{y}$$
$$= x^3 y^4\sqrt{xy}$$

In the final simplified form, we place the variable factors with exponents first and the radical factor second.

Practice Problem 5   Simplify.   **(a)** $\sqrt{x^{11}}$ **(b)** $\sqrt{x^5 y^3}$

In summary, to simplify a square root radical, you factor each part of the expression under the radical sign and simplify the perfect squares.

EXAMPLE 6   Simplify.   **(a)** $\sqrt{12y^5}$ **(b)** $\sqrt{18x^3 y^7 w^{10}}$

Solution

(a) $\sqrt{12y^5} = \sqrt{4 \cdot 3 \cdot y^4 \cdot y}$

$$= 2y^2\sqrt{3y} \qquad \text{Remember, } y^4 = (y^2)^2 \text{ and } \sqrt{(y^2)^2} = y^2.$$

(b) $\sqrt{18x^3 y^7 w^{10}} = \sqrt{9 \cdot 2 \cdot x^2 \cdot x \cdot y^6 \cdot y \cdot w^{10}}$

$$= 3xy^3 w^5\sqrt{2xy}$$

Practice Problem 6   Simplify.   **(a)** $\sqrt{48x^{11}}$ **(b)** $\sqrt{121x^6 y^7 z^8}$

Throughout the world there has been a significant increase in the production and consumption of energy from alternative sources. These are primarily geothermal, solar, wind, and waste recycling energy production. For example, in 1990 1.7 quadrillion Btu (British thermal units) of energy from alternative sources were produced in the world. By 2000 this had increased to 3 quadrillion Btu (that is, 3,000,000,000,000,000 Btu). Some scientists predict that by 2010 this will increase to $\sqrt{192}$ quadrillion Btu. When simplified this can be written as $8\sqrt{3}$ quadrillion Btu. (*Source:* U.S. Energy Information Administration.)

Student Solutions Manual CD/ Video PH Math Tutor Center MathXL®Tutorials on CD MathXL® MyMathLab® Interactmath.com

Verbal and Writing Skills

1. Is $\sqrt{6^2} = \left(\sqrt{6}\right)^2$?

2. Why is it *not correct* to say
$$\sqrt{30} = \sqrt{(-6)(-5)} = \sqrt{-6}\sqrt{-5}\,?$$

3. Is $\sqrt{(-9)^2} = -\sqrt{9^2}$?

4. Is $\sqrt{2^4} = \left(\sqrt{2}\right)^4$?

Simplify. Leave the answer in exponent form.

5. $\sqrt{8^2}$ **6.** $\sqrt{2^2}$ **7.** $\sqrt{18^4}$ **8.** $\sqrt{13^4}$ **9.** $\sqrt{9^6}$

10. $\sqrt{7^{10}}$ **11.** $\sqrt{33^8}$ **12.** $\sqrt{39^8}$ **13.** $\sqrt{5^{140}}$ **14.** $\sqrt{7^{160}}$

Simplify. Assume that all variables represent positive numbers.

15. $\sqrt{x^{12}}$ **16.** $\sqrt{x^{30}}$ **17.** $\sqrt{t^{18}}$ **18.** $\sqrt{w^{22}}$ **19.** $\sqrt{y^{26}}$

20. $\sqrt{w^{14}}$ **21.** $\sqrt{36x^8}$ **22.** $\sqrt{64x^6}$ **23.** $\sqrt{144x^2}$ **24.** $\sqrt{196y^{10}}$

25. $\sqrt{x^6y^4}$ **26.** $\sqrt{x^2y^{18}}$ **27.** $\sqrt{16x^2y^{20}}$ **28.** $\sqrt{64x^2y^4}$ **29.** $\sqrt{100x^{12}y^8}$

30. $\sqrt{81x^6y^{10}}$ **31.** $\sqrt{24}$ **32.** $\sqrt{27}$ **33.** $\sqrt{45}$ **34.** $\sqrt{20}$

35. $\sqrt{18}$ **36.** $\sqrt{32}$ **37.** $\sqrt{72}$ **38.** $\sqrt{60}$ **39.** $\sqrt{90}$

40. $\sqrt{125}$ **41.** $\sqrt{128}$ **42.** $\sqrt{96}$ **43.** $\sqrt{8x^3}$ **44.** $\sqrt{12y^5}$

45. $\sqrt{27w^5}$ **46.** $\sqrt{25x^5}$ **47.** $\sqrt{32z^9}$

48. $\sqrt{49x^5}$ **49.** $\sqrt{28x^5y^7}$ **50.** $\sqrt{18x^2y^3}$

Mixed Practice

Simplify. Assume all variables represent positive numbers.

51. $\sqrt{48y^3w}$ **52.** $\sqrt{12x^2y^3}$ **53.** $\sqrt{75x^2y^3}$ **54.** $\sqrt{27a^8y^7}$

55. $\sqrt{81x^8}$ **56.** $\sqrt{49x^6}$ **57.** $\sqrt{y^7}$ **58.** $\sqrt{w^9}$

59. $\sqrt{x^5y^{11}}$ **60.** $\sqrt{x^3y^{13}}$ **61.** $\sqrt{135x^5y^7}$ **62.** $\sqrt{180a^5bc^2}$

63. $\sqrt{80a^3b^3c^6}$ **64.** $\sqrt{63a^4b^6c^3}$ **65.** $\sqrt{169a^3b^8c^5}$ **66.** $\sqrt{121a^8b^5c^{13}}$

To Think About

Simplify. Assume that all variables represent positive numbers.

67. $\sqrt{(x + 5)^2}$

68. $\sqrt{x^2 + 4x + 4}$

69. $\sqrt{16x^2 + 8x + 1}$

70. $\sqrt{4x^2 + 20x + 25}$

71. $\sqrt{x^2y^2 + 14xy^2 + 49y^2}$

72. $\sqrt{16x^2 + 16x + 4}$

Applications

73. ***Electron Microscope*** In a research experiment at Lucent Technologies, Maria found that the rectangular viewing area of her electron microscope measured $\sqrt{75}$ square millimeters. Write this answer in simplified radical form.

74. ***Science Research*** During a research project at Sandia Labs, Mai Li found the rectangular area of a metal mesh filter needed to be $\sqrt{320}$ square centimeters. Write this in simplified radical form.

75. ***Electron Microscope*** When Maria adjusted the microscope discussed in exercise 73 to a lower magnification, she increased the rectangular viewing area to $\sqrt{1200}$ square millimeters.
(a) Write this answer in simplified radical form.

(b) How many times larger is this area than the original area found in exercise 73?

76. ***Electron Microscope*** When Mai Li continued with her research at Sandia, she found that a filter whose area was $\sqrt{1280}$ square centimeters produced better results for her experiment.
(a) Write this answer in simplified radical form.

(b) How many times larger is the area of this filter than the area of the original filter found in exercise 74?

Cumulative Review

77. Solve. $3x - 2y = 14$
$\qquad y = x - 5$

78. Solve. $\frac{1}{3}(2x - 4) = \frac{1}{4}x - 3$

79. ***Nuclear Energy Consumption*** Surprisingly, Asia and Australia lead the world in nuclear energy consumption. In 1984, they used 310 million tons of equivalent energy. In 1994, they used 560 million tons. Assuming the same annual rate of increase in energy consumption, estimate how much energy will be used in 2007.

80. ***Nuclear Energy Consumption*** In 1993, the United States consumed 84 quadrillion Btu of energy. This was 265% of the energy consumed by China in that same year. How many quadrillion Btu of energy were consumed by China?

Asia and Australia Nuclear Energy Consumption

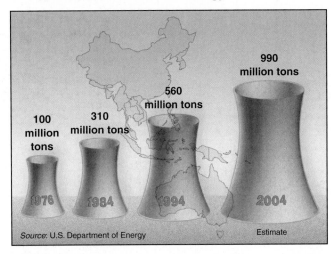

Source: U.S. Department of Energy

81. ***Nuclear Energy Consumption*** What is the expected percent of increase of consumption of nuclear energy for Asia and Australia from 1994 to 2004? Use the data in the Energy Consumption chart above to determine your answer.

9.3 ADDING AND SUBTRACTING RADICAL EXPRESSIONS

Student Learning Objectives

After studying this section, you will be able to:

 Add and subtract radical expressions with like radicals.

 Add and subtract radical expressions that must first be simplified.

1 Adding and Subtracting Radical Expressions with Like Radicals

Recall that when you simplify an algebraic expression you combine like terms. $8b + 5b = 13b$ because both terms contain the same variable b, and you can add the coefficients. This same idea is used when combining radicals. To add or subtract square root radicals, the numbers or the algebraic expressions under the radical signs must be the same. That is, the radicals must be **like radicals.**

EXAMPLE 1   Combine.   **(a)** $5\sqrt{2} - 8\sqrt{2}$ **(b)** $7\sqrt{a} + 3\sqrt{a} - 5\sqrt{a}$

Solution   First check to see if the radicands are the same. If they are, you can combine the radicals.

(a) $5\sqrt{2} - 8\sqrt{2} = (5 - 8)\sqrt{2} = -3\sqrt{2}$

(b) $7\sqrt{a} + 3\sqrt{a} - 5\sqrt{a} = (7 + 3 - 5)\sqrt{a} = 5\sqrt{a}$

Practice Problem 1   Combine.

(a) $7\sqrt{11} + 4\sqrt{11}$ **(b)** $4\sqrt{t} - 7\sqrt{t} + 6\sqrt{t} - 2\sqrt{t}$

NOTE TO STUDENT: *Fully worked-out solutions to all of the Practice Problems can be found at the back of the text starting at page SP-1*

Be careful. Not all the terms in an expression will have like radicals.

EXAMPLE 2   Combine. $5\sqrt{2a} + 3\sqrt{a} - 7\sqrt{2} + 3\sqrt{2a}$

Solution   The only terms that have the same radicand are $5\sqrt{2a}$ and $3\sqrt{2a}$. These terms may be combined. All other terms stay the same.

$$5\sqrt{2a} + 3\sqrt{a} - 7\sqrt{2} + 3\sqrt{2a} = 8\sqrt{2a} + 3\sqrt{a} - 7\sqrt{2}$$

Practice Problem 2   Combine. $3\sqrt{x} - 2\sqrt{xy} - 5\sqrt{y} + 7\sqrt{xy}$

2 Adding and Subtracting Radical Expressions That Must First Be Simplified

Sometimes it is necessary to simplify one or more radicals before their terms can be combined. Be sure to combine only like radicals.

EXAMPLE 3   Combine.   **(a)** $2\sqrt{3} + \sqrt{12}$ **(b)** $\sqrt{12} - \sqrt{27} + \sqrt{50}$

Solution

(a) $2\sqrt{3} + \sqrt{12} = 2\sqrt{3} + \sqrt{4 \cdot 3}$ Look for perfect square factors.

$\qquad\qquad\qquad = 2\sqrt{3} + 2\sqrt{3}$ Simplify and combine like radicals.

$\qquad\qquad\qquad = 4\sqrt{3}$

(b) $\sqrt{12} - \sqrt{27} + \sqrt{50} = \sqrt{4 \cdot 3} - \sqrt{9 \cdot 3} + \sqrt{25 \cdot 2}$

$\qquad\qquad\qquad\qquad = 2\sqrt{3} - 3\sqrt{3} + 5\sqrt{2}$ Only $2\sqrt{3}$ and $-3\sqrt{3}$ can be combined.

$\qquad\qquad\qquad\qquad = -\sqrt{3} + 5\sqrt{2}$ or $5\sqrt{2} - \sqrt{3}$

Practice Problem 3   Combine.

(a) $\sqrt{50} - \sqrt{18} + \sqrt{98}$ **(b)** $\sqrt{12} + \sqrt{18} - \sqrt{50} + \sqrt{27}$

EXAMPLE 4   Combine. $\sqrt{2a} + \sqrt{8a} + \sqrt{27a}$

Solution

$$\sqrt{2a} + \sqrt{8a} + \sqrt{27a} = \sqrt{2a} + \sqrt{4 \cdot 2a} + \sqrt{9 \cdot 3a}$$ Look for perfect square factors.

$$= \sqrt{2a} + 2\sqrt{2a} + 3\sqrt{3a}$$ Be careful. $\sqrt{2a}$ and $\sqrt{3a}$ are **not** like radicals. The expressions under the radical signs must be the same.

$$= 3\sqrt{2a} + 3\sqrt{3a}$$

Practice Problem 4   Combine. $\sqrt{9x} + \sqrt{8x} - \sqrt{4x} + \sqrt{50x}$

Special care should be taken if the original radical has a numerical coefficient. If the radical can be simplified, the two resulting numerical coefficients should be multiplied.

EXAMPLE 5   Combine. $2\sqrt{20} + 3\sqrt{45} - 4\sqrt{80}$

Solution

$$2\sqrt{20} + 3\sqrt{45} - 4\sqrt{80} = 2 \cdot \sqrt{4} \cdot \sqrt{5} + 3 \cdot \sqrt{9} \cdot \sqrt{5} - 4 \cdot \sqrt{16} \cdot \sqrt{5}$$

$$= 2 \cdot 2 \cdot \sqrt{5} + 3 \cdot 3 \cdot \sqrt{5} - 4 \cdot 4 \cdot \sqrt{5}$$

$$= 4\sqrt{5} + 9\sqrt{5} - 16\sqrt{5}$$

$$= -3\sqrt{5}$$

Practice Problem 5   Combine. $\sqrt{27} - 4\sqrt{3} + 2\sqrt{75}$

EXAMPLE 6   Combine. $3a\sqrt{8a} + 2\sqrt{50a^3}$

Solution

$$3a\sqrt{8a} + 2\sqrt{50a^3} = 3a\sqrt{4}\sqrt{2a} + 2\sqrt{25a^2}\sqrt{2a}$$

$$= 3a \cdot 2 \cdot \sqrt{2a} + 2 \cdot 5a \cdot \sqrt{2a}$$

$$= 6a\sqrt{2a} + 10a\sqrt{2a}$$

$$= 16a\sqrt{2a}$$

(*Note:* If you are unsure of the last step, show the use of the distributive property in performing the addition.)

$$6a\sqrt{2a} + 10a\sqrt{2a} = (6 + 10)a\sqrt{2a}$$

$$= 16a\sqrt{2a}$$

Practice Problem 6   Combine. $2a\sqrt{12a} + 3\sqrt{27a^3}$

Verbal and Writing Skills

1. List the steps involved in simplifying an expression such as $\sqrt{75x} + \sqrt{48x} + \sqrt{16x^2}$.

2. Julio said that, since $\sqrt{8}$ and $\sqrt{2}$ are not like radicals, $\sqrt{8} + \sqrt{2}$ cannot be simplified. What do you think?

Combine, if possible. Do not use a calculator or a table of square roots.

3. $3\sqrt{5} - \sqrt{5} + 4\sqrt{5}$

4. $\sqrt{7} + 2\sqrt{7} - 6\sqrt{7}$

5. $\sqrt{2} + 8\sqrt{3} - 5\sqrt{3} + 4\sqrt{2}$

6. $\sqrt{5} - \sqrt{6} + 3\sqrt{5} - 2\sqrt{6}$

7. $\sqrt{5} + \sqrt{45}$

8. $\sqrt{7} - \sqrt{63}$

9. $\sqrt{80} + 4\sqrt{20}$

10. $\sqrt{44} + 2\sqrt{99}$

11. $2\sqrt{8} - 3\sqrt{2}$

12. $2\sqrt{27} - 4\sqrt{3}$

13. $-\sqrt{2} + \sqrt{18} + \sqrt{98}$

14. $-\sqrt{54} + \sqrt{24} + 5\sqrt{6}$

15. $\sqrt{25} + \sqrt{72} + 3\sqrt{12}$

16. $5\sqrt{24} + \sqrt{81} + 3\sqrt{18}$

17. $2\sqrt{48} - 3\sqrt{8} + \sqrt{50}$

18. $\sqrt{50} + \sqrt{28} - \sqrt{9}$

19. $8\sqrt{2x} + \sqrt{50x}$

20. $6\sqrt{5y} - \sqrt{20y}$

21. $1.2\sqrt{3x} - 0.5\sqrt{12x}$

22. $-1.5\sqrt{2a} + 0.2\sqrt{8a}$

23. $\sqrt{20y} + 2\sqrt{45y} - \sqrt{5y}$

24. $\sqrt{72w} - 3\sqrt{2w} - \sqrt{50w}$

Mixed Practice

Combine if possible. Do not use a calculator or a table of square roots.

25. $\sqrt{50} - 3\sqrt{8}$

26. $\sqrt{75} - 5\sqrt{27}$

27. $3\sqrt{28x} - 5x\sqrt{63x}$

28. $-2b\sqrt{45b} + 5\sqrt{80b}$

29. $5\sqrt{8x^3} - 3x\sqrt{50x}$

30. $-2\sqrt{27y^3} + y\sqrt{12y}$

31. $5x\sqrt{48} - 2x\sqrt{75}$

32. $2x\sqrt{45} - 7x\sqrt{80}$

33. $2\sqrt{6y^3} - 2y\sqrt{54}$

34. $3\sqrt{20x^3} - 4x\sqrt{5x^2}$

35. $5x\sqrt{8x} - 24\sqrt{50x^3}$

36. $2x\sqrt{50x} - 3\sqrt{18x^3}$

Applications

▲ **37.** *Mars Mission* A planned unmanned mission to Mars may possibly require a small vehicle to make scientific measurements as it travels along the perimeter of a region like the one shown below. If the region is rectangular and the dimensions are as labeled, find the exact distance to be traveled by the vehicle in one trip around the perimeter.

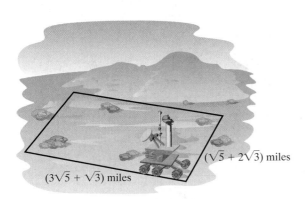

$(\sqrt{5} + 2\sqrt{3})$ miles

$(3\sqrt{5} + \sqrt{3})$ miles

▲ **38.** *Mars Mission* Suppose that on the second day of the mission to Mars, the vehicle must travel along the perimeter of a right triangle with dimensions as labeled below. Find the exact distance to be traveled by the vehicle in one trip around the perimeter.

$\sqrt{14}$ miles

$(2 - \sqrt{3})$ miles

$(2 + \sqrt{3})$ miles

39. *Underwater Mapping* A new underwater submersible is off the Great Barrier Reef on an underwater mapping mission. As the submersible approaches an underwater mountain, its mapping equipment estimates the height of the mountain as $42\sqrt{3} + 16\sqrt{2}$ meters. To assure that the sub does not hit the underwater mountain, the submersible controllers order it to rise $50\sqrt{3} + 70\sqrt{2}$ meters. If the estimates of the height were correct, how far above the mountain does the sub pass?

▲ **40.** *Underwater Search* The submersible has come upon a debris field, quite possibly from the wreck of an ancient Spanish galleon. The controllers program the sub to perform a search in a rectangular grid pattern, which is $23\sqrt{7} + 3\sqrt{11}$ meters long by $2\sqrt{7} + 18\sqrt{11}$ meters wide. Find the exact perimeter of the search grid.

Cumulative Review

41. Solve and graph on a number line. $7 - 3x \le 11$

42. Graph on a coordinate system. $3y - 2x \le 6$

43. *Music Store Purchases* Enrico went to the local music store and bought 10 CDs for $107.90. Some of the CDs were $7.99 and some were $11.99. How many of each did Enrico buy?

44. *Personal Investments* Shelly invested $8000 in two places. She invested in a certificate of deposit with a return of 7% simple interest, and she invested the rest of the money at 5% simple interest in a money market fund. At the end of the year she earned a total of $432 from these two places. How much did she invest in each place?

Student Learning Objectives

After studying this section, you will be able to:

1 Multiply monomial radical expressions.

2 Multiply a monomial radical expression by a polynomial.

3 Multiply two polynomial radical expressions.

NOTE TO STUDENT: Fully worked-out solutions to all of the Practice Problems can be found at the back of the text starting at page SP-1

1 **Multiplying Monomial Radical Expressions**

Recall the basic rule for multiplication of square root radicals,

$$\sqrt{a}\sqrt{b} = \sqrt{ab}.$$

We will use this concept to multiply square root radical expressions. Note that the direction "multiply" means to express the product in simplest form.

EXAMPLE 1 Multiply. $\sqrt{7}\sqrt{14x}$

Solution $\quad \sqrt{7}\sqrt{14x} = \sqrt{98x}$

We do *not* stop here, because the radical $\sqrt{98x}$ can be simplified.

$$\sqrt{98x} = \sqrt{49 \cdot 2x} = \sqrt{49}\sqrt{2x} = 7\sqrt{2x}$$

Practice Problem 1 Multiply. $\sqrt{3a}\sqrt{6a}$

EXAMPLE 2 Multiply.

(a) $\left(2\sqrt{3}\right)\left(5\sqrt{7}\right)$ **(b)** $\left(2a\sqrt{3a}\right)\left(3\sqrt{6a}\right)$

Solution

(a) $\left(2\sqrt{3}\right)\left(5\sqrt{7}\right) = 10\sqrt{21}$ $\quad$ Multiply the coefficients: $(2)(5) = 10$.
$\qquad\qquad\qquad\qquad\qquad\qquad$ Multiply the radicals: $\sqrt{3}\sqrt{7} = \sqrt{21}$.

(b) $\left(2a\sqrt{3a}\right)\left(3\sqrt{6a}\right) = 6a\sqrt{18a^2}$

$\qquad\qquad\qquad\qquad\quad = 6a\sqrt{9}\sqrt{2}\sqrt{a^2}$ $\quad$ Simplify $\sqrt{18a^2}$.

$\qquad\qquad\qquad\qquad\quad = 18a^2\sqrt{2}$ $\qquad$ Multiply $(6a)(3)(a)$.

Practice Problem 2 Multiply.

(a) $\left(2\sqrt{3}\right)\left(5\sqrt{5}\right)$ **(b)** $\left(4\sqrt{3x}\right)\left(2x\sqrt{6x}\right)$

EXAMPLE 3 Find the area of a farm field that measures $\sqrt{9400}$ feet long and $\sqrt{4800}$ feet wide. Express your answer as a simplified radical.

Solution We multiply.

$$\left(\sqrt{9400}\right)\left(\sqrt{4800}\right) = \left(10\sqrt{94}\right)\left(40\sqrt{3}\right)$$

$$= 400\sqrt{282} \text{ square feet.}$$

Practice Problem 3 Find the area of a computer chip that measures $\sqrt{180}$ millimeters long and $\sqrt{150}$ millimeters wide.

Multiplying a Monomial Radical Expression by a Polynomial

Recall that a binomial consists of two terms. $\sqrt{2} + \sqrt{5}$ is a binomial. We use the distributive property when multiplying a binomial by another factor.

EXAMPLE 4   Multiply and simplify. $\sqrt{5}\left(\sqrt{2} + 3\sqrt{7}\right)$

Solution
$$= \sqrt{5}\sqrt{2} + 3\sqrt{5}\sqrt{7}$$
$$= \sqrt{10} + 3\sqrt{35}$$

In similar fashion, we can multiply a trinomial by another factor.

Practice Problem 4   Multiply and simplify. $\sqrt{6}\left(\sqrt{3} + 2\sqrt{8}\right)$

We discover that $\sqrt{5} \cdot \sqrt{5} = 5$ and $\sqrt{3} \cdot \sqrt{3} = 3$. These are specific cases of the multiplication rule for square root radicals.

> For any nonnegative real number a,
> $$\sqrt{a} \cdot \sqrt{a} = a.$$

EXAMPLE 5   Multiply. $\sqrt{a}\left(3\sqrt{a} - 2\sqrt{5}\right)$

Solution
$$= 3\sqrt{a}\sqrt{a} - 2\sqrt{5}\sqrt{a} \quad \text{Use the distributive property.}$$
$$= 3a - 2\sqrt{5a}$$

Practice Problem 5   Multiply. $2\sqrt{x}\left(4\sqrt{x} - x\sqrt{2}\right)$

Be sure to simplify all radicals after multiplying.

 Multiplying Two Polynomial Radical Expressions

Recall the FOIL method used to multiply two binomials. The same method can be used for radical expressions.

Algebraic Expressions
$$(2x + y)(x - 2y) = 2x^2 - 4xy + xy - 2y^2$$
$$= 2x^2 - 3xy - 2y^2$$

Radical Expressions
$$\left(2\sqrt{3} + \sqrt{5}\right)\left(\sqrt{3} - 2\sqrt{5}\right) = 2\sqrt{9} - 4\sqrt{15} + \sqrt{15} - 2\sqrt{25}$$
$$= 2(3) - 3\sqrt{15} - 2(5)$$
$$= 6 - 3\sqrt{15} - 10$$
$$= -4 - 3\sqrt{15}$$

Let's look at this procedure more closely.

EXAMPLE 6 Multiply. $\left(\sqrt{2} + 5\right)\left(\sqrt{2} - 3\right)$

Solution

$$\left(\sqrt{2} + 5\right)\left(\sqrt{2} - 3\right)$$

$= \sqrt{4} - 3\sqrt{2} + 5\sqrt{2} - 15$ Multiply to obtain the four products.

$= 2 + 2\sqrt{2} - 15$ Combine the middle terms.

$= -13 + 2\sqrt{2}$

Practice Problem 6 Multiply. $\left(\sqrt{2} + \sqrt{6}\right)\left(2\sqrt{2} - \sqrt{6}\right)$

Be especially watchful if there is a number outside the radical sign. You must multiply the numerical coefficients very carefully.

EXAMPLE 7 Multiply. $\left(\sqrt{2} - 3\sqrt{6}\right)\left(\sqrt{2} + \sqrt{6}\right)$

Solution

$$\left(\sqrt{2} - 3\sqrt{6}\right)\left(\sqrt{2} + \sqrt{6}\right)$$

$= \sqrt{4} + \sqrt{12} - 3\sqrt{12} - 3\sqrt{36}$ Multiply to obtain the four products.

$= 2 - 2\sqrt{12} - 18$ Combine the middle terms.

$= -16 - 4\sqrt{3}$ Simplify $-2\sqrt{12}$.

NOTE TO STUDENT: Fully worked-out solutions to all of the Practice Problems can be found at the back of the text starting at page SP-1

Practice Problem 7 Multiply. $\left(\sqrt{6} + \sqrt{5}\right)\left(\sqrt{2} + 2\sqrt{5}\right)$

If an expression with radicals is squared, write the expression as the product of two binomials and multiply.

EXAMPLE 8 Multiply. $\left(2\sqrt{3} - \sqrt{6}\right)^2$

Solution

$= \left(2\sqrt{3} - \sqrt{6}\right)\left(2\sqrt{3} - \sqrt{6}\right)$

$= 4\sqrt{9} - 2\sqrt{18} - 2\sqrt{18} + \sqrt{36}$

$= 12 - 4\sqrt{18} + 6$

$= 18 - 12\sqrt{2}$

Practice Problem 8 Multiply. $\left(3\sqrt{5} - \sqrt{10}\right)^2$

Multiply. Be sure to simplify any radicals in your answer. Do not use a calculator or a table of square roots.

1. $\sqrt{3}\sqrt{10}$

2. $\sqrt{5}\sqrt{7}$

3. $\sqrt{2}\sqrt{22}$

4. $\sqrt{3}\sqrt{15}$

5. $\sqrt{18}\sqrt{3}$

6. $\sqrt{24}\sqrt{2}$

7. $(3\sqrt{5})(2\sqrt{6})$

8. $(2\sqrt{7})(3\sqrt{3})$

9. $\sqrt{5}\sqrt{10a}$

10. $\sqrt{3b}\sqrt{12}$

11. $(2\sqrt{14x})(3\sqrt{7x})$

12. $(3\sqrt{10x})(4\sqrt{5x})$

13. $(-4\sqrt{8a})(-2\sqrt{3a})$

14. $(4\sqrt{12a})(2\sqrt{6a})$

15. $(-3\sqrt{ab})(2\sqrt{b})$

16. $(-2\sqrt{x})(5\sqrt{xy})$

17. $\sqrt{5}(\sqrt{2} + \sqrt{3})$

18. $\sqrt{10}(\sqrt{2} + \sqrt{3})$

19. $\sqrt{3}(2\sqrt{6} + 5\sqrt{15})$

20. $\sqrt{6}(5\sqrt{12} - 4\sqrt{3})$

21. $2\sqrt{x}(\sqrt{x} - 8\sqrt{5})$

22. $-3\sqrt{b}(2\sqrt{a} + 3\sqrt{b})$

23. $\sqrt{6}(\sqrt{2} - 3\sqrt{6} + 2\sqrt{10})$

24. $\sqrt{10}(\sqrt{5} - 3\sqrt{10} + 5\sqrt{2})$

25. $(2\sqrt{3} + \sqrt{6})(\sqrt{3} - 2\sqrt{6})$

26. $(3\sqrt{5} + \sqrt{3})(\sqrt{5} + \sqrt{3})$

27. $(5 + 3\sqrt{2})(3 + \sqrt{2})$

28. $(7 - 2\sqrt{3})(4 + \sqrt{3})$

29. $(2\sqrt{7} - 3\sqrt{3})(\sqrt{7} + \sqrt{3})$

30. $(5\sqrt{6} - \sqrt{2})(\sqrt{6} + 3\sqrt{2})$

31. $(\sqrt{3} + 2\sqrt{6})(2\sqrt{3} - \sqrt{6})$

32. $(\sqrt{2} + 3\sqrt{10})(2\sqrt{2} - \sqrt{10})$

33. $(3\sqrt{7} - \sqrt{8})(\sqrt{8} + 2\sqrt{7})$

34. $(\sqrt{12} - \sqrt{5})(\sqrt{5} + 2\sqrt{12})$

35. $(2\sqrt{5} - 3)^2$

36. $(3\sqrt{2} + 5)^2$

37. $(\sqrt{3} + 5\sqrt{2})^2$

38. $(\sqrt{3} - 2\sqrt{7})^2$

39. $(3\sqrt{5} - \sqrt{3})^2$

40. $(2\sqrt{3} + 5\sqrt{5})^2$

Mixed Practice

Multiply. Be sure to simplify any radicals in your answer.

41. $(3\sqrt{5})(2\sqrt{8})$

42. $(2\sqrt{6})(3\sqrt{18})$

43. $\sqrt{7}(\sqrt{2} + 3\sqrt{14} - \sqrt{6})$

44. $\sqrt{5}(\sqrt{3} + 2\sqrt{10} - \sqrt{7})$

45. $(5\sqrt{2} - 3\sqrt{7})(3\sqrt{2} + \sqrt{7})$

46. $(6\sqrt{5} + 7\sqrt{6})(2\sqrt{5} - \sqrt{6})$

47. $2\sqrt{a}(3\sqrt{b} + \sqrt{ab} - 2\sqrt{a})$

48. $3\sqrt{x}(\sqrt{y} + 2\sqrt{xy} - 4\sqrt{x})$

Multiply the following binomials. Be sure to simplify your answer. Do not use a calculator or a table of square roots.

49. $(3x\sqrt{y} + \sqrt{5})(3x\sqrt{y} - \sqrt{5})$

50. $(5a\sqrt{3} - 3\sqrt{2})(5a\sqrt{3} + 3\sqrt{2})$

To Think About

51. (a) If a can be any number, can you think of an example of when it is not true that $\sqrt{a} \cdot \sqrt{a} = a$?

(b) Why is it necessary to state the restriction "for any nonnegative real number a"?

52. Hank was trying to evaluate the expression $\left(\sqrt{a} + \sqrt{b}\right)^2$. He wrote $\left(\sqrt{a} + \sqrt{b}\right)^2 = \left(\sqrt{a}\right)^2 + \left(\sqrt{b}\right)^2 = a + b$. This was a serious error! What is wrong with his work? How should he have done the problem?

Applications

Ramp Construction *A wheelchair access ramp is constructed with the dimensions as listed on the diagram. The gray shaded triangle is a right triangle.*

▲ **53.** Find the area of the shaded portion of the wheelchair ramp.

▲ **54.** Find the perimeter of the shaded portion of the wheelchair ramp. Leave your answer as a radical in simplified form.

$\sqrt{290}$ ft

$\sqrt{2}$ feet

$\sqrt{288}$ feet

Aquarium Construction *The architect for the new science building at Devers Prep School designed a space to place a built-in aquarium as the centerpiece of the lobby. The width and height of the front of the aquarium are as shown.*

▲ **55.** Find the viewing area of the front of the aquarium.

▲ **56.** If a specially designed wooden frame is to be placed around the perimeter of the front face of the aquarium, how much wood is needed?

$\sqrt{12}$ feet

$\sqrt{27}$ feet

Cumulative Review

57. Factor. $36x^2 - 49y^2$

58. Factor. $6x^2 + 11x - 10$

59. ***Speed in Knots*** Ships often measure their speed in knots (or nautical miles per hour). The regular mile (statute mile) that we are more familiar with is 5280 feet. A nautical mile is about 6076 feet. Write a formula of the form $m = ak$, where m is the number of statute miles, k is the number of nautical miles, and a is a constant number (in this case a fraction). Use the formula to find out how fast a Coast Guard Cutter is going in miles per hour if it is traveling at 35 knots.

60. ***Car Insurance*** Phil and Melissa LaBelle found that the collision and theft portion of their car insurance could be reduced by 15% if they purchased an auto security device. They purchased and had such a device installed for $350. The yearly collision and theft portion of their car insurance is $280. How many years will it take for the security device to pay for itself?

How are you doing with your homework assignments in Sections 9.1 to 9.4? Do you feel you have mastered the material so far? Do you understand the concepts you have covered? Before you go further in the textbook, take some time to do each of the following problems.

Assume all variables are positive.

9.1

Find the square roots, if possible. (If it is impossible to obtain a real number for an answer, say so.) Approximate to the nearest thousandth.

1. $\sqrt{49}$ **2.** $\sqrt{\dfrac{25}{64}}$ **3.** $\sqrt{81}$

4. $-\sqrt{100}$ **5.** $\sqrt{5}$ **6.** $-\sqrt{121}$

9.2

Simplify.

7. $\sqrt{0.04}$ **8.** $\sqrt{0.36}$ **9.** $\sqrt{x^4}$

10. $\sqrt{25x^4y^6}$ **11.** $\sqrt{50}$ **12.** $\sqrt{x^5}$

13. $\sqrt{36x^3}$ **14.** $\sqrt{8a^4b^3}$

9.3

Combine, if possible.

15. $\sqrt{98} + \sqrt{128}$ **16.** $3\sqrt{2} - \sqrt{8} + \sqrt{18}$

17. $2\sqrt{28} + 3\sqrt{63} - \sqrt{49}$ **18.** $3\sqrt{8} + \sqrt{12} + \sqrt{50} - 4\sqrt{75}$

19. $5\sqrt{2x} - 7\sqrt{8x}$ **20.** $-4x\sqrt{75} - 2x\sqrt{27}$

9.4

Multiply. Simplify your answer.

21. $\left(3\sqrt{5}\right)\left(2\sqrt{10}\right)$ **22.** $\left(5x\sqrt{7}\right)\left(3x\sqrt{14}\right)$

23. $\sqrt{2}\left(\sqrt{6} + 3\sqrt{2}\right)$ **24.** $\left(\sqrt{7} + 4\right)^2$

25. $\left(\sqrt{11} - \sqrt{10}\right)\left(\sqrt{11} + \sqrt{10}\right)$ **26.** $\left(2\sqrt{3} - \sqrt{5}\right)^2$

27. $\left(\sqrt{x} + 3\right)\left(\sqrt{x} - 4\right)$

28. $\left(3\sqrt{7} + 4\sqrt{3}\right)\left(5\sqrt{7} - 2\sqrt{3}\right)$

Now turn to page SA-28 for the answer to each of these problems. Each answer also includes a reference to the objective in which the problem is first taught. If you missed any of these problems, you should stop and review the Examples and Practice Problems in the referenced objective. A little review now will help you master the material in the upcoming sections of the text.

1. _____

2. _____

3. _____

4. _____

5. _____

6. _____

7. _____

8. _____

9. _____

10. _____

11. _____

12. _____

13. _____

14. _____

15. _____

16. _____

17. _____

18. _____

19. _____

20. _____

21. _____

22. _____

23. _____

24. _____

25. _____

26. _____

27. _____

28. _____

1 Using the Quotient Rule for Square Roots to Simplify a Fraction Involving Radicals

Just as there is a multiplication rule for square roots, there is a quotient rule for square roots. The rules are similar.

QUOTIENT RULE FOR SQUARE ROOTS

For all positive numbers a and b,

$$\frac{\sqrt{a}}{\sqrt{b}} = \sqrt{\frac{a}{b}} \quad \text{and} \quad \sqrt{\frac{a}{b}} = \frac{\sqrt{a}}{\sqrt{b}}.$$

This can be used to divide square root radicals or to simplify square root radical expressions involving division. We will be using both parts of the quotient rule.

EXAMPLE 1 Simplify. **(a)** $\dfrac{\sqrt{75}}{\sqrt{3}}$ **(b)** $\sqrt{\dfrac{25}{36}}$

Solution

(a) We notice that 3 is a factor of 75. We use the quotient rule to rewrite the expression.

$$\frac{\sqrt{75}}{\sqrt{3}} = \sqrt{\frac{75}{3}}$$
$$= \sqrt{25} \quad \text{Divide.}$$
$$= 5 \quad \text{Simplify.}$$

(b) Since both 25 and 36 are perfect squares, we will rewrite this as the quotient of square roots.

$$\sqrt{\frac{25}{36}} = \frac{\sqrt{25}}{\sqrt{36}}$$
$$= \frac{5}{6}$$

Practice Problem 1 Simplify. **(a)** $\dfrac{\sqrt{98}}{\sqrt{2}}$ **(b)** $\sqrt{\dfrac{81}{100}}$

EXAMPLE 2 Simplify. $\sqrt{\dfrac{20}{x^6}}$

Solution

$$\sqrt{\frac{20}{x^6}} = \frac{\sqrt{20}}{\sqrt{x^6}} = \frac{\sqrt{4}\sqrt{5}}{x^3} = \frac{2\sqrt{5}}{x^3} \quad \text{Don't forget to simplify } \sqrt{20} \text{ as } 2\sqrt{5}.$$

Practice Problem 2 Simplify. $\sqrt{\dfrac{50}{a^4}}$

② Rationalizing the Denominator of a Fraction with a Square Root in the Denominator

Sometimes, when calculating with fractions that contain radicals, it is advantageous to have an integer in the denominator. If a fraction has a radical in the denominator, we **rationalize the denominator.** That is, we multiply to change the fraction to an equivalent one that has an integer in the denominator. Remember, we do not want to change the value of the fraction. Thus we will multiply the numerator and the denominator by the same number.

EXAMPLE 3   Simplify. $\dfrac{3}{\sqrt{2}}$

Solution   Think, "What times 2 will make a perfect square?"

$$\frac{3}{\sqrt{2}} = \frac{3}{\sqrt{2}} \times 1 = \frac{3}{\sqrt{2}} \times \frac{\sqrt{2}}{\sqrt{2}} = \frac{3\sqrt{2}}{\sqrt{4}} = \frac{3\sqrt{2}}{2}$$

Practice Problem 3   Simplify. $\dfrac{9}{\sqrt{7}}$

(Note that a fraction is not simplified unless the denominator is rationalized.)

When rationalizing a denominator containing a square root radical, we will want to use the smallest possible radical that will yield the square root of a perfect square. Very often we will not use the radical that is in the denominator.

EXAMPLE 4   Simplify. **(a)** $\dfrac{\sqrt{7}}{\sqrt{8}}$ **(b)** $\dfrac{3}{\sqrt{x^3}}$

Solution

(a) Think, "What times 8 will make a perfect square?"

$$\frac{\sqrt{7}}{\sqrt{8}} = \frac{\sqrt{7}}{\sqrt{8}} \times \frac{\sqrt{2}}{\sqrt{2}} = \frac{\sqrt{14}}{\sqrt{16}} = \frac{\sqrt{14}}{4}$$

(b) Think, "What times x^3 will give an even exponent?"

$$\frac{3}{\sqrt{x^3}} \times \frac{\sqrt{x}}{\sqrt{x}} = \frac{3\sqrt{x}}{\sqrt{x^4}} = \frac{3\sqrt{x}}{x^2}$$

We could have simplified $\sqrt{x^3}$ before we rationalized the denominator.

$$\frac{3}{\sqrt{x^3}} = \frac{3}{x\sqrt{x}} \times \frac{\sqrt{x}}{\sqrt{x}} = \frac{3\sqrt{x}}{x\sqrt{x^2}} = \frac{3\sqrt{x}}{(x)(x)} = \frac{3\sqrt{x}}{x^2}$$

Practice Problem 4   Simplify. **(a)** $\dfrac{\sqrt{2}}{\sqrt{12}}$ **(b)** $\dfrac{6a}{\sqrt{a^7}}$

EXAMPLE 5 Simplify. $\dfrac{\sqrt{2}}{\sqrt{27x}}$

Solution Since it is not apparent what we should multiply 27 by to obtain a perfect square, we will begin by simplifying the denominator.

$$\frac{\sqrt{2}}{\sqrt{27x}} = \frac{\sqrt{2}}{3\sqrt{3x}}$$

Now it is easy to see that we multiply by $\dfrac{\sqrt{3x}}{\sqrt{3x}}$ to rationalize the denominator.

$$= \frac{\sqrt{2}}{3\sqrt{3x}} \times \frac{\sqrt{3x}}{\sqrt{3x}}$$

$$= \frac{\sqrt{6x}}{3\sqrt{9x^2}}$$

$$= \frac{\sqrt{6x}}{9x}$$

NOTE TO STUDENT: Fully worked-out solutions to all of the Practice Problems can be found at the back of the text starting at page SP-1

Practice Problem 5 Simplify. $\dfrac{\sqrt{2x}}{\sqrt{8x}}$

③ Rationalizing the Denominator of a Fraction with a Binomial Denominator Containing at Least One Square Root

Sometimes the denominator of a fraction is a binomial with a square root radical term. $\dfrac{1}{5 - 3\sqrt{2}}$ is such a fraction. How can we eliminate the radical term in the denominator? Recall that when we multiply $(a + b)(a - b)$, we obtain the square of a minus the square of b: $a^2 - b^2$. For example, $(x + 3)(x - 3) = x^2 - 9$. The resulting terms are perfect squares. We can use this idea to eliminate the radical in $5 - 3\sqrt{2}$.

$$\left(5 - 3\sqrt{2}\right)\left(5 + 3\sqrt{2}\right) = 5^2 + 15\sqrt{2} - 15\sqrt{2} - \left(3\sqrt{2}\right)^2 = 5^2 - \left(3\sqrt{2}\right)^2 = 25 - 18 = 7$$

Expressions like $\left(5 - 3\sqrt{2}\right)$ and $\left(5 + 3\sqrt{2}\right)$ are called **conjugates.**

EXAMPLE 6 Simplify. **(a)** $\dfrac{2}{\sqrt{3} - 4}$ **(b)** $\dfrac{\sqrt{x}}{\sqrt{5} + \sqrt{3}}$

Solution

(a) The conjugate of $\sqrt{3} - 4$ is $\sqrt{3} + 4$.

$$\frac{2}{\left(\sqrt{3} - 4\right)} \cdot \frac{\left(\sqrt{3} + 4\right)}{\left(\sqrt{3} + 4\right)} = \frac{2\sqrt{3} + 8}{\left(\sqrt{3}\right)^2 + 4\sqrt{3} - 4\sqrt{3} - 4^2}$$

$$= \frac{2\sqrt{3} + 8}{3 - 16}$$

$$= \frac{2\sqrt{3} + 8}{-13}$$

$$= -\frac{2\sqrt{3} + 8}{13}$$

(b) The conjugate of $\sqrt{5} + \sqrt{3}$ is $\sqrt{5} - \sqrt{3}$.

$$\frac{\sqrt{x}}{\left(\sqrt{5} + \sqrt{3}\right)} \cdot \frac{\left(\sqrt{5} - \sqrt{3}\right)}{\left(\sqrt{5} - \sqrt{3}\right)} = \frac{\sqrt{5x} - \sqrt{3x}}{\left(\sqrt{5}\right)^2 - \sqrt{15} + \sqrt{15} - \left(\sqrt{3}\right)^2}$$

$$= \frac{\sqrt{5x} - \sqrt{3x}}{5 - 3}$$

$$= \frac{\sqrt{5x} - \sqrt{3x}}{2}$$

Be careful not to combine $\sqrt{5x}$ and $\sqrt{3x}$ in the numerator. They are not like radicals.

Practice Problem 6   Simplify.

(a) $\dfrac{4}{\sqrt{3} + \sqrt{5}}$ **(b)** $\dfrac{\sqrt{a}}{\sqrt{10} - 3}$

Note that in Example 7 the numerator is a binomial.

EXAMPLE 7   Rationalize the denominator. $\dfrac{\sqrt{3} + \sqrt{2}}{\sqrt{3} - \sqrt{2}}$

Solution   The conjugate of $\sqrt{3} - \sqrt{2}$ is $\sqrt{3} + \sqrt{2}$.

$$\frac{\left(\sqrt{3} + \sqrt{2}\right)}{\left(\sqrt{3} - \sqrt{2}\right)} \cdot \frac{\left(\sqrt{3} + \sqrt{2}\right)}{\left(\sqrt{3} + \sqrt{2}\right)}$$

$$= \frac{\sqrt{9} + \sqrt{6} + \sqrt{6} + \sqrt{4}}{\left(\sqrt{3}\right)^2 - \left(\sqrt{2}\right)^2} \qquad \text{Multiply.}$$

$$= \frac{3 + 2\sqrt{6} + 2}{3 - 2} \qquad \text{Simplify and combine like terms.}$$

$$= \frac{5 + 2\sqrt{6}}{1}$$

$$= 5 + 2\sqrt{6}$$

Practice Problem 7   Rationalize the denominator. $\dfrac{\sqrt{3} + \sqrt{5}}{\sqrt{3} - \sqrt{5}}$

Note that in Examples 6 and 7 we used a raised dot $\cdot$ to indicate multiplication instead of the $\times$. You need to be familiar with both forms of multiplication notation.

Simplify. Be sure to rationalize all denominators. Do not use a calculator or a table of square roots.

1. $\dfrac{\sqrt{12}}{\sqrt{3}}$ **2.** $\dfrac{\sqrt{3}}{\sqrt{27}}$ **3.** $\dfrac{\sqrt{7}}{\sqrt{63}}$ **4.** $\dfrac{\sqrt{52}}{\sqrt{13}}$ **5.** $\dfrac{\sqrt{98}}{\sqrt{2}}$ **6.** $\dfrac{\sqrt{5}}{\sqrt{80}}$

7. $\dfrac{\sqrt{6}}{\sqrt{x^4}}$ **8.** $\dfrac{\sqrt{12}}{\sqrt{x^2}}$ **9.** $\dfrac{\sqrt{18}}{\sqrt{a^4}}$ **10.** $\dfrac{\sqrt{24}}{\sqrt{b^6}}$ **11.** $\dfrac{3}{\sqrt{7}}$ **12.** $\dfrac{2}{\sqrt{10}}$

13. $\dfrac{5}{\sqrt{13}}$ **14.** $\dfrac{6}{\sqrt{17}}$ **15.** $\dfrac{x\sqrt{x}}{\sqrt{2}}$ **16.** $\dfrac{\sqrt{3y}}{\sqrt{6}}$ **17.** $\dfrac{\sqrt{8}}{\sqrt{x}}$ **18.** $\dfrac{\sqrt{18}}{\sqrt{y}}$

19. $\dfrac{3}{\sqrt{12}}$ **20.** $\dfrac{4}{\sqrt{20}}$ **21.** $\dfrac{6}{\sqrt{a}}$ **22.** $\dfrac{8}{\sqrt{b}}$ **23.** $\dfrac{x}{\sqrt{2x^5}}$ **24.** $\dfrac{y}{\sqrt{5x^3}}$

25. $\dfrac{\sqrt{18}}{\sqrt{2x^3}}$ **26.** $\dfrac{\sqrt{24}}{\sqrt{6x^7}}$ **27.** $\sqrt{\dfrac{3}{5}}$ **28.** $\sqrt{\dfrac{7}{11}}$ **29.** $\sqrt{\dfrac{13}{14}}$ **30.** $\sqrt{\dfrac{15}{17}}$

31. $\dfrac{9}{\sqrt{32x}}$ **32.** $\dfrac{3}{\sqrt{50x}}$

New Directions

Rationalize the denominator. Simplify your answer. Do not use a calculator or a table of square roots.

33. $\dfrac{4}{\sqrt{3}-1}$ **34.** $\dfrac{5}{\sqrt{5}+1}$ **35.** $\dfrac{6}{\sqrt{5}+\sqrt{2}}$ **36.** $\dfrac{5}{\sqrt{11}-\sqrt{6}}$ **37.** $\dfrac{\sqrt{6}}{\sqrt{6}-\sqrt{3}}$

38. $\dfrac{\sqrt{3}}{\sqrt{5}+\sqrt{3}}$ **39.** $\dfrac{\sqrt{7}}{\sqrt{8}+\sqrt{7}}$ **40.** $\dfrac{\sqrt{5}}{\sqrt{5}+\sqrt{6}}$ **41.** $\dfrac{3x}{2\sqrt{2}-\sqrt{5}}$ **42.** $\dfrac{4x}{2\sqrt{7}+2\sqrt{6}}$

43. $\dfrac{\sqrt{x}}{\sqrt{6}+\sqrt{2}}$ **44.** $\dfrac{\sqrt{x}}{\sqrt{7}+\sqrt{2}}$ **45.** $\dfrac{\sqrt{5}-\sqrt{2}}{\sqrt{5}+\sqrt{2}}$ **46.** $\dfrac{\sqrt{7}+\sqrt{3}}{\sqrt{7}-\sqrt{3}}$

47. $\dfrac{4\sqrt{7}+3}{\sqrt{5}-\sqrt{2}}$ **48.** $\dfrac{4\sqrt{3}+2\sqrt{5}}{\sqrt{5}-\sqrt{3}}$ **49.** $\dfrac{4\sqrt{3}+2}{\sqrt{8}-\sqrt{6}}$

50. $\dfrac{3\sqrt{5}+4}{\sqrt{15}-\sqrt{3}}$ **51.** $\dfrac{x-25}{\sqrt{x}+5}$ **52.** $\dfrac{x-36}{\sqrt{x}-6}$

To Think About

53. **(a)** Multiply $\left(x + \sqrt{2}\right)$ by its conjugate.

(b) Use the result of part (a) to show that while $x^2 - 2$ cannot be factored by earlier methods, it can be factored using radicals.

(c) Factor using radicals. $x^2 - 12$

 54. When rationalized, $\dfrac{3}{\sqrt{11}}$ becomes $\dfrac{3\sqrt{11}}{11}$. Evaluate each expression with a calculator. What do you notice? Why?

Applications

▲ **55.** *Geometry* A pyramid has a volume V and a height h. The length of one side of the square base is $s = \sqrt{\dfrac{3V}{h}}$.

(a) Simplify this expression by rationalizing the denominator.

(b) Find the length of the side of a pyramid with a volume of 24 cubic feet and a height of 8 feet.

▲ **56.** *Geometry* A cylinder has a volume V and a height h. The radius of the cylinder is $r = \sqrt{\dfrac{V}{\pi h}}$.

(a) Simplify this expression by rationalizing the denominator.

(b) Find the radius of a cylinder of volume 81 cubic feet and height π feet.

▲ **57.** *Relay Tower* A reflector sheet for a microwave relay tower is shaped like a rectangle. The area of the rectangle is 3 square meters. Ideally the length should be exactly $\left(\sqrt{5} + 2\right)$ meters long. What should be the width of the rectangle? Express the answer exactly using radical expressions.

▲ **58.** *Computer Chip* A computer chip for a new computer is shaped like a rectangle. The area of the rectangle is 4 square millimeters. Ideally the width should be exactly $\left(\sqrt{5} - \sqrt{3}\right)$ millimeters. What should be the length of the rectangle? Express the answer exactly using radical expressions.

▲ **59.** *Relay Tower* An engineer needs the approximate dimensions of the rectangle in exercise 57. Approximate the length and the width of that rectangle to the nearest thousandth.

▲ **60.** *Computer Chip* An engineer needs the approximate dimensions of the computer chip in exercise 58. Approximate the length and the width of that computer chip to the nearest thousandth.

Cumulative Review

61. Solve for x (round to two decimal places). $3(2x - 6) = 5x - 7(x - 2) + 27$

62. Evaluate $3x^2 - x \div y + y^2 - y$ for $x = 12$ and $y = -3$.

63. *Caffeine in Coffee* Dr. Frank Day, a chemist, has determined that the types of percolated coffee he buys have 70% more caffeine per cup than the types of instant coffee he usually buys. He determined that the percolated coffee has 120 mg of caffeine per cup. Approximately how much caffeine is there per cup of instant coffee?

64. *Medical Insurance Costs* Wally and Mary Coyle are considering two family medical insurance plans that are quite similar. The Gold Plan costs $5000 per year and covers all doctor visits and all prescription drugs after a $50 deductible per doctor visit and a $30 deductible per prescription. The Master Plan costs $6000 per year and covers all doctor visits and all prescription drugs. Wally and Mary have discovered that half of all their doctor visits result in a written prescription. How many doctor visits per year would they need to average for the Master Plan to be more economical?

Student Learning Objectives

After studying this section, you will be able to:

 1 Use the Pythagorean theorem to solve applied problems.

 2 Solve radical equations.

1 Using the Pythagorean Theorem to Solve Applied Problems

In ancient Greece, mathematicians studied a number of properties of right triangles. They proved that the square of the longest side of a right triangle is equal to the sum of the squares of the other two sides. This property is called the Pythagorean theorem, in honor of the Greek mathematician Pythagoras (ca. 590 B.C.). The shorter sides, a and b, are referred to as the **legs** of the right triangle. The longest side, c, is called the **hypotenuse** of the right triangle. We state the theorem as follows.

> **PYTHAGOREAN THEOREM**
>
> In any right triangle, if c is the length of the hypotenuse and a and b are the lengths of the two legs, then
> $$c^2 = a^2 + b^2.$$

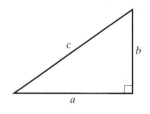

If we know any two sides of a right triangle, we can find the third side using this theorem.

Recall that in Section 9.1 we showed that if $s^2 = 4$ then $s = 2$ or -2. In general, if $x^2 = a$, then $x = \pm\sqrt{a}$. This is sometimes called "taking the square root of each side of an equation." The abbreviation $\pm\sqrt{a}$ means $+\sqrt{a}$ or $-\sqrt{a}$. It is read "plus or minus $\sqrt{a}$."

Notice that the hypotenuse is always opposite the right angle. The smallest side is opposite the smallest angle.

 EXAMPLE 1 The ramp to Tony Pitkin's barn rises 5 feet over a horizontal distance of 12 feet. How long is the ramp? That is, find the length of the hypotenuse of a right triangle whose legs are 5 feet and 12 feet.

Solution

1. ***Understand the problem.***
   Draw and label a diagram.

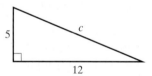

2. ***Write an equation.*** Write the Pythagorean Theorem. Substitute the known values into the equation.
$$c^2 = a^2 + b^2$$
$$c^2 = 5^2 + 12^2$$

3. ***Solve and state the answer.***
$$c^2 = 25 + 144$$
$$c^2 = 169 \qquad \text{Take the square root of each side of the equation.}$$
$$c = \pm\sqrt{169}$$
$$= \pm 13$$

The hypotenuse is 13 feet. We do not use -13 because length is not negative. Thus the length of the ramp is 13 feet.

4. ***Check.*** Substitute the values for a, b, and c in the Pythagorean Theorem and evaluate. Does $13^2 = 5^2 + 12^2$? Yes. The solution checks. ✓

NOTE TO STUDENT: Fully worked-out solutions to all of the Practice Problems can be found at the back of the text starting at page SP-1

Practice Problem 1 Find the length of the hypotenuse of a right triangle with legs of 9 centimeters and 12 centimeters.

Sometimes you will need to find one of the legs of a right triangle given the hypotenuse and the other leg.

 EXAMPLE 2   Find the unknown leg of a right triangle that has a hypotenuse of 10 yards and one leg of $\sqrt{19}$ yards.

Solution   Draw a diagram. The diagram is in the margin on the right.

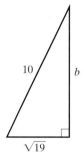

$$c^2 = a^2 + b^2 \qquad \text{Write the Pythagorean Theorem.}$$
$$10^2 = \left(\sqrt{19}\right)^2 + b^2 \quad \text{Substitute the known values into the equation.}$$
$$100 = 19 + b^2$$
$$81 = b^2$$
$$\pm 9 = b \qquad\qquad \text{Take the square root of each side of the equation.}$$

The leg is 9 yards.

▲ **Practice Problem 2**   The hypotenuse of a right triangle is $\sqrt{17}$ meters. One leg is 1 meter long. Find the length of the other leg.

Sometimes the answer to a problem will be an irrational number such as $\sqrt{33}$. It is best to leave the answer in radical form unless you are asked for an approximate answer. To find the approximate value, use a calculator or the square root table. $\sqrt{33}$ is an exact answer.

▲ **EXAMPLE 3**   A 25-foot ladder is placed against a building. The foot of the ladder is 8 feet from the wall. At approximately what height does the top of the ladder touch the building? Round to the nearest tenth.

Solution

$$c^2 = a^2 + b^2$$
$$25^2 = 8^2 + b^2$$
$$625 = 64 + b^2$$
$$561 = b^2$$
$$\pm\sqrt{561} = b$$

We want only the positive value for the distance, so $b = \sqrt{561}$. Using a calculator, we have 561 $\boxed{\sqrt{}}$ 23.68543856. Rounding, we obtain $b \approx 23.7$.

If you do not have a calculator, you will need to use the square root table in Appendix A. However, this square root table goes only as far as $\sqrt{200}$! To approximate larger radicals with the square root table, you will have to write the radical in a different fashion.

$b = \sqrt{3}\sqrt{187}$ Since $\sqrt{ab} = \sqrt{a}\sqrt{b}$ and $3 \cdot 187 = 561$.
$b \approx (1.732)(13.675)$ Replace each radical by a decimal approximation from the table.
$b \approx 23.7$ Multiply and round to the nearest tenth.

The ladder touches the building at a height of approximately 23.7 feet.

▲ **Practice Problem 3**   A support line is placed 3 meters away from the base of an 8-meter pole. If the support line is attached to the top of the pole and pulled tight (assume that it is a straight line), how long is the support line from the ground to the pole? Round to the nearest tenth.

2 Solving Radical Equations

A **radical equation** is an equation with a variable in one or more of the radicands. A square root radical equation can be simplified by squaring each side of the equation. The apparent solution to a radical equation *must* be checked by substitution into the *original* equation.

For some square root radical equations, you will need to *isolate the radical before* you square each side of the equation.

EXAMPLE 4   Solve and check. $1 + \sqrt{5x - 4} = 5$

Solution

$1 + \sqrt{5x - 4} = 5$	We want to isolate the radical first.
$\sqrt{5x - 4} = 4$	Subtract 1 from each side to isolate the radical.
$\left(\sqrt{5x - 4}\right)^2 = (4)^2$	Square each side.
$5x - 4 = 16$	Simplify and solve for x.
$5x = 20$	
$x = 4$	

Check.

$$1 + \sqrt{5(4) - 4} \overset{?}{=} 5$$
$$1 + \sqrt{20 - 4} \overset{?}{=} 5$$
$$1 + \sqrt{16} \overset{?}{=} 5$$
$$1 + 4 = 5 \ \checkmark$$

Thus 4 is the solution.

Practice Problem 4   Solve and check. $\sqrt{3x - 2} - 7 = 0$

NOTE TO STUDENT: Fully worked-out solutions to all of the Practice Problems can be found at the back of the text starting at page SP-1

Some problems will have an apparent solution, but that solution does not always check. An apparent solution that does not satisfy the original equation is called an **extraneous root.**

EXAMPLE 5   Solve and check. $\sqrt{x + 3} = -7$

Solution

$\sqrt{x + 3} = -7$	
$\left(\sqrt{x + 3}\right)^2 = (-7)^2$	Square each side.
$x + 3 = 49$	Simplify and solve for x.
$x = 46$	

Check.

$\sqrt{46 + 3} \overset{?}{=} -7$	
$\sqrt{49} \overset{?}{=} -7$	
$7 \neq -7!$	Does not check! The apparent solution is an extraneous root.

There is no solution to Example 5.

Practice Problem 5   Solve and check. $\sqrt{5x + 4} + 2 = 0$

Wait a minute! Is the original equation in Example 5 possible? No. $\sqrt{x+3} = -7$ is an impossible statement. We defined the $\sqrt{\ }$ symbol to mean the positive square root of the radicand. Thus, it cannot be equal to a negative number. If we had noticed this in the first step, we could have written down "no solution" immediately.

When you square both sides of an equation you may obtain a quadratic equation. In such cases, you may obtain two apparent solutions, and you must check both of them.

EXAMPLE 6   Solve and check. $\sqrt{3x+1} = x+1$.

Solution

$$\sqrt{3x+1} = x+1$$

$$\left(\sqrt{3x+1}\right)^2 = (x+1)^2 \qquad \text{Square each side.}$$

$$3x+1 = x^2 + 2x + 1 \quad \text{Simplify and set the equation equal to 0.}$$

$$0 = x^2 - x$$

$$0 = x(x-1) \qquad \text{Factor the quadratic equation.}$$

$$x = 0 \qquad x - 1 = 0 \qquad \text{Set each factor equal to 0 and solve.}$$

$$x = 1$$

Check.   If $x = 0$: $\sqrt{3(0)+1} \overset{?}{=} 0+1$ If $x = 1$: $\sqrt{3(1)+1} \overset{?}{=} 1+1$

$$\sqrt{0+1} \overset{?}{=} 0+1 \qquad\qquad\qquad \sqrt{3+1} \overset{?}{=} 2$$

$$\sqrt{1} \overset{?}{=} 1 \qquad\qquad\qquad\qquad \sqrt{4} \overset{?}{=} 2$$

$$1 = 1 \ \checkmark \qquad\qquad\qquad\qquad 2 = 2 \ \checkmark$$

Thus 0 and 1 are both solutions.

It is possible in these cases for both apparent solutions to check, for both to be extraneous, or for one to check and the other to be extraneous.

Practice Problem 6   Solve and check. $-2 + \sqrt{6x-1} = 3x - 2$

EXAMPLE 7   Solve and check. $\sqrt{2x-1} = x-2$

Solution

$$\sqrt{2x-1} = x-2$$

$$\left(\sqrt{2x-1}\right)^2 = (x-2)^2 \qquad \text{Square each side.}$$

$$2x - 1 = x^2 - 4x + 4$$

$$0 = x^2 - 6x + 5 \qquad \text{Simplify and set the equation to 0.}$$

$$0 = (x-5)(x-1) \quad \text{Solve for } x.$$

$$x - 5 = 0 \qquad x - 1 = 0$$

$$x = 5 \qquad\qquad x = 1$$

Check. If $x = 5$: $\sqrt{2(5) - 1} \overset{?}{=} 5 - 2$ If $x = 1$: $\sqrt{2(1) - 1} \overset{?}{=} 1 - 2$

$$\sqrt{10 - 1} \overset{?}{=} 3 \qquad\qquad \sqrt{2 - 1} \overset{?}{=} -1$$

$$\sqrt{9} \overset{?}{=} 3 \qquad\qquad\quad \sqrt{1} \overset{?}{=} -1$$

$$3 = 3 \; \checkmark \qquad\qquad\quad 1 \neq -1$$

It does not check. In this case 1 is an extraneous root.

Thus only 5 is a solution to this equation.

NOTE TO STUDENT: Fully worked-out solutions to all of the Practice Problems can be found at the back of the text starting at page SP-1

Practice Problem 7 Solve and check. $2 - x + \sqrt{x + 4} = 0$

Example 8 is somewhat different in format. Take a minute to see how it compares to Examples 6 and 7.

EXAMPLE 8   Solve and check. $\sqrt{3x + 3} = \sqrt{5x - 1}$

Solution   Here there are two radicals. Each radical is already isolated.

$$\sqrt{3x + 3} = \sqrt{5x - 1}$$

$$\left(\sqrt{3x + 3}\right)^2 = \left(\sqrt{5x - 1}\right)^2 \quad \text{Square each side.}$$

$$3x + 3 = 5x - 1 \qquad \text{Simplify and solve for } x.$$

$$3 = 2x - 1$$

$$4 = 2x$$

$$2 = x$$

Check. $\sqrt{3(2) + 3} \overset{?}{=} \sqrt{5(2) - 1}$

$$\sqrt{6 + 3} \overset{?}{=} \sqrt{10 - 1}$$

$$\sqrt{9} = \sqrt{9} \; \checkmark$$

Thus 2 is a solution.

Practice Problem 8 Solve and check. $\sqrt{2x + 1} = \sqrt{x - 10}$

9.6 EXERCISES

| Student Solutions Manual | CD/ Video | PH Math Tutor Center | MathXL®Tutorials on CD | MathXL® | MyMathLab® | Interactmath.com |

Use the Pythagorean Theorem to find the length of the third side of each right triangle. Leave any irrational answers in radical form.

▲ **1.**

▲ **2.**

▲ **3.** ▲ **4.**

 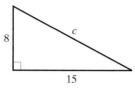

Exercises 5–14 refer to a right triangle with legs a and b and hypotenuse c. Find the exact length of the missing side.

▲ **5.** $a = 7; b = 7$
Find c.

▲ **6.** $a = 8; b = 4$
Find c.

▲ **7.** $a = \sqrt{14}, b = 7$
Find c.

▲ **8.** $a = \sqrt{21}, b = 5$
Find c.

▲ **9.** $c = 20, b = 18$
Find a.

▲ **10.** $c = 17, b = 16$
Find a.

▲ **11.** $c = \sqrt{82}, a = 5$
Find b.

▲ **12.** $a = \sqrt{5}, c = 7$
Find b.

 ▲ **13.** $c = 12.96, b = 8.35$
Find a to the nearest hundredth.

 ▲ **14.** $a = 7.61, b = 5.38$
Find c to the nearest hundredth.

Applications

Draw a diagram and use the Pythagorean Theorem to solve. Round to the nearest tenth.

▲ **15.** *Tenting* A guy rope is attached to the top of a tent pole. The tent pole is 8 feet tall. If the guy rope is pegged into the ground 5 feet from the tent, how long is the guy rope?

▲ **16.** *Baseball* A baseball diamond is a square. Each side of the square is 90 feet long. How far is it from home plate to second base? *Hint:* Draw the diagonal.

▲ **17.** *Flying a Kite* A kite is flying on a string that is 100 feet long and is fastened to the ground at the other end. Assume that the string is a straight line. How high is the kite if it is flying above a point that is 20 feet away from where it is fastened on the ground?

▲ **18.** *Boat Mooring* A boat must be moored 20 feet away from a dock. The boat will be 10 feet below the level of the dock at low tide. What is the minimum length of rope needed to go from the boat to the dock at low tide?

Radio Tower Construction *To estimate the cost of erecting a radio tower, we need to know how many feet of support cable are needed.*

▲ **19.** Ground anchor #2 is exactly 54 feet from the base of the tower (dashed line in diagram). The bottom ground support cable from level 6 to ground anchor #2 is attached to the tower at a height of 130 feet. The top ground support cable from level 6 to ground anchor #2 is attached to the tower at a height of 135 feet. How long is each support cable to level 6? (Round to the nearest hundredth of a foot.)

▲ **20.** Ground anchor #1 is exactly 28 feet from the base of the tower (dashed line in diagram). The bottom ground support cable from level 1 to ground anchor #1 is attached to the tower at a height 42 feet. The top ground support cable from level 1 to ground anchor #1 is attached to the tower at a height 47 feet. How long is each support cable to level 1? (Round to the nearest hundredth of a foot.)

Solve for the variable. Check your solutions.

21. $\sqrt{x + 2} = 3$

22. $\sqrt{x - 2} = 6$

23. $\sqrt{2x + 7} = 5$

24. $\sqrt{5x - 1} = 3$

25. $\sqrt{2x + 2} = \sqrt{3x - 5}$

26. $\sqrt{5x - 5} = \sqrt{4x + 1}$

27. $\sqrt{2x} - 5 = 4$

28. $\sqrt{13x} - 2 = 5$

29. $\sqrt{2x + 15} = x$

30. $\sqrt{7x - 10} = x$

31. $\sqrt{5y + 1} = y + 1$

32. $\sqrt{2y + 9} = y + 3$

33. $\sqrt{x + 3} = 3x - 1$

34. $\sqrt{2x + 3} = 2x - 9$

35. $\sqrt{3y + 1} - y = 1$

36. $\sqrt{3y - 8} + 2 = y$

37. $\sqrt{6y + 1} - 3y = y$

38. $\sqrt{2x + 3} - 2 = x$

To Think About

Solve for x.

39. $\sqrt{12x + 1} = 2\sqrt{6x} - 1$

40. $\sqrt{x - 2} + 3 = \sqrt{4x + 1}$

Cumulative Review

41. Solve. $\dfrac{5x}{x - 4} = 5 + \dfrac{4x}{x - 4}$

42. Multiply mentally. $(5x - 7)^2$

43. $f(x) = 2x^2 - 3x + 6$
Find $f(-2)$.

44. Is $x^2 + y^2 = 9$ a relation or a function?

Operating a Bird Sanctuary *Peter J. Dubacher and his wife, Betty Ann, have run a bird sanctuary for 25 years. This sanctuary costs $40,000 per year to run. Peter and Betty Ann care for 1700 birds, including 15 eagles, 2 peregrine falcons, and 19 emus.*

45. If the number of birds has grown by 50 every year for the last 25 years, how many birds did they start with in year 1?

46. If you average the cost to care for the birds over the 25 years, how much was spent per bird per year?

Student Learning Objectives

After studying this section, you will be able to:

1 Solve problems involving direct variation.

2 Solve problems involving inverse variation.

3 Identify the graphs of variation equations.

NOTE TO STUDENT: Fully worked-out solutions to all of the Practice Problems can be found at the back of the text starting at page SP-1

1 Solving Problems Involving Direct Variation

Often in daily life there is a relationship between two measurable quantities. For example, we turn up the thermostat and the heating bill increases. We say that the heating bill varies directly as the temperature on the thermostat. If y **varies directly as** x, then $y = kx$, where k is a constant. The constant is often called the **constant of variation.** Consider the following example.

EXAMPLE 1 Cliff works part-time in a local supermarket while going to college. His salary varies directly as the number of hours worked. Last week he earned $33.60 for working 7 hours. This week he earned $52.80. How many hours did he work?

Solution Let S = his salary,

h = the number of hours he worked, and

k = the constant of variation.

Since his salary varies directly as the number of hours he worked, we write

$$S = k \cdot h.$$

We can find the constant k by substituting the known values of $S = 33.60$ and $h = 7$.

$$33.60 = k \cdot 7 = 7k$$

$$\frac{33.60}{7} = \frac{7k}{7} \qquad \text{Solve for } k.$$

$$4.80 = k \qquad \text{The constant of variation is 4.80.}$$

$$S = 4.80h \qquad \text{Replace } k \text{ in the variation equation by 4.80.}$$

How many hours did he work to earn $52.80?

$$S = 4.80h \qquad \text{The direct variation equation with } k = 4.80.$$

$$52.80 = 4.80h \qquad \text{Substitute 52.80 for } S \text{ and solve for } h.$$

$$\frac{52.80}{4.80} = \frac{4.80h}{4.80}$$

$$11 = h$$

Cliff worked 11 hours this week.

Practice Problem 1 The *change* in temperature measured on the Celsius scale varies directly as the change measured on the Fahrenheit scale. The change in temperature from freezing water to boiling water is 100° on the Celsius scale and 180° on the Fahrenheit scale. If the Fahrenheit temperature drops 20°, what will be the change in temperature on the Celsius scale?

SOLVING A DIRECT VARIATION PROBLEM

1. Write the direct variation equation.

2. Solve for the constant k by substituting in known values.

3. Replace k in the direct variation equation by the value obtained in step 2.

4. Solve for the desired value.

EXAMPLE 2 If y varies directly as x, and $y = 20$ when $x = 3$, find the value of y when $x = 21$.

Solution

Step 1 $y = kx$

Step 2 To find k, we substitute $y = 20$ and $x = 3$.

$$20 = k(3)$$
$$20 = 3k$$
$$\frac{20}{3} = k$$

Step 3 We now write the variation equation with k replaced by $\frac{20}{3}$.

$$y = \frac{20}{3}x$$

Step 4 We replace x by 21 and find y.

$$y = \left(\frac{20}{\cancel{3}}\right)(\cancel{21}^{7}) = (20)(7) = 140$$

Thus $y = 140$ when $x = 21$.

Practice Problem 2 If y varies directly as x, and $y = 18$ when $x = 5$, find the value of y when $x = \frac{20}{23}$.

A number of real-life situations can be described by direct variation equations. The following table shows some of the more common forms of the direct variation equation. In each case, $k =$ the constant of variation.

SAMPLE DIRECT VARIATION SITUATIONS

Verbal Description	*Variation Equation*
y varies directly as x	$y = kx$
b varies directly as the square of c	$b = kc^2$
l varies directly as the cube of m	$l = km^3$
V varies directly as the square root of h	$V = k\sqrt{h}$

EXAMPLE 3 In a certain class of racing cars, the maximum speed varies directly as the square root of the horsepower of the engine. If a car with 225 horsepower can achieve a maximum speed of 120 mph, what speed could it achieve with 256 horsepower?

Solution Let $V =$ the maximum speed,
$h =$ the horsepower of the engine, and
$k =$ the constant of variation.

Step 1 Since the maximum speed (V) varies directly as the square root of the horsepower of the engine,

$$V = k\sqrt{h}.$$

Step 2 $120 = k\sqrt{225}$ Substitute known values of V and h.

$120 = k \cdot 15$ Solve for k.

$8 = k$

Step 3 Now we can write the direct variation equation with the known value for k.

$$V = 8\sqrt{h}$$

Step 4 $V = 8\sqrt{256}$ Substitute the value of $h = 256$.

$\quad\quad\quad = (8)(16)$ Solve for V.

$\quad\quad\quad = 128$

Thus a car with 256 horsepower could achieve a maximum speed of 128 mph.

NOTE TO STUDENT: *Fully worked-out solutions to all of the Practice Problems can be found at the back of the text starting at page SP-1*

Practice Problem 3 A certain type of car has a stopping distance that varies directly as the square of its speed. If this car is traveling at a speed of 20 mph on an ice-covered road, it can stop in 60 feet. If the car is traveling at 40 mph on an icy road, what will its stopping distance be?

② Solving Problems Involving Inverse Variation

If one variable is a constant multiple of the reciprocal of another, the two variables are said to **vary inversely.** If y varies inversely as x, we express this by the equation $y = \dfrac{k}{x}$, where k is the constant of variation. Inverse variation problems can be solved by a four-step procedure similar to that used for direct variation problems.

SOLVING AN INVERSE VARIATION PROBLEM

1. Write the inverse variation equation.
2. Solve for the constant k by substituting in known values.
3. Replace k in the inverse variation equation by the value obtained in step 2.
4. Solve for the desired value.

EXAMPLE 4   If y varies inversely as x, and $y = 12$ when $x = 7$, find the value of y when $x = \dfrac{2}{3}$.

Solution

Step 1 $y = \dfrac{k}{x}$

Step 2 $12 = \dfrac{k}{7}$ Substitute known values of x and y to find k.

$\quad\quad\quad 84 = k$

Step 3 $y = \dfrac{84}{x}$

Step 4 To find y when $x = \dfrac{2}{3}$, we substitute.

$$y = \dfrac{84}{\dfrac{2}{3}}$$

$$= \dfrac{84}{1} \div \dfrac{2}{3}$$

$$= \dfrac{\overset{42}{\cancel{84}}}{1} \cdot \dfrac{3}{\cancel{2}}$$

$$= 42 \cdot 3 = 126$$

Thus $y = 126$ when $x = \dfrac{2}{3}$.

Practice Problem 4 If y varies inversely as x, and $y = 8$ when $x = 15$, find the value of y when $x = \dfrac{3}{5}$.

EXAMPLE 5   A car manufacturer is thinking of reducing the size of the wheel used in a subcompact car. The number of times a car wheel must turn to cover a given distance varies inversely as the radius of the wheel. (Notice that this says that the smaller the wheel, the more times it must turn to cover a given distance.) A wheel with a radius of 0.35 meter must turn 400 times to cover a specified distance on a test track. How many times would it have to turn if the radius were reduced to 0.30 meter (see the sketch)?

0.35 m

Test distance

0.30 m

Solution   Let n = the number of times the car wheel turns,

r = the radius of the wheel, and

k = the constant of variation.

Step 1   Since the number of turns varies inversely as the radius, we can write the following.

$$n = \frac{k}{r} \quad \text{Write the variation equation.}$$

Step 2 $400 = \dfrac{k}{0.35}$ Substitute known values of n and r to find k.

$140 = k$

Step 3 $n = \dfrac{140}{r}$ Use the variation equation where k is known.

How many times must the wheel turn if the radius is 0.30 meter?

Step 4 $n = \dfrac{140}{0.30}$ Substitute 0.30 for r.

$n = 466\dfrac{2}{3}$

The wheel would have to turn $466\dfrac{2}{3}$ times to cover the same distance if the radius were only 0.30 meter.

Practice Problem 5   Over the last three years, the market research division of a calculator company found that the volume of sales of scientific calculators varies inversely as the price of the calculator. One year 120,000 calculators were sold at $30 each. How many calculators were sold the next year when the price was $24 for each calculator?

The following table contains various forms of the inverse variation equation. In each case, k = the constant of variation.

SAMPLE INVERSE VARIATION SITUATIONS

Verbal Description	*Variation Equation*
y varies inversely as x	$y = \dfrac{k}{x}$
b varies inversely as the square of c	$b = \dfrac{k}{c^2}$
l varies inversely as the cube of m	$l = \dfrac{k}{m^3}$
d varies inversely as the square root of t	$d = \dfrac{k}{\sqrt{t}}$

EXAMPLE 6   The illumination of a light source varies inversely as the square of the distance from the source. The illumination measures 25 candlepower when a certain light is 4 meters away. Find the illumination when the light is 8 meters away (see figure).

Solution   Let I = the measurement of illumination,
d = the distance from the light source, and
k = the constant of variation.

Step 1   Since the illumination varies inversely as the square of the distance,

$$I = \frac{k}{d^2}.$$

Step 2   We evaluate the constant by substituting the given values.

$$25 = \frac{k}{4^2} \quad \text{Substitute } I = 25 \text{ and } d = 4.$$

$$25 = \frac{k}{16} \quad \text{Simplify and solve for } k.$$

$$400 = k$$

Step 3   We may now write the variation equation with the constant evaluated.

$$I = \frac{400}{d^2}$$

Step 4 $I = \dfrac{400}{8^2}$ Substitute a distance of 8 meters.

$$= \frac{400}{64} \quad \text{Square 8.}$$

$$= \frac{25}{4} = 6.25$$

The illumination is 6.25 candlepower when the light source is 8 meters away.

Practice Problem 6   If the amount of power in an electrical circuit is held constant, the resistance in the circuit varies inversely as the square of the amount of current. If the amount of current is 0.01 ampere, the resistance is 800 ohms. What is the resistance if the amount of current is 0.02 ampere?

NOTE TO STUDENT: Fully worked-out solutions to all of the Practice Problems can be found at the back of the text starting at page SP-1

3 Identifying the Graphs of Variation Equations

The graphs of direct variation equations are shown in the following chart. Notice that as *x* increases, *y* also increases.

Variation Statement	Equation	Graph
1. *y* varies directly as *x*	$y = kx$	
2. *y* varies directly as x^2	$y = kx^2$	
3. *y* varies directly as x^3	$y = kx^3$	
4. *y* varies directly as the square root of *x*	$y = k\sqrt{x}$	

The graphs of inverse variation equations are shown in the following chart. Notice that as *x* increases, *y* decreases.

Variation Statement	Equation	Graph
5. *y* varies inversely as *x*	$y = \dfrac{k}{x}$	
6. *y* varies inversely as x^2	$y = \dfrac{k}{x^2}$	

9.7 EXERCISES

Student Solutions Manual · CD/Video · PH Math Tutor Center · MathXL®Tutorials on CD · MathXL® · MyMathLab® · Interactmath.com

1. If y varies directly as x, and $y = 9$ when $x = 2$, find y when $x = 16$.

2. If y varies directly as x, and $y = 5$ when $x = 3$, find y when $x = 18$.

3. If y varies directly as the cube of x, and $y = 12$ when $x = 2$, find y when $x = 7$.

4. If y varies directly as the square root of x, and $y = 7$ when $x = 4$, find y when $x = 25$.

5. If y varies directly as the square of x, and $y = 900$ when $x = 25$, find y when $x = 30$.

6. If y varies directly as the square of x, and $y = 8000$ when $x = 16$, find y when $x = 5$.

Applications

7. *Food Nutrition Facts* There are 200 calories in 50 grams of Swiss cheese. The number of calories varies directly with the weight of the cheese. How many calories will 70 grams of cheese have?

8. *Stretching a Spring* Hooke's Law states that the force needed to stretch a spring varies directly with the amount the spring is stretched. If a force of 50 pounds is needed to stretch a spring 5 inches, how far will a force of 120 pounds stretch the same spring?

9. *Storage Cube* The time it takes to fill a storage cube with sand varies directly as the cube of the side of the box. A storage cube that is 2.0 meters on each side (inside dimensions) can be filled in 7 minutes by a sand loader. How long will it take to fill a storage cube that is 4.0 meters on each side (inside dimensions)?

10. *Weight on the Moon* The weight of an object on the surface of the moon varies directly as the weight of the object on the surface of earth. An astronaut with his protective suit weighs 80 kilograms on earth's surface; on the moon and wearing the same suit, he will weigh 12.8 kilograms. If his moon rover vehicle weighs 900 kilograms on earth, how much will it weigh on the moon?

2.0 m
2.0 m
2.0 m

11. If y varies inversely as x and $y = 18$ when $x = 5$, find y when $x = 8$.

12. If y varies inversely as x and $y = 21$ when $x = 4$, find y when $x = 18$.

13. If y varies inversely as x, and $y = \frac{1}{4}$ when $x = 8$, find y when $x = 1$.

14. If y varies inversely as x, and $y = \frac{1}{5}$ when $x = 20$, find y when $x = 2$.

15. If y varies inversely as the square of x, and $y = 30$ when $x = 2$, find y when $x = 9$.

16. If y varies inversely as the cube of x, and $y = \frac{1}{9}$ when $x = 3$, find y when $x = \frac{1}{2}$.

Applications

17. *Ice Cube Melting* The amount of time in minutes that it takes for an ice cube to melt varies inversely as the temperature of the water that the ice cube is placed in. When an ice cube is placed in 60°F water, it takes 2.3 minutes to melt. How long would it take for an ice cube of this size to melt if it were placed in 40°F water?

18. *Textile Mills* The textile mills in Lowell, Massachusetts, were operated with water power. The Merrimack River and its canals powered huge turbines that generated the electricity for the mills. In each mill room, a single driveshaft supplied power to all machines by means of belts that ran to pulleys on the machines. The driveshaft turned at a constant speed, but the pulleys were different diameters, which provided for the different speeds of each machine. The speed of each machine varied inversely as the diameter of its pulley. If a machine with a 60-centimeter pulley turned at 1000 rpm, what was the speed of a machine with an 80-centimeter pulley?

19. *Variation in Weight on Earth* The weight of an object near earth's surface varies inversely as the square of its distance from the center of earth. An object weighs 1000 pounds on earth's surface. This is approximately 4000 miles from the center of earth. How much will the object weigh 6000 miles from the center of earth?

20. *Radio Waves* The wavelength, w, of a radio wave varies inversely as the frequency, f. A radio wave with a frequency of 1600 kilohertz has a length of 225 meters. What is the length of a radio wave with a frequency of 900 kilohertz?

21. *Distance Vision in a Boat* A person on the mast of a sailing ship 81 meters above the water can see approximately 12.3 kilometers out to sea. In general, the distance, d, you can see varies directly as the square root of your height, h, above the water. In this case, the equation is $d = \left(\frac{15}{11}\right)\sqrt{h}$, where d is measured in kilometers and h is measured in meters.
(a) Find the distances you can see when $h = 121$, 36, and 4. Round to the nearest tenth.

(b) Graph the equation using values of h of 121, 81, 36, and 4.

Height above the Water (in Meters)

22. *Computer Files* Suppose a computer has a hard drive that can hold 2000 files if each file is 100,000 bytes long. The number of files, f, the hard drive can hold varies inversely as the size, s, of a file. In this case, the equation is $f = \dfrac{200,000,000}{s}$, where f is the number of files and s is the size of a file in bytes.
(a) Find the number of files that can be held if $s = 400,000$, $s = 200,000$, and $s = 50,000$.

(b) Graph the equation using the values $s = 400,000$, $s = 200,000$, $s = 100,000$ and $s = 50,000$.

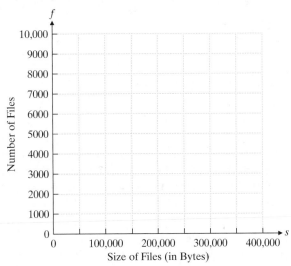
Size of Files (in Bytes)

To Think About

Write an equation to describe the variation and solve.

23. ***Planetary Motion*** Kepler's Third Law of Planetary Motion states that the square of the time it takes a planet to complete one revolution of the sun varies directly with the cube of the distance between the planet and the sun. Astronomers use 1 astronomical unit (au) as a measure of the distance from the earth to the sun.

   (a) Using this astronomical unit, find the constant k for the direct variation equation that represents Kepler's Third Law.

   (b) Mars is 1.5 times as far from the sun as the earth. Find the approximate length of the Martian year in terms of earth years.

24. ***Sailboat Design*** An important element of sailboat design is the calculation of the force of wind on the sail. The force varies directly with the square of the wind speed and directly with the area of the sail. For a certain class of racing boat, the force on 100 square feet of sail is 280 pounds when the wind speed is 6 mph. What would be the force on the sail if it were rigged for a storm (the amount of sail exposed to the wind is reduced by lashing part of the sail to the boom) so that only 60 square feet of sail were exposed and the wind speed were 40 mph?

Cumulative Review

25. Solve for a. $\dfrac{80{,}000}{320} = \dfrac{120{,}000}{320 + a}$

26. Graph the equation $S = kh$, where $k = 4.80$. Use h for the horizontal axis and S for the vertical. Assume that h and S are nonnegative.

27. Factor. $12x^2 - 20x - 48$

28. Solve. $2x - (5 - 3x) = 6(x - 2)$

▲ 29. ***Geometry*** A triangle has a perimeter of 69 feet. One side is 5 feet longer than the second side. The third side is twice the length of the second side. How long is each side?

30. ***Financial Investments*** Elizabeth invested $5000 for one year. Part was invested at 6% simple interest. The rest was invested at 10% simple interest. At the end of one year Elizabeth earned $428 in interest. How much did she invest at each rate?

Putting Your Skills to Work

The Mathematics of the International Space Station

Orbiting 250 miles above the earth, the International Space Station depends entirely on solar panels to produce the necessary operating electricity to maintain the station and to perform scientific experiments. The Space Station orbits the earth in approximately 90 minutes. During each orbit the station is in sunlight for approximately 60 minutes and in darkness for approximately 30 minutes. The electricity generated during sunlight hours is stored in batteries to provide power when the station is in darkness.

As of November 2003 the station has four huge solar panels each measuring 112 feet long by 39 feet wide. These panels produce a maximum of 3750 watts of electrical power. As the space station expands eventually there will be 24 solar panels. (*Source:* NASA)

Problems for Individual Investigation and Analysis

1. What is the area of one solar panel?

2. What is the maximum number of watts of electrical power produced by one solar panel?

Problems for Group Investigation and Cooperative Learning

3. How many watts of electrical power are produced by one square foot of the solar panel? Round your answer to the nearest hundredth.

4. If the four solar panels now installed were extended to each be 150 feet long, how much would the maximum number of watts of electrical power be increased? Use your answer obtained in 3 above. Round your answer to the nearest hundredth.

5. If the width of a solar panel remains constant, the electrical power that can be produced by the solar panel varies directly as the square root of the length of the panel. If a panel that is 130 feet long produces 4000 watts, how much would be produced by a panel of the same width that is 190 feet long? Round your answer to the nearest whole number.

6. If the length of a solar panel remains constant, the electrical power than can be produced by the solar panel varies directly as the square root of the width of the panel. If a panel that is 30 feet wide produces 4400 watts, how much would be produced by a panel of the same width that is 50 feet wide? Round your answer to the nearest whole number.

Topic	Procedure	Examples
Square roots, p. 518.	1. The square roots of perfect squares up to $\sqrt{225} = 15$ (or so) should be memorized. Approximate values of others can be found in tables or by calculator. 2. Negative numbers do not have real number square roots.	$\sqrt{196} = 14 \qquad \sqrt{169} = 13$ $\sqrt{200} \approx 14.142 \qquad \sqrt{13} \approx 3.606$ $\sqrt{-5}$ This is not a real number.
Simplifying radicals, p. 523.	Assuming that a and b are nonnegative, $$\sqrt{ab} = \sqrt{a}\sqrt{b}$$ Square root radicals are simplified by taking the square roots of all factors that are perfect squares and leaving all other factors under the radical sign.	$\sqrt{48} = \sqrt{16}\sqrt{3} = 4\sqrt{3}$ $\sqrt{12x^2y^3z} = \sqrt{2^2x^2y^2}\sqrt{3yz}$ $= 2xy\sqrt{3yz}$ $\sqrt{a^{11}} = \sqrt{a^{10}\cdot a} = \sqrt{a^{10}}\sqrt{a} = a^5\sqrt{a}$
Adding and subtracting radicals, p. 528.	Simplify all radicals and then add or subtract like radicals.	$\sqrt{12} + \sqrt{18} + \sqrt{27} + \sqrt{50}$ $= \sqrt{4}\sqrt{3} + \sqrt{9}\sqrt{2} + \sqrt{9}\sqrt{3} + \sqrt{25}\sqrt{2}$ $= 2\sqrt{3} + 3\sqrt{2} + 3\sqrt{3} + 5\sqrt{2}$ $= 5\sqrt{3} + 8\sqrt{2}$
Multiplying radicals, p. 532.	Square root radicals are multiplied like polynomials. Assuming that a and b are nonnegative, apply the rule $\sqrt{a}\sqrt{b} = \sqrt{ab}$ and simplify all results.	$(2\sqrt{6} + \sqrt{3})(\sqrt{6} - 3\sqrt{3})$ $= 2\sqrt{36} - 6\sqrt{18} + \sqrt{18} - 3\sqrt{9}$ $= 2(6) - 5\sqrt{18} - 3(3)$ $= 12 - 5\sqrt{9}\sqrt{2} - 9$ $= 12 - 5(3)\sqrt{2} - 9$ $= 3 - 15\sqrt{2}$
Dividing radicals, p. 538.	Division of radical expressions is accomplished by writing the division problem as a fraction and rationalizing the denominator.	Divide. $\sqrt{x} \div \sqrt{3}$ $\dfrac{\sqrt{x}}{\sqrt{3}} \cdot \dfrac{\sqrt{3}}{\sqrt{3}} = \dfrac{\sqrt{3x}}{\sqrt{9}} = \dfrac{\sqrt{3x}}{3}$
Conjugates, p. 540.	Note that the product of the sum and difference of two square root radicals is equal to the difference of their squares and thus contains no radicals. This sum and difference are called conjugates.	$(3\sqrt{6} + \sqrt{10})$ and $(3\sqrt{6} - \sqrt{10})$ are conjugates. $(3\sqrt{6} + \sqrt{10})(3\sqrt{6} - \sqrt{10})$ $= 9\sqrt{6}^2 - \sqrt{10}^2$ $= 9 \cdot 6 - 10 = 54 - 10 = 44$
Rationalizing denominators, p. 539.	To simplify a fraction with a square root radical in the denominator, the radical (in the denominator) must be removed. 1. Multiply the denominator by itself if it is a monomial. Multiply the numerator by the same quantity. 2. Multiply the denominator by its conjugate if it is a binomial. Multiply the numerator by the same quantity.	$\dfrac{\sqrt{10}}{\sqrt{6}} = \dfrac{\sqrt{10}\cdot\sqrt{6}}{\sqrt{6}\cdot\sqrt{6}} = \dfrac{2\sqrt{15}}{6} = \dfrac{\sqrt{15}}{3}$ $\dfrac{\sqrt{3}}{\sqrt{3} - \sqrt{2}} = \dfrac{\sqrt{3}(\sqrt{3} + \sqrt{2})}{(\sqrt{3} - \sqrt{2})(\sqrt{3} + \sqrt{2})}$ $= \dfrac{\sqrt{9} + \sqrt{6}}{3 - 2} = 3 + \sqrt{6}$
Radicals with fractions, p. 538.	Note that $\sqrt{\dfrac{5}{3}}$ is the same as $\dfrac{\sqrt{5}}{\sqrt{3}}$ and so must be rationalized.	$\sqrt{\dfrac{5}{3}} = \dfrac{\sqrt{5}}{\sqrt{3}} \cdot \dfrac{\sqrt{3}}{\sqrt{3}} = \dfrac{\sqrt{15}}{3}$
Pythagorean Theorem, p. 544.	In any right triangle with hypotenuse of length c and legs of lengths a and b, $$c^2 = a^2 + b^2.$$	Find c to the nearest tenth if $a = 7$ and $b = 9$. $c^2 = 7^2 + 9^2 = 49 + 81 = 130$ $c = \sqrt{130} \approx 11.4$

Topic	Procedure	Examples
Radical equations, p. 546.	To solve an equation containing a square root radical: 1. Isolate the radical by itself on one side of the equation. 2. Square both sides. 3. Solve the resulting equation. 4. Check all apparent solutions. Extraneous roots may have been introduced in step 2.	Solve. $\sqrt{3x+2} - 4 = 8$ $$\sqrt{3x+2} = 12$$ $$(\sqrt{3x+2})^2 = 12^2$$ $$3x + 2 = 144$$ $$3x = 142$$ $$x = \frac{142}{3}$$ Check: $\sqrt{3\left(\frac{142}{3}\right) + 2} \stackrel{?}{=} 12$ $$\sqrt{144} = 12 \checkmark$$ So $x = \frac{142}{3}$.
Variation: direct, p. 552.	If y varies directly with x, there is a constant of variation k such that $y = kx$. Once k is determined, the value of y or x can easily be computed.	y varies directly with x. When $x = 2$, $y = 7$. $$y = kx$$ $$7 = k(2) \quad \text{Substituting.}$$ $$k = \frac{7}{2} \quad \text{Solving.}$$ $$y = \frac{7}{2}x$$ What is y when $x = 8$? $$y = \frac{7}{2}x = \frac{7}{2} \cdot 8 = 28$$
Variation: inverse, p. 554.	If y varies inversely with x, there is a constant of variation k such that $$y = \frac{k}{x}.$$	y varies inversely with x. When x is 5, y is 12. What is y when x is 15? $$y = \frac{k}{x}$$ $$12 = \frac{k}{5} \quad \text{Substituting.}$$ $$k = 60 \quad \text{Solving.}$$ $$y = \frac{60}{x} \quad \text{Substituting.}$$ $$x = 15 \quad y = \frac{60}{15} = 4$$

Chapter 9 Review Problems

Section 9.1

Simplify or evaluate, if possible.

1. $\sqrt{36}$

2. $\sqrt{25}$

3. $\sqrt{169}$

4. $\sqrt{196}$

5. $-\sqrt{81}$

6. $\sqrt{64}$

7. $\sqrt{49}$

8. $-\sqrt{121}$

9. $-\sqrt{144}$

10. $\sqrt{289}$

11. $\sqrt{0.04}$

12. $\sqrt{0.49}$

13. $\sqrt{\dfrac{1}{25}}$

14. $\sqrt{\dfrac{36}{49}}$

Approximate, using the square root table in Appendix A or a calculator. Round to the nearest thousandth, if necessary.

15. $\sqrt{105}$

16. $\sqrt{198}$

17. $\sqrt{77}$

18. $\sqrt{88}$

Section 9.2

Simplify.

19. $\sqrt{27}$ **20.** $\sqrt{52}$ **21.** $\sqrt{28}$

22. $\sqrt{125}$ **23.** $\sqrt{40}$ **24.** $\sqrt{80}$

25. $\sqrt{x^8}$ **26.** $\sqrt{y^{10}}$ **27.** $\sqrt{x^5 y^6}$

28. $\sqrt{a^3 b^4}$ **29.** $\sqrt{16x^3 y^5}$ **30.** $\sqrt{98x^4 y^6}$

31. $\sqrt{12x^5}$ **32.** $\sqrt{27x^7}$ **33.** $\sqrt{75x^{10}}$

34. $\sqrt{125x^{12}}$ **35.** $\sqrt{120a^3 b^4 c^5}$ **36.** $\sqrt{121a^6 b^4 c}$

37. $\sqrt{56x^7 y^9}$ **38.** $\sqrt{99x^{13} y^7}$

Section 9.3

Simplify.

39. $\sqrt{63} - \sqrt{7} + \sqrt{28}$ **40.** $5\sqrt{6} - \sqrt{24} + 2\sqrt{54}$

41. $x\sqrt{3} + 3x\sqrt{3} + \sqrt{27x^2}$ **42.** $a\sqrt{2} + \sqrt{12a^2} + a\sqrt{98}$

43. $5\sqrt{5} - 6\sqrt{20} + 2\sqrt{10}$ **44.** $3\sqrt{6} - 5\sqrt{18} + 3\sqrt{24}$

Section 9.4

Simplify.

45. $\left(2\sqrt{x}\right)\left(3\sqrt{x^3}\right)$ **46.** $\left(-5\sqrt{a}\right)\left(2\sqrt{ab}\right)$

47. $\left(\sqrt{2a^3}\right)\left(\sqrt{8b^2}\right)$ **48.** $\left(5x\sqrt{x}\right)\left(-3x^2\sqrt{x}\right)$

49. $\sqrt{7}\left(\sqrt{28} - 3\sqrt{7}\right)$ **50.** $\sqrt{8}\left(5\sqrt{2} - 3\sqrt{8}\right)$

51. $\sqrt{2}\left(\sqrt{5} - \sqrt{3} - 2\sqrt{2}\right)$

52. $\sqrt{5}\left(\sqrt{6} - 2\sqrt{5} + \sqrt{10}\right)$

53. $\left(\sqrt{11} + 2\right)\left(2\sqrt{11} - 1\right)$

54. $\left(\sqrt{10} + 3\right)\left(3\sqrt{10} - 1\right)$

55. $\left(2 + 3\sqrt{6}\right)\left(4 - 2\sqrt{3}\right)$

56. $\left(5 - \sqrt{2}\right)\left(3 - \sqrt{12}\right)$

57. $\left(2\sqrt{3} + 3\sqrt{6}\right)^2$

58. $\left(5\sqrt{2} - 2\sqrt{6}\right)^2$

Section 9.5

Rationalize the denominator.

59. $\dfrac{1}{\sqrt{3x}}$

60. $\dfrac{2y}{\sqrt{5}}$

61. $\dfrac{x^2 y}{\sqrt{8}}$

62. $\dfrac{3ab}{\sqrt{2b}}$

63. $\sqrt{\dfrac{5}{6}}$

64. $\sqrt{\dfrac{6}{11}}$

65. $\dfrac{\sqrt{a^5}}{\sqrt{2a}}$

66. $\dfrac{\sqrt{x^3}}{\sqrt{3x}}$

67. $\dfrac{3}{\sqrt{5} + \sqrt{2}}$

68. $\dfrac{2}{\sqrt{6} - \sqrt{3}}$

69. $\dfrac{1 - \sqrt{5}}{2 + \sqrt{5}}$

70. $\dfrac{1 - \sqrt{3}}{3 + \sqrt{3}}$

Section 9.6

In exercises 71–77, use the Pythagorean Theorem to find the length of the missing side of a right triangle with legs a and b and hypotenuse c.

▲ **71.** $a = 5, b = 8$

▲ **72.** $c = \sqrt{11}, b = 3$

▲ **73.** $c = 5, a = 3.5$

▲ **74.** If $a = 2.400$ and $b = 2.000$, find c and round to the nearest thousandth.

▲ **75.** *Flagpole* A flagpole is 24 meters tall. A man stands 18 meters from the base of the pole. How far is it from the feet of the man to the top of the pole?

▲ **76.** *Distances Between Towns and Cities* A city is 20.0 miles east of a major airport. A small town is directly south of the city. The town is 50.0 miles from the airport. How far is it from the town to the city? (Round to the nearest tenth of a mile.)

▲ **77.** *Diving Pool* The diagonal measure of a swimming pool is $\sqrt{455}$ feet. If the pool is square, how long is each side? (Round to the nearest tenth of a foot.)

Solve. Be sure to verify your answers.

78. $\sqrt{x - 4} = 5$

79. $\sqrt{3x + 1} = 8$

80. $\sqrt{1 - 3x} = \sqrt{5 + x}$

81. $\sqrt{-5 + 2x} = \sqrt{1 + x}$ **82.** $\sqrt{10x + 9} = -1 + 2x$ **83.** $\sqrt{2x - 5} = 10 - x$

84. $6 - \sqrt{5x - 1} = x + 1$ **85.** $4x + \sqrt{x + 2} = 5x - 4$

Section 9.7

86. If y varies directly with the square root of x, and $y = 35$ when $x = 25$, find the value of y when $x = 121$.

87. If y varies inversely with the square of x, and $y = \dfrac{6}{5}$ when $x = 5$, find the value of y with $x = 15$.

88. If y varies inversely with the cube of x, and $y = 4$ when $x = 2$, find the value of y when $x = 4$.

89. *Illumination from a Light Source* The intensity of illumination from a light source varies inversely as the square of the distance form the light source. If an object is 4 meters from a light source, what is the intensity of the illumination compared to the illumination at the light source?

90. *Skid Marks of a Car* When an automobile driver slams on the brakes, the length of skid marks on the road varies directly with the square of the speed of the car. At 30 mph a certain car had skid marks 40 feet long. How long will the skid marks be if the car travels at 55 mph?

91. *Horsepower of a Boat* The horsepower that is needed to drive a racing boat through water varies directly with the cube of the speed of the boat. What will happen to the horsepower requirement if someone wants to double the maximum speed of a given boat?

Mixed Practice

Find the square root.

92. $\sqrt{\dfrac{36}{121}}$

93. $\sqrt{0.0004}$

Simplify.

94. $\sqrt{98x^6 y^3}$

Combine.

95. $3\sqrt{27} - 2\sqrt{75} + \sqrt{48}$

Multiply and simplify your answer.

96. $\left(2\sqrt{3} + \sqrt{5}\right)\left(3\sqrt{3} - \sqrt{10}\right)$

97. $\left(3\sqrt{5} + 2\right)^2$

Rationalize the denominator.

98. $\dfrac{5}{\sqrt{12}}$

99. $\dfrac{\sqrt{3} + \sqrt{6}}{2\sqrt{3} + \sqrt{2}}$

100. $\sqrt{2x - 3} = 9$

101. $\sqrt{10x + 5} = 2x + 1$

Remember to use your Chapter Test Prep Video CD to see the worked-out solutions to the test problems you want to review.

Evaluate.

1. $\sqrt{121}$

2. $\sqrt{\dfrac{9}{100}}$

Simplify.

3. $\sqrt{48x^2y^7}$

4. $\sqrt{100x^3yz^4}$

Combine and simplify.

5. $8\sqrt{3} + 5\sqrt{27} - 5\sqrt{48}$

6. $\sqrt{4a} + \sqrt{8a} + \sqrt{36a} + \sqrt{18a}$

Multiply.

7. $\left(2\sqrt{a}\right)\left(3\sqrt{b}\right)\left(2\sqrt{ab}\right)$

8. $\sqrt{5}\left(\sqrt{10} + 2\sqrt{3} - 3\sqrt{5}\right)$

9. $\left(2\sqrt{3} + 5\right)^2$

10. $\left(4\sqrt{2} - \sqrt{5}\right)\left(3\sqrt{2} + \sqrt{5}\right)$

In questions 11–14, simplify.

11. $\sqrt{\dfrac{x}{5}}$

12. $\dfrac{3}{\sqrt{12}}$

13. $\dfrac{\sqrt{3} + 4}{5 + \sqrt{3}}$

14. $\dfrac{3a}{\sqrt{5} + \sqrt{2}}$

15. Use the table of square roots or a calculator to approximate $\sqrt{156}$. Round your answer to the nearest hundredth.

Find the missing side by using the Pythagorean Theorem.

▲ **16.**

▲ **17.**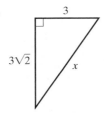

Solve. Verify your solutions.

18. $6 - \sqrt{2x + 1} = 0$

19. $x = 5 + \sqrt{x + 7}$

Solve.

20. The intensity of illumination from a light source varies inversely with the square of the distance from the light source. A photoelectric cell is placed 8 inches from a light source. Then it is moved so that it receives only one-fourth as much illumination. How far is it from the light source?

21. Commission on the sale of office supplies varies directly with the amount of the sale. If a salesman earns \$23.40 on sales of \$780, what does he earn on sales of \$2859?

22. The area of an equilateral triangle varies directly as the square of its perimeter. If the perimeter is 12 centimeters, the area is 6.93 square centimeters. What is the area of an equilateral triangle whose perimeter is 21 centimeters? Round to the nearest tenth.

1.	
2.	
3.	
4.	
5.	
6.	
7.	
8.	
9.	
10.	
11.	
12.	
13.	
14.	
15.	
16.	
17.	
18.	
19.	
20.	
21.	
22.	

1. _____

2. _____

3. _____

4. _____

5. _____

6. _____

7. _____

8. _____

9. _____

10. _____

11. _____

12. _____

13. _____

14. _____

15. _____

16. _____

17. _____

18. _____

19. _____

20. _____

21. _____

22. _____

23. _____

24. _____

25. _____

26. _____

Approximately one-half of this test covers the content of Chapters 0–8. The remainder covers the content of Chapter 9.

In questions 1–3, simplify.

1. $\dfrac{28}{42}$

2. $3\dfrac{1}{4} + 5\dfrac{2}{3}$

3. $6\dfrac{1}{4} \div 6\dfrac{2}{3}$

4. Write 0.07% as a decimal.

Simplify.

5. $-11 - 16 + 8 + 4 - 13 + 31$

6. $(-3)^2 \cdot 4 - 8 \div 2 - (3 - 2)^3$

7. $4a^2bc^3 - 2abc^2 - 7a^2bc^3$

8. $(x + 3)(x - 2)(x + 2)$

9. $(4x - 5)^2$

10. $(3x - 11)(3x + 11)$

11. $\dfrac{4x}{x + 2} + \dfrac{4}{x + 2}$

12. $\dfrac{x^2 - 5x + 6}{x^2 - 5x - 6} \div \dfrac{x^2 - 4}{x^2 + 2x + 1}$

Simplify, if possible.

13. $\sqrt{98x^5y^6}$

14. $\sqrt{50} - \sqrt{98} + \sqrt{162}$

15. $\left(\sqrt{6} - \sqrt{3}\right)^2$

16. $\dfrac{\sqrt{3} + \sqrt{2}}{\sqrt{3} - \sqrt{2}}$

17. $\left(3\sqrt{6} - \sqrt{2}\right)\left(\sqrt{6} + 4\sqrt{2}\right)$

18. $\dfrac{1}{3 - \sqrt{2}}$

19. $-\sqrt{25}$

20. $\sqrt{-16}$

Use the Pythagorean Theorem to find the missing side of each triangle. Be sure to simplify your answer.

21. $a = 4, b = \sqrt{7}$

▲ 22. $a = 19, c = 21$

In questions 23 and 24, solve for the variable. Be sure to verify your answer(s).

23. $\sqrt{5x + 6} = x$

24. $\sqrt{3y - 2} + 2 = y$

25. If y varies inversely as x, and $y = 4$ when $x = 5$, find the value of y when $x = 2$.

26. The surface area of a balloon varies directly with the square of the cross-sectional radius. When the radius is 2 meters, the surface area is 1256 square meters. Find the surface area if the radius is 0.5 meter.

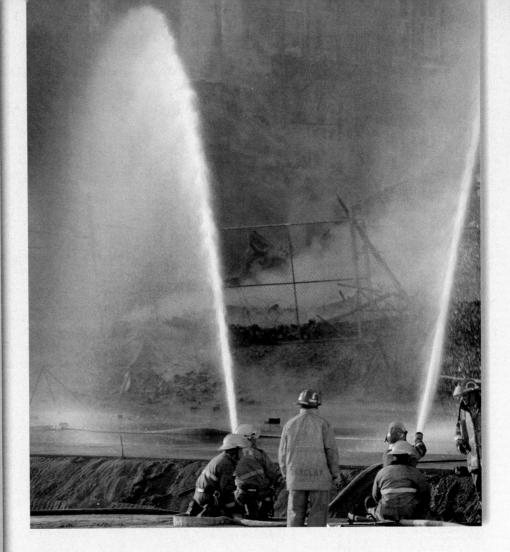

Firefighters need to use mathematics to determine water flow in fighting large fires. Calculations are made about the angle of the hose and the pressure needed to pump the water to the necessary height. Firefighters need to pass a mathematics exam in order to be appointed to city fire departments. Do you think you could handle the mathematics equations that are needed? Turn to page 607 to find out.

Quadratic Equations

1 Writing a Quadratic Equation in Standard Form

In Section 5.7 we introduced quadratic equations. A **quadratic equation** is a polynomial equation of degree two. For example,

$$3x^2 + 5x - 7 = 0, \quad \frac{1}{2}x^2 - 5x = 0, \quad 2x^2 = 5x + 9, \quad \text{and} \quad 3x^2 = 8$$

are all quadratic equations.

> The **standard form of a quadratic equation** is $ax^2 + bx + c = 0$, where a, b, and c are real numbers, and $a \neq 0$.

You will need to be able to recognize the real numbers that represent a, b, and c in specific equations. It will be easier to do so if these equations are placed in standard form. Note that it is usually easier to work with quadratic equations if a is positive, so in cases where a is negative, we recommend that you multiply the entire equation by -1. This is a suggestion and not a requirement.

EXAMPLE 1 Place each quadratic equation in standard form and identify the real numbers a, b, and c.

(a) $5x^2 - 6x + 3 = 0$ This equation is in standard form.

Solution $ax^2 + bx + c = 0$ Match each term to the standard form.

$$a = 5, \quad b = -6, \quad c = 3$$

(b) $2x^2 + 5x = 4$ The right-hand side is not zero. It is not in standard form.

Solution $2x^2 + 5x - 4 = 0$ Add -4 to each side of the equation.

$ax^2 + bx + c = 0$ Match each term to the standard form.

$$a = 2, \quad b = 5, \quad c = -4$$

(c) $-2x^2 + 15x + 4 = 0$ It is easier to work with quadratic equations if the first term is not negative.

Solution $2x^2 - 15x - 4 = 0$ Multiply each term on both sides of the equation by -1.

$ax^2 + bx + c = 0$

$$a = 2, \quad b = -15, \quad c = -4$$

(d) $7x^2 - 9x = 0$ This equation is in standard form.

Solution $ax^2 + bx + c = 0$ Note that the constant term is missing, so we know $c = 0$.

$$a = 7, \quad b = -9, \quad c = 0$$

Practice Problem 1 Place each quadratic equation in standard form and identify the real numbers a, b, and c. If $a < 0$, you can multiply each term of the equation by -1 to obtain an equivalent equation. This is not required, but most students find it helpful to have the first term be a positive number.

(a) $2x^2 + 12x - 9 = 0$ **(b)** $7x^2 = 6x - 8$

(c) $-x^2 - 6x + 3 = 0$ **(d)** $10x^2 - 12x = 0$

② Solving Quadratic Equations of the Form $ax^2 + bx = 0$ by Factoring

Notice that the terms in a quadratic equation of the form $ax^2 + bx = 0$ both have x as a factor. To solve such equations, begin by factoring out the x and any common numerical factor. Thus you may be able to factor out an x or an expression such as $5x$. Remember that you want to remove the *greatest common factor*. So if $5x$ is a common factor of each term, be sure to factor it out. Then use the zero factor property discussed in Section 5.7. This property states that if $a \cdot b = 0$, then either $a = 0$ or $b = 0$. Once each factor is set equal to 0, solve for x.

EXAMPLE 2　Solve. $7x^2 + 9x - 2 = -8x - 2$

Solution

$7x^2 + 9x - 2 = -8x - 2$	The equation is not in standard form.
$7x^2 + 9x - 2 + 8x + 2 = 0$	Add $8x + 2$ to each side.
$7x^2 + 17x = 0$	Collect like terms.
$x(7x + 17) = 0$	Factor.
$x = 0 \qquad 7x + 17 = 0$	Set each factor equal to zero.
$\qquad\qquad 7x = -17$	
$\qquad\qquad x = -\dfrac{17}{7}$	

One of the solutions to the preceding equation is 0. This will always be true of equations of this form. One root will be zero. The other root will be a nonzero real number.

Practice Problem 2　Solve. $2x^2 - 7x - 6 = 4x - 6$

③ Solving Quadratic Equations of the Form $ax^2 + bx + c = 0$ by Factoring

If an equation you are trying to solve contains fractions, clear the fractions by multiplying each term by the least common denominator of all the fractions in the equation. Sometimes an equation does not look like a quadratic equation. However, it takes the quadratic form once the fractions have been cleared. A note of caution: Always check a possible solution in the original equation. Any apparent solution that would make the denominator of any fraction in the original equation 0 is not a valid solution.

EXAMPLE 3　Solve and check. $8x - 6 + \dfrac{1}{x} = 0$

Solution

$8x - 6 + \dfrac{1}{x} = 0$	The equation has a fractional term.
$x(8x) - (x)(6) + x\left(\dfrac{1}{x}\right) = x(0)$	Multiply each term by the LCD which is x.
$8x^2 - 6x + 1 = 0$	Simplify. The equation is now in standard form.
$(4x - 1)(2x - 1) = 0$	Factor.
$4x - 1 = 0 \qquad 2x - 1 = 0$	Set each factor equal to 0.
$4x = 1 \qquad\quad 2x = 1$	Solve each equation for x.
$x = \dfrac{1}{4} \qquad\quad x = \dfrac{1}{2}$	

Check. Checking fractional roots is more difficult, but you should be able to do it if you work carefully.

$$\text{If } x = \frac{1}{4}: \quad 8\left(\frac{1}{4}\right) - 6 + \frac{1}{\frac{1}{4}} = 2 - 6 + 4 = 0 \quad \left(\text{since } 1 \div \frac{1}{4} = 1 \cdot \frac{4}{1} = 4 \right)$$

$$0 = 0 \; \checkmark$$

$$\text{If } x = \frac{1}{2}: \quad 8\left(\frac{1}{2}\right) - 6 + \frac{1}{\frac{1}{2}} = 4 - 6 + 2 = 0 \quad \left(\text{since } 1 \div \frac{1}{2} = 1 \cdot \frac{2}{1} = 2 \right)$$

$$0 = 0 \; \checkmark$$

Both roots check, so $\frac{1}{2}$ and $\frac{1}{4}$ are the two roots that satisfy the original equation

$$8x - 6 + \frac{1}{x} = 0.$$

It is also correct to write the answers in decimal form as 0.5 and 0.25. This will speed up the checking process, especially if you use a scientific calculator when performing the check.

NOTE TO STUDENT: Fully worked-out solutions to all of the Practice Problems can be found at the back of the text starting at page SP-1

Practice Problem 3 Solve and check. $3x + 10 - \dfrac{8}{x} = 0$

If a quadratic equation is given to us in standard form, we examine it to see if it can be factored. If it is possible to factor the quadratic equation, then we use the zero factor property to find the solutions.

EXAMPLE 4 Solve and check. $6x^2 + 7x - 10 = 0$

Solution

$$(6x - 5)(x + 2) = 0 \qquad \text{Factor the quadratic equation.}$$

$$6x - 5 = 0 \qquad x + 2 = 0 \qquad \text{Set each factor equal to zero and solve for } x.$$

$$6x = 5 \qquad\qquad x = -2$$

$$x = \frac{5}{6}$$

Thus the solutions to the equation are $\dfrac{5}{6}$ and -2.

Check each solution in the original equation.

$$6\left(\frac{5}{6}\right)^2 + 7\left(\frac{5}{6}\right) - 10 \overset{?}{=} 0 \qquad\qquad 6(-2)^2 + 7(-2) - 10 \overset{?}{=} 0$$

$$6\left(\frac{25}{36}\right) + 7\left(\frac{5}{6}\right) - 10 \overset{?}{=} 0 \qquad\qquad 6(4) + 7(-2) - 10 \overset{?}{=} 0$$

$$\frac{25}{6} + \frac{35}{6} - 10 \overset{?}{=} 0 \qquad\qquad 24 + (-14) - 10 \overset{?}{=} 0$$

$$\frac{60}{6} - 10 \overset{?}{=} 0 \qquad\qquad 10 - 10 = 0 \; \checkmark$$

$$10 - 10 = 0 \; \checkmark$$

Practice Problem 4 Solve and check. $2x^2 + 9x - 18 = 0$

It is important to place a quadratic equation in standard form before factoring. This will sometimes involve removing parentheses and collecting like terms.

EXAMPLE 5 Solve. $x + (x - 6)(x - 2) = 2(x - 1)$

Solution

$$x + x^2 - 6x - 2x + 12 = 2x - 2 \qquad \text{Remove parentheses.}$$
$$x^2 - 7x + 12 = 2x - 2 \qquad \text{Collect like terms.}$$
$$x^2 - 9x + 14 = 0 \qquad \text{Add } -2x + 2 \text{ to each side.}$$
$$(x - 7)(x - 2) = 0 \qquad \text{Factor.}$$
$$x = 7, \quad x = 2 \qquad \text{Use the zero property.}$$

The solutions to the equation are 7 and 2. The check is left to the student.

Practice Problem 5 Solve. $-4x + (x - 5)(x + 1) = 5(1 - x)$

EXAMPLE 6 Solve. $\dfrac{8}{x + 1} - \dfrac{x}{x - 1} = \dfrac{2}{x^2 - 1}$

Solution The LCD $= (x + 1)(x - 1)$. We multiply each term by the LCD.

$$(x + 1)(x - 1)\left(\frac{8}{x + 1}\right) - (x + 1)(x - 1)\left(\frac{x}{x - 1}\right) = (x + 1)(x - 1)\left[\frac{2}{(x + 1)(x - 1)}\right]$$

$$8(x - 1) - x(x + 1) = 2 \qquad \text{Simplify.}$$
$$8x - 8 - x^2 - x = 2 \qquad \text{Remove parentheses.}$$
$$-x^2 + 7x - 8 = 2 \qquad \text{Collect like terms.}$$
$$-x^2 + 7x - 8 - 2 = 0 \qquad \text{Add } -2 \text{ to each side.}$$
$$-x^2 + 7x - 10 = 0 \qquad \text{Simplify.}$$
$$x^2 - 7x + 10 = 0 \qquad \text{Multiply each term by } -1.$$
$$(x - 5)(x - 2) = 0 \qquad \text{Factor and solve for } x.$$
$$x = 5, \quad x = 2 \qquad \text{Use the zero factor property.}$$

Check. The check is left to the student.

Practice Problem 6 Solve. $\dfrac{10x + 18}{x^2 + x - 2} = \dfrac{3x}{x + 2} + \dfrac{2}{x - 1}$

④ **Solving Applied Problems Involving Quadratic Equations**

A manager of a truck delivery service is determining the possible routes he may have to send trucks on during a given day. His delivery service must have the capability to send trucks between any two of the following Texas cities: Austin, Dallas, Tyler, Lufkin, and Houston. He made a rough map of the cities.

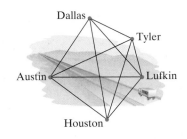

What is the maximum number of routes he may have to send trucks on in a given day, if there is a route between every two cities? One possible way to figure this out is to draw a straight line between every two cities. Let us assume that no three cities lie on a straight line. Therefore, the route between any two cities is distinct from all other routes. The following figure shows that we can draw exactly 10 lines, so the answer is that there are 10 truck routes. What if the manager has to service 18 cities? How many truck routes will there be then?

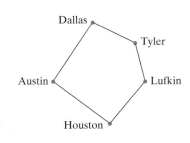

As the number of cities increases, it becomes quite difficult (and time consuming) to use a drawing to determine the number of routes. In a more advanced math course, it can be proved that the maximum number of truck routes (t) can be found by using the number of cities (n) in the following equation.

$$t = \frac{n^2 - n}{2}$$

Thus, to find the number of truck routes for 18 cities, we merely substitute 18 for n.

$$t = \frac{(18)^2 - (18)}{2}$$

$$= \frac{324 - 18}{2}$$

$$= \frac{306}{2}$$

$$= 153$$

Thus 153 truck routes are needed to service 18 cities if no three of those cities lie on a straight line.

EXAMPLE 7   A truck delivery company can handle a maximum of 36 truck routes in one day, and every two cities have a distinct truck route between them. How many separate cities can the truck company service?

Solution $t = \dfrac{n^2 - n}{2}$

$36 = \dfrac{n^2 - n}{2}$	Substitute 36 for t, the number of truck routes.
$72 = n^2 - n$	Multiply both sides of the equation by 2.
$0 = n^2 - n - 72$	Add -72 to each side to obtain the standard form of a quadratic equation.
$0 = (n - 9)(n + 8)$	Factor.
$n - 9 = 0 \qquad n + 8 = 0$	Set each factor equal to zero.
$n = 9 \qquad\qquad n = -8$	Solve for n.

We reject -8 as a meaningless solution for this problem. We cannot have a negative number of cities. Thus our answer is that the company can service 9 cities.

Check.   If we service $n = 9$ cities, does that really result in a maximum number of $t = 36$ truck routes?

$t = \dfrac{n^2 - n}{2}$	The equation in which $t =$ the number of truck routes and $n =$ the number of cities to be connected.
$36 \overset{?}{=} \dfrac{(9)^2 - (9)}{2}$	Substitute $t = 36$ and $n = 9$.
$36 \overset{?}{=} \dfrac{81 - 9}{2}$	
$36 \overset{?}{=} \dfrac{72}{2}$	
$36 = 36 \quad \checkmark$	

NOTE TO STUDENT: Fully worked-out solutions to all of the Practice Problems can be found at the back of the text starting at page SP-1

Practice Problem 7   How many cities can the company service with 28 truck routes?

Write in standard form. Determine the values of a, b, and c.

1. $x^2 + 8x + 7 = 0$

2. $x^2 + 6x - 5 = 0$

3. $8x^2 - 11x = 0$

4. $4x^2 + 20x = 0$

5. $x^2 + 15x - 7 = 12x + 8$

6. $3x^2 + 10x + 8 = -17x - 2$

Using the factoring method, solve for the roots of each quadratic equation. Be sure to place your equation in standard form before factoring.

7. $27x^2 - 9x = 0$

8. $6x + 18x^2 = 0$

9. $5x^2 - 7x = 3x$

10. $8x^2 - 9x = 7x$

11. $11x^2 - 13x = 8x - 3x^2$

12. $60x^2 + 14x = 9x^2 - 20x$

13. $x^2 - 3x - 28 = 0$

14. $x^2 + 3x - 40 = 0$

15. $2x^2 + 11x - 6 = 0$

16. $2x^2 - 9x - 5 = 0$

17. $3x^2 - 16x + 5 = 0$

18. $5x^2 - 21x + 4 = 0$

19. $x^2 = 3x + 18$

20. $x^2 = 10x - 24$

21. $12x^2 + 17x + 6 = 0$

22. $14x^2 + 29x + 12 = 0$

23. $m^2 + 13m + 24 = 2m - 6$

24. $p^2 + 15p + 25 = 6p + 5$

25. $8y^2 = 14y - 3$

26. $6y^2 = 11y - 3$

27. $25x^2 - 60x + 36 = 0$

Mixed Practice

28. $9x^2 - 24x + 16 = 0$

29. $(x - 5)(x + 2) = -10$

30. $(x - 6)(x + 1) = 8$

31. $3x^2 + 8x - 10 = -6x - 10$ **32.** $5x^2 - 7x + 12 = -8x + 12$ **33.** $y(y - 7) = 2(5 - 2y)$

34. $x(x - 4) = 3(x - 2)$ **35.** $(x - 6)(x - 4) = 3$ **36.** $(y + 6)(y - 1) = 8$

37. $2x(x + 3) = (3x + 1)(x + 1)$ **38.** $2(2x + 1)(x + 1) = x(7x + 1)$

Find the LCD and multiply each term by the LCD. Then solve. Check your solutions.

39. $x + \dfrac{8}{x} = 6$ **40.** $x - 1 = \dfrac{20}{x}$ **41.** $\dfrac{x}{4} - \dfrac{7}{x} = -\dfrac{3}{4}$ **42.** $\dfrac{x}{3} + \dfrac{2}{3} = \dfrac{5}{x}$

43. $\dfrac{3y - 8}{2} = \dfrac{2}{y + 1}$ **44.** $\dfrac{2x + 5}{2} = \dfrac{2}{x - 1}$ **45.** $\dfrac{4}{x} + \dfrac{3}{x + 5} = 2$

46. $\dfrac{7}{x} + \dfrac{6}{x - 1} = 2$ **47.** $\dfrac{24}{x^2 - 4} = 1 + \dfrac{2x - 6}{x - 2}$ **48.** $\dfrac{5x + 7}{x^2 - 9} = 1 + \dfrac{2x - 8}{x - 3}$

To Think About

49. Why can a quadratic equation in standard form with $c = 0$ always be solved?

50. Martha solved $(x + 3)(x - 2) = 14$ as follows.

$$x + 3 = 14 \qquad x - 2 = 14$$
$$x = 11 \qquad x = 16$$

Josette said this had to be wrong because these values do not check. Explain what is wrong with Martha's method.

51. The equation $ax^2 - 7x + c = 0$ has the roots $-\dfrac{3}{2}$ and 6. Find the values of a and c.

52. The equation $2x^2 + bx - 15 = 0$ has the roots -3 and $\dfrac{5}{2}$. Find the value of b.

Applications

Truck Routes   *The maximum number of truck routes, t, that are needed to provide direct service to n cities, where no three cities lie in a straight line, is given by the equation* $t = \dfrac{n^2 - n}{2}$.

Use this equation to find the solutions to exercises 53–58.

53. How many truck routes will be needed to run service between 20 cities?

54. How many truck routes will be needed to run service between 22 cities?

55. A company can handle a maximum of 15 truck routes in one day. How many separate cities can be serviced?

56. A company can handle a maximum of 21 truck routes in one day. How many separate cities can be serviced?

57. A company can handle a maximum of 45 truck routes in one day. How many separate cities can be serviced?

58. A company can handle a maximum of 55 truck routes in one day. How many separate cities can be serviced?

Business Revenue   *The revenue for producing the label for the new Starblazer video game is given by the formula* $R = -2.5n^2 + 80n - 280$, *where R is the revenue (in dollars) and n is the number of labels produced (in hundreds).*

59. If the revenue for producing labels is $320, what are the two values of the two different numbers of labels that may be produced?

60. If the revenue for producing labels is $157.50, what are the two values of the different numbers of labels that may be produced?

61. What is the average of the two answers you obtained in problem 59? Use the revenue equation to find the revenue for producing that number of labels. What is significant about that answer?

62. What is the average of the two answers you obtained in problem 60? Use the revenue equation to find the revenue for producing that number of labels. What is significant about that answer?

63. Find the revenue for producing 1500 and 1700 labels. Compare the two costs with the results from 59 and 61. What can you conclude?

64. Find the revenue for producing 1400 and 1800 labels. Compare the two costs with the results from 60 and 62. What can you conclude?

Cumulative Review

65. Simplify. $\dfrac{\dfrac{1}{x} + \dfrac{3}{x - 2}}{\dfrac{4}{x^2 - 4}}$

66. Add the fractions. $\dfrac{2x}{3x - 5} + \dfrac{2}{3x^2 - 11x + 10}$

Allergens in the Air   *Cars kick up a lot of dust. In Los Angeles, the dust consists of 20 known allergens, including molds, pollen, and animal dander. In two residential areas of Los Angeles, road dust contributed 12% of the allergens in the air. In contrast, in downtown Los Angeles, road dust is responsible for about 2% of the airborne allergens.*

67. If a 36-mg sample of dust were taken from the two residential areas of Los Angeles mentioned, how many milligrams of it would be road dust?

68. A sample of dust containing 2.6 mg of road dust was collected from downtown Los Angeles. What was the size of the sample of dust?

1 Solving Quadratic Equations Using the Square Root Property

Recall that the quadratic equation $x^2 = a$ has two possible solutions, $x = \sqrt{a}$ or $x = -\sqrt{a}$, where a is a nonnegative real number. That is $x = \pm\sqrt{a}$. This basic idea is called the **square root property.**

> **SQUARE ROOT PROPERTY**
>
> If $x^2 = a$, then $x = \sqrt{a}$ or $x = -\sqrt{a}$, for all nonnegative real numbers a.

EXAMPLE 1 Solve.

(a) $x^2 = 49$ **(b)** $x^2 = 20$ **(c)** $5x^2 = 125$

Solution

(a) $x^2 = 49$
$x = \pm\sqrt{49}$
$x = \pm 7$

(b) $x^2 = 20$
$x = \pm\sqrt{20}$
$x = \pm 2\sqrt{5}$

(c) $5x^2 = 125$ Divide both sides by 5 before taking the square root.

$x^2 = 25$
$x = \pm\sqrt{25}$
$x = \pm 5$

Practice Problem 1 Solve.

(a) $x^2 = 1$ **(b)** $x^2 - 50 = 0$ **(c)** $3x^2 = 81$

NOTE TO STUDENT: Fully worked-out solutions to all of the Practice Problems can be found at the back of the text starting at page SP-1

EXAMPLE 2 Solve. $3x^2 + 5x = 18 + 5x + x^2$

Solution Simplify the equation by placing all the variable terms on the left and the constants on the right.

$$3x^2 + 5x = 18 + 5x + x^2$$
$$2x^2 = 18$$
$$x^2 = 9$$
$$x = \pm 3$$

Practice Problem 2 Solve. $4x^2 - 5 = 319$

    The square root property can also be used if the term that is squared is a binomial.

    Sometimes when we use the square root property we obtain an irrational number. This will occur if the number on the right side of the equation is not a perfect square.

EXAMPLE 3   Solve. $(3x + 1)^2 = 8$

Solution $(3x + 1)^2 = 8$

$\qquad 3x + 1 = \pm\sqrt{8}$ Take the square root of both sides.

$\qquad 3x + 1 = \pm 2\sqrt{2}$ Simplify as much as possible.

Now we must solve the two equations expressed by the plus or minus statement.

$$3x + 1 = +2\sqrt{2} \qquad\qquad 3x + 1 = -2\sqrt{2}$$
$$3x = -1 + 2\sqrt{2} \qquad\qquad 3x = -1 - 2\sqrt{2}$$
$$x = \frac{-1 + 2\sqrt{2}}{3} \qquad\qquad x = \frac{-1 - 2\sqrt{2}}{3}$$

The roots of this quadratic equation are irrational numbers. They are

$$\frac{-1 + 2\sqrt{2}}{3} \quad \text{and} \quad \frac{-1 - 2\sqrt{2}}{3}.$$

We cannot simplify these roots further, so we leave them in this form.

Practice Problem 3   Solve. $(2x - 3)^2 = 12$

② Solving Quadratic Equations by Completing the Square

If a quadratic equation is not in a form where we can use the square root property, we can rewrite the equation by a method called **completing the square** to make it so. That is, we rewrite the equation with a perfect square on the left side so that we have an equation of the form $(ax + b)^2 = c$.

EXAMPLE 4   Solve. $x^2 + 12x = 4$

Solution   Write out the equation $x^2 + 12x = 4$. Think of a way to determine what number to add to each side.

$$x^2 + 12x \qquad\quad = 4$$

$$(x + 6)(x + 6) = 4 + 36 \qquad \text{We need a +36 to complete the square.}$$

$$x^2 + 12x + 36 = 4 + 36 \qquad \text{Add +36 to both sides of the equation.}$$

$$(x + 6)^2 = 40 \qquad\qquad \text{Write the left side as a perfect square.}$$

$$x + 6 = \pm\sqrt{40} \qquad\qquad \text{Solve for } x.$$

$$x + 6 = +2\sqrt{10} \qquad x + 6 = -2\sqrt{10}$$

$$x = -6 + 2\sqrt{10} \qquad\quad x = -6 - 2\sqrt{10}$$

The two roots are $-6 + 2\sqrt{10}$ and $-6 - 2\sqrt{10}$.

Practice Problem 4   Solve. $x^2 + 10x = 3$

Completing the square is a little more difficult when the coefficient of x is an odd number. Let's see what happens.

EXAMPLE 5 Solve. $x^2 - 3x - 1 = 0$

Solution Write the equation so that all the x-terms are on the left and the constants are on the right.

$$x^2 - 3x = 1$$

$$(x \quad)(x \quad)$$

The missing number when added to itself must be -3.

Since $-\dfrac{3}{2} + \left(-\dfrac{3}{2}\right) = -\dfrac{6}{2} = -3$, the missing number is $-\dfrac{3}{2}$.

$$\left(x - \dfrac{3}{2}\right)\left(x - \dfrac{3}{2}\right)$$

Complete the pattern.

$$x^2 - 3x + \dfrac{9}{4} = 1 + \dfrac{9}{4}$$

Complete the square by adding $\left(-\dfrac{3}{2}\right)\left(-\dfrac{3}{2}\right) = \dfrac{9}{4}$ to each side.

$$\left(x - \dfrac{3}{2}\right)^2 = \dfrac{13}{4}$$

Solve for x.

$$x - \dfrac{3}{2} = \pm\sqrt{\dfrac{13}{4}}$$

$$x - \dfrac{3}{2} = \pm\dfrac{\sqrt{13}}{2}$$

$$x - \dfrac{3}{2} = +\dfrac{\sqrt{13}}{2} \qquad x - \dfrac{3}{2} = -\dfrac{\sqrt{13}}{2}$$

$$x = \dfrac{3}{2} + \dfrac{\sqrt{13}}{2} \qquad x = \dfrac{3}{2} - \dfrac{\sqrt{13}}{2}$$

$$x = \dfrac{3 + \sqrt{13}}{2} \qquad x = \dfrac{3 - \sqrt{13}}{2}$$

The two roots are $\dfrac{3 \pm \sqrt{13}}{2}$.

NOTE TO STUDENT: *Fully worked-out solutions to all of the Practice Problems can be found at the back of the text starting at page SP-1*

Practice Problem 5 Solve. $x^2 - 5x = 7$

If a, the coefficient of the squared variable, is not 1, we divide all terms of the equation by a so that the coefficient of the squared variable will be 1.

Let us summarize for future reference the steps we have performed in order to solve a quadratic equation by completing the square.

COMPLETING THE SQUARE

1. Put the equation in the form $ax^2 + bx = c$. If $a \ne 1$, divide each term by a.

2. Square $\dfrac{b}{2}$ and add the result to both sides of the equation.

3. Factor the left side (a perfect square trinomial).

4. Use the square root property.

5. Solve the equations.

6. Check the solutions in the original equation.

EXAMPLE 6   Solve. $4x^2 + 4x - 3 = 0$

Solution

Step 1 $4x^2 + 4x = 3$ Place the constant on the right. This puts the equation in the form $ax^2 + bx = c$.

$x^2 + x = \dfrac{3}{4}$ Divide all terms by 4.

Step 2 $x^2 + 1x = \dfrac{3}{4}$ Take one-half the coefficient of x and square it. $\left(\dfrac{1}{2}\right)^2 = \dfrac{1}{4}$.

$x^2 + x + \dfrac{1}{4} = \dfrac{3}{4} + \dfrac{1}{4}$ Add $\dfrac{1}{4}$ to each side.

Step 3 $\left(x + \dfrac{1}{2}\right)^2 = 1$ Factor the left side.

Step 4 $x + \dfrac{1}{2} = \pm\sqrt{1}$ Use the square root property.

Step 5 $x + \dfrac{1}{2} = \pm 1$

$x + \dfrac{1}{2} = 1 \qquad\qquad x + \dfrac{1}{2} = -1$

$x = \dfrac{2}{2} - \dfrac{1}{2} \qquad\qquad x = -\dfrac{2}{2} - \dfrac{1}{2}$

$x = \dfrac{1}{2} \qquad\qquad x = -\dfrac{3}{2}$

Thus the two roots are $\dfrac{1}{2}$ and $-\dfrac{3}{2}$.

Step 6 $4\left(\dfrac{1}{2}\right)^2 + 4\left(\dfrac{1}{2}\right) - 3 \stackrel{?}{=} 0$ Check.

$4\left(\dfrac{1}{4}\right) + 4\left(\dfrac{1}{2}\right) - 3 \stackrel{?}{=} 0$

$1 + 2 - 3 = 0 \quad\checkmark$

$4\left(-\dfrac{3}{2}\right)^2 + 4\left(-\dfrac{3}{2}\right) - 3 \stackrel{?}{=} 0$

$4\left(\dfrac{9}{4}\right) + 4\left(-\dfrac{3}{2}\right) - 3 \stackrel{?}{=} 0$

$9 - 6 - 3 = 0 \quad\checkmark$

Both values check.

Practice Problem 6   Solve. $8x^2 + 2x - 1 = 0$

This method of completing the square will enable us to solve any quadratic equation that has real number roots. It is usually faster, however, to factor the quadratic equation if the polynomial is factorable.

Student Solutions Manual CD/Video PH Math Tutor Center MathXL®Tutorials on CD MathXL® MyMathLab® Interactmath.com

Solve using the square root property.

1. $x^2 = 64$

2. $x^2 = 36$

3. $x^2 = 98$

4. $x^2 = 72$

5. $x^2 - 28 = 0$

6. $x^2 - 45 = 0$

7. $5x^2 = 45$

8. $3x^2 = 48$

9. $6x^2 = 120$

10. $3x^2 = 150$

11. $3x^2 - 375 = 0$

12. $4x^2 - 252 = 0$

13. $5x^2 + 13 = 73$

14. $6x^2 + 5 = 53$

15. $13x^2 + 17 = 82$

16. $11x^2 + 14 = 47$

17. $(x - 7)^2 = 16$

18. $(x + 2)^2 = 36$

19. $(x + 4)^2 = 6$

20. $(x - 5)^2 = 36$

21. $(2x + 5)^2 = 2$

22. $(3x - 4)^2 = 6$

23. $(3x - 1)^2 = 7$

24. $(2x + 3)^2 = 10$

25. $(7x + 2)^2 = 12$

26. $(8x + 5)^2 = 20$

27. $(4x - 5)^2 = 54$

28. $(6x - 7)^2 = 72$

Solve by completing the square.

29. $x^2 - 6x = 11$

30. $x^2 - 2x = 4$

31. $x^2 + 6x - 7 = 0$

32. $x^2 - 12x - 13 = 0$

33. $x^2 - 12x - 5 = 0$

34. $x^2 + 16x + 30 = 0$

35. $x^2 - 7x = 0$

36. $x^2 - 9x = 0$

37. $5x^2 - 25x = 0$

38. $2x^2 + 5x = 3$

39. $2x^2 - 7x = 9$

40. $2x^2 + 13x = 7$

To Think About

41. Let b represent a constant value. Solve $x^2 + bx - 7 = 0$ for x by completing the square.

42. Solve $x^2 + bx + c = 0$ for x in terms of b and c.

Cumulative Review

43. Solve. $3a - 5b = 8$
$5a - 7b = 8$

44. Solve. $\dfrac{x - 4}{x - 2} - \dfrac{1}{x} = \dfrac{x - 3}{x^2 - 2x}$

45. *Hiking Knife* While hiking in the Grand Canyon, Will had to use his hunting knife to clear a fallen branch from the path. Find the pressure on the blade of his knife if the knife is 8 in. long and sharpened to an edge 0.01 in. wide and he applies a force of 16 pounds. Use the formula $P = \dfrac{F}{A}$, where P is the pressure on the blade of the knife, F is the force in pounds placed on the knife, and A is the area exposed to an object by the sharpened edge of the knife.

46. *Tire Defects* Darlene inspected a sample of 100 new steel-belted radial tires. 13 of the tires had defects in workmanship, while 9 of the tires had defects in materials. Of those defective tires, 6 had defects in both workmanship and materials. How many of the 100 tires had any kind of defect? (Assume that the only defects found were those in materials or workmanship.)

47. *Class Survey* A Sociology class took a survey of the class members. There were 48 students in the class, with an equal number of men and women. Twenty-one students said that they were single. If 7 of the single students were women, what percent of the men in the class were single? (Round to the nearest tenth of a percent.)

1 Solving a Quadratic Equation Using the Quadratic Formula

To find solutions of a quadratic equation, you can factor or you can complete the square when the trinomial is not factorable. If we use the method of completing the square on the general equation $ax^2 + bx + c = 0$, we can derive a general formula for the solution of any quadratic equation. This is called the **quadratic formula.**

QUADRATIC FORMULA

The roots of any quadratic equation of the form $ax^2 + bx + c = 0$, where a, b, and c are real numbers and $a \neq 0$, are

$$x = \frac{-b \pm \sqrt{b^2 - 4ac}}{2a}.$$

All you will need to know are the values of a, b, and c to solve a quadratic equation using the quadratic formula.

EXAMPLE 1 Solve using the quadratic formula. $3x^2 + 10x + 7 = 0$

Solution In our given equation, $3x^2 + 10x + 7 = 0$, we have

$a = 3$ (the coefficient of x^2), $\quad b = 10$ (the coefficient of x), and $c = 7$ (the constant term).

$$x = \frac{-b \pm \sqrt{b^2 - 4ac}}{2a} = \frac{-10 \pm \sqrt{(10)^2 - 4(3)(7)}}{2(3)} \qquad \text{Write the quadratic formula and substitute values for } a, b, \text{ and } c.$$

$$= \frac{-10 \pm \sqrt{100 - 84}}{6} \qquad \text{Simplify.}$$

$$= \frac{-10 \pm \sqrt{16}}{6} = \frac{-10 \pm 4}{6}$$

$$x = \frac{-10 + 4}{6} = \frac{-6}{6} = -1 \qquad \text{Using the positive sign.}$$

$$x = \frac{-10 - 4}{6} = \frac{-14}{6} = -\frac{7}{3} \qquad \text{Using the negative sign.}$$

Thus the two solutions are -1 and $-\frac{7}{3}$.

[*Note:* Here we obtain rational roots. We would obtain the same answer by factoring $3x^2 + 10x + 7 = 0$ into $(3x + 7)(x + 1) = 0$ and setting each factor $= 0$.]

NOTE TO STUDENT: Fully worked-out solutions to all of the Practice Problems can be found at the back of the text starting at page SP-1

Practice Problem 1 Solve using the quadratic formula. $x^2 - 7x + 6 = 0$

Often the roots will be irrational numbers. Sometimes these roots can be simplified. You should always leave your answer in simplest form.

EXAMPLE 2 Solve. $2x^2 - 4x - 9 = 0$

Solution $\qquad a = 2, \quad b = -4, \quad c = -9$

$$x = \frac{-b \pm \sqrt{b^2 - 4ac}}{2a} = \frac{-(-4) \pm \sqrt{(-4)^2 - 4(2)(-9)}}{2(2)}$$

$$= \frac{4 \pm \sqrt{16 + 72}}{4} = \frac{4 \pm \sqrt{88}}{4} \qquad \text{Do not stop here!}$$

Notice that we can simplify $\sqrt{88}$.

$$x = \frac{4 \pm 2\sqrt{22}}{4} = \frac{2(2 \pm \sqrt{22})}{4} = \frac{\overset{1}{\cancel{2}}(2 \pm \sqrt{22})}{\underset{2}{\cancel{4}}} = \frac{2 \pm \sqrt{22}}{2}$$

Be careful. Here we were able to divide the numerator and denominator by 2 because 2 was a factor of every term.

Practice Problem 2   Solve. $3x^2 - 8x + 3 = 0$

A quadratic equation *must* be written in the standard form $ax^2 + bx + c = 0$ *before* the quadratic formula can be used. Several algebraic steps may be needed to accomplish this objective. Also, since it is much easier to use the formula if a, b, and c are integers, we will avoid using fractional values.

EXAMPLE 3   Solve. $x^2 = 5 - \frac{3}{4}x$

Solution   First we obtain an equivalent equation that does not have fractions.

$$x^2 = 5 - \frac{3}{4}x$$

$$4(x^2) = 4(5) - 4\left(\frac{3}{4}x\right) \qquad \text{Multiply each term by the LCD of 4.}$$

$$4x^2 = 20 - 3x \qquad \text{Simplify.}$$

$$4x^2 + 3x - 20 = 0 \qquad \text{Add } 3x - 20 \text{ to each side.}$$

$$a = 4, \quad b = 3, \quad c = -20 \qquad \text{Substitute } a = 4, b = 3, \text{ and } c = -20 \text{ into the quadratic formula.}$$

$$x = \frac{-3 \pm \sqrt{(3)^2 - 4(4)(-20)}}{2(4)}$$

$$x = \frac{-3 \pm \sqrt{9 + 320}}{8}$$

$$x = \frac{-3 \pm \sqrt{329}}{8}$$

Practice Problem 3   Solve. $x^2 = 7 - \frac{3}{5}x$

② Finding a Decimal Approximation for the Real Roots of a Quadratic Equation

EXAMPLE 4   Find the roots of $3x^2 - 5x = 7$. Approximate to the nearest thousandth.

Solution

$$3x^2 - 5x = 7$$

$$3x^2 - 5x - 7 = 0 \qquad \text{Place the equation in standard form.}$$

$$a = 3, \quad b = -5, \quad c = -7$$

$$x = \frac{-(-5) \pm \sqrt{(-5)^2 - 4(3)(-7)}}{2(3)} = \frac{5 \pm \sqrt{25 + 84}}{6}$$

$$= \frac{5 \pm \sqrt{109}}{6}$$

Using most scientific calculators, we can find $\dfrac{5 + \sqrt{109}}{6}$ with these keystrokes.

$$5 \boxed{+} 109 \boxed{\sqrt{}} \boxed{=} \boxed{\div} 6 \boxed{=} \; 2.5733844$$

Rounding to the nearest thousandth, we have $x \approx 2.573$.

To find $\dfrac{5 - \sqrt{109}}{6}$, we use the following keystrokes.

$$5 \boxed{-} 109 \boxed{\sqrt{}} \boxed{=} \boxed{\div} 6 \boxed{=} \; -0.9067178$$

Rounding to the nearest thousandth, we have $x \approx -0.907$.

If you do not have a calculator with a square root key, look up $\sqrt{109}$ in the square root table. $\sqrt{109} \approx 10.440$. Using this result, we have

$$x \approx \frac{5 + 10.440}{6} = \frac{15.440}{6} \approx 2.573 \quad \text{(rounded to the nearest thousandth)}$$

and

$$x \approx \frac{5 - 10.440}{6} = \frac{-5.440}{6} \approx -0.907 \quad \text{(rounded to the nearest thousandth)}.$$

If you have a graphing calculator, there is an alternate method for finding the decimal approximation of the roots of a quadratic equation. In Section 10.4, you will cover graphing quadratic equations. You will discover that finding the roots of a quadratic equation of the form $ax^2 + bx + c = 0$ is equivalent to finding the values of x on the graph of $y = ax^2 + bx + c$, where $y = 0$. These may be approximated quite quickly with a graphing calculator.

NOTE TO STUDENT: *Fully worked-out solutions to all of the Practice Problems can be found at the back of the text starting at page SP-1*

Practice Problem 4 Find the roots of $2x^2 = 13x + 5$. Approximate to the nearest thousandth.

3 Determining Whether a Quadratic Equation Has No Real Solutions

EXAMPLE 5 Solve. $2x^2 + 5 = -3x$

Solution The equation is not in standard form.

$2x^2 + 5 = -3x$ Add $3x$ to both sides.

$2x^2 + 3x + 5 = 0$ We can now find a, b, and c.

$a = 2, \quad b = 3, \quad c = 5$ Substitute these values into the quadratic formula.

$$x = \frac{-3 \pm \sqrt{(3)^2 - 4(2)(5)}}{2(2)} = \frac{-3 \pm \sqrt{9 - 40}}{4} = \frac{-3 \pm \sqrt{-31}}{4}$$

There is no real number that is $\sqrt{-31}$. Since we are using only real numbers in this book, we say there is no solution to the problem. (*Note:* In more advanced math courses, these types of numbers, called complex numbers, will be studied.)

Practice Problem 5 Solve. $5x^2 + 2x = -3$

We can tell whether the roots of any given quadratic equation are real numbers. Look at the quadratic formula.

$$\frac{-b \pm \sqrt{b^2 - 4ac}}{2a}$$

The expression under the radical sign is called the **discriminant.**

$$b^2 - 4ac \text{ is the discriminant.}$$

If the discriminant is a negative number, the roots are not real numbers, and there is no real number solution to the equation.

EXAMPLE 6   Determine whether $3x^2 = 5x - 4$ has real number solution(s).

Solution   First, we place the equation in standard form. Then we need only check the discriminant.

$$3x^2 = 5x - 4$$
$$3x^2 - 5x + 4 = 0$$
$a = 3, \quad b = -5, \quad c = 4$ Substitute the values for a, b, and c into the discriminant and evaluate.

$$b^2 - 4ac = (-5)^2 - 4(3)(4) = 25 - 48 = -23$$

The discriminant is negative. Thus $3x^2 = 5x - 4$ has no real number solution(s).

Practice Problem 6   Determine whether each equation has real number solution(s).

(a) $5x^2 = 3x + 2$ **(b)** $2x^2 - 5 = 4x$

Verbal and Writing Skills

1. In the equation $3x^2 + 4x - 7 = 0$, what can we learn from the discriminant?

2. In the equation $7x^2 - 6x + 2 = 0$, what can we learn from the discriminant?

3. Is the equation $4x^2 = -5x + 6$ in standard form? If not, place it in standard form and find a, b, and c.

4. Is the equation $8x = -9x^2 + 2$ in standard form? If not, place it in standard form and find a, b, and c.

Solve using the quadratic formula. If there are no real roots, say so.

5. $x^2 + 3x - 10 = 0$

6. $x^2 - 5x + 4 = 0$

7. $x^2 - 3x - 8 = 0$

8. $x^2 + 9x + 7 = 0$

9. $4x^2 - 5x - 6 = 0$

10. $3x^2 + 7x - 6 = 0$

11. $2x^2 = 3x + 20$

12. $9x - 2 = 3x^2$

13. $6x^2 - 3x = 1$

14. $3 = 7x - 4x^2$

15. $x + \dfrac{3}{2} = 3x^2$

16. $\dfrac{2}{3}x^2 - x = \dfrac{5}{2}$

17. $\dfrac{x}{2} + \dfrac{5}{x} = \dfrac{7}{2}$

18. $\dfrac{x}{5} - \dfrac{3}{5} = \dfrac{2}{x}$

19. $5x^2 + 6x + 2 = 0$

20. $3x^2 - 4x + 2 = 0$

21. $3y^2 = 8y - 3$

22. $7y^2 = 10y - 2$

23. $\dfrac{d^2}{2} + \dfrac{5d}{6} - 2 = 0$

24. $\dfrac{k^2}{4} - \dfrac{5k}{6} = \dfrac{2}{3}$

25. $4x^2 + 8x + 11 = 0$

26. $5x^2 + 7x + 9 = 0$

27. $x^2 - 10x + 25 = 0$

28. $4x^2 - 12x + 9 = 0$

Use the quadratic formula to find the roots. Find a decimal approximation to the nearest thousandth.

29. $x^2 + 5x - 2 = 0$

30. $x^2 + 3x - 7 = 0$

31. $2x^2 - 7x - 5 = 0$

32. $3x^2 - 2x - 3 = 0$

33. $5x^2 + 10x + 1 = 0$

34. $2x^2 + 9x + 2 = 0$

Mixed Practice

Solve for the variable. Choose the method you feel will work best.

35. $6x^2 - 13x + 6 = 0$

36. $6x^2 - 7x - 3 = 0$

37. $3(x^2 + 1) = 10x$

38. $3 + x(x + 2) = 18$

39. $(t + 5)(t - 3) = 7$

40. $(b + 4)(b - 7) = 4$

41. $y^2 - \dfrac{2}{5}y = 2$

42. $y^2 - 3 = \dfrac{8}{3}y$

43. $3x^2 - 13 = 0$

44. $5x^2 + 9x = 0$

45. $x(x - 2) = 7$

46. $(3x + 1)^2 = 12$

To Think About

▲ **47.** *Swimming Pool Design* John and Chris Maney have designed a new in-ground swimming pool for their backyard. The pool is rectangular and measures 30 feet by 20 feet. They have also designed a tile border to go around the pool. In the diagram below, this border is x feet wide. The tile border covers 216 square feet in area. What is the width x of this tile border? (*Hint:* Find an expression for the area of the large rectangle and subtract an expression for the area of the small rectangle.)

▲ **48.** *Lynn Campus Entrance* The Alumni Association of North Shore Community College has designed a new garden for the front entrance of the Lynn campus. It is rectangular and measures 40 feet by 30 feet. The association has designed a cement walkway to go around this rectangular garden. In the diagram below, this cement walkway is x feet wide. The entire cement walkway covers 456 square feet in area. What is the width x of this cement walkway? (*Hint:* Find an expression for the area of the large rectangle and subtract an expression for the area of the small rectangle.)

Cumulative Review

49. Simplify. $\dfrac{3x}{x-2} + \dfrac{4}{x+2} - \dfrac{x+22}{x^2-4}$

50. Give the equation of the line passing through $(-8, 2)$ and $(5, -3)$ in **(a)** slope–intercept form and **(b)** standard form with integer coefficients.

51. Divide. $(x^3 + 8x^2 + 17x + 6) \div (x + 3)$

52. Solve for g. $2p = 3gf - 7p + 20$

▲ **53.** *Kaleidoscopes* A very large kaleidoscope is produced for a children's science museum. It is so big that it must be supported by a specially made stand. The kaleidoscope is in the shape of a cylinder, measures 5.35 feet in length, and has a radius of 2.40 feet. To prevent contamination by dust, the museum keeps the interior of the cylinder in a vacuum. Find the volume of air in cubic feet that must be removed from the cylinder to establish a vacuum. Use $\pi \approx 3.14$. Round to the nearest hundredth.

▲ **54.** *Archery Target* An archery bull's-eye is set up for the purpose of training Olympic athletes. There are 5 concentric circles within the largest circle. Each circle is 2 inches from the adjacent circle. The smallest circle has an area of 12.56 square inches. What is the area between the largest circle and the circle that is closest to the largest circle? Use $\pi \approx 3.14$. Round to the nearest hundredth.

How are you doing with your homework assignments in Sections 10.1 to 10.3? Do you feel you have mastered the material so far? Do you understand the concepts you have covered? Before you go further in the textbook, take some time to do each of the following problems.

10.1

Solve by factoring.

1. $x^2 - 13x - 48 = 0$

2. $5x^2 + 7x = 14x$

3. $5x^2 = 22x - 8$

4. $-2x + 1 = 8x^2$

5. $x(x + 9) = 4(x + 6)$

6. $\dfrac{5}{x + 2} = \dfrac{2x - 1}{5}$

Simplify all answers in problems 7–14.

10.2

Solve by taking the square root of each side of the equation.

7. $x^2 - 18 = 0$

8. $3x^2 + 1 = 76$

Solve by completing the square.

9. $x^2 + 8x + 5 = 0$

10. $2x^2 + 3x - 7 = 0$

10.3

Solve if possible using the quadratic formula.

11. $2x^2 + 4x - 5 = 0$

12. $2x^2 = 7x - 4$

13. $3x^2 + 8x + 1 = 0$

14. $5x^2 + 3 = 4x$

Now turn to page SA-30 for the answer to each of these problems. Each answer also includes a reference to the objective in which the problem is first taught. If you missed any of these problems, you should stop and review the Examples and Practice Problems in the referenced objective. A little review now will help you master the material in the upcoming sections of the text.

1. _____
2. _____
3. _____
4. _____
5. _____
6. _____
7. _____
8. _____
9. _____
10. _____
11. _____
12. _____
13. _____
14. _____

591

Student Learning Objectives

After studying this section, you will be able to:

1 Graph equations of the form $y = ax^2 + bx + c$ using ordered pairs.

2 Graph equations of the form $y = ax^2 + bx + c$ using the vertex formula.

1 Graphing Equations of the Form $y = ax^2 + bx + c$ Using Ordered Pairs

Recall that the graph of a linear equation is always a straight line. What is the graph of a quadratic equation? What does the graph look like if one of the variable terms is squared? We will use point plotting to look for a pattern. We begin with a table of values.

EXAMPLE 1   Graph. **(a)** $y = x^2$ **(b)** $y = -x^2$

Solution

(a) $y = x^2$

x	y
−3	9
−2	4
−1	1
0	0
1	1
2	4
3	9

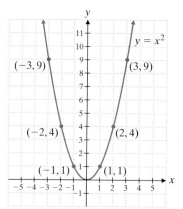

(b) $y = -x^2$

x	y
−3	−9
−2	−4
−1	−1
0	0
1	−1
2	−4
3	−9

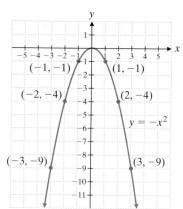

NOTE TO STUDENT: Fully worked-out solutions to all of the Practice Problems can be found at the back of the text starting at page SP-1

Practice Problem 1   Graph. **(a)** $y = 2x^2$ **(b)** $y = -2x^2$

(a) $y = 2x^2$

x	y
−2	
−1	
0	
1	
2	

(b) $y = -2x^2$

x	y
−2	
−1	
0	
1	
2	

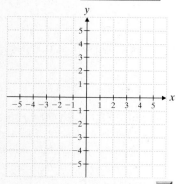

TO THINK ABOUT: When the Coefficient of x^2 Is Negative   What happens to the graph when the coefficient of x^2 is negative? What happens when the coefficient of x^2 is an integer greater than 1 or less than -1? Graph $y = 3x^2$ and $y = -3x^2$. What happens when $-1 < a < 1$? Graph $y = \frac{1}{3}x^2$ and $y = -\frac{1}{3}x^2$.

The curves that you have been graphing are called **parabolas.** What happens to the graph of a parabola as the quadratic equation that describes it changes?

In the preceding examples, the equations we worked with were relatively simple. They provided us with an understanding of the general shape of the graph of a quadratic equation. We have seen how changing the equation affects the graph of that equation. Let's look at an equation that is slightly more difficult to graph.

The **vertex** is the lowest point on a parabola that opens upward, or the highest point on a parabola that opens downward.

EXAMPLE 2   Graph $y = x^2 - 2x$. Identify the coordinates of the vertex.

Solution

x	y
-2	8
-1	3
0	0
1	-1
2	0
3	3
4	8

vertex

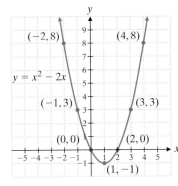

The vertex is at $(1, -1)$.

Notice that the x-intercepts are at $(0, 0)$ and $(2, 0)$. The x-coordinates are the solutions to the equation $x^2 - 2x = 0$.

$$x^2 - 2x = 0$$
$$x(x - 2) = 0$$
$$x = 0, \qquad x = 2$$

Practice Problem 2   Graph. For each table of values, use x-values from -4 to 2. Identify the coordinates of the vertex.
(a) $y = x^2 + 2x$ **(b)** $y = -x^2 - 2x$

(a) $y = x^2 + 2x$

x	y
-4	
-3	
-2	
-1	
0	
1	
2	

(b) $y = -x^2 - 2x$

x	y
-4	
-3	
-2	
-1	
0	
1	
2	

Graphing Calculator

Graphing Quadratic Equations

Graph $y = x^2 - 4x$ on a graphing calculator using an appropriate window.
Display:

Next you can use the Trace and Zoom features of your calculator to determine the vertex. Some calculators have a feature that will calculate the maximum or minimum point on the graph. Use the feature that calculates the minimum point to find the vertex.
Display:

Thus the vertex is at $(2, -4)$.

② Graphing Equations of the Form $y = ax^2 + bx + c$ Using the Vertex Formula

Finding the vertex is the key to graphing a parabola. Let's take a look at some of the previous graphs and identify the vertex of each parabola.

Equation:
$$y = x^2 \qquad y = x^2 - 3 \qquad y = (x - 1)^2 \qquad\qquad y = (x + 2)^2 \qquad\qquad y = x^2 - 2x$$
$$y = x^2 - 2x + 1 \qquad y = x^2 + 4x + 4$$

Vertex:
$$(0, 0) \qquad (0, -3) \qquad (1, 0) \qquad\qquad (-2, 0) \qquad\qquad (1, -1)$$

Notice that the x-coordinate of the vertex of the parabola for each of the first two equations is 0. You may also notice that these two equations do not have middle terms. That is, for $y = ax^2 + bx + c$, $b = 0$. This is not true of the last three equations. As you may have already guessed, there is a relationship between the quadratic equation and the x-coordinate of the vertex of the parabola it describes.

VERTEX OF A PARABOLA

The x-coordinate of the vertex of the parabola described by $y = ax^2 + bx + c$ is

$$x = \frac{-b}{2a}.$$

Once we have the x-coordinate, we can find the y-coordinate using the equation. We can plot the point and draw a general sketch of the curve. The sign of the squared term will tell us whether the graph opens upward or downward.

EXAMPLE 3 $y = -x^2 - 2x + 3$. Determine the vertex and the x-intercepts. Then sketch the graph.

Solution We will first determine the coordinates of the vertex. We begin by finding the x-coordinate.

$$x = \frac{-b}{2a} = \frac{-(-2)}{2(-1)} = \frac{2}{-2} = -1$$

$$y = -(-1)^2 - 2(-1) + 3 \qquad \text{Determine } y \text{ when } x \text{ is } -1.$$
$$= 4 \qquad\qquad\qquad\qquad \text{The point } (-1, 4) \text{ is the vertex of the parabola.}$$

Now we will determine the x-intercepts, those points where the graph crosses the x-axis. This occurs when $y = 0$. These are the solutions to the equation $-x^2 - 2x + 3 = 0$. Note that this equation is factorable.

$$-x^2 - 2x + 3 = 0 \qquad \text{Find the solutions to the equation.}$$
$$(-x + 1)(x + 3) = 0 \qquad \text{Factor.}$$
$$-x + 1 = 0 \qquad x + 3 = 0 \qquad \text{Solve for } x.$$
$$x = 1 \qquad\qquad x = -3 \qquad \text{The } x\text{-intercepts are at } (1, 0) \text{ and at } (-3, 0).$$

We now have enough information to sketch the graph. Since $a = -1$, the parabola opens downward and the vertex $(-1, 4)$ is the highest point. The graph crosses the x-axis at $(-3, 0)$ and at $(1, 0)$.

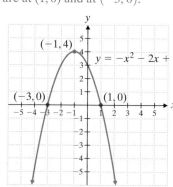

Practice Problem 3 $y = x^2 - 4x - 5$. Determine the vertex and the *x*-intercepts. Then sketch the graph.

NOTE TO STUDENT: Fully worked-out solutions to all of the Practice Problems can be found at the back of the text starting at page SP-1

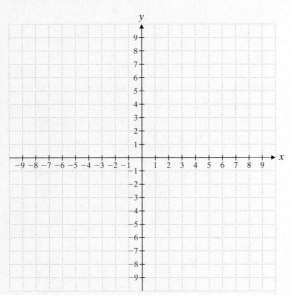

EXAMPLE 4   A person throws a ball into the air and its distance from the ground in feet is given by the equation $h = -16t^2 + 64t + 5$, where *t* is measured in seconds.

(a) Graph the equation $h = -16t^2 + 64t + 5$.

(b) What physical significance does the vertex of the graph have?

(c) After how many seconds will the ball hit the ground?

Solution

(a) Since the equation is a quadratic equation, the graph is a parabola. *a* is negative. Thus the parabola opens downward and the vertex is the highest point. We begin by finding the coordinates of the vertex. Our equation has ordered pairs (t, h) instead of (x, y). The *t*-coordinate of the vertex is

$$t = \frac{-b}{2a} = \frac{-64}{2(-16)} = \frac{-64}{-32} = 2.$$

Substitute $t = 2$ into the equation to find the *h*-coordinate of the vertex.

$$h = -16(2)^2 + 64(2) + 5 = -64 + 128 + 5 = 69$$

The vertex (t, h) is $(2, 69)$.

To sketch this graph, it would be helpful to know the *h*-intercept, that is, the point where $t = 0$.

$$h = -16(0)^2 + 64(0) + 5$$
$$= 5$$

The *h*-intercept is at $(0, 5)$.

The *t*-intercept is where $h = 0$. We will use the quadratic formula to find *t* when *h* is 0, that is, to find the roots of the quadratic equation $-16t^2 + 64t + 5 = 0$.

$$-16t^2 + 64t + 5 = 0$$

$$t = \frac{-64 \pm \sqrt{(64)^2 - 4(-16)(5)}}{2(-16)}$$

$$t = -\frac{64 \pm \sqrt{4096 + 320}}{-32} = -\frac{64 \pm \sqrt{4416}}{-32} = \frac{8 \pm \sqrt{69}}{4}$$

Using a calculator or a square root table, we have the following.

$$t = \frac{8 + 8.307}{4} \approx 4.077, \qquad t \approx 4.1 \quad \text{(to the nearest tenth)}$$

$$t = \frac{8 - 8.307}{4} \approx -0.077$$

We will not use negative t-values. We are studying the height of the ball from the time it is thrown ($t = 0$ seconds) to the time that it hits the ground (approximately $t = 4.1$ seconds). It is not useful to find points with negative values of t because they have no meaning in this situation.

(b) The vertex represents the greatest height of the ball. The ball is 69 feet above the ground 2 seconds after it is thrown.

(c) The ball will hit the ground approximately 4.1 seconds after it is thrown. This is the point at which the graph intersects the t-axis, the positive root of the equation.

NOTE TO STUDENT: Fully worked-out solutions to all of the Practice Problems can be found at the back of the text starting at page SP-1

Practice Problem 4 A ball is thrown upward with a speed of 32 feet per second from a distance 10 feet above the ground. The height of the ball in feet is given by the equation $h = -16t^2 + 32t + 10$, where t is measured in seconds.

(a) Graph the equation $h = -16t^2 + 32t + 10$.

(b) What is the greatest height of the ball?

(c) After how many seconds will the ball hit the ground?

10.4 EXERCISES

Student Solutions Manual | CD/Video | PH Math Tutor Center | MathXL®Tutorials on CD | MathXL® | MyMathLab® | Interactmath.com

Verbal and Writing Skills

1. How many x-intercepts can a parabola have?

2. How many vertices can a parabola have?

3. For a parabola in the form $y = ax^2 + bx + c$, what happens to the equation $x = -\dfrac{b}{2a}$ if $b = 0$?

4. For a parabola in the form $y = ax^2 + bx + c$, what can be said about the parabola if $a < 0$?

Graph each equation.

5. $y = x^2 + 2$

6. $y = x^2 - 4$

7. $y = -\frac{1}{3}x^2$

8. $y = -\frac{1}{2}x^2$

9. $y = 2x^2 - 3$

10. $y = 3x^2 + 2$

11. $y = (x - 2)^2$

12. $y = (x + 1)^2$

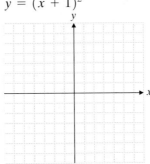

13. $y = -\frac{1}{2}x^2 + 4$

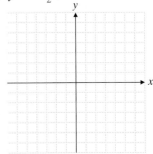

14. $y = -\frac{1}{3}x^2 + 6$

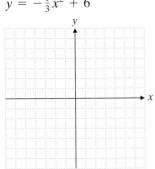

15. $y = \frac{1}{2}(x - 3)^2$

16. $y = \frac{1}{2}(x + 2)^2$

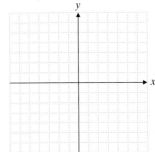

Graph each equation. Identify the coordinates of the vertex.

17. $y = x^2 + 4x$

18. $y = -x^2 + 4x$

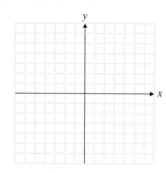

19. $y = -x^2 - 4x$

20. $y = x^2 - 4x$

21. $y = -2x^2 + 8x$

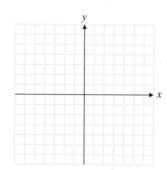

22. $y = x^2 + 6x$

Determine the vertex and the x-intercepts. Then sketch the graph.

23. $y = x^2 + 2x - 3$

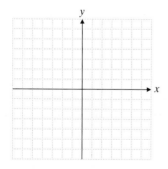

24. $y = x^2 - 4x + 3$

25. $y = -x^2 - 6x - 5$

26. $y = -x^2 + 6x - 8$

27. $y = x^2 + 6x + 9$

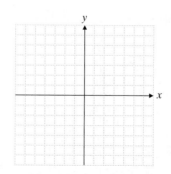

28. $y = -x^2 + 2x - 1$

Applications

29. ***Miniature Rockets*** A miniature rocket is launched so that its height h in meters after t seconds is given by the equation $h = -4.9t^2 + 39.2t + 4$.
    (a) Draw a graph of the equation.
    (b) How high is the rocket after 2 seconds?
    (c) What is the maximum height the rocket will attain?
    (d) How long after the launch will the rocket strike the earth?

30. ***Miniature Rockets*** A miniature rocket is launched so that its height h in meters after t seconds is given by the equation $h = -4.9t^2 + 19.6t + 10$.
    (a) Draw a graph of the equation.
    (b) How high is the rocket after 3 seconds?
    (c) What is the maximum height the rocket will attain?
    (d) How long after the launch will the rocket strike the earth?

31. ***Mosquito Population*** The number of mosquitoes in some areas varies with the amount of rain. The number, N, of mosquitoes measured in millions in Hamilton, Massachusetts, in May can be predicted by the equation $N = 9x - x^2$, where x is the number of inches of rain during that month. (*Source*: Massachusetts Environmental Protection Agency.)
    (a) Graph the equation.
    (b) How many inches of rain produce the maximum number of mosquitoes in May?
    (c) What is the maximum number of mosquitoes?

Use your graphing calculator to find the vertex, the y-intercept, and the x-intercepts of the parabolas described by the following equations. Round to the nearest thousandth when necessary.

32. $y = 3x^2 + 7x - 9$

33. $y = x^2 + 10.6x - 212.16$

34. $y = -156x^2 - 289x + 1133$

35. $y = -301x^2 - 167x + 1528$

Cumulative Review

36. If y varies inversely as x, and $y = \dfrac{1}{3}$ when $x = 21$, find y when $x = 7$.

37. If y varies directly as x^2, and $y = 12$ when $x = 2$, find y when $x = 5$.

38. *Football* A football player picks up a ball from the field and runs 60 yards. He runs the first half of this distance at 20 feet per second. He runs the second half at 24 feet per second. How long does it take him to run the 60 yards?

39. *Velocity of Sound* The equation $V = aT + b$ can be used to find the velocity of sound (V) in dry air in terms of the temperature (T) in degrees Celsius. At 40°C, sound travels at approximately 355 mph. At 49°C, sound travels at approximately 370 mph. Write a linear formula for the velocity of sound, based on temperature Celsius, in the form $V = aT + b$.

10.5 FORMULAS AND APPLIED PROBLEMS

1 Solving Applied Problems Using Quadratic Equations

We can use what we have learned about quadratic equations to solve word problems. Some word problems will involve geometry.

Recall that the Pythagorean Theorem states that in a right triangle $a^2 + b^2 = c^2$, where a and b are legs and c is the hypotenuse. We will use this theorem to solve word problems involving right triangles.

Student Learning Objective

After studying this section, you will be able to:

 Solve applied problems using quadratic equations

EXAMPLE 1 The hypotenuse of a right triangle is 25 meters in length. One leg is 17 meters longer than the other. Find the length of each leg.

Solution

1. **Understand the problem.** Draw a picture.
2. **Write an equation.** The problem involves a right triangle. Use the Pythagorean theorem.

$$a^2 + b^2 = c^2$$
$$x^2 + (x + 17)^2 = 25^2$$

3. **Solve and state the answer.**

$$x^2 + (x + 17)^2 = 25^2$$
$$x^2 + x^2 + 34x + 289 = 625$$
$$2x^2 + 34x - 336 = 0$$
$$x^2 + 17x - 168 = 0$$
$$(x + 24)(x - 7) = 0$$
$$x + 24 = 0 \qquad x - 7 = 0$$
$$x = -24 \qquad x = 7$$

(Note that $x = -24$ is not a valid solution for this particular word problem.)

One leg is 7 meters in length. The other leg is $x + 17 = 7 + 17 = 24$ meters in length.

4. **Check.** Check the conditions in the original problem.
   Do the two legs differ by 17 meters?

$$24 - 7 \overset{?}{=} 17 \qquad 17 = 17 \checkmark$$

   Do the sides of the triangle satisfy the Pythagorean theorem?

$$7^2 + 24^2 \overset{?}{=} 25^2$$
$$49 + 576 \overset{?}{=} 625$$
$$625 = 625 \checkmark$$

 Practice Problem 1 The hypotenuse of a right triangle is 30 meters in length. One leg is 6 meters shorter than the other leg. Find the length of each leg.

NOTE TO STUDENT: Fully worked-out solutions to all of the Practice Problems can be found at the back of the text starting at page SP-1

Sometimes it may help to use two variables in the initial part of the problem. Then one variable can be eliminated by substitution.

> **EXAMPLE 2** The ski club is renting a bus to travel to Mount Snow. The members agreed to share the cost of $180 equally. On the day of the ski trip, three members were sick with the flu and could not go. This raised by $10 the share of each person going on the trip. How many people originally planned to attend?

Solution

1. **Understand the problem.**

   Let s = the number of students in the ski club

   c = the cost for each student in the original group

2. **Write an equation(s).**

   number of students $\times$ cost per student = total cost

   $$s \cdot c = 180 \qquad \textbf{(1)}$$

   If three people are sick, the number of students drops by three, but the cost for each increases by $10. The total is still $180. Therefore, we have the following.

   $$(s - 3)(c + 10) = 180 \qquad \textbf{(2)}$$

3. **Solve and state the answer.**

   $$c = \frac{180}{s} \qquad \text{Solve equation \textbf{(1)} for } c.$$

   $$(s - 3)\left(\frac{180}{s} + 10\right) = 180 \qquad \text{Substitute } \frac{180}{s} \text{ for } c \text{ in equation \textbf{(2)}.}$$

   $$(s - 3)\left(\frac{180 + 10s}{s}\right) = 180 \qquad \text{Add } \frac{180}{s} + 10. \text{ Use the LCD which is } s.$$

   $$(s - 3)(180 + 10s) = 180s \qquad \text{Multiply both sides of the equation by } s.$$

   $$10s^2 + 150s - 540 = 180s \qquad \text{Use FOIL to multiply the binomials.}$$

   $$10s^2 - 30s - 540 = 0$$

   $$s^2 - 3s - 54 = 0$$

   $$(s - 9)(s + 6) = 0$$

   $$s = 9 \qquad s = -6$$

   The number of students originally in the ski club was 9.

4. **Check.** Would the cost increase by $10 if the number of students dropped from nine to six? Nine people in the club would mean that each would pay $20 ($180 \div 9 = 20$).

   If the number of students was reduced by three, there were six people who took the trip. Their cost was $30 ($180 \div 6 = 30$). The increase is $10.

   $$20 + 10 \stackrel{?}{=} 30$$

   $$30 = 30 \quad \checkmark$$

NOTE TO STUDENT: *Fully worked-out solutions to all of the Practice Problems can be found at the back of the text starting at page SP-1*

Practice Problem 2 Several friends from the University of New Mexico decided to charter a sport fishing boat for the afternoon while on a vacation at the coast. They agreed to share the rental cost of $240 equally. At the last minute, two of them could not go on the trip. This raised the cost for each person in the group by $4. How many people originally planned to charter the boat?

EXAMPLE 3   Minette is fencing a garden that borders the back of a large barn. She has 120 feet of fencing. She would like a rectangular garden that measures 1350 square feet in area. She wants to use the back of the barn, so she needs to use fencing on three sides only. What dimensions should she use for her garden?

Solution

1. **Understand the problem.** Draw a picture.

    Let x = the width of the garden in feet and

    y = the length of the garden in feet.

2. **Write an equation(s).** Look at the drawing. The 120 feet of fencing will be needed for the width twice (two sides) and the length once (one side).

$$120 = 2x + y \quad \textbf{(1)}$$

    The area formula is $A = (\text{width})(\text{length})$.

$$1350 = x(y) \quad \textbf{(2)}$$

3. **Solve and state the answer.**

$$y = 120 - 2x \qquad \text{Solve for } y \text{ in equation } \textbf{(1)}.$$
$$1350 = x(120 - 2x) \quad \text{Substitute for } y \text{ in equation } \textbf{(2)}.$$
$$1350 = 120x - 2x^2 \quad \text{Solve for } x.$$
$$2x^2 - 120x + 1350 = 0$$
$$x^2 - 60x + 675 = 0$$
$$(x - 15)(x - 45) = 0$$
$$x = 15 \qquad\qquad x = 45$$

First Solution: If the width is 15 feet, the length is 90 feet.

Check: $15 + 15 + 90 = 120$ ✓

$(15)(90) = 1350$ ✓

Second Solution: If the width is 45 feet, the length is 30 feet.

Check: $45 + 45 + 30 = 120$ ✓

$(45)(30) = 1350$ ✓

Width = x Length = y

First solution
15 ft
90 ft
15 ft

Thus we see that Minette has two choices that satisfy the given requirements (see the figures). She would probably choose the shape that is the most practical for her. (If her barn is shorter than 90 feet in length, she could not use the first solution!)

Remark: Is there another way to do this problem? Yes. Use only the variable x to represent the length and width. If you let x = the width and $120 - 2x$ = the length, you can immediately write equation **(3)**, which is

$$1350 = x(120 - 2x). \quad \textbf{(3)}$$

However, many students find that this is difficult to do, and they would prefer to use two variables. Work these problems in the way that is easiest for you.

Second solution
45 ft
45 ft
30 ft

▲ **Practice Problem 3** An Arizona rancher has a small fenced-in area against a canyon wall. Therefore, he only needs three sides of fencing to make a rectangular holding area for cattle. He has 160 feet of fencing available. He wants to make the rectangular area so that it has an area of 3150 square feet. What dimensions should he use for the rectangular holding area?

10.5 EXERCISES

Student Solutions Manual | CD/Video | PH Math Tutor Center | MathXL®Tutorials on CD | MathXL® | MyMathLab® | Interactmath.com

Applications

Solve.

▲ **1.** *Geometry* The base of a triangle is 5 cm shorter than the altitude. The area of the triangle is 88 sq cm. Find the lengths of the base and altitude.

▲ **2.** *Geometry* The area of a rectangle is 96 sq meters. If the length is 4 more than the width, what are the dimensions of the rectangle?

▲ **3.** *Garden Expansion* Joe wishes to expand the size of his front garden, which is square. Joe measures his front yard and decides that he can increase the width by 7 feet and the length by 11 feet. If the area of the newly expanded garden is 396 sq ft, what were the dimensions of the old garden?

▲ **4.** *Addition to a House* Mike and Norma have a home with a very small rectangular room on the first floor that they use for a living room. This room is 2 feet longer than it is wide. They have decided to put on an addition to the house that will give them a larger rectangular room to meet the need for an adequate living room for their family. The length of the living room will increase by 10 feet. The width of the living room will increase by 8 feet. The area of the new expanded living room will be 252 square feet. What were the dimensions of the present living room before the addition is built?

5. *Gift Expense* To help out Nina and Tom with their new twins, the employees at Tom's office decided to give the new parents a gift certificate worth $420 to furnish the babies' room. The employees planned to share the cost equally. At the last minute, seven people from the executive staff chipped in, lowering the cost per person by $5. How many people originally planned to give the gift?

6. *Septic Systems* Everyone in the School Street Townhouse Association agreed to equally share the cost of a new $6000 septic system. However, before the system was installed, five additional units were built and the five new families were asked to contribute their share. This lowered the contribution of each family by $200. How many families originally agreed to share the cost of the new septic system?

7. *Music Cost* A fraternity was sponsoring a spring fling, for which the $400 DJ cost was to be split evenly among those members attending. If 20 more members than were expected came, reducing the cost per person by $1, find the number of fraternity members that attended.

8. *Beach House Rental* The Cornell University *Wild Roses* Ultimate team took a spring break trip to Savannah, Georgia, where they all agreed to chip in and rent a beach house for the week. The cost of the rental for the week was $1500, and each girl was assessed an equal amount. Just before they left New York, three more girls decided to go. The captain said that each girl would now pay $25 less. How many people were planning on going on Spring Break?

▲ 9. *Fenway Park* A security fence encloses a rectangular area on one side of Fenway Park in Boston. Three sides of fencing are used, since the fourth side of the area is formed by a building. The enclosed area measures 1250 square feet. Exactly 100 feet of fencing is used to fence in three sides of this rectangle. What are the possible dimensions that could have been used to construct this area?

▲ 10. *Garden Fencing* Henry Redbone needs to fence in his garden, to protect it from the rabbits who love to feast on his lettuce. He plans on planting the garden next to the barn, so that he only needs to fence in three sides of the garden. If Henry has 80 feet of fencing available, and he wants the garden to have an area of 750 square feet, what are possible dimensions for the garden?

11. *Experimental Jet* A pilot is testing a new experimental craft. The cruising speed is classified information. In a test the jet traveled 2400 miles. The pilot revealed that if he had increased his speed 200 mph, the total trip would have taken 1 hour less. Determine the cruising speed of the jet.

▲ 12. *Gymnasium Floor Replacement* A gymnasium floor is being covered by square shock-absorbing tiles. The old gym floor required 864 square tiles. The new tiles are 2 inches larger in both length and width than the old tiles. The new flooring will require only 600 tiles. What is the length of a side of one of the new shock-absorbing tiles? (*Hint:* Since the area is the same in each case, write an expression for area with old tiles and one with new tiles. Set the two expressions equal to each other.)

Use the following information to solve exercises 13–18.

Business Expansion *Two young college graduates opened a chain of Quick Print Centers in southern California. The chain expanded rapidly during the early 1990s and many new stores were opened. The number of Quick Print Centers in operation during any year from 1990 to 2005 is given by the equation* $y = -2.5x^2 + 22.5x + 50$, *where x is the number of years since 1990.*

13. How many Quick Print Centers were in operation in 1994?

14. How many Quick Print Centers were in operation in 1993?

15. How many more Quick Print Centers were in operation in 1999 than in 2000?

16. How many fewer Quick Print Centers were in operation in 1998 than in 1996?

17. During what two years were there 100 Quick Print Centers in operation?

18. During what two years were there 85 Quick Print Centers in operation?

Cannon Shell Path Big Bertha, *an infamous World War I cannon, fired shells over an amazingly long distance. The path of a shell from this monstrous cannon can be modeled by the formula* $y = -0.018x^2 + 1.336x$, *where x is the horizontal distance the cannon shell traveled in miles and y is the height in miles the shell would reach on its flight.*

19. How far could *Big Bertha* send a shell? Round to the nearest whole number.

20. (a) How far into its journey would the shell achieve maximum height? Round to nearest whole number.

    (b) What was the maximum height in miles? Round to the nearest whole number.

To Think About

The sum of all the counting numbers from 1 to 6 is $1 + 2 + 3 + 4 + 5 + 6 = 21$. A shortcut to the answer is $6(7)\left(\dfrac{1}{2}\right) = 21$. In general, the sum of the counting numbers from 1 to n, which we can write as $1 + 2 + \cdots + n$, can be computed by the formula $\dfrac{n(n + 1)}{2} = s$.

21. If the sum of the first n counting numbers is 91, find n.

22. If the sum of the first n counting numbers is 136, find n.

Cumulative Review

23. Graph on a number line the solution to $3x + 11 \geq 9x - 4$.

24. Rationalize the denominator. $\dfrac{\sqrt{3}}{\sqrt{5} + 2}$

▲ **25.** *Recycling Bin* A recycling bin that has a length of 3 feet, a width of 2 feet, and a height of 2 feet has three separate sections inside for sorting. Section 1 is for paper, section 2 is for glass and plastic, and section 3 is for metal. Sections 1 and 2 have identical measurements, but section 3 has $\dfrac{2}{3}$ the volume of section 1 or 2. If all three sections have the same width and height, how long is section 3?

Putting Your Skills to Work

Firefighter Mathematics

A streaming column of water is pointed upward by a firefighter in order to bring a blaze under control. You have no doubt seen a picture of that action in newspapers and magazines. You recognize the flow of water looks very much the parabolas that we graphed in Section 10.4. In fact, the flow of water from a firefighter's hose can be modeled by a quadratic equations.

Problems for Individual Investigation and Analysis

1. For a certain angle and a certain water pressure, the path of the water from a hose at a fire scene can be modeled by the quadratic equation $h = -16t^2 + 80t + 4$, where h is the height of the water above ground in feet and t is the time in seconds after the water has left the hose. Find the value of h in the equation for $t = 0, 1, 2, 3, 4, 5$ seconds.

t	0	1	2	3	4	5
h						

2. Graph the equation by plotting the ordered pairs obtained in Problem 1.

Problems for Group Investigation and Cooperative Learning

3. How high does the water from the hose travel after 2 seconds?

4. What is the maximum height of the water?

5. How long after the water leaves the hose does it land on the ground? Round your answer to the nearest hundredth of a second.

6. If the water pressure in the hose is increased to that the path of water from the hose is given by the equation

   $h = -16t^2 + 90t + 5$, where h is the height in feet and t is the time in seconds, what is the maximum height of the water? At what time t does the water reach its maximum height? Round your answers to the nearest tenth.

Topic	Procedure	Examples
Solving quadratic equations by factoring, p. 571.	1. Clear the equation of fractions, if necessary. 2. Write as $ax^2 + bx + c = 0$. 3. Factor. 4. Set each factor equal to 0. 5. Solve the resulting equations.	Solve. $x + \dfrac{1}{2} = \dfrac{3}{2x}$ Multiply each term by LCD = $2x$. $$2x(x) + 2x\left(\dfrac{1}{2}\right) = 2x\left(\dfrac{3}{2x}\right)$$ $$2x^2 + x = 3$$ $$2x^2 + x - 3 = 0$$ $$(2x + 3)(x - 1) = 0$$ $2x + 3 = 0 \qquad x - 1 = 0$ $2x = -3 \qquad\qquad x = 1$ $$x = -\dfrac{3}{2}$$
Solving quadratic equations: taking the square root of each side of the equation, p. 578.	Begin solving quadratic equations of the form $ax^2 - c = 0$ by solving for x^2. Then use the property that if $x^2 = a$, $x = \pm\sqrt{a}$. This amounts to taking the square root of each side of the equation.	Solve. $2x^2 + 1 = 99$ $$2x^2 = 99 - 1$$ $$2x^2 = 98$$ $$\dfrac{2x^2}{2} = \dfrac{98}{2}$$ $$x^2 = 49$$ $$x = \pm\sqrt{49} = \pm 7$$
Solving quadratic equations: completing the square, p. 579.	1. Put the equation in the form $ax^2 + bx = c$. If $a \neq 1$, divide each term by a. 2. Square $\dfrac{b}{2}$ and add the result to both sides of the equation. 3. Factor the left side (a perfect square trinomial). 4. Use the square root property. 5. Solve the equations. 6. Check the solutions in the original equation.	Solve. $5x^2 - 3x - 7 = 0$ $$5x^2 - 3x = 7$$ $$x^2 - \dfrac{3}{5}x = \dfrac{7}{5}$$ $$x^2 - \dfrac{3}{5}x + \dfrac{9}{100} = \dfrac{7}{5} + \dfrac{9}{100}$$ $$\left(x - \dfrac{3}{10}\right)^2 = \dfrac{149}{100}$$ $$x - \dfrac{3}{10} = \pm\dfrac{\sqrt{149}}{10}$$ $$x = \dfrac{3}{10} \pm \dfrac{\sqrt{149}}{10}$$ $$x = \dfrac{3 \pm \sqrt{149}}{10}$$
Solving quadratic equations: quadratic formula, p. 584.	1. Put the equation in standard form: $ax^2 + bx + c = 0$. 2. Carefully determine the values of a, b, and c. 3. Substitute these into the formula $$x = \dfrac{-b \pm \sqrt{b^2 - 4ac}}{2a}.$$ 4. Simplify.	Solve. $3x^2 - 4x - 8 = 0$ $$a = 3,\ b = -4,\ c = -8$$ $$x = \dfrac{-(-4) \pm \sqrt{(-4)^2 - 4(3)(-8)}}{2(3)}$$ $$= \dfrac{4 \pm \sqrt{16 + 96}}{6}$$ $$= \dfrac{4 \pm \sqrt{112}}{6} = \dfrac{4 \pm \sqrt{16 \cdot 7}}{6}$$ $$= \dfrac{4 \pm 4\sqrt{7}}{6} = \dfrac{2(2 \pm 2\sqrt{7})}{2 \cdot 3} = \dfrac{2 \pm 2\sqrt{7}}{3}$$

Topic	Procedure	Examples
The discriminant, p. 587.	The *discriminant* in the quadratic formula is $b^2 - 4ac$. If it is negative the quadratic equation has no real solutions (since it is not possible to take the square root of a negative number and get a real number).	What kinds of solutions does this equation have? $$3x^2 + 2x + 1 = 0$$ $$b^2 - 4ac = 2^2 - 4 \cdot 3 \cdot 1$$ $$= 4 - 12 = -8$$ This equation has no real solutions.
Properties of parabolas, p. 594.	**1.** The graph of $y = ax^2 + bx + c$ is a parabola. It opens up if $a > 0$ and down if $a < 0$. **2.** The vertex of a parabola is its lowest (if $a > 0$) or highest (if $a < 0$) point. Its x-coordinate is $x = \dfrac{-b}{2a}$.	The vertex of $y = 3x^2 + 4x - 11$ has x-coordinate $$x = \frac{-b}{2a} = \frac{-4}{6} = -\frac{2}{3}.$$ The parabola opens upward.
Graphing parabolas, p. 594.	Graph the vertex of a parabola and several other points to get a good idea of its graph. Usually, we find the y-intercept and the x-intercepts if any exist. The following procedure is helpful. **1.** Determine if $a < 0$ or $a > 0$. **2.** Find the y-intercept. **3.** Find the x-intercepts. **4.** Find the vertex. **5.** Plot one or two extra points if necessary. **6.** Draw the graph.	Graph, $y = x^2 - 2x - 8$ $a = 1, b = -2, c = -8; a > 0$, so the parabola opens upward. Let $x = 0$. $y = 0^2 - 2(0) - 8 = -8$ The y-intercept is $(0, -8)$. $$0 = x^2 - 2x - 8$$ $$0 = (x - 4)(x + 2)$$ $$x = 4, \quad x = -2$$ The x-intercepts are $(4, 0)$ and $(-2, 0)$. The x-coordinate of the vertex is $$x = \frac{-b}{2a} = -\frac{-(-2)}{2} = 1.$$ If $x = 1$, $y = (1)^2 - 2(1) - 8 = -9$. The vertex is at $(1, -9)$.
Formulas and applied problems, p. 601.	Some word problems require the use of a quadratic equation. You may want to follow this four-step procedure for problem solving. **1.** Understand the problem. **2.** Write an equation. **3.** Solve and state the answer. **4.** Check.	The area of a rectangle is 48 square inches. The length is three times the width. Find the dimensions of the rectangle. **1.** Draw a picture. **2.** $(3w)w = 48$ **3.** $3w^2 = 48$ $w^2 = 16$ $w = 4$ inches, $l = 12$ inches **4.** $(4)(12) = 48$

Section 10.1

Write in standard form. Determine the values of a, b, and c. Do not solve.

1. $6x^2 = 5x - 8$

2. $9x^2 + 3x = -5x^2 + 16$

3. $(x - 4)(2x - 1) = x^2 + 7$

4. $2x(5x - 2) = x(3 - x)$

5. $\dfrac{3}{x^2} - \dfrac{6}{x} - 2 = 0$

6. $\dfrac{x}{x + 2} - \dfrac{3}{x - 2} = 5$

Place in standard form. Solve by factoring.

7. $x^2 + 26x + 25 = 0$

8. $x^2 + 16x + 64 = 0$

9. $x^2 + 6x - 55 = 0$

10. $4x^2 - 16x + 15 = 0$

11. $9x^2 - 24x + 16 = 0$

12. $15x^2 = 26x - 8$

13. $10x^2 = 19x - 7$

14. $3x^2 = -14x - 11$

15. $x^2 + \dfrac{1}{6}x - 2 = 0$

16. $\dfrac{1}{2}x^2 = \dfrac{3}{4}x - \dfrac{1}{4}$

17. $x^2 + \dfrac{2}{15}x = \dfrac{1}{15}$

18. $1 + \dfrac{13}{12x} - \dfrac{1}{3x^2} = 0$

19. $x^2 + 8x - 2 = 3x - 2$

20. $x^2 - 6x + 5 = 2x + 5$

21. $1 + \dfrac{2}{3x + 4} = \dfrac{3}{3x + 2}$

22. $2 - \dfrac{5}{x + 1} = \dfrac{3}{x - 1}$

23. $5 + \dfrac{24}{2 - x} = \dfrac{24}{2 + x}$

24. $\dfrac{4}{9} = \dfrac{5x}{3} - x^2$

Section 10.2

Solve by taking the square root of each side.

25. $x^2 - 8 = 41$ **26.** $x^2 + 11 = 92$ **27.** $x^2 - 5 = 17$ **28.** $x^2 + 11 = 50$

29. $2x^2 - 1 = 15$ **30.** $3x^2 + 4 = 154$ **31.** $3x^2 + 6 = 60$ **32.** $2x^2 - 5 = 43$

33. $(x - 4)^2 = 7$ **34.** $(x - 2)^2 = 3$ **35.** $(4x + 3)^2 = 24$ **36.** $(2x + 7)^2 = 45$

Solve by completing the square.

37. $x^2 + 10x - 11 = 0$ **38.** $x^2 + 12x + 11 = 0$

39. $2x^2 - 8x - 90 = 0$ **40.** $-5x^2 + 30x - 35 = 0$

41. $3x^2 + 6x - 6 = 0$ **42.** $2x^2 + 10x - 3 = 0$

Section 10.3

Solve using the quadratic formula.

43. $x^2 + 4x - 6 = 0$ **44.** $x^2 + 4x - 8 = 0$ **45.** $2x^2 - 7x + 4 = 0$

46. $2x^2 + 5x - 6 = 0$ **47.** $3x^2 - 8x - 4 = 0$ **48.** $4x^2 - 2x - 11 = 0$

49. $3x^2 - 5x = 4$ **50.** $4x^2 + 3x = 2$

Mixed Practice

Solve by any method. If there is no real number solution, so state.

51. $2x^2 - 9x + 10 = 0$

52. $4x^2 - 4x - 3 = 0$

53. $25x^2 + 10x + 1 = 0$

54. $2x^2 - 11x + 12 = 0$

55. $3x^2 - 6x + 2 = 0$

56. $5x^2 - 7x = 8$

57. $4x^2 + 4x = x^2 + 5$

58. $5x^2 + 7x + 1 = 0$

59. $x^2 = 9x + 3$

60. $3x^2 = 6 - 7x$

61. $-27x^2 + 3x = -12x^2 - 12$

62. $2x^2 - 1 = 35$

63. $\dfrac{(y - 2)^2}{20} + 3 + y = 0$

64. $\dfrac{(y + 2)^2}{5} + 2y = -9$

65. $3x^2 + 1 = 6 - 8x$

66. $2x^2 + 10x = 2x - 7$

67. $2y - 10 = 10y(y - 2)$

68. $\dfrac{3y - 2}{4} = \dfrac{y^2 - 2}{y}$

69. $\dfrac{y^2 + 5}{2y} = \dfrac{2y - 1}{3}$

70. $\dfrac{5x^2}{2} = x - \dfrac{7x^2}{2}$

Section 10.4

Graph each quadratic equation. Label the vertex.

71. $y = 2x^2$

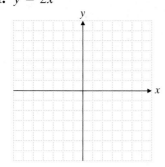

72. $y = x^2 + 4$

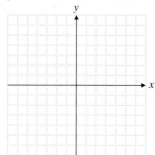

73. $y = x^2 - 3$

74. $y = -\dfrac{1}{2}x^2$

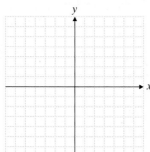

75. $y = x^2 - 3x - 4$

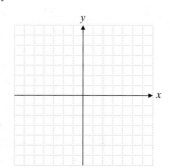

76. $y = \dfrac{1}{2}x^2 - 2$

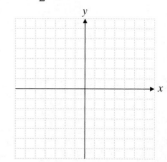

77. $y = -2x^2 + 12x - 17$

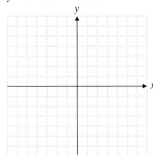

78. $y = -3x^2 - 2x + 4$

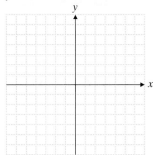

Section 10.5

Solve.

▲ **79.** *Dog Pen Fencing* Jon is building a dog pen against the house. He has 11 feet of fencing and will pen in an area of 15 square feet. What are the dimensions of the pen?

▲ **80.** *Geometry* The hypotenuse of a right triangle is 20 centimeters. One leg of the right triangle is 4 centimeters shorter than the other. What is the length of each leg of the right triangle?

▲ **81.** *TV Screen Dimensions* A television designer believes that the dimensions (length and width) of a TV screen are as important to sales as the size (measured by the diagonal) of the TV. If the diagonal of a newly designed TV is 30 inches and the length is 6 inches more than the width, what are the dimensions of the TV?

82. *Space Probe* Suppose an advanced alien civilization sends a space ship to orbit around out planet. They send a probe to earth to do research. The equation for the height of the probe is $h = -225t^2 + 291,600$, where h is the distance in feet from the surface of the earth t seconds after it leaves the mother ship. How long will it take the probe to reach earth from the mother ship?

83. *Golf Club* Last year the golf club on campus raised $720 through dues assigned equally to each member. This year there were four fewer members. As a result, the dues for each member went up by $6. How many members were in the golf club last year?

84. *Automobile Trip* Alice drove 90 miles to visit her cousin. Her average speed on the trip home was 15 mph faster than her average speed on the trip there. Her total travel time for both trips was 3.5 hours. What was her average rate of speed on each trip?

Use the following information to solve exercises 85–88.

Lacorazza Pizza Chain   The Lacorazza family opened a chain of pizza stores in New England in 1980. The number of pizza stores in operation during any year is given by the equation $y = -0.05x^2 + 2.5x + 60$, where x is the number of years since 1980.

85. According to the equation, how many pizza stores will still be in operation in 2030?

86. How many fewer pizza stores will still be in operation in 2040 than in 2030?

87. During what two years will there be exactly 80 pizza stores in operation?

88. During what two years will there be exactly 90 pizza stores in operation?

Remember to use your Chapter Test Prep Video CD to see the worked-out solutions to the test problems you want to review.

Solve by any desired method. If there is no real number solution, say so.

1. $5x^2 + 7x = 4$

2. $3x^2 + 13x = 10$

3. $2x^2 = 2x - 5$

4. $4x^2 - 19x = 4x - 15$

5. $12x^2 + 11x = 5$

6. $18x^2 + 32 = 48x$

7. $2x^2 - 11x + 3 = 5x + 3$

8. $5x^2 + 7 = 52$

9. $2x(x - 6) = 6 - x$

10. $x^2 - x = \dfrac{3}{4}$

Graph each quadratic equation. Locate the vertex.

11. $y = 3x^2 - 6x$

12. $y = -x^2 + 8x - 12$

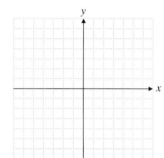

Solve.

▲ **13.** The hypotenuse of a right triangle is 15 meters in length. One leg is 3 meters longer than the other leg. Find the length of each leg.

14. When an object is thrown upward, its height (S) in meters is given (approximately) by the quadratic equation

$$S = -5t^2 + vt + h,$$

where v = the initial upward velocity in meters per second,
t = the time of flight in seconds, and
h = the height above level ground from which the object is thrown.

Suppose that a ball is thrown upward with a velocity of 33 meters/second at a height of 14 meters above the ground. How long after it is thrown will it strike the ground?

1. _____

2. _____

3. _____

4. _____

5. _____

6. _____

7. _____

8. _____

9. _____

10. _____

11. _____

12. _____

13. _____

14. _____

Approximately one-half of this test covers Chapters 0–9. The remainder covers the content of Chapter 10.

1. Simplify. $3x\{2y - 3[x + 2(x + 2y)]\}$

2. Evaluate. $\dfrac{3x - 2xy}{5x}$ for $x = -1; y = 2$

3. Solve for x. $\dfrac{1}{2}(x - 2) = \dfrac{1}{3}(x + 10) - 2x$

4. Factor completely. $16x^4 - 1$

5. Multiply. $(x - 3)(2x + 5)(x + 2)$

6. Solve for x. $\dfrac{3x}{x^2 - 4} = \dfrac{2}{x + 2} + \dfrac{4}{2 - x}$

7. Graph the line $y = -\dfrac{3}{4}x + 2$. Identify the slope and the y-intercept.

8. Find the equation of a line with a slope of $\dfrac{4}{3}$ that passes through the point $(-6, -2)$.

9. Solve for (x, y).
$$3x + 2y = 5$$
$$7x + y = 19$$

10. Simplify. $(-3x^3y^2)^4$

1. _____

2. _____

3. _____

4. _____

5. _____

6. _____

7. _____

8. _____

9. _____

10. _____

11. _____

11. Simplify. Assume that all variables are positive. $\sqrt{18x^5y^6z^3}$

12. _____

12. Multiply. $\left(\sqrt{2} + \sqrt{3}\right)\left(2\sqrt{2} - 4\sqrt{3}\right)$

13. _____

13. Rationalize the denominator. $\dfrac{\sqrt{5} - 3}{\sqrt{5} + 2}$

14. _____

14. Solve for t. $3(t + b) = 2a - 5t$

15. _____

15. Solve for b. $H = 25b^2 - 6$

16. _____

For exercises 16–22, solve.

16. $2x^2 + 3x = 35$ **17.** $(2x + 1)^2 = 20$

17. _____

18. $\dfrac{60}{R} = \dfrac{60}{R + 8} + 10$ **19.** $3x^2 + 11x + 2 = 0$

18. _____

20. $6x^2 + 11x - 10 = 0$ **21.** $x^2 = 72$

19. _____

22. $3x^2 + 4 = 79$

20. _____

23. Graph $y = x^2 + 6x + 10$. Locate the vertex.

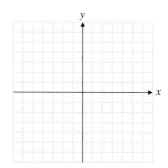

21. _____

22. _____

23. _____

24. _____

▲ **24.** The length of a rectangle is 2 meters less than three times its width. The area is 96 square meters. Find its length and width.

The following questions cover the content of Chapters 0–10. Follow the directions for each question and simplify your answers.

1. Simplify.
$-2x + 3y\{7 - 2[x - (4x + y)]\}$.

2. Evaluate if $x = -2$ and $y = 3$.
$2x^2 - 3xy - 4y$

3. Simplify. $(-3x^2y)(-6x^3y^4)$

4. Combine like terms.
$5x^2y - 6xy + 8xy - 3x^2y - 10xy$

5. Solve for x.
$\dfrac{1}{2}(x + 4) - \dfrac{2}{3}(x - 7) = 4x$

6. Solve for b. $p = 2a + 2b$

7. Solve for x and graph the resulting inequality on a number line.
$5x + 3 - (4x - 2) \le 6x - 8$

8. Multiply. $(2x + y)(x^2 - 3xy + 2y^2)$

9. Factor completely. $4x^2 - 18x - 10$

10. Factor completely. $3x^3 - 9x^2 - 30x$

11. Combine. $\dfrac{2}{x - 3} - \dfrac{3}{x^2 - x - 6} + \dfrac{4}{x + 2}$

12. Simplify. $\dfrac{\dfrac{3}{x} + \dfrac{5}{2x}}{1 + \dfrac{2}{x + 2}}$

13. Solve for x. $\dfrac{2}{x + 2} = \dfrac{4}{x - 2} + \dfrac{3x}{x^2 - 4}$

14. Find the slope of the line and then graph the line. $5x - 2y - 3 = 0$

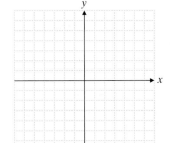

15. Find the equation of the line that has a slope of $-\dfrac{3}{4}$ and passes through the point $(-2, 5)$.

| 1. |
| 2. |
| 3. |
| 4. |
| 5. |
| 6. |
| 7. |
| 8. |
| 9. |
| 10. |
| 11. |
| 12. |
| 13. |
| 14. |
| 15. |

16. _____

17. _____

18. _____

19. _____

20. _____

21. _____

22. _____

23. _____

24. _____

25. _____

26. _____

27. _____

28. _____

29. _____

30. _____

▲ **16.** Two quarter-circles are attached to a rectangle. The dimensions are indicated in the figure. Find the area of the region. Use 3.14 as an approximation for π.

17. Solve for x and y.

$$2x + 7y = 4$$
$$-3x - 5y = 5$$

18. Solve for a and b.

$$a - \frac{3}{4}b = \frac{1}{4}$$
$$\frac{3}{2}a + \frac{1}{2}b = -\frac{9}{2}$$

19. Simplify and combine.
$$\sqrt{45x^3} + 2x\sqrt{20x} - 6\sqrt{5x^3}$$

20. Multiply and simplify.
$$\sqrt{6}\left(3\sqrt{2} - 2\sqrt{6} + 4\sqrt{3}\right)$$

21. Simplify by rationalizing the denominator. $\dfrac{\sqrt{3} + \sqrt{7}}{\sqrt{5} - \sqrt{7}}$

22. Solve for x. $12x^2 - 5x - 2 = 0$

23. Solve for y. $2y^2 = 6y - 1$

24. Solve for x. $4x^2 + 3 = 19$

▲ **25.** In the right triangle shown, sides b and c are given. Find the length of side a.

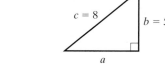

26. A number is tripled and then increased by 6. The result is 21. What is the original number?

▲ **27.** A rectangular region has a perimeter of 38 meters. The length of the rectangle is 2 meters shorter than double the width. What are the dimensions of the rectangle?

28. Margaret invested $7000 for one year. She placed part of the money in a tax-free bond earning 10% simple interest. She placed the rest in a money market account earning 14% simple interest. At the end of one year she had earned $860 in interest. How much did she invest at each interest rate?

29. A benefit concert was held on campus to raise scholarship money. A total of 360 tickets were sold. Admission prices were $5 for reserved seats and $3 for general admission. Total receipts were $1480. How many reserved-seat tickets were sold? How many general admission tickets were sold?

▲ **30.** The area of a triangle is 68 square meters. The altitude of the triangle is one meter longer than double the base of the triangle. Find the altitude and base of the triangle.

Beginning Algebra Glossary

Absolute value of a number (1.1) The absolute value of a number x is the distance between 0 and the number x on the number line. It is written as $|x|$. $|x| = x$ if $x \geq 0$, but $|x| = -x$ if $x < 0$.

Altitude of a geometric figure (1.8) The height of the geometric figure. In the three figures shown the altitude is labeled a.

Altitude of a trapezoid   Altitude of a parallelogram

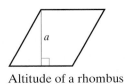

Altitude of a rhombus

Altitude of a triangle (1.8) The height of any given triangle. In the three triangles shown the altitude is labeled a.

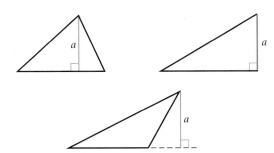

Associative property of addition (1.1) If a, b, and c are real numbers, then

$$a + (b + c) = (a + b) + c.$$

This property states that if three numbers are added it does not matter *which two numbers* are added first, the result will be the same.

Associative property of multiplication (1.3) If a, b, and c are real numbers, then

$$a \times (b \times c) = (a \times b) \times c.$$

This property states that if three numbers are multiplied it does not matter *which two numbers* are multiplied first; the result will be the same.

Base (1.4) The number or variable that is raised to a power. In the expression 2^6, the number 2 is the base.

Base of a triangle (1.8) The side of a triangle that is perpendicular to the altitude.

Binomial (4.3) A polynomial of two terms. The expressions $a + 2b$, $6x^3 + 1$, and $5a^3b^2 + 6ab$ are all binomials.

Circumference of a circle (1.8) The distance around a circle. The circumference of a circle is given by the formula $C = \pi d$ or $C = 2\pi r$, where d is the diameter of the circle and r is the radius of the circle.

Coefficient (4.1) A coefficient is a factor or a group of factors in a product. In the term $4xy$ the coefficient of y is $4x$, but the coefficient of xy is 4. In the term $-5x^3y$ the coefficient of x^3y is -5.

Commutative property for addition (1.1) If a and b are any real numbers, then $a + b = b + a$.

Commutative property for multiplication (1.3) If a and b are any real numbers, then $ab = ba$.

Complex fraction (6.4) A fraction that contains at least one fraction in the numerator or in the denominator or both. These three fractions are complex fractions:

$$\frac{7 + \dfrac{1}{x}}{x^2 + 2}, \qquad \frac{1 + \dfrac{1}{5}}{2 - \dfrac{1}{7}}, \qquad \text{and} \qquad \frac{\dfrac{1}{3}}{4}.$$

Constant (2.3) Symbol or letter that is used to represent exactly one single quantity during a particular problem or discussion.

Coordinates of a point (7.1) An ordered pair of numbers (x, y) that specifies the location of a point in a rectangular coordinate system.

Degree of a polynomial (4.3) The degree of the highest-degree term of a polynomial. The degree of the polynomial $5x^3 + 2x^2 - 6x + 8$ is 3. The degree of the polynomial $5x^2y^2 + 3xy + 8$ is 4.

Degree of a term of a polynomial (4.3) The sum of the exponents of the variables in the term. The degree of $3x^3$ is 3. The degree of $4x^5y^2$ is 7.

Denominator (0.1) and (6.1) The bottom number or algebraic expression in a fraction. The denominator of

$$\frac{3x - 2}{x + 4}$$

is $x + 4$. The denominator of $\dfrac{3}{7}$ is 7. The denominator of a fraction may not be zero.

Dependent equations (8.1) Two equations are dependent if every value that satisfies one equation satisfies the other. A system of two dependent equations in two variables will not have a unique solution.

Diagonal of a four-sided figure (1.8) A line connecting two nonadjacent corners of the figure. In each of the figures shown, line *AC* is a diagonal.

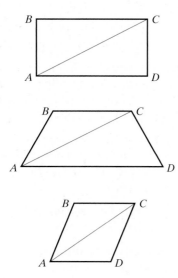

Difference (3.1) The result of subtracting one number or expression from another. The mathematical expression $x - 6$ can be written in words as the difference between x and 6.

Difference-of-two-squares polynomial (5.5) A polynomial of the form $a^2 - b^2$ that may be factored by using the formula

$$a^2 - b^2 = (a + b)(a - b).$$

Distributive property (1.6) For all real numbers $a, b,$ and $c, a(b + c) = ab + ac.$

Dividend (0.4) The number that is to be divided by another. In the problem $30 \div 5 = 6$, the three parts are as follows:

5 is the divisor
30 is the dividend
6 is the quotient

Divisor (0.4) The number you divide into another.

Domain of a relation (7.6) In any relation, the set of values that can be used for the independent variable is called its domain. This is the set of all the first coordinates of the ordered pairs that define the relation.

Equilateral triangle (1.8) A triangle with three sides equal in length and three angles that measure 60°. Triangle *ABC* is an equilateral triangle.

Even integers (1.3) Integers that are exactly divisible by 2, such as $\ldots, -4, -2, 0, 2, 4, 6, \ldots.$

Exponent (1.4) The number that indicates the power of a base. If the number is a positive integer it indicates how many times the base is multiplied. In the expression 2^6, the exponent is 6.

Expression (4.3) A mathematic expression is any quantity using numbers and variables. Therefore, $2x, 7x + 3,$ and $5x^2 + 6x$ are all mathematical expressions.

Extraneous solution (6.5) and (9.6) An obtained solution to an equation that when substituted back into the original equation, does *not* yield an identity. $x = 2$ is an extraneous solution to the equation

$$\frac{x}{x - 2} - 4 = \frac{2}{x - 2}.$$

An extraneous solution is also called an extraneous root.

Factor (0.1) and (5.1) When two or more numbers, variables, or algebraic expressions are multiplied, each is called a factor. If we write $3 \cdot 5 \cdot 2$, the factors are $3, 5,$ and 2. If we write $2xy$, the factors are $2, x,$ and y. In the expression $(x - 6)(x + 2)$, the factors are $(x - 6)$ and $(x + 2)$.

Fractions

Algebraic fractions (6.1) The indicated quotient of two algebraic expressions.

$$\frac{x^2 + 3x + 2}{x - 4} \quad \text{and} \quad \frac{y - 6}{y + 8}$$

are algebraic fractions. In these fractions the value of the denominator cannot be zero.

Numerical fractions (0.1) A set of numbers used to describe parts of whole quantities. A numerical fraction can be represented by the quotient of two integers for which the denominator is not zero. The numbers $\frac{1}{5}, -\frac{2}{3}, \frac{8}{2}, -\frac{4}{31}, \frac{8}{1},$ and $-\frac{12}{1}$ are all numerical fractions. The set of rational numbers can be represented by numerical fractions.

Function (7.6) A relation in which no two different ordered pairs have the same first coordinate.

Hypotenuse of a right triangle (9.6) The side opposite the right angle in any right triangle. The hypotenuse is always the longest side of a right triangle. Side *AB* is the hypotenuse of triangle below.

Identity (2.1) A statement that is always true. The equations $5 = 5, 7 + 4 = 7 + 4,$ and $x + 8 = x + 8$ are examples of identities.

Imaginary number (9.1) A complex number that is not a real number. Imaginary numbers can be created by taking the square root of a negative real number. The numbers $\sqrt{-9}$, $\sqrt{-7}$, and $\sqrt{-12}$ are all imaginary numbers.

Improper fraction (0.1) A numerical fraction whose numerator is larger than or equal to its denominator. $\frac{8}{3}$, $\frac{5}{2}$, and $\frac{7}{7}$ are improper fractions.

Inconsistent system of equations (8.1) A system of equations that does not have a solution.

Independent equations (8.1) Two equations that are not dependent are said to be independent.

Inequality (2.6), (2.7), and (7.5) A mathematical relationship between quantities that are not equal. $x \le -3$, $w > 5$, and $x < 2y + 1$ are mathematical inequalities.

Integers (1.1) The set of numbers $\ldots, -5, -4, -3, -2, -1, 0, 1, 2, 3, 4, 5, \ldots$.

Intercepts of an equation (7.2) The point or points where the graph of the equation crosses the x-axis or the y-axis or both. (*See x*-intercept.)

Irrational number (9.1) A real number that cannot be expressed in the form $\frac{a}{b}$, where a and b are integers and $b \ne 0$. $\sqrt{2}$, π, $5 + 3\sqrt{2}$, and $-4\sqrt{7}$ are irrational numbers.

Isosceles triangle (1.8) A triangle with two equal sides and two equal angles. Triangle ABC is an isosceles triangle. Angle BAC is equal to angle ACB. Side AB is equal in length to side BC.

Least common denominator of numerical fractions (0.2) The smallest whole number that is exactly divisible by all denominators of a group of fractions. The least common denominator (LCD) of $\frac{1}{6}$, $\frac{2}{3}$, and $\frac{3}{5}$ is 30. The least common denominator is also called the lowest common denominator.

Leg of a right triangle (9.6) One of the two shorter sides of a right triangle. Side AC and side BC are legs of triangle ABC.

Like terms (1.7) Terms that have identical variables and exponents. In the expression $5x^3 + 2xy^2 + 6x^2 - 3xy^2$, the term $2xy^2$ and the term $-3xy^2$ are like terms.

Linear equation in two variables (7.2) An equation of the form $Ax + By = C$, where A, B, and C are real numbers. The graph of a linear equation in two variables is a straight line.

Line of symmetry of a parabola (10.4) A line that can be drawn through a parabola such that, if the graph were folded on this line, the two halves of the curve would correspond exactly. The line of symmetry through the parabola formed by $y = ax^2 + bx + c$ is given by $x = \frac{-b}{2a}$. The line of symmetry of a parabola always passes through the vertex of the parabola.

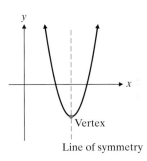

Mixed number (0.1) A number that consists of an integer written next to a proper fraction. $2\frac{1}{3}$, $4\frac{6}{7}$, and $3\frac{3}{8}$ are all mixed numbers. Mixed numbers are sometimes called mixed fractions or mixed numerals.

Monomial (4.3) A polynomial of one term. The expressions $3xy$, $5a^2b^3cd$, and -6 are all monomials.

Natural numbers (0.1) The set of numbers $1, 2, 3, 4, 5, \ldots$. This set is also called the set of counting numbers.

Numeral (0.1) The symbol used to describe a number.

Numerator (0.1) The top number or algebraic expression in a fraction. The numerator of $\frac{x + 3}{5x - 2}$ is $x + 3$. The numerator of $\frac{12}{13}$ is 12.

Numerical coefficient (4.1) The number that is multiplied by a variable or a group of variables. The numerical coefficient in $5x^3y^2$ is 5. The numerical coefficient in $-6abc$ is -6. The numerical coefficient in x^2y is 1. A numerical coefficient of 1 is not usually written.

Odd integers (1.3) Integers that are not exactly divisible by 2, such as $-3, -1, 1, 3, 5, 7, 9, \ldots$.

Opposite of a number (1.1) Two numbers that are the same distance from zero on the number line but lie on different sides of it are considered opposites. The opposite of -6 is 6. The opposite of $\frac{22}{7}$ is $-\frac{22}{7}$.

Ordered pair (7.1) A pair of numbers presented in a specified order. An ordered pair is often used to specify a location on a graph. Every point in a rectangular coordinate system can be represented by an ordered pair (x, y).

Origin (7.1) The point $(0, 0)$ in a rectangular coordinate system.

Parabola (10.4) A curve created by graphing a quadratic equation. The curves shown are all parabolas. The graph of the equation $y = ax^2 + bx + c$, where $a \neq 0$, will always be a parabola.

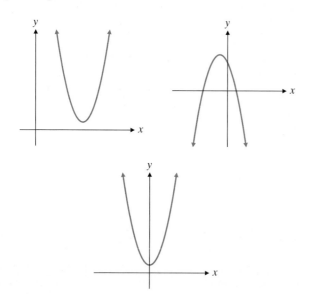

Parallel lines (7.3) and (8.1) Two straight lines that never intersect. The graph of an inconsistent system of two linear equations in two variables will result in parallel lines.

Parallelogram (1.8) A four-sided figure with opposite sides parallel. Figure $ABCD$ is a parallelogram.

Percent (0.5) Hundredths or "per one hundred"; indicated by the % symbol. Thirty-seven hundredths $\left(\dfrac{37}{100} \right) = 37\%$ (thirty-seven percent).

Perfect square number (5.5) A number that is the square of an integer. The numbers 1, 4, 9, 16, 25, 36, 49, 64, 81, 100, 121, 144, . . . are perfect square numbers.

Perfect square trinomial (5.5) A polynomial of the form $a^2 + 2ab + b^2$ or $a^2 - 2ab + b^2$ that may be factored using one of the following formulas:

$$a^2 + 2ab + b^2 = (a + b)^2$$

or

$$a^2 - 2ab + b^2 = (a - b)^2.$$

Perimeter (1.8) The distance around any plane figure. The perimeter of this triangle is 13. The perimeter of this rectangle is 20.

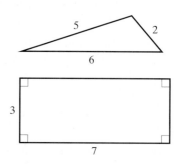

Pi (1.8) An irrational number, denoted by the symbol π, that is approximately equal to 3.141592654. In most cases 3.14 can be used as a sufficiently accurate approximation for π.

Polynomial (4.3) Expressions that contain terms with nonnegative integer exponents. The expressions $5ab + 6$, $x^3 + 6x^2 + 3$, -12, and $x + 3y - 2$ are all polynomials. The expressions $x^{-2} + 2x^{-1}$, $2\sqrt{x} + 6$, and $\dfrac{5}{x} + 2x^2$ are not polynomials.

Prime number (0.1) Any natural number greater than 1 whose only natural number factors are 1 and itself. The first eight prime numbers are 2, 3, 5, 7, 11, 13, 17, and 19.

Prime polynomial (5.6) A prime polynomial is a polynomial that cannot be factored by the methods of elementary algebra. $x^2 + x + 1$ is a prime polynomial.

Principal (3.4) In monetary problems, the principal is the original amount of money invested or borrowed.

Principal square root (9.1) For any given nonnegative number N, the principal square root of N (written $\sqrt{N}$) is the nonnegative number a if and only if $a^2 = N$. The principal square root of 25 $\left(\sqrt{25} \right)$ is 5.

Proper fraction (0.1) A numerical fraction whose numerator is less than its denominator; $\dfrac{3}{7}, \dfrac{2}{5}$, and $\dfrac{8}{9}$ are proper fractions.

Proportion (6.6) A proportion is an equation stating that two ratios are equal.

$$\frac{a}{b} = \frac{c}{d} \quad \text{where } b, d \neq 0$$

is a proportion.

Pythagorean Theorem (9.6) In any right triangle, if c is the length of the hypotenuse and a and b are the lengths of the two legs, then $c^2 = a^2 + b^2$.

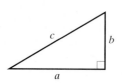

Quadratic equation (5.7) A quadratic equation is a polynomial equation with one variable that contains at least one term with the variable squared, but no term with the variable raised to a higher power. $5x^2 + 6x - 3 = 0$, $x^2 = 7$, and $5x^2 = 2x$ are all quadratic equations.

Quadratic formula (10.3) If $ax^2 + bx + c = 0$ and $a \neq 0$, then the solutions to the quadratic equation are given by

$$x = \frac{-b \pm \sqrt{b^2 - 4ac}}{2a}.$$

Quotient (0.4) The result of dividing one number or expression by another. In the problem $12 \div 4 = 3$, the quotient is 3.

Radical (9.1) An expression composed of a radical sign and a radicand. The expressions $\sqrt{5x}$, $\sqrt{\frac{3}{5}}$, $\sqrt{5x + b}$, and $\sqrt{10}$ are called radicals.

Radical equation (9.6) An equation that contains one or more radicals. $\sqrt{x + 4} = 12$, $\sqrt{x} = 3$, and $\sqrt{3x} + 4 = x + 2$ are all radical equations.

Radical sign (9.1) The symbol $\sqrt{}$ used to indicate the square root of a number.

Radicand (9.1) The expression beneath the radical sign. The radicand in $\sqrt{5ab}$ is $5ab$.

Range of a relation (7.6) In any relation, the set of values that represents the dependent variable is called its range. This is the set of all the second coordinates of the ordered pairs that define the relation.

Ratio (6.6) The ratio of one number a to another number b is the quotient $a \div b$ or $\frac{a}{b}$.

Rational numbers (1.1) and (9.1) A number that can be expressed in the form $\frac{a}{b}$, where a and b are integers and $b \neq 0$. $\frac{7}{3}$, $-\frac{2}{5}$, $\frac{7}{-8}$, $\frac{5}{1}$, 1.62, and 2.7156 are rational numbers.

Rationalizing the denominator (9.5) The process of transforming a fraction that contains a radical in the denominator to an equivalent fraction that does not contain a radical in the denominator.

Original Fraction	Equivalent Fraction with Rationalized Denominator
$\dfrac{3}{\sqrt{2}}$	$\dfrac{3\sqrt{2}}{2}$

Rationalizing the expression (9.5) The process of transforming a radical that contains a fraction to an equivalent expression that does not contain a fraction inside a radical.

Original Radical	Equivalent Expression after It Has Been Rationalized
$\sqrt{\dfrac{3}{5}}$	$\dfrac{\sqrt{15}}{5}$

Real number (9.1) Any number that is rational or irrational. 2, 7, $\sqrt{5}$, $\frac{3}{8}$, π, $-\frac{7}{5}$, and $-3\sqrt{5}$ are all real numbers.

Rectangle (1.8) A four-sided figure with opposite sides parallel and all interior angles measuring 90°. The opposite sides of a rectangle are equal.

Relation (7.6) A relation is any set of ordered pairs.

Rhombus (1.8) A parallelogram with four equal sides. Figure $ABCD$ is a rhombus.

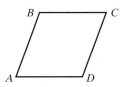

Right angle (1.8) An angle that measures 90°. Right angles are usually labeled in a sketch by using a small square to indicate that it is a right angle. Here angle ABC is a right angle.

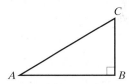

Right triangle (1.8) A triangle that contains a right angle.

Root of an equation (2.1) and (5.7) A value of the variable that makes an equation into a true statement. The root of an equation is also called the solution of an equation.

Scientific notation (4.2) A positive number is written in scientific notation if it is in the form $a \times 10^n$, where $1 \leq a < 10$ and n is an integer.

Slope-intercept form (7.3) The equation of a line that has slope m and the y-intercept at $(0, b)$ is given by $y = mx + b$.

Slope of a line (7.3) The ratio of change in y over the change in x for any two different points on a nonvertical line. The slope m is determined by

$$m = \frac{y_2 - y_1}{x_2 - x_1},$$

where $x_2 \neq x_1$ for any two points (x_1, y_1) and (x_2, y_2) on a nonvertical line.

Solution of an equation (2.1) A number that, when substituted into a given equation, yields an identity. The solution of an equation is also called the root of an equation.

Solution of a linear inequality (2.7) The possible values that make a linear inequality true.

Solution of an inequality in two variables (7.5) The set of all possible ordered pairs that when substituted into the inequality will yield a true statement.

Solution to a system of two equations in two variables (8.2) An ordered pair that can be substituted into each equation to obtain an identity in each case.

Square (1.8) A rectangle with four equal sides.

Square root (9.1) For any given nonnegative number N, the square root of N is the number a if $a^2 = N$. One square root of 16 is 4 since $(4)^2 = 16$. Another square root of 16 is -4 since $(-4)^2 = 16$. When we write $\sqrt{16}$, we want only the positive root, which is 4.

Standard form of a quadratic equation (5.7) A quadratic equation that is in the form $ax^2 + bx + c = 0$.

Subscript of a variable (3.2) A small number or letter written slightly below and to the right of a variable. In the expression $5 = 2(x - x_0)$, the subscript of x is 0. In the expression $t_f = 5(t_a - b)$ the subscript of the first t is f. The subscript of the second t is a. A subscript is used to indicate a different value of the variable.

System of equations (8.1) A set of two or more equations that must be considered together.

Term (1.7) A number, a variable, or a product of numbers and variables. For example, in the expression $a^3 - 3a^2b + 4ab^2 + 6b^3 + 8$, there are five terms. They are a^3, $-3a^2b$, $4ab^2$, $6b^3$, and 8. The terms of a polynomial are separated by plus and minus signs.

Trapezoid (1.8) A four-sided figure with two sides parallel. The parallel sides are called the bases of the trapezoid. Figure $ABCD$ is a trapezoid.

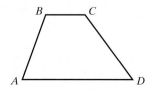

Trinomial (4.3) A polynomial of three terms. The expressions $x^2 + 6x - 8$ and $a + 2b - 3c$ are trinomials.

Variable (1.4) A letter that is used to represent a number or a set of numbers.

Variation (9.7) An equation relating values of one variable to those of other variables. An equation of the form $y = kx$, where k is a constant, indicates *direct variation*. An equation of the form $y = \dfrac{k}{x}$, where k is a constant, indicates *inverse variation*. In both cases, k is called the *constant of variation*.

Vertex of a parabola (10.4) The lowest point on a parabola opening upward or the highest point on a parabola opening downward. The x-coordinate of the vertex of the parabola formed by the equation $y = ax^2 + bx + c$ is given by $= \dfrac{-b}{2a}$.

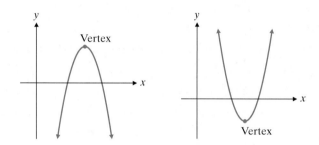

Vertical line test (7.6) If a vertical line can intersect the graph of a relation more than once, the relation is not a function.

Whole numbers (0.1) The set of numbers 0, 1, 2, 3, 4, 5, ...

x-intercept (7.2) The ordered pair $(a, 0)$ is the x-intercept of a line if the line crosses the x-axis at $(a, 0)$. The x-intercept of line l on the following graph is $(0, 4)$.

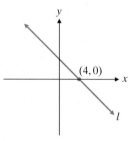

y-intercept (7.2) The ordered pair $(0, b)$ is the y-intercept of a line if the line crosses the y-axis at $(0, b)$. The y-intercept of line p on the following graph is $(0, 3)$.

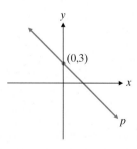

Appendix A Table of Square Roots

x	$\sqrt{x}$	x	$\sqrt{x}$	x	$\sqrt{x}$	x	$\sqrt{x}$	x	$\sqrt{x}$
1	1.000	41	6.403	81	9.000	121	11.000	161	12.689
2	1.414	42	6.481	82	9.055	122	11.045	162	12.728
3	1.732	43	6.557	83	9.110	123	11.091	163	12.767
4	2.000	44	6.633	84	9.165	124	11.136	164	12.806
5	2.236	45	6.708	85	9.220	125	11.180	165	12.845
6	2.449	46	6.782	86	9.274	126	11.225	166	12.884
7	2.646	47	6.856	87	9.327	127	11.269	167	12.923
8	2.828	48	6.928	88	9.381	128	11.314	168	12.961
9	3.000	49	7.000	89	9.434	129	11.358	169	13.000
10	3.162	50	7.071	90	9.487	130	11.402	170	13.038
11	3.317	51	7.141	91	9.539	131	11.446	171	13.077
12	3.464	52	7.211	92	9.592	132	11.489	172	13.115
13	3.606	53	7.280	93	9.644	133	11.533	173	13.153
14	3.742	54	7.348	94	9.695	134	11.576	174	13.191
15	3.873	55	7.416	95	9.747	135	11.619	175	13.229
16	4.000	56	7.483	96	9.798	136	11.662	176	13.266
17	4.123	57	7.550	97	9.849	137	11.705	177	13.304
18	4.243	58	7.616	98	9.899	138	11.747	178	13.342
19	4.359	59	7.681	99	9.950	139	11.790	179	13.379
20	4.472	60	7.746	100	10.000	140	11.832	180	13.416
21	4.583	61	7.810	101	10.050	141	11.874	181	13.454
22	4.690	62	7.874	102	10.100	142	11.916	182	13.491
23	4.796	63	7.937	103	10.149	143	11.958	183	13.528
24	4.899	64	8.000	104	10.198	144	12.000	184	13.565
25	5.000	65	8.062	105	10.247	145	12.042	185	13.601
26	5.099	66	8.124	106	10.296	146	12.083	186	13.638
27	5.196	67	8.185	107	10.344	147	12.124	187	13.675
28	5.292	68	8.246	108	10.392	148	12.166	188	13.711
29	5.385	69	8.307	109	10.440	149	12.207	189	13.748
30	5.477	70	8.367	110	10.488	150	12.247	190	13.784
31	5.568	71	8.426	111	10.536	151	12.288	191	13.820
32	5.657	72	8.485	112	10.583	152	12.329	192	13.856
33	5.745	73	8.544	113	10.630	153	12.369	193	13.892
34	5.831	74	8.602	114	10.677	154	12.410	194	13.928
35	5.916	75	8.660	115	10.724	155	12.450	195	13.964
36	6.000	76	8.718	116	10.770	156	12.490	196	14.000
37	6.083	77	8.775	117	10.817	157	12.530	197	14.036
38	6.164	78	8.832	118	10.863	158	12.570	198	14.071
39	6.245	79	8.888	119	10.909	159	12.610	199	14.107
40	6.325	80	8.944	120	10.954	160	12.649	200	14.142

Unless the value of $\sqrt{x}$ ends in 000, all values are rounded to the nearest thousandth.

Appendix B Metric Measurement and Conversion of Units

Student Learning Objectives

After studying this section, you will be able to:

1 Convert from one metric unit of measurement to another.

2 Convert between metric and U.S. units of measure.

3 Convert from one type of unit of measure to another.

1 Converting from One Metric Unit of Measurement to Another

Metric measurements are becoming more common in the United States. The metric system is used in many parts of the world and in the sciences. The basic unit of length in the metric system is the meter. For smaller measurements, centimeters are commonly used. There are 100 centimeters in 1 meter. For larger measurements, kilometers are commonly used. There are 1000 meters in 1 kilometer. The following tables give the metric units of measurement for length.

METRIC LENGTH

*1 kilometer (km)	=	1000 meters
1 hectometer (hm)	=	100 meters
1 dekameter (dam)	=	10 meters
*1 meter (m)	=	1 meter
1 decimeter (dm)	=	0.1 meter
*1 centimeter (cm)	=	0.01 meter
*1 millimeter (mm)	=	0.001 meter

The four most common units of metric measurement are indicated by an asterisk.

METRIC WEIGHT

1 kilogram (kg)	=	1000 grams
1 gram (g)	=	1 gram
1 milligram (mg)	=	0.001 gram

METRIC VOLUME

1 kiloliter (kL)	=	1000 liters
1 liter (L)	=	1 liter
1 milliliter (mL)	=	0.001 liter

One way to convert from one metric unit to another is to multiply by 1. We know mathematically that to multiply by 1 will yield an equivalent expression since $(a)(1) = a$. In our calculations when solving word problems, we treat dimension symbols much as we treat variables in algebra.

EXAMPLE 1 Fred drove a distance of 5.4 kilometers. How many meters is that?

Solution $5.4 \text{ km} \cdot 1 = 5.4 \text{ km} \cdot \dfrac{1000 \text{ m}}{1 \text{ km}} = \dfrac{(5.4)(1000)}{1} \cdot \dfrac{\text{km}}{\text{km}} \cdot \text{m}$

$= 5400 \text{ m}$

The distance is 5400 meters.

Practice Problem 1 June's room is 3.4 meters wide. How many centimeters is that?

NOTE TO STUDENT: Fully worked-out solutions to all of the Practice Problems can be found at the back of the text starting at page SP-1

EXAMPLE 2   A chemist measured 67 milliliters of a solution. How many liters is that?

Solution $67 \text{ mL} \cdot \dfrac{0.001 \text{ L}}{1 \text{ mL}} = 0.067 \text{ L}$

The chemist measured 0.067 liters of the solution.

Practice Problem 2 The container has 125 liters of water. How many kiloliters is that?

2 Converting between Metric and U.S. Units of Measure

The most common relationships between the U.S. and metric systems are listed in the following table. Most of these values are approximate.

METRIC CONVERSION RATIOS

LENGTH:	1 inch = 2.54 centimeters
	39.37 inches = 1 meter
	1 mile = 1.61 kilometers
	0.62 mile = 1 kilometer
WEIGHT:	1 pound = 454 grams
	2.20 pounds = 1 kilogram
	1 ounce = 28.35 grams
	0.0353 ounce = 1 gram
LIQUID CAPACITY:	1 quart = 946 milliliters
	1.06 quarts = 1 liter
	1 gallon = 3.785 liters

EXAMPLE 3   A box weighs 190 grams. How many ounces is that? Round to the nearest hundredth.

Solution $190 \text{ g} \cdot \dfrac{0.0353 \text{ oz}}{1 \text{ g}} = 6.707 \text{ oz}$

$\approx 6.71 \text{ oz}$ (rounded to the nearest hundredth)

The box weighs approximately 6.71 ounces.

Practice Problem 3   A bag of groceries weighs 5.72 pounds. How many kilograms is that?

EXAMPLE 4 Juanita drives 23 kilometers to work each day. How many miles is the trip? Round to the nearest mile.

Solution $23 \text{ km} \cdot \dfrac{0.62 \text{ mi}}{1 \text{ km}} = 14.26 \text{ mi} \approx 14 \text{ mi}$ (rounded to the nearest mile)

Juanita drives approximately 14 miles to work each day.

Practice Problem 4 Carlos installed an electrical connection that is 8.00 centimeters long. How many inches long is the connection? Round to the nearest hundredth.

NOTE TO STUDENT: Fully worked-out solutions to all of the Practice Problems can be found at the back of the text starting at page SP-1

EXAMPLE 5 Anita purchased 42.0 gallons of gasoline for her car last month. How many liters did she purchase? Round to the nearest liter.

Solution $42.0 \text{ gal} \cdot \dfrac{3.785 \text{ L}}{1 \text{ gal}} = 158.97 \text{ L} \approx 159 \text{ L}$ (rounded to the nearest liter)

Anita purchased approximately 159 liters of gasoline last month.

Practice Problem 5 Warren purchased a 3-liter bottle of Coca-Cola. How many quarts of Coca-Cola is that? Round to the nearest hundredth of a quart.

3 Converting from One Type of Unit of Measure to Another

Sometimes you will need to convert from one type of unit of measure to another. For example, you may need to convert days to minutes or miles per hour to feet per second. Recall the U.S. units of measure.

Length	Time
12 inches = 1 foot	60 seconds = 1 minute
3 feet = 1 yard	60 minutes = 1 hour
5280 feet = 1 mile	24 hours = 1 day
1760 yards = 1 mile	7 days = 1 week
Weight	**Volume**
16 ounces = 1 pound	2 cups = 1 pint
2000 pounds = 1 ton	2 pints = 1 quart
	4 quarts = 1 gallon

EXAMPLE 6 A car was traveling at 50.0 miles per hour. How many feet per second was the car traveling? Round to the nearest tenth of a foot per second.

Solution $\dfrac{50 \text{ mi}}{\text{hr}} \cdot \dfrac{5280 \text{ ft}}{1 \text{ mi}} \cdot \dfrac{1 \text{ hr}}{60 \text{ min}} \cdot \dfrac{1 \text{ min}}{60 \text{ sec}} = \dfrac{(50)(5280) \text{ ft}}{(60)(60) \text{ sec}} = \dfrac{73.333\ldots \text{ ft}}{\text{sec}}$

The car was traveling at 73.3 feet per second, rounded to the nearest tenth of a foot per second.

Practice Problem 6 A speeding car was traveling at 70.0 miles per hour. How many feet per second was the car traveling? Round to the nearest tenth of a foot per second.

1. How many meters are in 34 km?

2. How many meters are in 128 km?

3. How many centimeters are in 57 m?

4. How many centimeters are in 46 m?

5. How many millimeters are in 25 cm?

6. How many millimeters are in 63 cm?

7. How many meters are in 563 mm?

8. How many meters are in 831 mm?

9. How many milligrams are in 29.4 g?

10. How many milligrams are in 75.2 g?

11. How many kilograms are in 98.4 g?

12. How many kilograms are in 62.7 g?

13. How many milliliters are in 7 L?

14. How many milliliters are in 12 L?

15. How many kiloliters are in 4 mL?

16. How many kiloliters are in 3 mL?

Use the table of metric conversion ratios to find each of the following. Round all answers to the nearest tenth.

17. How many inches are in 4.2 cm?

18. How many inches are in 3.8 cm?

19. How many kilometers are in 14 mi?

20. How many kilometers are in 13 mi?

21. How many meters are in 110 in.?

22. How many meters are in 150 in.?

23. How many centimeters are in 7 in.?

24. How many centimeters are in 9 in.?

25. A box weighing 2.4 lb would weigh how many grams?

26. A box weighing 1.6 lb would weigh how many grams?

27. A man weighs 78 kg. How many pounds does he weigh?

28. A woman weighs 52 kg. How many pounds does she weigh?

29. Ferrante purchased 3 qt of milk. How many liters is that?

30. Wong Tin purchased 1 gal of milk. How many liters is that?

Answer the following questions. Round all answers to the nearest hundredth.

31. How many inches are in 3050 miles?

32. How many inches are in 4500 miles?

33. A truck traveled at a speed of 40 feet per second. How many miles per hour is that?

34. A car traveled at a speed of 55 feet per second. How many miles per hour is that?

35. How many years are in 3,500,000 seconds? (Use the approximate value that 365 days = 1 year.)

36. How many years are in 2,800,000 seconds? (Use the approximate value that 365 days = 1 year.)

Appendix C Interpreting Data from Tables, Charts, and Graphs

1 Tables

A table is a device used to organize information into categories. Using it you can readily find details about each category.

EXAMPLE 1

Table of Nutritive Values of Certain Popular "Fast Foods"

Type of Sandwich	Calories	Protein (g)	Fat (g)	Cholesterol (g)	Sodium (mg)
Burger King Whopper	630	27	38	90	880
McDonald's Big Mac	500	25	26	100	890
Wendy's Bacon Cheeseburger	440	22	25	65	870
Burger King BK Broiler (chicken)	280	20	10	50	770
McDonald's McChicken	415	19	20	50	830
Wendy's Grilled Chicken	290	24	7	60	670

Source: U.S. Government Agencies and Food Manufacturers

(a) Which food item has the least amount of fat per serving?

(b) Which beef item has the least amount of calories per serving?

(c) How much more protein does a Burger King Whopper have than a McDonald's Big Mac?

Solution

(a) The least amount of fat, 7 grams, is in the Wendy's Grilled Chicken Sandwich.

(b) The least amount of calories, 440, for a beef sandwich is the Wendy's Bacon Cheeseburger.

(c) The Burger King Whopper has 2 grams of protein more than the McDonald's Big Mac.

Practice Problem 1

(a) Which sandwich has the lowest level of sodium?

(b) Which sandwich has the highest level of cholesterol?

NOTE TO STUDENT: Fully worked-out solutions to all of the Practice Problems can be found at the back of the text starting at page SP-1

2 Charts

A chart is a device used to organize information in which not every category is the same. Example 2 illustrates a chart containing different types of data.

EXAMPLE 2   The following chart shows how people in Topsfield indicated they spent their free time.

Survey of Use of Leisure Time

Category	Activity	Hours spent per week
Single men	Gym	6
	Outdoor sports	4
	Dating	7
	Watching pro sports	12
	Reading & TV	3
Single women	Gym	4
	Outdoor sports	2
	Dating	7
	Time with friends	10
	Reading & TV	9
Couples	Time with family	21
	Time as a couple	8
	Time with friends	4
	Reading & TV	9
Children	Watching TV	28
	Playing outside	8
	Reading	1

Use the chart to answer the following questions about people in Topsfield.

(a) What is the average amount of time a couple spends together as a family during the week?

(b) How much more time do children spend watching TV than playing outside?

(c) What activity do single men spend most of their time doing?

Solution

(a) The average amount of time a couple spends together as a family is 21 hours per week.

(b) Children spend 20 more hours per week watching TV than playing outside.

(c) Single men spend more time per week watching pro sports (12 hr) than any other activity.

Practice Problem 2

(a) What two categories do single women spend the most time doing?

(b) What do couples spend most of their time doing?

(c) What is the most significant numerical difference in terms of number of hours per week spent by single women versus single men?

Pictographs

A pictograph uses a visually appropriate symbol to represent an amount of items. A pictograph is used in Example 3.

EXAMPLE 3 Consider the following pictograph.

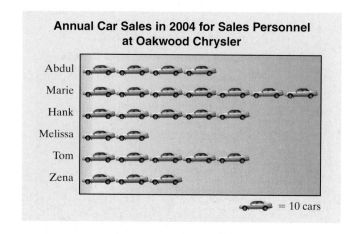

Annual Car Sales in 2004 for Sales Personnel at Oakwood Chrysler

= 10 cars

(a) How many cars did Melissa sell in 2004?
(b) Who sold the greatest number of cars?
(c) How many more cars did Tom sell than Zena?

Solution

(a) Melissa sold $2 \times 10 = 20$ cars.

(b) Marie sold the greatest number of cars.

(c) Tom sold $5 \times 10 = 50$ cars. Zena sold $3 \times 10 = 30$ cars. Now $50 - 30 = 20$. Therefore, Tom sold 20 cars more than Zena.

NOTE TO STUDENT: Fully worked-out solutions to all of the Practice Problems can be found at the back of the text starting at page SP-1

Practice Problem 3

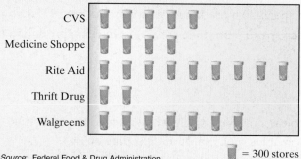

Approximate number of chain pharmacy stores in the United States in 2004

Source: Federal Food & Drug Administration

= 300 stores

(a) Approximately how many stores does Walgreens have?
(b) Approximately how many more stores does Rite Aid have than CVS?
(c) What is the combined number of Thrift Drug and Medicine Shoppe stores?

Bar Graphs

A bar graph is helpful for making comparisons and noting changes or trends. A scale is provided so that the height of the bar graph indicates a specific number. A bar graph is displayed in Example 4. A bar graph may be represented horizontally or vertically. In either case the basic concepts of interpreting a bar graph are the same.

EXAMPLE 4   The approximate population of California by year is given in the following bar graph.

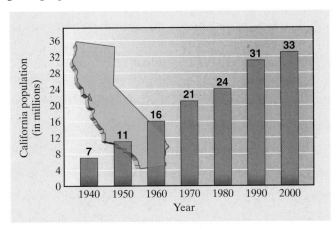

(a) What was the approximate population of California in 2000?

(b) How much greater was the population of California in 1980 than in 1970?

Solution

(a) The approximate population of California in 2000 was about 33 million.

(b) In 1980 it was 24 million. In 1970 it was 21 million. The population was approximately 3 million people more in California in 1980 than in 1970.

Practice Problem 4 The following bar graph depicts the number of fatal accidents for U.S. air carriers for scheduled flight service for aircraft with 30 seats or more.

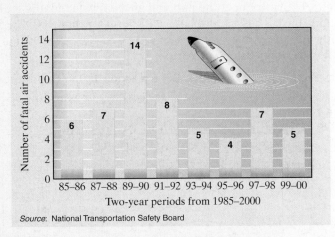

Source: National Transportation Safety Board

(a) What two-year period had the greatest number of fatal accidents?

(b) What was the increase in the number of fatal accidents from the 1987–1988 period to the 1989–1990 period?

(c) What was the decrease in the number of fatal accidents from the 1991–1992 period to the 1993–1994 period?

5 Line Graphs

A line graph is often used to display data when significant changes or trends are present. In a line graph, only a few points are actually plotted from measured values. The points are then connected by straight lines in order to show a trend. The intervening values between points may not exactly lie on the line. A line graph is displayed in Example 5.

EXAMPLE 5 The following line graph shows the number of customers per month coming to a restaurant in a tourist vacation community.

(a) What month had the greatest number of customers?

(b) How many customers came to the restaurant in April?

(c) How many more customers came in July than in August?

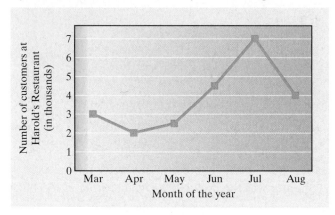

Solution

(a) More customers came during the month of July.

(b) Approximately 2000 customers came in April.

(c) In July there were 7000 customers, while in August there were 4000 customers. Thus, there were 3000 more customers in July than in August.

Practice Problem 5 The quality of the air is measured by the Pollutant Standards Index (PSI). To meet the national air quality standards set by the U.S. government, the air in a city cannot have a PSI greater than 100. The following line graph indicates the number of days that the PSI was greater than 100 in the city of Baltimore during an 11-year period.

(a) What was the number of days the PSI exceeded 100 in Baltimore in 1993?

(b) In what year did Baltimore have the fewest days in which the PSI exceeded 100?

Source: US Environmental Protective Agency

NOTE TO STUDENT: *Fully worked-out solutions to all of the Practice Problems can be found at the back of the text starting at page SP-1*

6 Pie Graphs and Circle Graphs

A pie graph or a circle graph indicates how a whole quantity is divided into parts. These graphs help you to visualize the size of the relative proportions of parts. Each piece of the pie or circle is called a sector. Example 6 uses a pie graph.

EXAMPLE 6   Together, the Great Lakes form the largest body of fresh water in the world. The total area of these five lakes is about 290,000 square miles, almost all of which is suitable for boating. The percentage of this total area taken up by each of the Great Lakes is shown in the pie graph.

(a) What percentage of the area is taken up by Lake Michigan?

(b) What lake takes up the largest percentage of area?

(c) How many square miles are taken up by Lake Huron and Lake Michigan together?

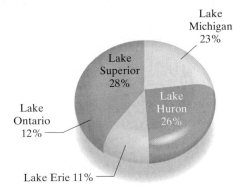

Solution

(a) Lake Michigan takes up 23% of the area.

(b) Lake Superior takes up the largest percentage.

(c) If we add 26 + 23, we get 49. Thus Lake Huron and Lake Michigan together take up 49% of the total area. 49% of 290,000 = (0.49)(290,000) = 142,100 square miles.

Practice Problem 6   Seattle receives on average about 37 inches of rain per year. However, the amount of rainfall per month varies significantly. The percent of rainfall that occurs during each quarter of the year is shown by the circle graph.

(a) What percent of the rain in Seattle falls between April and June?

(b) Forty percent of the rainfall occurs in what three-month period?

(c) How many inches of rain fall in Seattle from January to March?

C EXERCISES

| Student Solutions Manual | CD/ Video | PH Math Tutor Center | MathXL®Tutorials on CD | MathXL® | MyMathLab® | Interactmath.com |

In each case study carefully the appropriate visual display, then answer the questions. Consider the following table in answering exercises 1–10.

Table of Facts of the Rocky Mountain States

State	Area in Square Miles	Date Admitted to the Union	Estimated 2002 Population	Number of Representatives in U.S. Congress	Popular Name
Colorado	104,247	1876	4,507,000	7	Centennial State
Idaho	83,557	1890	1,341,000	2	Gem State
Montana	147,138	1889	909,000	1	Treasure State
Nevada	110,540	1864	2,173,000	3	Silver State
Utah	84,916	1896	2,316,000	3	Beehive State
Wyoming	97,914	1890	500,000	1	Equality State

1. What is the area of Utah in square miles?

2. What is the area of Montana in square miles?

3. What is the 2002 estimated population of Colorado?

4. What is the 2002 estimated population of Nevada?

5. How many representatives in the U.S. Congress come from Idaho?

6. How many representatives in the U.S. Congress come from Wyoming?

7. What is the popular name for Montana?

8. What is the popular name for Utah?

9. Which of these six states was the first one to be admitted to the Union?

10. In what year did two of these six states both get admitted to the Union?

Use this pictograph to answer exercises 11–16.

11. How many homes were built in Tarrant County in the year 2004?

12. How many homes were built in Essex County in the year 2004?

13. In what county were the most homes built?

14. How many more homes were built in Tarrant County than Waverly County?

15. How many homes were built in Essex County and Northface County combined?

16. How many homes were built in DuPage County and Waverly County combined?

Use this pictograph to answer exercises 17–20.

17. How many apartment units are rented for under $250 per month?

18. How many apartment units are rented for $800–$1249 per month?

19. How many more apartment units are available in the $500–$799 range than in the $800–$1249 range?

20. How many more apartment units are available in the $800–$1249 range than in the $1250 and up range?

Number of new homes built in 2004 in each of five counties

Approximate number of apartments in U.S. in 2004

Source: U.S. Bureau of the Census

Use this bar graph to answer exercises 21–26.

21. What was the population in Texas in 1950?

22. What was the population in Texas in 1990?

23. Between what two years was the increase in population the greatest?

24. Between what two years was the increase in population the smallest?

25. How many more people lived in Texas in 1970 than in 1950?

26. How many more people lived in Texas in 1980 than in 1960?

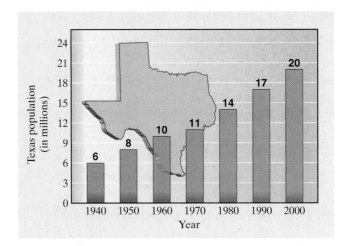

Use this bar graph to answer questions 27–30.

27. According to the bar graph, how many people watched a sports event at least once in the last 12 months?

28. According to the bar graph, how many people were involved in gardening at least once in the last 12 months?

29. What two activities were the most common?

30. What two activities were the least common?

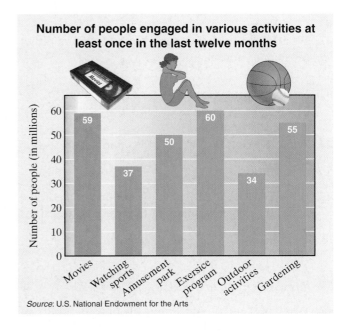

Number of people engaged in various activities at least once in the last twelve months

Source: U.S. National Endowment for the Arts

Use this line graph to answer exercises 31–36.

31. What was the profit in 2001?

32. What was the profit in 2000?

33. How much greater was the profit in 2004 than 2003?

34. In what year did the smallest profit occur?

35. Between what two years did the profit decrease the most?

36. Between what two years did the profit increase the most?

Profit of Wentworth Construction Company

Use this line graph to answer exercises 37–40.

37. How many people in the U.S. are in the age group of 5 to 24 years?

38. How many people in the U.S. are in the age group of 25 to 44 years?

39. Thirty-three million people are in what age group?

40. Fifty-one million people are in what age group?

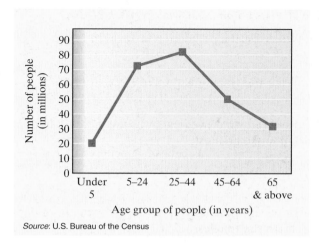

Source: U.S. Bureau of the Census

Use this circle graph to answer exercises 41–46.

41. What percent of the world's population is either Hindu or Buddhist?

42. What percent of the world's population is either Moslem or nonreligious?

43. What percent of the world's population is *not* Moslem?

44. What percent of the world's population is *not* Christian?

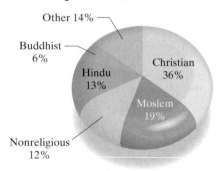

Religious faith distribution in the world

Source: United Nations Statistical Division

45. If there are approximately 6.5 billion people in the world in 2006, how many of them would we expect to be Hindu?

46. If there are approximately 6.9 billion people in the world in 2010, how many of them would we expect to be Moslem?

Use this circle graph to answer exercises 47–50.

47. What percent of the family income is spent for federal, state, and local taxes?

48. What percent of the family income is spent for food, medical care, and housing?

49. If the average two-income family earns $52,000 per year, how much is spent on transportation?

50. If the average two-income family earns $52,000 per year, how much is spent on recreation?

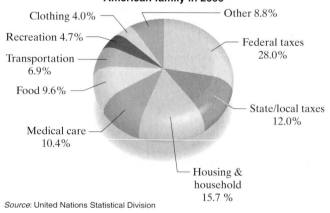

Distribution of spending by "average" two-income American family in 2000

Source: United Nations Statistical Division

Appendix D Inductive and Deductive Reasoning

① Using Inductive Reasoning to Reach a Conclusion

When we reach a conclusion based on specific observations, we are using **inductive reasoning.** Much of our early learning is based on simple cases of inductive reasoning. If a child touches a hot stove or other appliance several times and each time he gets burned, he is likely to conclude, "If I touch something that is hot, I will get burned." This is inductive reasoning. The child has thought about several actions and their outcomes and has made a conclusion or generalization.

The next few examples show how inductive reasoning can be used in mathematics.

Student Learning Objectives

After studying this section, you will be able to:

① Use inductive reasoning to reach a conclusion.

② Use deductive reasoning to reach a conclusion.

EXAMPLE 1 Find the next number in the sequence 10, 13, 16, 19, 22, 25, 28, ...

Solution We observe a pattern that each number is 3 more than the preceding number: $10 + 3 = 13$; $13 + 3 = 16$, and so on. Therefore, if we add 3 to 28, we conclude that the next number in the sequence is 31.

Practice Problem 1 Find the next number in the sequence 24, 31, 38, 45, 52, 59, 66, ...

NOTE TO STUDENT: Fully worked-out solutions to all of the Practice Problems can be found at the back of the text starting at page SP-1

EXAMPLE 2 Find the next number in the sequence 1, 8, 27, 64, 125, ...

Solution The sequence can be written as $1^3, 2^3, 3^3, 4^3, 5^3, \ldots$. Each successive integer is cubed. The next number would be 6^3 or 216.

Practice Problem 2 Find the next number in the sequence.

$$3, 8, 15, 24, 35, 48, 63, 80, \ldots$$

EXAMPLE 3 Guess the next seven digits in the following irrational number:

$$5.636336333633336\ldots$$

Solution Between 6's there are the digits 3, 33, 333, 3333, and so on. The pattern is that the number of 3's keeps increasing by 1 each time. Thus the next seven digits are 3333363.

Practice Problem 3 Guess the next seven digits in the following irrational number:

$$6.1213314441\ldots$$

EXAMPLE 4 Find the next two figures that would appear in the sequence.

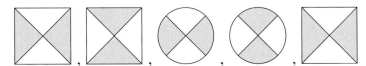

Solution We notice an alternating pattern: square, square, circle, circle, square . . . . We would next expect a square followed by a circle.

We notice a shading pattern of horizontal, vertical, horizontal, vertical, horizontal . . . . We would next expect vertical, then horizontal. Thus, the next two figures are

Practice Problem 4   Find the next two figures that would appear in the sequence.

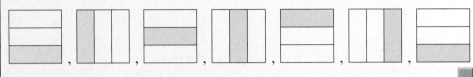

How accurate is inductive reasoning? Do we always come to the right conclusion? Conclusions arrived at by inductive reasoning are always tentative. They may require further investigation. When we use inductive reasoning we are using specific data to reach a general conclusion. However, we may have reached the wrong conclusion. To illustrate:

Take the sequence of numbers 20, 10, 5. What is the next number? You might say 2.5. In the sequence 20, 10, 5, each number appears to be one-half of the preceding number. Thus we would predict 20, 10, 5, 2.5, ...

But wait! There is another possibility. Maybe the sequence is 20, 10, 5, −5, −10, −20, −25, −35, −40, ...

To know for sure which answer is correct, we would need more information, such as more numbers in the sequence to verify the pattern. **You should always treat inductive reasoning conclusions as tentative, requiring further verification.**

2 Using Deductive Reasoning to Reach a Conclusion

Deductive reasoning requires us to take general facts, postulates, or accepted truths and use them to reach a specific conclusion. Suppose we know the following rules or "facts" of algebra:

> For all numbers a, b, and c, and for all numbers $d \neq 0$:
>
> **1.** *Addition principle of equations:*   If $a = b$, then $a + c = b + c$.
>
> **2.** *Division principle of equations:*   If $a = b$, then $\dfrac{a}{d} = \dfrac{b}{d}$.
>
> **3.** *Multiplication principle of equations:*   If $a = b$, then $ac = bc$.
>
> **4.** *Distributive property:* $a(b + c) = ab + ac$.

EXAMPLE 5   Use deductive reasoning and the four properties listed in the preceding box to justify each step in solving the equation.

$$2(7x - 2) = 38$$

Solution

Statement	Reason
1. $14x - 4 = 38$	**1.** Distributive property: $a(b + c) = ab + ac$ Here we distributed the 2.
2. $14x = 42$	**2.** Addition principle of equations: If $a = b$, then $a + c = b + c$ Here we added 4 to each side of the equation.

3. $\dfrac{14x}{14} = \dfrac{42}{14}$

   3. Division principle of equations:

   If $a = b$, then $\dfrac{a}{d} = \dfrac{b}{d}$

   $x = 3$ Here we divided each side by 14.

Practice Problem 5   Use deductive reasoning and the four properties listed in the preceding box to justify each step in solving the equation.

$$\frac{1}{6}x = \frac{1}{3}x + 4$$

Sometimes we need to make conclusions about angles and lines in geometry. The following properties are useful. We will refer to angles 1, 2, 3, 4.

1. If two lines intersect, the opposite angles are equal. Here $\angle 1 = \angle 2$ and $\angle 4 = \angle 3$.

2. If two lines intersect, the adjacent angles are supplementary (they add up to 180°). Here, $\angle 4 + \angle 1 = 180°$, and $\angle 1 + \angle 3 = 180°$, also $\angle 3 + \angle 2 = 180°$, and $\angle 4 + \angle 2 = 180°$.

3. If a transversal (intersecting line) crosses two parallel lines, the alternate interior angles are equal. In the figure at right, line P is a transversal. If line M is parallel to line N, then $\angle 3 = \angle 4$ and $\angle 1 = \angle 2$.

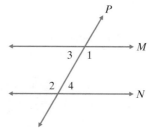

4. If two alternate interior angles on each side of a transversal cutting two straight lines are equal, the two straight lines are parallel. If $\angle 3 = \angle 4$, then line M is parallel to line N.

5. Supplements of equal angles are equal. In the figure at right, if $\angle 1$ and $\angle 2$ are supplementary, and $\angle 3$ and $\angle 4$ are supplementary, then if $\angle 1 = \angle 4$, then $\angle 2 = \angle 3$.

6. Transitive property of equality:

   If $a = b$ and $b = c$, then $a = c$.

We will now use these facts to prove some geometric conclusions.

EXAMPLE 6   Prove that line M is parallel to line N if $\angle 1 = \angle 3$.

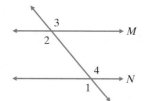

Solution

Statement	Reason
1. $\angle 3 = \angle 2$ and $\angle 4 = \angle 1$	**1.** If two lines intersect, the opposite angles are equal.
2. $\angle 2 = \angle 4$	**2.** Transitive property of equality.

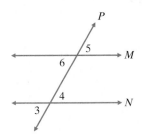

3. Therefore, line *M* is parallel to line *N*. (We write this as *M* ∥ *N*.)

3. If two alternate interior angles on each side of a transversal cutting two straight lines are equal, the straight lines are parallel.

Practice Problem 6 Refer to the diagram in the margin. Line *M* is parallel to line *N*. ∠5 = 26°. Prove ∠4 = 26°.

Now let us see if we can use deductive reasoning to solve the following problem.

EXAMPLE 7 For next semester, four professors from the psychology department have expressed the following desires for freshman courses. Each professor will teach only *one* freshman course next semester. The four freshman courses are General Psychology, Social Psychology, Psychology of Adjustment, and Educational Psychology.

1. Professors *A* and *B* don't want to teach General Psychology.
2. Professor *C* wants to teach Social Psychology.
3. Professor *D* will be happy to teach any course.
4. Professor *B* wants to teach Psychology of Adjustment.

Which professor will teach Educational Psychology, if all professors are given the courses they desire?

Solution Let us organize the facts by listing the four freshman courses and each professor: *A*, *B*, *C*, and *D*.

General Psychology	Social Psychology	Psychology of Adjustment	Educational Psychology
A	*A*	*A*	*A*
B	*B*	*B*	*B*
C	*C*	*C*	*C*
D	*D*	*D*	*D*

Step	Reason
1. We cross off Professors *A* and *B* from the General Psychology list.	**1.** Professors *A* and *B* don't want to teach General Psychology.

General Psychology			
~~*A*~~	*A*	*A*	*A*
~~*B*~~	*B*	*B*	*B*
C	*C*	*C*	*C*
D	*D*	*D*	*D*

2. We cross Professor *C* off every list (except Social Psychology) and mark that he will teach it.	**2.** Professor *C* wants to teach Social Psychology.

	Social Psychology		
~~*A*~~			
~~*B*~~			
~~*C*~~	*C*	~~*C*~~	~~*C*~~
D			

Step	Reason
3. Professor *D* is thus the only person who can teach General Psychology. We cross him off every other list and mark that he will teach General Psychology.	**3.** Professor *D* is happy to teach any course.

A			
B			
C	C	*C*	*C*
D	*D*	*D*	*D*

Step	Reason
4. We cross out all courses for Professor *B* except Psychology of Adjustment.	**4.** Professor *B* wants to teach Psychology of Adjustment.

Psychology of Adjustment			
A	*A*	A	A
B	*B*	B	*B*
C	C	*C*	*C*
D	*D*	*D*	*D*

Step	Reason
5. Professor *A* will teach Educational Psychology.	**5.** He is the only professor left. All others are assigned.

Educational Psychology
A

Practice Problem 7 A Honda, Toyota, Mustang, and Camaro are parked side by side, but not in that order.

1. The Camaro is parked on the right end.
2. The Mustang is between the Honda and the Toyota.
3. The Honda is not next to the Camaro.

Which car is parked on the left end?

NOTE TO STUDENT: Fully worked-out solutions to all of the Practice Problems can be found at the back of the text starting at page SP-1

D EXERCISES

| Student Solutions Manual | CD/ Video | PH Math Tutor Center | MathXL®Tutorials on CD | MathXL® | MyMathLab® | Interactmath.com |

Find the next number in the sequence.

1. 2, 4, 6, 8, 10, 12, …

2. 0, 5, 10, 15, 20, 25, …

3. 7, 16, 25, 34, 43, …

4. 12, 25, 38, 51, 64, …

5. 1, 16, 81, 256, 625, …

6. 1, 6, 13, 22, 33, 46, …

7. −7, 3, −6, 4, −5, 5, −4, 6, …

8. 2, −4, 8, −16, 32, −64, 128, …

9. $5x, 6x − 1, 7x − 2, 8x − 3, 9x − 4, …$

10. $60x, 30x, 15x, 7.5x, 3.75x, 1.875x, …$

In exercises 11–16, guess the next seven digits in each irrational number.

11. 8.181181118…

12. 3.043004300043…

13. 12.98987987698765…

14. 7.6574839201102938…

15. 2.14916253649…

16. 6.112223333…

17. Find the next row in this triangular pattern.

```
              1
          1       1
       1      2      1
     1     3      3     1
   1    4      6     4    1
 1    5    10     10    5    1
```

18. Find the next two figures that would appear in the sequence.

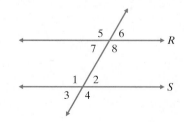

Use deductive reasoning and the four properties discussed in Example 5 to justify each step in solving each equation.

19. $12x − 30 = 6$

20. $3x + 7 = 4$

21. $4x − 3 = 3x − 5$

22. $2x − 9 = 4x + 5$

23. $8x − 3(x − 5) = 30$

24. $3x − 5(x − 1) = −5$

25. $\frac{1}{2}x + 6 = \frac{3}{2} + 6x$

26. $\frac{4}{5}x − 3 = \frac{1}{10}x + \frac{3}{5}$

In exercises 27–28, use deductive reasoning and the properties of geometry discussed in Example 6 to prove each statement.

27. If $\angle 1 = \angle 5$, prove that line P is parallel to line S.

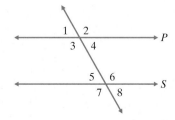

28. If line R is parallel to line S, prove that $\angle 6 = \angle 3$.

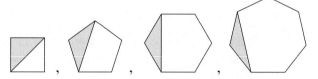

29. William, Brent, Charlie, and Dave competed in the Olympics. These four divers placed first, second, third, and fourth in the competition.
  1. James ranked between Brent and Dave.
  2. William did better than Brent.
  3. Brent did better than Dave.

Who finished in each of the four places?

30. The four floors of Hotel Royale are color-coded.
  1. The blue floor is directly below the green floor.
  2. The red floor is next to the yellow floor.
  3. The green floor is above the red floor.
  4. The blue floor is above the yellow floor.
  5. There is no floor below the red floor.

What is the order of the colors from top to bottom of Hotel Royale?

31. There are four people named Peter, Michael, Linda and Judy. Each of them has one occupation. They are a teacher, a butcher, a baker, and a candlestick maker.
  1. Linda is the baker.
  2. The teacher plans to marry the baker in October.
  3. Judy and Peter were married last year.

Who is the teacher?

32. A president of a company, a lawyer, a salesman, and a doctor are seated at the head table at a banquet.
  1. The lawyer prosecuted the doctor in a malpractice lawsuit and they should not sit together.
  2. The salesman should be at one end of the table.
  3. The lawyer should be at the far right.

From left to right how should they be seated?

33. John is considering purchasing a new car when he starts his new full-time job. The car with the best gas mileage is a Honda Civic, the car with the most conveniently located dealer is a Toyota Corolla, the car with the lowest price is a Ford Focus and the car that has best handling is the Dodge Neon. His next-door neighbor has a Honda and he does not want to copy his neighbor. If he does not buy a Toyota then he will definitely not go to graduate school. He told his best friend that he probably will not purchase a car based on its handling. He has definitely decided to go to graduate school. What car is he most likely to purchase?

34. Detective Smith recently found a stolen Corvette abandoned near Skull Rocks. A small empty boat was anchored 1 mile offshore with no person in the boat. Divers later found the body of a 35-year-old man $\frac{1}{2}$ mile further out to sea than the boat. If the man entered the water after 10:00 A.M. then the current of the outgoing tide was strong enough to take a dead body a distance of $\frac{1}{2}$ mile further out to sea. The coroner determined when the body was found at 4:00 P.M. that the man had been dead for seven hours. The detective determined that he died instantly when his head hit the rocks directly beneath the anchored boat. Could the man have acted alone based on the facts revealed so far?

35. Dr. Lowe is deciding whether he should do a root canal in order to save Fred's tooth or just remove the tooth and put in a false tooth. If Dr. Lowe works alone on the tooth the procedure will take 2 hours and 15 minutes. Fred wants to save the tooth but he does not want the dental bill for Tuesday's work to be more than $200. Dr. Lowe charges $100 per hour for dental procedures. If the bill is under $150 it is because Dr. Lowe's efficient dental assistant is helping with the oral surgery. The dental procedure is scheduled for Tuesday, when Dr. Lowe does not have any dental assistants available. Will Fred's tooth be saved or will he get a false tooth?

36. If Marcia passes the Bilingual Teacher's exam, she will stay in the Chicago area. If she devotes at least 50 hours of study time to practice her Spanish, she is confident that she can pass the exam. Either Marcia will stay in the Chicago area or she will move back to Massachusetts. If she did not pass the Bilingual Teacher's exam it is because she needed 12 hours of study time per month. Marcia started studying for the exam in April. She will be moving to Massachusetts. When was the Bilingual Teacher's exam given, if we know that Marcia took the exam?

Solutions to Practice Problems

Chapter 0

0.1 Practice Problems

1. (a) $\dfrac{10}{16} = \dfrac{2 \times 5}{2 \times 8} = \dfrac{5}{8}$ **(b)** $\dfrac{24}{36} = \dfrac{12 \times 2}{12 \times 3} = \dfrac{2}{3}$

(c) $\dfrac{36}{42} = \dfrac{6 \times 6}{6 \times 7} = \dfrac{6}{7}$

2. (a) $\dfrac{4}{12} = \dfrac{2 \times 2 \times 1}{2 \times 2 \times 3} = \dfrac{1}{3}$ **(b)** $\dfrac{25}{125} = \dfrac{5 \times 5 \times 1}{5 \times 5 \times 5} = \dfrac{1}{5}$

(c) $\dfrac{73}{146} = \dfrac{73 \times 1}{73 \times 2} = \dfrac{1}{2}$

3. (a) $\dfrac{18}{6} = \dfrac{3 \times 6}{6} = 3$ **(b)** $\dfrac{146}{73} = \dfrac{73 \times 2}{73} = 2$

(c) $\dfrac{28}{7} = \dfrac{7 \times 4}{7} = 4$

4. 56 out of 154 $= \dfrac{56}{154} = \dfrac{2 \times 7 \times 4}{2 \times 7 \times 11} = \dfrac{4}{11}$

5. (a) $\dfrac{12}{7} = 12 \div 7 = 7\overline{)12} = 1\dfrac{5}{7}$

$$\dfrac{7}{5} \quad \text{Remainder}$$

(b) $\dfrac{20}{5} = 20 \div 5 = 5\overline{)20} = 4$

$$\dfrac{20}{0} \quad \text{Remainder}$$

6. (a) $3\dfrac{2}{5} = \dfrac{(3 \times 5) + 2}{5} = \dfrac{15 + 2}{5} = \dfrac{17}{5}$

(b) $1\dfrac{3}{7} = \dfrac{(1 \times 7) + 3}{7} = \dfrac{7 + 3}{7} = \dfrac{10}{7}$

(c) $2\dfrac{6}{11} = \dfrac{(2 \times 11) + 6}{11} = \dfrac{22 + 6}{11} = \dfrac{28}{11}$

(d) $4\dfrac{2}{3} = \dfrac{(4 \times 3) + 2}{3} = \dfrac{12 + 2}{3} = \dfrac{14}{3}$

7. (a) $\dfrac{3}{8} = \dfrac{?}{24}$ Observe $8 \times 3 = 24$ **(b)** $\dfrac{5}{6} = \dfrac{?}{30}$

$$\dfrac{3 \times 3}{8 \times 3} = \dfrac{9}{24} \qquad\qquad \dfrac{5 \times 5}{6 \times 5} = \dfrac{25}{30}$$

(c) $\dfrac{2}{7} = \dfrac{?}{56}$

$$\dfrac{2 \times 8}{7 \times 8} = \dfrac{16}{56}$$

0.2 Practice Problems

1. (a) $\dfrac{3}{6} + \dfrac{2}{6} = \dfrac{3 + 2}{6} = \dfrac{5}{6}$ **(b)** $\dfrac{3}{11} + \dfrac{8}{11} = \dfrac{11}{11} = 1$

(c) $\dfrac{1}{8} + \dfrac{2}{8} + \dfrac{1}{8} = \dfrac{1 + 2 + 1}{8} = \dfrac{4}{8} = \dfrac{1}{2}$

(d) $\dfrac{5}{9} + \dfrac{8}{9} = \dfrac{5 + 8}{9} = \dfrac{13}{9} = 1\dfrac{4}{9}$

2. (a) $\dfrac{11}{13} - \dfrac{6}{13} = \dfrac{11 - 6}{13} = \dfrac{5}{13}$

(b) $\dfrac{8}{9} - \dfrac{2}{9} = \dfrac{8 - 2}{9} = \dfrac{6}{9} = \dfrac{2}{3}$

3. Find LCD of $\dfrac{1}{8}$ and $\dfrac{5}{12}$.

$$8 = 2 \cdot 2 \cdot 2$$
$$12 = \quad 2 \cdot 2 \cdot 3$$
$$2 \cdot 2 \cdot 2 \cdot 3 = 24 \qquad \text{LCD} = 24$$

4. Find the LCD using prime factors.

$$\dfrac{8}{35} \text{ and } \dfrac{6}{15}$$
$$35 = 7 \cdot 5$$
$$15 = \quad 5 \cdot 3$$
$$7 \cdot 5 \cdot 3 \qquad \text{LCD} = 105$$

5. Find LCD of $\dfrac{5}{12}$ and $\dfrac{7}{30}$.

$$12 = 3 \cdot 2 \cdot 2$$
$$30 = 3 \quad 2 \cdot 5$$
$$3 \cdot 2 \cdot 2 \cdot 5 \qquad \text{LCD} = 60$$

6. Find LCD of $\dfrac{1}{18}, \dfrac{2}{27}$ and $\dfrac{5}{12}$.

$$12 = 2 \cdot 2 \cdot 3$$
$$18 = 2 \quad 3 \cdot 3$$
$$27 = \quad 3 \cdot 3 \cdot 3$$
$$2 \cdot 2 \cdot 3 \cdot 3 \cdot 3 \qquad \text{LCD} = 108$$

7. Add $\dfrac{1}{8} + \dfrac{5}{12}$

First find the LCD.
$$8 = 2 \cdot 2 \cdot 2$$
$$12 = 2 \cdot 2 \quad \cdot 3$$
$$2 \cdot 2 \cdot 2 \cdot 3 \qquad \text{LCD} = 24$$
Then change to equivalent fractions and add.
$$\dfrac{1}{8} \times \dfrac{3}{3} + \dfrac{5}{12} \times \dfrac{2}{2} = \dfrac{3}{24} + \dfrac{10}{24} = \dfrac{3 + 10}{24} = \dfrac{13}{24}$$

8. $\dfrac{3}{5} + \dfrac{4}{25} + \dfrac{1}{10}$

First find the LCD.
$$5 = \quad 5$$
$$10 = \quad 2 \cdot 5$$
$$25 = \quad 5 \cdot 5$$
$$2 \cdot 5 \cdot 5 \qquad \text{LCD} = 50$$
Then change to equivalent fractions and add.
$$\dfrac{3}{5} \times \dfrac{10}{10} + \dfrac{4}{25} \times \dfrac{2}{2} + \dfrac{1}{10} \times \dfrac{5}{5} = \dfrac{30}{50} + \dfrac{8}{50} + \dfrac{5}{50}$$
$$= \dfrac{30 + 8 + 5}{50} = \dfrac{43}{50}$$

9. Add $\dfrac{1}{49} + \dfrac{3}{14}$

First find the LCD.
$$14 = 2 \cdot 7$$
$$49 = \quad 7 \cdot 7$$
$$2 \cdot 7 \cdot 7 \qquad \text{LCD} = 98$$
Then change to equivalent fractions and add.
$$\dfrac{1}{49} \times \dfrac{2}{2} + \dfrac{3}{14} \times \dfrac{7}{7} = \dfrac{2}{98} + \dfrac{21}{98} = \dfrac{2 + 21}{98} = \dfrac{23}{98}$$

10. $\dfrac{1}{12} - \dfrac{1}{30}$

First find the LCD.

$12 = 2 \cdot 2 \cdot 3$

$30 = \mid 2 \cdot 3 \cdot 5$

$\qquad 2 \cdot 2 \cdot 3 \cdot 5 \qquad$ LCD $= 60$

Then change to equivalent fractions and subtract.

$\dfrac{1}{12} \times \dfrac{5}{5} - \dfrac{1}{30} \times \dfrac{2}{2} = \dfrac{5}{60} - \dfrac{2}{60} = \dfrac{5-2}{60} = \dfrac{3}{60} = \dfrac{1}{20}$

11. $\dfrac{2}{3} + \dfrac{3}{4} - \dfrac{3}{8}$

First find the LCD.

$3 = \qquad\qquad\qquad 3$

$4 = \quad 2 \cdot 2$

$8 = \quad 2 \cdot 2 \cdot 2$

$\qquad 2 \cdot 2 \cdot 2 \cdot 3 \qquad$ LCD $= 24$

Then change to equivalent fractions and add and subtract.

$\dfrac{2}{3} \times \dfrac{8}{8} + \dfrac{3}{4} \times \dfrac{6}{6} - \dfrac{3}{8} \times \dfrac{3}{3}$

$\qquad = \dfrac{16}{24} + \dfrac{18}{24} - \dfrac{9}{24} = \dfrac{16 + 18 - 9}{24} = \dfrac{25}{24} = 1\dfrac{1}{24}$

12. (a) $1\dfrac{2}{3} + 2\dfrac{4}{5} = \dfrac{5}{3} + \dfrac{14}{5} = \dfrac{5}{3} \times \dfrac{5}{5} + \dfrac{14}{5} \times \dfrac{3}{3} = \dfrac{25}{15} + \dfrac{42}{15}$

$\qquad\qquad = \dfrac{25 + 42}{15} = \dfrac{67}{15} = 4\dfrac{7}{15}$

(b) $5\dfrac{1}{4} - 2\dfrac{2}{3} = \dfrac{21}{4} - \dfrac{8}{3} = \dfrac{21}{4} \times \dfrac{3}{3} - \dfrac{8}{3} \times \dfrac{4}{4} = \dfrac{63}{12} - \dfrac{32}{12}$

$\qquad\qquad = \dfrac{63 - 32}{12} = \dfrac{31}{12} = 2\dfrac{7}{12}$

13.

```
         6½
  ┌─────────────┐
4⅘│             │4⅘
  └─────────────┘
         6½
```

$4\dfrac{1}{5} + 4\dfrac{1}{5} + 6\dfrac{1}{2} + 6\dfrac{1}{2}$

$= \dfrac{21}{5} + \dfrac{21}{5} + \dfrac{13}{2} + \dfrac{13}{2}$

$\qquad$ LCD $= 10$

$\dfrac{21}{5} \times \dfrac{2}{2} + \dfrac{21}{5} \times \dfrac{2}{2} + \dfrac{13}{2} \times \dfrac{5}{5} + \dfrac{13}{2} \times \dfrac{5}{5}$

$= \dfrac{42}{10} + \dfrac{42}{10} + \dfrac{65}{10} + \dfrac{65}{10} = \dfrac{42 + 42 + 65 + 65}{10} = \dfrac{214}{10} = 21\dfrac{2}{5}$

The perimeter is $21\dfrac{2}{5}$ cm.

0.3 Practice Problems

1. (a) $\dfrac{2}{7} \times \dfrac{5}{11} = \dfrac{2 \cdot 5}{7 \cdot 11} = \dfrac{10}{77}$

(b) $\dfrac{1}{5} \times \dfrac{7}{10} = \dfrac{1 \times 7}{5 \times 10} = \dfrac{7}{50}$

(c) $\dfrac{9}{5} \times \dfrac{1}{4} = \dfrac{9 \times 1}{5 \times 4} = \dfrac{9}{20}$

(d) $\dfrac{8}{9} \times \dfrac{3}{10} = \dfrac{8 \times 3}{9 \times 10} = \dfrac{24}{90} = \dfrac{4}{15}$

2. (a) $\dfrac{3}{5} \times \dfrac{4}{3} = \dfrac{3 \cdot 4}{5 \cdot 3} = \dfrac{4}{5}$

(b) $\dfrac{9}{10} \times \dfrac{5}{12} = \dfrac{3 \cdot 3}{2 \cdot 5} \times \dfrac{5}{2 \cdot 2 \cdot 3} = \dfrac{3}{8}$

3. (a) $4 \times \dfrac{2}{7} = \dfrac{4}{1} \times \dfrac{2}{7} = \dfrac{4 \cdot 2}{1 \cdot 7} = \dfrac{8}{7} = 1\dfrac{1}{7}$

(b) $12 \times \dfrac{3}{4} = \dfrac{12}{1} \times \dfrac{3}{4} = \dfrac{2 \cdot 2 \cdot 3}{1} \times \dfrac{3}{2 \cdot 2} = 9$

4. Multiply. $5\dfrac{3}{5}$ times $3\dfrac{3}{4}$

$5\dfrac{3}{5} \times 3\dfrac{3}{4} = \dfrac{28}{5} \times \dfrac{15}{4} = \dfrac{2 \cdot 2 \cdot 7}{5} \times \dfrac{3 \cdot 5}{2 \cdot 2} = \dfrac{21}{1} = 21$

21 square miles

5. $3\dfrac{1}{2} \times \dfrac{1}{14} \times 4 = \dfrac{7}{2} \times \dfrac{1}{14} \times \dfrac{4}{1} = \dfrac{7}{2} \times \dfrac{1}{2 \cdot 7} \times \dfrac{2 \cdot 2}{1} = 1$

6. (a) $\dfrac{2}{5} \div \dfrac{1}{3} = \dfrac{2}{5} \times \dfrac{3}{1} = \dfrac{6}{5}$

(b) $\dfrac{12}{13} \div \dfrac{4}{3} = \dfrac{2 \cdot 2 \cdot 3}{13} \times \dfrac{3}{2 \cdot 2} = \dfrac{9}{13}$

7. (a) $\dfrac{3}{7} \div 6 = \dfrac{3}{7} \div \dfrac{6}{1} = \dfrac{3}{7} \times \dfrac{1}{6} = \dfrac{3}{42} = \dfrac{1}{14}$

(b) $8 \div \dfrac{2}{3} = \dfrac{8}{1} \times \dfrac{3}{2} = 12$

8. (a) $\dfrac{\frac{3}{11}}{\frac{5}{7}} = \dfrac{3}{11} \div \dfrac{5}{7} = \dfrac{3}{11} \times \dfrac{7}{5} = \dfrac{21}{55}$

(b) $\dfrac{\frac{12}{5}}{\frac{8}{15}} = \dfrac{12}{5} \div \dfrac{8}{15} = \dfrac{2 \cdot 2 \cdot 3}{5} \times \dfrac{3 \cdot 5}{2 \cdot 2 \cdot 2} = \dfrac{9}{2} = 4\dfrac{1}{2}$

9. (a) $1\dfrac{2}{5} \div 2\dfrac{1}{3} = \dfrac{7}{5} \div \dfrac{7}{3} = \dfrac{7}{5} \times \dfrac{3}{7} = \dfrac{3}{5}$

(b) $4\dfrac{2}{3} \div 7 = \dfrac{14}{3} \times \dfrac{1}{7} = \dfrac{2 \cdot 7}{3} \times \dfrac{1}{7} = \dfrac{2}{3}$

(c) $\dfrac{1\frac{1}{5}}{1\frac{2}{7}} = 1\dfrac{1}{5} \div 1\dfrac{2}{7} = \dfrac{6}{5} \div \dfrac{9}{7} = \dfrac{6}{5} \times \dfrac{7}{9} = \dfrac{2 \cdot 3}{5} \times \dfrac{7}{3 \cdot 3} = \dfrac{14}{15}$

10. $64 \div 5\dfrac{1}{3} = 64 \div \dfrac{16}{3} = 64 \times \dfrac{3}{16} = 12$ jars

11. $25\dfrac{1}{2} \times 5\dfrac{1}{4} = \dfrac{51}{2} \times \dfrac{21}{4} = \dfrac{1071}{8} = 133\dfrac{7}{8}$ miles

0.4 Practice Problems

1. (a) $0.9 = \dfrac{9}{10} =$ nine tenths   **(b)** $\dfrac{9}{100} =$ nine hundredths.

(c) $0.731 = \dfrac{731}{1000} =$ seven hundred thirty-one thousandths

(d) $1.371 = 1\dfrac{371}{1000} =$ one and three hundred seventy-one thousandths

(e) $0.0005 = \dfrac{5}{10,000} =$ five ten-thousandths

2. (a) $\dfrac{3}{8}$ $\quad$
```
   0.375
8)3.000
  2 4
   60
   56
   40
   40
    0
```

(b) $\dfrac{7}{200}$ $\quad$
```
     0.035
200)7.000
    6 00
    1 000
    1 000
        0
```

(c) $\dfrac{33}{20}$ $\quad$
```
    1.65
20)33.00
   20
   13 0
   12 0
    1 00
    1 00
       0
```

3. (a) $\dfrac{1}{6}$

$$\begin{array}{r} 0.166 = 0.1\overline{6} \\ 6\overline{)1.000} \\ \underline{6} \\ 40 \\ \underline{36} \\ 40 \\ \underline{36} \\ 4 \end{array}$$

(b) $\dfrac{5}{11}$

$$\begin{array}{r} 0.4545 = .\overline{45} \\ 11\overline{)5.0000} \\ \underline{4\,4} \\ 60 \\ \underline{55} \\ 50 \\ \underline{44} \\ 60 \\ \underline{55} \\ 5 \end{array}$$

4. (a) 0.8 $\dfrac{8}{10} = \dfrac{2 \cdot 2 \cdot 2}{2 \cdot 5} = \dfrac{4}{5}$

(b) 0.88 $\dfrac{88}{100} = \dfrac{11 \cdot 2 \cdot 2 \cdot 2}{5 \cdot 5 \cdot 2 \cdot 2} = \dfrac{22}{25}$

(c) 0.45 $\dfrac{45}{100} = \dfrac{5 \cdot 3 \cdot 3}{5 \cdot 5 \cdot 2 \cdot 2} = \dfrac{9}{20}$

(d) 0.148 $\dfrac{148}{1000} = \dfrac{2 \cdot 2 \cdot 37}{5 \cdot 5 \cdot 5 \cdot 2 \cdot 2 \cdot 2} = \dfrac{37}{250}$

(e) 0.612 $\dfrac{612}{1000} = \dfrac{17 \cdot 3 \cdot 3 \cdot 2 \cdot 2}{5 \cdot 5 \cdot 5 \cdot 2 \cdot 2 \cdot 2} = \dfrac{153}{250}$

(f) 0.016 $\dfrac{16}{1000} = \dfrac{2 \cdot 2 \cdot 2 \cdot 2}{5 \cdot 5 \cdot 5 \cdot 2 \cdot 2 \cdot 2} = \dfrac{2}{125}$

5. (a)
$$\begin{array}{r} 3.12 \\ 5.08 \\ 1.42 \\ \hline 9.62 \end{array}$$

(b)
$$\begin{array}{r} 152.003 \\ -136.118 \\ \hline 15.885 \end{array}$$

(c)
$$\begin{array}{r} 1.1 \\ 3.16 \\ 5.123 \\ \hline 9.383 \end{array}$$

(d)
$$\begin{array}{r} 1.0052 \\ -0.1234 \\ \hline 0.8818 \end{array}$$

6. (a)
$$\begin{array}{r} 0.061 \\ 5.0008 \\ 1.3 \\ \hline 6.3618 \end{array}$$

(b)
$$\begin{array}{r} 18.000 \\ -\;0.126 \\ \hline 17.874 \end{array}$$

7.
$$\begin{array}{r} 0.5 \\ \times\;0.3 \\ \hline 0.15 \end{array}$$

8.
$$\begin{array}{r} 0.12 \\ \times\;\;0.4 \\ \hline 0.048 \end{array}$$

9. (a)
$$\begin{array}{r} 1.23 \\ \times\;0.005 \\ \hline 0.00615 \end{array}$$

(b)
$$\begin{array}{r} 0.00002 \\ \times\;\;0.003 \\ \hline 0.00000006 \end{array}$$

10.
$$\begin{array}{r} \$5.26 \\ 6\overline{)31.56} \\ \underline{30} \\ 15 \\ \underline{12} \\ 36 \\ \underline{36} \\ 0 \end{array}$$

$5.26 for each box of paper

11.
$$\begin{array}{r} 30{,}000. \\ .06.\overline{)1800.00.} \end{array}$$

12.
$$\begin{array}{r} 4.9.\overline{)0.0.1764} \\ 0.0036 \\ 49\overline{)0.1764} \\ \underline{147} \\ 294 \\ \underline{294} \\ 0 \end{array}$$

13. (a) $0.0016 \times 100 = 0.16$
Move decimal point 2 places to the right.
(b) $2.34 \times 1000 = 2340$
Move decimal point 3 places to the right.
(c) $56.75 \times 10{,}000 = 567{,}500$
Move decimal point 4 places to the right.

14. (a) $\dfrac{5.82}{10}$ (Move decimal point 1 place to the left.) 0.582

(b) $\dfrac{123.4}{1000}$ (Move decimal point 3 places to the left.) 0.1234

(c) $\dfrac{0.00614}{10{,}000}$ (Move decimal point 4 places to the left.)
0.000000614

0.5 Practice Problems

1. (a) $0.92 = 92\%$ **(b)** $0.418 = 41.8\%$ **(c)** $0.7 = 70\%$
2. (a) $0.0019 = 0.19\%$ **(b)** $0.0736 = 7.36\%$
(c) $0.0003 = 0.03\%$
3. (a) 304%   **(b)** 518.6%   **(c)** 210%
4. (a) 0.07   **(b)** 0.093   **(c)** 0.002
5. (a) 1.31   **(b)** 3.016   **(c)** 0.0004
6. Change % to decimal and multiply.
(a) $0.18 \times 50 = 9$
(b) $0.04 \times 64 = 2.56$
(c) $1.56 \times 35 = 54.6$
7. $4.2\% \times 38{,}000 = 0.042 \times 38{,}000$
(a) $1596   **(b)** $38{,}000 + 1596 = \$39{,}596$
8. $\dfrac{37}{148}$ reduces to $\dfrac{37 \cdot 1}{37 \cdot 2 \cdot 2} = \dfrac{1}{4} = 0.25 = 25\%$

9. (a) $\dfrac{24}{48} = \dfrac{2 \cdot 2 \cdot 2 \cdot 3}{2 \cdot 2 \cdot 2 \cdot 2 \cdot 3} = \dfrac{1}{2} = 0.5 = 50\%$

(b) $\dfrac{4}{25} = 0.16 = 16\%$

10. $\dfrac{430}{1256} = \dfrac{215}{628} \approx 0.342 \approx 34\%$

0.6 Practice Problems

1. $100{,}000 \times 400 = 40{,}000{,}000$

2. $12\dfrac{1}{2} \to 10$; $9\dfrac{3}{4} \to 10$; First room $10 \times 10 = 100$ square feet.

$14\dfrac{1}{4} \to 15$; $18\dfrac{1}{2} \to 20$; Second room $15 \times 20 = 300$ square feet.

Both rooms $100 + 300 = 400$ square feet.

3. (a) $422.8 \to 400$ miles $19.3 \to 20$ gallons of gas

$\dfrac{400}{20} = 20$ miles/gallon

(b) $1.69\dfrac{9}{10} \to \$2.00$ per gallon $3862 \to 4000$ miles

$\dfrac{4000 \text{ miles}}{20 \text{ mi/gallon}} = 200$ gallons

200 gallons $\times \$2.00$ per gallon $= \$400.00$

4. (a) $\begin{array}{ll} 3{,}580{,}000{,}000 & \text{estimate} \to 4{,}000{,}000{,}000 \text{ miles} \\ 43{,}300 & \text{estimate} \to 40{,}000 \text{ mi/hour} \end{array}$

100,000 hours

(b) 24 hours/day estimate $\to \dfrac{100{,}000}{20} = 5000$ days

5. 56.93% estimate $\to 60\% \to 0.6$

$293{,}567.12$ estimate $\to 300{,}000$
$0.6 \times 300{,}000 = \$180{,}000$

0.7 Practice Problems

1.

Mathematics Blueprint for Problem Solving

Gather the Facts	What Am I Solving for?	What Must I Calculate?	Key Points to Remember
Living Room measures $16\frac{1}{2}$ ft $\times$ $10\frac{1}{2}$ ft.	Area of room in square feet.	Multiply $16\frac{1}{2}$ ft by $10\frac{1}{2}$ ft to get the area in square feet.	9 sq feet = 1 square yard
The carpet costs $20.00 per square yard.	Area of room in square yards. Cost of the carpet.	Divide the number of square feet by 9 to get the number of square yards. Multiply the number of square yards by $20.00.	

$16\frac{1}{2} \times 10\frac{1}{2} = 173\frac{1}{4}$ square feet

$173\frac{1}{4} \div 9 = 19\frac{1}{4}$ square yards

$19\frac{1}{4} \times 20 = \$385.00$ total cost of carpet

Check:

Estimate area of room $16 \times 10 = 160$ square feet
Estimate area in square yards $160 \div 10 = 16$
Estimate the cost $16 \times 20 = \$320.00$
This is close to our answer of $385.00. Our answer seems reasonable.

2. (a) $\dfrac{10}{55} \approx 0.181 \approx 18\%$ **(b)** $\dfrac{3,660,000}{13,240,000} \approx 0.276 \approx 28\%$

   (c) $\dfrac{15}{55} \approx 0.273 \approx 27\%$ **(d)** $\dfrac{3,720,000}{13,240,000} \approx 0.281 \approx 28\%$

   (e) We notice that 18% of the company's sales force is located in the Northwest, and they were responsible for 28% of the sales volume. The percent of sales compared to the percent of sales force is about 150%. 27% of the company's sales force is located in the Southwest, and they were responsible for 28% of the sales volume. The percent of sales compared to the percent of sales force is approximately 100%. It would appear that the Northwest sales force is more effective.

Chapter 1

1.1 Practice Problems

1.

	Number	Integer	Rational Number	Irrational Number	Real Number
(a)	$-\frac{2}{5}$		X		X
(b)	$1.515151\ldots$		X		X
(c)	-8	X	X		X
(d)	π			X	X

2. (a) Population growth of 1,259 is +1,259.
   (b) Depreciation of 763 is −763.00.
   (c) Wind chill factor of minus 10 is −10.

3. (a) The additive inverse of $+\dfrac{2}{5}$ is $-\dfrac{2}{5}$.
   (b) The additive inverse of -1.92 is $+1.92$.
   (c) The opposite of a loss of 12 yards on a football play is a gain of 12 yards on the play.

4. (a) $|-7.34| = 7.34$
   (b) $\left|\dfrac{5}{8}\right| = \dfrac{5}{8}$ **(c)** $\left|\dfrac{0}{2}\right| = \dfrac{0}{2} = 0$

1.1 Practice Problems (*continued*)

5. (a) $37 + 19$
   $37 + 19 = 56$
   $37 + 19 = +56$
   (b) $-23 + (-35)$
   $23 + 35 = 58$
   $-23 + (-35) = -58$

6. $-\dfrac{3}{5} + \left(-\dfrac{4}{7}\right)$
   $-\dfrac{21}{35} + \left(-\dfrac{20}{35}\right)$
   $-\dfrac{21}{35} + \left(-\dfrac{20}{35}\right) = -\dfrac{41}{35}$

7. $-12.7 + (-9.38)$
   $12.7 + 9.38 = 22.08$
   $-12.7 + (-9.38) = -22.08$

8. $-7 + (-11) + (-33)$
   $= -18 + (-33)$
   $= -51$

9. $-9 + 15$
   $15 - 9 = 6$
   $-9 + 15 = 6$

10. $-\dfrac{5}{12} + \dfrac{7}{12} + \left(-\dfrac{11}{12}\right)$
   $= \dfrac{2}{12} + \left(-\dfrac{11}{12}\right) = -\dfrac{9}{12} = -\dfrac{3}{4}$

11. $-6.3 + (-8.0) + 3.5$
   $= -14.3 + 3.5$
   $= -10.8$

12. -6 $+ 5 + (-7) + (-2) + 5 + 3$
   -6 $+ 5$
   -7 $+ 5$
   -2 $+ 3$
   $\overline{-15}$ $\overline{13}$
   $-15 + 13 = -2$

13. (a) $-2.9 + (-5.7) = -8.6$ **(b)** $\dfrac{2}{3} + \left(-\dfrac{1}{4}\right)$
   $= \dfrac{8}{12} + \left(-\dfrac{3}{12}\right) = \dfrac{5}{12}$

1.2 Practice Problems

1. $9 - (-3) = 9 + (+3) = 12$
2. $-12 - (-5) = -12 + (+5) = -7$
3. (a) $\dfrac{5}{9} - \dfrac{7}{9} = \dfrac{5}{9} + \left(-\dfrac{7}{9}\right) = -\dfrac{2}{9}$

(b) $-\dfrac{5}{21} - \left(-\dfrac{3}{7}\right) = -\dfrac{5}{21} + \left(+\dfrac{3}{7}\right) = -\dfrac{5}{21} + \left(+\dfrac{9}{21}\right) = +\dfrac{4}{21}$

4. $-17.3 - (-17.3)$

$\quad = -17.3 + 17.3$

$\quad = 0$

5. (a) $-21 - 9$ **(b)** $17 - 36$

$\quad = -21 + (-9)$ $= 17 + (-36)$

$\quad = -30$ $= -19$

(c) $12 - (-15)$ **(d)** $\dfrac{3}{5} - 2$

$\quad = 12 + (+15)$ $= \dfrac{3}{5} + (-2)$

$\quad = 27$ $= \dfrac{3}{5} + \left(-\dfrac{10}{5}\right) = -\dfrac{7}{5}$ or $-1\dfrac{2}{5}$

6. $350 - (-186)$

$\quad = 350 + 186$

$\quad = 536$ The helicopter is 536 feet from the sunken vessel.

1.3 Practice Problems

1. (a) $(-6)(-2) = 12$

(b) $(7)(9) = 63$

(c) $\left(-\dfrac{3}{5}\right)\left(\dfrac{2}{7}\right) = -\dfrac{6}{35}$

(d) $\left(\dfrac{5}{6}\right)(-7) = \left(\dfrac{5}{6}\right)\left(-\dfrac{7}{1}\right) = -\dfrac{35}{6}$

2. $(-5)(-2)(-6)$

$\quad = (+10)(-6) = -60$

3. (a) positive; $-2(-3) = 6$ **(b)** negative; $(-1)(-3)(-2)$

$\qquad\qquad\qquad\qquad\qquad\qquad\qquad\qquad = 3(-2)$

$\qquad\qquad\qquad\qquad\qquad\qquad\qquad\qquad = -6$

(c) positive; $-4\left(-\dfrac{1}{4}\right)(-2)(-6)$

$\qquad\qquad = 1(-2)(-6)$

$\qquad\qquad = -2(-6)$

$\qquad\qquad = +12$ or 12

4. (a) $-36 \div (-2) = 18$

(b) $-49 \div 7 = -7$

(c) $\dfrac{50}{-10} = -5$

(d) $\dfrac{-39}{13} = -3$

5. (a) $-12.6 \div (-1.8) = 7$

(b) $0.45 \div (-0.9) = -0.5$

6. $-\dfrac{5}{16} \div \left(-\dfrac{10}{13}\right) = \left(-\dfrac{5}{16}\right)\left(-\dfrac{13}{10}\right) = \left(-\dfrac{\overset{1}{\cancel{5}}}{16}\right)\left(-\dfrac{13}{\underset{2}{\cancel{10}}}\right) = \dfrac{13}{32}$

7. (a) $\dfrac{-12}{-\dfrac{4}{5}} = -12 \div \left(-\dfrac{4}{5}\right) = -12\left(-\dfrac{5}{4}\right) = \left(-\dfrac{\overset{3}{\cancel{12}}}{1}\right)\left(-\dfrac{5}{\underset{1}{\cancel{4}}}\right) = 15$

(b) $\dfrac{-\dfrac{2}{9}}{\dfrac{8}{13}} = -\dfrac{2}{9} \div \dfrac{8}{13} = -\dfrac{\overset{1}{\cancel{2}}}{9}\left(\dfrac{13}{\underset{4}{\cancel{8}}}\right) = -\dfrac{13}{36}$

8. (a) $6(-10) = -60$ yards

(b) $7(15) = 105$ yards

(c) $-60 + 105 = 45$ yards

1.4 Practice Problems

1. (a) $6(6)(6)(6) = 6^4$

(b) $-2(-2)(-2)(-2)(-2) = (-2)^5$

(c) $108(108)(108) = 108^3$

(d) $-11(-11)(-11)(-11)(-11)(-11) = (-11)^6$

(e) $(w)(w)(w) = w^3$

(f) $(z)(z)(z)(z) = z^4$

2. (a) $3^5 = (3)(3)(3)(3)(3) = 243$

(b) $2^2 + 3^3$

$\quad 2^2 = (2)(2) = 4$

$\quad 3^3 = (3)(3)(3) = 27$

$\quad 4 + 27 = 31$

3. (a) $(-3)^3 = -27$

(b) $(-2)^6 = 64$

(c) $-2^4 = -(2^4) = -16$

(d) $-(3^6) = -729$

4. (a) $\left(\dfrac{1}{3}\right)^3 = \left(\dfrac{1}{3}\right)\left(\dfrac{1}{3}\right)\left(\dfrac{1}{3}\right) = \dfrac{1}{27}$

(b) $(0.3)^4 = (0.3)(0.3)(0.3)(0.3) = 0.0081$

(c) $\left(\dfrac{3}{2}\right)^4 = \left(\dfrac{3}{2}\right)\left(\dfrac{3}{2}\right)\left(\dfrac{3}{2}\right)\left(\dfrac{3}{2}\right) = \dfrac{81}{16}$

(d) $(3)^4(4)^2$

$\quad 3^4 = (3)(3)(3)(3) = 81$

$\quad 4^2 = (4)(4) = 16$

$\quad (81)(16) = 1296$

(e) $4^2 - 2^4 = 16 - 16 = 0$

1.5 Practice Problems

1. $25 \div 5 \cdot 6 + 2^3$

$\quad = 25 \div 5 \cdot 6 + 8$

$\quad = 5 \cdot 6 + 8$

$\quad = 30 + 8$

$\quad = 38$

2. $(-4)^3 - 2^6$

$\quad = -64 - 64$

$\quad = -128$

3. $6 - (8 - 12)^2 + 8 \div 2$

$\quad = 6 - (-4)^2 + 8 \div 2$

$\quad = 6 - (16) + 8 \div 2$

$\quad = 6 - 16 + 4$

$\quad = -10 + 4$

$\quad = -6$

4. $\left(-\dfrac{1}{7}\right)\left(-\dfrac{14}{5}\right) + \left(-\dfrac{1}{2}\right) \div \left(\dfrac{3}{4}\right)$

$\quad = \left(-\dfrac{1}{7}\right)\left(-\dfrac{14}{5}\right) + \left(-\dfrac{1}{2}\right) \times \left(\dfrac{4}{3}\right)$

$\quad = \dfrac{2}{5} + \left(-\dfrac{2}{3}\right)$

$\quad = \dfrac{2 \cdot 3}{5 \cdot 3} + \left(-\dfrac{2 \cdot 5}{3 \cdot 5}\right)$

$\quad = \dfrac{6}{15} + \left(-\dfrac{10}{15}\right) = -\dfrac{4}{15}$

1.6 Practice Problems

1. (a) $3(x + 2y) = 3x + 6y$

(b) $-a(a - 3b) = -a(a) + (-a)(-3b) = -a^2 + 3ab$

2. (a) $-(-3x + y) = (-1)(-3x + y) = (-1)(-3x) + (-1)(y)$

$\qquad = 3x - y$

3. (a) $\dfrac{3}{5}(a^2 - 5a + 25) = \left(\dfrac{3}{5}\right)(a^2) + \left(\dfrac{3}{5}\right)(-5a) + \left(\dfrac{3}{5}\right)(25)$

$\qquad\qquad = \dfrac{3}{5}a^2 - 3a + 15$

(b) $2.5(x^2 - 3.5x + 1.2)$

$\quad = (2.5)(x^2) + (2.5)(-3.5x) + (2.5)(1.2)$

$\quad = 2.5x^2 - 8.75x + 3$

4. $-4x(x - 2y + 3) = (-4)(x)(x) - (-4)(x)(2)(y)$

$\qquad\qquad\qquad\qquad\qquad + (-4)(x)(3)$

$\qquad\qquad\qquad = -4x^2 + 8xy - 12x$

5. $(3x^2 - 2x)(-4) = (3x^2)(-4) - (2x)(-4) = -12x^2 + 8x$

6. $400(6x + 9y) = 400(6x) + 400(9y)$

$\qquad\qquad\qquad = 2400x + 3600y$

1.7 Practice Problems

1. (a) $5a$ and $8a$ are like terms.

     $2b$ and $-4b$ are like terms.

(b) y^2 and $-7y^2$ are like terms. These are the only like terms.

2. (a) $16y^3 + 9y^3 = (16 + 9)y^3 = 25y^3$

(b) $5a + 7a + 4a = (5 + 7 + 4)a = 16a$

3. $-8y^2 - 9y^2 + 4y^2 = (-8 - 9 + 4)y^2 = -13y^2$

4. (a) $1.3x + 3a - 9.6x + 2a = -8.3x + 5a$

(b) $5ab - 2ab^2 - 3a^2b + 6ab = 5ab + 6ab - 2ab^2 - 3a^2b$

                            $= 11ab - 2ab^2 - 3a^2b$

(c) $7x^2y - 2xy^2 - 3x^2y - 4xy^2 + 5x^2y$

      $= 7x^2y - 3x^2y + 5x^2y - 2xy^2 - 4xy^2 = 9x^2y - 6xy^2$

5. $5xy - 2x^2y + 6xy^2 - xy - 3xy^2 - 7x^2y$

     $= 5xy - xy - 2x^2y - 7x^2y + 6xy^2 - 3xy^2$

     $= 4xy - 9x^2y + 3xy^2$

6. $\dfrac{1}{7}a^2 + 2a^2 = \dfrac{1}{7}a^2 + \dfrac{2}{1}a^2 = \dfrac{1}{7}a^2 + \dfrac{2 \cdot 7}{1 \cdot 7}a^2$

            $= \dfrac{1}{7}a^2 + \dfrac{14}{7}a^2 = \dfrac{15}{7}a^2$

$-\dfrac{5}{12}b - \dfrac{1}{3}b = -\dfrac{5}{12}b - \dfrac{1 \cdot 4}{3 \cdot 4}b = -\dfrac{5}{12}b - \dfrac{4}{12}b$

                $= -\dfrac{9}{12}b = -\dfrac{3}{4}b$

Thus, our solution is $\dfrac{15}{7}a^2 - \dfrac{3}{4}b$.

7. $5a(2 - 3b) - 4(6a + 2ab) = 10a - 15ab - 24a - 8ab$

                             $= -14a - 23ab$

1.8 Practice Problems

1. $4 - \dfrac{1}{2}x = 4 - \dfrac{1}{2}(-8)$

     $= 4 + 4$

     $= 8$

2. (a) $4x^2 = 4(-3)^2 = 4(9) = 36$

(b) $(4x)^2 = [4(-3)]^2 = [-12]^2 = 144$

3. $2x^2 - 3x = 2(-2)^2 - 3(-2)$

           $= 2(4) - 3(-2)$

           $= 8 + 6$

           $= 14$

4. Area of a triangle is

$A = \dfrac{1}{2}ba$

altitude $= 3$ meters (m)

base $= 7$ meters (m)

$A = \dfrac{1}{2}(7\,\text{m})(3\,\text{m})$

   $= \dfrac{1}{2}(7)(3)(m)(m)$

   $= \left(\dfrac{7}{2}\right)(3)(m)^2$

   $= \dfrac{21}{2}(\text{m})^2$

   $= 10.5$ square meters

5. Area of a circle is

$A = \pi r^2$

$r = 3$ meters

$A \approx 3.14(3\,\text{m})^2$

   $= 3.14(9)(\text{m})^2$

   $= 28.26$ square meters

6. Formula $C = \dfrac{5}{9}(F - 32)$

             $= \dfrac{5}{9}(68 - 32)$

             $= \dfrac{5}{9}(36)$

             $= 5(4)$

             $= 20°$ Celsius

7. Use the formula.

$k = 1.61\ (r)$ Replace r by 35.

$k = 1.61\ (35)$

$k = 56.35$ The truck is violating the minimum speed limit.

1.9 Practice Problems

1. $5[4x - 3(y - 2)]$

   $= 5[4x - 3y + 6]$

   $= 20x - 15y + 30$

2. $-3[2a - (3b - c) + 4a]$

     $= -3[2a - 3b + c + 4a]$

     $= -3[6a - 3b + c]$

     $= -18a + 9b - 3c$

3. $3[4x - 2(1 - x)] - [3x + (x - 2)]$

   $= 3[4x - 2 + 2x] - [3x + x - 2]$

   $= 12x - 6 + 6x - 3x - x + 2$

   $= 14x - 4$

4. $-2\{5x - 3x[2x - (x^2 - 4x)]\}$

   $= -2\{5x - 3x[2x - x^2 + 4x]\}$

   $= -2\{5x - 3x[6x - x^2]\}$

   $= -2\{5x - 18x^2 + 3x^3\}$

   $= -10x + 36x^2 - 6x^3$

   $= -6x^3 + 36x^2 - 10x$

Chapter 2

2.1 Practice Problems

1. $x + 14 = 23$

$x + 14 + (-14) = 23 + (-14)$

$x + 0 = 9$

     $x = 9$

Check. $9 + 14 = 23$

               $23 = 23$

2. $17 = x - 5$

   $\underline{+\ 5 \qquad\quad +\ 5}$ **Check.** $17 \overset{?}{=} 22 - 5$

       $22 = x$ $17 = 17$

3. $0.5 - 1.2 = x - 0.3$

       $-0.7 = x - 0.3$

     $\underline{+0.3 \qquad\ +\ 0.3}$ **Check.** $0.5 - 1.2 \overset{?}{=} -0.4 - 0.3$

       $-0.4 = x$ $-0.7 = -0.7$

4. $x + 8 = -22 + 6$

      $x = -2$

Check. $-2 + 8 \overset{?}{=} -22 + 6$

                 $6 \neq -16$ This is not true.

Thus $x = -2$ is not a solution. Solve to find the solution.

   $x + 8 = -22 + 6 = -16$

         $x = -16 - 8$

         $x = -24$

5. $\dfrac{1}{20} - \dfrac{1}{2} = x + \dfrac{3}{5}$ *Check.*

$\dfrac{1}{20} - \dfrac{1 \cdot 10}{2 \cdot 10} = x + \dfrac{3 \cdot 4}{5 \cdot 4}$ $\dfrac{1}{20} - \dfrac{1}{2} \overset{?}{=} -1\dfrac{1}{20} + \dfrac{3}{5}$

$\dfrac{1}{20} - \dfrac{10}{20} = x + \dfrac{12}{20}$ $\dfrac{1}{20} - \dfrac{10}{20} \overset{?}{=} -\dfrac{21}{20} + \dfrac{12}{20}$

$-\dfrac{9}{20} = x + \dfrac{12}{20}$ $-\dfrac{9}{20} = -\dfrac{9}{20}$ ✓

$-\dfrac{9}{20} - \dfrac{12}{20} = x + \dfrac{12}{20} - \dfrac{12}{20}$

$-\dfrac{21}{20} = -1\dfrac{1}{20} = x$

2.2 Practice Problems

1. $(8)\dfrac{1}{8}x = -2(8)$ **2.** $\dfrac{9x}{9} = \dfrac{72}{9}$ **3.** $\dfrac{6x}{6} = \dfrac{50}{6}$

$\quad x = -16$ $\quad x = 8$ $\quad x = \dfrac{25}{3}$

4. $\dfrac{-27x}{-27} = \dfrac{54}{-27}$ **5.** $\dfrac{-x}{-1} = \dfrac{36}{-1}$ **6.** $\dfrac{-51}{-6} = \dfrac{-6x}{-6}$

$\quad x = -2$ $\quad x = -36$ $\dfrac{17}{2} = x$

7. $16.2 = 5.2x - 3.4x$
$16.2 = 1.8x$
$\dfrac{16.2}{1.8} = \dfrac{1.8x}{1.8}$
$9 = x$

2.3 Practice Problems

1. $9x + 2 = 38$ *Check.* $9(4) + 2 \overset{?}{=} 38$
$\quad\; -2 \quad -2$ $36 + 2 \overset{?}{=} 38$
$\quad \dfrac{9x}{9} = \dfrac{36}{9}$ $38 = 38$ ✓
$\quad\; x = 4$

2. $13x = 2x - 66$ *Check.* $13(-6) \overset{?}{=} 2(-6) - 66$
$\; -2x \quad -2x$ $-78 \overset{?}{=} -12 - 66$
$\; \dfrac{11x}{11} = \dfrac{-66}{11}$ $-78 = -78$ ✓
$\quad x = -6$

3. $3x + 2 = 5x + 2$ *Check.* $3(0) + 2 \overset{?}{=} 5(0) + 2$
$\quad\; -2 \qquad -2$ $2 = 2$ ✓
$\;\; 3x = 5x$
$-5x \quad -5x$
$-2x = 0$
$\quad x = 0$

4. $-z + 8 - z = 3z + 10 - 3$
$-2z + 8 = 3z + 7$
$-3z \qquad -3z$
$-5z + 8 = 7$
$\quad -8 = -8$
$\dfrac{-5z}{-5} = \dfrac{-1}{-5}$
$\quad z = \dfrac{1}{5}$

5. $4x - (x + 3) = 12 - 3(x - 2)$
$4x - x - 3 = 12 - 3x + 6$
$3x - 3 = -3x + 18$
$+3x \qquad\quad +3x$
$6x - 3 = 18$
$\quad +3 \qquad +3$
$\dfrac{6x}{6} = \dfrac{21}{6}$
$x = \dfrac{21}{6} = \dfrac{7}{2}$ or $3\dfrac{1}{2}$

Check. $4\left(\dfrac{7}{2}\right) - \left(\dfrac{7}{2} + 3\right) \overset{?}{=} 12 - 3\left(\dfrac{7}{2} - 2\right)$
$14 - \dfrac{13}{2} \overset{?}{=} 12 - 3\left(\dfrac{3}{2}\right)$
$\dfrac{28}{2} - \dfrac{13}{2} \overset{?}{=} \dfrac{24}{2} - \dfrac{9}{2};\quad \dfrac{15}{2} = \dfrac{15}{2}$ ✓

6. $4(-2x - 3) = -5(x - 2) + 2$
$-8x - 12 = -5x + 10 + 2$
$-8x - 12 = -5x + 12$
$+5x \qquad\quad +5x$
$-3x - 12 = 12$
$\quad +12 = +12$
$\dfrac{-3x}{-3} = \dfrac{24}{-3}$
$\quad x = -8$

7. $0.3x - 2(x + 0.1) = 0.4(x - 3) - 1.1$
$0.3x - 2x - 0.2 = 0.4x - 1.2 - 1.1$
$-1.7x - 0.2 = 0.4x - 2.3$
$-0.4x \qquad -0.4x$
$-2.1x - 0.2 = -2.3$
$\quad + 0.2 \quad +0.2$
$\dfrac{-2.1x}{-2.1} = \dfrac{-2.1}{-2.1}$
$\quad x = 1$

8. $5(2z - 1) + 7 = 7z - 4(z + 3)$
$10z - 5 + 7 = 7z - 4z - 12$
$10z + 2 = 3z - 12$
$-3z \qquad -3z$
$7z + 2 = -12$
$\quad -2 \qquad -2$
$\dfrac{7z}{7} = \dfrac{-14}{7}$
$z = -2$

Check. $5[2(-2) - 1] + 7 \overset{?}{=} 7(-2) - 4[-2 + 3]$
$5(-5) + 7 \overset{?}{=} -14 - 4(1)$
$-25 + 7 \overset{?}{=} -18$
$-18 = -18$ ✓

2.4 Practice Problems

1. $\dfrac{3}{8}x - \dfrac{3}{2} = \dfrac{1}{4}x$
$3x - 12 = 2x$
$-2x \qquad -2x$
$x - 12 = 0$
$\quad +12 \quad +12$
$x = 12$

2. $\dfrac{5x}{4} - 1 = \dfrac{3x}{4} + \dfrac{1}{2}$ *Check.* $\dfrac{5(3)}{4} - 1 \overset{?}{=} \dfrac{3(3)}{4} + \dfrac{1}{2}$
$5x - 4 = 3x + 2$ $\dfrac{15}{4} - 1 \overset{?}{=} \dfrac{9}{4} + \dfrac{1}{2}$
$-3x \qquad -3x$ $\dfrac{15}{4} - \dfrac{4}{4} \overset{?}{=} \dfrac{9}{4} + \dfrac{2}{4}$
$2x - 4 = +2$
$\quad +4 = +4$ $\dfrac{11}{4} = \dfrac{11}{4}$ ✓
$\dfrac{2x}{2} = \dfrac{6}{2}$
$x = 3$

3.
$$\frac{x+6}{9} = \frac{x}{6} + \frac{1}{2}$$
$$\frac{x}{9} + \frac{6}{9} = \frac{x}{6} + \frac{1}{2}$$
$$18\left(\frac{x}{9}\right) + 18\left(\frac{6}{9}\right) = 18\left(\frac{x}{6}\right) + 18\left(\frac{1}{2}\right)$$
$$2x + 12 = 3x + 9$$
$$\underline{-2x \qquad\qquad -2x}$$
$$12 = x + 9$$
$$\underline{-9 \qquad -9}$$
$$3 = x$$

4. $\dfrac{1}{3}(x-2) = \dfrac{1}{4}(x+5) - \dfrac{5}{3}$
$$\frac{1}{3}x - \frac{2}{3} = \frac{1}{4}x + \frac{5}{4} - \frac{5}{3}$$
$$4x - 8 = \quad 3x + 15 - 20$$
$$4x - 8 = \quad 3x - 5$$
$$\underline{-3x \qquad = -3x}$$
$$x - 8 = \qquad -5$$
$$\underline{+8 \qquad\quad +8}$$
$$x \quad = \qquad 3$$

Check. $\dfrac{1}{3}(3-2) \stackrel{?}{=} \dfrac{1}{4}(3+5) - \dfrac{5}{3}$
$$\frac{1}{3}(1) \stackrel{?}{=} \frac{1}{4}(8) - \frac{5}{3}$$
$$\frac{1}{3} \stackrel{?}{=} 2 - \frac{5}{3}$$
$$\frac{1}{3} \stackrel{?}{=} \frac{6}{3} - \frac{5}{3}$$
$$\frac{1}{3} = \frac{1}{3} \checkmark$$

5. $2.8 = 0.3(x - 2) + 2(0.1x - 0.3)$
$$2.8 = 0.3x - 0.6 + 0.2x - 0.6$$
$$10(2.8) = 10(0.3x) - 10(0.6) + 10(0.2x) - 10(0.6)$$
$$28 = 3x - 6 + 2x - 6$$
$$28 = 5x - 12$$
$$\underline{+12 \qquad\quad +12}$$
$$40 = 5x$$
$$8 = x$$

2.5 Practice Problems

1.
$$d = rt$$
$$\frac{4565}{8.3} = \frac{8.3r}{8.3}$$
$$550 \text{ mph} = r$$

2. Solve for m. $\dfrac{E}{c^2} = \dfrac{mc^2}{c^2}$ $\dfrac{E}{c^2} = m$

3. Solve for y. $8 - 2y + 3x = 0$
$$\underline{\qquad -3x = \quad -3x}$$
$$8 - 2y \quad = \quad -3x$$
$$\underline{-8 \qquad\qquad -8}$$
$$\frac{-2y}{-2} = \frac{-3x - 8}{-2}$$
$$y = \frac{3x+8}{2} \text{ or } y = \frac{3}{2}x + 4$$

4. Solve for d. $C = \pi d$
$$\frac{C}{\pi} = \frac{\pi d}{\pi}$$
$$\frac{C}{\pi} = d$$

2.6 Practice Problems

1. (a) $7 > 2$ **(b)** $-3 > -4$ **(c)** $-1 < 2$
   (d) $-8 < -5$ **(e)** $0 > -2$ **(f)** $\dfrac{2}{5} > \dfrac{3}{8}$

2. (a) $x > 5$; x is greater than 5

   (b) $x \le -2$; x is less than or equal to -2

   (c) $3 > x$; 3 is greater than x (or x is less than 3)

   (d) $x \ge -\dfrac{3}{2}$; x is greater than or equal to $-\dfrac{3}{2}$

3. (a) $t \le 180$ **(b)** $d < 15{,}000$

2.7 Practice Problems

1. (a) $9 + 4 > 6 + 4$
$$13 > 10$$
   (b) $-2 - 3 < 5 - 3$
$$-5 < 2$$
   (c) $1(2) > -3(2)$
$$2 > -6$$
   (d) $\dfrac{10}{5} < \dfrac{15}{5}$
$$2 < 3$$

2. (a) $7 > 2$ **(b)** $-3 < -1$
$$-14 < -4 \qquad\qquad 3 > 1$$
   (c) $-10 \ge -20$ **(d)** $-15 \le -5$
$$1 \le 2 \qquad\qquad 3 \ge 1$$

3. $8x - 2 \quad < 3$
$$\underline{+2 \qquad +2}$$
$$\frac{8x}{8} < \frac{5}{8}$$
$$x < \frac{5}{8}$$

4. $4 - 5x > \quad 7$
$$\underline{-4 \qquad\qquad -4}$$
$$\frac{-5x}{-5} < \frac{3}{-5}$$
$$x < \frac{-3}{5}$$

5. $\dfrac{1}{2}x + 3 < \dfrac{2}{3}x$
$$3x + 18 < \quad 4x$$
$$\underline{-4x \qquad\qquad -4x}$$
$$-x + 18 < \quad 0$$
$$\underline{\quad -18 \qquad -18}$$
$$\frac{-x}{-1} > \frac{-18}{-1}$$
$$x > 18$$

6. $\dfrac{1}{2}(3 - x) \le 2x + 5$
$$\frac{3}{2} - \frac{1}{2}x \le 2x + 5$$
$$3 - x \le 4x + 10$$
$$-5x \le 7$$
$$x \ge -\frac{7}{5}$$

7. $2000n - 700{,}000 \ge 2{,}500{,}000$
$$2{,}000n \ge 3{,}200{,}000$$
$$n \ge 1{,}600$$

Chapter 3

3.1 Practice Problems

1. (a) $x + 4$ **(b)** $3x$ **(c)** $x - 8$ **(d)** $\frac{1}{4}x$

2. (a) $3x + 8$ **(b)** $3(x + 8)$ **(c)** $\frac{1}{3}(x + 4)$

3. Let $a = $ Ann's hours per week.
Then $a - 17 = $ Marie's hours per week.

4.

$l = 2w + 5$
w

width $= w$
length $= 2w + 5$

5. 1st angle $= s - 16$
2nd angle $= s$
3rd angle $= 2s$

$2s$
s $s - 16$

6. Let $x = $ the number of students in the fall.

$\frac{2}{3}x = $ the number of students in the spring.

$\frac{1}{5}x = $ the number of students in the summer.

3.2 Practice Problems

1. $\frac{3}{4}x = -81$

$x = -108$

2. $3x - 2 = 49$

$3x = 51$

$x = 17$

3. $x + (3x - 12) = 24$

$4x - 12 = 24$

$4x = 36$

$x = 9$

First number is 9. Second number $= 3(9) - 12 = 15$.

4. Let rainfall in Canada $= x$
Then rainfall in Texas $= 3x - 14$

$3x - 14 = \quad 43$

$\underline{+14 \qquad +14}$

$\dfrac{3x}{3} = \dfrac{57}{3}$

$x = \quad 19$

19 inches in Canada
43 inches in Texas

5. (a) $\dfrac{220}{4} = 55$ mph   **(b)** $\dfrac{225}{4.5} = 50$ mph

(c) The trip leaving the city by 5 mph.

6. $\dfrac{x + x + 78 + 80 + 100 + 96}{6} = 90$

$354 + 2x = 540$

$2x = 186$

$x = 93$

She needs a 93 on the final exam.

3.3 Practice Problems

1. short piece $= x$
long piece $= x + 17$
$x + (x + 17) = 89$

$2x + 17 = 89$

$2x = 72$

$x = 36$ feet

Therefore,   Short piece   36 feet
          Long piece $36 + 17 = 53$ feet

2. Family 1 $= x + 360$
Family 2 $= x$
Family 3 $= 2x - 200$
$(x) + (x + 360) + (2x - 200) = 3960$

$4x + 160 = 3960$

$x = 950$

Therefore,   Family #2 $= \$950.00$
          Family #3 $= 2(950) - 200 = \$1700.00$
          Family #1 $= 950 + 360 = \$1310.00$

3. Let the second side $= x$ Therefore the
          first side $= x - 30$ second side $= 300$ meters
          third side $= \frac{1}{2}x$ first side $= 270$ meters
$x + x - 30 + \frac{1}{2}x = 720$ third side $= 150$ meters

$5x - 60 = 1440$

$5x = 1500$

$x = 300$

3.4 Practice Problems

1. $3(25) + (0.20)m = 350$

$75 + 0.20m = 350$

$0.20m = 275$

$m = 1375$ miles

2. $0.38x = 4560$

$x = \$12,000.00$

3. $x + 0.07x = 13,910$

$1.07x = 13,910$

$x = \$13,000.00$

4. $x = 7000(0.12)(1)$

$x = \$840.00$

5. $0.09x + 0.07(8000 - x) = 630$

$0.09x + 560 - 0.07x = 630$

$0.02x + 560 = 630$

$0.02x = 70$

$x = \$3500$

Therefore, she invested \$3500.00 at 9%.
$(8000 - 3500) = \$4500.00$ at 7%

6. $x = $ Dimes
$x + 5 = $ Quarters
$0.10x + 0.25(x + 5) = 5.10$

$0.10x + 0.25x + 1.25 = 5.10$

$0.35x + 1.25 = 5.10$

$0.35x = 3.85$

$x = 11$

Therefore, she has 11 dimes and 16 quarters.

7.   Nickels $= 2x$
Dimes $= x$
Quarters $= x + 4$
$0.05(2x) + 0.10(x) + 0.25(x + 4) = 2.35$

$0.1x + 0.10x + 0.25x + 1 = 2.35$

$0.45x + 1 = 2.35$

$0.45x = 1.35$

$x = 3$

Therefore, the boy has 3 dimes, 6 nickels, and 7 quarters.

3.5 Practice Problems

1. $A = \dfrac{1}{2}ab$

$= \dfrac{1}{2}(20)(14)$

$= (10)(14)$

$= 140$ square inches

2. $A = lw$

$120 = l(8)$

$\dfrac{120}{8} = l$

$l = 15$ yards

3. $A = \dfrac{1}{2}a(b_1 + b_2)$

$256 = \dfrac{1}{2}a(12 + 20)$

$256 = \dfrac{1}{2}(a)(32)$

$256 = 16a$

$\dfrac{256}{16} = a$

$a = 16$ feet

4. $C = 2\pi r$

$C = 2(3.14)(15)$

$C = 94.2$

$C \approx 94$ meters

5. $P = a + b + c$

$P = 15 + 15 + 15$

$P = 45$ cm

6.

$132° + x + x = 180°$

$132° + 2x = 180°$

$2x = 48°$

$x = 24°$

7. Surface Area $= 4\pi r^2$

$= 4(3.14)(5)^2$

$= 314$ m^2

8. $V = \pi r^2 h$

$V = 3.14(3)^2(4) \approx 113$ ft^3

9. Calculate the area of the pool.

$A = lw$

$A = 12 \times 8 = 96$ sq. ft.

Now add 6 feet to the length and 6 feet to the width of the pool and calculate the area.

$A = lw$

$A = 18 \times 14 = 252$ sq. ft.

Now subtract the areas.

$252 - 96 = 156$ sq. ft. at \$12 a square foot.

$156 \times 12 = \$1872$

10. $V = lwh$

$V = (6)(5)(8) = 240$ ft^3

Weight $= 240$ ft$^3 \times \dfrac{62.4 \text{ lb}}{1 \text{ ft}^3} = 14{,}976$ lb $\approx 15{,}000$ lb

3.6 Practice Problems

1. (a) height ≤ 6

(b) speed > 65

(c) area ≥ 560

(d) profit margin ≥ 50

2. $\dfrac{1050 + 1250 + 950 + x}{4} \geq 1100$

$\dfrac{x + 3250}{4} \geq 1100$

$x + 3250 \geq 4400$

$x \geq 1150$

The rope must hold at least 1150 pounds.

3. $1400 + 0.02x > 2200$

$0.02x > 800$

$x > 40{,}000$

Rita must sell more than \$40,000 worth of products each month.

Chapter 4

4.1 Practice Problems

1. (a) $a^7 \cdot a^5 = a^{7+5} = a^{12}$ **(b)** $w^{10} \cdot w = w^{10+1} = w^{11}$

2. (a) $x^3 \cdot x^9 = x^{3+9} = x^{12}$ **(b)** $3^7 \cdot 3^4 = 3^{7+4} = 3^{11}$

(c) $a^3 \cdot b^2 = a^3 \cdot b^2$ (cannot be simplified)

3. (a) $(-a^8)(a^4) = (-1 \cdot 1)(a^8 \cdot a^4)$

$= -1(a^8 \cdot a^4)$

$= -1a^{12} = -a^{12}$

(b) $(3y^2)(-2y^3) = (3)(-2)(y^2 \cdot y^3) = -6y^5$

(c) $(-4x^3)(-5x^2) = (-4)(-5)(x^3 \cdot x^2) = 20x^5$

4. $(2xy)\left(-\dfrac{1}{4}x^2y\right)(6xy^3) = (2)\left(-\dfrac{1}{4}\right)(6)(x \cdot x^2 \cdot x)(y \cdot y \cdot y^3)$

$= -3x^4y^5$

5. (a) $\dfrac{10^{13}}{10^7} = 10^{13-7} = 10^6$ **(b)** $\dfrac{x^{11}}{x} = x^{11-1} = x^{10}$

(c) $\dfrac{y^{18}}{y^8} = y^{18-8} = y^{10}$

6. (a) $\dfrac{c^3}{c^4} = \dfrac{1}{c^{4-3}} = \dfrac{1}{c}$ **(b)** $\dfrac{10^{31}}{10^{56}} = \dfrac{1}{10^{56-31}} = \dfrac{1}{10^{25}}$

(c) $\dfrac{z^{15}}{z^{21}} = \dfrac{1}{z^{21-15}} = \dfrac{1}{z^6}$

7. (a) $\dfrac{-7x^7}{-21x^9} = \dfrac{1}{3x^{9-7}} = \dfrac{1}{3x^2}$ **(b)** $\dfrac{15x^{11}}{-3x^4} = -5x^{11-4} = -5x^7$

(c) $\dfrac{23x^8}{46x^9} = \dfrac{1}{2x^{9-8}} = \dfrac{1}{2x}$

8. (a) $\dfrac{x^7y^9}{y^{10}} = \dfrac{x^7}{y}$ **(b)** $\dfrac{12x^5y^6}{-24x^3y^8} = -\dfrac{x^2}{2y^2}$

9. (a) $\dfrac{10^7}{10^7} = 1$

(b) $\dfrac{12a^4}{15a^4} = \dfrac{4}{5}\left(\dfrac{a^4}{a^4}\right) = \dfrac{4}{5}(1) = \dfrac{4}{5}$

10. (a) $\dfrac{-20a^3b^8c^4}{28a^3b^7c^5} = -\dfrac{5a^0b}{7c} = -\dfrac{5(1)b}{7c} = -\dfrac{5b}{7c}$

(b) $\dfrac{5x^0y^6}{10x^4y^8} = \dfrac{5(1)y^6}{10x^4y^8} = \dfrac{1}{2x^4y^2}$

11. $\dfrac{(-6ab^5)(3a^2b^4)}{16a^5b^7} = \dfrac{-18a^3b^9}{16a^5b^7} = -\dfrac{9b^2}{8a^2}$

12. (a) $(a^4)^3 = a^{4\cdot3} = a^{12}$

(b) $(10^5)^2 = 10^{5\cdot2} = 10^{10}$ **(c)** $(-1)^{15} = -1$

13. (a) $(3xy)^3 = (3)^3x^3y^3 = 27x^3y^3$

(b) $(yz)^{37} = y^{37}z^{37}$

(c) $(-3x^3)^2 = (-3)^2(x^3)^2 = 9x^6$

14. (a) $\left(\dfrac{x}{5}\right)^3 = \dfrac{x^3}{5^3} = \dfrac{x^3}{125}$

(b) $\left(\dfrac{4a}{b}\right)^6 = \dfrac{4^6a^6}{b^6} = \dfrac{4096a^6}{b^6}$

15. $\left(\dfrac{-2x^3y^0z}{4xz^2}\right)^5 = \left(\dfrac{-x^2}{2z}\right)^5 = \dfrac{(-1)^5(x^2)^5}{2^5z^5} = -\dfrac{x^{10}}{32z^5}$

4.2 Practice Problems

1. (a) $x^{-12} = \dfrac{1}{x^{12}}$ **(b)** $w^{-5} = \dfrac{1}{w^5}$ **(c)** $z^{-2} = \dfrac{1}{z^2}$

2. (a) $4^{-3} = \dfrac{1}{4^3} = \dfrac{1}{64}$ **(b)** $2^{-4} = \dfrac{1}{2^4} = \dfrac{1}{16}$

3. (a) $\dfrac{3}{w^{-4}} = 3w^4$ **(b)** $\dfrac{x^{-6}y^4}{z^{-2}} = \dfrac{y^4z^2}{x^6}$ **(c)** $x^{-6}y^{-5} = \dfrac{1}{x^6y^5}$

4. (a) $(2x^4y^{-5})^{-2} = 2^{-2}x^{-8}y^{10} = \dfrac{y^{10}}{2^2x^8} = \dfrac{y^{10}}{4x^8}$

(b) $\dfrac{y^{-3}z^{-4}}{y^2z^{-6}} = \dfrac{z^6}{y^2y^3z^4} = \dfrac{z^6}{y^5z^4} = \dfrac{z^2}{y^5}$

5. (a) $78{,}200 = 7.82 \times 10{,}000 = 7.82 \times 10^4$
(b) $4{,}786{,}000 = 4.786 \times 1{,}000{,}000 = 4.786 \times 10^6$
6. (a) $0.98 = 9.8 \times 10^{-1}$
(b) $0.000092 = 9.2 \times 10^{-5}$
7. (a) $1.93 \times 10^6 = 1.93 \times 1{,}000{,}000 = 1{,}930{,}000.0$

(b) $8.562 \times 10^{-5} = 8.562 \times \dfrac{1}{100{,}000} = 0.00008562$

8. $30{,}900{,}000{,}000{,}000{,}000 = 3.09 \times 10^{16}$ meters
9. (a) $(56{,}000)(1{,}400{,}000{,}000) = (5.6 \times 10^4)(1.4 \times 10^9)$
$= (5.6)(1.4)(10^4)(10^9)$
$= 7.84 \times 10^{13}$

(b) $\dfrac{0.000111}{0.00000037} = \dfrac{1.11 \times 10^{-4}}{3.7 \times 10^{-7}} = \dfrac{1.11}{3.7} \times \dfrac{10^{-4}}{10^{-7}}$

$= \dfrac{1.11}{3.7} \times \dfrac{10^7}{10^4} = 0.3 \times 10^3 = 3.0 \times 10^2$

10. $159 \text{ parsecs} = (159 \text{ parsecs})\dfrac{(3.09 \times 10^{13} \text{ kilometers})}{1 \text{ parsec}}$

$= 491.31 \times 10^{13}$ kilometers

$d = r \times t$

$491.31 \times 10^{13} \text{ km} = \dfrac{50{,}000 \text{ km}}{1 \text{ hr}} \times t$

$4.9131 \times 10^{15} \text{ km} = \dfrac{5 \times 10^4 \text{ km}}{1 \text{ hr}} \times t$

$\dfrac{4.9131 \times 10^{15} \text{ km}}{\dfrac{5 \times 10^4 \text{ km}}{1 \text{ hr}}} = t$

$\dfrac{4.9131 \times 10^{15} \text{ km } (1 \text{ hr})}{5.0 \times 10^4 \text{ km}} = t$

$0.98262 \times 10^{11} \text{ hr} = t$

9.83×10^{10} hours

4.3 Practice Problems

1. (a) This polynomial is of degree 5. It has two terms, so it is a binomial.
(b) This polynomial is of degree 7, since the sum of the exponents is $3 + 4 = 7$. It has one term, so it is a monomial.
(c) This polynomial is of degree 3. It has three terms, so it is a trinomial.
2. $(-8x^3 + 3x^2 + 6) + (2x^3 - 7x^2 - 3)$
$= [-8x^3 + 2x^3] + [3x^2 - 7x^2] + [6 - 3]$
$= [(-8 + 2)x^3] + [(3 - 7)x^2] + [6 - 3]$
$= -6x^3 - 4x^2 + 3$

3. $\left(-\dfrac{1}{3}x^2 - 6x - \dfrac{1}{12}\right) + \left(\dfrac{1}{4}x^2 + 5x - \dfrac{1}{3}\right)$

$= \left[-\dfrac{1}{3}x^2 + \dfrac{1}{4}x^2\right] + [-6x + 5x] + \left[-\dfrac{1}{12} - \dfrac{1}{3}\right]$

$= \left[\left(-\dfrac{1}{3} + \dfrac{1}{4}\right)x^2\right] + [(-6 + 5)x] + \left[-\dfrac{1}{12} - \dfrac{1}{3}\right]$

$= \left[\left(-\dfrac{4}{12} + \dfrac{3}{12}\right)x^2\right] + [-x] + \left[-\dfrac{1}{12} - \dfrac{4}{12}\right]$

$= -\dfrac{1}{12}x^2 - x - \dfrac{5}{12}$

4. $(3.5x^3 - 0.02x^2 + 1.56x - 3.5) + (-0.08x^2 - 1.98x + 4)$
$= 3.5x^3 + [-0.02 - 0.08]x^2 + [1.56 - 1.98]x + [-3.5 + 4]$
$= 3.5x^3 - 0.1x^2 - 0.42x + 0.5$
5. $(5x^3 - 15x^2 + 6x - 3) - (-4x^3 - 10x^2 + 5x + 13)$
$= (5x^3 - 15x^2 + 6x - 3) + (4x^3 + 10x^2 - 5x - 13)$
$= [5 + 4]x^3 + [-15 + 10]x^2 + [6 - 5]x + [-3 - 13]$
$= 9x^3 - 5x^2 + x - 16$
6. $(x^3 - 7x^2y + 3xy^2 - 2y^3) - (2x^3 + 4xy - 6y^3)$
$= [1 - 2]x^3 - 7x^2y + 3xy^2 - 4xy + [-2 + 6]y^3$
$= -x^3 - 7x^2y + 3xy^2 - 4xy + 4y^3$

7. (a) 1974 is 4 years later than 1970, so $x = 4$.
$0.03(4) + 5.4 = 0.12 + 5.4$
$= 5.52$
We estimate that the average truck in 1974 obtained 5.52 miles per gallon.
(b) 2006 is 36 years later than 1970, so $x = 36$.
$0.03(36) + 5.4 = 1.08 + 5.4$
$= 6.48$
We predict that the average truck in 2006 will obtain 6.48 miles per gallon.

4.4 Practice Problems

1. $4x^3(-2x^2 + 3x) = 4x^3(-2x^2) + 4x^3(3x)$
$= 4(-2)(x^3)(x^2) + 4(3)(x^3)(x)$
$= -8x^5 + 12x^4$
2. (a) $-3x(x^2 + 2x - 4) = -3x^3 - 6x^2 + 12x$
(b) $6xy(x^3 + 2x^2y - y^2) = 6x^4y + 12x^3y^2 - 6xy^3$
3. $(-6x^3 + 4x^2 - 2x)(-3xy) = 18x^4y - 12x^3y + 6x^2y$
4. $(5x - 1)(x - 2) = 5x^2 - 10x - x + 2 = 5x^2 - 11x + 2$
5. $(8a - 5b)(3a - b) = 24a^2 - 8ab - 15ab + 5b^2$
$= 24a^2 - 23ab + 5b^2$
6. $(3a + 2b)(2a - 3c) = 6a^2 - 9ac + 4ab - 6bc$
7. $(3x - 2y)(3x - 2y) = 9x^2 - 6xy - 6xy + 4y^2$
$= 9x^2 - 12xy + 4y^2$
8. $(2x^2 + 3y^2)(5x^2 + 6y^2) = 10x^4 + 12x^2y^2 + 15x^2y^2 + 18y^4$
$= 10x^4 + 27x^2y^2 + 18y^4$
9. $A = (\text{length})(\text{width}) = (7x + 3)(2x - 1)$
$= 14x^2 - 7x + 6x - 3$
$= 14x^2 - x - 3$
There are $(14x^2 - x - 3)$ square feet in the room.

4.5 Practice Problems

1. $(6x + 7)(6x - 7) = (6x)^2 - (7)^2 = 36x^2 - 49$
2. $(3x + 5y)(3x - 5y) = (3x)^2 - (5y)^2 = 9x^2 - 25y^2$
3. (a) $(4a - 9b)^2 = (4a)^2 - 2(4a)(9b) + (9b)^2$
$= 16a^2 - 72ab + 81b^2$
(b) $(5x + 4)^2 = (5x)^2 + 2(5x)(4) + (4)^2 = 25x^2 + 40x + 16$
4. $(4x^3 - 2x^2 + x)(x^2 + 3x - 2)$

$\begin{array}{r} 4x^3 - 2x^2 + x \\ x^2 + 3x - 2 \\ \hline -8x^3 + 4x^2 - 2x \\ 12x^4 - 6x^3 + 3x^2 \\ 4x^5 - 2x^4 + x^3 \\ \hline 4x^5 - 10x^4 - 13x^3 + 7x^2 - 2x \end{array}$

5. $(2x^2 + 5x + 3)(x^2 - 3x - 4)$
$= 2x^2(x^2 - 3x - 4) + 5x(x^2 - 3x - 4) + 3(x^2 - 3x - 4)$
$= 2x^4 - 6x^3 - 8x^2 + 5x^3 - 15x^2 - 20x + 3x^2 - 9x - 12$
$= 2x^4 - x^3 - 20x^2 - 29x - 12$
6. $(3x - 2)(2x + 3)(3x + 2) = (3x - 2)(3x + 2)(2x + 3)$
$= [(3x)^2 - 2^2](2x + 3)$
$= (9x^2 - 4)(2x + 3)$

$\begin{array}{r} 9x^2 - 4 \\ 2x + 3 \\ \hline 27x^2 + 0x - 12 \\ 18x^3 + 0x^2 - 8x \\ \hline 18x^3 + 27x^2 - 8x - 12 \end{array}$

Thus we have
$(3x - 2)(2x + 3)(3x + 2) = 18x^3 + 27x^2 - 8x - 12.$

4.6 Practice Problems

1. $\dfrac{15y^4 - 27y^3 - 21y^2}{3y^2} = \dfrac{15y^4}{3y^2} - \dfrac{27y^3}{3y^2} - \dfrac{21y^2}{3y^2} = 5y^2 - 9y - 7$

2.

$$x + 4)\overline{x^3 + 10x^2 + 31x + 35} \quad \begin{array}{c} x^2 + 6x + 7 \end{array}$$

$$\underline{x^3 + 4x^2}$$
$$6x^2 + 31x$$
$$\underline{6x^2 + 24x} \qquad \text{Ans. } x^2 + 6x + 7 + \dfrac{7}{x + 4}$$
$$7x + 35$$
$$\underline{7x + 28}$$
$$7$$

3.

$$x - 1)\overline{2x^3 - x^2 + 0x + 1} \quad \begin{array}{c} 2x^2 + x + 1 \end{array}$$

$$\underline{2x^3 - 2x^2}$$
$$x^2 + 0x$$
$$\underline{x^2 - x} \qquad \text{Ans. } 2x^2 + x + 1 + \dfrac{2}{x - 1}$$
$$x + 1$$
$$\underline{x - 1}$$
$$2$$

4.

$$4x - 3)\overline{20x^3 - 11x^2 - 11x + 6} \quad \begin{array}{c} 5x^2 + x - 2 \end{array}$$

$$\underline{20x^3 - 15x^2}$$
$$4x^2 - 11x$$
$$\underline{4x^2 - 3x} \qquad \text{Ans. } 5x^2 + x - 2$$
$$-8x + 6$$
$$\underline{-8x + 6}$$
$$0$$

Check. $(4x - 3)(5x^2 + x - 2)$
$= 20x^3 + 4x^2 - 8x - 15x^2 - 3x + 6$
$= 20x^3 - 11x^2 - 11x + 6$

Chapter 5

5.1 Practice Problems

1. (a) $7(3a - b)$
  (b) $5xy + 8x$
    $= x(5y + 8)$
2. $4a(3a + 4b^2 - 3ab)$
3. (a) $8(2a^3 - 3b^3)$
  (b) $r^3(s^2 - 4rs + 7r^2)$
4. $9ab^2c(2a^2 - 3bc - 5ac)$
5. $6xy^2(5x^2 - 4x + 1) = 30x^3y^2 - 24x^2y^2 + 6xy^2$
6. $3(a + 5b) + x(a + 5b)$
    $= (a + 5b)(3 + x)$
7. $(9y^2 - 2)(8y - 1)$
8. $\pi b^2 - \pi a^2 = \pi(b^2 - a^2)$

5.2 Practice Problems

1. $(2x - 7)(3y - 8)$
2. $3x(2x - 5) + 2(2x - 5)$
    $= (2x - 5)(3x + 2)$
3. $a(x + 2) + 4b(x + 2)$
    $= (x + 2)(a + 4b)$
4. $6a^2 + 3ac + 10ab + 5bc$
    $= 3a(2a + c) + 5b(2a + c)$
    $= (2a + c)(3a + 5b)$
5. $6xy + 14x - 15y - 35$
    $= 2x(3y + 7) - 5(3y + 7)$
    $= (3y + 7)(2x - 5)$
6. $3x + 6y - 5ax - 10ay$
    $= 3(x + 2y) - 5a(x + 2y)$
    $= (3 - 5a)(x + 2y)$

7. $10ad + 27bc - 6bd - 45ac$
    $= 10ad - 6bd - 45ac + 27bc$
    $= 2d(5a - 3b) - 9c(5a - 3b)$
    $= (2d - 9c)(5a - 3b)$
    Check: $10ad - 6bd - 45ac + 27bc$

5.3 Practice Problems

1. $(x + 6)(x + 2)$
2. $(x + 2)(x + 15)$
3. $(x - 9)(x - 2)$
4. $x^2 - 11x + 24 = (x - 8)(x - 3)$ or $(x - 3)(x - 8)$
5. $x^2 - 5x - 24 = (x - 8)(x + 3)$ or $(x + 3)(x - 8)$
6. $y^2 + 17y - 60 = (y + 20)(y - 3)$ or $(y - 3)(y + 20)$
7. $(x + 5)(x - 12)$
8. $(a^2 + 7)(a^2 - 6)$
9. $3x^2 + 45x + 150$
    $= 3(x^2 + 15x + 50)$
    $= 3(x + 5)(x + 10)$
10. $4x^2 - 8x - 140$
    $= 4(x^2 - 2x - 35)$
    $= 4(x + 5)(x - 7)$
11. $x(x + 1) - 4(5)$
    $= x^2 + x - 20$
    $= (x + 5)(x - 4)$

5.4 Practice Problems

1. $(2x + 5)(x + 1)$
2. $(9x - 1)(x - 7)$
3. $(3x - 7)(x + 2)$
4. $2x^2 + 7x + 5 = 2x^2 + 5x + 2x + 5$
    $= x(2x + 5) + 1(2x + 5)$
    $= (2x + 5)(x + 1)$
5. $9x^2 - 63x - x + 7$
    $= 9x(x - 7) - 1(x - 7)$
    $= (x - 7)(9x - 1)$
6. $3x^2 + 6x - 2x - 4$
    $= 3x(x + 2) - 2(x + 2)$
    $= (x + 2)(3x - 2)$
7. $8x^2 + 8x - 6 = 2(4x^2 + 4x - 3)$
    $= 2[4x^2 - 2x + 6x - 3]$
    $= 2[2x(2x - 1) + 3(2x - 1)]$
    $= 2(2x - 1)(2x + 3)$
8. $24x^2 - 38x + 10$
    $= 2(12x^2 - 19x + 5)$
    $= 2(12x^2 - 4x - 15x + 5)$
    $= 2[4x(3x - 1) - 5(3x - 1)]$
    $= 2(4x - 5)(3x - 1)$

5.5 Practice Problems

1. $64x^2 - 1 = (8x - 1)(8x + 1)$
2. $(6x + 7)(6x - 7)$
3. $(10x + 9y)(10x - 9y)$
4. $(x^4 + 1)(x^4 - 1)$
    $= (x^4 + 1)(x^2 + 1)(x^2 - 1)$
    $= (x^4 + 1)(x^2 + 1)(x + 1)(x - 1)$
5. $x^2 + 10x + 25 = (x + 5)(x + 5)$ or $(x + 5)^2$
6. $25x^2 - 30x + 9 = (5x - 3)(5x - 3) = (5x - 3)^2$
7. (a) $25x^2 + 60xy + 36y^2 = (5x + 6y)^2$
  (b) $64x^6 - 48x^3 + 9 = (8x^3 - 3)^2$
8. $9x^2 + 15x + 4 = (3x + 1)(3x + 4)$
9. $20x^2 - 45$
    $= 5(4x^2 - 9)$
    $= 5(2x - 3)(2x + 3)$
10. $3(25x^2 - 20x + 4)$
    $= 3(5x - 2)(5x - 2)$
    $= 3(5x - 2)^2$

5.6 Practice Problems

1. (a) $3(x^2 - 12x + 36)$
$= 3(x - 6)(x - 6)$
$= 3(x - 6)^2$

(b) $9y^2(x^4 - 1)$
$= 9y^2(x^2 + 1)(x^2 - 1)$
$= 9y^2(x^2 + 1)(x + 1)(x - 1)$

(c) $5x(x^2 - 3xy + 2x - 6y)$
$= 5x[x(x - 3y) + 2(x - 3y)]$
$= 5x(x + 2)(x - 3y)$

(d) $-(4x^2 - 12x + 9)$
$= -(2x - 3)(2x - 3)$
$= -(2x - 3)^2$

2. $x^2 - 9x - 8$
The factors of -8 are
$(-2)(4) = -8$
$(2)(-4) = -8$
$(-8)(1) = -8$
$(-1)(8) = -8$
None of these pairs will add up to be the coefficient of the middle term. Thus the polynomial cannot be factored. It is prime.

3. $25x^2 + 82x + 4$
Check to see if this is a perfect square trinomial.
$2[(5)(2)] = 2(10) = 20$
This is not the coefficient of the middle term. The grouping number equals 100. No factors add to 82. It is prime.

5.7 Practice Problems

1. $10x^2 - x - 2 = 0$
$(5x + 2)(2x - 1) = 0$ $5x + 2 = 0$ $2x - 1 = 0$
$5x = -2$ $2x = 1$
$x = -\dfrac{2}{5}$ $x = \dfrac{1}{2}$

Check. $10\left(-\dfrac{2}{5}\right)^2 - \left(-\dfrac{2}{5}\right) - 2 \stackrel{?}{=} 0$ $10\left(\dfrac{1}{2}\right)^2 - \dfrac{1}{2} - 2 \stackrel{?}{=} 0$
$10\left(\dfrac{4}{25}\right) + \dfrac{2}{5} - 2 \stackrel{?}{=} 0$ $10\left(\dfrac{1}{4}\right) - \dfrac{1}{2} - 2 \stackrel{?}{=} 0$
$\dfrac{8}{5} + \dfrac{2}{5} - 2 \stackrel{?}{=} 0$ $\dfrac{5}{2} - \dfrac{1}{2} - 2 \stackrel{?}{=} 0$
$\dfrac{10}{5} - \dfrac{10}{5} \stackrel{?}{=} 0$ $\dfrac{4}{2} - 2 \stackrel{?}{=} 0$
$0 = 0$ $2 - 2 \stackrel{?}{=} 0$
$0 = 0$

Thus $-\dfrac{2}{5}$ and $\dfrac{1}{2}$ are both roots for the equation.

2. $3x^2 + 11x - 4 = 0$
$(3x - 1)(x + 4) = 0$ $3x - 1 = 0$ $x + 4 = 0$
$3x = 1$ $x = -4$
$x = \dfrac{1}{3}$

3. $7x^2 + 11x = 0$ $x = 0$ $7x + 11 = 0$
$x(7x + 11) = 0$ $7x = -11$
$x = \dfrac{-11}{7}$

4. $x^2 - 6x + 4 = -8 + x$ $x - 3 = 0$ $x - 4 = 0$
$x^2 - 7x + 12 = 0$ $x = 3$ $x = 4$
$(x - 3)(x - 4) = 0$

5. $\dfrac{2x^2 - 7x}{3} = 5$
$2x^2 - 7x = 15$ $2x + 3 = 0$ $x - 5 = 0$
$2x^2 - 7x - 15 = 0$ $x = -\dfrac{3}{2}$ $x = 5$
$(2x + 3)(x - 5) = 0$

6. Let w = width, then
$3w + 2$ = length.
$(3w + 2)w = 85$ $3w + 17 = 0$ $w - 5 = 0$
$3w^2 + 2w = 85$ $w = -\dfrac{17}{3}$ $w = 5$
$3w^2 + 2w - 85 = 0$
$(3w + 17)(w - 5) = 0$
The only valid answer is width = 5 meters.
length = $3(5) + 2 = 17$ meters

7. Let b = base.
$b - 3$ = altitude.
$\dfrac{b(b - 3)}{2} = 35$
$b^2 - 3b = 70$
$b^2 - 3b - 70 = 0$
$(b + 7)(b - 10) = 0$
$b + 7 = 0$
$b = -7$
This is not a valid answer.
$b - 10 = 0$
$b = 10$
Thus the base = 10 centimeters.
altitude = $10 - 3 = 7$ centimeters

8. $-5t^2 + 45 = 0$
$-5(t^2 - 9) = 0$
$t^2 - 9 = 0$
$(t + 3)(t - 3) = 0$
$t = 3$ $t = -3$ $t = -3$ is not a valid answer
Thus it will be 3 seconds before he breaks the water's surface.

Chapter 6

6.1 Practice Problems

1. $\dfrac{28}{63} = \dfrac{7 \times 2 \times 2}{7 \times 3 \times 3} = \dfrac{4}{9}$

2. $\dfrac{12x - 6}{14x - 7} = \dfrac{6(2x - 1)}{7(2x - 1)} = \dfrac{6}{7}$

3. $\dfrac{4x^2 - 9}{2x^2 - x - 3} = \dfrac{(2x - 3)(2x + 3)}{(2x - 3)(x + 1)} = \dfrac{2x + 3}{x + 1}$

4. $\dfrac{x^3 - 16x}{x^3 - 2x^2 - 8x} = \dfrac{x(x^2 - 16)}{x(x^2 - 2x - 8)} = \dfrac{x(x + 4)(x - 4)}{x(x - 4)(x + 2)} = \dfrac{x + 4}{x + 2}$

5. $\dfrac{2x - 5}{5 - 2x} = \dfrac{-1(-2x + 5)}{(5 - 2x)} = \dfrac{-1(5 - 2x)}{(5 - 2x)} = -1$

6. $\dfrac{4x^2 + 3x - 10}{25 - 16x^2} = \dfrac{(4x - 5)(x + 2)}{(5 + 4x)(5 - 4x)} = \dfrac{(4x - 5)(x + 2)}{-1(4x - 5)(5 + 4x)}$
$= \dfrac{x + 2}{-1(5 + 4x)} = -\dfrac{x + 2}{5 + 4x}$

7. $\dfrac{4x^2 - 9y^2}{4x^2 + 12xy + 9y^2} = \dfrac{(2x + 3y)(2x - 3y)}{(2x + 3y)(2x + 3y)} = \dfrac{2x - 3y}{2x + 3y}$

8. $\dfrac{25a^2 - 16b^2}{10a^2 + 3ab - 4b^2} = \dfrac{(5a + 4b)(5a - 4b)}{(5a + 4b)(2a - b)} = \dfrac{5a - 4b}{2a - b}$

6.2 Practice Problems

1. $\dfrac{6x^2 + 7x + 2}{x^2 - 7x + 10} \cdot \dfrac{x^2 + 3x - 10}{2x^2 + 11x + 5}$
$= \dfrac{(2x + 1)(3x + 2)}{(x - 5)(x - 2)} \cdot \dfrac{(x + 5)(x - 2)}{(2x + 1)(x + 5)} = \dfrac{3x + 2}{x - 5}$

2. $\dfrac{2y^2 - 6y - 8}{y^2 - y - 2} \cdot \dfrac{y^2 - 5y + 6}{2y^2 - 32}$
$= \dfrac{2(y^2 - 3y - 4)}{(y - 2)(y + 1)} \cdot \dfrac{(y - 3)(y - 2)}{2(y^2 - 16)}$
$= \dfrac{2(y - 4)(y + 1)}{(y - 2)(y + 1)} \cdot \dfrac{(y - 3)(y - 2)}{2(y - 4)(y + 4)} = \dfrac{y - 3}{y + 4}$

3. $\dfrac{x^2 + 5x + 6}{x^2 + 8x} \div \dfrac{2x^2 + 5x + 2}{2x^2 + x}$
$= \dfrac{(x + 2)(x + 3)}{x(x + 8)} \cdot \dfrac{x(2x + 1)}{(2x + 1)(x + 2)} = \dfrac{x + 3}{x + 8}$

4. $\dfrac{x + 3}{x - 3} \div (9 - x^2) = \dfrac{x + 3}{x - 3} \cdot \dfrac{1}{(3 + x)(3 - x)}$
$= \dfrac{1}{(x - 3)(3 - x)}$

6.3 Practice Problems

1. $\dfrac{2s+t}{2s-t} + \dfrac{s-t}{2s-t} = \dfrac{2s+t+s-t}{2s-t} = \dfrac{3s}{2s-t}$

2. $\dfrac{b}{(a-2b)(a+b)} - \dfrac{2b}{(a-2b)(a+b)} = \dfrac{b-2b}{(a-2b)(a+b)}$

$= \dfrac{-b}{(a-2b)(a+b)}$

3. $\dfrac{7}{6x+21}$, $\dfrac{13}{10x+35}$ $\quad 6x+21 = 3(2x+7)$
$10x+35 = 5(2x+7)$
$\text{LCD} = 3\cdot5\cdot(2x+7) = 15(2x+7)$

4. (a) $\dfrac{3}{50xy^2z}$, $\dfrac{19}{40x^3yz}$ $\quad 50xy^2z = 2\cdot5^2\cdot x\cdot y^2\cdot z$
$40x^3yz = 2^3\cdot5\cdot x^3\cdot y\cdot z$
$\text{LCD} = 2^3\cdot5^2\cdot x^3\cdot y^2\cdot z$
$\text{LCD} = 200x^3y^2z$

(b) $\dfrac{2}{x^2+5x+6}$, $\dfrac{6}{3x^2+5x-2}$
$x^2+5x+6 = (x+3)(x+2)$
$3x^2+5x-2 = (3x-1)(x+2)$
$\text{LCD} = (x+2)(x+3)(3x-1)$

5. $\dfrac{7}{a} + \dfrac{3}{abc} = \dfrac{7bc+3}{abc}$ $\quad\text{LCD} = abc$

6. $\dfrac{2a-b}{a^2-4b^2} + \dfrac{2}{a+2b}$ $\quad\text{LCD} = (a+2b)(a-2b)$

$= \dfrac{2a-b}{(a+2b)(a-2b)} + \dfrac{2(a-2b)}{(a+2b)(a-2b)}$

$= \dfrac{2a-b}{(a+2b)(a-2b)} + \dfrac{2a-4b}{(a+2b)(a-2b)}$

$= \dfrac{2a-b+2a-4b}{(a+2b)(a-2b)}$

$= \dfrac{4a-5b}{(a+2b)(a-2b)}$

7. $\dfrac{7a}{a^2+2ab+b^2} + \dfrac{4}{a^2+ab}$

$= \dfrac{7a}{(a+b)(a+b)} + \dfrac{4}{a(a+b)}$ $\quad\text{LCD} = a(a+b)^2$

$= \dfrac{7a^2}{a(a+b)(a+b)} + \dfrac{4(a+b)}{a(a+b)(a+b)} = \dfrac{7a^2+4a+4b}{a(a+b)^2}$

8. $\dfrac{x+7}{3x-9} - \dfrac{x-6}{x-3} = \dfrac{x+7}{3(x-3)} - \dfrac{x-6}{x-3}$ $\quad\text{LCD} = 3(x-3)$

$= \dfrac{x+7-3(x-6)}{3(x-3)} = \dfrac{x+7-3x+18}{3(x-3)} = \dfrac{-2x+25}{3(x-3)}$

9. $\dfrac{x-2}{x^2-4} - \dfrac{x+1}{2x^2+4x} = \dfrac{x-2}{(x+2)(x-2)} - \dfrac{x+1}{2x(x+2)}$
$\text{LCD} = 2x(x+2)(x-2)$

$= \dfrac{2x(x-2)}{2x(x+2)(x-2)} - \dfrac{(x-2)(x+1)}{2x(x+2)(x-2)}$

$= \dfrac{2x^2-4x-(x^2-x-2)}{2x(x+2)(x-2)} = \dfrac{2x^2-4x-x^2+x+2}{2x(x+2)(x-2)}$

$= \dfrac{x^2-3x+2}{2x(x+2)(x-2)} = \dfrac{(x-2)(x-1)}{2x(x+2)(x-2)} = \dfrac{x-1}{2x(x+2)}$

6.4 Practice Problems

1. $\dfrac{\dfrac{1}{a}+\dfrac{1}{b}}{\dfrac{2}{ab^2}} = \dfrac{\dfrac{b+a}{ab}}{\dfrac{2}{ab^2}}$

$= \dfrac{b+a}{ab} \div \dfrac{2}{ab^2} = \dfrac{b+a}{ab} \cdot \dfrac{ab^2}{2} = \dfrac{ab^2(a+b)}{2ab} = \dfrac{b(a+b)}{2}$

2. $\dfrac{\dfrac{1}{a}+\dfrac{1}{b}}{\dfrac{1}{a}-\dfrac{1}{b}} = \dfrac{\dfrac{b+a}{ab}}{\dfrac{b-a}{ab}} = \dfrac{b+a}{ab} \cdot \dfrac{ab}{b-a} = \dfrac{b+a}{b-a}$

3. $\dfrac{\dfrac{x}{x^2+4x+3}+\dfrac{2}{x+1}}{x+1} = \dfrac{\dfrac{x}{(x+1)(x+3)}+\dfrac{2}{x+1}}{x+1}$

$= \dfrac{\dfrac{x+2(x+3)}{(x+1)(x+3)}}{x+1} = \dfrac{\dfrac{x+2x+6}{(x+1)(x+3)}}{(x+1)}$

$= \dfrac{3x+6}{(x+1)(x+3)}\cdot\dfrac{1}{(x+1)} = \dfrac{3(x+2)}{(x+1)^2(x+3)}$

4. $\dfrac{\dfrac{6}{x^2-y^2}}{\dfrac{1}{x-y}+\dfrac{3}{x+y}} = \dfrac{\dfrac{6}{(x+y)(x-y)}}{\dfrac{(x+y)+3(x-y)}{(x+y)(x-y)}}$

$= \dfrac{6}{(x+y)(x-y)} \div \dfrac{x+y+3x-3y}{(x+y)(x-y)}$

$= \dfrac{6}{(x+y)(x-y)} \cdot \dfrac{(x+y)(x-y)}{4x-2y}$

$= \dfrac{6}{(x+y)(x-y)} \cdot \dfrac{(x+y)(x-y)}{2(2x-y)} = \dfrac{3}{2x-y}$

5. $\dfrac{\dfrac{2}{3x^2}-\dfrac{3}{y}}{\dfrac{5}{xy}-4}$ $\quad\text{LCD} = 3x^2y$

$\dfrac{3x^2y\left(\dfrac{2}{3x^2}\right) - 3x^2y\left(\dfrac{3}{y}\right)}{3x^2y\left(\dfrac{5}{xy}\right) - 3x^2y(4)} = \dfrac{2y-3x^2(3)}{3x(5)-12x^2y}$

$= \dfrac{2y-9x^2}{15x-12x^2y}$

6. $\dfrac{\dfrac{6}{x^2-y^2}}{\dfrac{7}{x-y}+\dfrac{3}{x+y}} = \dfrac{\dfrac{6}{(x+y)(x-y)}}{\dfrac{7}{x-y}+\dfrac{3}{x+y}}$
$\text{LCD} = (x+y)(x-y)$

$\dfrac{(x+y)(x-y)\left(\dfrac{6}{(x+y)(x-y)}\right)}{(x+y)(x-y)\left(\dfrac{7}{x-y}\right)+(x+y)(x-y)\left(\dfrac{3}{x+y}\right)}$

$= \dfrac{6}{7(x+y)+3(x-y)} = \dfrac{6}{7x+7y+3x-3y}$

$= \dfrac{6}{10x+4y}$

$= \dfrac{3}{5x+2y}$

6.5 Practice Problems

1. $\dfrac{3}{x} + \dfrac{4}{5} = \dfrac{-2}{x}$ $\quad\text{LCD} = 5x$

$= 15 + 4x = -10$
$4x = -10 - 15$
$4x = -25$
$x = -\dfrac{25}{4}$

Check.

$\dfrac{3}{-25} + \dfrac{4}{5} \overset{?}{=} \dfrac{-2}{-25}$
$\quad\dfrac{4}{\ } \qquad\qquad \dfrac{4}{\ }$

$\dfrac{12}{-25} + \dfrac{4}{5} \overset{?}{=} \dfrac{-2}{-25}$
$\qquad\qquad\qquad\quad \dfrac{4}{\ }$

$-\dfrac{12}{25} + \dfrac{20}{25} \overset{?}{=} \dfrac{8}{25}$

$\dfrac{8}{25} = \dfrac{8}{25} \quad\checkmark$

2. $\dfrac{4}{2x+1} = \dfrac{6}{2x-1}$ LCD $= (2x+1)(2x-1)$

$$(2x+1)(2x-1)\left[\dfrac{4}{2x+1}\right] = (2x+1)(2x-1)\left[\dfrac{6}{2x-1}\right]$$

$$4(2x-1) = 6(2x+1)$$
$$8x - 4 = 12x + 6$$
$$-4x = 10$$
$$x = -\dfrac{5}{2}$$

Check.

$$\dfrac{4}{2\left(-\dfrac{5}{2}\right)+1} \overset{?}{=} \dfrac{6}{2\left(-\dfrac{5}{2}\right)-1}$$

$$\dfrac{4}{-5+1} \overset{?}{=} \dfrac{6}{-5-1}$$

$$\dfrac{4}{-4} \overset{?}{=} \dfrac{6}{-6}$$

$$-1 = -1 \ \checkmark$$

3. $\dfrac{x-1}{x^2-4} = \dfrac{2}{x+2} + \dfrac{4}{x-2}$

$$\dfrac{x-1}{(x+2)(x-2)} = \dfrac{2}{x+2} + \dfrac{4}{x-2} \qquad \text{LCD} = (x+2)(x-2)$$

$$(x+2)(x-2)\left[\dfrac{x-1}{(x+2)(x-2)}\right]$$

$$= (x+2)(x-2)\left[\dfrac{2}{x+2}\right] + (x+2)(x-2)\left[\dfrac{4}{x-2}\right]$$

$$x - 1 = 2(x-2) + 4(x+2)$$
$$x - 1 = 2x - 4 + 4x + 8$$
$$x - 1 = 6x + 4$$
$$-5x = 5$$
$$x = -1$$

Check.

$$\dfrac{-1-1}{(-1)^2-4} \overset{?}{=} \dfrac{2}{-1+2} + \dfrac{4}{-1-2}$$

$$\dfrac{-2}{-3} \overset{?}{=} \dfrac{2}{1} + \dfrac{4}{-3}$$

$$\dfrac{2}{3} \overset{?}{=} \dfrac{6}{3} - \dfrac{4}{3}$$

$$\dfrac{2}{3} = \dfrac{2}{3} \ \checkmark$$

4. $\dfrac{2x}{x+1} = \dfrac{-2}{x+1} + 1$ LCD $= (x+1)$

$$(x+1)\left[\dfrac{2x}{x+1}\right] = (x+1)\left[\dfrac{-2}{x+1}\right] + (x+1)[1]$$

$$2x = -2 + x + 1$$
$$2x = x - 1$$
$$x = -1$$

Check.

$$\dfrac{2(-1)}{-1+1} \overset{?}{=} \dfrac{-2}{-1+1} + 1$$

$$\dfrac{-2}{0} \overset{?}{=} \dfrac{-2}{0} + 1$$

These expressions are not defined; therefore, there is no solution to this problem.

6.6 Practice Problems

1. $\dfrac{8}{420} = \dfrac{x}{315}$

$$8(315) = 420x$$
$$2520 = 420x$$
$$x = 6$$

It would take Brenda 6 hours to drive 315 miles.

2. $\dfrac{\dfrac{5}{8}}{30} = \dfrac{2\dfrac{1}{2}}{x}$

$$\dfrac{5}{8}x = 30\left(2\dfrac{1}{2}\right)$$

$$\dfrac{5}{8}x = 75$$

$$x = 120$$

Therefore $2\dfrac{1}{2}$ inches would represent 120 miles.

3. $\dfrac{13}{x} = \dfrac{16}{18}$

$$13(18) = 16x$$
$$234 = 16x$$
$$x = 14\dfrac{5}{8} \text{ cm}$$

4. $\dfrac{6}{7} = \dfrac{x}{38.5}$

$$6(38.5) = 7x$$
$$231 = 7x$$
$$x = 33 \text{ feet}$$

5. Train A time $= \dfrac{180}{x+10}$ Train B time $= \dfrac{150}{x}$

$$\dfrac{180}{x+10} = \dfrac{150}{x}$$

$$180x = 150(x+10)$$
$$180x = 150x + 1500$$
$$30x = 1500$$
$$x = 50$$

Train B travels 50 kilometers per hour. Train A travels $50 + 10 = 60$ kilometers per hour.

6.

	Number of Hours	Part of the Job Done in One Hour
John	6 hours	$\dfrac{1}{6}$
Dave	7 hours	$\dfrac{1}{7}$
John & Dave Together	x	$\dfrac{1}{x}$

$$\dfrac{1}{6} + \dfrac{1}{7} = \dfrac{1}{x} \qquad \text{LCD} = 42x$$

$$7x + 6x = 42$$
$$13x = 42$$
$$x = 3\dfrac{3}{13} \qquad \dfrac{3}{13}\text{ hour} \times \dfrac{60 \text{ min}}{1 \text{ hour}} = \dfrac{180}{13}\text{ min} = 13.846 \text{ min}$$

Thus, doing the job together will take 3 hours and 14 minutes.

Chapter 7

7.1 Practice Problems

1. Point B is 3 units to the right on the x-axis and 4 units up from the point where we stopped on the x-axis.

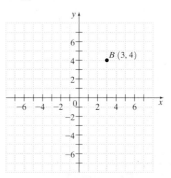

2. (a) Begin by counting 2 squares to the left starting at the origin. Since the y-coordinate is negative, count 4 units down from the point where we stopped on the x-axis. Label the point I.

(b) Begin by counting 4 squares to the left of the origin. Then count 5 units up because the y-coordinate is positive. Label the point J.

(c) Begin by counting 4 units to the right of the origin. Then count 2 units down because the y-coordinate is negative. Label the point K.

3. The points are plotted in the figure.

4. Move along the x-axis to get as close as possible to B. We end up at 2. Thus the first number of the ordered pair is 2. Then count 7 units upward on a line parallel to the y-axis to reach B. So the second number of the ordered pair is 7. Thus, $B = (2, 7)$.

5. $A = (-2, -1); B = (-1, 3); C = (0, 0); D = (2, -1); E = (3, 1)$

6. (a) Replace x by 0 in the equation.

$$3(0) - 4y = 12$$
$$0 - 4y = 12$$
$$y = -3 \qquad (0, -3)$$

(b) Replace the variable y by 3.

$$3x - 4(3) = 12$$
$$3x - 12 = 12$$
$$3x = 24$$
$$x = 8 \qquad (8, 3)$$

(c) Replace the variable y by -6.

$$3x - 4(-6) = 12$$
$$3x + 24 = 12$$
$$3x = -12$$
$$x = -4 \qquad (-4, -6)$$

7. (a)

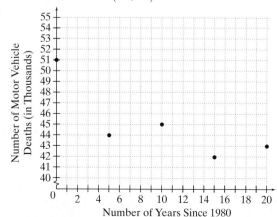

(b) 1980 was a significant high in motor vehicle deaths. During 1985–2000, the number of motor vehicle deaths was relatively stable.

7.2 Practice Problems

1. Graph $x + y = 10$.

Let $x = 0$.
$$0 + y = 10$$
$$y = 10$$

Let $x = 5$.
$$5 + y = 10$$
$$y = 5$$

Let $x = 2$.
$$2 + y = 10$$
$$y = 8$$

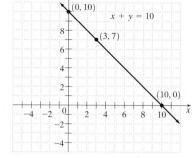

Plot the ordered pairs.
$(0, 10)$, $(5, 5)$, and $(2, 8)$

2.
$$7x + 3 = -2y + 3$$
$$7x + 3 - 3 = -2y + 3 - 3$$
$$7x = -2y$$
$$7x + 2y = -2y + 2y$$
$$7x + 2y = 0$$

Let $x = 0$.
$$7(0) + 2y = 0$$
$$2y = 0$$
$$y = 0$$

Let $x = -2$.
$$7(-2) + 2y = 0$$
$$-14 + 2y = 0$$
$$2y = 14$$
$$y = 7$$

Let $x = 2$.
$$7(2) + 2y = 0$$
$$14 + 2y = 0$$
$$2y = -14$$
$$y = -7$$

Graph the ordered pairs. $(0, 0)$, $(2, -7)$, and $(-2, 7)$

3. $2y - x = 6$

Find the two intercepts.

Let $x = 0$.   Let $y = 0$.
$$2y - 0 = 6 \qquad 2(0) - x = 6$$
$$2y = 6 \qquad\quad -x = 6$$
$$y = 3 \qquad\quad x = -6$$

Find the third point.

Let $y = 1$.
$$2(1) - x = 6$$
$$2 - x = 6$$
$$-x = 4$$
$$x = -4$$

Graph the ordered pairs.
$(0, 3)$, $(-6, 0)$, and $(-4, 1)$

4. $2y - 3 = 0$

Solve for y.
$$2y = 3$$
$$y = \frac{3}{2}$$

This line is parallel to the x-axis. It is a horizontal line $1\frac{1}{2}$ units above the x-axis.

5. $x + 3 = 0$

Solve for x.
$$x = -3$$

This line is parallel to the y-axis. It is a vertical line 3 units to the left of the y-axis.

7.3 Practice Problems

1. $m = \dfrac{y_2 - y_1}{x_2 - x_1} = \dfrac{-1 - 1}{-4 - 6} = \dfrac{-2}{-10} = \dfrac{1}{5}$

2. $m = \dfrac{y_2 - y_1}{x_2 - x_1} = \dfrac{1 - 0}{-1 - 2} = \dfrac{1}{-3} = -\dfrac{1}{3}$

3. (a) $m = \dfrac{4 - 0}{-4 - (-4)} = \dfrac{4}{0}$

$\dfrac{4}{0}$ is undefined. Therefore there is no slope and the line is a vertical line through $x = -4$.

(b) $m = \dfrac{-11 - (-11)}{3 - (-7)} = \dfrac{0}{10} = 0$

$m = 0$. The line is a horizontal line through $y = -11$.

4. Solve for y.

$4x - 2y = -5$
$-2y = -4x - 5$
$y = \dfrac{-4x - 5}{-2}$
$y = 2x + \dfrac{5}{2}$ Slope $= 2$ y-intercept $= \left(0, \dfrac{5}{2}\right)$

5. (a) $y = mx + b$

$m = -\dfrac{3}{7}$ y-intercept $= \left(0, \dfrac{2}{7}\right)$

$y = -\dfrac{3}{7}x + \dfrac{2}{7}$

(b) $y = -\dfrac{3}{7}x + \dfrac{2}{7}$

$7(y) = 7\left(-\dfrac{3}{7}x\right) + 7\left(\dfrac{2}{7}\right)$

$7y = -3x + 2$

$3x + 7y = 2$

6. y-intercept $= (0, -1)$. Thus the coordinates of the y-intercept for this line are $(0, -1)$. Plot the point. Slope is $\dfrac{\text{rise}}{\text{run}}$. Since the slope for this line is $\dfrac{3}{4}$, we will go up (rise) 3 units and go over (run) 4 units to the right from the point $(0, -1)$. This is the point $(4, 2)$.

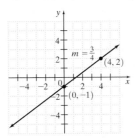

7. $y = -\dfrac{2}{3}x + 5$

The y-intercept is $(0, 5)$ since $b = 5$. Plot the point $(0, 5)$. The slope is $-\dfrac{2}{3} = \dfrac{-2}{3}$. Begin at $(0, 5)$, go down 2 units and to the right 3 units. This is the point $(3, 3)$. Draw a line that connects the points $(0, 5)$ and $(3, 3)$.

8. (a) Parallel lines have the same slope. Line j has a slope of $\dfrac{1}{4}$.

(b) Perpendicular lines have slopes whose product is -1.

$m_1 m_2 = -1$

$\dfrac{1}{4} m_2 = -1$

$4\left(\dfrac{1}{4}\right) m_2 = -1(4)$

$m_2 = -4$

Thus line k has a slope of -4.

9. (a) The slope of line n is $\dfrac{1}{4}$. The slope of a line that is parallel to line n is $\dfrac{1}{4}$.

(b) $m_1 m_2 = -1$

$\dfrac{1}{4} m_2 = -1$

$m_2 = -4$

The slope of a line that is perpendicular to n is -4.

7.4 Practice Problems

1. $y = mx + b$

$12 = -\dfrac{3}{4}(-8) + b$

$12 = 6 + b$

$6 = b$

The equation of the line is $y = -\dfrac{3}{4}x + 6$.

2. Find the slope.

$m = \dfrac{y_2 - y_1}{x_2 - x_1} = \dfrac{1 - 5}{-1 - 3} = \dfrac{-4}{-4} = 1$

Using either of the two points given, substitute x and y values into the equation $y = mx + b$.

$m = 1$ $x = 3$ and $y = 5$.

$y = mx + b$
$5 = 1(3) + b$
$5 = 3 + b$
$2 = b$

The equation of the line is $y = x + 2$.

3. The y-intercept is $(0, 1)$. Thus $b = 1$. Look for another point in the line. We choose $(6, 2)$. Count the number of vertical units from 1 to 2 (rise). Count the number of horizontal units from 0 to 6 (run).

$m = \dfrac{1}{6}$ Now we can write the equation of the line.

$y = mx + b$

$y = \dfrac{1}{6}x + 1$

7.5 Practice Problems

1. Graph $x - y \geq -10$.

Begin by graphing the line $x - y = -10$. You may use any method discussed previously. Since there is an equal sign in the inequality, we will draw a solid line to indicate that the line is part of the solution set. The easiest test point is $(0, 0)$. Substitute $x = 0$, $y = 0$ in the inequality.

$x - y \geq -10$

$0 - 0 \geq -10$

$0 \geq -10$ True

Therefore, shade the side of the line that includes the point $(0, 0)$.

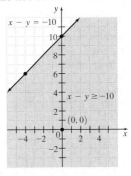

2. Step 1 Graph $y = \frac{1}{2}x$. Since $>$ is used, the line should be a dashed line.

Step 2 We see the line passes through $(0, 0)$.

Step 3 Choose another test point. We will choose $(-1, 1)$.

$$y > \frac{1}{2}x$$

$$1 > \frac{1}{2}(-1)$$

$$1 > -\frac{1}{2} \quad \text{true}$$

Shade the region that includes $(-1, 1)$, that is, the region above the line.

3. Step 1 Graph $y = -3$. Since $\geq$ is used, the line should be solid.
Step 2 Test $(0, 0)$ in the inequality.
$$y \geq -3$$
$$0 \geq -3 \quad \text{true}$$
Shade the region that includes $(0, 0)$, that is, the region above the line $y = -3$.

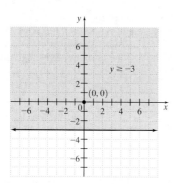

7.6 Practice Problems

1. The domain is $\{-3, 3, 0, 20\}$. The range is $\{-5, 5\}$.
2. (a) Look at the ordered pairs. No two ordered pairs have the same first coordinate. Thus this set of ordered pairs defines a function.
   (b) Look at the ordered pairs. Two different ordered pairs, $(60, 30)$ and $(60, 120)$, have the same first coordinate. Thus this relation is not a function.
3. (a) Looking at the table, we see that no two different ordered pairs have the same first coordinate. The cost of gasoline is a function of the distance traveled.
   Note that cost depends on distance. Thus distance is the independent variable. Since a negative distance does not make sense, the domain is {all nonnegative real numbers}.
   The range is {all nonnegative real numbers}.
   (b) Looking at the table, we see two ordered pairs, $(5, 20)$ and $(5, 30)$, have the same first coordinate. Thus this relation is not a function.

4. Construct a table, plot the ordered pairs and connect the points.

x	$y = x^2 - 2$	y
-2	$y = (-2)^2 - 2 = 2$	2
-1	$y = (-1)^2 - 2 = -1$	-1
0	$y = 0 - 2 = -2$	-2
1	$y = (1)^2 - 2 = -1$	-1
2	$y = (2)^2 - 2 = 2$	2

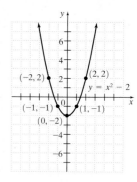

5. Select values of y and then substitute it into the equation to obtain x.

y	$x = y^2 - 1$	x	y
-2	$x = (-2)^2 - 1 = 3$	3	-2
-1	$x = (-1)^2 - 1 = 0$	0	-1
0	$x = (0)^2 - 1 = -1$	-1	0
1	$x = (1)^2 - 1 = 0$	0	1
2	$x = (2)^2 - 1 = 3$	3	2

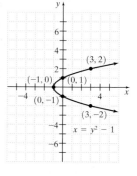

6. $y = \frac{6}{x}$

x	$y = \frac{6}{x}$	y
-3	$y = \frac{6}{-3} = -2$	-2
-2	$y = \frac{6}{-2} = -3$	-3
-1	$y = \frac{6}{-1} = -6$	-6
0	We cannot divide by 0.	
1	$y = \frac{6}{1} = 6$	6
2	$y = \frac{6}{2} = 3$	3
3	$y = \frac{6}{3} = 2$	2

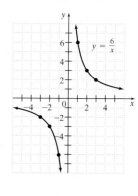

7. (a) The graph of a vertical line is not a function.
   (b) This curve is a function. Any vertical line will cross the curve in only one location.
   (c) This curve is not the graph of a function. There exist vertical lines that will cross the curve in more than one place.
8. $f(x) = -2x^2 + 3x - 8$
   (a) $f(2) = -2(2)^2 + 3(2) - 8$
      $f(2) = -2(4) + 3(2) - 8$
      $f(2) = -8 + 6 - 8$
      $f(2) = -10$
   (b) $f(-3) = -2(-3)^2 + 3(-3) - 8$
      $f(-3) = -2(9) + 3(-3) - 8$
      $f(-3) = -18 - 9 - 8$
      $f(-3) = -35$

(c) $f(0) = -2(0)^2 + 3(0) - 8$
$f(0) = -2(0) + 3(0) - 8$
$f(0) = 0 + 0 - 8$
$f(0) = -8$

Chapter 8

8.1 Practice Problems

1. $x + y = 12$
$-x + y = 4$
Graph the equations on the same coordinate plane.
The lines intersect at the point $x = 4$, $y = 8$. Thus the solution to the system of equations is $(4, 8)$.

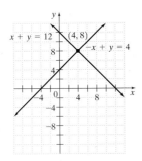

2. $4x + 2y = 8$
$-6x - 3y = 6$
Graph both equations on the same coordinate plane.
These lines are parallel. They do not intersect. Hence there is no solution to this system of equations.

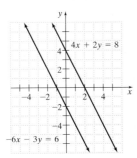

3. $3x - 9y = 18$
$-4x + 12y = -24$
Graph both equations on the same coordinate plane.
Notice both equations represent the same line. Thus there is an infinite number of solutions to this system.

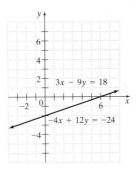

4. (a) Bill Tupper's Electrical Service charges $100 for a house call and $30 per hour. Thus we obtain the first equation,

$y = 100 + 30x$. Wire for Hire charges $50 for a house call and $40 per hour. Thus we obtain the second equation, $y = 50 + 40x$.

(b) Bill Tupper's Electrical Service $y = 100 + 30x$
Let $x = 0$ $y = 100 + 30(0) = 100$
Let $x = 4$ $y = 100 + 30(4) = 220$
Let $x = 8$ $y = 100 + 30(8) = 340$

x	y
0	100
4	220
8	340

Wire for Hire $y = 50 + 40x$
Let $x = 0$ $y = 50 + 40(0) = 50$
Let $x = 4$ $y = 50 + 40(4) = 210$
Let $x = 8$ $y = 50 + 40(8) = 370$

x	y
0	50
4	210
8	370

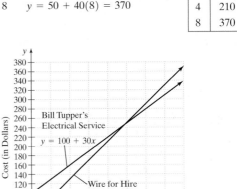

(c) We see that the graphs of the two lines intersect at $(5, 250)$. Thus the two companies will charge the same if 5 hours of electrical repairs are required.

(d) We draw a dashed line at $x = 6$. We see the line representing Wire for Hire is higher than the line representing Bill Tupper's Electrical Service after 5 hours. Thus the cost would be less if we use Bill Tupper's Electrical Service for 6 hours of work.

8.2 Practice Problems

1. $5x + 3y = 19$
$2x - y = 12$

Step 1   Solve the second equation for y.
$2x - y = 12$
$-y = -2x + 12$
$y = 2x - 12$

Step 2   Now substitute $y = 2x - 12$ into the first equation.
$5x + 3y = 19$
$5x + 3(2x - 12) = 19$

Step 3   Solve this equation.
$5x + 6x - 36 = 19$
$11x - 36 = 19$
$11x = 55$
$x = 5$

Step 4   Now obtain the value for the second variable.
$2x - y = 12$
$2(5) - y = 12$
$10 - y = 12$
$-y = 2$
$y = -2$
The solution is $(5, -2)$.

Step 5   Check.

$5x + 3y = 19$ $2x - y = 12$
$5(5) + 3(-2) \overset{?}{=} 19$ $2(5) - (-2) \overset{?}{=} 12$
$25 - 6 \overset{?}{=} 19$ $10 + 2 \overset{?}{=} 12$
$19 = 19$ ✓ $12 = 12$ ✓

2. $\dfrac{x}{3} - \dfrac{y}{2} = 1$

$x + 4y = -8$

Clear the first equation of fractions. We observe that the LCD is 6.

$6\left(\dfrac{x}{3}\right) - 6\left(\dfrac{y}{2}\right) = 6(1)$

$2x - 3y = 6$

Step 1   Solve the second equation for x.

$x + 4y = -8$

$x = -4y - 8$

Step 2   Substitute $x = -4y - 8$ into the equation $2x - 3y = 6$.

$2(-4y - 8) - 3y = 6$

Step 3   Solve this equation.

$-8y - 16 - 3y = 6$

$-11y - 16 = 6$

$-11y = 22$

$y = -2$

Step 4   Obtain the value of the second variable.

$x + 4y = -8$

$x + 4(-2) = -8$

$x - 8 = -8$

$x = 0$

The solution is $(0, -2)$.

Step 5   Check.

$\dfrac{x}{3} - \dfrac{y}{2} = 1$ $x + 4y = -8$

$\dfrac{0}{3} - \dfrac{(-2)}{2} \overset{?}{=} 1$ $0 + 4(-2) \overset{?}{=} -8$

$0 + 1 \overset{?}{=} 1$ $-8 = -8$ ✓

$1 = 1$ ✓

8.3 Practice Problems

1. $3x + y = 7$

$5x - 2y = 8$

Multiply the first equation by 2.

$2(3x) + 2(y) = 2(7)$

$6x + 2y = 14$

$\underline{+\ 5x - 2y = \ \ 8}$

$11x \qquad = 22$

$x = \ \ 2$

$5(2) - 2y = 8$

$10 - 2y = 8$

$-2y = -2$

$y = 1$

The solution is $(2, 1)$.

2. $4x + 5y = 17$

$3x + 7y = 12$

Multiply the first equation by -3 and the second equation by 4.

$-3(4x) + (-3)(5y) = (-3)(17)$

$4(3x) + 4(7y) = 4(12)$

$-12x - 15y = -51$

$\underline{12x + 28y = \ \ \ 48}$

$13y = -3$

$y = -\dfrac{3}{13}$

Substitute $y = -\dfrac{3}{13}$ into one of the original equations.

$4x + 5\left(-\dfrac{3}{13}\right) = 17$

$4x - \dfrac{15}{13} = 17$

$13(4x) - 13\left(\dfrac{15}{13}\right) = 13(17)$

$52x - 15 = 221$

$52x = 236$

$x = \dfrac{59}{13}$

The solution is $\left(\dfrac{59}{13}, -\dfrac{3}{13}\right)$.

3. Multiply the first equation by 12.

$12\left(\dfrac{2}{3}x\right) - 12\left(\dfrac{3}{4}y\right) = 12(3)$

$8x - 9y = 36$

We now have an equivalent system without fractions.

$8x - 9y = 36$

$-2x + \ y = \ 6$

Eliminate the x-variable.

$4(-2x) + 4(y) = 4(6)$

$-8x + 4y = 24$

$\underline{8x - 9y = 36}$

$-5y = \ \ 60$

$y = -12$

Substitute $y = -12$ into one of the original equations.

$-2x + (-12) = 6$

$-2x - 12 = 6$

$-2x = 18$

$x = -9$

The solution is $(-9, -12)$.

4. Multiply each term of each equation by 10.

$10(0.2x) + 10(0.3y) = 10(-0.1)$

$10(0.5x) - 10(0.1y) = 10(-1.1)$

$2x + 3y = -1$

$5x - y = -11$

Eliminate the y-variable.

$2x + 3y = -\ 1$

$\underline{15x - 3y = -33}$

$17x \qquad = -34$

$x = \ -2$

Substitute $x = -2$ into one of the equations.

$2(-2) + 3y = -1$

$-4 + 3y = -1$

$3y = 3$

$y = 1$

The solution is $(-2, 1)$.

8.4 Practice Problems

1. (a) $3x + 5y = 1485$

$x + 2y = 564$

Solve for x in the second equation and solve using the substitution method.

$x = -2y + 564$

$3(-2y + 564) + 5y = 1485$

$-6y + 1692 + 5y = 1485$

$-y = -207$

$y = 207$

Substitute $y = 207$ into the second equation and solve for x.

$x + 2(207) = 564$

$x + 414 = 564$

$x = 150$

The solution is $(150, 207)$.

(b) $7x + 6y = 45$

$6x - 5y = -2$

Using the addition method, multiply the first equation by -6 and the second equation by 7.

$-6(7x) + (-6)(6y) = (-6)(45)$

$7(6x) - 7(5y) = 7(-2)$

$-42x - 36y = -270$

$\underline{42x - 35y = \ \ -14}$

$-71y = -284$

$y = 4$

Substitute $y = 4$ into one of the original equations and solve for x.

$7x + 6(4) = 45$

$7x + 24 = 45$

$7x = 21$

$x = 3$

The solution is $(3, 4)$.

2. $4x + 2y = 2$
$-6x - 3y = 6$
Use the addition method. Multiply the first equation by 3 and the second equation by 2.
$$\begin{aligned} 12x + 6y &= 6 \\ -12x - 6y &= 12 \\ \hline 0 &= 18 \end{aligned}$$
The statement $0 = 18$ is inconsistent. There is no solution to this system of equations.

3. $3x - 9y = 18$
$-4x + 12y = -24$
Use the addition method. Multiply the first equation by 4 and the second equation by 3.
$$\begin{aligned} 12x - 36y &= 72 \\ -12x + 36y &= -72 \\ \hline 0 &= 0 \end{aligned}$$
The statement $0 = 0$ is true and can identify the equations as dependent. There are an infinite number of solutions to the system.

8.5 Practice Problems

1. Let x = number of gallons/hour pumped by the smaller pump and y = number of gallons/hour pumped by the larger pump.
1st day $8x + 5y = 49,000$
2nd day $5x + 3y = 30,000$
Use the addition method.
Multiply the first equation by 5 and the second equation by -8.
$$\begin{aligned} 40x + 25y &= 245,000 \\ -40x - 24y &= -240,000 \\ \hline y &= 5,000 \end{aligned}$$
Substitute $y = 5,000$ into one of the original equations and solve for x.
$$\begin{aligned} 8x + 5(5000) &= 49,000 \\ 8x + 25,000 &= 49,000 \\ 8x &= 24,000 \\ x &= 3,000 \end{aligned}$$
The smaller pump removes 3000 gallons per hour.
The larger pump removes 5000 gallons per hour.

2. Let x = cost/gallon of unleaded premium gasoline and y = cost/gallon of unleaded regular gasoline.
Last week's purchase $7x + 8y = 27.22$
This week's purchase $8x + 4y = 22.16$
Use the addition method. Multiply the first equation by 8 and the second equation by -7.
$$\begin{aligned} 56x + 64y &= 217.76 \\ -56x - 28y &= -155.12 \\ \hline 36y &= 62.64 \\ y &= 1.74 \end{aligned}$$
Substitute $y = 1.74$ into one of the original equations and solve for x.
$$\begin{aligned} 7x + 8(1.74) &= 27.22 \\ 7x + 13.92 &= 27.22 \\ 7x &= 13.30 \\ x &= 1.90 \end{aligned}$$
Unleaded premium gasoline costs \$1.90/gallon.
Unleaded regular gasoline costs \$1.74/gallon.

3. Let x = the amount of 20% solution and y = the amount of 80% solution.
Total amount of 65% solution
$x + y = 4000$
Amount of acid in 65% solution
$0.2x + 0.8y = 0.65(4000)$
Use the substitution method. Solve for x in the first equation.
$x = 4000 - y$
Substitute $x = 4000 - y$ into the second equation.
$$\begin{aligned} 0.2(4000 - y) + 0.8y &= 0.65(4000) \\ 800 - 0.2y + 0.8y &= 2600 \\ 800 + 0.6y &= 2600 \\ 0.6y &= 1800 \\ y &= 3000 \end{aligned}$$
Now substitute $y = 3000$ into the first equation.

$$\begin{aligned} x + y &= 4000 \\ x + 3000 &= 4000 \\ x &= 1000 \end{aligned}$$
Therefore he will need 1000 liters of 20% solution and 3000 liters of 80% solution.

4. To go downstream with the current, we have as the rate of travel
$$r = \frac{d}{t} = \frac{72}{3} = 24 \text{ miles/hour}$$
To go upstream against the current, we have as the rate of travel
$$r = \frac{d}{t} = \frac{72}{4} = 18 \text{ miles/hour}$$
Let x = the speed of the boat in miles/hour and y = the speed of the current in miles/hour.
Thus we have the equations $x + y = 24$ and $x - y = 18$. Use the addition method to solve.
$$\begin{aligned} x + y &= 24 \\ x - y &= 18 \\ \hline 2x &= 42 \\ x &= 21 \end{aligned}$$
Substitute $x = 21$ into one of the original equations and solve for y.
$$\begin{aligned} 21 + y &= 24 \\ y &= 3 \end{aligned}$$
Thus the boat speed is 21 miles/hour and the speed of the current is 3 miles/hour.

Chapter 9

9.1 Practice Problems

1. **(a)** $\sqrt{64} = 8$ because $8^2 = 64$.
   (b) $-\sqrt{121} = -11$ because $(11)^2 = 121$.

2. **(a)** $\sqrt{\dfrac{9}{25}} = \dfrac{\sqrt{9}}{\sqrt{25}} = \dfrac{3}{5}$ **(b)** $-\sqrt{\dfrac{121}{169}} = \dfrac{-\sqrt{121}}{\sqrt{169}} = -\dfrac{11}{13}$

3. **(a)** $-\sqrt{0.0036} = -0.06$ **(b)** $\sqrt{2500} = 50$ **(c)** $\sqrt{196} = 14$

4. **(a)** 1.732   **(b)** 5.916   **(c)** 11.269

9.2 Practice Problems

1. **(a)** 6   **(b)** 13^2 **(c)** 18^6
2. **(a)** y^9 **(b)** x^{15}
3. **(a)** $25y^2$ **(b)** $x^8 y^{11}$ **(c)** $11x^6 y^3$
4. **(a)** $\sqrt{49} \cdot \sqrt{2} = 7\sqrt{2}$ **(b)** $\sqrt{4} \cdot \sqrt{3} = 2\sqrt{3}$
   (c) $\sqrt{25} \cdot \sqrt{3} = 5\sqrt{3}$
5. **(a)** $\sqrt{x^{11}} = \sqrt{x^{10}} \cdot \sqrt{x} = x^5 \sqrt{x}$
   (b) $\sqrt{x^5 y^3} = \sqrt{x^4 y^2} \cdot \sqrt{xy} = x^2 y \sqrt{xy}$
6. **(a)** $\sqrt{48x^{11}} = \sqrt{16x^{10}} \cdot \sqrt{3x} = 4x^5 \sqrt{3x}$
   (b) $\sqrt{121x^6 y^7 z^8} = \sqrt{121x^6 y^6 z^8} \cdot \sqrt{y} = 11x^3 y^3 z^4 \sqrt{y}$

9.3 Practice Problems

1. **(a)** $11\sqrt{11}$ **(b)** $\sqrt{t}$
2. $3\sqrt{x} - 2\sqrt{xy} + 7\sqrt{xy} - 5\sqrt{y}$
   $= 3\sqrt{x} + 5\sqrt{xy} - 5\sqrt{y}$
3. **(a)** $\sqrt{50} - \sqrt{18} + \sqrt{98}$
   $= \sqrt{25} \cdot \sqrt{2} - \sqrt{9} \cdot \sqrt{2} + \sqrt{49} \cdot \sqrt{2}$
   $= 5\sqrt{2} - 3\sqrt{2} + 7\sqrt{2} = 2\sqrt{2} + 7\sqrt{2} = 9\sqrt{2}$
   (b) $\sqrt{12} + \sqrt{18} - \sqrt{50} + \sqrt{27}$
   $= \sqrt{4} \cdot \sqrt{3} + \sqrt{9} \cdot \sqrt{2} - \sqrt{25} \cdot \sqrt{2} + \sqrt{9} \cdot \sqrt{3}$
   $= 2\sqrt{3} + 3\sqrt{3} + 3\sqrt{2} - 5\sqrt{2} = 5\sqrt{3} - 2\sqrt{2}$
4. $\sqrt{9x} + \sqrt{8x} - \sqrt{4x} + \sqrt{50x}$
   $= 3\sqrt{x} + 2\sqrt{2x} - 2\sqrt{x} + 5\sqrt{2x}$
   $= \sqrt{x} + 7\sqrt{2x}$
5. $\sqrt{27} - 4\sqrt{3} + 2\sqrt{75}$
   $= \sqrt{9}\sqrt{3} - 4\sqrt{3} + 2\sqrt{25}\sqrt{3}$
   $= 3\sqrt{3} - 4\sqrt{3} + 10\sqrt{3} = 9\sqrt{3}$
6. $2a\sqrt{12a} + 3\sqrt{27a^3}$
   $= 2a\sqrt{4}\sqrt{3a} + 3\sqrt{9a^2}\sqrt{3a}$
   $= 2a \cdot 2 \cdot \sqrt{3a} + 3 \cdot 3a \cdot \sqrt{3a}$
   $= 4a\sqrt{3a} + 9a\sqrt{3a}$
   $= 13a\sqrt{3a}$

9.4 Practice Problems

1. $\sqrt{3a}\sqrt{6a} = \sqrt{18a^2} = 3a\sqrt{2}$

2. (a) $(2\sqrt{3})(5\sqrt{5}) = 10\sqrt{15}$

   (b) $(4\sqrt{3x})(2x\sqrt{6x}) = 8x\sqrt{18x^2} = 24x^2\sqrt{2}$

3. $(\sqrt{180})(\sqrt{150})$

   $= (6\sqrt{5})(5\sqrt{6})$

   $= 30\sqrt{30}$ square millimeters

4. $\sqrt{6}(\sqrt{3} + 2\sqrt{8})$

   $= (\sqrt{6})(\sqrt{3}) + (\sqrt{6})(2\sqrt{8})$

   $= \sqrt{18} + 2\sqrt{48}$

   $= \sqrt{9} \cdot \sqrt{2} + 2\sqrt{16} \cdot \sqrt{3}$

   $= 3\sqrt{2} + 8\sqrt{3}$

5. $2\sqrt{x}(4\sqrt{x} - x\sqrt{2})$

   $= 8\sqrt{x^2} - 2x\sqrt{2x}$

   $= 8x - 2x\sqrt{2x}$

6. $(\sqrt{2} + \sqrt{6})(2\sqrt{2} - \sqrt{6})$

   $= 2\sqrt{4} - \sqrt{12} + 2\sqrt{12} - \sqrt{36}$

   $= 2(2) + \sqrt{12} - 6 = 4 + 2\sqrt{3} - 6 = -2 + 2\sqrt{3}$

7. $(\sqrt{6} + \sqrt{5})(\sqrt{2} + 2\sqrt{5})$

   $= \sqrt{12} + 2\sqrt{30} + \sqrt{10} + 2\sqrt{25}$

   $= 2\sqrt{3} + 2\sqrt{30} + \sqrt{10} + 2(5)$

   $= 2\sqrt{3} + 2\sqrt{30} + \sqrt{10} + 10$

8. $(3\sqrt{5} - \sqrt{10})^2$

   $= (3\sqrt{5} - \sqrt{10})(3\sqrt{5} - \sqrt{10})$

   $= 9\sqrt{25} - 3\sqrt{50} - 3\sqrt{50} + \sqrt{100}$

   $= 45 - 6\sqrt{25}\sqrt{2} + 10 = 55 - 30\sqrt{2}$

9.5 Practice Problems

1. (a) $\dfrac{\sqrt{98}}{\sqrt{2}} = \sqrt{\dfrac{98}{2}} = \sqrt{49} = 7$

   (b) $\sqrt{\dfrac{81}{100}} = \dfrac{\sqrt{81}}{\sqrt{100}} = \dfrac{9}{10}$

2. $\sqrt{\dfrac{50}{a^4}} = \dfrac{5\sqrt{2}}{a^2}$

3. $\dfrac{9}{\sqrt{7}} \cdot \dfrac{\sqrt{7}}{\sqrt{7}} = \dfrac{9\sqrt{7}}{7}$

4. (a) $\dfrac{\sqrt{2}}{\sqrt{12}} \cdot \dfrac{\sqrt{3}}{\sqrt{3}} = \dfrac{\sqrt{6}}{\sqrt{36}} = \dfrac{\sqrt{6}}{6}$

   (b) $\dfrac{6a}{\sqrt{a^7}} \cdot \dfrac{\sqrt{a}}{\sqrt{a}} = \dfrac{6a\sqrt{a}}{\sqrt{a^8}} = \dfrac{6a\sqrt{a}}{a^4} = \dfrac{6\sqrt{a}}{a^3}$

5. $\dfrac{\sqrt{2x}}{\sqrt{8x}} = \dfrac{\sqrt{2x}}{2\sqrt{2x}} \cdot \dfrac{\sqrt{2x}}{\sqrt{2x}} = \dfrac{\sqrt{4x^2}}{2\sqrt{4x^2}} = \dfrac{2x}{4x} = \dfrac{1}{2}$

6. (a) $\dfrac{4}{\sqrt{3} + \sqrt{5}} \cdot \dfrac{\sqrt{3} - \sqrt{5}}{\sqrt{3} - \sqrt{5}}$

   $= \dfrac{4(\sqrt{3} - \sqrt{5})}{3 - 5} = \dfrac{4(\sqrt{3} - \sqrt{5})}{-2}$

   $= \dfrac{2(\sqrt{3} - \sqrt{5})}{-1} = -2(\sqrt{3} - \sqrt{5})$

   (b) $\dfrac{\sqrt{a}}{(\sqrt{10} - 3)} \cdot \dfrac{(\sqrt{10} + 3)}{(\sqrt{10} + 3)}$

   $= \dfrac{\sqrt{a}(\sqrt{10} + 3)}{\sqrt{100} + 3\sqrt{10} - 3\sqrt{10} - 9}$

   $= \dfrac{\sqrt{a}(\sqrt{10} + 3)}{10 - 9} = \sqrt{a}(\sqrt{10} + 3)$

7. $\dfrac{\sqrt{3} + \sqrt{5}}{\sqrt{3} - \sqrt{5}} \cdot \dfrac{\sqrt{3} + \sqrt{5}}{\sqrt{3} + \sqrt{5}} = \dfrac{8 + 2\sqrt{15}}{3 - 5} = \dfrac{8 + 2\sqrt{15}}{-2}$

   $= \dfrac{8}{-2} + \dfrac{2\sqrt{15}}{-2} = -4 - \sqrt{15}$

9.6 Practice Problems

1. $c^2 = a^2 + b^2$

   $c^2 = 9^2 + 12^2$

   $c^2 = 81 + 144$

   $c^2 = 225$

   $c = \pm\sqrt{225}$

   $c = \pm 15$

   $c = +15$ centimeters because length is not negative.

2. $(\sqrt{17})^2 = (1)^2 + b^2$

   $17 = 1 + b^2$

   $16 = b^2$

   $\sqrt{16} = b$

   $4 = b$

   The leg is 4 meters long.

3. $c^2 = 3^2 + 8^2$

   $c^2 = 9 + 64$

   $c^2 = 73$

   $c = \sqrt{73}$

   $c \approx 8.5$ meters

4. $\sqrt{3x - 2} - 7 = 0$ ***Check.***

   $\sqrt{3x - 2} = 7$ $\sqrt{3(17) - 2} - 7 \overset{?}{=} 0$

   $(\sqrt{3x - 2})^2 = (7)^2$ $\sqrt{51 - 2} - 7 \overset{?}{=} 0$

   $3x - 2 = 49$ $\sqrt{49} - 7 \overset{?}{=} 0$

   $3x = 51$ $7 - 7 \overset{?}{=} 0$

   $x = 17$ $0 = 0$ ✓

5. $\sqrt{5x + 4} + 2 = 0$ ***Check.***

   $\sqrt{5x + 4} = -2$ $\sqrt{5(0) + 4} + 2 \overset{?}{=} 0$

   $(\sqrt{5x + 4})^2 = (-2)^2$ $\sqrt{4} + 2 \overset{?}{=} 0$

   $5x + 4 = 4$ $2 + 2 \overset{?}{=} 0$

   $5x = 0$ $4 \neq 0$

   $x = 0$

No, this does not check. There is no solution.

6. $-2 + \sqrt{6x - 1} = 3x - 2$

   $\sqrt{6x - 1} = 3x$

   $(\sqrt{6x - 1})^2 = (3x)^2$

   $6x - 1 = 9x^2$

   $0 = 9x^2 - 6x + 1$ Factor.

   $0 = (3x - 1)^2$

   $0 = 3x - 1$

   $x = \dfrac{1}{3}$ is a double root.

   Check.

   $-2 + \sqrt{6\left(\dfrac{1}{3}\right) - 1} \overset{?}{=} 3\left(\dfrac{1}{3}\right) - 2$

   $-2 + \sqrt{1} \overset{?}{=} 1 - 2$

   $-2 + 1 \overset{?}{=} 1 - 2$

   $-1 = -1$ ✓

7. $2 - x + \sqrt{x + 4} = 0$

   $\sqrt{x + 4} = x - 2$

   $(\sqrt{x + 4})^2 = (x - 2)^2$

   $x + 4 = x^2 - 4x + 4$

   $0 = x^2 - 5x$

   $0 = x(x - 5)$

   $x = 0$ $x - 5 = 0$

               $x = 5$

   Check.

   $2 - 0 + \sqrt{0 + 4} \overset{?}{=} 0$

   $2 + \sqrt{4} \overset{?}{=} 0$

   $2 + 2 \overset{?}{=} 0$

   $4 \neq 0$

   $x = 0$ does not check.

   $2 - 5 + \sqrt{5 + 4} \overset{?}{=} 0$

   $-3 + \sqrt{9} \overset{?}{=} 0$

   $-3 + 3 \overset{?}{=} 0$

   $0 = 0$ ✓

The only solution is $x = 5$.

8.
$$\sqrt{2x + 1} = \sqrt{x - 10}$$
$$\left(\sqrt{2x + 1}\right)^2 = \left(\sqrt{x - 10}\right)^2$$
$$2x + 1 = x - 10$$
$$x = -11$$

$x = -11$ produces a negative radicand and thus is extraneous. There is no solution.

9.7 Practice Problems

1. Change of $100°C$ = change of $180°F$.

Let C = the change in C,
f = the change in F, and
k = the constant of variation.

$$C = kf$$
$$100 = k \cdot 180$$
$$k = \frac{5}{9}$$

if $C = kf$ then

$$C = \frac{5}{9}f$$

if $f = -20$ then

$$C = \frac{5}{9}(-20)$$

$$C = -11\frac{1}{9}$$

The temperature drops $11\frac{1}{9}$ degrees Celsius.

2. $y = kx$

To find k we substitute $y = 18$ and $x = 5$.
$$18 = k(5)$$
$$\frac{18}{5} = k$$

We now write the variation equation with k replaced by $\frac{18}{5}$.

$$y = \frac{18}{5}x$$

Replace x by $\frac{20}{23}$ and solve for y.

$$y = \frac{18}{5} \cdot \frac{20}{23}$$

$$y = \frac{72}{23}$$

Thus $y = \frac{72}{23}$ when $x = \frac{20}{23}$.

3. Let d = the distance to stop the car,
s = the speed of the car, and
k = the constant of variation.

Since the distance varies directly as the square of the speed, we have

$$d = ks^2$$

To evaluate k we substitute the known distance and speed.
$$60 = k(20)^2$$
$$60 = k(400)$$
$$k = \frac{3}{20}$$

Now write the variation equation with the constant evaluated.

$$d = \frac{3}{20}s^2$$

$$d = \frac{3}{20}(40)^2 \quad \text{Substitute } s = 40.$$

$$d = \frac{3}{20}(1600)$$

$$d = 240$$

The distance to stop the car going 40 mph on an ice-covered road is 240 feet.

4.
$$y = \frac{k}{x}$$

$$8 = \frac{k}{15} \quad \text{Substitute known values of } x \text{ and } y \text{ to find } k.$$

$$120 = k$$

$$y = \frac{120}{x} \quad \text{Write the variation equation with } k \text{ replaced by 120.}$$

$$y = \frac{120}{\frac{3}{5}} \quad \text{Find } y \text{ when } x = \frac{3}{5}.$$

$$y = 200 \quad \text{Thus } y = 200 \text{ when } x = \frac{3}{5}.$$

5. Let V = the volume of sales,
p = the price of the calculator, and
k = the constant of variation.

Since volume varies inversely as the price, then

$$V = \frac{k}{p} \qquad \text{Evaluate } k \text{ by substituting known values for } V \text{ and } p.$$

$$120,000 = \frac{k}{30}$$

$$3,600,000 = k$$

$$V = \frac{3,600,000}{24} \quad \text{Write the variation equation with the evaluated constant and substitute } p = 24.$$

$$V = 150,000$$

Thus the volume increased to 150,000 calculators sold per year when the price was reduced to $24 per calculator.

6. Let R = the resistance in the circuit,
i = the amount of current, and
k = the constant of variation.

Since the resistance in the circuit varies inversely as the square of the amount of current,

$$R = \frac{k}{i^2}.$$

To evaluate k, substitute the known values for R and i.

$$800 = \frac{k}{(0.01)^2}$$

$$k = 0.08$$

$$R = \frac{0.08}{(0.02)^2} \quad \text{Write the variation equation with the evaluated constant of } k = 0.08 \text{ and substitute } i = 0.02.$$

$$R = \frac{0.08}{0.0004}$$

$$R = 200$$

Thus the resistance is 200 ohms if the amount of current is increased to 0.02 ampere.

Chapter 10

10.1 Practice Problems

1. (a) $2x^2 + 12x - 9 = 0$
$a = 2, b = 12, c = -9$

(b) $7x^2 = 6x - 8$
$7x^2 - 6x + 8 = 0$
$a = 7, b = -6, c = 8$

(c) $-x^2 - 6x + 3 = 0$
$x^2 + 6x - 3 = 0$
$a = 1, b = 6, c = -3$

(d) $10x^2 - 12x = 0$
$a = 10, b = -12, c = 0$

2.
$$2x^2 - 7x - 6 = 4x - 6$$
$$2x^2 - 11x = 0$$
$$x(2x - 11) = 0$$
$$x = 0 \qquad 2x - 11 = 0$$
$$x = \frac{11}{2}$$

The two roots are 0 and $\frac{11}{2}$.

3. $3x + 10 - \dfrac{8}{x} = 0$

$3x^2 + 10x - 8 = 0$

$(3x - 2)(x + 4) = 0$

$3x - 2 = 0 \qquad x + 4 = 0$

$x = \dfrac{2}{3} \qquad\qquad x = -4$

Check.

$3\left(\dfrac{2}{3}\right) + 10 - \dfrac{8}{\frac{2}{3}} \overset{?}{=} 0 \qquad 3(-4) + 10 - \dfrac{8}{-4} \overset{?}{=} 0$

$2 + 10 - 12 \overset{?}{=} 0 \qquad\qquad -12 + 10 + 2 \overset{?}{=} 0$

$2 - 2 \overset{?}{=} 0 \qquad\qquad\qquad -2 + 2 \overset{?}{=} 0$

$0 = 0 \; \checkmark \qquad\qquad\qquad 0 = 0 \; \checkmark$

Both roots check, so $x = \dfrac{2}{3}$ and $x = -4$ are the two roots that

satisfy the equation $3x + 10 - \dfrac{8}{x} = 0$.

4. $2x^2 + 9x - 18 = 0$

$(2x - 3)(x + 6) = 0$

$2x - 3 = 0 \qquad x + 6 = 0$

$x = \dfrac{3}{2} \qquad\qquad x = -6$

Check.

$2\left(\dfrac{3}{2}\right)^2 + 9\left(\dfrac{3}{2}\right) - 18 \overset{?}{=} 0 \qquad 2(-6)^2 + 9(-6) - 18 \overset{?}{=} 0$

$2\left(\dfrac{9}{4}\right) + \dfrac{27}{2} - 18 \overset{?}{=} 0 \qquad\qquad 2(36) - 54 - 18 \overset{?}{=} 0$

$\dfrac{9}{2} + \dfrac{27}{2} - 18 \overset{?}{=} 0 \qquad\qquad\qquad 72 - 54 - 18 \overset{?}{=} 0$

$\dfrac{36}{2} - 18 \overset{?}{=} 0 \qquad\qquad\qquad\qquad 18 - 18 \overset{?}{=} 0$

$18 - 18 \overset{?}{=} 0 \qquad\qquad\qquad\qquad 0 = 0 \; \checkmark$

$0 = 0 \; \checkmark$

Both roots check, so $x = \dfrac{3}{2}$ and $x = -6$ are the two roots that

satisfy the equation $2x^2 + 9x - 18 = 0$.

5. $-4x + (x - 5)(x + 1) = 5(1 - x)$

$-4x + x^2 - 4x - 5 = 5 - 5x$

$x^2 - 8x - 5 = 5 - 5x$

$x^2 - 3x - 10 = 0$

$(x + 2)(x - 5) = 0$

$x + 2 = 0 \qquad x - 5 = 0$

$x = -2 \qquad\quad x = 5$

Check. $\quad -4(-2) + (-2 - 5)(-2 + 1) \overset{?}{=} 5(1 + 2)$

$8 + (-7)(-1) \overset{?}{=} 5(3)$

$8 + 7 \overset{?}{=} 15$

$15 = 15 \; \checkmark$

$-4(5) + (5 - 5)(5 + 1) \overset{?}{=} 5(-4)$

$-20 + 0 \overset{?}{=} -20$

$-20 = -20 \; \checkmark$

Both roots check, so $x = -2$ and $x = 5$ are the two roots that
satisfy the equation $-4x + (x - 5)(x + 1) = 5(1 - x)$.

6. $\dfrac{10x + 18}{x^2 + x - 2} = \dfrac{3x}{x + 2} + \dfrac{2}{x - 1}$

$\dfrac{10x + 18}{(x - 1)(x + 2)} = \dfrac{3x}{x + 2} + \dfrac{2}{x - 1}$ Multiply by the LCD.

$10x + 18 = 3x(x - 1) + 2(x + 2)$

$10x + 18 = 3x^2 - 3x + 2x + 4$

$0 = 3x^2 - 11x - 14$

$0 = (3x - 14)(x + 1)$

$3x - 14 = 0 \qquad x + 1 = 0$

$x = \dfrac{14}{3} \qquad\qquad x = -1$

7. $t = $ the number of truck routes

$n = $ the number of cities they can service

$t = \dfrac{n^2 - n}{2}$

$28 = \dfrac{n^2 - n}{2}$

$56 = n^2 - n$

$0 = n^2 - n - 56$

$0 = (n + 7)(n - 8)$

$n + 7 = 0 \qquad n - 8 = 0$

$n = -7 \qquad\qquad n = 8$

We reject $n = -7$. There cannot be a negative number of cities.
Thus our answer is 8 cities.

10.2 Practice Problems

1. (a) $x^2 = 1$

$x = \pm\sqrt{1}$

$x = \pm 1$

(b) $x^2 - 50 = 0$

$x^2 = 50$

$x = \pm\sqrt{50}$

$x = \pm 5\sqrt{2}$

(c) $3x^2 = 81$

$x^2 = 27$

$x = \pm\sqrt{27} = \pm 3\sqrt{3}$

2. $4x^2 - 5 = 319$

$4x^2 = 324$

$x^2 = 81$

$x = \pm\sqrt{81} = \pm 9$

3. $(2x - 3)^2 = 12$

$\sqrt{(2x - 3)^2} = \pm\sqrt{12}$

$2x - 3 = \pm 2\sqrt{3}$

$2x = 3 \pm 2\sqrt{3}$

$x = \dfrac{3 \pm 2\sqrt{3}}{2}$

4. $x^2 + 10x = 3$

$x^2 + 10x + (5)^2 = 3 + 25$

$(x + 5)^2 = 28$

$\sqrt{(x + 5)^2} = \pm\sqrt{28}$

$x + 5 = \pm\sqrt{28}$

$x = -5 \pm 2\sqrt{7}$

5. $x^2 - 5x = 7$

$x^2 - 5x + \left(\dfrac{5}{2}\right)^2 = 7 + \left(\dfrac{5}{2}\right)^2$

$\left(x - \dfrac{5}{2}\right)^2 = 7 + \dfrac{25}{4}$

$\left(x - \dfrac{5}{2}\right)^2 = \dfrac{28}{4} + \dfrac{25}{4}$

$\left(x - \dfrac{5}{2}\right)^2 = \dfrac{53}{4}$

$x - \dfrac{5}{2} = \pm\sqrt{\dfrac{53}{4}}$

$x = \dfrac{5}{2} \pm \dfrac{\sqrt{53}}{2}$

$x = \dfrac{5 \pm \sqrt{53}}{2}$

6. $8x^2 + 2x - 1 = 0$

$x^2 + \dfrac{1}{4}x - \dfrac{1}{8} = 0$

$x^2 + \dfrac{1}{4}x = \dfrac{1}{8}$

$x^2 + \dfrac{1}{4}x + \dfrac{1}{64} = \dfrac{1}{8} + \dfrac{1}{64}$

$\left(x + \dfrac{1}{8}\right)^2 = \dfrac{9}{64}$

$\sqrt{\left(x + \dfrac{1}{8}\right)^2} = \pm\sqrt{\dfrac{9}{64}}$

$x + \dfrac{1}{8} = \pm\dfrac{3}{8}$

$x = -\dfrac{1}{8} \pm \dfrac{3}{8}$

$x = -\dfrac{1}{2} \quad \text{and} \quad x = \dfrac{1}{4}$

10.3 Practice Problems

1. $x^2 - 7x + 6 = 0$

$$x = \frac{-(-7) \pm \sqrt{(-7)^2 - 4(1)(6)}}{2(1)}$$

$$x = \frac{7 \pm \sqrt{49 - 24}}{2}$$

$$x = \frac{7 \pm \sqrt{25}}{2}$$

$$x = \frac{7 \pm 5}{2}$$

$$x = \frac{7 + 5}{2} \qquad x = \frac{7 - 5}{2}$$

$$x = 6 \quad \text{and} \quad x = 1$$

2. $3x^2 - 8x + 3 = 0$

$$x = \frac{-(-8) \pm \sqrt{(-8)^2 - 4(3)(3)}}{2(3)}$$

$$x = \frac{8 \pm \sqrt{64 - 36}}{6}$$

$$x = \frac{8 \pm \sqrt{28}}{6}$$

$$x = \frac{8 \pm 2\sqrt{7}}{6}$$

$$x = \frac{2(4 \pm \sqrt{7})}{6} = \frac{4 \pm \sqrt{7}}{3}$$

3.
$$x^2 = 7 - \frac{3}{5}x$$

$$5x^2 = 35 - 3x$$

$$5x^2 + 3x - 35 = 0$$

$$x = \frac{-3 \pm \sqrt{(3)^2 - 4(5)(-35)}}{2(5)}$$

$$x = \frac{-3 \pm \sqrt{9 + 700}}{10}$$

$$x = \frac{-3 \pm \sqrt{709}}{10}$$

4. $2x^2 = 13x + 5$

$$2x^2 - 13x - 5 = 0$$

$$x = \frac{-(-13) \pm \sqrt{(-13)^2 - 4(2)(-5)}}{2(2)}$$

$$x = \frac{13 \pm \sqrt{169 + 40}}{4}$$

$$x = \frac{13 \pm \sqrt{209}}{4}$$

$$x = \frac{13 + \sqrt{209}}{4} \qquad x = \frac{13 - \sqrt{209}}{4}$$

$$x \approx 6.864 \quad \text{and} \quad x \approx -0.364$$

5. $5x^2 + 2x = -3$

$$5x^2 + 2x + 3 = 0$$

$$x = \frac{-2 \pm \sqrt{(2)^2 - 4(5)(3)}}{2(5)}$$

$$x = \frac{-2 \pm \sqrt{4 - 60}}{10}$$

$$x = \frac{-2 \pm \sqrt{-56}}{10}$$

There is no real number that is $\sqrt{-56}$.
There is no solution to this problem.

6. (a) $5x^2 - 3x - 2 = 0$

$$b^2 - 4ac = (-3)^2 - 4(5)(-2)$$
$$= 9 + 40 = 49$$

Because $49 > 0$ there will be two real roots.

(b) $2x^2 - 4x - 5 = 0$

$$b^2 - 4ac = (-4)^2 - 4(2)(-5)$$
$$= 16 + 40 = 56$$

Because $56 > 0$ there will be two real roots.

10.4 Practice Problems

1. (a) $y = 2x^2$

x	y
-2	8
-1	2
0	0
1	2
2	8

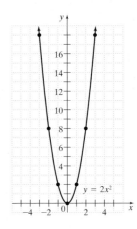

(b) $y = -2x^2$

x	y
-2	-8
-1	-2
0	0
1	-2
2	-8

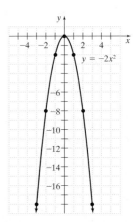

2. (a) $y = x^2 + 2x$

x	y	
-4	8	
-3	3	
-2	0	
-1	-1	vertex
0	0	
1	3	
2	8	

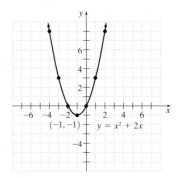

(b) $y = -x^2 - 2x$

x	y	
-4	-8	
-3	-3	
-2	0	
-1	1	vertex
0	0	
1	-3	
2	-8	

3. $y = x^2 - 4x - 5$

$$x = \frac{-b}{2a} = \frac{-(-4)}{2(1)} = 2$$

If $x = 2$, then
$$y = (2)^2 - 4(2) - 5$$
$$y = 4 - 8 - 5$$
$$y = -9$$
The point $(2, -9)$ is the vertex. x-intercepts:
$$x^2 - 4x - 5 = 0$$
$$(x + 1)(x - 5) = 0$$
$$x + 1 = 0 \qquad x - 5 = 0$$
$$x = -1 \qquad x = 5$$
The x-intercepts are the points $(-1, 0)$ and $(5, 0)$.

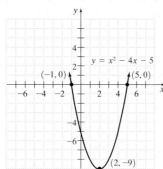

4. $h = -16t^2 + 32t + 10$

(a) Since the equation is a quadratic equation, the graph is a parabola. a is negative, therefore the parabola opens downward and the vertex is the highest point.
$$\text{Vertex} = t = \frac{-b}{2a} = \frac{-32}{2(-16)} = \frac{-32}{-32} = 1$$
If $t = 1$, then $h = -16(1)^2 + 32(1) + 10$
$$h = -16 + 32 + 10$$
$$h = 26$$
The vertex (t, h) is $(1, 26)$.
Now find the h-intercept. The point where $t = 0$.
$$h = -16t^2 + 32t + 10$$
$$h = 10$$
The h-intercept is at $(0, 10)$.

Table Values	
t	**h**
0	10
0.5	22
1	26
1.5	22
2	10
2.275	0

(b) The ball is 26 feet above the ground one second after it is thrown. This is the greatest height of the ball.

(c) The ball hits the ground when $h = 0$.
$$0 = -16t^2 + 32t + 10$$
$$t = \frac{-32 \pm \sqrt{(32)^2 - 4(-16)(10)}}{2(-16)}$$
$$t = \frac{-32 \pm \sqrt{1024 + 640}}{-32}$$
$$t = \frac{-32 \pm \sqrt{1664}}{-32}$$
$$t = \frac{4 \pm \sqrt{26}}{4}$$
$$t \approx \frac{4 - 5.099}{4} \quad \text{and} \quad t \approx \frac{4 + 5.099}{4}$$
$$t \approx -0.275 \qquad t \approx 2.275$$
In this situation it is not useful to find points for negative values of t. Therefore $t \approx 2.3$ seconds. The ball will hit the ground approximately 2.3 seconds after it is thrown.

10.5 Practice Problems

1. $a^2 + b^2 = c^2$ 　　　Let $a = x$
　　　　　　　　　　　　　$b = x - 6$
$$x^2 + (x - 6)^2 = 30^2$$
$$x^2 + x^2 - 12x + 36 = 900$$
$$2x^2 - 12x + 36 = 900$$
$$2x^2 - 12x - 864 = 0$$
$$x^2 - 6x - 432 = 0$$
$$(x + 18)(x - 24) = 0$$
$$x + 18 = 0 \qquad x - 24 = 0$$
$$x = -18 \qquad x = 24$$
One leg is 24 meters, the other leg is $x - 6$ or $24 - 6 = 18$ meters.

2. Let $c =$ cost for each student in the original group
　　　$s =$ number of students
Number of students $\times$ cost per student $=$ total cost
$$s \times c = 240$$
If 2 people cannot go, then the number of students drops by 2 but the cost of each increases by \$4.
$$(s - 2)(c + 4) = 240$$
If $s \times c = 240$, then
$$c = \frac{240}{s}$$
$$(s - 2)\left(\frac{240}{s} + 4\right) = 240 \quad \text{Multiply by the LCD which is } s.$$
$$(s - 2)(240 + 4s) = 240s$$
$$4s^2 + 232s - 480 = 240s$$
$$4s^2 - 8s - 480 = 0$$
$$s^2 - 2s - 120 = 0$$
$$(s - 12)(s + 10) = 0$$
$$s - 12 = 0 \qquad s + 10 = 0$$
$$s = 12 \qquad s = -10$$
The number of students originally going on the trip was 12.

Check. If 12 students originally planned the trip then
$$12 \times c = 240.$$
$$c = \$20 \text{ per student}$$
If 2 students dropped out, does this mean they must increase the cost of the trip by \$4?
$$10 \times c = 240$$
$$c = \$24.00 \quad \text{Yes.}$$

3. Let $x =$ width
　　　$y =$ length
$$160 = 2x + y$$
The area of a rectangle is (width)(length).
$$A = x(y) \quad \text{Substitute } y = 160 - 2x.$$
$$3150 = x(160 - 2x)$$
$$3150 = 160x - 2x^2$$
$$2x^2 - 160x + 3150 = 0$$
$$x^2 - 80x + 1575 = 0$$
$$(x - 45)(x - 35) = 0$$
$$x = 45 \quad \text{and} \quad x = 35$$
First solution:
If the width is 45 feet then the length is
$$160 = 2(45) + y$$
$$160 = 90 + y$$
$$y = 70 \text{ feet}$$

Check. 　　$45 + 45 + 70 \stackrel{?}{=} 160$
$$90 + 70 \stackrel{?}{=} 160$$
$$160 = 160 \quad \checkmark$$

Second solution:
If the width is 35 feet then the length is
$$160 = 2(35) + y$$
$$160 = 70 + y$$
$$y = 90 \text{ feet}$$

Check. 　　$35 + 35 + 90 \stackrel{?}{=} 160$
$$70 + 90 \stackrel{?}{=} 160$$
$$160 = 160 \quad \checkmark$$

Appendix B

Practice Problems

1. $3.4 \text{ m} \cdot 1 = 3.4 \text{ m} \cdot \dfrac{100 \text{ cm}}{1 \text{ m}} = \dfrac{(3.4)(100)}{1} \cdot \dfrac{\text{m}}{\text{m}} \cdot \text{cm} = 340 \text{ cm}$

2. $125 \text{ L} \cdot \dfrac{0.001 \text{ kL}}{1 \text{ L}} = \dfrac{(125)(0.001)}{1} \cdot \dfrac{\text{L}}{\text{L}} \cdot \text{kL} = 0.125 \text{ kL}$

3. $5.72 \text{ lb} \cdot \dfrac{1 \text{ kg}}{2.20 \text{ lb}} = 2.6 \text{ kg}$

4. $8 \text{ cm} \cdot \dfrac{1 \text{ in.}}{2.54 \text{ cm}} \approx 3.15 \text{ in.}$

5. $3 \text{ L} \cdot \dfrac{1.06 \text{ qt}}{1 \text{ L}} = 3.18 \text{ qt}$

6. $\dfrac{70 \text{ mi}}{\text{hr}} \cdot \dfrac{5280 \text{ ft}}{1 \text{ mi}} \cdot \dfrac{1 \text{ hr}}{60 \text{ min}} \cdot \dfrac{1 \text{ min}}{60 \text{ sec}} = \dfrac{(70)(5280) \text{ ft}}{(60)(60) \text{ sec}}$
   $\approx 102.7 \text{ ft/sec}$

Appendix C

Practice Problems

1. **(a)** The lowest level of sodium, 670 mg, is in the Wendy's Grilled Chicken Sandwich.
   (b) The highest level of cholesterol, 100 g, is in the McDonald's Big Mac.
2. **(a)** The two activities single women spend the most time doing are spending time with friends, and reading and watching TV.
   (b) Couples spend more time per week with their family (21 hr) than any other activity.
   (c) The greatest difference between single men and single women is the time spent reading and watching TV (6 hr).
3. **(a)** Walgreens has $7 \times 300 = 2100$ stores.
   (b) Rite Aid has $9 \times 300 = 2700$ stores. CVS has $5 \times 300 = 1500$ stores. Since $2700 - 1500 = 1200$, Rite Aid has 1200 more stores than CVS.
   (c) Thrift Drug has $2 \times 300 = 600$ stores, and Medicine Shoppe has $4 \times 300 = 1200$ stores. There are $600 + 1200 = 1800$ total Thrift Drug and Medicine Shoppe stores.
4. **(a)** The greatest number of fatal accidents, 14, occurred during the 1989–1990 period.
   (b) During the 1987–1988 period there were 7 accidents. During 1989–1990 there were 14 accidents. The increase in the number of fatal accidents from 1987–1988 to 1989–1990 was 7.
   (c) During the 1991–1992 period there were 8 accidents. During 1993–1994 there were 5 accidents. The decrease in the number of fatal accidents from 1991–1992 to 1993–1994 was 3.
5. **(a)** There were 12 days in 1993 in which the PSI exceeded 100.
   (b) The year with the fewest days in which the PSI exceeded 100 was 1992.
6. **(a)** 15% of Seattle's rainfall falls between April and June.
   (b) 40% of Seattle's rainfall occurs from October to December.
   (c) 35% of the total rainfall occurs from January to March. This is $35\% \times 37 = 0.35 \times 37 = 12.95$ inches.

Appendix D

Practice Problems

1. Each number is 7 more than the preceding number. The next number in the sequence is $66 + 7 = 73$.
2. The sequence could be written $2^2 - 1, 3^2 - 1, 4^2 - 1, 5^2 - 1, \ldots$. The next number would be $10^2 - 1 = 99$.
3. Between 1's there are one 2, two 3's, and three 4's. Next would come four 5's, a 1, and then five 6's. The next seven digits are 5555166.
4. In the first figure, the bottom rectangle is shaded and the figure is rotated 90° clockwise to get the second figure. Then, in the third figure, the middle rectangle is shaded. This is rotated 90° clockwise to get the fourth figure. Then, the top rectangle is shaded and the figure is rotated. In the last figure, the bottom rectangle is shaded again and the sequence is repeated.

The next two figures are

5.

Statement	Reason
1. $x = 2x + 24$	**1.** Multiplication principle: If $a = b$, then $ac = bc$. Here we multiplied by 6.
2. $-x = 24$	**2.** Addition principle: If $a = b$, then $a + c = b + c$. Here we added $-2x$ to each side of the equation.
3. $x = -24$	**3.** Multiplication principle: If $a = b$, then $ac = bc$. Here we multiplied each side by -1.

6.

Statement	Reason
1. $\angle 6 = \angle 5 = 26°$	**1.** If two lines intersect, the opposite angles are equal.
2. $\angle 4 = \angle 6 = 26°$	**2.** If a transversal crosses two parallel lines, the alternate interior angles are equal.

7. Let us organize the facts by listing the cars and each possible position: 1, 2, 3, 4, in which position 1 means far left and position 4 means far right.

Honda	Toyota	Mustang	Camaro
1	1	1	1
2	2	2	2
3	3	3	3
4	4	4	4

Step	Reason
1. We indicate that the Camaro is in position 4 and cross out the other 4's since no other car can be in position 4. We also cross out positions 1, 2, and 3 for the Camaro. 	Honda	Toyota	Mustang	Camaro	 1, 1, 1, 1̶ 2, 2, 2, 2̶ 3, 3, 3, 3̶ 4̶, 4̶, 4̶, [4]	**1.** The Camaro is parked on the right end.
2. The Mustang cannot be in position 1 since it's between two cars. It also cannot be in position 3 as that would place it next to the Camaro. Therefore, the Mustang must be in position 2. Cross off the 2's in all other columns.	**2.** The Mustang is between the Honda and the Toyota.
3. The Honda must be in position 1 since it cannot be in position 3, next to the Camaro. This leaves position 3 for the Toyota.	**3.** The Honda is not next to the Camaro.

The Honda, in position 1, is the car parked on the left end.

Answers to Selected Exercises

Chapter 0

0.1 Exercises

1. 12   **3.** When two or more numbers are multiplied, each number that is multiplied is called a factor. In 2 × 3, 2 and 3 are factors.

5.   **7.** $\frac{3}{4}$ **9.** $\frac{1}{3}$ **11.** 5   **13.** $2\frac{5}{6}$ **15.** $12\frac{1}{3}$ **17.** $5\frac{3}{7}$ **19.** $\frac{16}{5}$

21. $\frac{33}{5}$ **23.** 8   **25.** 12   **27.** 21   **29.** 12   **31.** $22\frac{16}{17}$ **33.** $\frac{33}{160}$ **35.** $\frac{1}{4}$ **37.** $\frac{1}{2}$

39. Aaron Dunsay $\frac{3}{5}$, Paul Banks $\frac{2}{5}$, Tom Re $\frac{4}{5}$ **41.** $\frac{13}{16}$

0.2 Exercises

1. Answers may vary. A sample answer is: 8 is exactly divisible by 4.
3. 105   **5.** 20   **7.** 54   **9.** 90

11. $\frac{5}{8}$ **13.** $\frac{2}{7}$ **15.** $\frac{29}{24}$ or $1\frac{5}{24}$ **17.** $\frac{31}{63}$ **19.** $\frac{11}{15}$ **21.** $\frac{17}{36}$ **23.** $\frac{3}{2}$ or $1\frac{1}{2}$ **25.** $\frac{1}{2}$ **27.** $7\frac{11}{15}$ **29.** $1\frac{35}{72}$ **31.** $4\frac{11}{12}$

33. $6\frac{13}{28}$ **35.** $5\frac{19}{24}$ **37.** $4\frac{3}{7}$ **39.** $8\frac{7}{8}$ **41.** $\frac{29}{18}$ or $1\frac{11}{18}$ **43.** $\frac{83}{14}$ or $5\frac{13}{14}$ **45.** $\frac{10}{21}$ **47.** $\frac{27}{10}$ or $2\frac{7}{10}$ **49.** $10\frac{7}{24}$ miles

51. $4\frac{1}{12}$ hours   **53.** A=12 inches, B=$15\frac{7}{8}$ inches   **55.** $1\frac{5}{8}$ inches   **57.** $\frac{9}{11}$ **58.** $\frac{133}{5}$

0.3 Exercises

1. First, change each number to an improper fraction. Look for a common factor in the numerator and denominator to divide by, and, if one is found, peform the division. Multiply the numerators. Multiply the denominators.

3. $\frac{20}{7}$ or $2\frac{6}{7}$ **5.** $\frac{17}{30}$ **7.** $\frac{6}{25}$ **9.** $\frac{12}{5}$ or $2\frac{2}{5}$ **11.** $\frac{18}{5}$ or $3\frac{3}{5}$ **13.** $\frac{3}{5}$ **15.** $\frac{1}{7}$ **17.** 14   **19.** $\frac{7}{6}$ or $1\frac{1}{6}$ **21.** $\frac{15}{14}$ or $1\frac{1}{14}$ **23.** $\frac{8}{35}$

25. $\frac{4}{3}$ or $1\frac{1}{3}$ **27.** $\frac{7}{12}$ **29.** $\frac{8}{15}$ **31.** 1   **33.** $\frac{5}{4}$ or $1\frac{1}{4}$ **35.** $\frac{219}{4}$ or $54\frac{3}{4}$ **37.** $\frac{17}{2}$ or $8\frac{1}{2}$ **39.** 28   **41.** $\frac{3}{16}$ **43.** $71\frac{1}{2}$ yards

45. $2\frac{4}{5}$ miles   **47.** $\frac{29}{31}$ **48.** $\frac{3}{7}$

0.4 Exercises

1. 10; 100; 10,000; and so on   **3.** three, left   **5.** 0.625   **7.** 0.2   **9.** $0.\overline{63}$ **11.** $\frac{4}{5}$ **13.** $\frac{5}{8}$ **15.** $\frac{13}{5}$ or $2\frac{3}{5}$ **17.** 2.09   **19.** 10.82

21. 261.208   **23.** 131.79   **25.** 30.282   **27.** 0.10575   **29.** 219.7   **31.** 2.64   **33.** 261.5   **35.** 0.257   **37.** 3450   **39.** 0.0076
41. 73,600   **43.** 0.73892   **45.** 1.425   **47.** 11.7257   **49.** 2.12   **51.** 768.3   **53.** 24.13 cm   **55.** $212.75

57. No, there were 0.3 milligrams of copper in excess.   **59.** $\frac{2}{3}$ **60.** $\frac{1}{6}$ **61.** $\frac{93}{100}$ **62.** $\frac{11}{10}$ or $1\frac{1}{10}$

How Am I Doing? Sections 0.1–0.4

1. $\frac{3}{11}$ (obj. 0.1.2)   **2.** $\frac{2}{5}$ (obj. 0.1.2)   **3.** $3\frac{3}{4}$ (obj. 0.1.3)   **4.** $\frac{33}{7}$ (obj. 0.1.3)   **5.** 6 (obj. 0.1.4)   **6.** 35 (obj. 0.1.4)

7. 120 = LCD (obj. 0.2.2)   **8.** $\frac{5}{7}$ (obj. 0.2.1)   **9.** $\frac{19}{42}$ (obj. 0.2.3)   **10.** $8\frac{5}{12}$ (obj. 0.2.4)   **11.** $\frac{7}{18}$ (obj. 0.2.3)

12. $\frac{4}{21}$ (obj. 0.2.3)   **13.** $1\frac{33}{40}$ (obj. 0.2.4)   **14.** $\frac{10}{9}$ or $1\frac{1}{9}$ (obj. 0.3.1)   **15.** $\frac{21}{2}$ or $10\frac{1}{2}$ (obj. 0.3.1)   **16.** $\frac{7}{2}$ or $3\frac{1}{2}$ (obj. 0.3.2)

17. $\frac{28}{39}$ (obj. 0.3.2)   **18.** 0.875 (obj. 0.4.2)   **19.** $0.\overline{5}$ (obj. 0.4.2)   **20.** 0.015 (obj. 0.4.2)   **21.** 19.651 (obj. 0.4.4)

22. 2.0664 (obj. 0.4.5)   **23.** 0.45 (obj. 0.4.6)   **24.** 2.246 (obj. 0.4.4)

0.5 Exercises

1. Answers may vary. Sample answers are below. 19% means 19 out of 100 parts. Percent means per 100. 19% is really a fraction with a denominator of 100. In this case it would be $\frac{19}{100}$.
3. 28%   **5.** 56.8%   **7.** 7.6%   **9.** 239%   **11.** 0.03   **13.** 0.004   **15.** 2.5   **17.** 0.074   **19.** 5.2   **21.** 13   **23.** 72.8
25. 150%   **27.** 5%   **29.** 250%   **31.** 85%   **33.** \$4.92 tip \$37.72 total   **35.** approximately 21%   **37.** 540
39. \$1 an hour raise; \$13.50 per hour   **41.** \$29,640 sales commission; \$35,040 annual salary   **43.** \$2.97   **45.** \$538.13
46. \$6220.50   **47.** 22 miles per gallon   **48.** 4.0 inches

0.6 Exercises

1. 210,000   **3.** 18,000,000   **5.** 240   **7.** 20,000   **9.** 0.1   **11.** \$4000.00   **13.** \$400   **15.** \$6400   **17.** 25 miles per gallon
19. \$1666.67   **21.** \$60,000,000,000   **23.** \$4000   **25.** 10%   **27.** \$240   **29.** 2.1   **30.** 56.25%   **31.** \$6.73   **32.** 270 tires

0.7 Exercises

1. \$269.17   **3. (a)** $82\frac{1}{3}$ feet   **(b)** 90 feet; \$17.90   **5.** jog, $2\frac{2}{3}$ miles; walk $3\frac{1}{9}$ miles rest, $4\frac{4}{9}$ minutes; walk, $1\frac{7}{9}$ miles

7. Betty; Melinda increases each activity by $\frac{2}{3}$ by day 3 but Betty increases each activity by $\frac{7}{9}$ by day 3.   **9.** $4\frac{1}{2}$ miles

11. \$98,969.00   **13. (a)** \$36,000   **(b)** \$11,160   **15.** 18%   **17.** 69%

Putting Your Skills to Work

1. 12,700 sq mi   **2.** 9400 sq mi   **3.** About 4 times   **4.** About 319 times

Chapter 0 Review Problems

1. $\frac{3}{4}$ **2.** $\frac{3}{10}$ **3.** $\frac{18}{41}$ **4.** $\frac{3}{5}$ **5.** $\frac{23}{5}$ **6.** $6\frac{4}{5}$ **7.** $6\frac{1}{2}$ **8.** 15   **9.** 5   **10.** 40   **11.** 22   **12.** $\frac{17}{20}$ **13.** $\frac{29}{24}$ or $1\frac{5}{24}$
14. $\frac{4}{15}$ **15.** $\frac{13}{30}$ **16.** $\frac{173}{30}$ or $5\frac{23}{30}$ **17.** $\frac{79}{20}$ or $3\frac{19}{20}$ **18.** $2\frac{29}{36}$ **19.** $\frac{23}{12}$ or $1\frac{11}{12}$ **20.** $\frac{30}{11}$ or $2\frac{8}{11}$ **21.** $\frac{21}{2}$ or $10\frac{1}{2}$ **22.** $\frac{19}{8}$ or $2\frac{3}{8}$
23. $\frac{20}{7}$ or $2\frac{6}{7}$ **24.** $\frac{1}{16}$ **25.** $\frac{24}{5}$ or $4\frac{4}{5}$ **26.** $\frac{3}{20}$ **27.** 6   **28.** 7.201   **29.** 7.737   **30.** 29.561   **31.** 4.436   **32.** 0.03745
33. 362,341   **34.** 0.07956   **35.** 10.368   **36.** 0.00186   **37.** 0.07132   **38.** 1.3075   **39.** 90   **40.** 1.82   **41.** 0.5   **42.** 0.375
43. $\frac{9}{25}$ **44.** 0.014   **45.** 0.361   **46.** 0.0002   **47.** 1.253   **48.** 510   **49.** 3.96   **50.** 64%   **51.** 12.5%   **52.** 60%
53. 16,324,803   **54.** 75%   **55.** 400,000,000,000   **56.** 2500   **57.** 300,000   **58.** 19   **59.** \$12,000   **60.** 30   **61.** \$240
62. \$300   **63.** \$349.07   **64.** 8%   **65.** 1840 miles; 1472 miles   **66.** 1500 miles; 1050 miles   **67.** $7\frac{1}{2}$ cubic inches   **68.** 200 boxes
69. \$462.80   **70.** 46 packages   **71.** \$1585.50   **72.** \$371.20

How Am I Doing? Chapter 0 Test

1. $\frac{8}{9}$ (obj. 0.1.2)   **2.** $\frac{4}{3}$ (obj. 0.1.2)   **3.** $\frac{45}{7}$ (obj. 0.1.3)   **4.** $11\frac{2}{3}$ (obj. 0.1.3)   **5.** $\frac{15}{8}$ or $1\frac{7}{8}$ (obj. 0.2.3)   **6.** $\frac{39}{8}$ or $4\frac{7}{8}$ (obj. 0.2.4)
7. $\frac{5}{6}$ (obj. 0.2.4)   **8.** $\frac{4}{3}$ or $1\frac{1}{3}$ (obj. 0.3.1)   **9.** $\frac{5}{24}$ (obj. 0.3.1)   **10.** $\frac{7}{2}$ or $3\frac{1}{2}$ (obj. 0.3.2)   **11.** $\frac{65}{8}$ or $8\frac{1}{8}$ (obj. 0.3.1)
12. $\frac{43}{22}$ or $1\frac{21}{22}$ (obj. 0.3.2)   **13.** 14.64 (obj. 0.4.4)   **14.** 3.9897 (obj. 0.4.4)   **15.** 1.312 (obj. 0.4.5)   **16.** 73.85 (obj. 0.4.5)
17. 230 (obj. 0.4.6)   **18.** 263,259 (obj. 0.4.7)   **19.** 7.3% (obj. 0.5.2)   **20.** 1.965 (obj. 0.5.3)   **21.** 6.3 (obj. 0.5.4)
22. 0.336 (obj. 0.5.4)   **23.** 6% (obj. 0.5.5)   **24.** 30% (obj. 0.5.5)   **25.** 18 (obj. 0.3.2)   **26.** 100 (obj. 0.6.1)   **27.** 700 (obj. 0.6.1)
28. 65% (obj. 0.7.1)   **29.** 60 tiles (obj. 0.7.1)

Chapter 1

1.1 Exercises

1. Whole number, rational number, real number   **3.** Irrational number, real number   **5.** Rational number, real number
7. Rational number, real number   **9.** Irrational number, real number   **11.** $-20,000$ **13.** $-37\frac{1}{2}$ **15.** $+7$ **17.** -8 **19.** 2.73
21. 1.3   **23.** $\frac{5}{6}$ **25.** -11 **27.** -31 **29.** $\frac{1}{4}$ **31.** $-\frac{7}{13}$ **33.** $\frac{1}{35}$ **35.** -3.8 **37.** 0.4   **39.** -14.16 **41.** -6 **43.** -5

45. $-\dfrac{4}{15}$ **47.** -8 **49.** -3 **51.** 59 **53.** $\dfrac{7}{18}$ **55.** $\dfrac{2}{5}$ **57.** 0.76 **59.** 12 **61.** 0 **63.** 15.94 **65.** \$167 profit

67. $-\$3800$ **69.** 3 yard gain **71.** 3500 **73.** \$32,000,000 **75.** 18 **77.** 6 yards **79.** $\dfrac{2}{3}$ **80.** $\dfrac{8}{27}$ **81.** $\dfrac{1}{12}$ **82.** $\dfrac{25}{34}$

83. 1.52 **84.** 0.65 **85.** 1.141 **86.** 0.26

1.2 Exercises

1. First change subtracting -3 to adding a positive three. Then use the rules for addition of two real numbers with different signs.

Thus, $-8 - (-3) = -8 + 3 = -5$. **3.** -17 **5.** -5 **7.** -11 **9.** 8 **11.** 5 **13.** 0 **15.** -3 **17.** $-\dfrac{2}{5}$ **19.** $\dfrac{27}{20}$ or $1\dfrac{7}{20}$

21. $-\dfrac{19}{12}$ or $-1\dfrac{7}{12}$ **23.** -0.9 **25.** 4.47 **27.** $-\dfrac{17}{5}$ or $-3\dfrac{2}{5}$ **29.** $\dfrac{40}{7}$ or $5\dfrac{5}{7}$ **31.** -53 **33.** -73 **35.** 7.1 **37.** $8\dfrac{3}{4}$ **39.** $-6\dfrac{1}{6}$

41. $-\dfrac{21}{20}$ or $-1\dfrac{1}{20}$ **43.** -8.5 **45.** $-5\dfrac{4}{5}$ **47.** 8.162 **49.** -5.047 **51.** 7 **53.** -48 **55.** -2 **57.** 11 **59.** -62

61. 1.6 **63.** 38 **65.** \$149 **67.** 6051 meters **69.** -21 **70.** -51 **71.** -19 **72.** $-8°C$ **73.** $6\dfrac{2}{3}$ miles were snow covered

1.3 Exercises

1. To multiply two real numbers, multiply the absolute values. The sign of the result is positive if both numbers have the same sign, but negative if the two numbers have opposite signs. **3.** -20 **5.** 0 **7.** 24 **9.** 0.264 **11.** -1.75 **13.** $-\dfrac{3}{2}$ or $-1\dfrac{1}{2}$ **15.** $\dfrac{9}{11}$ **17.** $-\dfrac{5}{26}$

19. 4 **21.** 6 **23.** 20 **25.** -12 **27.** -130 **29.** -0.6 **31.** -0.9 **33.** $-\dfrac{3}{10}$ **35.** $\dfrac{20}{3}$ or $6\dfrac{2}{3}$ **37.** $\dfrac{7}{10}$ **39.** 14

41. $-\dfrac{5}{4}$ or $-1\dfrac{1}{4}$ **43.** $\dfrac{15}{16}$ **45.** -24 **47.** 24 **49.** -16 **51.** -0.00018 **53.** $-\dfrac{8}{35}$ **55.** $\dfrac{2}{27}$ **57.** 9 **59.** -2 **61.** 17

63. -72 **65.** -1 **67.** He gave \$4.40 to each boy and to himself. **69.** \$328.50 **71.** 20 yards **73.** 70 yards

75. The Panthers gained 20 yards **77.** The Panthers would have gained 105 fewer yards. **78.** -6.69 **79.** $-\dfrac{11}{6}$ or $-1\dfrac{5}{6}$

80. -15 **81.** -88 **82.** 266 square yards

1.4 Exercises

1. The base is 4 and the exponent is 4. Thus you multiply $(4)(4)(4)(4) = 256$.
3. The answer is negative. When you raise a negative number to an odd power the result is always negative.
5. If you have parentheses surrounding the -2, then the base is -2 and the exponent is 4. The result is 16. If you do not have parentheses, then the base is 2. You evaluate to obtain 16 and then take the negative of 16, which is -16. Thus $(-2)^4 = 16$ but $-2^4 = -16$. **7.** 5^7 **9.** w^2

11. p^4 **13.** $(3q)^3$ or 3^3q^3 **15.** 27 **17.** 81 **19.** 216 **21.** -27 **23.** 16 **25.** -25 **27.** $\dfrac{1}{16}$ **29.** $\dfrac{8}{125}$ **31.** 1.21

33. 0.0016 **35.** 256 **37.** -256 **39.** 161 **41.** 116 **43.** -91 **45.** 23 **47.** -576 **49.** -512 **51.** 16,777,216

53. -7 **55.** -19 **56.** $-\dfrac{5}{3}$ or $-1\dfrac{2}{3}$ **57.** -8 **58.** 2.52 **59.** \$1672

1.5 Exercises

1. $3(4) + 6(5)$ **3.** (a) 90 (b) 42 **5.** 12 **7.** 13 **9.** -29 **11.** 24 **13.** 21 **15.** 13 **17.** -6 **19.** 42 **21.** $\dfrac{9}{4}$ or $2\dfrac{1}{4}$

23. 0.848 **25.** $\dfrac{3}{10}$ **27.** 5 **29.** $-\dfrac{23}{2}$ or $-11\dfrac{1}{2}$ **31.** 7.56 **33.** $\dfrac{1}{4}$ **35.** $3(-2) + 9(-1) + 5(0) + 1(1)$ **37.** 1 above par

39. 0.125 **40.** $-\dfrac{19}{12}$ or $-1\dfrac{7}{12}$ **41.** -1 **42.** $\dfrac{72}{125}$ **43.** 45 ounces **44.** 0.684 **45.** 0.54

How Am I Doing? Sections 1.1–1.5

1. -9 (obj. 1.1.4) **2.** $-\dfrac{41}{24}$ or $-1\dfrac{17}{24}$ (obj. 1.1.3) **3.** 1.24 (obj. 1.1.3) **4.** -7.1 (obj. 1.1.4) **5.** 11 (obj. 1.2.1)

6. $-\dfrac{17}{15}$ or $-1\dfrac{2}{15}$ (obj. 1.2.1) **7.** 12.3 (obj. 1.2.1) **8.** 10 (obj. 1.2.1) **9.** -96 (obj. 1.3.1) **10.** $\dfrac{10}{11}$ (obj. 1.3.1)

11. -0.9 (obj. 1.3.3) **12.** $-\dfrac{10}{17}$ (obj. 1.3.3) **13.** 0.343 (obj. 1.4.2) **14.** 256 (obj. 1.4.2) **15.** -256 (obj. 1.4.2)

16. $\dfrac{8}{27}$ (obj. 1.4.2) **17.** 141 (obj. 1.4.2) **18.** 7 (obj. 1.5.1) **19.** 10 (obj. 1.5.1) **20.** 11 (obj. 1.5.1)

21. -3.3 (obj. 1.5.1) **22.** $-\dfrac{9}{20}$ (obj. 1.5.1)

1.6 Exercises

1. variable   **3.** Here we are multiplying 4 by x by x. Since we know from the definition of exponents that x multiplied by x is x^2, this gives us an answer of $4x^2$.

5. Yes, $a(b - c)$ can be written as $a[b + (-c)]$

$$3(10 - 2) = (3 \times 10) - (3 \times 2)$$
$$3 \times 8 = 30 - 6$$
$$24 = 24$$

7. $3x - 6y$ **9.** $-8a + 6b$ **11.** $9x + 3y$ **13.** $-10a - 15b$ **15.** $-x + 3y$ **17.** $-81x + 45y - 72$ **19.** $-10x + 2y - 12$

21. $10x^2 - 20x + 15$ **23.** $\dfrac{x^2}{5} + 2xy - \dfrac{4x}{5}$ **25.** $5x^2 + 10xy + 5xz$ **27.** $-4x + 6$ **29.** $18x^2 + 3xy - 3x$

31. $-3x^2y - 2xy^2 + xy$ **33.** $6x^2y + 9xy^2 - 6xy$ **35.** $9.43x^2 - 5.29x + 0.92$ **37.** $0.36x^3 + 0.09x^2 - 0.15x$
39. $0.3x^2 + 0.4xy - 2.5x$ **41.** $800(5x + 14y) = 4000x + 11{,}200y$ square feet   **43.** $8xy - 20y$ dollars
45. $4x(3000 - 2y) = 12{,}000 - 8xy$ square feet   **46.** -16 **47.** 64 **48.** 14 **49.** 4 **50.** 10 **51.** 56% **52.** 14 days

1.7 Exercises

1. A term is a number, a variable, or a product of numbers and variables.
3. The two terms $5x$ and $-8x$ are like terms because they both have the variable x with the exponent of one.
5. The only like terms are $7xy$ and $-14xy$ because the other two have different exponents even though they have the same variables.
7. $-25b^2$ **9.** $18x^4 + 7x^2$ **11.** $-5x - 5y$ **13.** $7.1x - 3.5y$ **15.** $-2x - 8.7y$ **17.** $5p + q - 18$ **19.** $5bc - 6ac$

21. $x^2 - 10x + 3$ **23.** $-10y^2 - 16y + 12$ **25.** $-\dfrac{1}{15}x - \dfrac{2}{21}y$ **27.** $\dfrac{11}{20}a^2 - \dfrac{5}{6}b$ **29.** $-8ab - 3a - 6b$ **31.** $28a - 20b$

33. $-27ab - 11b^2$ **35.** $-8x^2 - 39y$ **37.** $32x + 23$ **39.** $7a + 9b$ **41.** $7a + 7b + 3$ centimeters   **43.** $36x - 20$ inches

45. $-\dfrac{2}{15}$ **46.** $-\dfrac{5}{6}$ **47.** $\dfrac{23}{50}$ **48.** $-\dfrac{15}{98}$ **49.** 0.2 liter

1.8 Exercises

1. -5 **3.** -11 **5.** $\dfrac{25}{2}$ or $12\dfrac{1}{2}$ **7.** -26 **9.** -1.3 **11.** $\dfrac{25}{4}$ or $6\dfrac{1}{4}$ **13.** 10 **15.** 3 **17.** -24 **19.** -20 **21.** 9 **23.** 39

25. -2 **27.** 15 **29.** 42 **31.** -9 **33.** 29 **35.** 49 **37.** 32 **39.** $-\dfrac{1}{2}$ **41.** 352 square feet   **43.** 1.24 square centimeters
45. 32 square inches   **47.** $56{,}000$ square feet   **49.** 50.24 square centimeters   **51.** $-78.5°C$ **53.** $\$2340.00$ **55.** $-58°F$ to $140°F$
57. 1.4 miles; air is thin, oxygen is lacking   **59.** 16 **60.** $-x^2 + 2x - 4y$ **61.** 6.2 minutes/song
62. $53.5 = 54$ papers each approximately

1.9 Exercises

1. $-(3x + 2y)$ **3.** distributive   **5.** $3x + 6y$ **7.** $5a + 3b$ **9.** $x - 7y$ **11.** $8x^3 - 4x^2 + 12x$ **13.** $-2x + 26y$
15. $4x - 6y - 3$ **17.** $15a - 60ab$ **19.** $12a^3 - 19a^2 - 22a$ **21.** $3a^2 + 16b + 12b^2$ **23.** $-7a + 8b$ **25.** $9b^2 + 36b - 12$
27. $12a^2 - 8b$ **29.** 219 unsuccessful; 8541 successful   **31.** $97.52°F$ **32.** $453{,}416$ square feet   **33.** $300{,}000$ sq feet; $\$16{,}500.00$
34. 11.375 sq ft   **35.** Great Danes weigh 264.4 to 330.75 kg on average.   **36.** Miniature Pinschers weigh on average 19.845 to 30.87 kg.
$\$1387.75$

Putting Your Skills to Work

1. about $21{,}190{,}964$ **2.** about $9{,}518{,}000$ **3.** $846{,}622{,}338$ people   **4.** $76{,}598{,}250$ people   **5.** $92{,}190{,}000$ **6.** $105{,}690{,}000$

Chapter 1 Review Problems

1. -8 **2.** -4.2 **3.** -9 **4.** 1.9 **5.** $-\dfrac{1}{3}$ **6.** $-\dfrac{7}{22}$ **7.** $\dfrac{1}{6}$ **8.** $\dfrac{22}{15}$ or $1\dfrac{7}{15}$ **9.** 8 **10.** 13 **11.** -33 **12.** 9.2

13. $-\dfrac{13}{8}$ or $-1\dfrac{5}{8}$ **14.** $\dfrac{1}{2}$ **15.** -22.7 **16.** -88 **17.** -3 **18.** 18 **19.** 32 **20.** $-\dfrac{2}{3}$ **21.** $-\dfrac{25}{7}$ or $-3\dfrac{4}{7}$ **22.** -72

23. 30 **24.** -30 **25.** -4 **26.** 16 **27.** -29 **28.** 1 **29.** $-\dfrac{1}{2}$ **30.** $-\dfrac{4}{7}$ **31.** -30 **32.** -5 **33.** -9.1 **34.** 0.9

35. 10.1 **36.** -1.2 **37.** 1.9 **38.** -1.3 **39.** 24 yards   **40.** $-22°F$ **41.** 7363 feet   **42.** $2\dfrac{1}{4}$ point loss   **43.** -243

44. -128 **45.** 625 **46.** $\dfrac{8}{27}$ **47.** -81 **48.** 0.36 **49.** $\dfrac{25}{36}$ **50.** $\dfrac{27}{64}$ **51.** -44 **52.** 30 **53.** 1 **54.** $15x - 35y$

55. $6x^2 - 14xy + 8x$ **56.** $-7x^2 + 3x - 11$ **57.** $-6xy^2 - 3xy + 3y^2$ **58.** $-5a^2b + 3bc$ **59.** $-3x - 4y$ **60.** $-5x^2 - 35x - 9$

61. $10x^2 - 8x - \dfrac{1}{2}$ **62.** -55 **63.** 1 **64.** -4 **65.** -15 **66.** 10 **67.** -16 **68.** $\dfrac{32}{5}$ **69.** $\$810$ **70.** $86°F$

71. $\$2119.50$ **72.** $\$8580.00$ **73.** $100{,}000$ sq ft   **74.** 10.45 sq feet $\$689.70$ **75.** $-2x + 42$ **76.** $-17x - 18$ **77.** $-2 + 10x$
$\$200{,}000$

78. $-12x^2 + 63x$ **79.** $5xy^3 - 6x^3y - 13x^2y^2 - 6x^2y$ **80.** $x - 10y + 35 - 15xy$ **81.** $10x - 22y - 36$

82. $-10a + 25ab - 15b^2 - 10ab^2$ **83.** $-3x - 9xy + 18y^2$ **84.** $10x + 8xy - 32y$ **85.** -2.3 **86.** 8 **87.** $-\frac{22}{15}$ or $-1\frac{7}{15}$

88. $-\frac{1}{8}$ **89.** -1 **90.** -0.5 **91.** $\frac{3}{2}$ or $1\frac{1}{2}$ **92.** 6 **93.** 240 **94.** -25.42 **95.** \$600   **96.** 0.0081

97. -0.0625 **98.** 10 **99.** $-4.9x + 4.1y$ **100.** $-\frac{1}{9}$ **101.** $-\frac{2}{3}$ **102.** The dog doesn't have a fever. 101.48° is below normal.

103. $3y^2 + 12y - 7x - 28$ **104.** $-12x + 6y + 12xy$

How Am I Doing? Chapter 1 Test

1. -0.3 (obj. 1.1.4)   **2.** 2 (obj. 1.2.1)   **3.** $-\frac{14}{3}$ or $-4\frac{2}{3}$ (obj. 1.3.1)   **4.** -70 (obj. 1.3.1)   **5.** 4 (obj. 1.3.3)   **6.** -3 (obj. 1.3.3)

7. -64 (obj. 1.4.2)   **8.** 2.56 (obj. 1.4.2)   **9.** $\frac{16}{81}$ (obj. 1.4.2)   **10.** 6.8 (obj. 1.5.1)   **11.** -25 (obj. 1.5.1)

12. $-5x^2 - 10xy + 35x$ (obj. 1.6.1)   **13.** $6a^2b^2 + 4ab^3 - 14a^2b^3$ (obj. 1.6.1)   **14.** $2a^2b + \frac{15}{2}ab$ (obj. 1.7.2)

15. $-1.8x^2y - 4.7xy^2$ (obj. 1.7.2)   **16.** $5a + 30$ (obj. 1.7.2)   **17.** $14x - 16y$ (obj. 1.7.2)   **18.** 122 (obj. 1.8.1)   **19.** 37 (obj. 1.8.1)

20. $\frac{13}{6}$ or $2\frac{1}{6}$ (obj. 1.8.1)   **21.** 96.6 kilometers/hr (obj. 1.8.2)   **22.** 22,800 sq ft (obj. 1.8.2)   **23.** \$23.12 (obj. 1.8.2)   **24.** 3 cans (obj. 1.8.2)

25. $3x - 6xy - 21y^2$ (obj. 1.9.1)   **26.** $-3a - 9ab + 3b^2 - 3ab^2$ (obj. 1.9.1)

Chapter 2

2.1 Exercises

1. equals, equal   **3.** solution   **5.** answers vary   **7.** $x = 4$ **9.** $x = 12$ **11.** $x = 17$ **13.** $x = -5$ **15.** $x = -13$ **17.** $x = 62$
19. $x = 15$ **21.** $x = 21$ **23.** $x = 0$ **25.** $x = -7$ **27.** $x = 21$ **29.** no, $x = 9$ **31.** no, $x = -13$ **33.** yes   **35.** yes

37. $x = -1.8$ **39.** $x = 2.5$ **41.** $x = 1$ **43.** $x = -\frac{1}{4}$ **45.** $x = -7$ **47.** $x = \frac{17}{6}$ or $2\frac{5}{6}$ **49.** $x = \frac{13}{12}$ or $1\frac{1}{12}$ **51.** $x = 7.2$

53. $x = 1.4906$ **55.** $-2x - 4y$ **56.** $-2y^2 - 4y + 4$ **57.** yes, this was $1.9\% < 3\%$ **58.** \$14.92   **59.** 117 feet

2.2 Exercises

1. 6 **3.** 7 **5.** $x = 36$ **7.** $x = -27$ **9.** $x = 80$ **11.** $x = -15$ **13.** $x = 4$ **15.** $x = 8$ **17.** $x = -\frac{8}{3}$ or $-2\frac{2}{3}$

19. $x = 50$ **21.** $x = 15$ **23.** $x = -7$ **25.** $x = 0.2$ or $\frac{1}{5}$ **27.** $x = 4$ **29.** no, $x = -7$ **31.** yes   **33.** $y = -0.03$

35. $t = \frac{8}{3}$ or $2\frac{2}{3}$ **37.** $y = -0.7$ **39.** $x = 3$ **41.** $x = -4$ **43.** $x = \frac{7}{9}$ **45.** $x = 1$ **47.** $x = 3$ **49.** $x = 27$

51. $x = -5.26$ **53.** To solve an equation, we are performing steps to get an equivalent equation that has the same solution. Now $a = b$ and $a(0) = b(0)$ are not equivalent equations because they do not have the same solution. So we must have the requirement that when we multiply both sides of the equation by c, it is absolutely essential that c is nonzero.

55. $48 - 6 = 42$ **56.** $-27 - 10 = -37$ **57.** $5 + 16 = 21$ **58.** $93\frac{1}{3}\%$ **59.** \$632   **60.** 104 calves   **61.** 27 earthquakes

2.3 Exercises

1. $x = 2$ **3.** $x = 6$ **5.** $x = -4$ **7.** $x = 13$ **9.** $x = 3.1$ **11.** $x = 40$ **13.** $x = -27$ **15.** $x = 8$ **17.** $x = 3$ **19.** $x = 7$
21. $x = -9$ **23.** yes   **25.** no, $x = -11$ **27.** $x = -1$ **29.** $x = 7$ **31.** $y = 1$ **33.** $x = 16$ **35.** $y = 3$ **37.** $x = 4$

39. $x = -\frac{1}{3}$ **41.** $x = 2.5$ or $2\frac{1}{2}$ **43.** $x = 6.5$ or $6\frac{1}{2}$ **45.** $x = 8$ **47.** $x = -3.2$ or $-3\frac{1}{5}$ **49.** $y = 2$ **51.** $x = -4$ **53.** $z = 5$

55. $a = -6.5$ **57.** $x = -\frac{2}{3}$ **59.** $x = -0.25$ or $-\frac{1}{4}$ **61.** $x = 8$ **63.** $x = -4.5$ **65.** $x = \frac{7}{6}$ or $1\frac{1}{6}$ **67.** $x = -4.23$

69. $14x^2 - 14xy$ **70.** $-10x - 60$ **71.** $\$844\frac{1}{4}$ or \$844.25   **72.** (a) \$91.00 (b) \$94.50

2.4 Exercises

1. $x = -1$ **3.** $x = 1$ **5.** $x = 1$ **7.** $x = 24$ **9.** $y = 20$ **11.** $x = 3$ **13.** $x = \frac{7}{3}$ or $2\frac{1}{3}$ **15.** $x = -3.5$ **17.** yes   **19.** no

21. $x = 1$ **23.** $x = 8$ **25.** $x = 2$ **27.** $x = -3$ **29.** $y = 4$ **31.** $x = -22$ **33.** $x = 2$ **35.** $x = -12$ **37.** $x = -\frac{5}{3}$ or $-1\frac{2}{3}$

39. $x = 0.4$ **41.** no solution   **43.** infinite number of solutions   **45.** $x = 0$ **47.** no solution   **49.** $\frac{27}{14}$ or $1\frac{13}{14}$ **50.** $\frac{19}{20}$

51. $-\frac{52}{3}$ or $-17\frac{1}{3}$ **52.** $\frac{22}{5}$ or $4\frac{2}{5}$ **53.** 264 pairs   **54.** 3173 seats   **55.** \$226.08   **56.** \$108

How Am I Doing? Sections 2.1–2.4

1. $x = -9$ (obj. 2.1.1)   **2.** $x = -10.9$ (obj. 2.1.1)   **3.** $x = 9$ (obj. 2.2.2)   **4.** $x = -8$ (obj. 2.2.2)   **5.** $x = \dfrac{2}{3}$ (obj. 2.3.1)

6. $x = \dfrac{7}{26}$ (obj. 2.3.2)   **7.** $x = \dfrac{3}{5}$ (obj. 2.3.3)   **8.** $x = 3.75$ (obj. 2.3.3)   **9.** $x = -1.6$ or $-1\dfrac{3}{5}$ (obj. 2.3.3)   **10.** $x = \dfrac{8}{9}$ (obj. 2.4.1)

11. $x = -3$ (obj. 2.4.1)   **12.** $x = \dfrac{9}{11}$ (obj. 2.4.1)   **13.** $x = 5$ (obj. 2.4.1)   **14.** $x = -\dfrac{1}{3}$ (obj. 2.4.1)

2.5 Exercises

1. Multiply each term by 5. Then add -160 to each side. Then divide each side by 9. We would obtain $\dfrac{5F - 160}{9} = C$.

3. (a) 10 meters **(b)** 16 meters   **5. (a)** $y = -\dfrac{4}{3}x + 6$ **(b)** $y = 10$ **7.** $b = \dfrac{2A}{h}$ **9.** $P = \dfrac{I}{rt}$ **11.** $m = \dfrac{y - b}{x}$ **13.** $y = \dfrac{2}{3}x - 2$

or $y = \dfrac{18 - 4x}{3}$

15. $x = -\dfrac{3}{2}y + 6$ **17.** $y = \dfrac{c - ax}{b}$ **19.** $r^2 = \dfrac{A}{\pi}$ **21.** $g = \dfrac{2s}{t^2}$ **23.** $t = \dfrac{A - P}{pr}$ **25.** $h = \dfrac{S - 2\pi r^2}{2\pi r}$ **27.** $h = \dfrac{V}{\pi r^2}$

29. $L = \dfrac{V}{WH}$ **31.** $r^2 = \dfrac{3V}{\pi h}$ **33.** $w = \dfrac{P - 2L}{2}$ **35.** $a^2 = c^2 - b^2$ **37.** $C = \dfrac{5F - 160}{9}$ **39.** $R = \dfrac{E^2}{P}$ **41.** $S = \dfrac{360A}{\pi r^2}$

43. 0.8 mile   **45.** 30 feet   **47. (a)** $x = \dfrac{V - 7050}{1100}$ **(b)** 2002   **49.** A doubles   **51.** A increases 4 times   **53.** 31.2   **54.** 0.096

55. 1250 feet of fencing   **56.** 2.5%   **57.** 39,000 square feet   **58.** $10\dfrac{7}{12}$ hours

2.6 Exercises

1. Yes, both statements imply 5 is to the right of -6 on the number line.   **3.** $>$ **5.** $<$ **7.** $>$ **9.** $<$ **11.** $<$
13. $<$ **15.** 0.0247   **17.**   **19.**   **21.**

23.   **25.**   **27.** $x \geq -\dfrac{2}{3}$ **29.** $x < -20$

31. $x > 2.8$ **33.** $W > 175$ **35.** $h \geq 37$ **37.** $h \geq 48$ **39.**
41. 6.08   **42.** 15%   **43.** 2%   **44.** 37.5%   **45.** \$15   **46.** 495 feet

2.7 Exercises

1. $x \leq -3$ **3.** $x \leq 5$

5. $x > -9$ **7.** $x \geq 8$

9. $x < -12$ **11.** $x < \dfrac{7}{2}$ **13.** $x \geq -2$

15. $x < \dfrac{3}{2}$ **17.** $x > -6$ **19.** $x > \dfrac{1}{3}$

21. $3 > 1$ Adding any number to both sides of an inequality doesn't reverse the direction.   **23.** $x > 3$ **25.** $x \geq 4$ **27.** $x < -1$

29. $x \leq 14$ **31.** $x < -3$ **33.** $x > -\dfrac{4}{11}$ **35.** 76 or greater   **37.** 8 days or more   **39.** 260 feet   **40.** width = 6.5 inches
height = 3.6 inches

41. 63.6 sq inches   **42.** 15,600 sq feet

Putting Your Skills to Work

1. 8.4 gallons per hour   **2.** 252 gallons   **3.** 32.3 gallons per hour   **4.** 55 miles per hour   **5.** $x = \dfrac{y + 0.1}{0.5}$ or $2y + 0.2 = x$ **6.** 35 mph

Chapter 2 Review Problems

1. -7 **2.** -3 **3.** 2   **4.** -3 **5.** 2.75 or $2\dfrac{3}{4}$ **6.** -13 **7.** 40.4   **8.** -7 **9.** -2 **10.** -27 **11.** 20   **12.** -9 **13.** -3

14. -1 **15.** 4   **16.** 3   **17.** 3   **18.** -7 **19.** 3   **20.** $-\dfrac{7}{2}$ or $-3\dfrac{1}{2}$ or -3.5 **21.** -3 **22.** 2.1   **23.** 5   **24.** 1   **25.** 20

26. $\dfrac{2}{3}$ **27.** 5   **28.** $\dfrac{35}{11}$ or $3\dfrac{2}{11}$ **29.** 4   **30.** -17 **31.** $\dfrac{2}{5}$ or 0.4   **32.** 32   **33.** $\dfrac{26}{7}$ or $3\dfrac{5}{7}$ **34.** -1 **35.** $\dfrac{1}{5}$ **36.** $\dfrac{3}{4}$ **37.** -17

38. -2.5 **39.** 23   **40.** -32 **41.** $-3\dfrac{2}{5}$ or $-\dfrac{17}{5}$ or -3.4 **42.** 9   **43.** $-\dfrac{16}{13}$ or $-1\dfrac{3}{13}$ **44.** 9.75 or $9\dfrac{3}{4}$ **45.** $y = 3x - 10$

46. $y = \dfrac{-5x - 7}{2}$ **47.** $r = \dfrac{A - P}{Pt}$ **48.** $h = \dfrac{A - 4\pi r^2}{2\pi r}$ **49.** $P = \dfrac{3H - a - 3}{2}$ **50.** $y = \dfrac{c - ax}{b}$ **51. (a)** $T = \dfrac{1000c}{WR}$

(b) $T = 6000$ **52. (a)** $y = \dfrac{5}{3}x - 4$ **(b)** $y = 11$ **53. (a)** $h = \dfrac{V}{lw}$ **(b)** $h = 6$ **54.** $x \le 2$

55. $x \ge 1$ **56.** $x < -4$ **57.** $x > -3$

58. $x \ge 6$ **59.** $x < 5$ **60.** $x < 10$

61. $x > -3$ **62.** $x > \dfrac{17}{2}$ **63.** $x > \dfrac{7}{5}$

64. $h \le 32$ hours   **65.** $n \le 17$ **66.** $-\dfrac{7}{3}$ or $-2\dfrac{1}{3}$ **67.** $-\dfrac{2}{7}$ **68.** 0   **69.** 4   **70.** 4   **71.** -5 **72.** $d = \dfrac{6c - H}{5}$

73. $b = \dfrac{4H - 3c}{2}$ **74.** $x < 2$ **75.** $x \le -8$

76. $x \ge \dfrac{19}{7}$ **77.** $x \ge -15$

How Am I Doing? Chapter 2 Test

1. $x = 2$ (obj. 2.3.1) **2.** $x = \dfrac{1}{3}$ (obj. 2.3.2) **3.** $y = -\dfrac{7}{2}$ or $-3\dfrac{1}{2}$ or -3.5 (obj. 2.3.3) **4.** $y = 8.4$ or $8\dfrac{2}{5}$ or $\dfrac{42}{5}$ (obj. 2.4.1) **5.** $x = 1$ (obj. 2.3.3)

6. $x = -1.2$ (obj. 2.3.3)   **7.** $y = 7$ (obj. 2.4.1)   **8.** $y = \dfrac{7}{3}$ or $2\dfrac{1}{3}$ (obj. 2.3.3)   **9.** $x = 13$ (obj. 2.3.3)   **10.** $x = 20$ (obj. 2.3.3)

11. $x = 10$ (obj. 2.3.3)   **12.** $x = -4$ (obj. 2.3.3)   **13.** $x = 12$ (obj. 2.3.3)   **14.** $x = -\dfrac{1}{5}$ or -0.2 (obj. 2.4.1)   **15.** $x = 3$ (obj. 2.4.1)

16. $x = 2$ (obj. 2.4.1)   **17.** $x = -2$ (obj. 2.4.1)   **18.** $w = \dfrac{A - 2P}{3}$ (obj. 2.5.1)   **19.** $w = \dfrac{6 - 3x}{4}$ (obj. 2.5.1)

20. $a = \dfrac{2A - hb}{h}$ (obj. 2.5.1)   **21.** $y = \dfrac{10ax - 5}{8ax}$ (obj. 2.5.1)   **22.** $B = \dfrac{3V}{h}$ (obj. 2.5.1)   **23.** $B = 30$ sq. in. (obj. 2.5.1)

24. $x \le -3$ (obj. 2.7.1)   **25.** $x > -\dfrac{5}{4}$ (obj. 2.7.1)

26. $x < 2$ (obj. 2.7.1)   **27.** $x \ge \dfrac{1}{2}$ (obj. 2.7.1)

Cumulative Test for Chapters 0–2

1. $\dfrac{4}{21}$ **2.** $\dfrac{79}{20}$ or $3\dfrac{19}{20}$ **3.** $\dfrac{32}{15}$ or $2\dfrac{2}{15}$ **4.** 0.6888   **5.** 0.12   **6.** 35.69   **7.** 30   **8.** $12ab - 28ab^2$ **9.** $25x^2$

10. $6x - 40 + 48y - 24xy$ **11.** $x = \dfrac{40}{11}$ or $3\dfrac{7}{11}$ **12.** $x = 4$ **13.** $y = 1$ **14.** $a = \dfrac{6 + 8b}{3}$ **15.** $b = \dfrac{3H - 8a}{2}$ **16.** $t = \dfrac{I}{Pr}$

17. $a = \dfrac{2A}{h} - b$ **18.** $x < 5$ **19.** $x \le 3$

20. $x > -15$ **21.** $x \le -1$ **22.** $x \le 1$ **23.** 92

Chapter 3

3.1 Exercises

1. $x + 5$ **3.** $x - 6$ **5.** $\dfrac{1}{8}x$ or $\dfrac{x}{8}$ **7.** $2x$ **9.** $3 + \dfrac{1}{2}x$ **11.** $2x + 9$ **13.** $\dfrac{1}{3}(x + 7)$ **15.** $\dfrac{1}{3}x - 2x$ **17.** $3x - 7$

19. $x =$ value of a share of AT&T stock   **21.** $2w + 7 =$ length   **23.** $x =$ number of boxes sold by Keiko
$x + 74.50 =$ value of a share of IBM stock $w =$ width $x - 43 =$ number of boxes sold by Sarah
$x + 53 =$ number of boxes sold by Imelda

25. 1st angle $= s - 16$ **27.** $v =$ value of exports of Canada   **29.** 1st angle $= 3x$ **31.** $A =$ area of Minnesota
2nd angle $= s$ $2v =$ value of exports of Japan   2nd angle $= x$ $\dfrac{1}{2}A =$ area of Kentucky
3rd angle $= 2s$ 3rd angle $= x - 14$
$\dfrac{2}{5}A =$ area of Maine

33. $x =$ points for an arrow in the blue ring.   **35.** $x =$ men aged 16–24   **37.** $x = 7$ **38.** $x = -\dfrac{5}{2}$ or $-2\dfrac{1}{2}$
$3x - 6 =$ points awarded for an arrow in the gold ring. $x + 51 =$ men aged 25–34
$x - 60 =$ men aged 35–44
$x - 132 =$ men aged 45 and above

39. $x = 12$ **40.** $w = 7$

3.2 Exercises

1. 1261   **3.** 2368   **5.** 182   **7.** 43   **9.** 9   **11.** -5 **13.** 12   **15.** 15 used bikes   **17.** 34,415 wildfires   **19.** 6 CDs
21. 4 items   **23.** 320.2 miles per hour   **25.** 550 lb   **27.** 5 miles   **29.** She traveled 52 mph on the mountain road. It was 12 mph faster
on the highway route.   **31.** 105 degrees F   **33.** 9 miles per gallon   **35.** (a) $F - 40 = \dfrac{x}{4}$ (b) 200 chirps (c) 77°F
36. $10x^3 - 30x^2 - 15x$ **37.** $-2a^2b + 6ab - 10a^2$ **38.** $-5x - 6y$ **39.** $-4x^2y - 7xy^2 - 8xy$ **40.** 3000 apples   **41.** 60 people

3.3 Exercises

1. The long piece is 32 meters long.   **3.** Dave earned \$34,300 per year.   **5.** David worked 30 hours.   **7.** 11 students born in Oklahoma.
The short piece is 15 meters long.   Elsie earned \$36,900 per year.   Sarah worked 45 hours.   19 students born in Texas.
Kate worked 25 hours.   2 students born in Arizona.

9. Length is 13 meters.   **11.** Length is 41 centimeters.   **13.** width = 32.5 cm   **15.** The cheetah can run 70 mph.
Width is 8 meters.   Width is 7 centimeters.   length = 62.5 cm   The jackal can run 35 mph.
The elk can run 45 mph.

17. Length A = 20 inches   **19.** (a) 18 hours (b) 21 hours (c) 19 mph   **21.** original square was 11 m $\times$ 11 m
Length B = 15 inches
Length C = 10 inches
Length D = 13 inches

23. $-8x^3 + 12x^2 - 32x$ **24.** $5a^2b + 30ab - 10a^2$ **25.** $-19x + 2y - 2$ **26.** $9x^2y - 6xy^2 + 7xy$ **27.** 82%   **28.** 20%

How Am I Doing? Sections 3.1–3.3

1. $3x - 40$ (obj. 3.1.2)   **2.** $\dfrac{1}{2}x + 12$ (obj. 3.1.2)   **3.** $\dfrac{5x}{12}$ or $\dfrac{5}{12}x$ (obj. 3.1.2)   **4.** $\dfrac{1}{4}(x + 10)$ or $\dfrac{x + 10}{4}$ (obj. 3.1.2)
5. -26 (obj. 3.2.1)   **6.** 104 (obj. 3.2.1)   **7.** 40 months (obj. 3.2.2)   **8.** 5 hours (obj. 3.2.2)
9. The fourth person earns \$60,000 per year.   **10.** The long piece is 62 feet.   **11.** Sunday = 50 pages
The fifth person earns \$62,500 per year. (obj. 3.2.2)   The short piece is 22 feet. (obj. 3.3.1)   Monday = 75 pages
Tuesday = 135 pages (obj. 3.3.1)

12. 1st field = 360 hours
2nd field = 120 hours
3rd field = 30 hours (obj. 3.3.1)

3.4 Exercises

1. 11 bags   **3.** 7 hours   **5.** 12 weeks   **7.** \$360   **9.** \$22,000   **11.** \$6000   **13.** \$3000 @ 7%   **15.** Conservative fund = \$250,000
\$2000 @ 5%   growth fund = \$150,000
17. \$12,000   **19.** 13 quarters   **21.** 18 nickels   **23.** 30 regular stamps   **25.** \$10 bills = 8   **27.** \$925,000 worth of furniture
9 nickels   6 dimes   22 breast cancer stamps   \$20 bills = 16
9 quarters   \$100 bills = 11
29. more than 225 miles   **31.** 12   **32.** -17 **33.** -28 **34.** -25 **35.** \$95.20   **36.** \$19,040

3.5 Exercises

1. distance around   **3.** surface   **5.** 180°   **7.** 156 in.^{2} **9.** 98 in.^{2} **11.** 10 feet   **13.** 9 inches   **15.** 153.86 sq ft
17. 9520.48 miles   **19.** 24 inches   **21.** 134,400 sq meters   **23.** 5 centimeters   **25.** 200.96 cm   **27.** 74°   **29.** equal angles = 80°
3rd angle = 20°
31. 71° each   **33.** 1st angle = 46°   **35.** 52°   **37.** 6 feet   **39.** 2512 in.^{3} **41.** Yes   **43.** $V = 452.16$ cm^3 **45.** 163.28 ft^2
2nd angle = 23° $S = 376.8$ cm^2
3rd angle = 111°

47. (a) 22.77 sq. yd (b) \$56.93   **49.** (a) 73.96 inches (b) 181.93 cubic inches   **51.** 9.3 feet wide and 34.2 feet long
53. no; the perimeter may be 19 feet or it may be 15.5 feet   **55.** 6.28 ft   **56.** $3x^2 + 8x - 24$ **57.** $-11x - 17$
58. $-4x - 9$ **59.** $22x - 84$

3.6 Exercises

1. $x > 67,000$ **3.** $x \le 120$ **5.** $h \le 1500$ **7.** $x \ge 93$ **9.** It must be less than or equal to 140 cm.   **11.** 22 or more deliveries
13. Depth can be no more than 7.5 feet.   **15.** $x \le 384.6$ miles   **17.** $F < 230°$ **19.** The budget will exceed \$307.1 billion after 2005.

21. It will be 8 hours or less   **23.** 455 or more discs would need to be sold.   **25.** $x < -4$ **26.** $x \le 3$ **27.** $x \ge 4\dfrac{2}{3}$ **28.** $x > -4$
29. (a) 2920 gallons (b) \$24.82

Putting Your Skills to Work

1. $a = \dfrac{7463}{4} = 1865.75$ **2.** The herd is approximately 32,097 for 1995.
The herd is approximately 11,571 for 1984.
$b = 15,303$
$y = 1866x + 15,303$

3. The equation is fairly accurate.   **4.** $a = 1651$ approximately
The equation predicts 47,025 and the measured value is 3650 lower. $y = 1651x + 15,303$

Chapter 3 Review Problems

1. $x + 19$ **2.** $\frac{2}{3}x$ **3.** $\frac{1}{2}x$ or $\frac{x}{2}$ **4.** $x - 18$ **5.** $3(x + 4)$ **6.** $2x - 3$ **7.** $r =$ the number of retired people; $4r =$ the number of working people; $0.5r =$ the number of unemployed people   **8.** $3w + 5 =$ the length; $w =$ the width
9. $b =$ the number of degrees in angle B; $2b =$ the number of degrees in angle A; $b - 17 =$ the number of degrees in angle C
10. $a =$ the number of students in algebra; $a + 29 =$ the number of students in biology; $0.5a =$ the number of students in geology
11. $x = 5$ **12.** $x = 7$ **13.** \$40   **14.** 16 years old   **15.** 6.6 hours; 6 hours   **16.** 90   **17.** 23 feet
18. 1st angle $= 32°$; 2nd angle $= 96°$; 3rd angle $= 52°$ **19.** 31.25 yd; 18.75 yd   **20.** Jon $=$ \$30,000; Lauren $=$ \$18,000
21. 310 kilowatt-hours   **22.** 280 miles   **23.** \$3000   **24.** \$200   **25.** \$7000 at 12%; \$2000 at 8%
26. \$2000 at 4.5%; \$3000 at 6%   **27.** 18 nickels; 6 dimes; 9 quarters   **28.** 7 nickels; 8 dimes; 10 quarters
29. S.A. $= 3215.36$ sq. in.   **30.** $P = 24$ feet   **31.** $71°$ **32.** 42 sq. mi.   **33.** 1920 ft^3 **34.** 113.04 cm^3 **35.** 254.34 cm^2
36. 12 in.   **37.** \$3440   **38.** \$23,864   **39.** 200 miles or less   **40.** more than \$300,000   **41.** CP distance $= 630$ miles
UP distance $= 960$ miles
42. less than \$16.15   **43.** $24\frac{4}{5}$ min   **44.** 4 ounces   **45.** more than 3 years   **46.** more than 60 months
47. The height is 4 feet. The width is 10 feet.   **48.** It will take 10 years.   **49.** They made 27 field goals and 9 free throws.
50. The width is 16 feet and the length is 22 feet.   **51.** Carl $= 12$ hours   **52.** \$13,600 @ 9.75%
Ryan $= 16$ hours   \$15,600 @ 8.5%
53. first angle measures $60°$; second angle measures $40°$; third angle measures $80°$;   **54.** \$1224   **55.** 94   **56.** No, she traveled 9000 miles for business. Her correct deduction was \$2970.   **57.** It can travel 640 miles in one hour. It would take approximately 169 minutes.
58. 12 seconds   **59.** 5 Explorers and 10 Caravans   **60.** smallest angle $= 32°$ **61.** 183.26 kg
largest angle $= 96°$
third angle $= 52°$

How Am I Doing? Chapter 3 Test

1. 35 (obj. 3.2.1)   **2.** 36 (obj. 3.2.1)   **3.** -4 (obj. 3.2.1)   **4.** first side $= 20$ m; second side $= 30$ m; third side $= 16$ m (obj. 3.2.2)
5. width $= 20$ m; length $= 47$ m (obj. 3.2.2)   **6.** 1st pollutant $= 8$ ppm; 2nd pollutant $= 4$ ppm; 3rd pollutant $= 3$ ppm (obj. 3.4.2)
7. 15 months (obj. 3.4.1)   **8.** \$21,000 (obj. 3.4.2)   **9.** \$1400 at 14%; \$2600 at 11% (obj. 3.4.3)
10. 16 nickels; 7 dimes; 8 quarters (obj. 3.4.4)   **11.** 213.52 in. (obj. 3.5.1)   **12.** 192 square inches (obj. 3.5.1)   **13.** 4187 in.^{3} (obj. 3.5.2)
14. 96 cm^2 (obj. 3.5.1)   **15.** \$450 (obj. 3.5.3)   **16.** at least an 82 (obj. 3.6.2)   **17.** more than \$110,000 (obj. 3.6.2)
18. 600 hours (obj. 3.6.2)

Cumulative Test for Chapters 0–3

1. 3.69   **2.** $\frac{31}{24}$ or $1\frac{7}{24}$ **3.** $-23y + 18$ **4.** 21   **5.** $\frac{2H - 3a}{5} = b$ **6.** $x \geq -3$;   **7.** -7
8. Psychology students $= 50$; World Hist. students $= 84$ **9.** width $= 7$ cm; length $= 32$ cm   **10.** \$135,000
11. \$3000 at 15%; \$4000 at 7%   **12.** 7 nickels; 12 dimes; 4 quarters   **13.** $A = 162.5$ m^2; \$731.25
14. $V = 113.04$ cubic inches; 169.56 pounds   **15.** 25 days

Chapter 4

4.1 Exercises

1. When multiplying exponential expressions with the same base, keep the base the same and add the exponents.
3. $\frac{2^2}{2^3} = \frac{2 \cdot 2}{2 \cdot 2 \cdot 2} = \frac{1}{2}$ **5.** $6; x; y$: 11 and 1   **7.** $2^2 a^3 b$ **9.** $-3a^2 b^2 c^3$ **11.** 7^{10} **13.** 5^{26} **15.** x^{12} **17.** w^{32} **19.** $-20x^6$
$= \frac{1}{2^{3-2}}$

21. $50x^3$ **23.** $24x^5 y^4$ **25.** $\frac{2}{15}x^3 y^5$ **27.** $-2.75x^3 yz$ **29.** 0   **31.** $80x^3 y^7$ **33.** $-24x^4 y^7$ **35.** 0   **37.** $-24a^4 b^3 x^2 y^5$

39. $-30x^3 y^4 z^6$ **41.** y^7 **43.** $\frac{1}{y^3}$ **45.** $\frac{1}{11^{12}}$ **47.** 3^4 **49.** $\frac{a^8}{4}$ **51.** $\frac{x^7}{y^6}$ **53.** $2x^4$ **55.** $-\frac{x^3}{2y^2}$ **57.** $\frac{f^2}{30g^5}$ **59.** $-3x^3$ **61.** $\frac{y^2}{16x^3}$

63. $\frac{3a}{4}$ **65.** $\frac{17a^2 b}{9c^3}$ **67.** $-27x^5 yz^3$ **69.** x^{12} **71.** $x^7 y^{14}$ **73.** $r^6 s^{12}$ **75.** $27a^9 b^6 c^3$ **77.** $9a^8$ **79.** $\frac{x^7}{128m^{28}}$ **81.** $\frac{25x^2}{49y^4}$

83. $81a^8 b^{12}$ **85.** $-8x^9 z^3$ **87.** $\frac{9}{x}$ **89.** $27a^4 b^7$ **91.** $\frac{64}{y^{10}}$ **93.** $\frac{16x^4}{y^{12}}$ **95.** $\frac{a^4 b^8}{c^{12} d^{16}}$ **97.** $-3x^3 y^4 z^7$ **99.** -11 **100.** -46

101. $-\frac{7}{4}$ **102.** 5   **103.** approximately 17.6%   **104.** approximately 47.6% of rain forest has been lost

105. 34,500 sq km of rain forest lost

4.2 Exercises

1. $\dfrac{1}{x^4}$ **3.** $\dfrac{1}{81}$ **5.** y^8 **7.** $\dfrac{z^6}{x^4 y^5}$ **9.** $\dfrac{y^6}{x^5}$ **11.** $\dfrac{x^9}{8}$ **13.** $\dfrac{3}{x^2}$ **15.** $\dfrac{1}{16x^4 y^2}$ **17.** $\dfrac{3xz^3}{y^2}$ **19.** $3x$ **21.** $\dfrac{wy^3}{x^5 z^2}$ **23.** $\dfrac{1}{8}$ **25.** $\dfrac{z^8}{9y^4}$

27. $\dfrac{1}{x^6 y}$ **29.** 1.2378×10^5 **31.** 6.3×10^{-2} **33.** 8.8961×10^{11} **35.** 1.963×10^{-8} **37.** $302{,}000$ **39.** 0.00047 **41.** $983{,}000$

43. 2.37×10^{-5} mph   **45.** 0.000007 meters   **47.** 4.368×10^{19} **49.** 1.0×10^1 **51.** 8.1×10^{-11} **53.** 4.5×10^5

55. 2.33×10^4 dollars   **57.** 1.90×10^{11} hours   **59.** 6.6×10^{-5} miles   **61.** $\$1.6 \times 10^8$ **63.** 159.3% **65.** -0.8 **66.** -1 **67.** $-\dfrac{1}{28}$

68. Candidate 1 shook 2524 hands.   **69.**   Mario $= \$36{,}094$
Candidate 2 shook 1016 hands.   Alfonso $= \$27{,}352$
Gina $= \$48{,}554$

4.3 Exercises

1. A polynomial in x is the sum of a finite number of terms of the form ax^n, where a is any real number and n is a whole number. An example is $3x^2 - 5x - 9$.   **3.** The degree of a polynomial in x is the largest exponent of x in any of the terms of the polynomial.

5. degree 4; monomial   **7.** degree 5; trinomial   **9.** degree 5; binomial   **11.** $5x - 28$ **13.** $-2x^2 + 2x - 1$ **15.** $\dfrac{5}{6}x^2 + \dfrac{1}{2}x - 9$

17. $3.4x^3 + 2.2x^2 - 13.2x - 5.4$ **19.** $5x - 24$ **21.** $\dfrac{1}{15}x^2 - \dfrac{1}{14}x + 11$ **23.** $-2x^3 + 5x - 10$ **25.** $-4.7x^4 - 0.7x^2 - 1.6x + 0.4$

27. $-8x + 7$ **29.** $-3x^2 y + 6xy^2 - 4$ **31.** $x^4 - 3x^3 - 4x^2 - 24$ **33.** 5.4 **35.** 2020 **37.** $727{,}000$ **39.** $434{,}400$ **41.** $3x^2 + 12x$

43. $x = \dfrac{3y - 2}{8}$ **44.** $B = \dfrac{3A}{2CD}$ **45.** $\dfrac{2A}{h} - c = b$ **46.** $d = \dfrac{5xy}{B}$ **47.** approximately 696.8 billion dollars   **48.** 300% increase

How Am I Doing? Sections 4.1–4.3

1. $-15x^3 y^6$ (obj. 4.1.1)   **2.** $-\dfrac{7y^3}{5x^7}$ (obj. 4.1.2)   **3.** $\dfrac{4x^5}{y^9}$ (obj. 4.1.2)   **4.** $81x^{20} y^4$ (obj. 4.1.3)   **5.** $\dfrac{x^6}{16y^8}$ (obj. 4.2.1)   **6.** $\dfrac{y^{15}}{3x^3}$ (obj. 4.2.1)

7. 5.874×10^4 (obj. 4.2.2)   **8.** 9.362×10^{-5} (obj. 4.2.2)   **9.** $2.3x^2 - 0.4x - 5.4$ (obj. 4.3.2)   **10.** $7x^2 + 17x - 26$ (obj. 4.3.3)

11. $\dfrac{1}{6}x^3 + \dfrac{3}{8}x^2 + 3x$ (obj. 4.3.3)   **12.** $\dfrac{5}{16}x^2 - \dfrac{3}{10}x - \dfrac{3}{8}$ (obj. 4.3.2)

4.4 Exercises

1. $-12x^4 + 2x^2$ **3.** $21x^3 - 9x^2$ **5.** $-4x^6 + 10x^4 - 2x^3$ **7.** $x + \dfrac{3}{2}x^2 + \dfrac{5}{2}x^3$ **9.** $-15x^4 y^2 + 6x^3 y^2 - 18x^2 y^2$

11. $9x^4 y + 3x^3 y - 24x^2 y$ **13.** $3x^4 - 9x^3 + 15x^2 - 6x$ **15.** $-2x^3 y^3 + 12x^2 y^3 - 16xy$ **17.** $-28x^5 y + 12x^4 y + 8x^3 y - 4x^2 y$
19. $-6c^2 d^5 + 8c^2 d^3 - 12c^2 d$ **21.** $12x^7 - 6x^5 + 18x^4 + 54x^3$ **23.** $-35x^7 + 56x^6 - 21x^5$ **25.** $x^2 + 13x + 30$ **27.** $x^2 + 8x + 12$
29. $x^2 - 6x - 16$ **31.** $x^2 - 9x + 20$ **33.** $-20x^2 - 7x + 6$ **35.** $7x^2 + 14xy - 4x - 8y$ **37.** $15x^2 - 5xy + 6x - 2y$
39. $20y^2 - 7y - 3$ **41.** The signs are incorrect.   **43.** $20x$ **45.** $20x^2 - 23xy + 6y^2$ **47.** $49x^2 - 28x + 4$ **49.** $16a^2 + 16ab + 4b^2$
The result is $-3x + 6$.

51. $0.8x^2 + 11.94x - 0.9$ **53.** $\dfrac{1}{4}x^2 + \dfrac{1}{24}x - \dfrac{1}{12}$ **55.** $2ax - 10ab - 3bx + 15b^2$ **57.** $35x^2 + 78xy + 27y^2$

59. $20y^2 - 22yz + 6z^2$ **61.** $28x^2 - 30x - 18$ **62.** $x = -10$ **63.** $w = -\dfrac{25}{7}$ or $-3\dfrac{4}{7}$ **64.** 7

65. 24 tiger's eye marbles   **66.** 9 twenties   **67.** 18.4 hrs   **68.** 20.8 hrs   **69.** 35.2 hrs   **70.** 36.4 hrs
48 cat's eye marbles   8 tens
23 fives

4.5 Exercises

1. binomial   **3.** The middle term is missing. The answer should be $16x^2 - 56x + 49$.   **5.** $y^2 - 49$ **7.** $x^2 - 81$ **9.** $49x^2 - 16$
11. $4x^2 - 49$ **13.** $25x^2 - 9y^2$ **15.** $0.36x^2 - 9$ **17.** $9y^2 + 6y + 1$ **19.** $25x^2 - 40x + 16$ **21.** $49x^2 + 42x + 9$

23. $9x^2 - 42x + 49$ **25.** $\dfrac{4}{9}x^2 + \dfrac{1}{3}x + \dfrac{1}{16}$ **27.** $81x^2 y^2 + 72xyz + 16z^2$ **29.** $49x^2 - 9y^2$ **31.** $49c^2 - 84cd + 36d^2$

33. $81a^2 - 100b^2$ **35.** $25x^2 + 90xy + 81y^2$ **37.** $x^3 + 3x^2 - 13x + 6$ **39.** $4x^4 - 7x^3 + 2x^2 - 3x - 1$
41. $a^4 + a^3 - 13a^2 + 17a - 6$ **43.** $3x^3 - 2x^2 - 25x + 24$ **45.** $3x^3 - 13x^2 - 6x + 40$ **47.** $2a^3 + 3a^2 - 50a - 75$
49. $24x^3 + 14x^2 - 11x - 6$ **51.** $\$11{,}000$ @ 7%   **52.** width $= 7$ m   **53.** $-28.12°$F   **54.** 2.0625×10^5 meters
$\$7{,}000$ @ 11%   length $= 10$ m

4.6 Exercises

1. $5x^3 - 3x + 4$ **3.** $2y^2 - 3y - 1$ **5.** $9x^4 - 4x^2 - 7$ **7.** $8x^4 - 9x + 6$ **9.** $3x + 5$ **11.** $x - 3 - \dfrac{32}{x - 5}$

13. $3x^2 - 4x + 8 - \dfrac{10}{x + 1}$ **15.** $2x^2 - 3x - 2 - \dfrac{5}{2x + 5}$ **17.** $2x^2 + 3x - 1$ **19.** $6y^2 + 3y - 8 + \dfrac{7}{2y - 3}$

21. $y^2 - 4y - 1 - \dfrac{9}{y+3}$ **23.** $y^3 + 2y^2 - 5y - 10 - \dfrac{25}{y-2}$ **25.** $2y^2 + \dfrac{1}{2}y + \dfrac{7}{8} - \dfrac{\frac{49}{8}}{4y-1}$ **27.** 77,000 gallons

28. 1965: $0.75   **29.** 200 cats   **30.** pages 519 and 520   **31. (a)** 4.2
1970: $0.86   **(b)** 7.4
1971: $0.82   **(c)** 76.2%
                                                                (d) 13.0

Putting Your Skills to Work

1. $13,526   **2.** $568,082   **3.** 35.4%   **4.** $136,777   **5.** $45,272   **6.** $43,912

Chapter 4 Review Problems

1. $-18a^7$ **2.** 5^{23} **3.** $6x^4y^6$ **4.** $-14x^4y^9$ **5.** $\dfrac{1}{7^{12}}$ **6.** $\dfrac{1}{x^5}$ **7.** y^{14} **8.** $\dfrac{1}{9^{11}}$ **9.** $-\dfrac{3}{5x^5y^4}$ **10.** $-\dfrac{2a}{3b^6}$ **11.** x^{24} **12.** b^{30}

13. $9a^6b^4$ **14.** $81x^{12}y^4$ **15.** $\dfrac{25a^2b^4}{c^6}$ **16.** $\dfrac{y^9}{64w^{15}z^6}$ **17.** $\dfrac{b^5}{a^3}$ **18.** $\dfrac{m^8}{p^5}$ **19.** $\dfrac{2y^3}{x^6}$ **20.** $\dfrac{x^8}{9y^6}$ **21.** $\dfrac{y^8}{25x^4}$ **22.** $\dfrac{3y^2}{x^3}$ **23.** $\dfrac{4w^2}{x^5y^6z^8}$

24. $\dfrac{b^5c^3d^4}{27a^2}$ **25.** 1.563402×10^{11} **26.** 1.79632×10^5 **27.** 7.8×10^{-3} **28.** 6.173×10^{-5} **29.** 120,000   **30.** 6,034,000

31. 3,000,000   **32.** 0.25   **33.** 0.0000432   **34.** 0.000000006   **35.** 2.0×10^{13} **36.** 9.36×10^{19} **37.** 9.6×10^{-10} **38.** 7.8×10^{-11}
39. 3.504×10^8 kilometers   **40.** 7.94×10^{14} cycles   **41.** 6×10^9 **42.** $11.4x^2 - 7.3x + 5.3$ **43.** $6.7x^2 - 11x + 3$

44. $-x^3 + 2x^2 - x + 8$ **45.** $7x^3 - 3x^2 - 6x + 4$ **46.** $\dfrac{1}{10}x^2y - \dfrac{13}{21}x + \dfrac{5}{12}$ **47.** $\dfrac{1}{4}x^2 - \dfrac{1}{4}x + \dfrac{1}{10}$ **48.** $-x^2 - 2x + 10$

49. $-6x^2 - 6x - 3$ **50.** $15x^2 + 2x - 1$ **51.** $28x^2 - 29x + 6$ **52.** $20x^2 + 48x + 27$ **53.** $10x^3 - 30x^2 + 15x$
54. $-18x^4 + 6x^3 - 3x^2 + 12x$ **55.** $-6x^3y^3 + 9x^3y^2 + 12x^2y^2$ **56.** $5a^2 - 8ab - 21b^2$ **57.** $8x^4 - 10x^2y - 12x^2 + 15y$
58. $-15x^6y^2 - 9x^4y + 6x^2y$ **59.** $9x^2 - 12x + 4$ **60.** $25x^2 - 9$ **61.** $49x^2 - 36y^2$ **62.** $25a^2 - 20ab + 4b^2$
63. $64x^2 + 144xy + 81y^2$ **64.** $4x^3 + 27x^2 + 5x - 3$ **65.** $2x^3 - 7x^2 - 42x + 72$ **66.** $2y^2 + 3y + 4$ **67.** $6x^3 + 7x^2 - 18x$

68. $4x^2y - 6x + 8y$ **69.** $53x^3 - 12x^2 + 19x + 13$ **70.** $3x - 2$ **71.** $3x - 7$ **72.** $3x^2 + 2x + 4 + \dfrac{9}{2x-1}$

73. $2x^2 - 5x + 13 - \dfrac{27}{x+2}$ **74.** $4x + 1$ **75.** $4x - 5 + \dfrac{11}{2x+1}$ **76.** $x^2 + 3x + 8$ **77.** $2x^2 + 4x + 5 + \dfrac{11}{x-2}$ **78.** 4^{29}

79. 6^{15} **80.** $125x^8y^{14}$ **81.** $-\dfrac{2x^5}{y^3}$ **82.** $-64x^{15}y^{18}$ **83.** 7.86×10^{-5} **84.** $-4x^3 - 12x + 17$ **85.** $-12x^5 - 30x^4 + 12x^2$

86. $14x^3 + 15x^2 - 23x + 6$ **87.** $2x^2 + 5x + 3 + \dfrac{4}{3x-2}$ **88.** $46.07 per person   **89.** 1.263×10^9 people   **90.** 2.733×10^{-23} gram
91. 3.3696×10^{31} joules   **92.** $3xy + 2x$ **93.** $2x^2 - 4y^2$

How Am I Doing? Chapter 4 Test

1. 3^{34} (obj. 4.1.1)   **2.** $\dfrac{1}{25^{16}}$ (obj. 4.1.7)   **3.** 8^{24} (obj. 4.1.3)   **4.** $12x^4y^{10}$ (obj. 4.1.1)   **5.** $-\dfrac{7x^3}{5}$ (obj. 4.1.2)   **6.** $-125x^3y^{18}$ (obj. 4.1.3)

7. $\dfrac{49a^{14}b^4}{9}$ (obj. 4.1.3)   **8.** $\dfrac{3x^4}{4}$ (obj. 4.1.3)   **9.** $\dfrac{1}{64}$ (obj. 4.2.1)   **10.** $\dfrac{6c^5}{a^4b^3}$ (obj. 4.2.1)   **11.** $3xy^7$ (obj. 4.2.1)

12. 5.482×10^{-4} (obj. 4.2.2)   **13.** 582,000,000 (obj. 4.2.2)   **14.** 2.4×10^{-6} (obj. 4.2.2)   **15.** $-2x^2 + 5x$ (obj. 4.3.2)
16. $-11x^3 - 4x^2 + 7x - 8$ (obj. 4.3.3)   **17.** $-21x^5 + 28x^4 - 42x^3 + 14x^2$ (obj. 4.4.1)   **18.** $15x^4y^3 - 18x^3y^2 + 6x^2y$ (obj. 4.4.1)
19. $10a^2 + 7ab - 12b^2$ (obj. 4.4.2)   **20.** $6x^3 - 11x^2 - 19x - 6$ (obj. 4.5.3)   **21.** $49x^4 + 28x^2y^2 + 4y^4$ (obj. 4.5.2)
22. $25s^2 - 121t^2$ (obj. 4.5.1)   **23.** $12x^4 - 14x^3 + 25x^2 - 29x + 10$ (obj. 4.5.3)   **24.** $3x^4 + 4x^3y - 15x^2y^2$ (obj. 4.5.1)
25. $3x^3 - x + 5$ (obj. 4.6.1)   **26.** $2x^2 - 7x + 4$ (obj. 4.6.2)   **27.** $2x^2 + 6x + 12$ (obj. 4.6.2)   **28.** 3.044×10^9 barrels per year (obj. 4.2.2)
29. 4.18×10^6 miles (obj. 4.2.2)

Cumulative Test for Chapters 0–4

1. $-\dfrac{11}{24}$ **2.** -0.74 **3.** $-\dfrac{6}{7}$ **4.** 110.55   **5.** $2x^2 - 13x$ **6.** 35   **7.** $x = -\dfrac{9}{2}$ **8.** $x = 15$ **9.** $x > -1$ **10.** $d = \dfrac{3y-8x}{11}$

11. 12,400 employees   **12.** $199.20   **13.** $20x^2 + 9x - 18$ **14.** $9x^2 - 30x + 25$ **15.** $6x^3 - 17x^2 - 26x - 8$ **16.** $-20x^5y^8$

17. $-\dfrac{2x^3}{3y^9}$ **18.** $16x^{12}y^8$ **19.** $\dfrac{9z^8}{w^2x^3y^4}$ **20.** 1.36×10^{15} **21.** 5.6×10^{-4} **22.** 4.0×10^{-35} **23.** $5x^3 + 7x^2 - 6x + 50$

24. $-36x^3y^2 + 18x^2y^3 - 48xy^4$ **25.** $6x^3 - 19x^2 + 18x - 5$ **26.** $x + 5 + \dfrac{3}{x-3}$

Chapter 5

5.1 Exercises

1. factors   **3.** no, $6a^3 + 3a^2 - 9a$ has a common factor of $3a$.   **5.** $3a(a + 1)$ **7.** $7ab(3 - 2b)$ **9.** $2\pi r(h + r)$
11. $5x(x^2 + 5x - 3)$ **13.** $4(3ab - 7bc + 5ac)$ **15.** $8x^2(2x^3 + 3x - 4)$ **17.** $7x(2xy - 5y - 9)$ **19.** $9x(6x - 5y + 2)$
21. $y(3xy - 2a + 5x - 2)$ **23.** $8xy(3x - 5y)$ **25.** $7x^2y^2(x + 3)$ **27.** $9xy(x - 2y - 3)$ **29.** $(x + 2y)(7a - b)$
31. $(x - 4)(3x - 2)$ **33.** $(2a - 3c)(6b - 5d)$ **35.** $(b - a^2)(7c - 5d + 2f)$ **37.** $(xy - 3)(2a - 4 - z)$ **39.** $(a - 3b)(4a^3 + 1)$
41. $(2a + 3)(1 - 7x)$ **43.** $A = 2.786(a + b + c + d)$ sq. in.   **45.** 17, 19, 21   **47.** 2.24 feet per minute   **48.** 23,220,000 tons
49. 38,700,000 tons   **50.** approximately 815 pounds per person   **51.** approximately 445 pounds per person

5.2 Exercises

1. We must remove a common factor of 5 from the last two terms. This will give us $3x(x - 2y) + 5(x - 2y)$. Then our final answer is
$(x - 2y)(3x + 5)$.   **3.** $(a + 4)(b - 3)$ **5.** $(x - 4)(x^2 + 3)$ **7.** $(a + 3b)(2x - y)$ **9.** $(3a + b)(x - 2)$ **11.** $(a + 2b)(5 + 6c)$
13. $(a - b)(5 - 2x)$ **15.** $(y - 2)(y - 3)$ **17.** $(7 + y)(2 - y)$ **19.** $(3x + y)(2a - 1)$ **21.** $(2x - 3)(x + 4)$
23. $(t - 1)(t^2 + 1)$ **25.** $(4x + 3w)(7x + 2y^2)$ **27.** We must rearrange the terms in a different order so that the expression in the
parentheses is the same in each case. We use the order $6a^2 - 8ad + 9ab - 12bd$ to factor $2a(3a - 4d) + 3b(3a - 4d) = (3a - 4d)(2a + 3b)$.
29. 8 seconds   **30.** \$420   **31.** \$126.9 million   **32.** 390 floor seats; 510 balcony seats   **33.** 7,315,000 tons in 2000; 9,363,200 tons in 2005

5.3 Exercises

1. product, sum   **3.** $(x + 1)(x + 1)$ **5.** $(x + 2)(x + 10)$ **7.** $(x - 3)(x - 1)$ **9.** $(x - 7)(x - 4)$ **11.** $(x + 5)(x - 4)$
13. $(x - 14)(x + 1)$ **15.** $(x + 7)(x - 5)$ **17.** $(x - 6)(x + 4)$ **19.** $(x + 8)(x + 4)$ **21.** $(x - 6)(x - 4)$ **23.** $(x + 3)(x + 10)$
25. $(x - 5)(x - 1)$ **27.** $(a + 8)(a - 2)$ **29.** $(x - 4)(x - 8)$ **31.** $(x + 7)(x - 3)$ **33.** $(x + 7)(x + 8)$ **35.** $(x + 2)(x - 23)$
37. $(x + 12)(x - 3)$ **39.** $(x + 3y)(x - 5y)$ **41.** $(x - 7y)(x - 9y)$ **43.** $4(x + 5)(x + 1)$ **45.** $6(x + 1)(x + 2)$

47. $2(x - 3)(x - 7)$ **49.** $3(x + 4)(x - 6)$ **51.** $7(x + 5)(x - 2)$ **53.** $3(x - 1)(x - 5)$ **55.** $A = x^2 - \dfrac{\pi x^2}{4} = x^2\left(1 - \dfrac{\pi}{4}\right)$;

$A = 13.76$ sq in   **57.** $(12 + x)(10 - x)$ **59.** $t = \dfrac{A - P}{Pr}$ **60.** $x \geq -\dfrac{5}{3}$ **61.** 120 miles

62. \$130,000   **63.** 1:30 A.M.   **64.** 9:00 P.M.   **65.** 25°C   **66.** June

5.4 Exercises

1. $(4x + 1)(x + 3)$ **3.** $(5x + 2)(x + 1)$ **5.** $(3x - 7)(x + 1)$ **7.** $(2x + 1)(x - 3)$ **9.** $(3x + 1)(3x + 2)$
11. $(3x - 5)(5x - 3)$ **13.** $(2x - 5)(x + 4)$ **15.** $(4x - 1)(2x + 3)$ **17.** $(3x + 2)(2x - 3)$ **19.** $(3x + 5)(2x - 3)$
21. $(x - 2)(7x + 9)$ **23.** $(9y - 4)(y - 1)$ **25.** $(5a + 2)(a - 3)$ **27.** $(7x - 5)(2x + 1)$ **29.** $(5x - 2)(3x + 2)$
31. $(6x + 5)(2x + 3)$ **33.** $(6x + 1)(2x - 3)$ **35.** $(2x^2 - 1)(x^2 + 8)$ **37.** $(2x + 5y)(x + 3y)$ **39.** $(5x - 4y)(x + 4y)$
41. $2(2x + 3)(x + 7)$ **43.** $2(4x - 1)(x - 3)$ **45.** $5(2x + 1)(x - 3)$ **47.** $3x(2x - 5)(x + 4)$ **49.** $(5x - 2)(x + 1)$

51. $2(3x - 2)(2x - 5)$ **53.** $(6x - 1)(2x - 3)$ **55.** $2(2x - 1)(2x + 5)$ **57.** $x = \dfrac{1}{10}$ **58.** $x = \dfrac{5 + 3y}{7}$ **59.** 26.5%   **60.** 45.4%
61. China   **62.** from 1995 to 2000

How Am I Doing? Sections 5.1–5.4

1. $3(2xy - 5z + 7)$ (obj. 5.1.1)   **2.** $4x(5x - 8y + 3)$ (obj. 5.1.1)   **3.** $(4x - 5)(7 - b)$ (obj. 5.1.1)   **4.** $(8y + 3z)(2x - 5y)$ (obj. 5.1.1)
5. $(6 + x)(3 - y)$ (obj. 5.2.1)   **6.** $(3x + 4w)(5 - 3b)$ (obj. 5.2.1)   **7.** $(x - 5)(x^2 - 3)$ (obj. 5.2.1)   **8.** $(a + 3b)(7 + 2b)$ (obj. 5.2.1)
9. $(x - 7)(x - 8)$ (obj. 5.3.1)   **10.** $(x + 16)(x - 4)$ (obj. 5.3.1)   **11.** $(x + 8y)(x + 5y)$ (obj. 5.3.1)   **12.** $7(x + 5)(x - 7)$ (obj. 5.3.2)
13. $(5x - 2)(2x + 1)$ (obj. 5.4.1)   **14.** $(x - 7)(3x - 2)$ (obj. 5.4.1)   **15.** $(2x + 3y)(3x + 4y)$ (obj. 5.4.2)
16. $2x(7x + 4)(x - 2)$ (obj. 5.4.3)

5.5 Exercises

1. $(3x + 1)(3x - 1)$ **3.** $(9x - 4)(9x + 4)$ **5.** $(x + 7)(x - 7)$ **7.** $(2x - 5)(2x + 5)$ **9.** $(x + 5)(x - 5)$
11. $(1 - 4x)(1 + 4x)$ **13.** $(4x - 7y)(4x + 7y)$ **15.** $(6x - 13y)(6x + 13y)$ **17.** $(9x + 10)(9x - 10)$ **19.** $(5a + 9b)(5a - 9b)$
21. $(3x + 1)^2$ **23.** $(y - 5)^2$ **25.** $(6x - 5)^2$ **27.** $(7x + 2)^2$ **29.** $(x + 7)^2$ **31.** $(5x - 4)^2$ **33.** $(9x + 2y)^2$ **35.** $(5x - 3y)^2$
37. $(4a + 9b)^2$ **39.** $(7x - 3y)^2$ **41.** $(7x + 3)^2$ **43.** $(7x + 3)(7x - 3)$ **45.** $(x^2 + 6)(x^2 - 6)$ **47.** $(3x^2 - 2)^2$
49. No two binomials can be multiplied to obtain $9x^2 + 1$.   **51.** 49, one answer   **53.** $4(2x - 3)(2x + 3)$ **55.** $3(7x - y)(7x + y)$
57. $3(2x - 3)^2$ **59.** $2(7x + 3)^2$ **61.** $(x - 11)(x - 4)$ **63.** $(2x - 1)(x + 3)$ **65.** $(4x - 11)(4x + 11)$ **67.** $(3x + 7)^2$

69. $3(x + 5)(x - 3)$ **71.** $5(x - 4)(x + 4)$ **73.** $7(x + 3)^2$ **75.** $2(x - 9)(x - 7)$ **77.** $x^2 + 3x + 4 + \dfrac{-3}{x - 2}$
78. $2x^2 + x - 5$ **79.** 1.2 ounces of greens; 1.05 ounces of bulk vegetables; 0.75 ounce of fruit   **80.** 1.44 ounces of greens; 1.26 ounces of
bulk vegetables; 0.9 ounce of fruit   **81.** 3838 feet above sea level   **82.** 5 miles

5.6 Exercises

1. $a(6a + 2b - 3)$ **3.** $(4x - 5y)(4x + 5y)$ **5.** $(3x - 2y)^2$ **7.** $(x + 5)(x + 3)$ **9.** $(3x + 2)(5x - 1)$ **11.** $(a - 3c)(x + 3y)$
13. $(y + 7)^2$ **15.** $(2x - 3)^2$ **17.** $(2x - 3)(x - 4)$ **19.** $(x - 10y)(x + 7y)$ **21.** $(a + 3)(x - 5)$ **23.** $5x(3 - x)(3 + x)$
25. $5xy^3(x - 1)^2$ **27.** $3xy(3z + 2)(3z - 2)$ **29.** $3(x + 7)(x - 5)$ **31.** $3x(x + 4)(x - 3)$ **33.** $-1(2x^2 + 1)(x + 2)(x - 2)$
35. prime   **37.** $5(x^2 + 2xy - 6y)$ **39.** $3x(2x + y)(5x - 2y)$ **41.** $2(3x - 5)(4x - 3)$ **43.** prime   **45.** \$78.24
46. 372 live strains   **47.** \$74   **48.** \$86   **49.** $y = 0.8x + 57$ **50.** $y = -3.3x + 144$

5.7 Exercises

1. $x = -3, 7$ **3.** $x = -13, x = -3$ **5.** $x = \dfrac{3}{2}, 2$ **7.** $x = \dfrac{2}{3}, \dfrac{3}{2}$ **9.** $x = 0, -13$ **11.** $x = 3, -3$ **13.** $x = 0, 1$ **15.** $x = \dfrac{2}{3}, 2$

17. $x = -3, 2$ **19.** $x = -\dfrac{1}{2}$ **21.** $x = 0, -2$ **23.** $x = -3, -4$ **25.** $x = -\dfrac{3}{2}, 4$ **27.** You can always factor out x.

29. $L = 14$ m; $W = 10$ m   **31.** 66 groups   **33.** 10 students   **35.** 12 meters above ground after 2 sec   **37.** 5 additional helicopters

39. 2415 telephone calls   **41.** 18 people   **43.** $-10x^5y^4$ **44.** $12a^{10}b^{13}$ **45.** $-\dfrac{3a^4}{2b^2}$ **46.** $\dfrac{1}{3x^5y^4}$

Putting Your Skills to Work

1. 402   **2.** 114   **3.** Increased from 29% to 34%   **4.** Decreased from 34% to 32%   **5.** 354   **6.** 132   **7.** $y = 6(5x + 49)$
8. $y = 11(x + 9)$

Chapter 5 Review Problems

1. $4x^2(3x - 5y)$ **2.** $5x^3(2 - 7y)$ **3.** $7xy(x - 2y - 3x^2y^2)$ **4.** $25a^4b^4(2b - 1 + 3ab)$ **5.** $3a(a^2 + 2a - 3b + 4)$
6. $2(x - 2y + 3z + 6)$ **7.** $(a + 3b)(2a - 5)$ **8.** $3xy(5x^2 + 2y + 1)$ **9.** $(3x - 7)(a - 2)$ **10.** $(a + 5b)(a - 4)$
11. $(x^2 + 3)(y - 2)$ **12.** $3(2x - y)(5a + 7)$ **13.** $(5x - 1)(3x + 2)$ **14.** $(5w - 3)(6w + z)$ **15.** $(x - 7)(x + 5)$
16. $(x - 4)(x - 6)$ **17.** $(x + 6)(x + 8)$ **18.** $(x + 3y)(x + 5y)$ **19.** $(x^2 + 7)(x^2 + 6)$ **20.** $(x^2 - 7)(x^2 + 5)$
21. $6(x + 2)(x + 3)$ **22.** $3(x + 1)(x + 12)$ **23.** $2(x - 6)(x - 8)$ **24.** $4(x - 5)(x - 6)$ **25.** $(4x - 5)(x + 3)$
26. $(3x - 1)(4x + 5)$ **27.** $(5x + 4)(3x - 1)$ **28.** $(3x - 2)(2x - 3)$ **29.** $(2x - 3)(x + 1)$ **30.** $(3x - 4)(x + 2)$
31. $(10x - 1)(2x + 5)$ **32.** $(5x - 1)(4x + 5)$ **33.** $(3a - 2)(2a + 5)$ **34.** $(3a - 2)(2a - 5)$ **35.** $2(x - 1)(3x + 5)$
36. $2(x + 1)(3x - 5)$ **37.** $2(2x - 3)(x - 5)$ **38.** $4(x - 9)(x + 4)$ **39.** $2(3x + 2)(2x - 5)$ **40.** $3(3x - 2)(2x + 7)$
41. $(3x - 2y)(2x - 5y)$ **42.** $2(3x - y)(x - 5y)$ **43.** $(7x + y)(7x - y)$ **44.** $4(2x - 3y)(2x + 3y)$ **45.** $(5x + 3)^2$
46. $(7x - 2)^2$ **47.** $(5x - 6)(5x + 6)$ **48.** $(10x - 3)(10x + 3)$ **49.** $(y - 6x)(y + 6x)$ **50.** $(3y - 5x)(3y + 5x)$ **51.** $(6x + 1)^2$
52. $(5x - 2)^2$ **53.** $(4x - 3y)^2$ **54.** $(7x - 2y)^2$ **55.** $2(x - 3)(x + 5)$ **56.** $3(x - 5)(x + 5)$ **57.** $7(2x + 5)^2$ **58.** $8(3x - 4)^2$
59. $(2x + 3y)(2x - 3y)$ **60.** $(x + 3)^2$ **61.** $(x - 3)(x - 6)$ **62.** $(x + 15)(x - 2)$ **63.** $(x - 1)(6x + 7)$ **64.** $(5x - 2)(2x + 1)$
65. $12(2x - 5)$ **66.** $4xy(2xy - 1)$ **67.** $10x^2y^2(5x + 2)$ **68.** $13ab(2a^2 - b^2 + 4ab^3)$ **69.** $x(x - 8)^2$ **70.** $2(x + 10)^2$
71. $3(x - 3)^2$ **72.** $x(5x - 6)^2$ **73.** $(7x + 5)(x - 2)$ **74.** $(4x + 3)(x - 4)$ **75.** $xy(3x + 2y)(3x - 2y)$
76. $x^3a(3a + 4x)(a - 5x)$ **77.** $2(3a + 5b)(2a - b)$ **78.** $(11a + 3b)^2$ **79.** $(a - 1)(7 - b)$ **80.** $(4b - 5)(2 + 7c)$
81. $(6b - 7)(3 - 2x)$ **82.** $(b - 7)(5x + 4y)$ **83.** $x(2a - 1)(a - 7)$ **84.** $x(x + 4)(x - 4)(x + 1)(x - 1)$
85. $(x^2 + 9y^6)(x + 3y^3)(x - 3y^3)$ **86.** $(3x^2 - 5)(2x^2 + 3)$ **87.** $yz(14 - x)(2 - x)$ **88.** $x(3x + 2)(4x + 3)$
89. $(2w + 1)(8w - 5)$ **90.** $3(2w - 1)^2$ **91.** $2y(2y - 1)(y + 3)$ **92.** $(5y - 1)(2y + 7)$ **93.** $8y^8(y^2 - 2)$
94. $9(x^2 + 4)(x - 2)(x + 2)$ **95.** prime   **96.** prime   **97.** $4y(2y^2 - 5)(y^2 + 3)$ **98.** $3x(3y + 7)(y - 2)$ **99.** $(4x^2y - 7)^2$

100. $2xy(8x + 1)(8x - 1)$ **101.** $(2x + 5)(a - 2b)$ **102.** $(2x + 1)(x + 3)(x - 3)$ **103.** $-3, 6$ **104.** $-4, \dfrac{3}{2}$ **105.** $0, \dfrac{1}{6}$

106. $4, -3$ **107.** $-5, \dfrac{1}{2}$ **108.** $-8, -3$ **109.** $-5, -9$ **110.** $-\dfrac{3}{5}, 2$ **111.** -3 **112.** $-3, \dfrac{3}{4}$ **113.** $\dfrac{1}{5}, 2$

114. base $= 10$ cm; altitude $= 7$ cm   **115.** width $= 7$ feet, length $= 15$ feet   **116.** 6 seconds   **117.** 8 amperes, 12 amperes

How Am I Doing? Chapter 5 Test

1. $(x + 14)(x - 2)$ (obj. 5.3.1)   **2.** $(4x + 9)(4x - 9)$ (obj. 5.5.1)   **3.** $(5x + 1)(2x + 5)$ (obj. 5.4.2)   **4.** $(3a - 5)^2$ (obj. 5.5.2)
5. $x(7 - 9x + 14y)$ (obj. 5.1.1)   **6.** $(2x + 3b)(5y - 4)$ (obj. 5.4.2)   **7.** $2x(3x - 4)(x - 2)$ (obj. 5.6.1)
8. $c(5a - 1)(a - 2)$ (obj. 5.6.1)   **9.** $(9x + 10)(9x - 10)$ (obj. 5.5.1)   **10.** $(3x - 1)(3x - 4)$ (obj. 5.6.1)
11. $5(2x - 3)(2x + 3)$ (obj. 5.6.1)   **12.** prime (obj. 5.6.2)   **13.** $x(3x + 5)(x + 2)$ (obj. 5.6.1)   **14.** $-5y(2x - 3y)^2$ (obj. 5.6.1)
15. $(9x + 1)(9x - 1)$ (obj. 5.5.1)   **16.** $(9y^2 + 1)(3y - 1)(3y + 1)$ (obj. 5.5.1)   **17.** $(x + 3)(2a - 5)$ (obj. 5.6.1)
18. $(a + 2b)(w + 2)(w - 2)$ (obj. 5.6.1)   **19.** $3(x - 6)(x + 5)$ (obj. 5.6.1)   **20.** $x(2x + 5)(x - 3)$ (obj. 5.6.1)   **21.** $-5, -9$ (obj. 5.7.1)
22. $-\dfrac{7}{3}, -2$ (obj. 5.7.1)   **23.** $-\dfrac{5}{2}, 2$ (obj. 5.7.1)   **24.** $x = 7, -4$ (obj. 5.7.1)   **25.** width $= 7$ miles; length $= 13$ miles (obj. 5.7.2)

Cumulative Test for Chapters 0–5

1. 15%   **2.** 0.494   **3.** 5.8   **4.** $8x^4y^{10}$ **5.** 81   **6.** $27x^2 + 6x - 8$ **7.** $2x^3 - 12x^2 + 19x - 3$ **8.** $x \le -3$ **9.** 2   **10.** -15

11. $q = \dfrac{p + 3}{5}$ **12.** $(3x - 1)(2x - 1)$ **13.** $(3x + 4)(2x - 1)$ **14.** $(3x + 2)(3x - 1)$ **15.** $(11x + 8y)(11x - 8y)$

16. $-4(5x + 6)(4x - 5)$ **17.** prime   **18.** $x(4x + 5)^2$ **19.** $(9x^2 + 4b^2)(3x + 2b)(3x - 2b)$ **20.** $(2x + 3)(a - 2b)$

21. $3(5x - 3)(x + 1)$ **22.** $-8, 3$ **23.** $\dfrac{5}{3}, 2$ **24.** length $= 20$ ft
width $= 15$ ft

Chapter 6

6.1 Exercises

1. 3   **3.** $\dfrac{6}{x}$ **5.** $\dfrac{3x + 1}{1 - 3x}$ **7.** $\dfrac{a(a - 2b)}{2b}$ **9.** $\dfrac{x + 2}{x}$ **11.** $\dfrac{x - 5}{3x - 1}$ **13.** $\dfrac{x - 3}{x(x - 7)}$ **15.** $\dfrac{3x - 2}{x + 4}$ **17.** $\dfrac{3x - 5}{4x - 1}$ **19.** $\dfrac{x - 5}{x + 1}$

21. $-\dfrac{1}{2x}$ **23.** $\dfrac{-2x - 3}{x + 5}$ **25.** $\dfrac{4x + 5}{2x - 1}$ **27.** $\dfrac{2x - 3}{-x + 5}$ **29.** $\dfrac{a - b}{2a - b}$ **31.** $\dfrac{3x - 2y}{3x + 2y}$ **33.** $\dfrac{x(x + a)}{b(x + c)}$ **35.** $9x^2 - 42x + 49$

36. $49x^2 - 36y^2$ **37.** $2x^3 - 9x^2 - 2x + 24$ **38.** $2x^4 - 5x^3 - 10x^2 - 14x - 15$ **39.** $1\frac{5}{8}$ acre   **40.** 6 hrs, 25 min

41. 874,000,000 people   **42.** 358,000,000 people

6.2 Exercises

1. factor numerator and denominator completely and divide out common factors   **3.** $\dfrac{2(x+4)}{x-4}$ **5.** $\dfrac{3x^2}{4(x-3)}$ **7.** $\dfrac{x-2}{x+3}$ **9.** $\dfrac{(x+6)(x+2)}{x+5}$

11. $x+2$ **13.** $\dfrac{3(x+2y)}{4(x+3y)}$ **15.** $\dfrac{(x+5)(x-2)}{3x-1}$ **17.** $\dfrac{-5(x+3)}{2(x-2)}$ or $\dfrac{5(x+3)}{-2(x-2)}$ or $-\dfrac{5(x+3)}{2(x-2)}$ **19.** 1   **21.** $\dfrac{x-1}{x+3}$

23. The denominator cannot be zero. So x cannot be 2, −7, or 6.   **25.** $x=-8$ **26.** $7x^3 - 22x^2 + 2x + 3$ **27.** 5 milligrams

28. $79.5(8981)x = \$713,989.5x$ **29.** Harold's is 6 ft by 6 ft.
George's is 9 ft by 4 ft.

6.3 Exercises

1. The LCD would be a product that contains each factor. However, any repeated factor in any one denominator must be repeated the greatest number of times it occurs in any one denominator. So the LCD would be $(x+5)(x+3)^2$.

3. $\dfrac{3x+1}{x+5}$ **5.** $\dfrac{2x-5}{x+3}$ **7.** $\dfrac{2x-7}{5x+7}$ **9.** $3a^2b^2$ **11.** $90x^3y^5$ **13.** $15(x-3)$ **15.** x^2-9 **17.** $(x+5)(3x-1)^2$

19. $\dfrac{7+3a}{ab}$ **21.** $\dfrac{3x-13}{(x+7)(x-7)}$ **23.** $\dfrac{4y^2-4y}{(y+2)(y-2)}$ **25.** $\dfrac{43a+12}{5a(3a+2)}$ **27.** $\dfrac{4z+x}{6xyz}$ **29.** $\dfrac{9x+14}{2(x-3)}$ **31.** $\dfrac{x+10}{(x+5)(x-5)}$

33. $\dfrac{3a+17b}{10}$ **35.** $\dfrac{-4x+34}{(2x-3)(x+2)}$ **37.** $\dfrac{-7x}{(x+3)(x-1)(x-4)}$ **39.** $\dfrac{5x+19}{(x+2)(x+3)(x+5)}$ **41.** $\dfrac{4x-13}{(x-4)(x-3)}$

43. $\dfrac{11x}{y-2x}$ **45.** $\dfrac{-17y^2-10y}{(4y-1)(2y+1)(y-5)}$ **47.** $\dfrac{6}{y+3}$ **49.** $\dfrac{5x^2-5x+7}{(x-3)(x+2)}$ **51.** $x=-7$ **52.** $y=\dfrac{5ax+6bc}{2a}$

53. $x<\dfrac{3}{2}$ **54.** $81x^{12}y^{16}$ **55.** at least 17 days   **56.** 566,100 more people   **57. (a)** 900,000 fish   **(b)** three more months

58. more than 7 videos per month

How Am I Doing? Sections 6.1–6.3

1. $\dfrac{8}{x}$ (obj. 6.1.1)   **2.** $\dfrac{2x+3}{x-7}$ (obj. 6.1.1)   **3.** $\dfrac{y+3}{3x^2(3-y)}$ (obj. 6.1.1)   **4.** $\dfrac{x-4}{x+2}$ (obj. 6.1.1)   **5.** $\dfrac{3x}{2}$ (obj. 6.2.1)

6. $-\dfrac{x+5}{3(x-2)}$ or $\dfrac{x+5}{-3(x-2)}$ (obj. 6.2.1)   **7.** $\dfrac{2x-1}{2}$ (obj. 6.2.2)   **8.** $\dfrac{4a-3}{2a-5}$ (obj. 6.2.2)   **9.** $\dfrac{xy-3ax-3ay}{axy}$ (obj. 6.3.3)

10. $\dfrac{7}{2(x+2)}$ (obj. 6.3.3)   **11.** $\dfrac{x^2+2x-9}{(x+3)(x+7)}$ (obj. 6.3.3)   **12.** $\dfrac{-5x^2-23x-8}{(x+5)(x-2)(x-4)}$ (obj. 6.3.3)

6.4 Exercises

1. $\dfrac{3x}{2+5x}$ **3.** $y+x$ **5.** $\dfrac{x^2-2x}{4+5x}$ **7.** $\dfrac{2}{3x+10}$ **9.** $\dfrac{1}{xy}$ **11.** $\dfrac{2x-1}{x}$ **13.** $\dfrac{2}{x-1}$ **15.** $\dfrac{3a^2+9}{a^2+2}$ **17.** $\dfrac{3x+9}{2x-5}$

19. $\dfrac{y+1}{-y+1}$ **21.** No expression in any denominator can be zero because division by zero is undefined. So −3, 5, and 0 are not allowed.

23. $w=\dfrac{p-2l}{2}$ **24.** $x>-1$ **25.** \$1875   **26.** approximately 57%

6.5 Exercises

1. −12   **3.** 3.5 or $\dfrac{7}{2}$ **5.** 15   **7.** 3   **9.** −5   **11.** $-\dfrac{14}{3}$ or $-4\dfrac{2}{3}$ **13.** −2   **15.** 8   **17.** −0.5 or $-\dfrac{1}{2}$ **19.** no solution

21. −3   **23.** −5   **25.** no solution   **27.** 4   **29.** no solution   **31.** 2, 4   **33.** $(3x+4)(2x-3)$ **34.** $-\dfrac{10}{3}$ **35.** width = 7 m
length = 20 m

36. \$946.45   **37.** 21.9%   **38.** 90.9%   **39.** 10.3 million people   **40.** 3.5 million people

6.6 Exercises

1. 18   **3.** $\dfrac{204}{5}$ or $40\dfrac{4}{5}$ **5.** $\dfrac{56}{5}$ or 11.2   **7.** $\dfrac{91}{4}$ or 22.75   **9.** 110 miles   **11. (a)** 1043 Australian dollars   **(b)** 133 more Australian dollars

13. 56 mph   **15.** 29 miles   **17.** $18\dfrac{17}{20}$ inches   **19.** 200 meters   **21.** $9\dfrac{1}{3}$ in.   **23.** $38\dfrac{2}{5}$ m   **25.** 48 in.   **27.** 35 inches

29. 61.5 mph   **31.** commuter = 250 km/hr   **33. (a)** \$0.11   **(b)** \$0.09   **(c)** \$3.73   **35.** $2\dfrac{2}{9}$ hours or 2 hrs, 13 min   **37.** $3\dfrac{3}{7}$ hours or 3 hrs, 26 min
helicopter = 210 km/hr

39. 8.92465×10^{-4} **40.** 582,000,000   **41.** $\dfrac{w^8}{x^3y^2z^4}$ **42.** $\dfrac{27}{8}$ or $3\dfrac{3}{8}$

Putting Your Skills to Work

1. \$3000   **2.** \$1200   **3.** $P=\dfrac{24x}{x+3}$ **4.** \$18,857   **5.** \$15,000   **6.** \$21,000

Chapter 6 Review Problems

1. $\dfrac{x}{x-y}$ **2.** $-\dfrac{4}{5}$ **3.** $\dfrac{x+3}{x-4}$ **4.** $\dfrac{x+2}{x+4}$ **5.** $\dfrac{x+3}{x-7}$ **6.** $\dfrac{2(x+4)}{3}$ **7.** $\dfrac{x}{x-5}$ **8.** $\dfrac{2x+3}{2x}$ **9.** $\dfrac{2(x-4y)}{2x-y}$ **10.** $\dfrac{2-y}{3y-1}$

11. $\dfrac{x-2}{(5x+6)(x-1)}$ **12.** $4x+2y$ **13.** $\dfrac{x-25}{x+5}$ **14.** $\dfrac{2(y-3)}{3y}$ **15.** $\dfrac{(y+2)(2y+1)(y-2)}{(2y-1)(y+1)^2}$ **16.** $\dfrac{4y(3y-1)}{(3y+1)(2y+5)}$

17. $\dfrac{3y(4x+3)}{2(x-5)}$ **18.** $\dfrac{x+2}{2}$ **19.** $\dfrac{2(x+3y)}{x-4y}$ **20.** $\dfrac{3}{16}$ **21.** $\dfrac{9x+2}{x(x+1)}$ **22.** $\dfrac{5x^2+7x+1}{x(x+1)}$ **23.** $\dfrac{10x-22}{(x+2)(x-4)}$

24. $\dfrac{(x-1)(x-2)}{(x+3)(x-3)}$ **25.** $\dfrac{2xy+4x+5y+6}{2y(y+2)}$ **26.** $\dfrac{(2a+b)(a+4b)}{ab(a+b)}$ **27.** $\dfrac{3x-2}{3x}$ **28.** $\dfrac{2x^2+7x-2}{2x(x+2)}$ **29.** $\dfrac{3}{2(x-9)}$

30. $\dfrac{1-2x-x^2}{(x+5)(x+2)}$ **31.** $\dfrac{1}{11}$ **32.** $\dfrac{5}{3x^2}$ **33.** $w-2$ **34.** 1 **35.** $-\dfrac{y^2}{2}$ **36.** $\dfrac{x+2y}{y(x+y+2)}$ **37.** $\dfrac{-1}{a(a+b)}$ or $-\dfrac{1}{a(a+b)}$

38. $\dfrac{-3a-b}{b}$ or $-\dfrac{3a+b}{b}$ **39.** $\dfrac{5y(x+5y)^2}{x(x-6y)}$ **40.** $\dfrac{-3y}{2(x+2y)}$ or $-\dfrac{3y}{2(x+2y)}$ **41.** 2 **42.** 15 **43.** $\dfrac{2}{7}$ **44.** -8 **45.** -4

46. -2 **47.** $\dfrac{1}{2}$ **48.** $\dfrac{1}{5}$ **49.** no solution   **50.** $x=-4$; $x=6$ **51.** 6 **52.** 2 **53.** -2 **54.** no solution   **55.** 9 **56.** 0

57. 2.8 **58.** 1.3 **59.** 26.4 **60.** 2 **61.** 16 **62.** 9.1 **63.** 8.3 gallons   **64.** 167 cookies   **65.** 46 gallons   **66.** 91.5 miles

67. train $= 60$ mph, car $= 40$ mph   **68.** 3 hrs, 5 min   **69.** 182 ft   **70.** 1200 ft   **71.** 3 hrs, 20 min   **72.** 12 hrs   **73.** $-\dfrac{a+4}{16a^2}$ **74.** $\dfrac{6b^2}{b+5}$

75. $\dfrac{x-2y}{x+2y}$ **76.** $\dfrac{x-8y}{x+7y}$ **77.** $\dfrac{x+6}{x}$ **78.** $\dfrac{x+6}{x}$ **79.** $\dfrac{b^2-a^2}{ab(x+y)}$ **80.** $\dfrac{1}{3}$ **81.** $-\dfrac{6}{y^2}$ **82.** $\dfrac{y(x+3y)}{2(x+2y)}$ **83.** 23

84. -12 **85.** -2 **86.** 10.7 kg   **87.** 420 miles

How Am I Doing? Chapter 6 Test

1. $\dfrac{2}{3a}$ (obj. 6.1.1)   **2.** $\dfrac{2x^2(2-y)}{(y+2)}$ (obj. 6.1.1)   **3.** $\dfrac{5}{12}$ (obj. 6.2.1)   **4.** $\dfrac{1}{3y(x-y)}$ (obj. 6.2.1)   **5.** $\dfrac{2a+1}{a+2}$ (obj. 6.2.2)

6. $\dfrac{3a+4}{(a+1)(a-2)}$ (obj. 6.3.3)   **7.** $\dfrac{x-a}{ax}$ (obj. 6.3.3)   **8.** $-\dfrac{x+2}{x+3}$ (obj. 6.3.3)   **9.** $\dfrac{x}{4}$ (obj. 6.4.2)   **10.** $\dfrac{6x}{5}$ (obj. 6.4.1)

11. $\dfrac{2x-3y}{4x+y}$ (obj. 6.1.1)   **12.** $\dfrac{x}{(x+2)(x+4)}$ (obj. 6.3.3)   **13.** $-\dfrac{1}{5}$ (obj. 6.5.1)   **14.** $x=4$ (obj. 6.5.1)   **15.** no solution (obj. 6.5.2)

16. $x=\dfrac{47}{6}$ (obj. 6.5.1)   **17.** $x=\dfrac{45}{13}$ (obj. 6.6.1)   **18.** $x=37.2$ (obj. 6.6.1)   **19.** 151 flights (obj. 6.6.1)   **20.** $368 (obj. 6.6.1)

21. 102 feet (obj. 6.6.2)

Cumulative Test for Chapters 0–6

1. 0.006 cm   **2.** 35% **3.** $92.50   **4.** $x=6$ **5.** $h=\dfrac{A}{\pi r^2}$ **6.** $x>1.25$ [number line]   **7.** $x\le 26$

8. $(a+b)(3x-2y)$ **9.** $2a(4a+b)(a-5b)$ **10.** $-4x^3y^{12}$ **11.** $\dfrac{2x+5}{x+7}$ **12.** $\dfrac{(x-2)(3x+1)}{3x(x+5)}$ **13.** $\dfrac{1}{2}$ **14.** $\dfrac{11x-3}{2(x+2)(x-3)}$

15. $\dfrac{6c-28}{(c+2)(c-2)(c-3)}$ **16.** $x=-2$ **17.** $x=-\dfrac{9}{2}$ **18.** $\dfrac{x+8}{6x(x-3)}$ **19.** $\dfrac{3ab^2+2a^2b}{5b^2-2a^2}$ **20.** $x=\dfrac{7}{10}$ **21.** 208 miles

22. 484 phone calls

Chapter 7

7.1 Exercises

1. 0 **3.** The order in which you write the numbers matters. The graph of $(5,1)$ is not the same as the graph of $(1,5)$.

5.

7. $R:(-3,-5)$
$S:\left(-4\tfrac{1}{2},0\right)$
$X:(3,-5)$
$Y:\left(2\tfrac{1}{2},6\right)$

9. $(-4,-1)$
$(-3,-2)$
$(-2,-3)$
$(-1,-5)$
$(0,-3)$
$(2,-1)$

11. (a) $(0, 7)$ **(b)** $(2, 15)$ **13. (a)** $(-5, 22)$ **(b)** $(4, -14)$ **15. (a)** $(-3, -5)$ **(b)** $(5, 1)$ **17. (a)** $(-2, 0)$ **(b)** $(-4, 3)$

19. (a) $(1, 20)$ **(b)** $\left(\dfrac{9}{5}, 12\right)$ **21.** B5   **23.** E1   **25.** D3

27. (a)

Number of years since 1997

(b) The number of cellular users has significantly increased in this 5-year period.

29. (a)

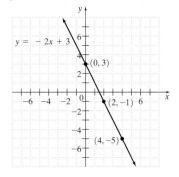

(b) An estimated 57 to 58 million people would buy books online in 2004.

31. 8   **32.** $\dfrac{x - y}{x - 3y}$ **33.** 1133.54 square yards   **34.** 922   **35.** 94.1%   **36. (a)** $55 per sq ft **(b)** $6160

7.2 Exercises

1. No, replacing x by -2 and y by 5 in the equation does not result in a true statement.   **3.** x-axis

5. $y = -2x + 1$
   $(0, 1)$
   $(-2, 5)$
   $(1, -1)$

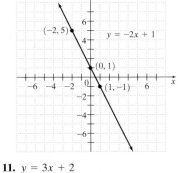

7. $y = x - 4$
   $(0, -4)$
   $(2, -2)$
   $(4, 0)$

9. $y = -2x + 3$

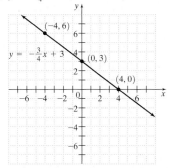

11. $y = 3x + 2$

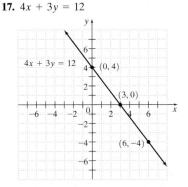

13. $3x - 2y = 0$

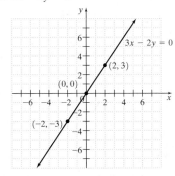

15. $y = -\dfrac{3}{4}x + 3$

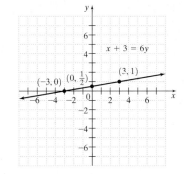

17. $4x + 3y = 12$

19. $y = 6 - 2x$

21. $x + 3 = 6y$

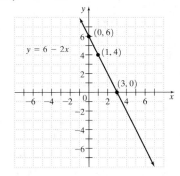

23. $y - 2 = 3y$

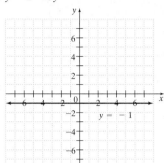

$y = -1$

25. $2x + 9 = 5x$

$x = 3$

27.

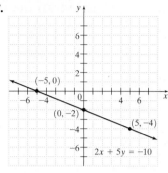

$(-5, 0)$

$(0, -2)$

$(5, -4)$

$2x + 5y = -10$

29. $C = 0; C = 120; C = 240; C = 360;$
$C = 480; C = 600$

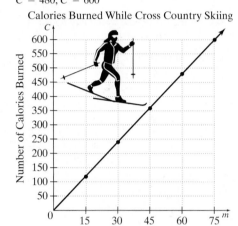

Calories Burned While Cross Country Skiing

Number of Calories Burned

Minutes Spent Skiing

31.

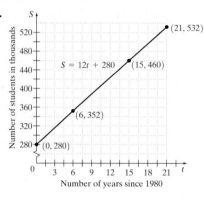

$(21, 532)$

$S = 12t + 280$ $(15, 460)$

$(6, 352)$

$(0, 280)$

Number of students in thousands

Number of years since 1980

33. 28 **34.** $x \geq -\dfrac{14}{3}$

$-6 \quad -5 \; -\frac{14}{3} \quad -4$

35. $W = 9.5$ meters **36.** 212 **37.** \$1.24 **38.** 76 milligrams **39.** one base = 4 in.
$L = 17$ meters other base = 5 in.

7.3 Exercises

1. No, division by zero is impossible, so the slope is undefined. **3.** 3 **5.** -1 **7.** $\dfrac{3}{5}$ **9.** $-\dfrac{1}{4}$ **11.** $-\dfrac{4}{3}$ **13.** $-\dfrac{16}{5}$

15. $m = 8; (0, 9)$ **17.** $m = -3; (0, 4)$ **19.** $m = -\dfrac{8}{7}\left(0, \dfrac{3}{4}\right)$ **21.** $m = -6; (0, 0)$ **23.** $m = -6; \left(0, \dfrac{4}{5}\right)$ **25.** $m = -\dfrac{5}{2}; \left(0, \dfrac{3}{2}\right)$

27. $m = \dfrac{7}{3}; \left(0, -\dfrac{4}{3}\right)$ **29.** $y = \dfrac{3}{5}x + 3$ **31.** $y = 6x - 3$ **33.** $y = -\dfrac{5}{4}x - \dfrac{3}{4}$

35.

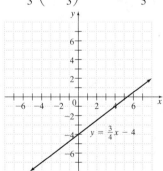

$y = \dfrac{3}{4}x - 4$

37.

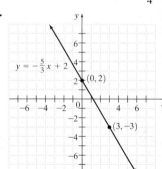

$y = -\dfrac{5}{3}x + 2$

$(0, 2)$

$(3, -3)$

39.

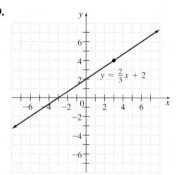

$y = \dfrac{2}{3}x + 2$

41.

43.
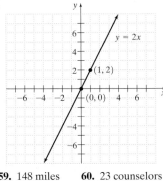

45. (a) $\dfrac{5}{6}$ **(b)** $-\dfrac{6}{5}$ **47. (a)** 6   **(b)** $-\dfrac{1}{6}$

49. (a) $\dfrac{1}{3}$ **(b)** -3 **51.** yes; $2x - 3y = 18$

53. (a) $y = 56x + 490$ **(b)** $m = 56$; $(0, 490)$
   (c) The amount of increase in income in U.S. federal budget in billions of dollars per year

55. $x > 4$

56. $x < \dfrac{12}{5}$

57. $x \le 24$

58. $x \le -22$
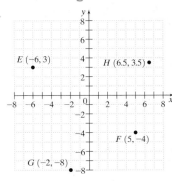

59. 148 miles   **60.** 23 counselors   **61.** 3.5%

How Am I Doing? Sections 7.1–7.3

1.   (obj. 7.1.1)
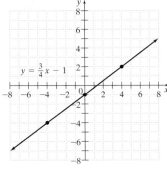

2. $A\ (3, 9)$
$B\ (7, -7)$ (obj. 7.1.2)
$C\ (-7, -8)$
$D\ (-4, 3)$

3. $(4, -25)$; $(0, 3)$; $(-2, 17)$ (obj. 7.1.3)
4. $(0, 6)$; $(-5, 0)$; $(-10, -6)$ (obj. 7.1.3)

5.   (obj. 7.2.1)

6.

(obj. 7.2.1)

7.
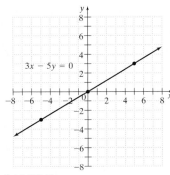

(obj. 7.2.2)

8.

(obj. 7.2.2)

9. $\dfrac{5}{8}$ (obj. 7.3.1)   **10.** slope $= 3$
y-intercept $= \left(0, -\dfrac{5}{3}\right)$
(obj. 7.3.2)

11. $-\dfrac{4}{3}$ (obj. 7.3.5)   **12.** $-\dfrac{11}{3}$ (obj. 7.3.5)

7.4 Exercises

1. $y = 4x + 12$ **3.** $y = -3x + 11$ **5.** $y = -3x + \dfrac{7}{2}$ **7.** $y = -\dfrac{2}{5}x - 1$ **9.** $y = -2x - 6$ **11.** $y = -4x + 2$

13. $y = 5x - 10$ **15.** $y = \dfrac{1}{3}x + \dfrac{1}{2}$ **17.** $y = -2x + 11$ **19.** $y = -3x + 3$ **21.** $y = -\dfrac{2}{3}x + 1$ **23.** $y = \dfrac{2}{3}x - 4$

25. $y = -\dfrac{2}{3}x$ **27.** $y = -2$ **29.** $y = -2$ **31.** $x = 4$ **33.** $y = \dfrac{3}{4}x - 4$ **35.** $y = -\dfrac{1}{2}x + 4$ **37.** $y = 2.4x + 227$

39. $\dfrac{1}{3}$ **40.** $\dfrac{(2x + 5)(x - 3)}{(2x - 5)^2}$ **41.** \$61.20   **42.** 290 minutes   **43.** 4.95%   **44.** \$12,640   **45.** width $= 550$ feet
length $= 1170$ feet

7.5 Exercises

1. No, all points in one region will be solutions to the inequality while all points in the other region will not be solutions. Thus testing any point will give the same result, as long as the point is not on the boundary line.

3.

5.

7.

9.

11.

13.

15.

17.

19.

21.

23. (a) $x < 3$ (b) $x < 3$
(c) $3x + 3$ will be greater than $5x - 3$ for those values of x where the graph of $y = 3x + 3$ lies above the graph of $y = 5x + 3$.

25. $x = \dfrac{3}{5}, -3$ **26.** $y = -\dfrac{7}{3}x + 7$

27. 9200 miles   **28.** 16

29. 90 ounces of zinc

7.6 Exercises

1. You can describe a function using a table of values, an algebraic equation, or a graph.   **3.** Possible values of the independent variable
5. If a vertical line can intersect the graph more than once, the relation is not a function. If no such line exists, then the relation is a function.

7. (a) Domain $= \left\{\dfrac{2}{5}, 3, -3\right\}$ (b) Not a function   **9.** (a) Domain $= \{5, 7\}$ (b) Not a function
Range $= \left\{4, \dfrac{2}{5}, -1\right\}$ Range $= \{0, 2.3, 3.1\}$

11. (a) Domain $= \{1, 9, 12, 14\}$ (b) function   **13.** (a) Domain $= \{3, 5, 7\}$ (b) Not a function
Range $= \{1, 3, 12\}$ Range $= \{75, 85, 95, 100\}$

15.

17.

19.

21.

23.

25.

27.

29.

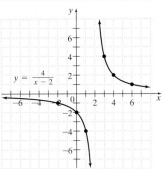

31. function   **33.** not a function   **35.** function
37. not a function   **39. (a)** 26 **(b)** 2 **(c)** −4
41. (a) 3 **(b)** 24 **(c)** 9
43. (a) $\frac{3}{5}$ **(b)** $\frac{3}{2}$ **(c)** −1   **45.** yes

47. 1980

49. $f(0) = 800$ $f(3) = 878$ $f(6) = 992$ $f(10) = 1200$
The curve slopes more steeply for larger values of x.
Growth rate is increasing as x gets larger.

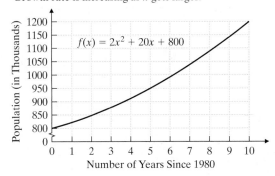

51. $-2x^2 - 14x + 6$ **52.** $-3x^2 - 7x + 11$

53. $4x^3 + 6x^2 - x + 6$ **54.** $2x + 4 + \dfrac{-1}{3x - 1}$

Putting Your Skills to Work

1. 213, 408 acres; 78,000,000 acres   **2.** 4337 million acres   **3.** 4181 million acres, 3947 million acres   **4.** $r(t) = 4337 - 78t$
5. (0, 4337)   **6.** $r(t) = 4337 - 78t$
  (5, 3947)
  (10, 3557) r is the dependent variable.
  (15, 3167) t is the independent variable.
  (20, 2777)

7. about 827 million acres
8. approximately 2045

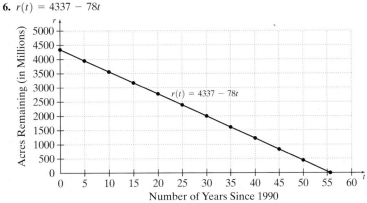

Chapter 7 Review Problems

1.

2.

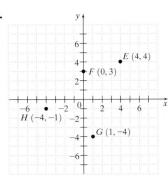

3. $(0, 7); (-1, 10)$ **4. (a)** $(1, 2)$ **(b)** $(-4, 4)$
5. (a) $(6, -1)$ **(b)** $(6, 3)$

6.

7.

8.
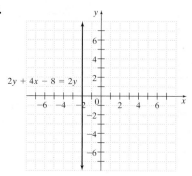

9. $m = -\dfrac{5}{6}$ **10.** $m = \dfrac{9}{11}; \left(0, \dfrac{5}{11}\right)$ **11.** $y = -\dfrac{1}{2}x + 3$ **12.** $m = -\dfrac{5}{3}$

13.

14.

15.
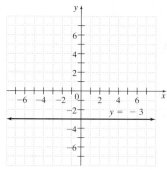

16. $y = 2x - 4$ **17.** $y = -6x + 14$ **18.** $y = -\dfrac{3}{2}x - 3$ **19.** $y = 7$ **20.** $-\dfrac{2}{3}$ **21.** $-\dfrac{4}{3}$ **22.** $y = \dfrac{2}{3}x - 3$ **23.** $y = -3x + 1$

24. $x = 5$ **25.**

26.

27.
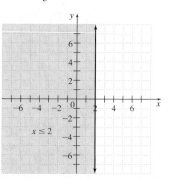

28. Domain: $\{-6, -5, 5\}$ **29.** Domain: $\{2, 5, 6, -2\}$ **30.** function   **31.** not a function   **32.** function
Range: $\{-6, 5\}$ Range: $\{-3, 4\}$
not a function   function

33.

34.

35.

36. (a) 7 (b) 31   **37.** (a) -1 (b) -5 **38.** (a) -11 (b) 1   **39.** (a) 1 (b) $\dfrac{1}{5}$ **40.** (a) 0 (b) 4   **41.** \$210   **42.** \$174

43. $y = 0.09x + 30$
$(0, 30)$ tells us that if Russ and Norma use no electricity, the minimum cost will be \$30.
44. $m = 0.09$; the electric bill increases 0.09 for each kilowatt-hour of use.   **45.** 1300 kilowatt-hours   **46.** 2400 kilowatt-hours
47. 42.8 quadrillion Btu   **48.** 43.6 quadrillion Btu
49. $m = 0.4$. This means that for each additional year the U.S. consumes 0.4 quadrillion Btu of petroleum.
50. y-intercept $= (0, 23.6)$. This means that when $x = 0$, which is the year 1960, a total of 23.6 quadrillion Btu of petroleum were used.
51. 2020   **52.** 2030

53.

54.

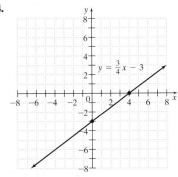

55. $-\dfrac{2}{5}$ **56.** $m = -\dfrac{7}{6}$

$y\text{-intercept} = \left(0, \dfrac{5}{3}\right)$

57. $y = \dfrac{2}{3}x - 7$

58.

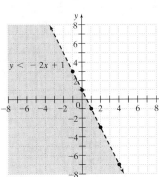

59. 22,000,000 in 1980
20,700,000 in 1990
19,400,000 in 2000

60.

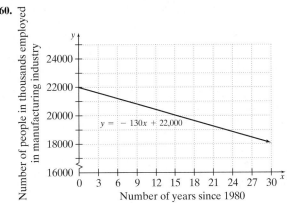

61. $m = -130$. The slope tells us that the number of people employed in manufacturing decreases each year by 130,000.

62. y-intercept is $(0, 22,000)$. This tells us in the year 1980, the number of manufacturing jobs was 22,000 thousands which is 22,000,000.

63. 2010   **64.** 2020

How Am I Doing? Chapter 7 Test

1.

(obj. 7.1.1)

2.

(obj. 7.2.1)

3.

(obj. 7.2.1)

4.

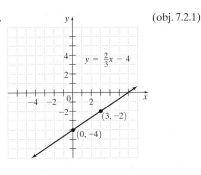

(obj. 7.2.1)

5. $m = -\dfrac{3}{2}; \left(0, \dfrac{5}{2}\right)$ (obj. 7.3.2)   **6.** $m = 1$ (obj. 7.3.1)   **7.** $y = \dfrac{1}{2}x - 4$ (obj. 7.3.3)   **8.** $m = 0$ (obj. 7.3.1)

9. $y = -\dfrac{3}{2}x + \dfrac{7}{2}$ (obj. 7.4.2)   **10.** $x = 2$; undefined (obj. 7.4.2)

11. (obj. 7.5.1)   **12.** 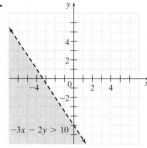 (obj. 7.5.1)

13. No, two different ordered pairs have the same first coordinate. (obj. 7.6.1)   **14.** Yes, any vertical line passes through no more than one point on the graph. (obj. 7.6.3)

15. 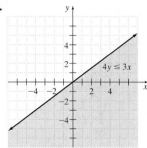 (obj. 7.6.2)   **16. (a)** -3 **(b)** -3 (obj. 7.6.4)   **17. (a)** -3 **(b)** $-\dfrac{1}{5}$ (obj. 7.6.4)

Cumulative Test for Chapters 0–7

1. $-27x^9y^{12}z^3$ **2.** $-\dfrac{2y^{10}}{3x^8}$ **3.** $x = 7$ **4.** $R = 40$ **5.** $x = 5$ **6.** $w = \dfrac{A - lh}{l}$ or $w = \dfrac{A}{l} - h$ **7.** $2(5a + 7b)(5a - 7b)$

8. $(2x - 3)(2x + 5)$ **9.** $2(x - 1)$ **10.** $y = 3x - 10$ **11.** $x = 7$ **12.** $y = -\dfrac{2}{3}x - 3$ **13.** $m = 0$ **14.** $m = \dfrac{3}{7}$

15.   **16.**   **17.**

18. no 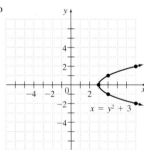   **19.** yes   **20. (a)** 1 **(b)** 10

Chapter 8

8.1 Exercises

1. The lines are parallel. One line intersects the y-axis at -5. The other line intersects the y-axis at 6. There is no solution. The system is inconsistent.   **3.** intersect, one   **5.** The lines intersect at the y-intercept. The y-intercept is the solution of the system.

7.

9.

11.

13.

15.

17.

19.

21.

23.

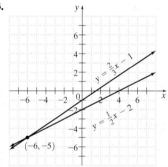

25. (a) Newtown: $y = 7x$ **(b)**
Tryus: $y = 36 + 3x$

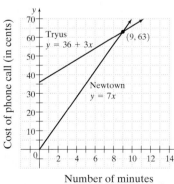

x	1	4	8	12
y = 7x	7	28	56	84

x	1	4	8	12
y = 36 + 3x	39	48	60	72

(c) 9 minutes   **(d)** Tryus

27. $(-33, -40)$ **29.** $(47.52, -61.38)$ **31.** $x \geq \dfrac{1}{8}$ **32.** $d = \dfrac{5y + 16}{x}$ **33.** $9x^2 - 42x + 49$ **34.** $(5x - 6)^2$ **35.** 90,000 people

36. 300,000 people   **37.** 547,500 people   **38.** 38.9%

8.2 Exercises

1. $(3, -1)$ **3.** $(1, 2)$ **5.** $(3, -5)$ **7.** $(0, 4)$ **9.** $(9, 5)$ **11.** $(2, 3)$ **13.** $\left(\dfrac{2}{3}, \dfrac{5}{3}\right)$ **15.** $(2, -3)$ **17.** $(-3, 6)$ **19.** $(3, -2)$

21. $(6, -5)$ **23.** $(1, 3)$ **25.** $(-4, -12)$ **27.** $(8, 1)$ **29.** 3, 7   **31.** both   **33.** no solution, parallel lines have no points in common

35. (a) $y = 90x + 50$ **(b)** 5 times plowing, $500   **(c)** 13 times, Adirondack Plowing   **36.** $m = -\dfrac{7}{11}; \left(0, \dfrac{19}{11}\right)$
$y = 80x + 100$

37.

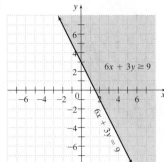

38. 398 meters/min   **39.** \$3,914,000   **40.** 8736 nights   **41.** \$162

8.3 Exercises

1. Eliminate x by multiplying the second equation by 2, then adding the equations.
3. Eliminate y by multiplying the first equation by 4, multiplying the second equation by 3, then adding the equations.   **5.** $(-1, -4)$
7. $(2, -1)$ **9.** $\left(-\dfrac{3}{2}, -2\right)$ **11.** $\left(3, \dfrac{2}{5}\right)$ **13.** $(1, 0)$ **15.** $\left(-\dfrac{1}{2}, \dfrac{1}{3}\right)$ **17.** $(-10, 4)$ **19.** $(-3, -2)$ **21.** $\left(\dfrac{1}{3}, 1\right)$

23. Multiply Equation 1 by 4   **25.** Multiply Equation 1 by 6.   **27.** $\left(2, \dfrac{1}{5}\right)$ **29.** $(3, -2)$ **31.** $\left(-2, \dfrac{4}{3}\right)$ **33.** $(-3, 3)$
                                      Multiply Equation 2 by 20

35. Multiply Equation 1 by 10.   **37.** Multiply Equation 1 by 10.   **39.** $(2, 3)$ **41.** $(3, -5)$ **43.** $\left(2, \dfrac{2}{5}\right)$ **45.** $(3, 5)$ **47.** $(8, 6)$
$5x - 3y = 1$ Multiply Equation 2 by 100.
$5x + 3y = 6$ $40x + 5y = 90$
                                      $20x - 5y = 100$
49. $(3, 5)$ **51.** $(1, 3)$ **53.** between 550 and 770 airplanes   **54.** 12.4% more   **55.** 99 mph or greater   **56.** 36 games
57. $3(5x - 1)(5x + 1)$ **58.** $(2x - 5)(3x - 5)$

How Am I Doing? Sections 8.1–8.3

1.

2.

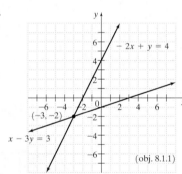

3. $(6, -4)$ (obj. 8.2.1)

4. $(3, -2)$ (obj. 8.2.1)   **5.** $(-2, 3)$ (obj. 8.2.2)   **6.** $(1, -2)$ (obj. 8.2.1)   **7.** $(7, 1)$ (obj. 8.3.1)   **8.** $(3, 3)$ (obj. 8.3.1)
9. $(-9, -12)$ (obj. 8.3.2)   **10.** $(1, -3)$ (obj. 8.3.3)

8.4 Exercises

1. Parallel lines, inconsistent   **3.** Intersect, independent, consistent   **5.** $(3, -7)$ **7.** $(24, 19)$ **9.** no solution
11. infinite number of solutions   **13.** $(5, 4)$ **15.** $(50, 40)$ **17.** $(3, 1)$ **19.** $\left(\dfrac{1}{2}, \dfrac{2}{3}\right)$ **21.** $\left(\dfrac{7}{3}, \dfrac{8}{3}\right)$ **23.** $(1, -2)$ **25.** $\left(\dfrac{3}{5}, \dfrac{3}{5}\right)$
27. no solution, inconsistent system   **29.** $(-4, -3)$ **31.** 2   **32.** 0.1   **33.** \$21,500   **34.** \$38,500   **35.** 1261 tickets
36. approximately 52.4 cents per mile

8.5 Exercises

1. 8 1st class   **3.** 5 @ \$35   **5.** 9 hrs day shift   **7.** original length = 10 ft   **9.** 8 qts of 50% antifreeze originally.
   6 coach class   7 @ \$40   14 hrs night shift   original width = 9 ft   She added 8 quarts of 80% solution.
                                                               new length = 14 ft
                                                               new width = 14 ft

11. 30 lb nuts   **13.** 50 km/hr = wind speed   **15.** Jewelry wants 10 \$10 bills and 7 \$20 bills.   **17.** \$0.69 per lb. = oranges
   20 lb raisins   550 km/hr = plane's speed   Cosmetics wants 18 \$10 bills and 3 \$20 bills.   \$0.79 per lb. = apples

19. 160 general admission   **21.** 2013; 5,680,000   **23.** $29,000 = heart kits   **25.** $\dfrac{2x^2 - 12}{x^2 - 9}$ **26.** 35   **27.** $-64x^9y^6$
130 student admission $36,000 = kidney kits

28. 35%   **29.** six inches of space

Putting Your Skills to Work

1. 372 million in 1980   **2.** 242 million in 1980   **3.** No. America experienced greater growth. It grew by 3 million more people in the period
480 million in 2000   347 million in 2000   1980 to 2000.   **4.** The difference will be 136 million people in 2020. yes
534 million in 2010   399.5 million in 2010

5.

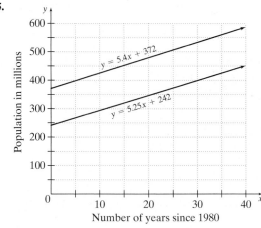

Number of years since 1980

6. They don't intersect on the graph paper. If they intersect it would have to be for values of x less than 0.
7. Yes, algebraically we get a solution at $x = -866.7$ and $y = -4308.2$.
8. This would have no meaning since y is negative and you cannot have negative population.

Chapter 8 Review Problems

1.

2.

3.

4.

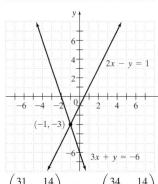

5. $(3, 3)$ **6.** $(3, 5)$ **7.** $(15, 21)$ **8.** $(8, 12)$ **9.** $(2, 1)$ **10.** $(-1, 0)$ **11.** $(0, -3)$

12. $(2, -5)$ **13.** $(4, -4)$ **14.** $(10, 2)$ **15.** $\left(\dfrac{4}{5}, -\dfrac{6}{5}\right)$ **16.** $\left(\dfrac{33}{13}, -\dfrac{28}{13}\right)$ **17.** $(6, -4)$

18. $(-1, -4)$ **19.** no solution, inconsistent system   **20.** $(9, 4)$ **21.** $\left(-3, -\dfrac{2}{3}\right)$

22. infinite number of solutions, dependent equations   **23.** $(-2, -2)$ **24.** $(-1, -2)$

25. $(4, -3)$ **26.** $\left(-\dfrac{24}{41}, \dfrac{2}{41}\right)$ **27.** infinite number of solutions, dependent equations

28. $(1960, 1040)$ **29.** $(8, -5)$ **30.** $(-4, -7)$ **31.** $\left(\dfrac{29}{28}, \dfrac{1}{7}\right)$ **32.** $(4, 10)$ **33.** $(10, 8)$

34. $(-12, -15)$ **35.** no solution inconsistent system   **36.** infinite solutions dependent equations

37. $\left(\dfrac{31}{41}, -\dfrac{14}{41}\right)$ **38.** $\left(\dfrac{34}{43}, -\dfrac{14}{43}\right)$ **39.** $(-5, 4)$ **40.** $(9, -8)$ **41.** $(2, 4)$ **42.** $(3, 1)$ **43.** $(3, 2)$ **44.** $(12, -12)$

45. 120 reserved seats; 640 general admission seats   **46.** 21 2-lb bags   **47.** 275 miles/hour in still air; 25 miles/hour wind speed
42 5-lb bags

48. 32 liters   **49.** 44 regular-sized cars; 42 compact cars   **50.** 8 tons of 15% salt; 16 tons of 30% salt

51. speed of the boat, 19 kilometers per hour; speed of the current, 4 kilometers per hour   **52.** $3 for a gram of copper; $4 hourly labor rate

53. 11 lower skydeck seats   **54.** 19 field level infield seats
12 upper skydeck seats   8 field level bases seats

How Am I Doing? Chapter 8 Test

1. $(-3, -4)$ (obj. 8.2.1)   **2.** $(-3, 4)$ (obj. 8.3.1)   **3.**

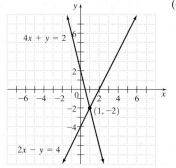

(obj. 8.1.1)

4. $(6, 2)$ (obj. 8.4.1)   **5.** $(4, 3)$ (obj. 8.4.1)   **6.** $(5, 1)$ (obj. 8.4.1)   **7.** $(3, 0)$ (obj. 8.4.1)
8. no solution, inconsistent system (obj. 8.4.1, 8.4.2)   **9.** infinite number of solutions, dependent equations (obj. 8.4.1, 8.4.2)
10. $(-2, 3)$ (obj. 8.4.1)   **11.** $(-5, 3)$ (obj. 8.4.1)   **12.** $(-3.5, -4.5)$ or $\left(-\dfrac{7}{2}, -\dfrac{9}{2}\right)$ (obj. 8.4.1)   **13.** $5, -3$ (obj. 8.5.1)
14. 16,000 of the \$12 tickets; 14,500 of the \$8 tickets (obj. 8.5.1)
15. A shirt costs \$20 and a pair of slacks costs \$24. (obj. 8.5.1)
16. 1000 booklets; \$450 total cost (obj. 8.5.1)   **17.** wind = 50 kilometers/hour; jet = 450 kilometers/hour (obj. 8.5.1)

Cumulative Test for Chapters 0–8

1. \$4.09   **2.** -8 **3.** $15x^3 - 19x^2 + 11x - 3$ **4.** $\dfrac{1}{(x + 3)(x - 4)}$ **5.** $x > 21$ **6.** $x = \dfrac{1}{2}, -\dfrac{2}{3}$ **7.** 3000   **8.** $x = \dfrac{5}{3}$
9. $5t^2(3t - 2)(t - 1)$ **10.** $x = 3$ **11.** $(8, -8)$ **12.** $(24, 8)$ **13.** infinite number of solutions; dependent equations   **14.** $(5, 1)$
15. inconsistent system; no solution   **16.** 36 kg of water   **17.** 1 mile/hour
18. the printer that broke prints 1200 labels/hour; the other printer prints 1800 labels/hour

Chapter 9

9.1 Exercises

1. The principal square root of N, where $N \geq 0$ is a nonnegative number a that has the property $a^2 = N$.   **3.** no; $(0.3)(0.3) = 0.09$
5. ± 3 **7.** ± 7 **9.** 4   **11.** 9   **13.** -6 **15.** 0.9   **17.** $\dfrac{6}{11}$ **19.** $\dfrac{7}{8}$ **21.** 20   **23.** -100 **25.** 13   **27.** $-\dfrac{1}{8}$ **29.** $\dfrac{3}{4}$
31. 0.06   **33.** 130   **35.** -17 **37.** 5.196   **39.** 8.602   **41.** -11.533 **43.** -13.964 **45.** 13 feet   **47.** 7.5 seconds
49. 9.5 seconds   **51.** 3   **53.** -6 **55.** 3   **57.** 3   **59.** no, a 4th power must be nonnegative.   **61.** $(2, 1)$
62. no solution, inconsistent system   **63.** each snowboard is \$250   **64.** 470 mph   **65.** % decrease = 11.3%
each pair of goggles is \$50   length of stay = 8.3 days
66. % of decrease = 15.9%
rate of decrease for U.S. is 4.6% greater

9.2 Exercises

1. yes, $\sqrt{6^2} = \sqrt{6 \cdot 6} = \sqrt{36} = 6$ **3.** no, $\sqrt{(-9)^2} = \sqrt{(-9)(-9)} = \sqrt{81} = 9$ **5.** 8   **7.** 18^2 **9.** 9^3 **11.** 33^4 **13.** 5^{70}
$\left(\sqrt{6}\right)^2 = \left(\sqrt{6}\right)\left(\sqrt{6}\right) = \sqrt{36} = 6$ but $-\sqrt{9^2} = -\sqrt{9 \cdot 9} = -\sqrt{81} = -9$
15. x^6 **17.** t^9 **19.** y^{13} **21.** $6x^4$ **23.** $12x$ **25.** $x^3 y^2$ **27.** $4xy^{10}$ **29.** $10x^6 y^4$ **31.** $2\sqrt{6}$ **33.** $3\sqrt{5}$ **35.** $3\sqrt{2}$
37. $6\sqrt{2}$ **39.** $3\sqrt{10}$ **41.** $8\sqrt{2}$ **43.** $2x\sqrt{2x}$ **45.** $3w^2\sqrt{3w}$ **47.** $4z^4\sqrt{2z}$ **49.** $2x^2 y^3 \sqrt{7xy}$ **51.** $4y\sqrt{3yw}$ **53.** $5xy\sqrt{3y}$
55. $9x^4$ **57.** $y^3\sqrt{y}$ **59.** $x^2 y^5 \sqrt{xy}$ **61.** $3x^2 y^3 \sqrt{15xy}$ **63.** $4abc^3 \sqrt{5ab}$ **65.** $13ab^4 c^2 \sqrt{ac}$ **67.** $x + 5$ **69.** $4x + 1$
71. $y(x + 7)$ **73.** $5\sqrt{3}$ square millimeters   **75.** (a) $20\sqrt{3}$ square millimeters   **77.** $(4, -1)$ **78.** -4 **79.** 885 million tons
(b) 4 times larger
80. approximately 31.7 quadrillion Btu   **81.** approximately 76.8%

9.3 Exercises

1. 1. Simplify each radical term.   **3.** $6\sqrt{5}$ **5.** $5\sqrt{2} + 3\sqrt{3}$ **7.** $4\sqrt{5}$ **9.** $12\sqrt{5}$ **11.** $\sqrt{2}$ **13.** $9\sqrt{2}$ **15.** $5 + 6\sqrt{2} + 6\sqrt{3}$
2. Combine like radicals.
17. $8\sqrt{3} - \sqrt{2}$ **19.** $13\sqrt{2x}$ **21.** $0.2\sqrt{3x}$ **23.** $7\sqrt{5y}$ **25.** $-\sqrt{2}$ **27.** $6\sqrt{7x} - 15x\sqrt{7x}$ **29.** $-5x\sqrt{2x}$ **31.** $10x\sqrt{3}$
33. $2y\sqrt{6y} - 6y\sqrt{6}$ **35.** $-110x\sqrt{2x}$ **37.** $\left(8\sqrt{5} + 6\sqrt{3}\right)$ miles   **39.** $8\sqrt{3} + 54\sqrt{2}$ meters   **41.** $x \geq -\dfrac{4}{3}$

$$\xleftarrow{\hspace{1cm}} \overset{\substack{-3 \quad -2 \quad -1}}{\underset{\substack{-\frac{4}{3}}}{\rule{3cm}{0.4pt}}} \bullet \xrightarrow{\hspace{1cm}}$$

42.

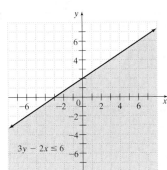

$3y - 2x \leq 6$

43. 3 CDs @ $7.99
7 CDs @ $11.99

44. $1600 @ 7%
$6400 @ 5%

9.4 Exercises

1. $\sqrt{30}$ **3.** $2\sqrt{11}$ **5.** $3\sqrt{6}$ **7.** $6\sqrt{30}$ **9.** $5\sqrt{2a}$ **11.** $42x\sqrt{2}$ **13.** $16a\sqrt{6}$ **15.** $-6b\sqrt{a}$ **17.** $\sqrt{10} + \sqrt{15}$
19. $6\sqrt{2} + 15\sqrt{5}$ **21.** $2x - 16\sqrt{5x}$ **23.** $2\sqrt{3} - 18 + 4\sqrt{15}$ **25.** $-6 - 9\sqrt{2}$ **27.** $21 + 14\sqrt{2}$ **29.** $5 - \sqrt{21}$
31. $-6 + 9\sqrt{2}$ **33.** $2\sqrt{14} + 34$ **35.** $29 - 12\sqrt{5}$ **37.** $53 + 10\sqrt{6}$ **39.** $48 - 6\sqrt{15}$ **41.** $12\sqrt{10}$ **43.** $\sqrt{14} + 21\sqrt{2} - \sqrt{42}$
45. $9 - 4\sqrt{14}$ **47.** $6\sqrt{ab} + 2a\sqrt{b} - 4a$ **49.** $9x^2y - 5$ **51.** **(a)** It is not true when a is negative.   **53.** 12 sq feet
(b) In order to deal with real numbers
55. 18 square feet   **57.** $(6x + 7y)(6x - 7y)$ **58.** $(2x + 5)(3x - 2)$ **59.** $m = \left(\dfrac{1519}{1320}\right)k$; approx. 40.3 mph   **60.** $8\dfrac{1}{3}$ years

How Am I Doing? Sections 9.1–9.4

1. 7 (obj. 9.1.1)   **2.** $\dfrac{5}{8}$ (obj. 9.1.1)   **3.** 9 (obj. 9.1.1)   **4.** -10 (obj. 9.1.1)   **5.** 2.236 (obj. 9.1.2)   **6.** -11 (obj. 9.1.2)
7. 0.2 (obj. 9.2.1)   **8.** 0.6 (obj. 9.2.1)   **9.** x^2 (obj. 9.2.1)   **10.** $5x^2y^3$ (obj. 9.2.1)   **11.** $5\sqrt{2}$ (obj. 9.2.2)   **12.** $x^2\sqrt{x}$ (obj. 9.2.2)
13. $6x\sqrt{x}$ (obj. 9.2.2)   **14.** $2a^2b\sqrt{2b}$ (obj. 9.2.2)   **15.** $15\sqrt{2}$ (obj. 9.3.2)   **16.** $4\sqrt{2}$ (obj. 9.3.2)   **17.** $13\sqrt{7} - 7$ (obj. 9.3.2)
18. $11\sqrt{2} - 18\sqrt{3}$ (obj. 9.3.2)   **19.** $-9\sqrt{2x}$ (obj. 9.3.2)   **20.** $-26x\sqrt{3}$ (obj. 9.3.2)   **21.** $30\sqrt{2}$ (obj. 9.4.1)   **22.** $105x^2\sqrt{2}$ (obj. 9.4.1)
23. $2\sqrt{3} + 6$ (obj. 9.4.2)   **24.** $23 + 8\sqrt{7}$ (obj. 9.4.3)   **25.** 1 (obj. 9.4.3)   **26.** $17 - 4\sqrt{15}$ (obj. 9.4.3)   **27.** $x - \sqrt{x} - 12$ (obj. 9.4.3)
28. $81 + 14\sqrt{21}$ (obj. 9.4.3)

9.5 Exercises

1. 2   **3.** $\dfrac{1}{3}$ **5.** 7   **7.** $\dfrac{\sqrt{6}}{x^2}$ **9.** $\dfrac{3\sqrt{2}}{a^2}$ **11.** $\dfrac{3\sqrt{7}}{7}$ **13.** $\dfrac{5\sqrt{13}}{13}$ **15.** $\dfrac{x\sqrt{2x}}{2}$ **17.** $\dfrac{2\sqrt{2x}}{x}$ **19.** $\dfrac{\sqrt{3}}{2}$ **21.** $\dfrac{6\sqrt{a}}{a}$ **23.** $\dfrac{\sqrt{2x}}{2x^2}$
25. $\dfrac{3\sqrt{x}}{x^2}$ **27.** $\dfrac{\sqrt{15}}{5}$ **29.** $\dfrac{\sqrt{182}}{14}$ **31.** $\dfrac{9\sqrt{2x}}{8x}$ **33.** $2\sqrt{3} + 2$ **35.** $2\left(\sqrt{5} - \sqrt{2}\right)$ or $2\sqrt{5} - 2\sqrt{2}$ **37.** $2 + \sqrt{2}$ **39.** $2\sqrt{14} - 7$
41. $x(2\sqrt{2} + \sqrt{5})$ or $2x\sqrt{2} + x\sqrt{5}$ **43.** $\dfrac{\sqrt{6x} - \sqrt{2x}}{4}$ **45.** $\dfrac{7 - 2\sqrt{10}}{3}$ **47.** $\dfrac{4\sqrt{35} + 3\sqrt{5} + 3\sqrt{2} + 4\sqrt{14}}{3}$ **49.** $5\sqrt{6} + 8\sqrt{2}$
51. $\sqrt{x} - 5$ **53.** **(a)** $x^2 - 2$ **55.** **(a)** $s = \dfrac{1}{h}\sqrt{3Vh}$ **57.** $\left(3\sqrt{5} - 6\right)$ meters   **59.** Length $= 4.236$ m
(b) $x^2 - 2 = \left(x + \sqrt{2}\right)\left(x - \sqrt{2}\right)$ **(b)** $s = 3$ ft   Width $= 0.708$ m
(c) $\left(x + 2\sqrt{3}\right)\left(x - 2\sqrt{3}\right)$
61. 7.38   **62.** 448   **63.** 71 mg   **64.** more than 15 doctor visits

9.6 Exercises

1. 13   **3.** $2\sqrt{14}$ **5.** $7\sqrt{2}$ **7.** $3\sqrt{7}$ **9.** $2\sqrt{19}$ **11.** $\sqrt{57}$ **13.** 9.91   **15.** 9.4 feet   **17.** 98.0 ft   **19.** top cable $= 145.4'$
bottom cable $= 140.77'$
21. 7   **23.** 9   **25.** 7   **27.** $\dfrac{81}{2}$ **29.** 5 only   **31.** 0, 3   **33.** 1 only   **35.** 0, 1   **37.** $\dfrac{1}{2}$ only   **39.** $\dfrac{2}{3}$ only   **41.** 5
42. $25x^2 - 70x + 49$ **43.** 20   **44.** relation   **45.** 450 birds   **46.** approximately $37.21

9.7 Exercises

1. $y = 72$ **3.** $y = \dfrac{1029}{2}$ **5.** $y = 1296$ **7.** $L = 280$ calories   **9.** $T = 56$ min   **11.** $y = \dfrac{45}{4}$ **13.** $y = 2$ **15.** $y = \dfrac{40}{27}$

17. 3.45 minutes   **19.** $444\dfrac{2}{9}$ lb   **21.** **(a)** 15 km, 8.2 km, and 2.7 km   **23.** **(a)** $k = 1$ **25.** 160
(b) 1.837 Earth years
(b)

$d = \dfrac{15}{11}\sqrt{h}$

Distance you can see in kilometers

Height above the water in meters

26.

27. $4(3x + 4)(x - 3)$ **28.** 7 **29.** 1st side = 21 feet **30.** $1800 @ 6%
2nd side = 16 feet $3200 @ 10%
3rd side = 32 feet

Putting Your Skills to Work

1. 4368 sq ft   **2.** 937.5 watts   **3.** 0.21 watt produced by one sq ft of solar panel.   **4.** 1244.88 watts   **5.** approximately 4836 watts
6. approximately 5680 watts

Chapter 9 Review Problems

1. 6   **2.** 5   **3.** 13   **4.** 14   **5.** −9   **6.** 8   **7.** 7   **8.** −11   **9.** −12   **10.** 17   **11.** 0.2   **12.** 0.7   **13.** $\dfrac{1}{5}$ **14.** $\dfrac{6}{7}$
15. 10.247   **16.** 14.071   **17.** 8.775   **18.** 9.381   **19.** $3\sqrt{3}$ **20.** $2\sqrt{13}$ **21.** $2\sqrt{7}$ **22.** $5\sqrt{5}$ **23.** $2\sqrt{10}$ **24.** $4\sqrt{5}$
25. x^4 **26.** y^5 **27.** $x^2y^3\sqrt{x}$ **28.** $ab^2\sqrt{a}$ **29.** $4xy^2\sqrt{xy}$ **30.** $7x^2y^3\sqrt{2}$ **31.** $2x^2\sqrt{3x}$ **32.** $3x^3\sqrt{3x}$ **33.** $5x^5\sqrt{3}$
34. $5x^6\sqrt{5}$ **35.** $2ab^2c^2\sqrt{30ac}$ **36.** $11a^3b^2\sqrt{c}$ **37.** $2x^3y^4\sqrt{14xy}$ **38.** $3x^6y^3\sqrt{11xy}$ **39.** $4\sqrt{7}$ **40.** $9\sqrt{6}$ **41.** $7x\sqrt{3}$
42. $8a\sqrt{2} + 2a\sqrt{3}$ **43.** $-7\sqrt{5} + 2\sqrt{10}$ **44.** $9\sqrt{6} - 15\sqrt{2}$ **45.** $6x^2$ **46.** $-10a\sqrt{b}$ **47.** $4ab\sqrt{a}$ **48.** $-15x^4$ **49.** −7
50. −4   **51.** $\sqrt{10} - \sqrt{6} - 4$ **52.** $\sqrt{30} - 10 + 5\sqrt{2}$ **53.** $20 + 3\sqrt{11}$ **54.** $27 + 8\sqrt{10}$ **55.** $8 - 4\sqrt{3} + 12\sqrt{6} - 18\sqrt{2}$
56. $15 - 3\sqrt{2} - 10\sqrt{3} + 2\sqrt{6}$ **57.** $66 + 36\sqrt{2}$ **58.** $74 - 40\sqrt{3}$ **59.** $\dfrac{\sqrt{3x}}{3x}$ **60.** $\dfrac{2y\sqrt{5}}{5}$ **61.** $\dfrac{x^2y\sqrt{2}}{4}$ **62.** $\dfrac{3a\sqrt{2b}}{2}$
63. $\dfrac{\sqrt{30}}{6}$ **64.** $\dfrac{\sqrt{66}}{11}$ **65.** $\dfrac{a^2\sqrt{2}}{2}$ **66.** $\dfrac{x\sqrt{3}}{3}$ **67.** $\sqrt{5} - \sqrt{2}$ **68.** $\dfrac{2\left(\sqrt{6} + \sqrt{3}\right)}{3}$ **69.** $-7 + 3\sqrt{5}$ **70.** $\dfrac{3 - 2\sqrt{3}}{3}$
71. $c = \sqrt{89}$ **72.** $a = \sqrt{2}$ **73.** $b = \dfrac{\sqrt{51}}{2}$ **74.** $c = 3.124$ **75.** 30 meters   **76.** 45.8 miles   **77.** 15.1 feet   **78.** $x = 29$
79. $x = 21$ **80.** $x = -1$ **81.** $x = 6$ **82.** $x = 4$ **83.** $x = 7$ **84.** $x = 2$ **85.** $x = 7$ **86.** $y = 77$ **87.** $y = \dfrac{2}{15}$
88. $y = \dfrac{1}{2}$ **89.** $\dfrac{1}{16}$ as bright   **90.** approximately 134 feet   **91.** It will be 8 times as much   **92.** $\dfrac{6}{11}$ **93.** 0.02   **94.** $7x^3y\sqrt{2y}$
95. $3\sqrt{3}$ **96.** $18 - 2\sqrt{30} + 3\sqrt{15} - 5\sqrt{2}$ **97.** $49 + 12\sqrt{5}$ **98.** $\dfrac{5\sqrt{3}}{6}$ **99.** $\dfrac{6 - \sqrt{6} + 6\sqrt{2} - 2\sqrt{3}}{10}$ **100.** $x = 42$
101. $x = -\dfrac{1}{2}, x = 2$

How Am I Doing? Chapter 9 Test

1. 11 (obj. 9.1.1)   **2.** $\dfrac{3}{10}$ (obj. 9.1.1)   **3.** $4xy^3\sqrt{3y}$ (obj. 9.2.2)   **4.** $10xz^2\sqrt{xy}$ (obj. 9.2.2)   **5.** $3\sqrt{3}$ (obj. 9.3.2)
6. $8\sqrt{a} + 5\sqrt{2a}$ (obj. 9.3.2)   **7.** $12ab$ (obj. 9.4.1)   **8.** $5\sqrt{2} + 2\sqrt{15} - 15$ (obj. 9.4.2)   **9.** $37 + 20\sqrt{3}$ (obj. 9.4.3)
10. $19 + \sqrt{10}$ (obj. 9.4.3)   **11.** $\dfrac{\sqrt{5x}}{5}$ (obj. 9.5.2)   **12.** $\dfrac{\sqrt{3}}{2}$ (obj. 9.5.2)   **13.** $\dfrac{17 + \sqrt{3}}{22}$ (obj. 9.5.3)   **14.** $a\left(\sqrt{5} - \sqrt{2}\right)$ (obj. 9.5.3)
15. 12.49 (obj. 9.1.2)   **16.** $x = \sqrt{133}$ (obj. 9.6.1)   **17.** $x = 3\sqrt{3}$ (obj. 9.6.1)   **18.** $x = \dfrac{35}{2}$ (obj. 9.6.2)   **19.** $x = 9$ (obj. 9.6.2)
20. 16 inches (obj. 9.7.2)   **21.** $85.77 (obj. 9.7.1)   **22.** $A = 21.2 \text{ cm}^2$ (obj. 9.7.1)

Cumulative Test for Chapters 0–9

1. $\dfrac{2}{3}$ **2.** $8\dfrac{11}{12}$ **3.** $\dfrac{15}{16}$ **4.** 0.0007   **5.** 3   **6.** 31   **7.** $-3a^2bc^3 - 2abc^2$ **8.** $x^3 + 3x^2 - 4x - 12$ **9.** $16x^2 - 40x + 25$
10. $9x^2 - 121$ **11.** $\dfrac{4x + 4}{x + 2}$ **12.** $\dfrac{(x - 3)(x + 1)}{(x - 6)(x + 2)}$ **13.** $7x^2y^3\sqrt{2x}$ **14.** $7\sqrt{2}$ **15.** $9 - 6\sqrt{2}$ **16.** $5 + 2\sqrt{6}$ **17.** $10 + 22\sqrt{3}$
18. $\dfrac{3 + \sqrt{2}}{7}$ **19.** −5   **20.** not a real number   **21.** $c = \sqrt{23}$ **22.** $b = 4\sqrt{5}$ **23.** $x = 6$ **24.** $y = 6$ **25.** $y = 10$ **26.** 78.5 m^2

Chapter 10

10.1 Exercises

1. $a = 1$ **3.** $a = 8$ **5.** $a = 1$ **7.** $x = 0$ **9.** $x = 0$ **11.** $x = 0$ **13.** $x = 7$ **15.** $x = \dfrac{1}{2}$ **17.** $x = \dfrac{1}{3}$
  $b = 8$ $b = -11$ $b = 3$ $x = \dfrac{1}{3}$ $x = 2$ $x = \dfrac{3}{2}$ $x = -4$ $x = -6$ $x = 5$
  $c = 7$ $c = 0$ $c = -15$

19. $x = 6$
$x = -3$

21. $x = -\dfrac{3}{4}$
$x = -\dfrac{2}{3}$

23. $m = -5$
$m = -6$

25. $y = \dfrac{1}{4}$
$y = \dfrac{3}{2}$

27. $x = \dfrac{6}{5}$

29. $x = 0$
$x = 3$

31. $x = 0$
$x = -\dfrac{14}{3}$

33. $y = -2$
$y = 5$

35. $x = 3$
$x = 7$

37. $x = 1$

39. $x = 4$
$x = 2$

41. $x = -7$
$x = 4$

43. $y = -\dfrac{4}{3}$
$y = 3$

45. $x = \dfrac{5}{2}$
$x = -4$

47. $x = -\dfrac{10}{3}$
$x = 4$

49. You can always factor out x.

51. $a = \dfrac{14}{9}$
$c = -14$

53. 190 truck routes

55. 6 cities

57. 10 cities

59. $n = 1200$
$n = 2000$

61. average $= 1600$
revenue $= \$360$

63. Revenue for $1500 = \$357.50$
Revenue for $1700 = \$357.50$
It appears the maximum revenue is at 1600 labels.

65. $\dfrac{(2x - 1)(x + 2)}{2x}$

66. $\dfrac{2x^2 - 4x + 2}{3x^2 - 11x + 10}$

67. approximately 4.32 mg

68. approximately 130 mg

10.2 Exercises

1. $x = \pm 8$
3. $x = \pm 7\sqrt{2}$
5. $x = \pm 2\sqrt{7}$
7. $x = \pm 3$
9. $x = \pm 2\sqrt{5}$
11. $x = \pm 5\sqrt{5}$
13. $x = \pm 2\sqrt{3}$
15. $x = \pm\sqrt{5}$

17. $x = 3, x = 11$
19. $x = -4 \pm \sqrt{6}$
21. $x = \dfrac{-5 \pm \sqrt{2}}{2}$
23. $x = \dfrac{1 \pm \sqrt{7}}{3}$
25. $x = \dfrac{-2 \pm 2\sqrt{3}}{7}$
27. $x = \dfrac{5 \pm 3\sqrt{6}}{4}$

29. $x = 3 \pm 2\sqrt{5}$
31. $x = 7, x = 1$
33. $x = 6 \pm \sqrt{41}$
35. $x = 0, x = 7$
37. $x = 0, x = 5$
39. $x = \dfrac{9}{2}, x = -1$

41. $x = \dfrac{-b \pm \sqrt{28 + b^2}}{2}$
43. $a = -4, b = -4$
44. $x = 1, x = 5$
45. 200 lb per sq. in.

46. A total of 16 tires were defective. 6 had both kinds of defects. 7 had only defects in workmanship. 3 had defects in materials.
47. 58.3%

10.3 Exercises

1. $\sqrt{b^2 - 4ac} = \sqrt{100} = 10$
Therefore, there are two rational roots.

3. No, standard form is $4x^2 + 5x - 6 = 0$.
Then $a = 4, b = 5$, and $c = -6$.

5. $x = 2, x = -5$

7. $x = \dfrac{3 \pm \sqrt{41}}{2}$
9. $x = 2, x = -\dfrac{3}{4}$
11. $x = 4, x = -\dfrac{5}{2}$
13. $x = \dfrac{3 \pm \sqrt{33}}{12}$
15. $x = \dfrac{1 \pm \sqrt{19}}{6}$

17. $x = 5, x = 2$
19. no real solution
21. $y = \dfrac{4 \pm \sqrt{7}}{3}$
23. $d = \dfrac{4}{3}, d = -3$
25. no real solution
27. $x = 5$

29. $x = 0.372$
$x = -5.372$
31. $x = 4.108, x = -0.608$
33. $x = -1.894, x = -0.106$
35. $x = \dfrac{3}{2}, x = \dfrac{2}{3}$
37. $x = 3 \quad x = \dfrac{1}{3}$

39. $t = -1 \pm \sqrt{23}$
41. $y = \dfrac{1 \pm \sqrt{51}}{5}$
43. $x = \dfrac{\pm\sqrt{39}}{3}$
45. $x = 1 \pm 2\sqrt{2}$
47. 2 feet
49. $\dfrac{3(x + 5)}{x + 2}$

50. (a) $y = -\dfrac{5x}{13} - \dfrac{14}{13}$
51. $x^2 + 5x + 2$
52. $\dfrac{9p - 20}{3f} = g$
53. 96.76 cubic feet
54. 138.16 sq inches
(b) $5x + 13y = -14$

How Am I Doing? Sections 10.1–10.3

1. $x = 16, x = -3$ (obj. 10.1.2)
2. $x = 0, x = \dfrac{7}{5}$ (obj. 10.1.3)
3. $x = \dfrac{2}{5}, x = 4$ (obj. 10.1.3)
4. $x = \dfrac{1}{4}, x = -\dfrac{1}{2}$ (obj. 10.1.3)

5. $x = -8, x = 3$ (obj. 10.1.3)
6. $x = 3, x = -\dfrac{9}{2}$ (obj. 10.1.3)
7. $x = \pm 3\sqrt{2}$ (obj. 10.2.1)
8. $x = \pm 5$ (obj. 10.2.1)

9. $x = -4 \pm \sqrt{11}$ (obj. 10.2.2)
10. $x = \dfrac{-3 \pm \sqrt{65}}{4}$ (obj. 10.2.2)
11. $x = \dfrac{-2 \pm \sqrt{14}}{2}$ (obj. 10.3.1)
12. $x = \dfrac{7 \pm \sqrt{17}}{4}$ (obj. 10.3.1)

13. $x = \dfrac{-4 \pm \sqrt{13}}{3}$ (obj. 10.3.1)
14. no real solution (obj. 10.3.3)

10.4 Exercises

1. It may have 2, 1, or none.
3. If $b = 0$, then the parabola always has a vertex at $(0, c)$ which is on the y-axis.

5.

$y = x^2 + 2$

7.

$y = -\dfrac{1}{3}x^2$

9.

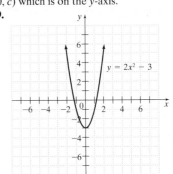

$y = 2x^2 - 3$

11.

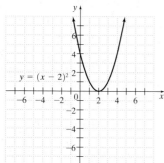

$y = (x - 2)^2$

13.

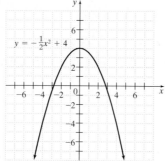

$y = -\frac{1}{2}x^2 + 4$

15.

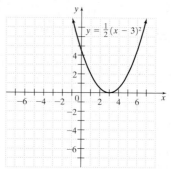

$y = \frac{1}{2}(x - 3)^2$

17.

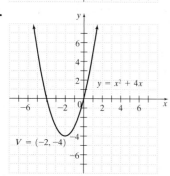

$y = x^2 + 4x$

$V = (-2, -4)$

19.

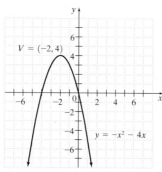

$V = (-2, 4)$

$y = -x^2 - 4x$

21.

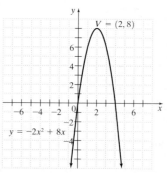

$V = (2, 8)$

$y = -2x^2 + 8x$

23.

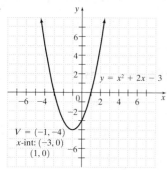

$y = x^2 + 2x - 3$

$V = (-1, -4)$

x-int: $(-3, 0)$

$(1, 0)$

25.

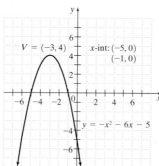

$V = (-3, 4)$ x-int: $(-5, 0)$

$(-1, 0)$

$y = -x^2 - 6x - 5$

27.

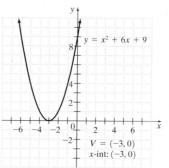

$y = x^2 + 6x + 9$

$V = (-3, 0)$

x-int: $(-3, 0)$

29. (a)

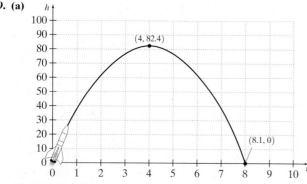

(4, 82.4)

(8.1, 0)

(b) 62.8 meters

(c) 82.4 meters

(d) about 8.1 seconds

31. (a)

Number of mosquitoes in millions

(4.5, 20.25)

$N = 9x - x^2$

Number of inches of rain

(b) 4.5 inches

(c) 20.25 million

33. vertex: $(-5.3, -240.25)$
y-intercept: $(0, -212.16)$
x-intercepts: $(10.2, 0)$ and $(-20.8, 0)$

35. vertex: $(-0.277, 1551.164)$
y-intercept: $(0, 1528)$
x-intercepts: $(1.993, 0)$ and $(-2.548, 0)$

36. $y = 1$ **37.** $y = 75$

38. 8.25 seconds   **39.** $V = \frac{5}{3}T + \frac{865}{3}$

10.5 Exercises

1. base = 11 cm
altitude = 16 cm

3. 11 ft by 11 ft

5. 21 people

7. 80 members were expected.
100 members attended.

9. width = 25 feet
length = 50 feet

11. 600 mph   **13.** 100   **15.** 25   **17.** 1994 and 1995   **19.** about 74 miles   **21.** $n = 13$ **23.**

$2 \quad \frac{5}{2} \quad 3$

24. $\sqrt{15} - 2\sqrt{3}$ **25.** 0.75 feet

Putting Your Skills to Work

1.

t	0	1	2	3	4	5
h	4	68	100	100	68	4

2.

3. 100 feet   **4.** 104 feet   **5.** 5.05 seconds   **6.** 131.6 feet, 2.8 seconds

Chapter 10 Review Problems

1. $a = 6, b = -5, c = 8$ **2.** $a = 14, b = 3, c = -16$ **3.** $a = 1, b = -9, c = -3$ **4.** $a = 11, b = -7, c = 0$
5. $a = -2, b = -6, c = 3$ **6.** $a = 4, b = 5, c = -14$ **7.** $x = -25, x = -1$ **8.** $x = -8$ **9.** $x = 5, x = -11$
10. $x = \dfrac{3}{2}, x = \dfrac{5}{2}$ **11.** $x = \dfrac{4}{3}$ **12.** $x = \dfrac{2}{5}, x = \dfrac{4}{3}$ **13.** $x = \dfrac{1}{2}, x = \dfrac{7}{5}$ **14.** $x = -\dfrac{11}{3}, x = -1$ **15.** $x = -\dfrac{3}{2}, x = \dfrac{4}{3}$
16. $x = \dfrac{1}{2}, x = 1$ **17.** $x = \dfrac{1}{5}, x = -\dfrac{1}{3}$ **18.** $x = \dfrac{1}{4}, x = -\dfrac{4}{3}$ **19.** $x = 0, x = -5$ **20.** $x = 0, x = 8$ **21.** $x = 0, x = -\dfrac{5}{3}$
22. $x = 0, x = 4$ **23.** $x = -\dfrac{2}{5}, x = 10$ **24.** $x = \dfrac{1}{3}, x = \dfrac{4}{3}$ **25.** $x = \pm 7$ **26.** $x = \pm 9$ **27.** $x = \pm\sqrt{22}$ **28.** $x = \pm\sqrt{39}$
29. $x = \pm 2\sqrt{2}$ **30.** $x = \pm 5\sqrt{2}$ **31.** $x = \pm 3\sqrt{2}$ **32.** $x = \pm 2\sqrt{6}$ **33.** $x = 4 \pm \sqrt{7}$ **34.** $x = 2 \pm \sqrt{3}$ **35.** $x = \dfrac{-3 \pm 2\sqrt{6}}{4}$
36. $x = \dfrac{-7 \pm 3\sqrt{5}}{2}$ **37.** $x = 1, x = -11$ **38.** $x = -1, x = -11$ **39.** $x = 9, x = -5$ **40.** $x = 3 \pm \sqrt{2}$ **41.** $x = -1 \pm \sqrt{3}$
42. $x = \dfrac{-5 \pm \sqrt{31}}{2}$ **43.** $x = -2 \pm \sqrt{10}$ **44.** $x = -2 \pm 2\sqrt{3}$ **45.** $x = \dfrac{7 \pm \sqrt{17}}{4}$ **46.** $x = \dfrac{-5 \pm \sqrt{73}}{4}$ **47.** $x = \dfrac{4 \pm 2\sqrt{7}}{3}$
48. $x = \dfrac{1 \pm 3\sqrt{5}}{4}$ **49.** $x = \dfrac{5 \pm \sqrt{73}}{6}$ **50.** $x = \dfrac{-3 \pm \sqrt{41}}{8}$ **51.** $x = \dfrac{5}{2}, x = 2$ **52.** $x = \dfrac{3}{2}, x = -\dfrac{1}{2}$ **53.** $x = -\dfrac{1}{5}$
54. $x = 4, x = \dfrac{3}{2}$ **55.** $x = \dfrac{3 \pm \sqrt{3}}{3}$ **56.** $x = \dfrac{7 \pm \sqrt{209}}{10}$ **57.** $x = \dfrac{-2 \pm \sqrt{19}}{3}$ **58.** $x = \dfrac{-7 \pm \sqrt{29}}{10}$ **59.** $x = \dfrac{9 \pm \sqrt{93}}{2}$
60. $x = -3, x = \dfrac{2}{3}$ **61.** $x = -\dfrac{4}{5}, x = 1$ **62.** $x = \pm 3\sqrt{2}$ **63.** $y = -8$ only   **64.** $y = -7$ only   **65.** $x = \dfrac{-4 \pm \sqrt{31}}{3}$
66. $x = \dfrac{-4 \pm \sqrt{2}}{2}$ **67.** $y = \dfrac{11 \pm \sqrt{21}}{10}$ **68.** $y = -4, y = 2$ **69.** $y = -3, y = 5$ **70.** $x = 0, x = \dfrac{1}{6}$

71.

72.

73.

74.

75.

76.

77.

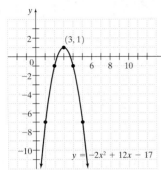

$y = -2x^2 + 12x - 17$

78.

$y = -3x^2 - 2x + 4$

79. 6 feet by 2.5 feet   **80.** 16 centimeters, 12 centimeters   **81.** width = 18 inches   length = 24 inches   **82.** $t = 36$ seconds   **83.** 24 members

84. going 45 mph, returning 60 mph   **85.** 60 pizza stores   **86.** 30 fewer pizza stores   **87.** 1990 and 2020   **88.** 2000 and 2010

How Am I Doing? Chapter 10 Test

1. $x = \dfrac{-7 \pm \sqrt{129}}{10}$ (obj. 10.3.1)   **2.** $x = -5,\ x = \dfrac{2}{3}$ (obj. 10.1.3)   **3.** no real solution (obj. 10.3.3)   **4.** $x = \dfrac{3}{4},\ x = 5$ (obj. 10.2.2)

5. $x = -\dfrac{5}{4},\ x = \dfrac{1}{3}$ (obj. 10.2.1)   **6.** $x = \dfrac{4}{3}$ (obj. 10.2.2)   **7.** $x = 0, x = 8$ (obj. 10.3.1)   **8.** $x = \pm 3$ (obj. 10.3.1)   **9.** $x = -\dfrac{1}{2},\ x = 6$ (obj. 10.3.1)

10. $x = \dfrac{3}{2},\ x = -\dfrac{1}{2}$ (obj. 10.3.1)   **11.**   (obj. 10.4.1)   **12.**   (obj. 10.4.1)

$y = 3x^2 - 6x$ $(1, -3)$

$y = -x^2 + 8x - 12$ $(4, 4)$

13. 9 meters, 12 meters (obj. 10.5.1)   **14.** 7 seconds (obj. 10.5.1)

Cumulative Test for Chapters 1–10

1. $-27x^2 - 30xy$ **2.** $-\dfrac{1}{5}$ **3.** 2   **4.** $(4x^2 + 1)(2x - 1)(2x + 1)$ **5.** $2x^3 + 3x^2 - 17x - 30$ **6.** $-\dfrac{12}{5}$

7.

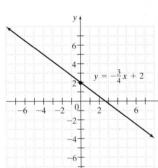

$y = -\dfrac{3}{4}x + 2$

slope $-\dfrac{3}{4}$; y-intercept = $(0, 2)$ **8.** $y = \dfrac{4}{3}x + 6$ **9.** $(3, -2)$ **10.** $81x^{12}y^8$ **11.** $3x^2y^3z\sqrt{2xz}$

12. $-2\sqrt{6} - 8$ **13.** $-5\sqrt{5} + 11$ **14.** $t = \dfrac{2a - 3b}{8}$

15. $b = \pm\dfrac{\sqrt{H + 6}}{5}$ **16.** $x = \dfrac{7}{2}, x = -5$ **17.** $x = \dfrac{-1 \pm 2\sqrt{5}}{2}$ **18.** $R = -12, R = 4$

19. $x = \dfrac{-11 \pm \sqrt{97}}{6}$ **20.** $x = \dfrac{2}{3}, x = -\dfrac{5}{2}$ **21.** $x = \pm 6\sqrt{2}$

22. $x = \pm 5$ **23.**   **24.** length 16 meters, width 6 meters

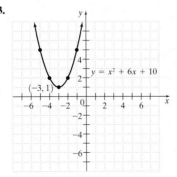

$y = x^2 + 6x + 10$ $(-3, 1)$

Practice Final Examination

1. $-2x + 21y + 18xy + 6y^2$ **2.** 14 **3.** $18x^5y^5$ **4.** $2x^2y - 8xy$ **5.** $x = \dfrac{8}{5}$ **6.** $b = \dfrac{p - 2a}{2}$

7. $x \geq 2.6$ 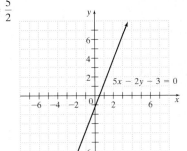   **8.** $2x^3 - 5x^2y + xy^2 + 2y^3$ **9.** $2(2x + 1)(x - 5)$ **10.** $3x(x - 5)(x + 2)$ **11.** $\dfrac{6x - 11}{(x + 2)(x - 3)}$

12. $\dfrac{11(x + 2)}{2x(x + 4)}$ **13.** $-\dfrac{12}{5}$ **14.** slope $= \dfrac{5}{2}$ **15.** $3x + 4y = 14$ or $y = -\dfrac{3}{4}x + \dfrac{7}{2}$

$5x - 2y - 3 = 0$

16. 38.13 sq in.   **17.** $x = -5, y = 2$ **18.** $a = -2, b = -3$ **19.** $x\sqrt{5x}$ **20.** $6\sqrt{3} - 12 + 12\sqrt{2}$ **21.** $-\dfrac{\sqrt{15} + \sqrt{35} + \sqrt{21} + 7}{2}$

22. $x = \dfrac{2}{3}, x = -\dfrac{1}{4}$ **23.** $y = \dfrac{3 \pm \sqrt{7}}{2}$ **24.** $x = \pm 2$ **25.** $\sqrt{39}$ **26.** 5 **27.** width 7 meters, length 12 meters

28. $3000 at 10%, $4000 at 14%   **29.** 200 reserved-seat tickets, 160 general admission tickets   **30.** base 8 meters, altitude 17 meters

Appendix B Exercises

1. 34,000 m   **3.** 5700 cm   **5.** 250 mm   **7.** 0.563 m   **9.** 29,400 mg   **11.** 0.0984 kg   **13.** 7000 mL   **15.** 0.000004 kL
17. 1.7 in.   **19.** 22.5 km   **21.** 2.8 m   **23.** 17.8 cm   **25.** 1089.6 g   **27.** 171.6 lb   **29.** 2.8 L   **31.** 193,248,000 in.
33. 27.27 miles per hour   **35.** 0.11 yr

Appendix C Exercises

1. 84,916 square miles   **3.** 4,507,000   **5.** 2   **7.** Treasure State   **9.** Nevada   **11.** 2000 homes   **13.** Dupage County
15. 2800 homes   **17.** 4,000,000 apartment units   **19.** 7,200,000 more apartment units   **21.** 8,000,000   **23.** It is a three-way tie. The increase was 3 million in three cases. It occurred between 1970 and 1980, and again between 1980 and 1990, and finally again between 1990 and 2000.   **25.** 3 million more   **27.** 37 million   **29.** Movies and an Exercise Program   **31.** $5.6 million   **33.** $2.5 million greater
35. Between 2001 and 2002   **37.** 74 million   **39.** an age group of 65 years and above   **41.** 19%   **43.** 83%   **45.** 754 million people
47. 40%   **49.** $3588

Appendix D Exercises

1. 14   **3.** 52   **5.** 1296   **7.** -3 **9.** $10x - 5$ **11.** 1111811   **13.** 9876549   **15.** 6481100   **17.** 1 6 15 20 15 6 1

19. $x = 3$, steps will vary   **21.** $x = -2$, steps will vary   **23.** $x = 3$, steps will vary   **25.** $x = \dfrac{9}{11}$, steps will vary   **27.** answers will vary

29. William was first, Brent was second, James was third, Dave was fourth   **31.** The teacher is Michael   **33.** Toyota Corolla
35. Unless Fred is willing to pay over $200 for Tuesday's procedure, he will probably get a false tooth.

Applications Index

Photo Credits

CHAPTER 0 **CO** © Paul Almasy/Corbis Bettmann   **p. 31** Stockbyte   **p. 40** Adam Jones/Photo Researchers, Inc.
p. 44 Peter Hvizdak/The Image Works

CHAPTER 1 **CO** SuperStock, Inc.   **p. 69** Peter Skinner/Photo Researchers, Inc.   **p. 90** Tony Freeman/PhotoEdit
p. 101 AP Wide World Photos   **p. 116** John Coletti/Stock Boston   **p. 124** Photo Researchers, Inc.   **p. 125** Elizabeth Crews/
The Image Works

CHAPTER 2 **CO** © Bill Schid/Corbis Bettmann   **p. 145** Mark Downy/Getty Images, Inc.
p. 145 C. Allan Morgan/Peter Arnold, Inc.   **p. 146** Jerry Wachter/Photo Researchers, Inc.   **p. 179** Jeffrey Dunn/Stock Boston
p. 182 Stephen J. Krasemann/Photo Researchers, Inc.

CHAPTER 3 **CO** © Kennan Ward/Corbis Bettmann   **p. 201** Ken Graham/Getty Images, Inc.
p. 205 NASA/Science Source/Photo Researchers, Inc.   **p. 206** Townsend P. Dickinson/The Image Works
p. 207 Bob Daemmrich/Stock Boston   **p. 208** Dennis MacDonald/PhotoEdit   **p. 212** Dave Bartruff/Corbis Bettmann
p. 226 AP Wide World Photos   **p. 237** George Ranalli/Photo Researchers, Inc.   **p. 234** NASA/Johnson Space Center

CHAPTER 4 **CO** Rob Crandall/Stock Boston   **p. 265** NASA/John F. Kennedy Space Center

CHAPTER 5 **CO** Jim Cummins/Getty Images, Inc.   **p. 306** Malcolm Fielding, Johnson Matthey PLC/Science Photo
Library/Photo Researchers, Inc.   **p. 308** Luis Veiga/Getty Images, Inc.   **p. 333** Alok Kavan/Photo Researchers, Inc.
p. 343 Philip H. Coblentz/World Travel Images, Inc.   **p. 347** AP Wide World Photos

CHAPTER 6 **CO** Philippe Hays/Peter Arnold, Inc.

CHAPTER 7 **CO** Ted Mead/Photolibrary.com   **p. 415** Bill Bachman/PhotoEdit   **p. 425** Warwick Kent/Photolibray.com
p. 427 Eye Wire Collection/Getty Images—Photodisc   **p. 436** Corbis/Bettmann   **p. 459** Jacques Jangoux/Getty Images,
Inc.—Stone

CHAPTER 8 **CO** © Jeremy Horner/Corbis Bettmann   **p. 508** Thomas Hoepker/Magnum Photos, Inc.

CHAPTER 9 **CO** NASA/Johnson Space Center   **p. 520** David Madison/Getty Images, Inc.—Stone
p. 525 Glen Allison/Getty Images, Inc.—Stone   **p. 531** World Perspective/Stone   **p. 552** Juan Silva Production/Getty Images, Inc.
p. 560 Alastair Black/Getty Images, Inc.—Stone   **p. 561** Photo Researchers, Inc.

CHAPTER 10 **CO** Jurgen Vogt, Getty Images, Inc.   **p. 583** David Muench/Stone   **p. 607** AP Wide World Photos

R EAD THIS LICENSE CAREFULLY BEFORE OPENING THIS PACKAGE. BY OPENING THIS PACKAGE, YOU ARE AGREEING TO THE TERMS AND CONDITIONS OF THIS LICENSE. IF YOU DO NOT AGREE, DO NOT OPEN THE PACKAGE. PROMPTLY RETURN THE UNOPENED PACKAGE AND ALL ACCOMPANYING ITEMS TO THE PLACE YOU OBTAINED THEM. *THESE TERMS APPLY TO ALL LICENSED SOFTWARE ON THE DISK EXCEPT THAT THE TERMS FOR USE OF ANY SHAREWARE OR FREEWARE ON THE DISKETTES ARE AS SET FORTH IN THE ELECTRONIC LICENSE LOCATED ON THE DISK:*

Single PC Site License

1. GRANT OF LICENSE and OWNERSHIP: The enclosed computer programs and any data ("Software") are licensed, not sold, to you by Pearson Education, Inc. publishing as Pearson Prentice Hall ("We" or the "Company") in consideration of your adoption of the accompanying Company textbooks and/or other materials, and your agreement to these terms. You own only the disk(s) but we and/or our licensors own the Software itself. This license allows instructors and students enrolled in the course using the Company textbook that accompanies this Software (the "Course") to use and display the enclosed copy of the Software on an unlimited number of computers, for academic use only, so long as you comply with the terms of this Agreement. You may make one copy for back up only. We reserve any rights not granted to you.

2. USE RESTRICTIONS: You may not sell or license copies of the Software or the Documentation to others. You may not transfer, distribute or make available the Software or the Documentation. You may not reverse engineer, disassemble, decompile, modify, adapt, translate or create derivative works based on the Software or the Documentation. You may be held legally responsible for any copying or copyright infringement that is caused by your failure to abide by the terms of these restrictions.

3. TERMINATION: This license is effective until terminated. This license will terminate automatically without notice from the Company if you fail to comply with any provisions or limitations of this license. Upon termination, you shall destroy the Documentation and all copies of the Software. All provisions of this Agreement as to limitation and disclaimer of warranties, limitation of liability, remedies or damages, and our ownership rights shall survive termination.

4. DISCLAIMER OF WARRANTY: THE COMPANY AND ITS LICENSORS MAKE NO WARRANTIES ABOUT THE SOFTWARE, WHICH IS PROVIDED "AS-IS." IF THE DISK IS DEFECTIVE IN MATERIALS OR WORKMANSHIP, YOUR ONLY REMEDY IS TO RETURN IT TO THE COMPANY WITHIN 30 DAYS FOR REPLACEMENT UNLESS THE COMPANY DETERMINES IN GOOD FAITH THAT THE DISK HAS BEEN MISUSED OR IMPROPERLY INSTALLED, REPAIRED, ALTERED OR DAMAGED. THE COMPANY DISCLAIMS ALL WARRANTIES, EXPRESS OR IMPLIED, INCLUDING WITHOUT LIMITATION, THE IMPLIED WARRANTIES OF MERCHANTABILITY AND FITNESS FOR A PARTICULAR PURPOSE. THE COMPANY DOES NOT WARRANT, GUARANTEE OR MAKE ANY REPRESENTATION REGARDING THE ACCURACY, RELIABILITY, CURRENTNESS, USE, OR RESULTS OF USE, OF THE SOFTWARE.

5. LIMITATION OF REMEDIES AND DAMAGES: IN NO EVENT, SHALL THE COMPANY OR ITS EMPLOYEES, AGENTS, LICENSORS OR CONTRACTORS BE LIABLE FOR ANY INCIDENTAL, INDIRECT, SPECIAL OR CONSEQUENTIAL DAMAGES ARISING OUT OF OR IN CONNECTION WITH THIS LICENSE OR THE SOFTWARE, INCLUDING, WITHOUT LIMITATION, LOSS OF USE, LOSS OF DATA, LOSS OF INCOME OR PROFIT, OR OTHER LOSSES SUSTAINED AS A RESULT OF INJURY TO ANY PERSON, OR LOSS OF OR DAMAGE TO PROPERTY, OR CLAIMS OF THIRD PARTIES, EVEN IF THE COMPANY OR AN AUTHORIZED REPRESENTATIVE OF THE COMPANY HAS BEEN ADVISED OF THE POSSIBILITY OF SUCH DAMAGES. SOME JURISDICTIONS DO NOT ALLOW THE LIMITATION OF DAMAGES IN CERTAIN CIRCUMSTANCES, SO THE ABOVE LIMITATIONS MAY NOT ALWAYS APPLY.

6. GENERAL: THIS AGREEMENT SHALL BE CONSTRUED IN ACCORDANCE WITH THE LAWS OF THE UNITED STATES OF AMERICA AND THE STATE OF NEW YORK, APPLICABLE TO CONTRACTS MADE IN NEW YORK, AND SHALL BENEFIT THE COMPANY, ITS AFFILIATES AND ASSIGNEES. This Agreement is the complete and exclusive statement of the agreement between you and the Company and supersedes all proposals, prior agreements, oral or written, and any other communications between you and the company or any of its representatives relating to the subject matter. If you are a U.S. Government user, this Software is licensed with "restricted rights" as set forth in subparagraphs (a)-(d) of the Commercial Computer-Restricted Rights clause at FAR 52.227-19 or in subparagraphs (c)(1)(ii) of the Rights in Technical Data and Computer Software clause at DFARS 252.227-7013, and similar clauses, as applicable.

Should you have any questions concerning this agreement or if you wish to contact the Company for any reason, please contact in writing: Customer Service Pearson Prentice Hall, 200 Old Tappan Road, Old Tappan NJ 07675.

Minimum System Requirements

Windows	Macintosh
Pentium II 300 MHz processor	Power PC G3 233 MHz or better
Windows 98 or later	Mac OS 9.x or 10.x
64 MB RAM	64 MB RAM
800 x 600 resolution	800 x 600 resolution
8x or faster CD-ROM drive	8x or faster CD-ROM drive
QuickTime 6.0 or later	QuickTime 6.0 or later

PROPERTIES OF REAL NUMBERS

	Addition	Multiplication
Commutative Properties	$a + b = b + a$	$ab = ba$
Associative Properties	$(a + b) + c = a + (b + c)$	$(ab)c = a(bc)$
Identity Properties	$a + 0 = 0 + a = a$	$a \cdot 1 = 1 \cdot a = a$
Inverse Properties	$a + (-a) = -a + a = 0$	$a \cdot \dfrac{1}{a} = \dfrac{1}{a} \cdot a = 1 \ (a \neq 0)$
Distributive Property		$a(b + c) = ab + ac$

PROPERTIES OF EXPONENTS

If $x, y \neq 0$, then

$$x^a \cdot x^b = x^{a+b}$$

$$\frac{x^a}{x^b} = x^{a-b} \text{ if } a \geq b$$

$$\frac{x^a}{x^b} = \frac{1}{x^{b-a}} \text{ if } a < b$$

$$x^0 = 1$$

$$(x^a)^b = x^{ab}$$

$$(xy)^a = x^a y^a$$

$$\left(\frac{x}{y}\right)^a = \frac{x^a}{y^a}$$

ABSOLUTE VALUE

Definition

$$|x| = \begin{cases} x \text{ if } x \geq 0 \\ -x \text{ if } x < 0 \end{cases}$$

INEQUALITIES

If $a < b$, then for all values of c, $a + c < b + c$ and $a - c < b - c$.

If $a < b$ when c is a **positive number** $(c > 0)$, then $ac < bc$ and $\dfrac{a}{c} < \dfrac{b}{c}$.

If $a < b$ when c is a **negative number** $(c < 0)$, then $ac > bc$ and $\dfrac{a}{c} > \dfrac{b}{c}$.

FACTORING AND MULTIPLYING FORMULAS

Perfect-Square Trinomials	$a^2 + 2ab + b^2 = (a + b)^2$ $a^2 - 2ab + b^2 = (a - b)^2$
Difference of Two Squares	$a^2 - b^2 = (a + b)(a - b)$
Sum of Two Squares	$a^2 + b^2$ *cannot be factored*

PROPERTIES OF LINES AND SLOPES

The **slope** of any nonvertical line passing through (x_1, y_1) and (x_2, y_2) is $m = \dfrac{y_2 - y_1}{x_2 - x_1}$ $(x_1 \neq x_2)$.

The **standard form** of the equation of a straight line is $Ax + By = C$.

The **slope–intercept** form of the equation of a straight line with slope m and y-intercept $(0, b)$ is $y = mx + b$.

A **horizontal line** has a slope of zero. The equation of a horizontal line can be written as $y = b$.

A **vertical line** has no slope (the slope is not defined for a vertical line).

The equation of a vertical line can be written as $x = a$.